AF737636

METHODS IN MOLECULAR BIOLOGY

Series Editor
John M. Walker
School of Life and Medical Sciences
University of Hertfordshire
Hatfield, Hertfordshire, AL10 9AB, UK

For further volumes:
http://www.springer.com/series/7651

Eukaryotic Genomic Databases

Methods and Protocols

Edited by

Martin Kollmar

Group Systems Biology of Motor Proteins, Department of NMR-Based Structural Biology, Max-Planck-Institute for Biophysical Chemistry, Göttingen, Germany

Humana Press

Editor
Martin Kollmar
Group Systems Biology of Motor Proteins
Department of NMR-Based Structural Biology
Max-Planck-Institute for Biophysical Chemistry
Göttingen, Germany

ISSN 1064-3745 ISSN 1940-6029 (electronic)
Methods in Molecular Biology
ISBN 978-1-4939-7736-9 ISBN 978-1-4939-7737-6 (eBook)
https://doi.org/10.1007/978-1-4939-7737-6

Library of Congress Control Number: 2018935916

© Springer Science+Business Media, LLC, part of Springer Nature 2018
Chapters 4 and 6 are licensed under the terms of the Creative Commons Attribution 4.0 International License (http://creativecommons.org/licenses/by/4.0/). For further details see license information in the chapters.
This work is subject to copyright. All rights are reserved by the Publisher, whether the whole or part of the material is concerned, specifically the rights of translation, reprinting, reuse of illustrations, recitation, broadcasting, reproduction on microfilms or in any other physical way, and transmission or information storage and retrieval, electronic adaptation, computer software, or by similar or dissimilar methodology now known or hereafter developed.
The use of general descriptive names, registered names, trademarks, service marks, etc. in this publication does not imply, even in the absence of a specific statement, that such names are exempt from the relevant protective laws and regulations and therefore free for general use.
The publisher, the authors and the editors are safe to assume that the advice and information in this book are believed to be true and accurate at the date of publication. Neither the publisher nor the authors or the editors give a warranty, express or implied, with respect to the material contained herein or for any errors or omissions that may have been made. The publisher remains neutral with regard to jurisdictional claims in published maps and institutional affiliations.

Printed on acid-free paper

This Humana Press imprint is published by the registered company Springer International Publishing AG part of Springer Nature.
The registered company address is: 233 Spring Street, New York, NY 10013, U.S.A.

Preface

Since genome sequencing became affordable, the number of available eukaryotic genomes has dramatically increased. Large-scale sequencing of thousands of species is under way (e.g., 1k fungi project, Y1000+ yeast project, Genomes 10k vertebrates project, B10K bird project, 5k insects project, 959 nematodes project, 10KP plant project). Species-dedicated communities complement genome-based knowledge by population genomics, transcriptomics, and various other types of data. This volume introduces databases containing genome-based data and providing access to genome-wide analyses. The primary audience involves geneticists and molecular biologists who want to keep their knowledge up to date on where and how to obtain eukaryotic genomics data. Many databases also contain more complex data and analyses that are often hidden to the novice or occasional user and are usually not described in documentations or online help pages. Chapters will describe database contents as well as typical use cases, written in the spirit of the series which aims to provide practical guidance and troubleshooting advice. The focus of this book is on databases from all taxa except plants, which are described in another volume.

Although most of the databases provide access to the same data analysis tools such as BLAST, JBrowse, and InterMine, and connect the genome annotations with the same feature databases such as GO, KEGG, Pfam, and phenotype ontologies, the described use cases and approaches to capture and combine these data are very different. Therefore, it is highly recommended to read and follow the protocols of multiple chapters even if the main interest is in a single species or taxon. The databases presented in this volume are under active curation and development. General technological developments and demands to integrate extended and new data types will lead to continuous changes to and major updates of these databases. However, the rationale how to access and combine the described genomic data will remain independent of whether the look and feel of tools will change. The clear guidance presented in the chapters should facilitate accessing eukaryotic genomic data and stimulate comparative genomic research.

Göttingen, Germany *Martin Kollmar*

Acknowledgment

Pedro Assis, Gail Binkley, Juan Carlos Chan, J. Michael Cherry, Anne Eagle, Ken Frazer, Sibyl Gao, Patrick Kalita, Kalpana Karra, Jonathan Knight, Ruben Lancaster, Kevin MacPherson, Prita Mani, Ryan Martin, Stuart Miyasato, Sierra Moxon, Hans-Michael Muller, Cecilia Nakamura, Paulo Nuin, Holly Paddock, Christian Pich, Matthew Russell, Kevin Schaper, Tim Schedl, Gary Schindelman, Xiang Shao, Travis Sheppard, Matt Simison, Amy Singer, Lincoln Stein, Paul W. Sternberg, Sabrina Toro, Daniel Wang, Shuai Weng, Edith Wong, Adam Wright, Karen Yook

Contents

Contributors

JULIE AGAPITE · *Biological Laboratories, Harvard University, Cambridge, MA, USA*
GIULIA ANTONAZZO · *Department of Physiology, Development and Neuroscience, University of Cambridge, Cambridge, UK*
HELEN ATTRILL · *Department of Physiology, Development and Neuroscience, University of Cambridge, Cambridge, UK*
KIMBERLY VAN AUKEN · *Division of Biology and Biological Engineering, California Institute of Technology, Pasadena, CA, USA*
PHILLIP BAKER · *Department of Biology, University of New Mexico, Albuquerque, NM, USA*
EVELINA Y. BASENKO · *Centre for Genomic Research, Institute of Integrative Biology, University of Liverpool, Liverpool, UK*
MATTHEW BERRIMAN · *Wellcome Trust Sanger Institute, Hinxton, Cambridge, UK*
JONATHAN BINKLEY · *Department of Genetics, Stanford University, Stanford, CA, USA*
BRUCE J. BOLT · *European Molecular Biology Laboratory, European Bioinformatics Institute (EMBL-EBI), Hinxton, Cambridge, UK*
YVONNE M. BRADFORD · *The Zebrafish Information Network, University of Oregon, Eugene, OR, USA*
KRIS BROLL · *Biological Laboratories, Harvard University, Cambridge, MA, USA*
NICHOLAS BROWN · *Department of Physiology, Development and Neuroscience, University of Cambridge, Cambridge, UK*
GREGORY W. BURNS · *Division of Animal Sciences, University of Missouri, Columbia, MO, USA*
KEVIN BURNS · *Xenbase Curation Team, Division of Developmental Biology, Cincinnati Children's Hospital, Cincinnati, OH, USA*
SCOTT CAIN · *Informatics and Bio-computing Platform, Ontario Institute for Cancer Research, Toronto, ON, Canada*
R. ANDREW CAMERON · *Department of Biological Sciences, Mellon Institute, Carnegie Mellon University, Pittsburgh, PA, USA; Division of Biology and Biological Engineering, California Institute of Technology, Pasadena, CA, USA*
GREGORY A. CARY · *Department of Biological Sciences, Mellon Institute, Carnegie Mellon University, Pittsburgh, PA, USA*
MEI-JU MAY CHEN · *Graduate Institute of Biomedical Electronics and Bioinformatics, National Taiwan University, Taipei, Taiwan*
WEN J. CHEN · *Division of Biology and Biological Engineering, California Institute of Technology, Pasadena, CA, USA*
CHRISTOPHER P. CHILDERS · *USDA, Agricultural Research Service, National Agricultural Library, Beltsville, MD, USA*
STANLEY CHU · *Xenbase Development Team, Department of Computer Science, University of Calgary, Calgary, AB, Canada; Xenbase Development Team, Department of Biological Science, University of Calgary, Calgary, AB, Canada*
RICHARD CRIPPS · *Department of Biology, University of New Mexico, Albuquerque, NM, USA*

MADELINE CROSBY · *Biological Laboratories, Harvard University, Cambridge, MA, USA*
KATHRYN CROUCH · *Wellcome Trust Centre for Molecular Parasitology, University of Glasgow, Glasgow, UK*
BRYON CZOCH · *Department of Biology, Indiana University, Bloomington, IN, USA*
PAUL DAVIS · *European Molecular Biology Laboratory, European Bioinformatics Institute, Cambridge, UK*
MELINDA R. DWINELL · *Genomic Sciences and Precision Medicine Center, Medical College of Wisconsin, Milwaukee, WI, USA; Department of Physiology, Medical College of Wisconsin, Milwaukee, WI, USA*
CHRISTINE G. ELSIK · *Division of Animal Sciences, University of Missouri, Columbia, MO, USA; Division of Plant Sciences, University of Missouri, Columbia, MO, USA; MU Informatics Institute, University of Missouri, Columbia, MO, USA*
DAVID EMMERT · *Biological Laboratories, Harvard University, Cambridge, MA, USA*
STACIA R. ENGEL · *Department of Genetics, Stanford University, Palo Alto, CA, USA*
KATHLEEN FALLS · *Biological Laboratories, Harvard University, Cambridge, MA, USA*
DAVID S. FASHENA · *The Zebrafish Information Network, University of Oregon, Eugene, OR, USA*
SILVIE FEXOVA · *Department of Physiology, Development and Neuroscience, University of Cambridge, Cambridge, UK*
MALCOLM E. FISHER · *Xenbase Curation Team, Division of Developmental Biology, Cincinnati Children's Hospital, Cincinnati, OH, USA*
JOSHUA FORTRIEDE · *Xenbase Curation Team, Division of Developmental Biology, Cincinnati Children's Hospital, Cincinnati, OH, USA*
PHANI GARAPATI · *Department of Physiology, Development and Neuroscience, University of Cambridge, Cambridge, UK*
SUSAN RUSSO GELBART · *Biological Laboratories, Harvard University, Cambridge, MA, USA*
OMID GHIASVAND · *Department of Biomedical Engineering, Medical College of Wisconsin and Marquette University, Milwaukee, WI, USA*
JOSH GOODMAN · *Department of Biology, Indiana University, Bloomington, IN, USA*
SIAN GRAMATES · *Biological Laboratories, Harvard University, Cambridge, MA, USA*
CHRISTIAN GROVE · *Division of Biology and Biological Engineering, California Institute of Technology, Pasadena, CA, USA*
GARY GRUMBLING · *Department of Biology, Indiana University, Bloomington, IN, USA*
DARREN E. HAGEN · *Division of Animal Sciences, University of Missouri, Columbia, MO, USA*
OMAR S. HARB · *Department of Biology, University of Pennsylvania, Philadelphia, PA, USA*
MIDORI A. HARRIS · *Department of Biochemistry, Cambridge Systems Biology Centre, University of Cambridge, Cambridge, UK*
TODD HARRIS · *Informatics and Bio-computing Platform, Ontario Institute for Cancer Research, Toronto, ON, Canada*
G. THOMAS HAYMAN · *Department of Biomedical Engineering, Medical College of Wisconsin and Marquette University, Milwaukee, WI, USA*
SAGE T. HELLERSTEDT · *Department of Genetics, Stanford University, Palo Alto, CA, USA*
VERONICA F. HINMAN · *Department of Biological Sciences, Mellon Institute, Carnegie Mellon University, Pittsburgh, PA, USA*

MATTHEW J. HOFFMAN · *Genomic Sciences and Precision Medicine Center, Medical College of Wisconsin, Milwaukee, WI, USA; Department of Physiology, Medical College of Wisconsin, Milwaukee, WI, USA*
DOUGLAS G. HOWE · *The Zebrafish Information Network, University of Oregon, Eugene, OR, USA*
KEVIN L. HOWE · *European Molecular Biology Laboratory, European Bioinformatics Institute (EMBL-EBI), Hinxton, Cambridge, UK*
CHRISTINA JAMES-ZORN · *Xenbase Curation Team, Division of Developmental Biology, Cincinnati Children's Hospital, Cincinnati, OH, USA*
TAMSIN JONES · *Department of Physiology, Development and Neuroscience, University of Cambridge, Cambridge, UK*
KAMRAN KARIMI · *Xenbase Development Team, Department of Computer Science, University of Calgary, Calgary, AB, Canada; Xenbase Development Team, Department of Biological Science, University of Calgary, Calgary, AB, Canada*
THOMAS KAUFMAN · *Department of Biology, Indiana University, Bloomington, IN, USA*
PAUL J. KERSEY · *European Molecular Biology Laboratory, European Bioinformatics Institute (EMBL-EBI), Hinxton, Cambridge, UK*
RANJANA KISHORE · *Division of Biology and Biological Engineering, California Institute of Technology, Pasadena, CA, USA*
JESSICA C. KISSINGER · *Center for Tropical and Emerging Global Diseases, University of Georgia, Athens, GA, USA; Institute of Bioinformatics, University of Georgia, Athens, GA, USA; Department of Genetics, University of Georgia, Athens, GA, USA*
MARTIN KOLLMAR · *Group Systems Biology of Motor Proteins, Department of NMR-Based Structural Biology, Max-Planck-Institute for Biophysical Chemistry, Göttingen, Germany*
OLIVIA W. LANG · *Department of Genetics, Stanford University, Palo Alto, CA, USA*
AOIFE LARKIN · *Department of Physiology, Development and Neuroscience, University of Cambridge, Cambridge, UK*
STANLEY J. F. LAULEDERKIND · *Department of Biomedical Engineering, Medical College of Wisconsin and Marquette University, Milwaukee, WI, USA*
MEIYEE LAW · *Mouse Genome Informatics, The Jackson Laboratory, Bar Harbor, ME, USA*
RAYMOND LEE · *Division of Biology and Biological Engineering, California Institute of Technology, Pasadena, CA, USA*
YU-YU LIN · *Graduate Institute of Biomedical Electronics and Bioinformatics, National Taiwan University, Taipei, Taiwan*
ANTONIA LOCK · *Department of Genetics, Evolution and Environment, University College London, London, UK*
VANEET LOTAY · *Xenbase Development Team, Department of Computer Science, University of Calgary, Calgary, AB, Canada; Xenbase Development Team, Department of Biological Science, University of Calgary, Calgary, AB, Canada*
JOHN MARTIN · *McDonnell Genome Institute, Washington University in St. Louis, St. Louis, MO, USA*
STEVEN J. MARYGOLD · *Department of Physiology, Development and Neuroscience, University of Cambridge, Cambridge, UK*
BEVERLEY MATTHEWS · *Biological Laboratories, Harvard University, Cambridge, MA, USA*
GILLIAN MILLBURN · *Department of Physiology, Development and Neuroscience, University of Cambridge, Cambridge, UK*

MAKEDONKA MITREVA · *McDonnell Genome Institute, Washington University in St. Louis, St. Louis, MO, USA; Division of Infectious Diseases, Department of Medicine, Washington University School of Medicine, St. Louis, MO, USA*
BENJAMIN MOORE · *European Molecular Biology Laboratory, European Bioinformatics Institute, Cambridge, UK*
ROBERT S. NASH · *Department of Genetics, Stanford University, Palo Alto, CA, USA*
VICTORIA NEWMAN · *European Molecular Biology Laboratory, European Bioinformatics Institute, Cambridge, UK*
MICHAEL PAULINI · *European Molecular Biology Laboratory, European Bioinformatics Institute, Cambridge, UK*
TROY PELLS · *Xenbase Development Team, Department of Computer Science, University of Calgary, Calgary, AB, Canada; Xenbase Development Team, Department of Biological Science, University of Calgary, Calgary, AB, Canada*
NORBERT PERRIMON · *Biological Laboratories, Harvard University, Cambridge, MA, USA*
EMILY PERRY · *European Molecular Biology Laboratory, European Bioinformatics Institute, Cambridge, UK*
VICTORIA PETRI · *Department of Biomedical Engineering, Medical College of Wisconsin and Marquette University, Milwaukee, WI, USA*
MONICA F. POELCHAU · *USDA, Agricultural Research Service, National Agricultural Library, Beltsville, MD, USA*
VIRGILIO PONFERRADA · *Xenbase Curation Team, Division of Developmental Biology, Cincinnati Children's Hospital, Cincinnati, OH, USA*
JEFF DE PONS · *Department of Biomedical Engineering, Medical College of Wisconsin and Marquette University, Milwaukee, WI, USA*
DANIELA RACITI · *Division of Biology and Biological Engineering, California Institute of Technology, Pasadena, CA, USA*
SRIDHAR RAMACHANDRAN · *The Zebrafish Information Network, University of Oregon, Eugene, OR, USA*
ALIX J. REY · *Department of Physiology, Development and Neuroscience, University of Cambridge, Cambridge, UK*
FAYE H. RODGERS · *Wellcome Trust Sanger Institute, Hinxton, Cambridge, UK*
DAVID S. ROOS · *Department of Biology, University of Pennsylvania, Philadelphia, PA, USA*
BRUCE A. ROSA · *McDonnell Genome Institute, Washington University in St. Louis, St. Louis, MO, USA*
KIM RUTHERFORD · *Department of Biochemistry, Cambridge Systems Biology Centre, University of Cambridge, Cambridge, UK*
LEYLA RUZICKA · *The Zebrafish Information Network, University of Oregon, Eugene, OR, USA*
GILBERTO DOS SANTOS · *Biological Laboratories, Harvard University, Cambridge, MA, USA*
ERIK SEGERDELL · *Xenbase Curation Team, Division of Developmental Biology, Cincinnati Children's Hospital, Cincinnati, OH, USA*
MYRIAM SHAFIE · *Wellcome Trust Sanger Institute, Hinxton, Cambridge, UK*
ACHCHUTHAN SHANMUGASUNDRAM · *Centre for Genomic Research, Institute of Integrative Biology, University of Liverpool, Liverpool, UK*

DAVID R. SHAW · *Mouse Genome Informatics, The Jackson Laboratory, Bar Harbor, ME, USA*
GAVIN SHERLOCK · *Department of Genetics, Stanford University, Stanford, CA, USA*
MARY SHIMOYAMA · *Department of Biomedical Engineering, Medical College of Wisconsin and Marquette University, Milwaukee, WI, USA*
FATIMA SILVA-FRANCO · *Centre for Genomic Research, Institute of Integrative Biology, University of Liverpool, Liverpool, UK*
DOMINIC SIMM · *Group Systems Biology of Motor Proteins, Department of NMR-Based Structural Biology, Max-Planck-Institute for Biophysical Chemistry, Göttingen, Germany; Theoretical Computer Science and Algorithmic Methods, Institute of Computer Science, Georg-August-University, Göttingen, Germany*
MAREK S. SKRZYPEK · *Department of Genetics, Stanford University, Stanford, CA, USA*
CERI E. VAN SLYKE · *The Zebrafish Information Network, University of Oregon, Eugene, OR, USA*
JENNIFER R. SMITH · *Department of Biomedical Engineering, Medical College of Wisconsin and Marquette University, Milwaukee, WI, USA*
HELEN SPARROW · *European Molecular Biology Laboratory, European Bioinformatics Institute, Cambridge, UK*
VICTOR STRELETS · *Department of Biology, Indiana University, Bloomington, IN, USA*
CHRISTOPHER TABONE · *Biological Laboratories, Harvard University, Cambridge, MA, USA*
ADITI TAYAL · *Division of Animal Sciences, University of Missouri, Columbia, MO, USA*
JYOTHI THOTA · *Department of Biomedical Engineering, Medical College of Wisconsin and Marquette University, Milwaukee, WI, USA*
JIM THURMOND · *Department of Biology, Indiana University, Bloomington, IN, USA*
VITOR TROVISCO · *Department of Physiology, Development and Neuroscience, University of Cambridge, Cambridge, UK*
MARY ANN TULI · *Division of Biology and Biological Engineering, California Institute of Technology, Pasadena, CA, USA*
MAREK A. TUTAJ · *Department of Biomedical Engineering, Medical College of Wisconsin and Marquette University, Milwaukee, WI, USA*
MONIKA TUTAJ · *Department of Biomedical Engineering, Medical College of Wisconsin and Marquette University, Milwaukee, WI, USA*
RAHUL TYAGI · *McDonnell Genome Institute, Washington University in St. Louis, St. Louis, MO, USA*
DEEPAK R. UNNI · *Division of Animal Sciences, University of Missouri, Columbia, MO, USA*
JOSE-MARIA URBANO · *Department of Physiology, Development and Neuroscience, University of Cambridge, Cambridge, UK*
PETER D. VIZE · *Xenbase Development Team, Department of Computer Science, University of Calgary, Calgary, AB, Canada*
DONG ZHUO WANG · *Xenbase Development Team, Department of Computer Science, University of Calgary, Calgary, AB, Canada; Xenbase Development Team, Department of Biological Science, University of Calgary, Calgary, AB, Canada*
SHUR-JEN WANG · *Department of Biomedical Engineering, Medical College of Wisconsin and Marquette University, Milwaukee, WI, USA*

YING WANG · *Xenbase Development Team, Department of Computer Science, University of Calgary, Calgary, AB, Canada; Xenbase Development Team, Department of Biological Science, University of Calgary, Calgary, AB, Canada*
SUSANNE WARRENFELTZ · *Center for Tropical and Emerging Global Diseases, University of Georgia, Athens, GA, USA*
MAGGIE WERNER-WASHBURNE · *Department of Biology, University of New Mexico, Albuquerque, NM, USA*
GARY WILLIAMS · *European Molecular Biology Laboratory, European Bioinformatics Institute, Cambridge, UK*
VALERIE WOOD · *Department of Biochemistry, Cambridge Systems Biology Centre, University of Cambridge, Cambridge, UK*
PINGLEI ZHOU · *Biological Laboratories, Harvard University, Cambridge, MA, USA*
AARON ZORN · *Xenbase Curation Team, Division of Developmental Biology, Cincinnati Children's Hospital, Cincinnati, OH, USA*
MARK ZYTKOVICZ · *Biological Laboratories, Harvard University, Cambridge, MA, USA*

Chapter 1

Identifying Sequenced Eukaryotic Genomes and Transcriptomes with diArk

Martin Kollmar and Dominic Simm

Abstract

The diArk Eukaryotic Genome Database is a manually curated and updated repository of available eukaryotic genome and transcriptome assemblies. diArk is a key resource for researchers interested in comparative eukaryotic genomics, and the entry point to browsing sequenced eukaryotes in general and to find the most closely related species to the own organism of interest in particular. The exponentially increasing number of sequenced species demands sophisticated search and data presentation tools. In this chapter we describe how to navigate the diArk database keeping a first-time user in mind.

Key words Eukaryotes, Sequenced genomes, Genome assembly, Transcriptome assembly

1 Introduction

The diArk Eukaryotic Genome Database (http://www.diark.org) was started in 2005 as a central repository for eukaryotes with genome assembly data available [1]. The species comprised human, the most widely used model organisms such as *Drosophila melanogaster*, *Caenorhabditis elegans*, and *Arabidopsis thaliana*, and multiple fungi. While dedicated databases were available for most of these species connecting the genomes with additional data and analyses, diArk filled the gap as central entry point allowing browsing through the sequenced eukaryotome and providing links to species databases and primary data repositories. Given the progress in sequencing strategy and technology development it was clear that genome assemblies will become available with increasing pace. Several genome sequencing initiatives had been started aiming to obtain genome assemblies of multiple related species. Large scale at that time meant the sequencing of 12 *Drosophila* species [2] or the sequencing of 29 low- and high-coverage mammalian genomes [3]. Soon after, size and speed of the projects increased by two

Martin Kollmar (ed.), *Eukaryotic Genomic Databases: Methods and Protocols*, Methods in Molecular Biology, vol. 1757,
https://doi.org/10.1007/978-1-4939-7737-6_1, © Springer Science+Business Media, LLC, part of Springer Nature 2018

orders of magnitude. Current large scale projects cover all major branches of eukaryotic life and new genomes are released on a daily basis. The most important ongoing whole-genome sequencing projects are the sequencing of vertebrates within Genome 10K [4], the sequencing of all 10,500 bird species within B10K [5], the sequencing of 5,000 arthropods (i5k) [6], the 959 Nematodes project [7, 8], the 1000 Fungal Genomes Project (1FKG; http://1000.fungalgenomes.org), and the sequencing of 1000 yeasts (y1000+; https://y1000plus.wei.wisc.edu/). In addition, transcriptomes have been sequenced and assembled from 1000 plants (1KP project) [9], more than 1200 insects (1KITE project; http://www.1kite.org), and hundreds of marine microorganisms within MMETSP (Marine Microbial Eukaryote Transcriptome Sequencing Project) [10]. Pilot data have been obtained for two further ambitious projects, the 1000 Plant & Animal Reference Genomes Project and Fish-T1K [11], but it is not clear whether these projects still continue. Most recently, plans to sequence the genomes of at least 10,000 plants have been announced (10KP) [12]. In addition to these massive sequencing efforts, there are dozens of small-scale sequencing projects dedicated to a few species, and hundreds of groups generate genome and transcriptome assemblies of just single species. Because most journals require only the raw sequencing data to be submitted to NCBI/ENA/DDBJ, genome and transcriptome assemblies and annotations are often stored at university servers or digital repositories.

At diArk, we manually track the availability of genome and transcriptome assemblies by browsing publications, genome databases of species and taxa from large-scale efforts and small research communities, and the NCBI genome assembly database [13]. While diArk includes almost all available genome assemblies, manually keeping track with the pace of transcriptome assembly generation is more difficult. In case of the 1KP and MMETSP projects hundreds of transcriptome assemblies were made available at once, and these assemblies have not yet been integrated into diArk. At diArk, we not only provide links to assembly data but also provide the genome and transcriptome data directly [14, 15]. For many species, updated and/or alternative genome assemblies are available. Alternative genome assemblies have been generated by using different software on the same sequencing read data, or by independent sequencing and assembling efforts. diArk provides metadata for each assembly such as date of generation, version, sequencing method, and assembly software used, and reports simple statistics such as number of entries (contigs, supercontigs, etc.), total sequence length, GC content, and N50. This chapter explains how to use diArk's search and filter tools, and how genome and transcriptome data are displayed.

2 Methods

2.1 diArk Website Navigation

diArk's home page offers several entry points to search for sequenced genomes and transcriptomes (Fig. 1). The main access to tools and information is via the dark-grey menu bar near the top of the page which is present on all diArk pages and facilitates quick navigation to the search tools, to extensive statistical analyses of the data, to help pages and other information. In addition to the menu items, the website provides two autocompletion search forms. An autocompletion search box, which can be added to the search plugins of the Firefox browser, is located on the right site of the menu bar and provides quick access to species summary pages (Fig. 1A; *see* Subheading 2.2). Right in the middle of the website is an autocompletion search field by which a search using the Fast Search functionalities is started (Fig. 1B; *see* Subheading 2.3). The home page also includes a quick view on a randomly selected species, a short list of the latest species and their genome/transcriptome assembly data added to diArk, a short list of the most recent publications about species' genome/transcriptome analyses, and a small box to the right with some metrics of diArk's content (Fig. 1C). More statistics can be obtained by clicking the info-icon in the header of this box, and by browsing the statistics subpage, which is accessible from the menu bar.

Because many species (especially fungi and yeasts) are known under multiple, synonymously used scientific names, and because many species were renamed recently because of better classification based on now available whole-genome data, we adhered to a few conventions from the beginning of the diArk project. First, diArk's reference scientific species names are the main species entries as given in the NCBI-Taxonomy database [16]. There is no notification by NCBI-Taxonomy, and thus we only adjust names as soon as we observe changes. Second, as an exception to this convention, we consistently use the fungi's teleomorph names as diArk's scientific reference names. Teleomorph and anamorph are the sexual and asexual states, respectively, of fungi, and both states were often given different scientific names in times of pure morphological classification. Many of these states were later shown to belong to the same holomorph (the whole fungus, including anamorph and teleomorph) but all scientific names remained in use. NCBI-Taxonomy does not have, to our knowledge, a convention on how to assign a main species name to these fungi. In most cases, the name of the most widely analysed state seems to be taken. Accordingly, in case the anamorph is the main experimental research state for a fungus, the corresponding genome/transcriptome assemblies at diArk refer to the less used teleomorph name. Third, different sequenced strains/breeds/cultivars/isolates of the same species (summarized as "strains" from here on) get different

Fig. 1 diArk home page. The dark-grey menu bar below the page header can be used to quickly navigate to search tools, statistical representations of sequenced genome/transcriptome data, help pages and more. (A) Use the autocompletion search box to access species summary pages. The menu bar and the search box are visible from all diArk pages. (B) Entering a search text in the autocompletion search form and selecting a species from the list of hits starts the Fast Search. (C) Some metrics on diArk's content are shown in the right box. More numbers will be shown upon clicking the info-icon

entries in diArk to facilitate distinguishing multiple genome/transcriptome assemblies of the same species. Sometimes, this information cannot be revealed and the respective assemblies are combined under a common species entry without strain information. These assemblies (and also multiple assemblies of identical strains) are numbered consecutively.

To facilitate the search for sequenced genomes/transcriptomes in case researchers only know a common name but not the scientific name (or one of the multiple used scientific names) the species are tagged by "alternative" and synonymously used scientific names and by common names, which are all fully available for the keyword search.

2.2 The Fastest Access to Sequenced Genome and Transcriptome Data

Species summary pages provide the most basic information about available genome and transcriptome assembly data. These pages are accessed by using the autocompletion search box in the menu bar which is present on all diArk pages (Fig. 1A). The query is automatically performed in all species name categories, diArk's main scientific names, alternative/synonymous scientific names, names of anamorphs, and common names. The search box can be added to the list of search plugins of the Firefox browser by clicking on the Firefox browser icon on the right site of the search box. This allows accessing diArk's species summary pages from anywhere.

1. To access a species summary page, first open the diArk home page (http://www.diark.org) (Fig. 1). Locate the autocompletion search box in the right top corner of the page.
2. To find a sequenced eukaryote of interest, start typing your query into the search box (e.g., type "dict"). The autocompletion function will list the first 10 hits for your query string in alphabetical order and summarize the remaining if more hits were found. Direct matches of the query string to diArk's main species names are shown in bold. In case the query string matched to a common name, a name of an anamorph or an alternative scientific name, diArk's main species name is shown in the list (e.g., type "dolphin," "chicken," or "duck").
3. Move the mouse to the name of your eukaryote of interest. For faster orientation, the currently activated species name is highlighted in yellow. Click on the name of the eukaryote of interest to open the species summary page (e.g., click on "Dictyostelium discoideum AX4").
4. On top of the species summary page is a header with diArk's main species name in bold and a brief taxonomic overview with the three most ancient and the three most recent taxa (Fig. 2A). The page lists species-related information such as full taxonomy, alternative and common species names, comments about specific strains sequenced, and provides links to NCBI taxonomy and Encyclopedia of Life pages (Fig. 2B). The project-related part of the page lists the type of data (genomic DNA or transcribed DNA) and the sequencing centers where the data have been obtained from (Fig. 2C). Via the links you can directly go to the sequencing centers' reference pages for the respective species. The projects also list and provide links to community-driven species home pages such as dictyBase, XenBase, or ZFIN. Using these links you can directly access more

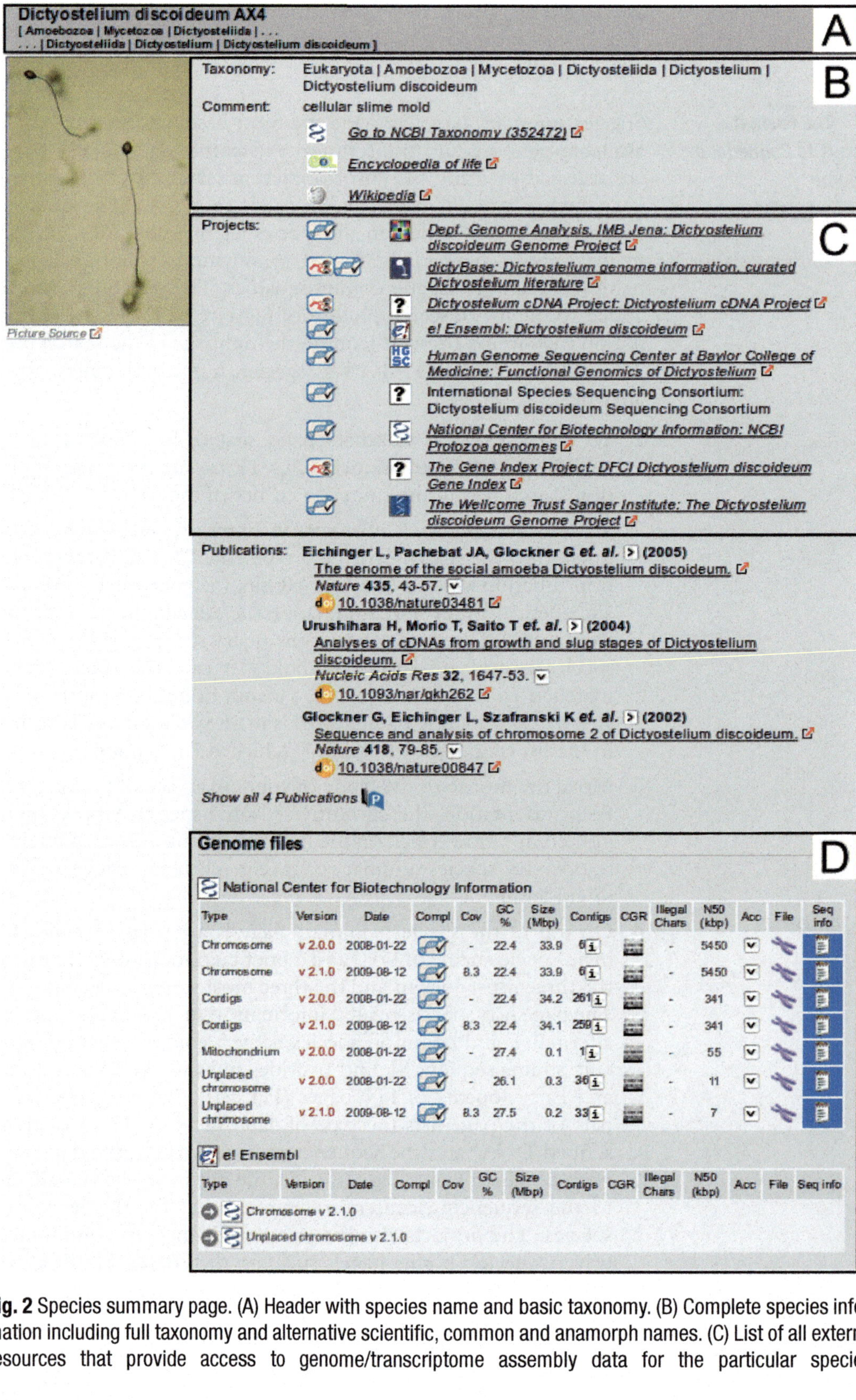

Fig. 2 Species summary page. (A) Header with species name and basic taxonomy. (B) Complete species information including full taxonomy and alternative scientific, common and anamorph names. (C) List of all external resources that provide access to genome/transcriptome assembly data for the particular species.

species-related data and analysis tools. If the genome and/or transcriptiome assemblies of the species have already been published, these data are listed in the Publications section.

5. The lower part of the species summary page contains the Genome files section which is sorted by projects (Fig. 2D). For each project the sequence data available is tabulated. The table provides a quick overview about assembly version, assembly release date, genome assembly completeness, sequencing coverage (if known), GC content, assembly size, contig number, presence of illegal characters in the data (not being g/G, a/A, t/T, c/C, or n/N), and the N50 of the dataset. These data allow an easy comparison if multiple versions of the same data, different assembly states (e.g., contigs vs. supercontigs vs. chromosome; Table 1), and alternative assemblies (indicated by "_assembly") are available. In addition, the Genome files tables contain icons that can be clicked for more detailed information. Click on the info-icon in the contigs column to view N50 (contigs sorted by lengths) and A50 (cumulated assembly length by contig number) plots (Fig. 3A). Clicking the Chaos Game Representation (CGR) fingerprint icon provides the CGR of the assembly and Frequency Chaos Game Representations (FCGRs) of the data in all resolutions up to 1024 × 1024 (Fig. 3B). Click the arrow-down button in the Acc column to see accession numbers, if available. The chromosome icon in the File column allows access to the assembly data in archived fasta format. Click the Seq info icon to get further information about the sequencing method (e.g., Sanger, Roche/454, Illumina, and PacBio) and the assembly software. The background color of the Seq info icon indicates the sequencing method with, for example, blue being used for Sanger, red for Roche/454, and yellow for Illumina sequencing.
6. In case multiple strains have been sequenced for the same species, only one can be selected from the search box list. The Genome files sections are collapsed if multiple species are shown, but can be opened by clicking *Show genomes*. If you want to get a quick overview about all sequenced strains, use the Taxonomy Search Module (*see* Subheading 2.4).

2.3 The Fast Search

The Fast Search is enabled by using the autocompletion search field in the middle of the home page or by selecting Search Database from the menu bar. The autocompletion function in the search

Fig. 2 (continued) Some of these external resources provide extensive additional data and multiple analyses tools. (D) List of assembly files available at diArk sorted by project. In case several projects host exactly the same data, these data are linked. The table contains multiple metrics for comparing different assemblies and evaluating the quality of each assembly. Multiple icons provide further detailed information and plots upon clicking

Table 1
Genome and transcriptome assembly file types

Chromosome	Every fasta-entry in this file is a chromosome. There are a few exceptions where chromosomes are split in two or more fasta-entries.
Uchromosome	These files contain contigs/supercontigs which could not be mapped to any (unknown chr.) or anchored (random chr.) to a certain chromosome.
Supercontigs	Every fasta-entry represents a supercontig which consists of sorted contigs separated by estimated or fixed numbers of "N" bases.
Usupercontigs	These files contain contigs which could not be mapped to supercontigs.
Ultracontigs	Every fasta-entry represents a number of supercontigs assembled to an ultracontig.
Contigs	Contigs (from "contiguous sequence") are the smallest pieces of an assembly and consist of overlapping sequence reads.
Ureads	These files contain the unplaced reads, reads that could not be assembled to contigs. These files are especially important for low-coverage genomes that in most cases end up with very short contigs. In these cases, small proteins or some exons can be reconstructed from the ureads-files.
Apicoplast	These files contain the apicoplast DNA. The apicoplast is a relict, nonphotosynthetic plastid found in Apicomplexa. It is proposed that it evolved via secondary endosymbiosis. The apicoplast is surrounded by four membranes within the outermost part of the endomembrane system.
Chloroplast	These files contain the chloroplast DNA. Chloroplasts are organelles found in plant cells and eukaryotic algae that conduct photosynthesis.
Kinetoplast	These files contain the kinetoplast DNA. A Kinetoplast is a disk-shaped mass of circular DNAs inside a large mitochondrion that contains many copies of the mitochondrial genome. Kinetoplasts are only found in protozoa of the class kinetoplastea. Kinetoplasts are usually adjacent to the organisms' flagellar basal body leading to the thought that they are tightly bound to the cytoskeleton.
Plastid	These files contain plastid DNA. Plastids are major double-membrane organelles found in the cells of plants, algae, and some other eukaryotic organisms.

(continued)

Table 1
(continued)

Mito	These files contain the mitochondrial DNA. Mitochondria are membrane-enclosed organelles found in most eukaryotic cells.
Nucleomorph	These files contain nucleomorph DNA. Nucleomorphs are small, vestigial eukaryotic nuclei found between the inner and outer pairs of membranes in certain plastids. So far, nucleomorphs have only been identified in cryptomonads, which belong to the Chromista supergroup, and in chlorarachniophytes, which are a subphylum of Rhizaria.
TSA	Transcriptome shotgun assembly.
_assembly	This extension is used to distinguish multiple assemblies of the same strain/breed/cultivar/isolate. Different assemblies can be based on using the same sequencing data but different assembly software and protocols, or are the result of sequencing and assembling the same strain/breed/cultivar/isolate multiple times.
_haploid	This extension to an assembly type indicates that the data (e.g., contigs, supercontigs) have been merged to generate an assembly representing a haploid state.
_diploid	This extension to an assembly type indicates that both alleles were assembled independently.

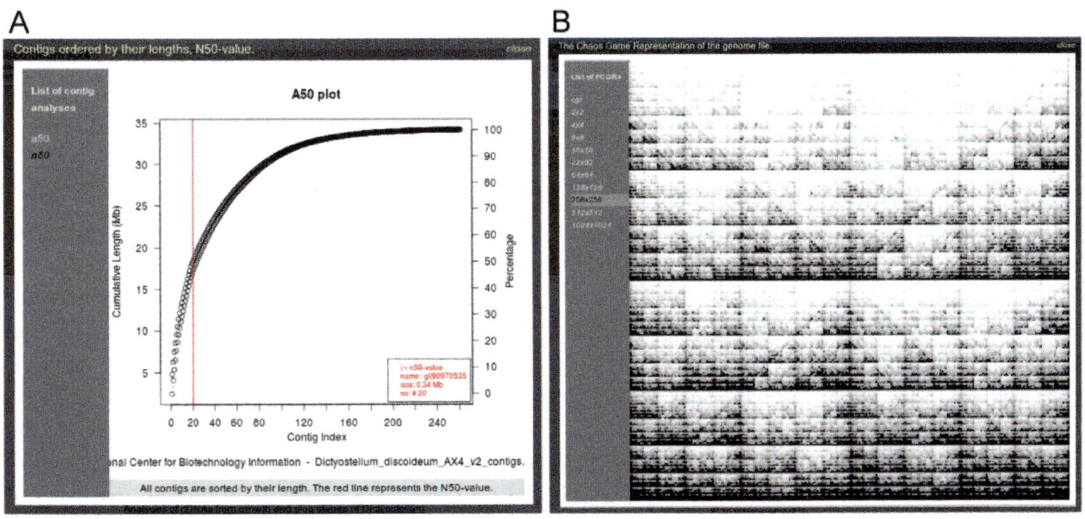

Fig. 3 Genome assembly analyses. (A) By clicking the info-icon in the *Contigs* column of the Genome files section of the species summary page, the user can inspect N50 and A50 plots of the assemblies. (B) A Chaos Game Representation (CGR) is a unique fingerprint for each genome dataset and the number of pixels in the graph is exactly the number of nucleotides in the dataset. Therefore, for comparing CGRs usually Frequency Chaos Game Representations (FCGRs) are taken, in which the pixels in a certain resolution-dependent section of the graph are summed

field works similar to that in the autocompletion search box in the menu bar (*see* Subheading 2.2). However, small pictures of the species, if available, allow easier identification of the species of interest, and additional information is provided for each species in terms of linked project pages and available genome assembly files. In contrast to the species summary pages, the Fast Search allows easy extension or filtering of the search space by selecting/deselecting species, and the Results tabs provide full information about species and assembly file related data.

1. Go to diArk's home page and enter "dict" in the autocompletion search field in the middle of the page.
2. Use the up- and down-keys and press enter, or use the mouse to select "**Dict**yostelium discoideum AX4 | Dd" and start the Fast Search.
3. The Fast Search contains a headline summarizing filter parameters (Fig. 4), which were preselected and combined from the extended Search Modules from the complex search (*see* Subheading 2.4). Below the headline are short sections separating filter for species names, several taxa and model organisms, and a few options to filter by sequencing type, completeness of sequencing, and availability of genome assembly files for download in diArk (Fig. 4A). The small exclamation mark icons next to model organism names indicate, that genome/transcriptome assemblies are available for multiple strains/breeds/cultivars/isolates and that only the most commonly used is selected (e.g., in the case of *Caenorhabditis elegans* the Bristol N2 strains, and for *Saccharomyces cerevisiae* the S288c strain). The Sequencing type filter allows selecting those species for which only EST, genome or RNAseq data or for which multiple data types are available.
4. Move further down the page to inspect the Search Results which are organized in multiple tabs (Fig. 4B). By default the **Species result** view is opened summarizing species related information, links to species sequencing centers and other resources with assembly data, and publications. This result tab is organized species by species. The **Projects result** view is organized by sequencing center allowing a resource-centric view on the data in case multiple species were selected. The **Publications result** view summarizes all publications related to the selected species. The **Genome Stats result** view provides a

Fig. 4 (continued) Below the species selection section there are three options to select certain sequencing types (genome, EST, transcriptome, and combinations of these), complete or incomplete genomes, and genome data available for download at diArk. (B) Data for the selected species can be browsed in seven Result tabs. By default, the Species result tab is opened which is identical to the species section on the species summary page (*see* Fig. 2B). Here, the References result tab is opened providing information about available analysis tools and data types at species home pages and repositories

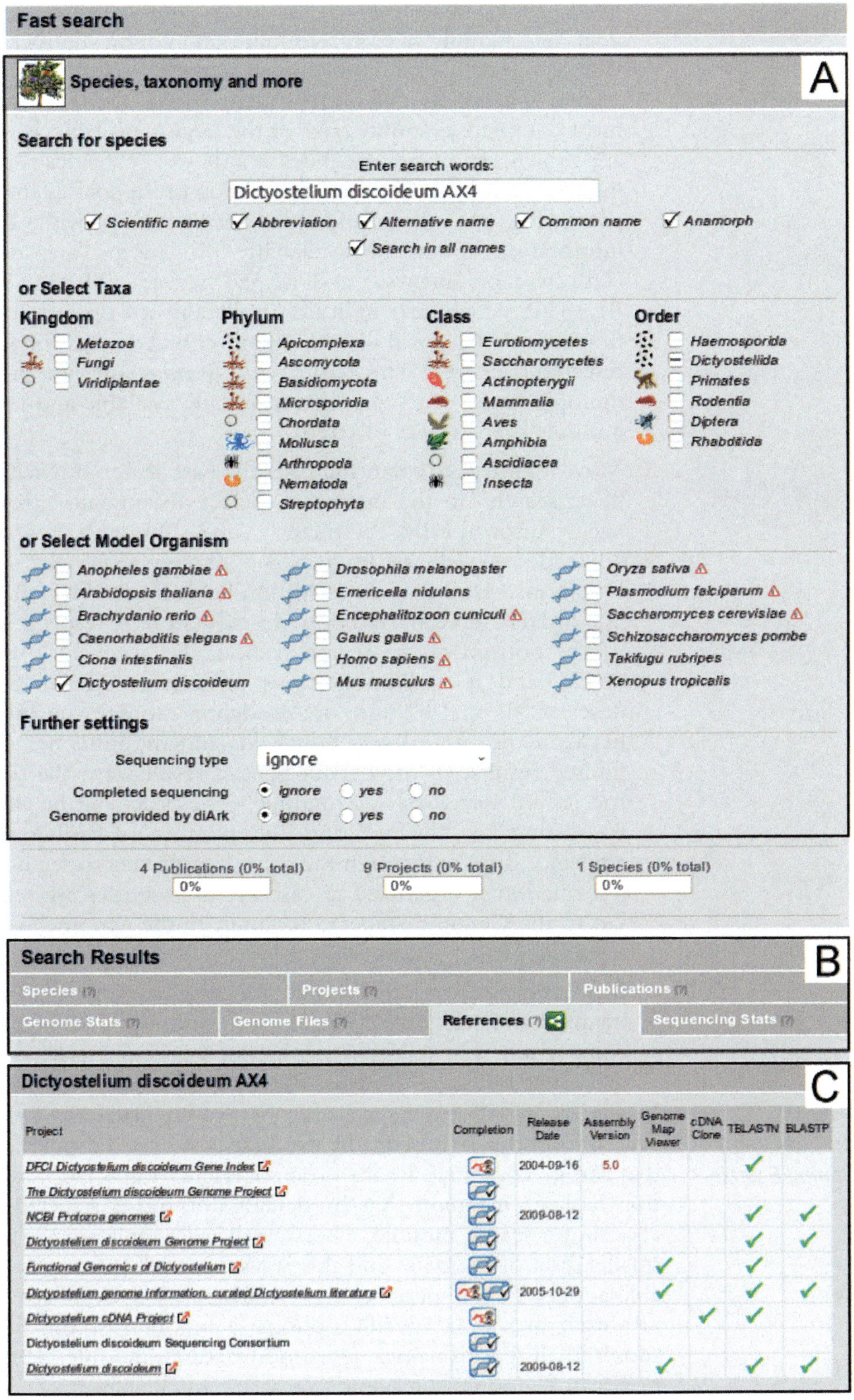

Fig. 4 The Fast Search. (A) The Fast Search offers a collection of the most commonly used search options. Species can be searched using an autocompletion search field and are selected by pressing the enter key or the left mouse button. Species and taxa can also be selected from the list of model organisms and taxa.

quick overview about chromosome numbers (if known), genome assembly size, GC content, and contig numbers. This view is especially useful for comparing multiple strains or closely related species. By default, the analysis data for the suspected most complete assembly (size of the largest available assembly) are shown. The **Genome Files result** view provides the same information as the Genome files section of the species summary pages (Fig. 2D). The **References result** view provides further information about the availability of genome map viewers (GBrowse or JBrowse) and BLAST servers (TBLASTN and BLASTP, which also indicates availability of protein annotations), and the possibility to obtain cDNA clones for further research (Fig. 4C). The **Sequencing Stats result** view provides multiple plots, which are obviously only available and useful if multiple species were selected.

5. Move up the page again and reset the Fast Search by clicking on "Fast search" in the menu bar. Select "Mammalia" from the taxa. Automatically, "Primates" and "Rodentia" are also selected, as well as human and mouse from the Model Organisms. Dashes in the selection fields from "Metazoa" and "Chordata" indicate selection of a subset of the respective taxa. At the bottom of the Search section, the number of selected species and related publications/projects are given. Select/deselect "Primates" and/or "Rodentia" to see the interplay between the various selection fields and the influence on the filtered results. In the default Species result view, the selected species are sorted by "Taxonomy." The order can be changed to sorting by "Name" and the number of visible species adjusted. Browse through the other Result tabs to see how the information is organized in case multiple species are selected. Go to the Genome Stats tab to compare the genome assembly sizes (Fig. 5).
6. Move up again to the Search for species input field, enter "danio" and select the "Brachydanio rerio str. Tuebingen" for comparison.

2.4 Complex Searches by Combining Search Modules

In the extended search, available by Search Database and then Search from the menu bar, the database is searched using modules that can be combined in any order. These modules work as selection baskets, meaning that by default nothing is selected when choosing a search module. Users are usually interested in only a small subset of the data, and this is faster to select from the available options than to deselect the rest. There are five different modules each providing specific options: a module for the full-text search in all species names, a taxonomy search module, a publication search module, a module to search sequencing project related data, and a module to filter by parameters related to genome/transcriptome assembly files. A search operation can consist of any combination of modules and their options. By adding

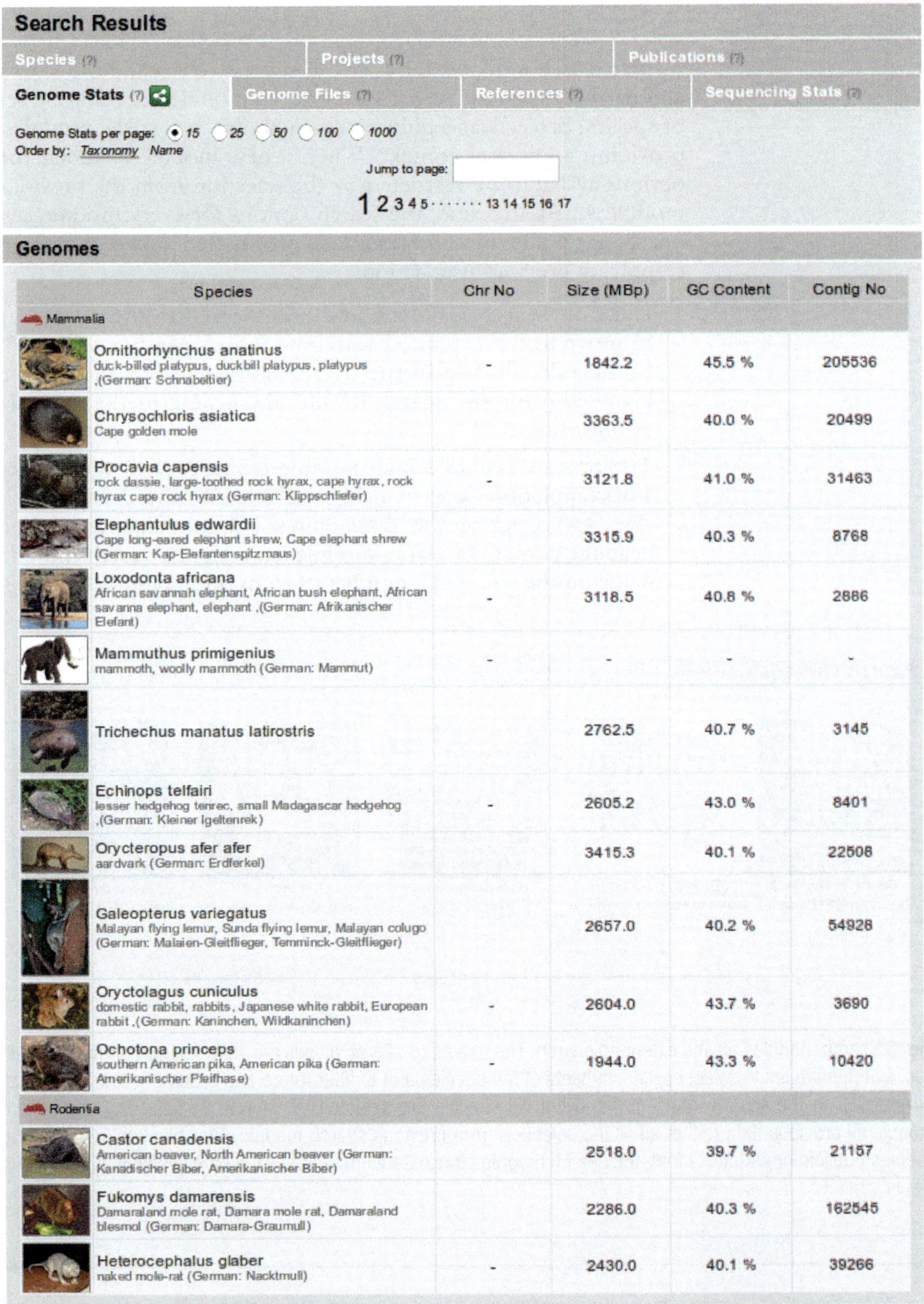

Species	Chr No	Size (MBp)	GC Content	Contig No
Mammalia				
Ornithorhynchus anatinus duck-billed platypus, duckbill platypus, platypus .(German: Schnabeltier)	-	1842.2	45.5 %	205536
Chrysochloris asiatica Cape golden mole		3363.5	40.0 %	20499
Procavia capensis rock dassie, large-toothed rock hyrax, cape hyrax, rock hyrax cape rock hyrax (German: Klippschliefer)	-	3121.8	41.0 %	31463
Elephantulus edwardii Cape long-eared elephant shrew, Cape elephant shrew (German: Kap-Elefantenspitzmaus)		3315.9	40.3 %	8768
Loxodonta africana African savannah elephant, African bush elephant, African savanna elephant, elephant .(German: Afrikanischer Elefant)	-	3118.5	40.8 %	2886
Mammuthus primigenius mammoth, woolly mammoth (German: Mammut)	-	-	-	-
Trichechus manatus latirostris		2762.5	40.7 %	3145
Echinops telfairi lesser hedgehog tenrec, small Madagascar hedgehog .(German: Kleiner Igeltenrek)	-	2605.2	43.0 %	8401
Orycteropus afer afer aardvark (German: Erdferkel)		3415.3	40.1 %	22508
Galeopterus variegatus Malayan flying lemur, Sunda flying lemur, Malayan colugo (German: Malaien-Gleitflieger, Temminck-Gleitflieger)		2657.0	40.2 %	54928
Oryctolagus cuniculus domestic rabbit, rabbits, Japanese white rabbit, European rabbit .(German: Kaninchen, Wildkaninchen)	-	2604.0	43.7 %	3690
Ochotona princeps southern American pika, American pika (German: Amerikanischer Pfeifhase)	-	1944.0	43.3 %	10420
Rodentia				
Castor canadensis American beaver, North American beaver (German: Kanadischer Biber, Amerikanischer Biber)		2518.0	39.7 %	21157
Fukomys damarensis Damaraland mole rat, Damara mole rat, Damaraland blesmol (German: Damara-Graumull)		2286.0	40.3 %	162545
Heterocephalus glaber naked mole-rat (German: Nacktmull)	-	2430.0	40.1 %	39266

Fig. 5 The Genome Stats result view. The Genome Stats result view shows the genome size and GC content for the selected species, here all mammals. To better evaluate these metrics the number of contigs for the underlying dataset is given. In most cases the sizes of the contig datasets are bigger than those of supercontigs and chromosomes because the latter do not contain those contigs that cannot be ordered into supercontigs or mapped to chromosomes

further search modules the user can successively refine the search and narrow down the result list. For each module the resulting set of species, projects and publications is shown below the modules, providing additional context. When a new module is added the options available are restricted by the selection from the previous module(s). At any time, the search options for every module can be changed and modifications are propagated down the chain reapplying previous user actions.

1. Click on Search Database and then Search in the menu bar to move to the extended search by Search Modules (Fig. 6). By default, all data in the database are selected and can be browsed using the Search Result tabs as described above (*see* Subheading 2.3).
2. The Species Names search module is identical to the species autocompletion search field available in the Fast Search (Fig. 4A). Click on the Taxonomy search module picture. The headline contains a search module icon and the search module name on the left, and a number of icons on the right (Fig. 7A).

Fig. 6 Search modules for the extended search. The extended search allows any combination of the available five search modules to select specific subsets of the species and to filter for certain assembly characteristics. When opening the search from the menu bar, all species are selected by default. By selecting any search module all previous data (either all or the selection from previous search modules) are blocked. By selecting species/publications/projects from the search modules these data immediately become available in the Result views

Fig. 7 (continued) contracting the module (arrow-up/arrow-down icon). (B) Section of the Taxonomy search module that allows selecting any combination of species from a full taxonomic tree. Here, "Dictyostelium discoideum AX4" was searched and selected in the autocompletion field, which opened the corresponding part of the tree up to the searched species. Species connected to internal nodes are automatically shown. The tree can be expanded and contracted using the arrow-down/arrow-up icons, and species and taxa are selected using the check boxes

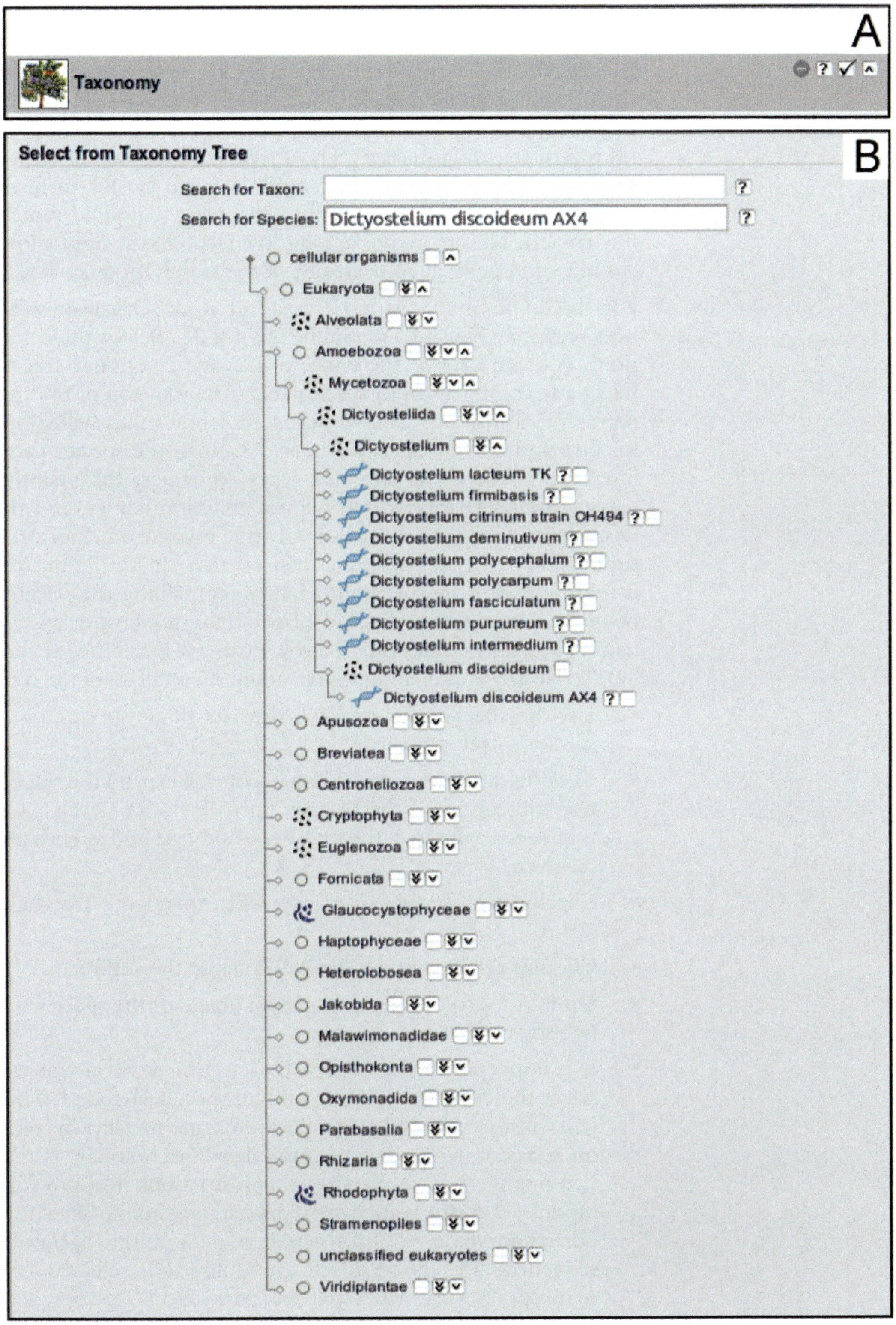

Fig. 7 The Taxonomy search module. (A) The header of the Taxonomy search module. The icons in the right top corner of the header are available in all search modules and allow removing the module (minus icon), getting explanations for the search options (help icon), activating/deactivating the module (check box), and expanding/

By clicking the minus icon, the respective search module and the respective applied search filter can be removed. The question mark icon allows opening a short help as separate window on top of the webinterface. The checkbox allows deselecting/selecting the respective search module from the list of filters while the selected options within the search module remain unchanged. The arrow-up/arrow-down icon to the right allows closing and opening, respectively, of the search module view.

3. You should be familiar with the Taxa and Model Organism selection sections (Fig. 4A; *see* Subheading 2.3). Below these sections, you can browse the entire eukaryotic taxonomic tree to select any combination of species (Fig. 7B). On top of the tree representation, there are two autocompletion input fields, one for taxa and one for species names. As usual, the species name input field accepts all different types of names, the scientific names, the alternatively used names, common names, and the anamorphs, while it selects the scientific reference name. After submission of the taxon/species name search, the taxonomy tree is reloaded with the specific branch(es) containing the selected taxon/species opened. Instead of searching for taxa or species names, taxa and species can be browsed and selected by expanding/collapsing and including/excluding subsections of the tree.
   - Use the checkboxes to select/deselect single species, or all species within a taxon.
   - Clicking the double arrow-down icon will expand the respective subsection of the tree by up to 5 nodes (levels). On mouse-over the icon the number of subtaxa and species will be given.
   - Clicking the arrow-down icon will expand the tree by 1 taxon.
   - Clicking the arrow-up icon will collapse the subtree.
   - On mouse-over a species name an image of the species will be shown.
   - It is important to know that by selecting a taxon you will select the complete subtree and all species included. Thus you will also select all taxa/species that are not shown based on restrictions in previous modules. The rationale is that you might want to change your restrictions in a previous module but still want to keep your taxonomic selection. For example, if you restricted your search to genomic sequences in the Projects module and selected the Arthropoda you will only see Arthropoda species with sequenced genomes. If you then change your selection in the Projects module to also include the cDNA/EST projects, your species selection will automatically expand to include all Arthropoda for which genomic sequences and/or cDNA/EST data are available.

4. Go to the header of the Taxonomy search module and click the minus icon to remove it from the current search, or click the Search menu item to reset the search. From the Search Modules select the Publications search module. The publications related to the species and sequencing projects can be searched in several ways. Full-text searches are provided for titles, authors, abstracts, and journals. In addition, the journal search input field is supplemented with an autocompletion function. Searches can be restricted by dates. By default, searches are unrestricted and include the full time span of publication dates. The option to select "All Publications" is useful as filter for selecting only species with published genome/transcriptome because unpublished genome data are often under embargo and should be used with respect to data sharing policies [17].
5. Remove the Publications search module and select the Projects search module. Here, you can select all species sequenced by a certain sequencing center or available from a specific species home page. The selection of species can be filtered by Sequencing type (*see* Subheading 2.3) and by analysis tools available at species home pages and data repositories, e.g., only those data resources can be selected which provide access to a BLASTP server or which allow ordering cDNA clones. Further down in the search module, there is a section to select one or more specific references. Reference is referred to sequencing centers, data repositories, and species home pages, in diArk. For example, the Joint Genome Institute is a sequencing center (= *reference* in diArk) and hosts many species-dedicated websites (=*projects* in diArk). Selection of a certain option in the table at the top of the module will automatically disable those references that do not provide the selected data and tools. If you are particularly interested in for example filtering for a certain Sequencing type, select the respective type and select "All Projects." Keep in mind that all search modules work as selection baskets. Thus, whatever filter you apply from the table at the top of the module, you have to "select" one (or all) of the references to include any data.
6. Reset the search, or close the Projects search module, and click on the Genome Files search module picture (Fig. 8). The Genome File search module allows filtering for assembly-related metadata and metrics. By default, all assemblies are included. By restricting the time span of the assembly release dates, for example, only assemblies of the current year can be selected. For several assemblies, release dates could not be revealed, and these assemblies can either be included or excluded in total. The data can further be filtered by assembly completeness, presence or absence of illegal characters, sequencing coverage, GC-content,

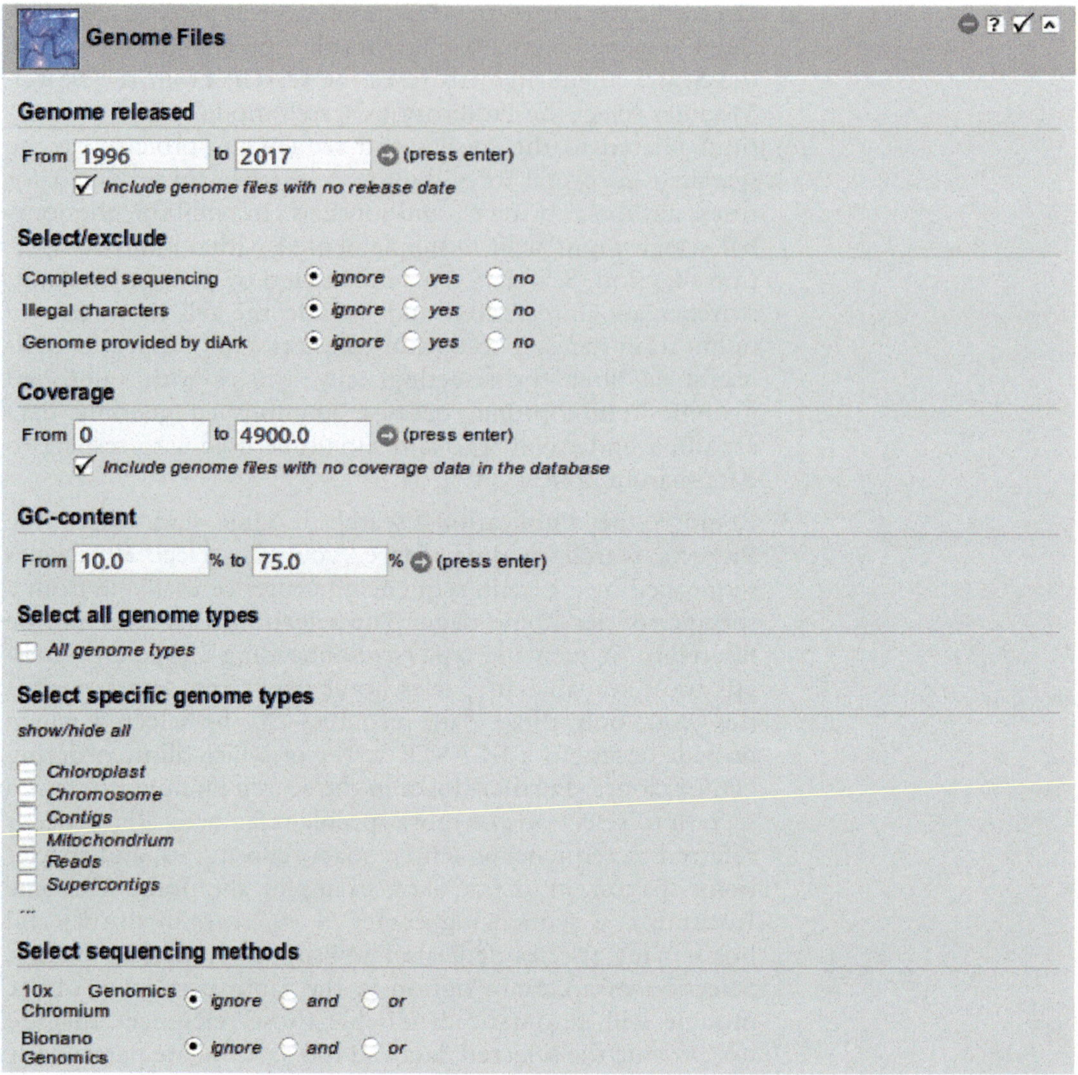

Fig. 8 The Genome Files search module. The Genome Files search module allows filtering the genome/transcriptome data for release dates, complete/incomplete sequencing, sequencing coverage, and GC content. These parameters can only restrict the space of selectable genome/transcriptome assembly files. Thus, either "all genome types" or at least one of the listed genome types need to be selected to inspect the respective genome/transcriptome assembly data via the Result views

genome type, sequencing method, and assembly software. In the Genome File search module, selection of any data is done via choosing one or more genome types. Thus, whatever filter you apply from the other options, you have to "select" one (or all) of the genome types.

7. Browse through the result tabs as explained above (*see* Subheading 2.3).

References

1. Odronitz F, Hellkamp M, Kollmar M (2007) diArk--a resource for eukaryotic genome research. BMC Genomics 8:103. https://doi.org/10.1186/1471-2164-8-103
2. Clark AG, Eisen MB, Smith DR et al (2007) Evolution of genes and genomes on the drosophila phylogeny. Nature 450:203–218. https://doi.org/10.1038/nature06341
3. Lindblad-Toh K, Garber M, Zuk O et al (2011) A high-resolution map of human evolutionary constraint using 29 mammals. Nature 478:476–482. https://doi.org/10.1038/nature10530
4. Genome 10K Community of Scientists (2009) Genome 10K: a proposal to obtain whole-genome sequence for 10 000 vertebrate species. J Hered 100:659–674. https://doi.org/10.1093/jhered/esp086
5. Zhang G, Rahbek C, Graves GR et al (2015) Genomics: bird sequencing project takes off. Nature 522:34. https://doi.org/10.1038/522034d
6. i5K Consortium (2013) The i5K initiative: advancing arthropod genomics for knowledge, human health, agriculture, and the environment. J Hered 104:595–600. https://doi.org/10.1093/jhered/est050
7. Kumar S, Schiffer PH, Blaxter M (2012) 959 nematode genomes: a semantic wiki for coordinating sequencing projects. Nucleic Acids Res 40:D1295–D1300. https://doi.org/10.1093/nar/gkr826
8. Kumar S, Koutsovoulos G, Kaur G, Blaxter M (2012) Toward 959 nematode genomes. WormBook 1:42–50. https://doi.org/10.4161/worm.19046
9. Matasci N, Hung L-H, Yan Z et al (2014) Data access for the 1,000 plants (1KP) project. GigaScience 3:17. https://doi.org/10.1186/2047-217X-3-17
10. Keeling PJ, Burki F, Wilcox HM et al (2014) The marine microbial eukaryote Transcriptome sequencing project (MMETSP): illuminating the functional diversity of eukaryotic life in the oceans through transcriptome sequencing. PLoS Biol 12:e1001889. https://doi.org/10.1371/journal.pbio.1001889
11. Sun Y, Huang Y, Li X et al (2016) Fish-T1K (Transcriptomes of 1,000 fishes) project: large-scale transcriptome data for fish evolution studies. GigaScience 5:18. https://doi.org/10.1186/s13742-016-0124-7
12. Normile D (2017) Plant scientists plan massive effort to sequence 10,000 genomes. In: Sci. AAAS. http://www.sciencemag.org/news/2017/07/plant-scientists-plan-massive-effort-sequence-10000-genomes. Accessed 28 Aug 2017
13. Kitts PA, Church DM, Thibaud-Nissen F et al (2016) Assembly: a resource for assembled genomes at NCBI. Nucleic Acids Res 44:D73–D80. https://doi.org/10.1093/nar/gkv1226
14. Hammesfahr B, Odronitz F, Hellkamp M, Kollmar M (2011) diArk 2.0 provides detailed analyses of the ever increasing eukaryotic genome sequencing data. BMC Res Notes 4:338. https://doi.org/10.1186/1756-0500-4-338
15. Kollmar M, Kollmar L, Hammesfahr B, Simm D (2015) diArk – the database for eukaryotic genome and transcriptome assemblies in 2014. Nucleic Acids Res 43:D1107–D1112. https://doi.org/10.1093/nar/gku990
16. Federhen S (2012) The NCBI taxonomy database. Nucleic Acids Res 40:D136–D143. https://doi.org/10.1093/nar/gkr1178
17. Kaye J, Heeney C, Hawkins N et al (2009) Data sharing in genomics — re-shaping scientific practice. Nat Rev Genet 10:331–335. https://doi.org/10.1038/nrg2573

Chapter 2

An Introduction to the Saccharomyces Genome Database (SGD)

Olivia W. Lang, Robert S. Nash, Sage T. Hellerstedt, Stacia R. Engel, and The SGD Project

Abstract

The *Saccharomyces* Genome Database (SGD) is a well-established, key resource for researchers studying *Saccharomyces cerevisiae*. In addition to updating and maintaining the official genomic sequence of this highly studied organism, SGD provides integrated data regarding gene functions and phenotypes, which are extracted from the published literature. The vast amount and variety of data housed in the database can prove challenging to navigate for the first-time user. Therefore, this chapter serves as an introduction describing how to search the database in order to discover new information. We introduce the different types of pages on the website, and describe how to manipulate the tables and diagrams therein to display, download, or analyze the data using various SGD tools.

Key words *Saccharomyces cerevisiae*, Genome database, Phenotype, Gene ontology, Yeast

1 Introduction

The *Saccharomyces* Genome Database (SGD; www.yeastgenome.org) is a public, encyclopedic resource for the budding yeast *Saccharomyces cerevisiae*. As the official keepers for the genome sequence and gene nomenclature of yeast, SGD is a core component of the yeast research community and an essential tool for experimental design and data analysis [1]. The SGD project extracts function and gene information from published literature through a process of human curation, and pools these facts with data from other repositories to create a resource for researchers that integrates sequence data with evidence-based annotations. It is this core activity of manual curation that is essential for converting the

The members of the SGD Project are listed in the acknowledgements section.

Martin Kollmar (ed.), *Eukaryotic Genomic Databases: Methods and Protocols*, Methods in Molecular Biology, vol. 1757, https://doi.org/10.1007/978-1-4939-7737-6_2, © Springer Science+Business Media, LLC, part of Springer Nature 2018

vast amount of published data into a coherent body of knowledge. All information in SGD can be traced back to its original source so that users can review the experiment details and interpretations from the authors.

The tools within SGD offer different ways for users to make both specific and broad queries. YeastMine offers a way of performing bulk queries of information from the database [2, 3] while SPELL searches SGD's curated expression datasets for genes with similar expression profiles [4]. Other tools available for exploring data at SGD include GO Term Finder [5, 6], GO Slim Mapper [5], YeastPathways [3], JBrowse genome browser [7, 8], and Variant Viewer [9]. Since users primarily use SGD as an encyclopedia for yeast genes, this chapter focuses on how to access the data at SGD using the website search and gene pages.

The majority of the manual curation efforts at SGD use controlled vocabularies to capture the main research findings from the scientific literature. SGD primarily uses the Gene Ontology (GO; www.geneontology.org; [10]) and the Ascomycete Phenotype Ontology (APO; [11]) to capture functional information about yeast genes. Three categories of Gene Ontology terms capture information on the function of a gene product: Molecular Function, Biological Process, and Cellular Component.

Many SGD users primarily use SGD to look up specific yeast genes' roles and functions. In order to successfully navigate the website and perform fruitful searches, it is important to have a basic understanding of how data are captured and displayed on the website. This chapter will explain how to use the website's search tool, and also how data are displayed in tables and network diagrams in pages across the SGD site (*see* **Note 1**).

2 Methods

2.1 Exploring SGD Through a Search Query: Finding a Gene Page

A great way to start exploring the kind of information captured by SGD is to learn to use SGD's website search tool. The search includes many features that help you narrow your results to find the information you are looking for. For this example, we will step through how to find cell-cycle-related protein-coding genes.

2.1.1 Navigating from SGD's Home Page

Open the SGD home page (www.yeastgenome.org) and browse the purple menu bar near the top of the page to find links to SGD tools and resources like the Genome Browser and SGD's instance of YeastMine (labeled "Gene Lists" in the Analyze section). The home page also includes a slideshow of images provided by members of the community, a list of upcoming meetings and conferences, and a news section for blog posts and announcements (Fig. 1).

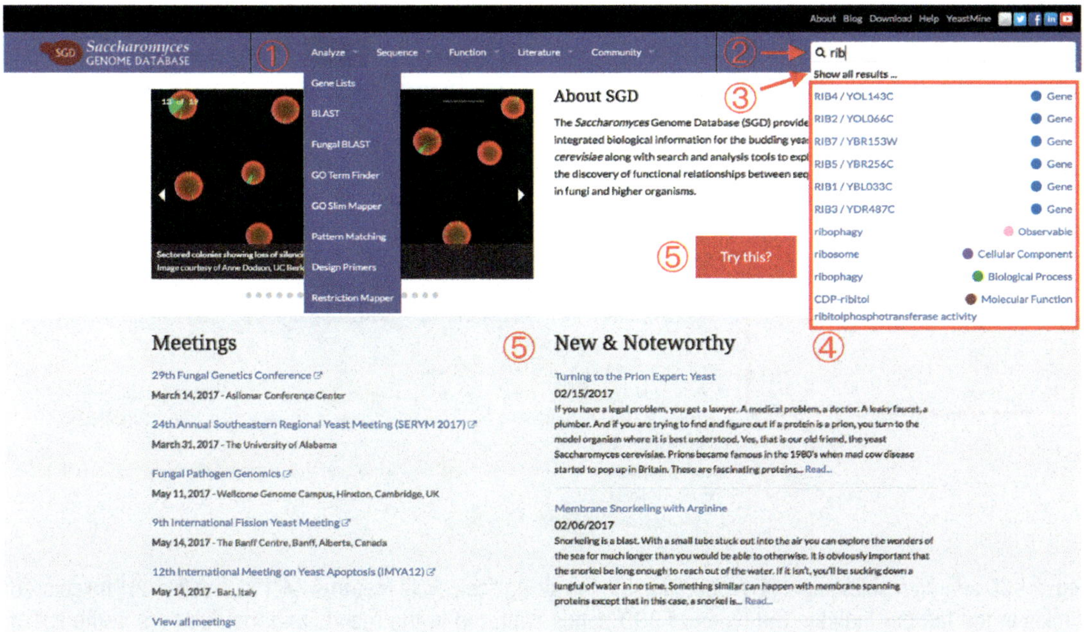

Fig. 1 SGD home page. (1) Browse the purple menu bar for links to SGD tools and external resources. (2) Type your query into the search box. (3) Clicking "Show all results" will show all results matching the search string and will not take you directly to an SGD page. (4) For quick-links that send you directly to SGD pages, select one of the options in the quick results box. (5) Not sure what to search for? Then click the "Try this?" button to get the results from one of a set of selected biological terms and phrases

2.1.2 Enter Search

Enter a search query, like "cell cycle," into the search box within the purple bar. As you type, you will notice results showing up below the search box. Clicking on one of these results will send you directly to the Gene/Term/Reference page. Keep in mind that if your search query is a unique gene name, PubMed Identifier, or Gene Ontology Identifier (prefixed with "GO:"), you will be sent directly to the corresponding gene, reference, or GO term page. If you do not want to be sent directly to an SGD data page, click on "Show all results ..." which pops up with the autofill options when you are typing in the search box (*see* **Note 2**).

2.1.3 Search Results

Filter your results to the data type you are looking for by selecting one of the categories on the left-hand side (Fig. 2). The number of results for a category is displayed next to the category name and colored circle. For this example, select the blue "Gene" category.

2.1.4 Drilling Down

After selecting a category, the left bar will change to display further filter options. For the Genes category, you can filter by "Feature Type" or by associated Gene Ontology terms. Five of the filter values are displayed by default, but you can click "Show more" and "Show all" to see more options. For this example, select the "ORF" feature type to show only the protein-coding genes (*see* **Note 3**).

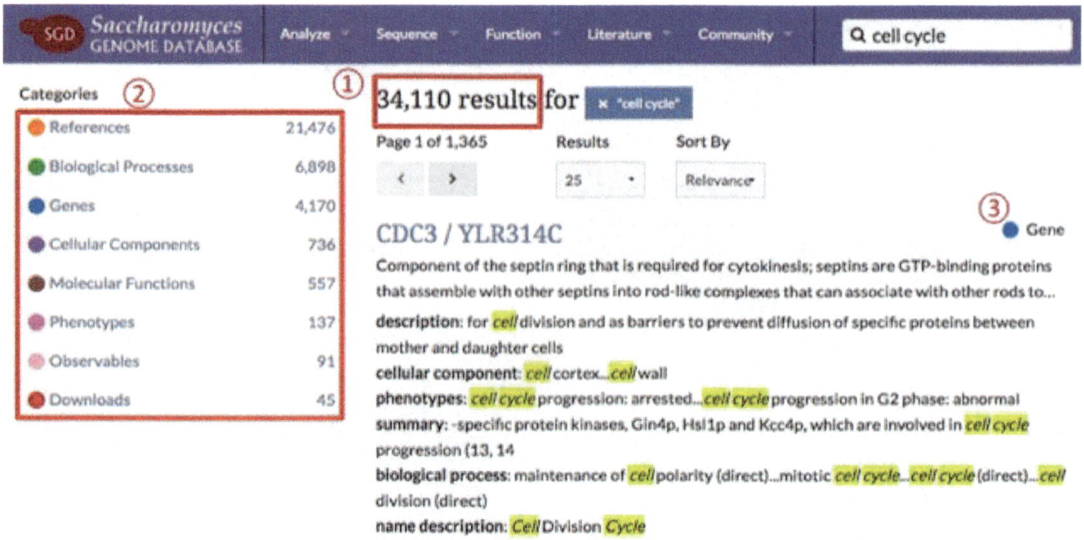

Fig. 2 SGD search results page. (1) A search using the string "cell cycle" returns 34,110 results. (2) The colored circles in the left bar indicate the types of SGD pages captured in the results and the numbers in the list of categories on the left-side bar indicate the number of results from each category. (3) Each result is an SGD page with the type of page indicated at the right and highlighted text showing how it relates to the search query

2.1.5 Editing Your Search

It is possible to edit your search by removing filters in a search query. As you progress through a search, blue rectangles are added to the top of the results page with text relating to the filter applied. These "breadcrumbs" can show you the search path to get to the current results page and by clicking the "x" you can remove the specific breadcrumb. For this example, you could remove the "ORF" breadcrumb to retrieve all the genes relating to the cell cycle.

2.1.6 Wrapped View

To download the list of genes from this search query, click the "Wrapped" button next to the highlighted "List" button. When the page is fully loaded, you will be able to download the list of genes by clicking the "Download" button next to the "Analyze" button. The function of the "Analyze" button will be explained in the section below (Fig. 3).

2.1.7 Saving Search Results in a URL

If you wish to save your search results for later, just copy and save the URL in your browser when you are on the search results page. This is especially useful for sharing search results with colleagues or bookmarking your results.

2.1.8 Selecting a Gene

To enter the SGD locus summary page, click the blue name of the gene either in the "List" or "Wrapped" view (*see* **Note 4**).

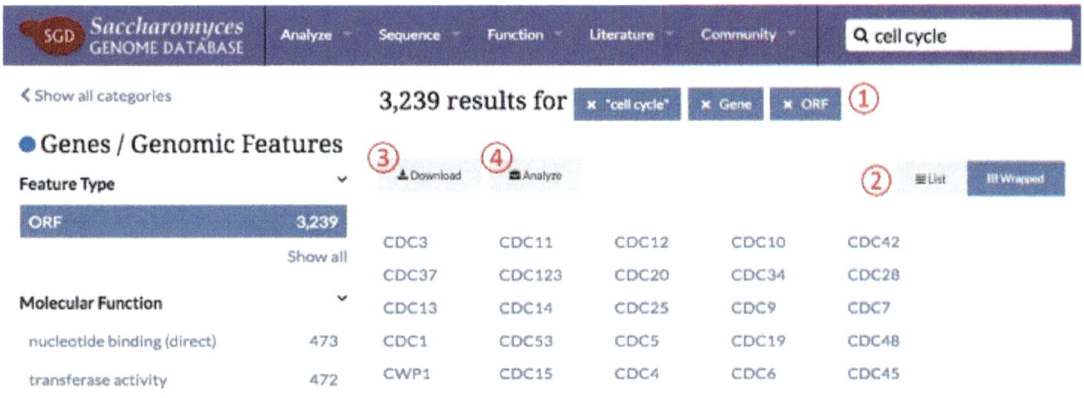

Fig. 3 With the selection of the "ORF" filter, the results of your query narrowed down to 3239 results. (1) Note the blue rectangles that appear near the number of search results with the addition of each new filter. These are called "breadcrumbs" because you can use them to trace back the filters applied to your query. You can also remove filters by clicking the "x" for each breadcrumb you wish to remove. (2) The results are displayed using the "Wrapped" view here. The "Wrapped" view includes the options of (3) downloading the results as a text file, or (4) sending the list to the "Analyze" tool

2.2 Annotation Tables at SGD

SGD displays much of its data in tabular form on various pages of the website. Tables are used to display data including ontology annotations, gene interactions, posttranslational modifications, and protein abundance. In this section, we will be exploring the annotations on the Gene Ontology Term page for "GTP-binding" (GO:0005525).

2.2.1 Navigating to the Gene Ontology Term Page

From the SGD home page, start typing in "GTP-binding" until you see the Molecular Function term pop-up in the list of quick-results. Click on the result to go straight to the GO term page (*see* **Note 5**).

2.2.2 Navigating to the GO Annotations Table

Scroll to the "Annotations" table by using the left-hand vertical menu or by scrolling past the term overview with description and past the figure of related ontology terms. Next to the "Annotations" header is a count of the number of annotations and the number of associated genes.

2.2.3 Reading GO Annotations

Each row represents an annotation that associates a gene to the "GTP-binding" molecular function term. For example, the first entry of this table was curated by SGD on March 23, 2011 to say: the study by Shin et al. (2011) [12] uses a direct assay to demonstrate that the FUN12 gene interacts selectively with guanosine triphosphate. If you need more information, the blue text within the tables often links to other SGD pages. Click on a gene if you wish to go to the SGD Locus Summary Page for that gene (Fig. 4).

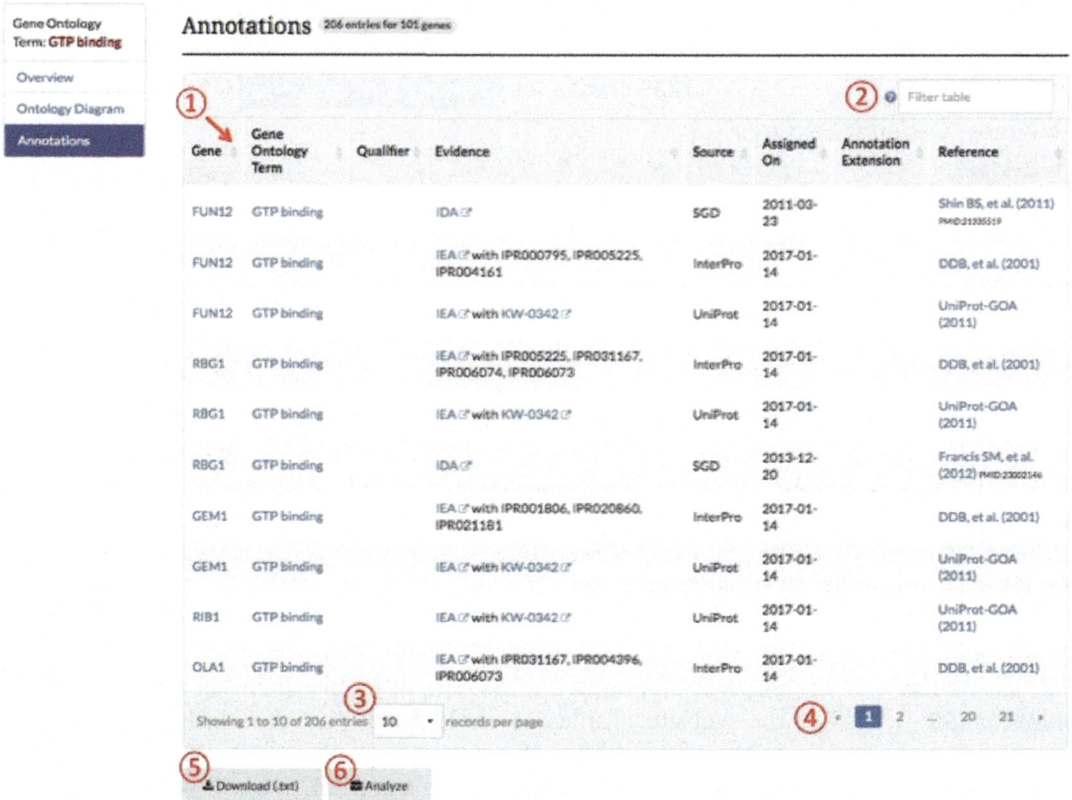

Gene	Gene Ontology Term	Qualifier	Evidence	Source	Assigned On	Annotation Extension	Reference
FUN12	GTP binding		IDA	SGD	2011-03-23		Shin BS, et al. (2011) PMID:21335519
FUN12	GTP binding		IEA with IPR000795, IPR005225, IPR004161	InterPro	2017-01-14		DDB, et al. (2001)
FUN12	GTP binding		IEA with KW-0342	UniProt	2017-01-14		UniProt-GOA (2011)
RBG1	GTP binding		IEA with IPR005225, IPR031167, IPR006074, IPR006073	InterPro	2017-01-14		DDB, et al. (2001)
RBG1	GTP binding		IEA with KW-0342	UniProt	2017-01-14		UniProt-GOA (2011)
RBG1	GTP binding		IDA	SGD	2013-12-20		Francis SM, et al. (2012) PMID:23002546
GEM1	GTP binding		IEA with IPR001806, IPR020860, IPR021181	InterPro	2017-01-14		DDB, et al. (2001)
GEM1	GTP binding		IEA with KW-0342	UniProt	2017-01-14		UniProt-GOA (2011)
RIB1	GTP binding		IEA with KW-0342	UniProt	2017-01-14		UniProt-GOA (2011)
OLA1	GTP binding		IEA with IPR031167, IPR004396, IPR006073	InterPro	2017-01-14		DDB, et al. (2001)

Fig. 4 An SGD data table. (1) Click the up/down arrows next to the column header to sort the table by the values within a given column (2) By typing into the "Filter table" search, the table will only display entries that have matching text. You can see more annotations by (3) displaying more annotations per page or (4) flipping through the pages of annotations. (5) The "Download" button saves the contents of the table into a text file, while the (6) "Analyze" button leads to the Analyze page where you can send the list of genes into one of SGD's tools

2.2.4 Sort Feature for the Table

You can sort the table by values within each of the header sections, like "Gene" or "Source," by clicking on the up and down arrows within the header boxes. This is commonly used to see which genes have been annotated with the GO term and how many of those annotations have been made.

2.2.5 Search the Table

The tables at SGD also have a search feature in the upper right side of the table. Click the blue circle with the "?" for help information on how to use it. Typing anything within the search box will filter the table to show only annotations that contain the matching text anywhere within the entry.

2.2.6 Downloading Annotations

Most tables at SGD have a "Download (.txt)" button where you can download the current contents of the table in the order that they are being displayed. For example, if you sorted the entries by

"Source" and entered text within the search box to filter the annotations displayed, the downloaded file will save the subset of features selected by the search all ordered by source.

2.2.7 Analyze Toolbox

Clicking on the "Analyze" toolbox below the table will send you to the analyze page where you can use the selected genes from the table as input for several of SGD's tools. You can send the list of genes to the "GO Term Finder" tool, the "GO Slim Mapper" tool, the gene expression tool, "SPELL," or the popular "YeastMine" tool (*see* **Note 1**).

2.3 Network Diagrams at SGD

Some SGD pages have interactive Network Diagrams that offer a visual approach to studying the data. These diagrams are used on several different pages (Gene Ontology, Phenotype, Interaction, Regulation, and Literature), and also for displaying the ontologies on GO term and APO term pages.

2.3.1 Go to a Protein Page

Go to the SGD gene page of your favorite gene and select the "Protein" tab at the top. The first section includes overview information about the protein product and the subsequent "Experimental Data" section includes information from published datasets.

2.3.2 Protein "Domains and Classification" Section

Scroll to the "Domains and Classifications" section to view the table, the "Domain Locations" diagram, and the "Shared Domains" network diagram. For more information on any of these figures or tables, click the blue "i" icon for details on how the data were generated or how the data are displayed (Fig. 5).

2.3.3 "Shared Domains" Network Section

Scroll to the "Shared Domains" network diagram and click on the blue "i" icon for information about the network and the blue "?" for information on how to interact with the network. You can find these blue icons all across the SGD website pages next to headers and within tables. Clicking on these icons will give you important information about how the data at SGD is captured and displayed. The date of the last update to the diagram is displayed in the network diagram on the lower right (Fig. 5).

2.3.4 Clicking on the Nodes

Clicking on the nodes will send you to an SGD page. There are two kinds of nodes: domain nodes (square) that send you to SGD domain pages, and gene nodes (round) that send you to the SGD locus summary page for the gene. The domain nodes are connected to genes that contain that domain. The gene nodes are connected to nodes representing domains that are found within that gene. The legend is displayed at the bottom left (Fig. 5).

2.3.5 Dragging Nodes/Groups of Nodes Around

You can move around the nodes of the graph by clicking on a node and dragging it to where you want it to be. For particularly large networks, it is useful to highlight a group of nodes and move the

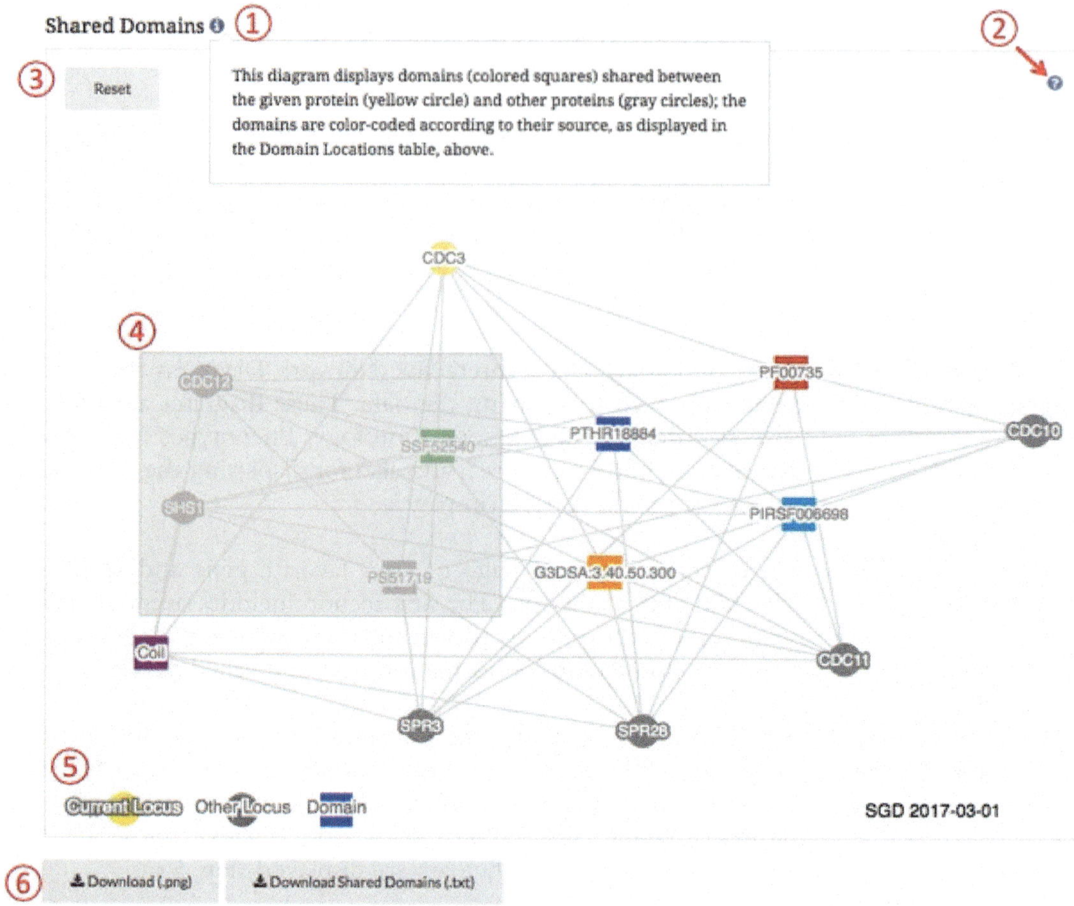

Fig. 5 An SGD network diagram. (1) Clicking the blue info icon displays an explanation of how the data are displayed. (2) Clicking the blue question mark displays help descriptions for how to interact with the diagram. (3) Clicking the "Reset" button returns the diagram to its original state. (4) Dragging your mouse across the figure will select a group of genes that you can manipulate together. (5) The legend describes the differences between the domain and gene nodes. (6) The Download options allow download of a .png or .txt file of the figure

group together. To do this, click in the empty space of the graph and drag the mouse to expand a rectangle that will enclose the nodes you wish to group. This selects the group of genes so you can click and drag any one of the highlighted nodes to move the entire group (Fig. 5).

2.3.6 Resetting the Diagram

At any time you wish to reset the network diagram to the original conformation, click the "Reset" button in the upper left corner of the diagram.

2.3.7 Downloading the Diagram

To download the network as an image, click "Download (.png)," or download the data displayed in the network diagram as a file using the "Download (.txt)" button.

3 Notes

1. For help on how to use specific tools, or other aspects of the SGD, please visit the SGD YouTube channel (youtube.com/c/SaccharomycesGenomeDatabase) for video tutorials or send an e-mail to the SGD help desk (sgd-helpdesk@lists.stanford.edu).
2. If you are searching for a specific phrase, you can use quotations as in a Google search to group words into a specific order (ex., "cell cycle").
3. When the "Genes" category is selected, the left bar changes to display the subcategory options for further filtering. The "Show more" button displays more filter options. You can return to view the other categories by clicking the "Show all categories" button above the left bar.
4. By clicking any of the blue gene names in the search results, you will be sent to an SGD Locus Summary page. The summary page includes an overview description and highlights from various topics of information. From here, you can look at the details of the information SGD captures by clicking any of the tabs at the top or any of the links to the right of the section header. For example, clicking the "Phenotypes" tab will take you to a page with phenotype annotations for the gene of interest.
5. Alternatively, you can enter in the term ID (GO:0005525).

Acknowledgments

This work was supported by a U41 grant from the National Human Genome Research Institute at the US National Institutes of Health (HG001315) to the *Saccharomyces* Genome Database project. The content is solely the responsibility of the authors and does not necessarily represent the official views of the National Human Genome Research Institute or the National Institutes of Health. Additional members of the SGD Project include: J. Michael Cherry, Gail Binkley, Felix Gondwe, Kalpana Karra, Kevin MacPherson, Stuart Miyasato, Travis Sheppard, Matt Simison, Marek Skrzypek, Shuai Weng, and Edith Wong.

References

1. Engel SR, Dietrich FS, Fisk DG, Binkley G, Balakrishnan R, Costanzo MC, Dwight SS, Hitz BC, Karra K, Nash RS, Weng S, Wong ED, Lloyd P, Skrzypek MS, Miyasato SR, Simison M, Cherry JM (2013) The reference genome sequence of Saccharomyces cerevisiae: then and now. G3 (Bethesda) 4:389–398
2. Engel SR, MacPherson KA (2016) Using model organism databases (MODs). Curr Protoc Essential Lab Tech 13:11.4.1–11.4.22
3. Cherry JM (2015) The Saccharomyces Genome Database: advanced searching methods and data mining. Cold Spring Harb Protoc. https://doi.org/10.1101/pdb.prot088906

4. Hibbs MA, Hess DC, Myers CL, Huttenhower C, Li K, Troyanskaya OG (2007) Exploring the functional landscape of gene expression: directed search of large microarray compendia. Bioinformatics 23(20):2692–2699
5. Skrzypek MS, Binkley J, Sherlock G (2016) How to use the Candida Genome Database. Methods Mol Biol 1356:3–15
6. Cherry JM (2015) The Saccharomyces Genome Database: gene product annotation of function, process, and component. Cold Spring Harb Protoc. https://doi.org/10.1101/pdb.prot088914
7. Skinner ME, Uzilov AV, Stein LD, Mungall CJ, Holmes IH (2009) JBrowse: a next-generation genome browser. Genome Res 19(9):1630–1638
8. Buels R, Yao E, Diesh CM, Hayes RD, Munoz-Torres M, Helt G, Goodstein DM, Elsik CG, Lewis SE, Stein L, Holmes IH (2016) JBrowse: a dynamic web platform for genome visualization and analysis. Genome Biol 17:66
9. Sheppard TK, Hitz BC, Engel SR, Song G, Balakrishnan R, Binkley G, Costanzo MC, Dalusag KS, Demeter J, Hellerstedt ST, Karra K, Nash RS, Paskov KM, Skrzypek MS, Weng S, Wong ED, Cherry JM (2016) The Saccharomyces Genome Database variant viewer. Nucleic Acids Res 44(D1):D698–D702
10. Ashburner M, Ball CA, Blake JA, Botstein D, Butler H, Cherry JM, Davis AP, Dolinski K, Dwight SS, Eppig JT, Harris MA, Hill DP, Issel-Tarver L, Kasarskis A, Lewis S, Matese JC, Richardson JE, Ringwald M, Rubin GM, Sherlock G (2000) Gene ontology: tool for the unification of biology. The Gene Ontology Consortium. Nat Genet 25(1):25–29
11. Engel SR, Balakrishnan R, Binkley G, Christie KR, Costanzo MC, Dwight SS, Fisk DG, Hirschman JE, Hitz BC, Hong EL, Krieger CJ, Livstone MS, Miyasato SR, Nash R, Oughtred R, Park J, Skrzypek MS, Weng S, Wong ED, Dolinski K, Botstein D, Cherry JM (2010) Saccharomyces genome database provides mutant phenotype data. Nucleic Acids Res 38:D433–D436. https://doi.org/10.1093/nar/gkp917
12 Shin B-S, Acker MG, Kim J-R, Maher KN, Arefin SM, Lorsch JR, Dever TE (2011) Structural integrity of alpha-helix H12 in translation initiation factor eIF5B is critical for 80S complex stability. RNA 17(4):687–696

Chapter 3

Using the *Candida* Genome Database

Marek S. Skrzypek, Jonathan Binkley, and Gavin Sherlock

Abstract

Studying *Candida* biology requires access to genomic sequence data in conjunction with experimental information that together provide functional context to genes and proteins, and aid in interpreting newly generated experimental data. The *Candida* Genome Database (CGD) curates the *Candida* literature, and integrates functional information about *Candida* genes and their products with a set of analysis tools that facilitate searching for sets of genes and exploring their biological roles. This chapter describes how the various types of information available at CGD can be searched, retrieved, and analyzed. Starting with the guided tour of the CGD Home page and Locus Summary page, this unit shows how to navigate the various assemblies of the *C. albicans* genome, how to use Gene Ontology tools to make sense of large-scale data, and how to access the microarray data archived at CGD, as well as visualize high-throughput sequencing data through the use of JBrowse.

Key words *Candida*, Genome database, Expression analysis, Gene ontology, GO slim, JBrowse

1 Introduction

The *Candida* Genome Database (CGD; http://www.candidagenome.org) was started in 2004 to serve as an online compendium of genomic, genetic, and molecular biology information about *Candida albicans*. The approach for CGD itself was derived from the *Saccharomyces* Genome Database (SGD), both in terms of the code base itself, which was modified from SGD's code, but also in terms of the staff who have worked at CGD over the last 13 years—many of them have also worked at SGD at various times. Thus CGD was able to provide a database with a similar look and feel to a community that was already used to SGD, and a similar outlook on the purpose of the database—that is, to serve the experimental scientists working at the bench, to better enable them to understand and interpret their own data, and to design experiments based on the collective knowledge of the community. At the outset, CGD was based on the then newly assembled genomic sequence of strain SC5314 [1], a clinical strain isolated from a patient with disseminated candidiasis [2] (originally from the

Martin Kollmar (ed.), *Eukaryotic Genomic Databases: Methods and Protocols*, Methods in Molecular Biology, vol. 1757, https://doi.org/10.1007/978-1-4939-7737-6_3, © Springer Science+Business Media, LLC, part of Springer Nature 2018

Squibb Collection, and first referenced in the literature in 1968 [3–5]). The primary goal of CGD was to couple the sequence data with the literature-derived experimental data in a single, easy to navigate web-based resource. Since then, the expanding scope of *Candida* research, facilitated by the progress in sequencing of genomes from other strains and species, has prompted the incorporation of similar data for *C. glabrata* CBS138, *C. parapsilosis* CDC317, and *C. dubliniensis* CD36. CGD has also become an archive that provides access to genomic sequences for other related strains and species, including *C. albicans* WO-1, *C. guilliermondii* ATCC_6260, *C. lusitaniae* ATCC_42720, *C. orthopsilosis* Co 90-125, *C. tropicalis* MYA-3404, *Debaryomyces hansenii* CBS767, and *Lodderomyces elongisporus* NRLL YB-4239.

At the core of CGD lies human curation, a process that involves manually extracting gene-specific experimental information from the published, peer-reviewed literature and associating those annotations with the relevant genomic features. Genes and their annotations are organized in such a fashion that the information is easily browsable, searchable and retrievable for further analysis and perusal. CGD curators also make sure that every piece of information is traceable to its original source, usually a publication in a scientific journal, thus providing access to all available experimental details and their interpretations. CGD also includes a rigorous analysis of orthology between species [6] and of protein domain structure, which allows consistent predictions of functions for genes that have not been experimentally characterized [7]. Thus, CGD provides a structured, unbiased and continuously updated collection of a large variety of experimental results and computational predictions that has become indispensible for *Candida* researchers.

In order to ensure a uniform representation of biological information across different organisms, most genome databases use controlled vocabularies to annotate various attributes of genes and gene products. The most widely used vocabulary for capturing the key aspects of gene product biology is the Gene Ontology (GO; http://www.geneontology.org/; [8]). GO is a system of standardized terms with defined relationships that describe a primary activity of the gene product (Molecular Function), a broader cellular role the gene product is involved in (Biological Process), and the predominant localization, such as a protein complex, a subcellular structure, or an organelle (Cellular Component). CGD uses GO as the main vocabulary to annotate genes. Another data type that CGD captures, mutant phenotypes, is curated using Ascomycete Phenotype Ontology (APO), a vocabulary developed at *Saccharomyces* Genome Database (SGD) [9] that we adapted to the specific needs of *Candida* biology.

The information in CGD is organized in a system of interlinked web pages, designed with the goal of making them intuitive, easily navigable and user-friendly. However, the sheer complexity of the data presented in CGD can make it difficult for newcomers to find the information they are looking for. This chapter provides help for navigating the site and highlights some of the more recently added features of CGD. We present an overview of the main entry point, the Home Page, and the central organizing principle of the database, the Locus Summary page. We show how to navigate between the *C. albicans* genome and the genomes of other *Candida* species. We also explain how to perform some of the most common types of analysis that utilize GO annotations. Finally, we show how to access and browse large-scale datasets collected at CGD, including archived microarray datasets, as well as some more recent high throughput sequencing data.

2 Methods

2.1 Exploring CGD Locus Summary Pages

The Locus Summary page (LSP) serves as the primary organizing principle for gene-specific data in CGD. There is an LSP for each protein- or RNA-coding gene, as well as for each of the other annotated chromosomal features (*see* **Note 1**). The LSP provides an overview of what is known about that particular entity and serves as an access point to more detailed information and analysis tools.

1. To access a Locus Summary page, first open the CGD home page (http://www.candidagenome.org) (Fig. 1). Take a look at the yellow menu bar at the top of the page. This bar is present on most CGD pages and facilitates quick navigation to various data search and analysis tools, genome browsers, and literature tools; it also offers links to bulk download tools as well as a variety of community-related information. Hover the mouse over each item to reveal a menu with available options.
2. To find an LSP for a particular feature of interest, enter your query into the "Search our site" box above the banner (*see* **Note 2**). After entering an identifier that unambiguously points to a specific entity (for example, enter orf19.922), clicking on the "Go" button will open the LSP for that entity. If your search produces multiple hits, such as a gene name that is used in several *Candida* species represented in CGD (enter *erg11*, as an example), you will see a "CGD Quick Search Result" page that lists the type and number of hits, and a species they come from. Positive hits are hyperlinked to either their respective LSPs, or to an intermediate list of individual hits. Click on an individual hit to open its LSP.
3. The Locus Summary page has several tabs, whose type and number may differ for different feature types. The Summary tab is open by default. The other tabs on an LSP for a

Fig. 1 Locus summary page for *C. albicans ERG2*—Basic Information section; red arrows point to the search box present on most CGD pages, and the tabs, that allow switching between different types of information

protein-coding gene are Locus History, Literature, Gene Ontology, Phenotype, Homologs, and Protein. Browse through the tabs to see what types of information they contain.

4. Go back to the Summary tab and look at the Basic Information section at the top of the page. It contains a headline-like Description that summarizes the most significant features of the locus. It also lists all the names associated with the locus including the standard name—typically the genetic name

under which the gene was first published. If there are other names by which the gene has been referred to in the literature, they are listed as aliases. The identifier assigned during genome sequencing is listed as Systematic Name along with the name of the reference strain used in the sequencing project (*see* **Note 3**). The Basic Information section allows easy navigation to genes in other organisms. Click on any hyperlinked name listed among "Orthologous genes in *Candida* species" to open its LSP in CGD. You can also click on the "View ortholog cluster" link, which will show the orthologs from 15 *Candida* species in their genomic context, a report produced by the Candida Gene Order Browser [6]. To explore orthologs from other fungal species, select a gene from "Ortholog(s) in non-CGD species" and you will access information at other resources: AspGD, Broad Institute, PomBase, and SGD, for genes from *A. nidulans, N. crassa, S. pombe, S. cerevisiae*, respectively (*see* **Note 4**). At the bottom of this section, there is a thumbnail showing the chromosomal location of the gene, hyperlinked to GBrowse [10, 11], which provides a graphical interface to inspect the genomic context of the gene.

5. The GO Annotations and Mutant Phenotypes sections show current information about the function of the gene. For more detailed information, including the references on which these annotations are based, open the Gene Ontology or Phenotype tab, respectively. GO annotations on the Summary tab are divided by the GO aspect (Molecular Function, Biological Process, Cellular Component) and by the annotation method. Manually curated annotations are assigned by a CGD curator based on reading of the scientific literature, while computational annotations are produced by transferring experiment-based GO annotations from orthologous genes in other species, or by predictions based on domain structure. Each annotation is accompanied by an evidence code that indicates the reason behind the annotation, for instance, IMP means Inferred from Mutant Phenotype. Click on any evidence code to see a table of all the evidence codes and their definitions. Annotations based on sequence similarity, genetic or physical interactions also list the source gene(s) and the organism. Click on any gene name to see a report for that gene in CGD or in an external database. Each GO term itself is hyperlinked to another CGD page that provides more information, including term definitions, a diagram depicting the relevant segment of the ontology, and a list of other genes in CGD that are also annotated with that term. Similarly, each mutant phenotype is hyperlinked to a page that lists all other genes in CGD that display the same phenotype.

6. The Sequence Information section shows the basic data about the gene (chromosomal coordinates, intron-exon structure), but also provides easy access to sequences and sequence analysis tools. Open the drop-down menu next to Retrieve Sequences to see available options that include retrieval of DNA sequence in several configurations and, for protein-coding genes, for the predicted protein sequence as well. Similarly, open the Sequence Analysis Tools drop-down menu to start BLAST searches, restriction analysis or primer design tools. More analysis tools are available from the banner on top of the page; click on Search, Sequence, Tools, or Download to see the options. In addition, at the bottom of the Sequence Information section there are links to sequence data available from external resources, such as GenBank, UniProt and others.

2.2 Explore *Candida* Genomes Using Genome Browsers

Genome browsers allow visualization of massive amounts of information in an easily understandable graphical format. The genomic sequence is overlaid with functional annotations, thus providing an overview of entire regions and, at the same time, a quick access to individual genome features. The genome browsers implemented in CGD are Generic Genome Browser, or GBrowse [10, 11], and JavaScript-based Genome Browser JBrowse [12].

2.2.1 Browsing the Chromosomal Features Using GBrowse

1. Access GBrowse from the menu bar at the top of most CGD pages. When you hover the cursor over the GBrowse option, a menu unfolds that lets you select one of the *Candida* species available in CGD (*see* **Note 5**).
2. In the Search panel of the main GBrowse window (Fig. 2), you can search for a particular gene of interest by entering its name in the Landmark or Region box. The search tool also works for other types of queries (*see* **Note 6**).
3. You can switch to a different Data Source by selecting one of the historic *C. albicans* assemblies, one of the other *Candida* species available in CGD, or you can change the browser to display protein data.
4. The Search panel also provides the ability to scroll up and down along the chromosome using the yellow <<, <, >, >> buttons, or zoom in and out using the yellow – or + buttons. You can also select a desired zoom level from a pull-down menu. In addition, the Search panel provides tools to generate a restriction map of the displayed region, download the sequence and save the current display as a snapshot.
5. The Overview and Region panels show schematic diagrams of the entire chromosome, or the selected region, with the highlight representing the currently displayed region. The diagrams can be used for navigation: a single click recenters the display on the clicked location; a click-and-drag specifies a new region to display.

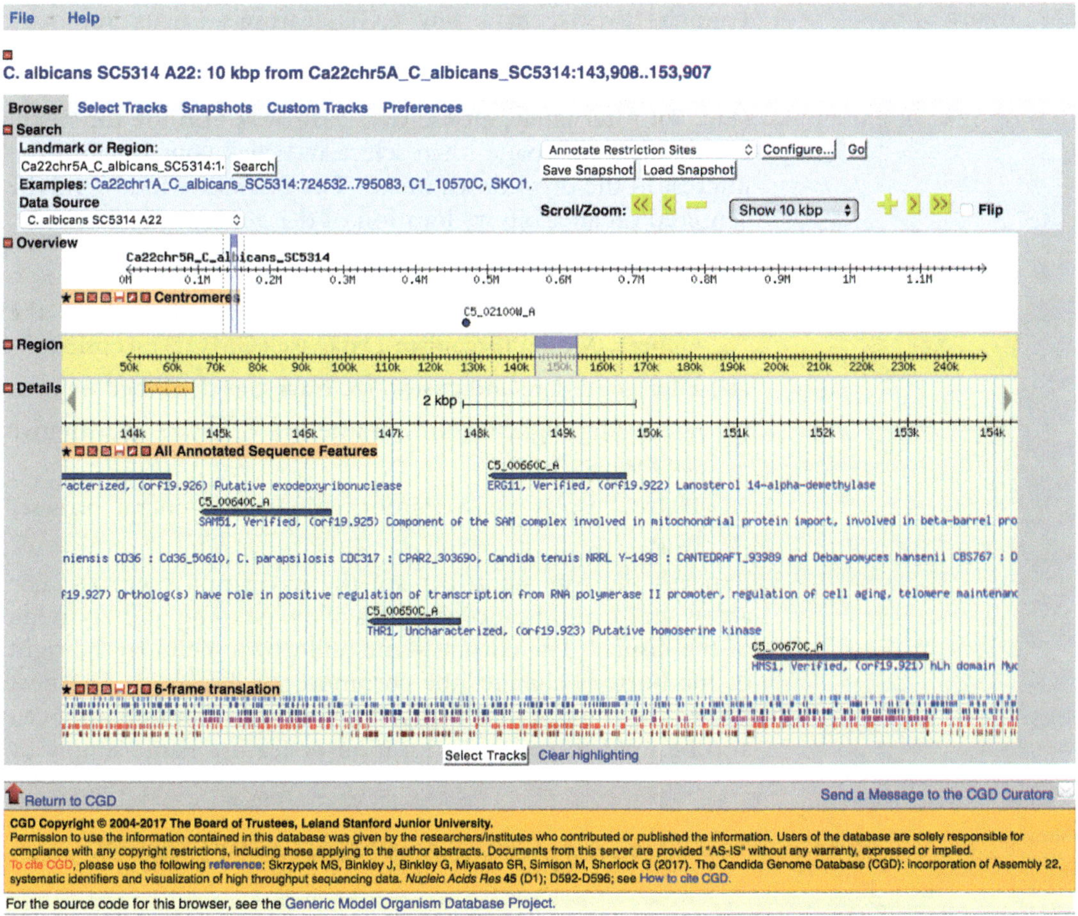

Fig. 2 GBrowse main window, showing genomic features for a region of chromosome 5

6. The Details panel with the default settings displays all annotated sequence features, including protein coding genes, tRNA and snRNA genes, transposons, and centromeres. The features are shown as horizontal bars with pointed ends and their color indicates their orientation, where applicable: features encoded on the Watson strand are red and those on the Crick strand are blue. Depending on the zoom level, each feature may be labeled with its name and description. Hovering over a feature generates a window with more information and clicking on a feature opens its Locus Summary page.
7. The content displayed in the Details panel is highly customizable. Click on Select Tracks tab on top, or Select Tracks button at the bottom to find the list of available datasets that can be overlaid on the sequence. Check or uncheck the desired boxes and hit Back to Browser to see the results (*see* **Note 7**)

2.2.2 Browse and Analyze Large-Scale Datasets Using JBrowse

1. Launch JBrowse from any CGD feature's Locus Summary page by clicking on the JBrowse logo about halfway down the page. The JBrowse window will be centered upon that feature (Fig. 3). Alternately, mouse-over "JBrowse" on the top menu bar of any CGD page, then select and click your organism of interest in the drop-down menu. The browser window will be centered on an arbitrary location of the genome (*see* **Note 8**).
2. The top menu bar of the browser has four items:
   - (a) JBrowse: Displays version information with links to the Generic Model Organism Database (GMOD) websites.
   - (b) File: Load data and customize tracks, as described below.
   - (c) View: Highlight features of interest, or resize quantitative data tracks.
   - (d) Help: Display version information and general browser help.
3. Beneath the top menu bar are the navigation controls. The arrow and magnification buttons allow panning and zooming of the display. The chromosome pull-down menu to the right of the buttons changes the chromosome displayed. The next box to the right of that displays the current location, but it can also be used to search by location or feature name simply by

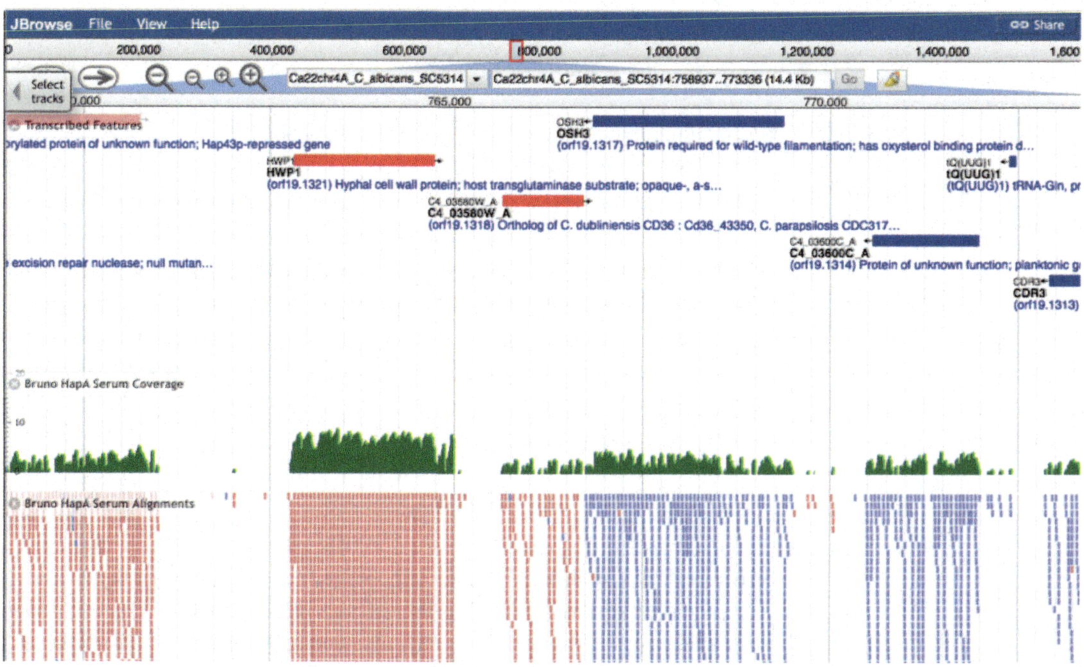

Fig. 3 JBrowse main window, showing short reads from an RNA-Seq experiment aligned against a region of the genome. Read color (red or blue) allows inference of the expressed strand, and corresponds to the direction of known genes

clicking and entering text (best search results will be obtained by entering the systematic name of the target feature). Just above the buttons and search box is a linear representation of the current chromosome, with a draggable red box over the current location. Change the current location by clicking and dragging the red box. Change the size of the region displayed by clicking on either side of the box and resizing it.

4. The main browser window displays data tracks of various types. A label in the far left of each track shows the track name, and has a pull-down menu with technical information about the track. You can change the relative position of the track in the window by clicking on the label and dragging it up or down to the desired position. The main types of tracks are:

   (a) Sequence Tracks display nucleotide sequence of both strands, as well as 6-frame translated amino acid sequences. Note that the window may need to be zoomed-in considerably for sequence tracks to be displayed.

   (b) Feature Tracks show all transcribed features annotated at CGD. Features encoded on the "W" (plus) strand are displayed in red, and features on the "C" (minus) strand in blue. Clicking on a feature brings up an information window, and clicking on the feature name within that window opens the CGD Locus Summary page for that feature.

   (c) Quantitative Tracks graphically display various types of numeric data—for example, expression levels, degree of conservation, or density of aligned high-throughput sequence reads. Most quantitative data are shown on a log scale, but this can be changed in the information pull-down menu at the far left of the track.

   (d) Alignment Tracks show the genomic alignments of sequence reads from a variety of high-throughput experiment types: DNA-Seq, RNA-Seq, ChIP-Seq, etc. (*see* **Note 9**).

   (e) Variation Tracks display small variations (SNPs or indels) between two different genomes (or between the two haplotypes of a diploid genome, such as that of *C. albicans* SC5314).

5. To choose which tracks to display, click on the "Select Tracks" tab on the upper-left side of the main JBrowse window. This brings up the Track Selection Menu (Fig. 4). The many available tracks are listed in the right panel of the menu. To select tracks, click the checkbox next to the item description. The left panel of the selection menu gives a number of filtering criteria to help narrow down the track choices. For example, you might filter by experimental technique (RNA-Seq, DNA-Seq,

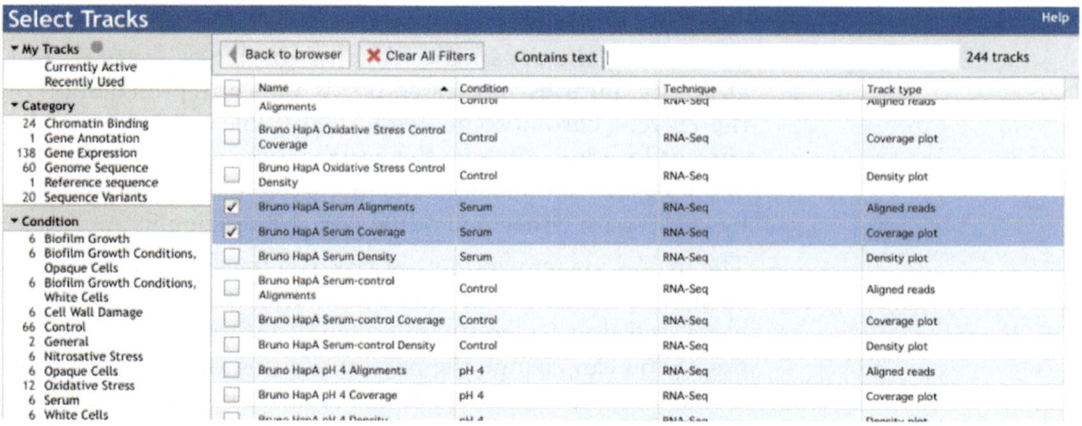

Fig. 4 JBrowse track selection window. Different datasets can be selected for display

etc.), experimental condition (oxidative stress, pH, control, etc.), or source publication (either by first author or PubMed ID). Select "Currently Active" under "My Tracks" to show only the tracks that are currently selected and displayed in the main window. The "Clear All Filters" button above the right panel removes any filters applied from the left panel (but does not deselect tracks), returning the display to all available tracks. You can also find tracks using the search box above the right panel: for example, searching *C. albicans* SC5314 tracks with the text "stress" shows selections from both oxidative stress and nitrosative stress experiments. Clicking the "Back to Browser" button above the right panel executes any track changes you made and returns to the main JBrowse window.

6. You can display all instances of a particular sequence or motif as a separate track. From the File menu in the top menu bar, select "Add sequence search track", and enter the query sequence (either as plain text or a regular expression). You can search using either nucleotide or amino acid sequence, and you can search on either or both strands.

7. You can combine information from separate tracks into a custom "combination" track. For example, you may want to create a Search Track for a given sequence motif, as above, but only display the instances that occur in genes included in the "Transcribed Features" track. From the File menu in the top menu bar, select "Add combination track". A new "empty" track is created. Click on the information labels of the tracks you wish to combine, drag them over to the information label of the new combination track, and release to combine (the label will appear red while you are dragging it, and turn green when it is in position over the combination track). After the different tracks have been added, a menu pops up to select

details of the combination (intersection, union, etc.). For the above example, you would select "Intersection".

8. You can load your own data for display as tracks in CGD JBrowse. Accepted data formats include BAM, GFF3, Wiggle/BigWig, or VCF. Data can be either in files on your local system or from a remote web server. From the File menu in the top menu bar, select "Open track file or URL", and add local files to the left box, or URLs to files from web servers to the right box. Note that in order to display correctly, the data must be annotated with the sequence coordinates of the current CGD genome version.

2.3 GO Term Finder and GO Slim Mapper, Tools for GO Analysis

The GO vocabularies are structured as hierarchies, with more general terms, so-called "parents", encompassing more specific "child" terms. It is a standard requirement for the curators to assign the most specific (granular) term that the evidence presented in the publication supports. Within the hierarchical structure of the ontology, annotation to a child term implies that annotation to its parent term is also correct. This feature of GO has to be taken into account during analysis of large-scale data, when it is necessary to identify common biological features in a set of genes that are, for instance, coregulated in a microarray experiment. Finding statistically significant similarities in GO annotations for multiple genes requires a tool, GO Term Finder, that is able to move up the GO hierarchy from the specific terms used to annotate the genes in a list to their GO parent terms that the genes may have in common. Given the complexity of GO, it is often desirable to group genes into broader categories, using only high-level terms. The GO Slims are such lists of high-level terms from each ontology branch (Molecular Function, Biological Process, and Cellular Component), carefully selected to cover most of the curated GO information. The tool, GO Slim Mapper, allows users to pick such GO Slim terms and then map a set of genes to them.

2.3.1 Using GO Term Finder

1. Open the GO Term Finder page by selecting it from the options in the Search or GO pull-down menus in the menu bar (*see* **Note 10**).
2. Select the species in Step 1; the default species is *Candida albicans*, but you can run this analysis for other CGD-curated species: *C. glabrata*, *C. parapsilosis* or *C. dubliniensis*.
3. In Step 2, enter a list of gene names. You can either type the name of the genes in the input box or upload a file that contains the list. Either genetic names (CGD Standard Names, e.g., '*AAF1*') or systematic names ('C3_06470W_A', or orf19 identifiers, e.g., 'orf19.7436') may be used (*see* **Note 11**). As an example, use these genes: *MOB2 GAL10 AGE3 ECM33 MNS1 SOG2 VRG4 DPM1 SAC1 CBK1*.

4. In Step 3, select one of the three branches of GO (biological process, molecular function, or cellular component) by checking the boxes. The tool only searches one of the three branches at a time.
5. Click the Search button after Step 3 to use the default settings or go further down to Steps 4 and 5 to specify and customize your background set and/or refine the types of annotations in your background set.
6. You may change your background set in Step 4. The default background set includes all the genes in the database that have at least one GO annotation. You can also customize the background set by choosing which feature type(s) it should include.
7. In Step 5 you can deselect specific types of GO annotations that should not be used for calculations. By default, annotations collected by all methods and with all types of evidence are included.
8. The results page displays the significant shared GO terms (or their parents) in both graphic and table form, within the set of genes entered on the previous page.
9. The graphic shows the GO tree that includes terms used directly or indirectly in annotations for the genes in your list. The terms are color-coded to indicate their statistical significance (*p*-value score). Genes associated with the GO terms are shown in gray boxes, with links to their respective Locus Summary pages.
10. The table below the graph lists each significant GO term, the number of times the GO term is used to annotate genes in the list and the number of times that the term is used to annotate genes in the background set (*see* **Note 12**).
11. Additional columns list the *p*-value, the False Discovery Rate (FDR), and a list of all the genes annotated, either directly or indirectly, to the term. FDR is an estimate of the percent chance that a particular GO term might actually be a false positive. It represents the fraction of the nodes with p-values as good or better than the node with this FDR that would be expected to be false positives.
12. The statistical significance of the association of a particular GO term with a group of genes in the list is indicated by the *p*-value: the probability of seeing at least *x* number of genes out of the total n genes in the list annotated to a particular GO term, given the proportion of genes in the whole genome that are annotated to that GO Term. The closer the *p*-value is to zero, the more significant the particular GO term association with the group of genes is (i.e., the less likely to occur by chance).

2.3.2 Using GO Slim Mapper

1. Select GO Slim Mapper from the options in the Search or GO pull-down menus in the menu bar. In Step1 select the species for your analysis (default is *Candida albicans*). In Step 2 you can type or paste your list of genes or upload them as a file. As an example, you can use the same gene list as previously: *MOB2 GAL10 AGE3 ECM33 MNS1 SOG2 VRG4 DPM1 SAC1 CBK1*.
2. In Step 3, use the pull-down menu to select the GO Set Name: GO Slim Component, GO Slim Function, or GO Slim Process. The list of terms from the selected set appears in the window in Step 4.
3. In Step 4 you can specify which particular term, or terms, you want to use. The default setting is "Select ALL Terms," but you can highlight only terms you are interested in.
4. In optional Step 5 you can exclude certain types of annotations, e.g., computational and high-throughput experiments.
5. Click Search to start the process; note that long lists of genes will take significant time to analyze.
6. Results appear in a table with three columns: the GO Slim terms chosen, with a link to graphical depiction of that branch of GO, the percentage of genes in your list annotated to each term, and the genes from your list that are annotated to that term. You can also download the results in a tab-delimited file.

2.4 Browsing Large-Scale Data and GeneXplorer

CGD has been collecting large-scale datasets from its inception. The datasets from authors' websites or from supplementary materials that accompany publications are archived in the Datasets page accessible from the Download menu in the banner, thus offering a convenient one-stop shop for datasets pertaining to *Candida* that can be then imported into one's own tools for analysis. The microarray results have been built into GeneXplorer, a web tool that allows browsing and analysis of the expression data at the CGD site, with the benefit of an instant access to functional annotations [13].

1. To access the archived datasets, click on the Download pull-down menu in the top banner and select Datasets. The resulting page lists all the publications from which the datasets are archived in CGD, ordered by year and author (Fig. 5). All the references have icons that lead to the abstract, full-text articles and to the original data website, as well as to the download directory at CGD. The microarray-containing references have a pair (or pairs) of green-blue icons. Click the one on the right that is labeled ".pcl" and you will get the entire datasets in the pcl format, a standard format that makes the dataset amenable to analysis by many popular software suites. Find a publication of interest and click the icon left to the ".pcl" icon to see the data in GeneXplorer (Fig. 5).

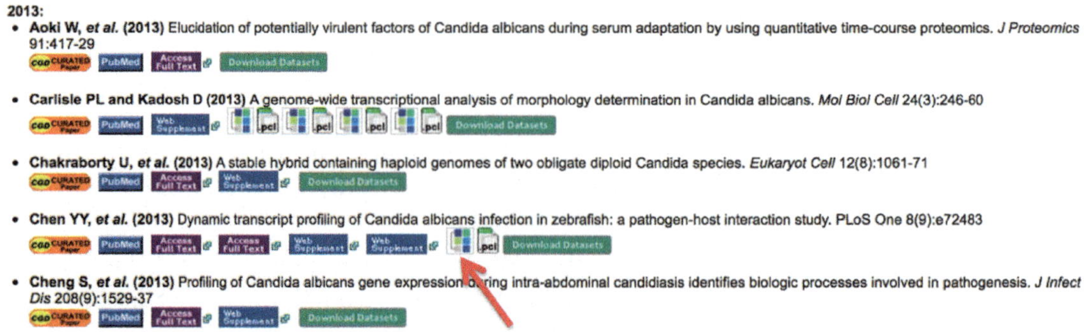

Fig. 5 Example entries in the Dataset Archive page. The red arrow points to the icon leading to the GeneXplorer window

Fig. 6 GeneXplorer page: red arrows indicate the zoom buttons, the gene search box, and the individual expression profile to click on to see genes similarly expressed to *KTR2*

2. The GeneXplorer page (Fig. 6) shows a heat map image of the clustered expression profiles from the selected publication, where increased or decreased gene expression is shown in shades of red or green, respectively. Zoom in and out using the + or – buttons above the cluster. Select any particular region by dragging over it. Click on a profile for a particular gene to see the expression patterns of the most similarly or dissimilarly expressed genes. Find a gene of interest by entering its name into the "Search for" box. Click on any gene name to go to its Locus Summary page.

3 Notes

1. Several entities other than protein-coding genes have their Locus Summary pages in CGD, including various RNA genes, such as tRNA, rRNA, snRNA, as well as centromeres, telomeres, repeated sequences, and other annotated chromosomal features.
2. The query entered into the search box may be a *Candida* gene or protein name (standard or systematic name, or an alias), author or colleague name, PubMed ID, or any keyword (such as a functional term or phenotype). It can even be a name of an ortholog from one of the non-CGD species. When there are multiple hits, a list of matches is displayed. The "Search our site" box is case-insensitive and accepts a wildcard character *. For example, enter "act*" to retrieve any piece of data starting with "act". Also, the search box has an autocomplete feature, which provides suggestions when you start typing your query.
3. For *C. albicans*, the systematic name shown on the Locus Summary page is always the name of the haplotype A allele, as denoted by an "A" suffix, with the corresponding haplotype B allele listed below. This new systematic name is based on the known chromosomal location and haplotype, and it consists of the chromosome (C1–C7 and CR for the eight nuclear chromosomes, CM for the mitochondrial chromosome), a unique number indicating the order of features along chromosomes, the strand (W for Watson or C for Crick) and the haplotype (A or B) [14]. Since the systematic names from previous assemblies of *C. albicans* genome (so called "orf19" names) continue to be widely used, the Assembly 19/21 identifier is also listed. The systematic Identifiers from earlier assemblies are included among the aliases.
4. The ortholog mappings among *Candida* strains, and between *Candida* strains and *S. cerevisiae*, are derived from the curated syntenic groupings at the Candida Gene Order Browser (CGOB) [15]. The ortholog mappings between *Candida* strains and *S. pombe*, *A. nidulans*, and *N. crassa* are made by pairwise comparisons using the InParanoid software [16].
5. Gbrowse is also accessible from any Locus Summary page. Click on a thumbnail in the Chromosomal Location section and a browser window will open, centered on the gene of the page and 10 kbp upstream and downstream. Note that in *C. albicans* there are two alleles and the two thumbnails lead to either haplotype A or haplotype B chromosomes.
6. The GBrowse search tool recognizes several types of landmarks, such as a chromosome ("Ca22chr1A_C_albicans_SC5314") or coordinates ("Ca22chr1A_C_albicans_

SC5314:724532.0.795083"), a gene name ("*ATP1*"), a systematic ORF name ("C1_04610W_A"), an alias ("orf19.6854"). You can also use keywords ("ATP synthase") and wildcards ("ATP*"). When the search results include multiple hits, the browser will show a diagram of all chromosomes with landmarks matching your query and a table with a list of hits. Click on one of the landmarks, either in the diagram or in the table, to see that region to the Details panel of the browser.

7. There is a set of icons to the left of the track name that provides additional controls. Hovering over each icon produces a pop-up window that explains its function.
8. CGD has aligned high-throughput sequence data to each of the two haplotypes of *C. albicans* SC5314 independently, and the two haplotypes are displayed in JBrowse separately. This is primarily to reduce alignment errors for short-read sequences, but it also allows sequence variation to be viewed in the context of either haplotype. By default, JBrowse opens with the "A" haplotype displayed.
9. The browser window may need to be zoomed in to view alignments, and scrolling may be necessary to view all the aligned reads. Browser speed and performance may be greatly diminished if more than one alignment track is displayed. For each alignment, CGD also provides two quantitative tracks, summarizing read coverage and read density. These may be preferred if more than one dataset are to be viewed or compared.
10. Links to both GO Term Finder and GO Slim Mapper appear at the bottom of many pages with results of searches that produce a list of genes. Selecting those links open the respective tool with the gene list already preloaded into the input box.
11. When a name is entered that is an alias for one gene or feature, the program will map the name to that gene. If the name is an alias for more than one gene but not a standard or systematic name for any genes, the program will present a list of possible mappings. The user can decide which gene was intended and edit the input.
12. Because the frequency of any given annotation within the background set is compared against the frequency of the annotation within the query set (input), the choice of background set affects the significance of the results that are returned by the tool. Please note that the specific background set of genes that was used in the absence of any user-defined set (the default background set) has changed over time.

Acknowledgments

This work was supported by National Institute of Dental and Craniofacial Research at the US National Institutes of Health (grant no. R01 DE015873).

References

1. Jones T, Federspiel NA, Chibana H et al (2004) The diploid genome sequence of *Candida albicans*. Proc Natl Acad Sci U S A 101(19):7329–7334. https://doi.org/10.1073/pnas.0401648101
2. Gillum AM, Tsay EY, Kirsch DR (1984) Isolation of the *Candida albicans* gene for orotidine-5′-phosphate decarboxylase by complementation of S. cerevisiae *ura3* and *E. coli pyrF* mutations. Mol Gen Genet 198(2):179–182
3. Aszalos A, Robison RS, Lemanski P et al (1968) Trienine, an antitumor triene antibiotic. J Antibiot (Tokyo) 21(10):611–615
4. Maestrone G, Semar R (1968) Establishment and treatment of cutaneous *Candida albicans* infection in the rabbit. Naturwissenschaften 55(2):87–88
5. Meyers E, Miraglia GJ, Smith DA et al (1968) Biological characterization of prasinomycin, a phosphorus-containing antibiotic. Appl Microbiol 16(4):603–608
6. Fitzpatrick DA, O'Gaora P, Byrne KP et al (2010) Analysis of gene evolution and metabolic pathways using the Candida Gene Order Browser. BMC Genomics 11:290. https://doi.org/10.1186/1471-2164-11-290
7. Inglis DO, Arnaud MB, Binkley J et al (2012) The *Candida* genome database incorporates multiple Candida species: multispecies search and analysis tools with curated gene and protein information for *Candida albicans* and *Candida glabrata*. Nucleic Acids Res 40:D667–D674. https://doi.org/10.1093/nar/gkr945
8. Ashburner M, Ball CA, Blake JA et al (2000) Gene ontology: tool for the unification of biology. The Gene Ontology Consortium. Nat Genet 25(1):25–29. https://doi.org/10.1038/75556
9. Engel SR, Balakrishnan R, Binkley G et al (2010) Saccharomyces Genome Database provides mutant phenotype data. Nucleic Acids Res 38(Database issue):D433–D436. https://doi.org/10.1093/nar/gkp917
10. Stein LD (2013) Using GBrowse 2.0 to visualize and share next-generation sequence data. Brief Bioinform 14(2):162–171. https://doi.org/10.1093/bib/bbt001
11. Stein LD, Mungall C, Shu S et al (2002) The generic genome browser: a building block for a model organism system database. Genome Res 12(10):1599–1610. https://doi.org/10.1101/gr.403602
12. Buels R, Yao E, Diesh CM et al (2016) JBrowse: a dynamic web platform for genome visualization and analysis. Genome Biol 17:66. https://doi.org/10.1186/s13059-016-0924-1
13. Rees CA, Demeter J, Matese JC et al (2004) GeneXplorer: an interactive web application for microarray data visualization and analysis. BMC Bioinformatics 5:141. https://doi.org/10.1186/1471-2105-5-141
14. Skrzypek MS, Binkley J, Binkley G et al (2017) The *Candida* Genome Database (CGD): incorporation of assembly 22, systematic identifiers and visualization of high throughput sequencing data. Nucleic Acids Res 45:D592–D596. https://doi.org/10.1093/nar/gkw924
15. Maguire SL, OhEigeartaigh SS, Byrne KP et al (2013) Comparative genome analysis and gene finding in *Candida* species using CGOB. Mol Biol Evol 30(6):1281–1291. https://doi.org/10.1093/molbev/mst042
16. Ostlund G, Schmitt T, Forslund K et al (2010) InParanoid 7: new algorithms and tools for eukaryotic orthology analysis. Nucleic Acids Res 38:D196–D203. https://doi.org/10.1093/nar/gkp931

Chapter 4

PomBase: The Scientific Resource for Fission Yeast

Antonia Lock, Kim Rutherford, Midori A. Harris, and Valerie Wood

Abstract

The fission yeast *Schizosaccharomyces pombe* has become well established as a model species for studying conserved cell-level biological processes, especially the mechanics and regulation of cell division. PomBase integrates the *S. pombe* genome sequence with traditional genetic, molecular, and cell biological experimental data as well as the growing body of large datasets generated by emerging high-throughput methods. This chapter provides insight into the curation philosophy and data organization at PomBase, and provides a guide to using PomBase for infrequent visitors and anyone considering exploring *S. pombe* in their research.

Key words *Schizosaccharomyces pombe*, Fission yeast, Biological database, Model organism, Gene ontology, GO slim, Annotation extensions

1 Introduction

PomBase (http://www.pombase.org/), funded by the Wellcome Trust, is the model organism database (MOD) for the fission yeast *Schizosaccharomyces pombe*. Its primary goals are:

- To support the exploratory and hypothesis-driven research needs of those using the model eukaryote fission yeast as an experimental system.
- To provide an integrated model of a eukaryotic cell.
- To promote and support the use of fission yeast as a model eukaryotic system, with particular relevance to human biology.
- To provide a community hub, and support for in-depth community led curation.

To accomplish these goals, PomBase integrates the *S. pombe* genome sequence and features with genome-wide datasets and detailed, comprehensive gene-oriented manual curation of published literature, and provides tools to interrogate these data [1, 2].

Martin Kollmar (ed.), *Eukaryotic Genomic Databases: Methods and Protocols*, Methods in Molecular Biology, vol. 1757,
https://doi.org/10.1007/978-1-4939-7737-6_4, © The Author(s) 2018

As an experimental organism, fission yeast is inexpensive to grow, proliferates rapidly and is amenable to genetic manipulation. Researchers typically study isogenic strains of *S. pombe* derived from a single isolate known as 968 h90. This facilitates the comparison of results across different laboratories. Aside from the PomBase database, several organism specific resources are available to fission yeast researchers, including the genome-wide deletion mutant collection [3, 4] and the Orfeome localization collection [5].

The sequence of the reference strain 972 h–(a naturally occurring 968 h90 derivative) was published in 2002 [6]. Fission yeast has a compact genome, 14 Mb in size, that consists of three chromosomes of 3.5–5.6 Mb plus a 19 kb mitochondrial genome. 5071 protein-coding genes, of which more than two thirds are conserved in human, are annotated in the reference genome, along with rRNAs, tRNAs, snRNAs, snoRNAs, and other noncoding RNAs.

The fission yeast community comprises ~2000 researchers working primarily or exclusively with *S. pombe* or other *Schizosaccharomyces* species. In addition, fission yeast is used extensively as a complementary organism by those studying conserved cellular mechanisms in vertebrate systems, including the cell cycle, cytokinesis, chromosome segregation, epigenetic mechanisms, DNA metabolism, and drug responses [7–10]. PomBase thus serves a large (~15,000 unique users per month) and varied user base with diverse experience and requirements.

2 Data Curation in PomBase

The most precise and reliable molecular data in PomBase are generated by manual curation of the fission yeast literature. Automated methods, such as annotation transfer based on sequence orthology, and high-throughput datasets supplement the body of manually curated data.

To enable the fission yeast community to contribute directly to PomBase, we have developed Canto [11], an intuitive web-based literature curation tool. Canto allows both professional curators and community researchers to use state-of-the-art annotation techniques to build complex connections among genes, ontology terms, and supporting metadata. Notably, the use of ontology terms and "annotation extensions" described below underlies the generation of comprehensive curated networks representing biological processes. By combining the topic-specific expertise of biological experts with PomBase curators' familiarity with ontologies and annotation practices, Canto usage yields literature curation of a particularly high standard of accuracy and specificity [12]. To date (August 2017) approximately 10,000 annotations have been submitted by community curators for over 500 publications.

3 PomBase Gene Page Organization

Like other model organism databases, PomBase organizes data into pages summarizing genes, publications, ontology terms, and others, of which the most intensively used are gene pages. Each gene page gathers all data relevant to the gene into one place, with a menu that shows available data types at a glance and facilitates navigation within the page (Fig. 1). Gene pages can be accessed directly by typing a gene name in the search field at the top right corner of each PomBase page, and selecting it from the drop-down list (e.g., *clp1*, *cdc2*, *cdc25*, *mde4*).

Curated data types include ontology-based annotations for gene function (Gene Ontology; GO), phenotypes, and modifications, genetic and physical interactions, qualitative and quantitative gene expression data, protein features, complementation,

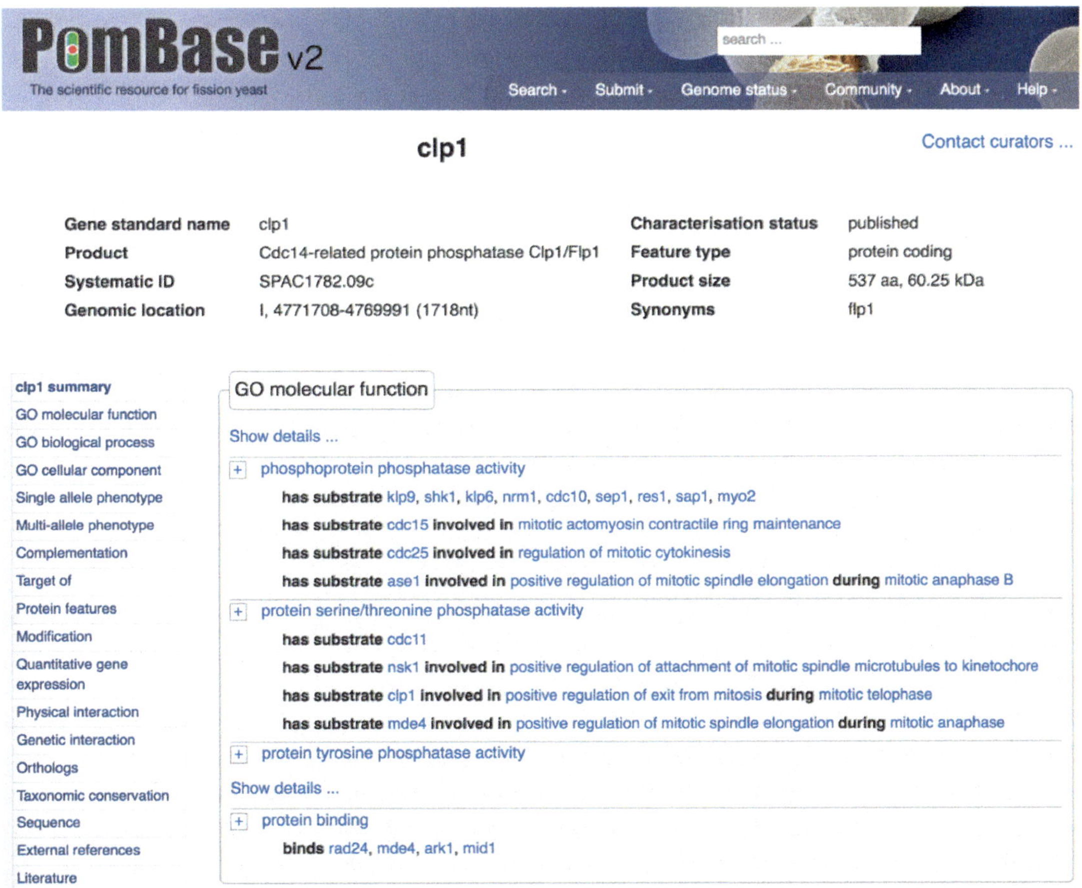

Fig. 1 The quick-links menu. The menu displayed on the left-hand side of gene pages provides an overview of the different data types available for specific genes, and enables rapid navigation between the different sections of the gene page

orthologs and taxonomic conservation. Gene pages also provide gene and protein sequences, and links to gene-specific entries in external databases, and a collection of literature relevant to each gene. We discuss several of these in depth below.

3.1 Gene Ontology Data

The Gene Ontology (GO) section of the gene pages shows a table of annotations using each of the main branches of GO: molecular function, biological process, and cellular component [13, 14]. By default, the tables display a nonredundant summary of annotated terms and extensions. Figure 2A shows a selection of molecular function and biological process annotations for the protein phosphatase *clp1*. An expanded view shows all annotations as well as supporting metadata such as references, evidence, term IDs, and qualifiers. Ontology terms, genes, and references in the annotation views link to additional PomBase pages. The biological process section also lists any GO slim terms (*see* below) applicable to the gene.

The *clp1* GO molecular function annotation shown in Fig. 2A also illustrates the usage of GO annotation extensions. PomBase was a pioneer in the implementation of annotation extensions [15], which allow curation of effector–target relationships (such as protein kinase substrates) or spatiotemporal information (such as where and when a function is executed). Extensions on the *clp1* "serine/threonine phosphatase activity" molecular function annotation indicate that Clp1 dephosphorylates different substrates to contribute to different regulatory processes (e.g., Clp1 dephosphorylates Mde4 to positively regulate spindle elongation during anaphase).

Figure 2B shows a summary of the relationships used at PomBase to curate annotation extensions, and then, as described below, to build networks using the resulting connections among gene products.

3.2 Phenotype Data

Phenotypes are curated by PomBase using the Fission Yeast Phenotype Ontology (FYPO), an ontology of over 6000 precomposed phenotype terms [16]. Fission yeast researchers typically study isogenic strains, making it possible to define "normal" and "abnormal" phenotypes in mutants compared to the behavior of the "wild type" reference strain.

PomBase curates single mutant allele and multiallele genotypes, which are displayed in separate gene page sections. The phenotype view is further split into population and cell level phenotypes and users can toggle between a summary view (Fig. 3A) and a detailed view (Fig. 3B). Gene deletion viability is indicated at the top of the single mutant phenotype section. The displayed phenotypes can be filtered by broad phenotypic categories (viability, cell cycle, morphology, etc.), improving the usability of the very long phenotype lists now present for many genes (green box, Fig. 3A).

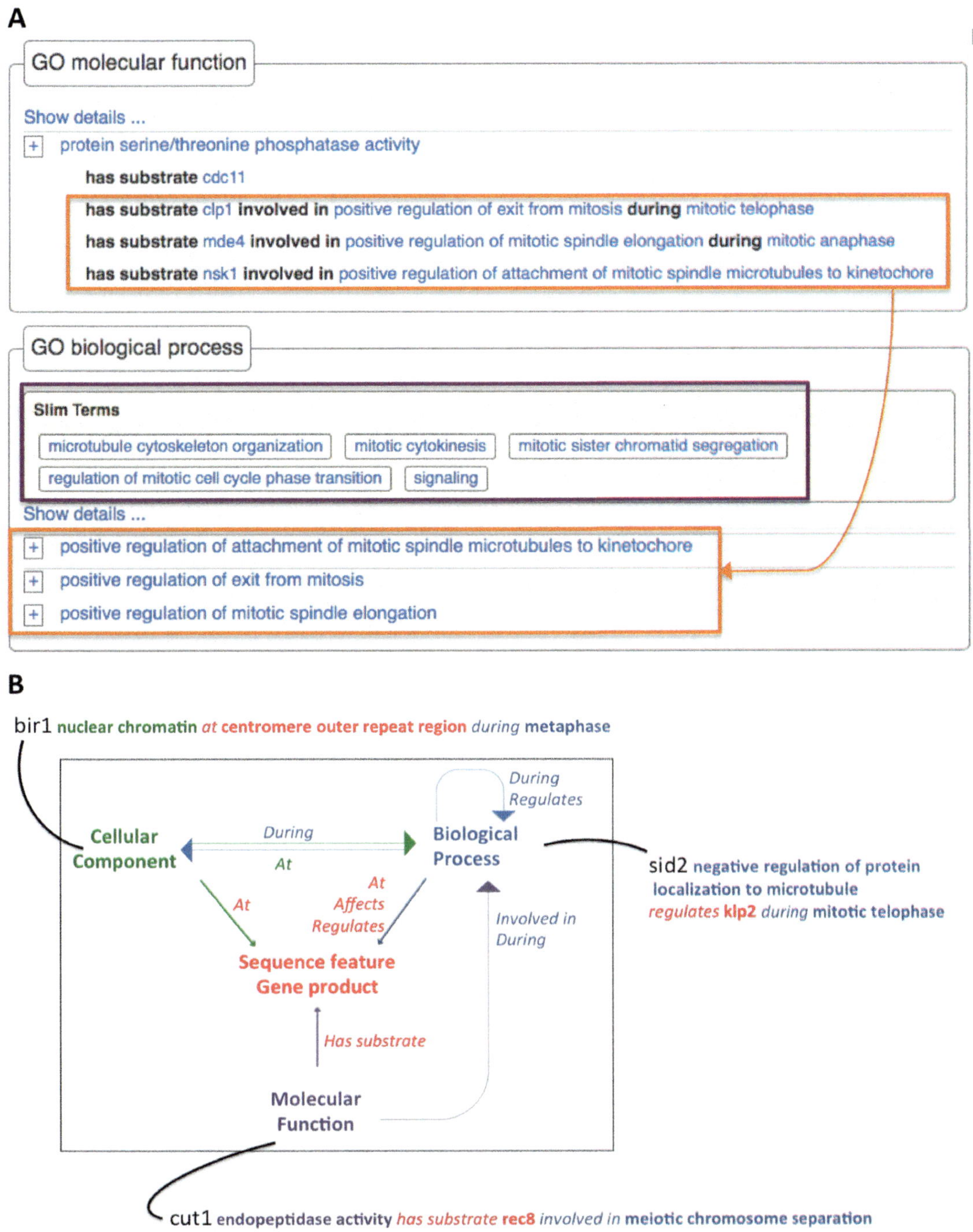

Fig. 2 GO annotations and extensions. (A) Summary view of selected annotations on the *clp1* gene page. The orange boxes highlight annotations representing Clp1's roles: Clp1 dephosphorylates the Nsk1 protein to positively regulate spindle attachment to the kinetochore. During anaphase, it dephosphorylates Mde4 to positively regulate spindle elongation. Clp1 also directly targets itself during telophase to promote mitotic exit. Processes linked to molecular functions are also shown in the biological process section. Biological process annotations that map to the PomBase GO slim are shown at the top of the biological process section Fig. 2 (continued) (purple box). (B) Relations used in GO annotation extensions, showing how each links one gene to other genes or additional ontology terms, with examples for each GO branch

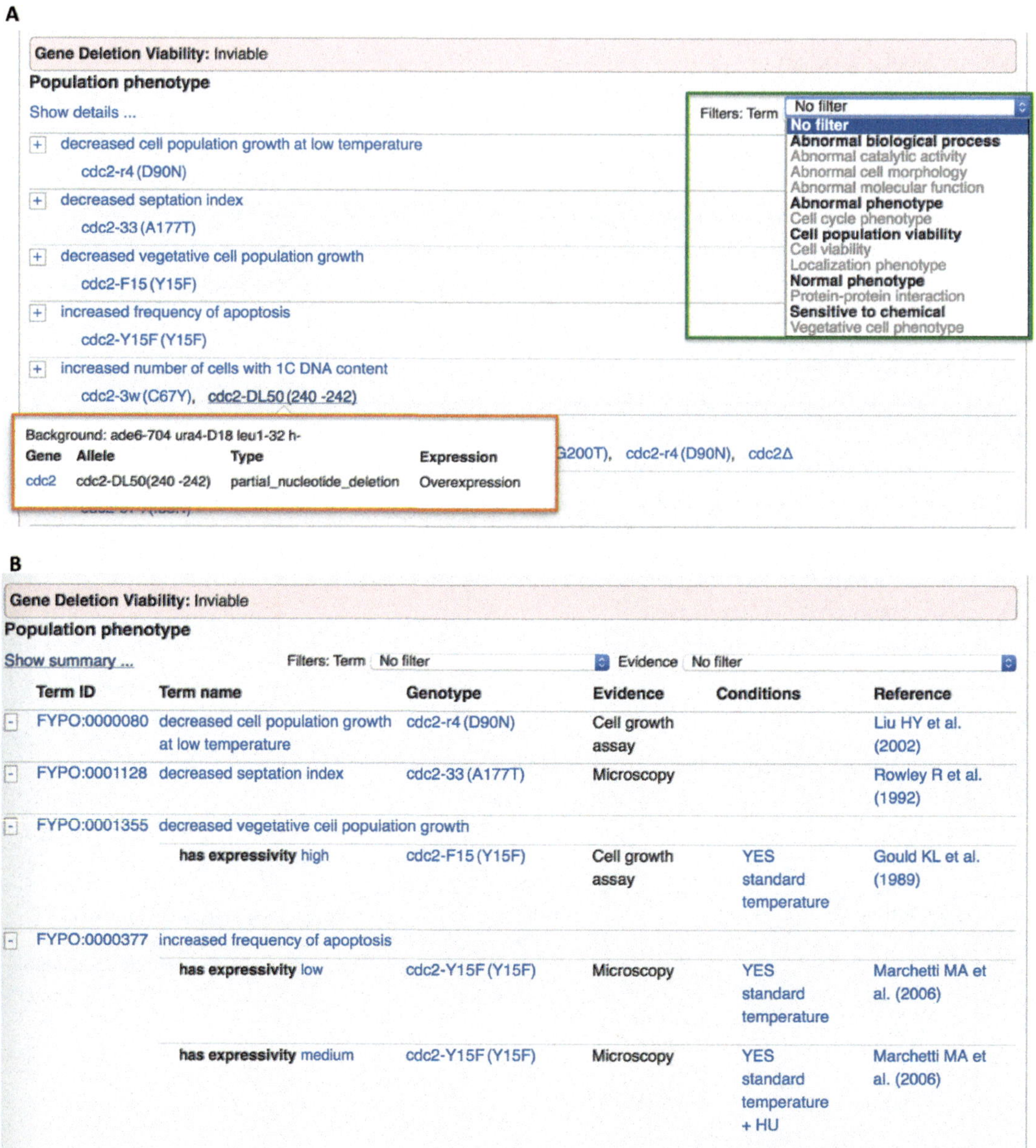

Fig. 3 The PomBase phenotype display. (A) Summary view (B) Detailed view of a subset of phenotypes associated with alleles of *cdc2*. In the summary view, redundant annotations (including annotation to the same phenotype term, but with different extensions or metadata) and metadata are hidden. The detailed view shows all annotations, plus the cited references, evidence, extensions indicating penetrance, expressivity, or affected gene products, and any curated experimental, conditions. Genotype details, including the type of mutation for each allele, expression level of the gene products, and any background genotype information, are provided in a mouse-over (shown in A, orange box). A drop-down menu enables filtering for subsets of phenotypes (shown in A, green box)

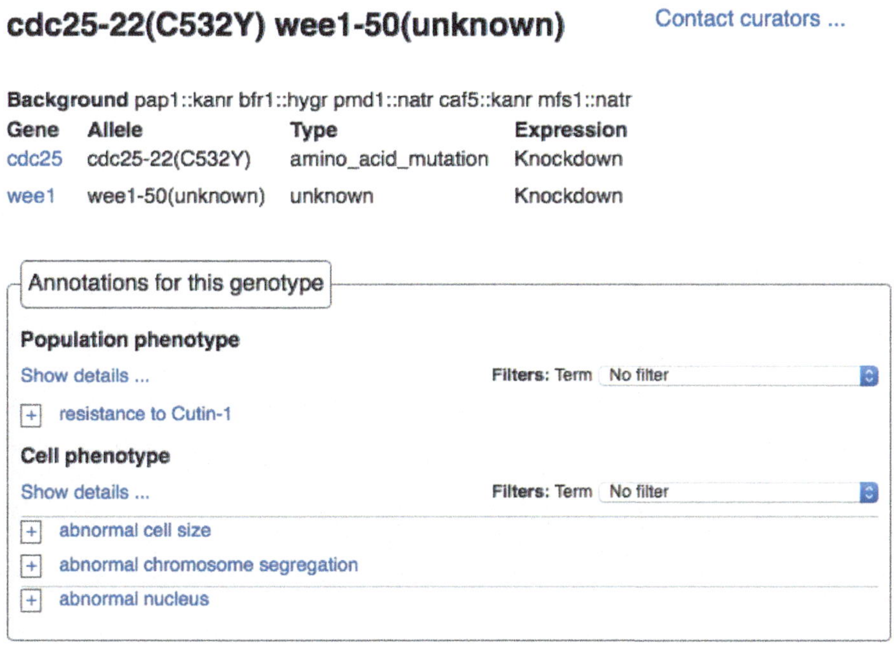

Fig. 4 Genotype pages. Each genotype page displays allele and expression details and all annotations associated with the genotype. In this example, the double mutant comprising *cdc25-22* (C532Y) and *wee1-50*, both at reduced expression levels, in the background *pap1::kanr bfr1::hygr pmd1::natr caf5::kanr mfs1::natr* has been associated with four different phenotype terms

Each phenotype annotation also links to a page dedicated to the genotype, which displays details (name, description, expression level) for the alleles that make up the genotype, any background alleles, and all annotated phenotypes (Fig. 4).

3.3 Targets

The "Target of" section provides information about gene products or mutations that affect the gene of interest, derived from the reciprocal of annotations specifying targeted genes, such as the substrates of molecular functions. Target annotations indicate the connecting ontology term and the specific relationship between the two genes. For example, Clp1 dephosphorylates (Fig. 2A) and Cdc2 phosphorylates (Fig. 6) the Mde4 protein. Since Mde4 is targeted by these proteins, *clp1* and *cdc2* are listed in the *mde4* "Target of" section (Fig. 5). Users can thus navigate entire biological pathways; downstream by a gene product's GO molecular function substrates, and upstream by effectors in the "target of" section. Reciprocal annotations are also generated from phenotype and protein modification annotations.

3.4 Taxonomic Conservation, Orthologs, and Disease Curation

To support the growing cohort of researchers using both fission yeast and other species, PomBase maintains manually curated inventories of orthologous proteins for fission yeast vs. human and fission yeast vs. budding yeast (*Saccharomyces cerevisiae*). Both are

Target of

Show details ...

Ontology	Relationship	Gene/genotype	Product
GO	substrate of	cdc2	cyclin-dependent protein kinase Cdk1/Cdc2
GO	substrate of	clp1	Cdc14-related protein phosphatase Clp1/Flp1
GO	regulated by	mei4	meiotic forkhead transcription factor Mei4
FYPO	affected by mutation in	cdc2	cyclin-dependent protein kinase Cdk1/Cdc2
FYPO	affected by mutation in	clp1	Cdc14-related protein phosphatase Clp1/Flp1
FYPO	affected by mutation in	mde4	microtubule-site clamp monopolin complex subunit Mde4

Fig. 5 The *mde4* "Target of" section. Because *cdc2* is annotated to a protein kinase molecular function term, with Mde4 specified as a substrate, *cdc2* is listed in the "target of" section for *mde4*. Reciprocal annotations are also generated from phenotype and protein modification annotations. For example, a mutation in *cdc2* has an effect on *mde4*, with phenotypic details available on the *cdc2* gene page, and a "target of" annotation using the "affected by mutation in" relationship on the *mde4* gene page

compiled by integrating published data and in-house analyses with the consensus from numerous orthology resources [17]. The human–fission yeast curated orthology dataset now identifies human orthologs for 69% of the fission yeast proteome.

Gene pages show any manually curated orthologous genes in human and budding yeast, and the basic gene search will retrieve available *S. pombe* orthologs using human standard gene names or budding yeast systematic (ORF) names. Where a fission yeast gene has a human ortholog that has been implicated in a disease, the PomBase gene page notes the disease and a source publication.

The taxonomic conservation section shows a broad domain kingdom or phylum level conservation for protein-coding genes. Taxon restrictions are also recorded where applicable. Other terms may also be assigned, such as whether the gene is conserved one-to-one. Classifiers are assigned manually from a small in-house controlled vocabulary (Table 1).

Taxonomic conservation can be used to retrieve high quality broad taxon classification specific datasets for analyses, or to provide functional clues for unstudied proteins based on presence or absence in particular kingdom or phyla.

4 Building Networks

The growing body of GO annotations with annotation extensions in PomBase creates connections between gene products, and provides rich biological context to their interactions. These connections can be exploited to reconstruct biological pathways. For

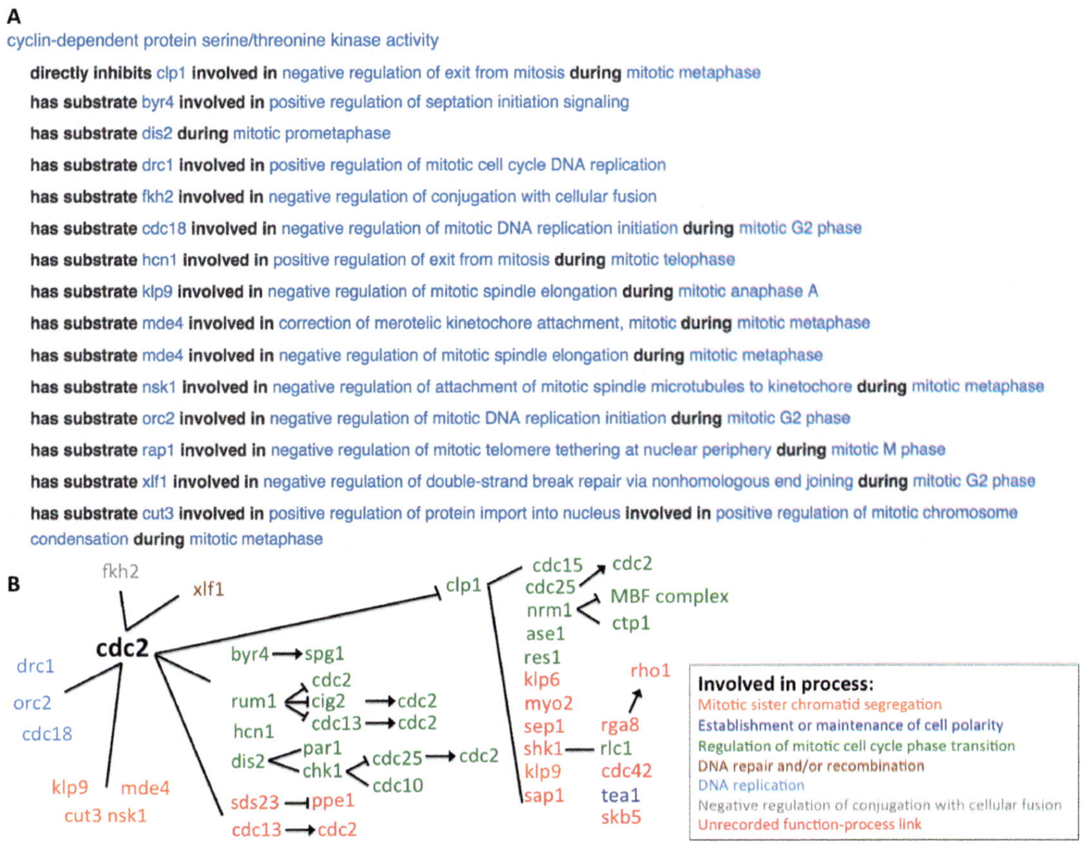

Fig. 6 *cdc2* function–process links and downstream signaling cascades. (A) Showing the subset of Cdc2 phosphorylation targets with function–process links. Biological processes that the enzyme–substrate interaction is part of, or happens during, are indicated using the "*involved in*" and "*during*" annotation extension relationships. (B) Targets downstream of Cdc2 can be accessed via the hyperlinked annotation extension substrates, enabling users to follow biological pathways. The capturing of targets makes it possible to reconstruct pathways for a systems level representation of gene networks

example, the highly conserved cyclin-dependent serine/threonine kinase Cdc2 (homolog of the mammalian *CDK1*) is known to directly phosphorylate over 140 different proteins. A number of these *cdc2*–substrate connections are linked to the biological processes that the interaction is regulating (Fig. 6A). Annotated substrates can be followed, in order to move down biological pathways (Fig. 6B).

PomBase will use the connections curated between gene products (enzyme–substrate links, and high confidence physical interaction data), and the links to the biological processes they are involved in, to automatically construct networks for biological processes. This approach will ultimately create a detailed and reliable curation-based network for a eukaryotic cell.

Table 1
Taxonomic conservation groups. Taxonomic conservation groups are assigned manually from a controlled set of terms at the kingdom/domain level. A gene may be annotated to multiple different orthologous groups. Taxon restrictions are recorded for where orthologs have not been identified outside of the noted taxa (fungi or eukaryotes). The absence of an ortholog in the *S. cerevisiae* reference sequence is also recorded. Copy number conservation is also documented, for example whether the gene is conserved one-to-one or whether orthologs cannot be distinguished. In some cases, faster evolving duplicates may be observed—this is where a copy of a gene appears to evolve faster than another copy of the gene

Orthologous groups	Conserved in archaea
	Conserved in bacteria
	Conserved in eukaryotes
	Conserved in fungi
	Conserved in metazoa
	Conserved in vertebrates
	Schizosaccharomyces specific
	Schizosaccharomyces pombe specific
Taxon restrictions	Conserved in fungi only
	Conserved in eukaryotes only
Others	No apparent *S.cerevisiae* ortholog
	Predominantly single copy (one-to-one)
	Orthologs cannot be distinguished
	Faster evolving duplicate

5 GO Slim Summary

PomBase maintains the fission yeast GO slim, a set of broad, high level GO biological process terms providing coverage for most gene products with process annotations (http://www.pombase.org/browse-curation/fission-yeast-go-slim-terms). Like other GO slim sets (*see* http://geneontology.org/page/go-slim-and-subset-guide), the fission yeast GO slim supports genome-level overview of GO annotation coverage, and can be used to summarize large-scale experimental results.

The PomBase GO slim terms encompass 99.5% of all genes with a biological process annotation. Uninformative (very high level grouping terms) are excluded from the PomBase GO-slim set. Table 2 shows the number of gene products annotated to each fission yeast GO slim term. Of the 5071 *S. pombe* proteins, 748 do not have a biological process annotation because their biological

Table 2
Fission yeast GO slim annotations. For each term in the fission yeast GO slim, the number of annotated genes is shown. Note that a gene may be annotated to more than one slim term

GO slim term	Number of genes
Actin cytoskeleton organization	89
Apoptotic process	8
Ascospore formation	74
Autophagy	49
Carbohydrate derivative metabolic process	276
Carbohydrate metabolic process	138
Cell adhesion	20
Cell wall organization or biogenesis	104
Cellular amino acid metabolic process	190
Chromatin organization	278
Cofactor metabolic process	177
Conjugation with cellular fusion	100
Cytoplasmic translation	485
Detoxification	59
DNA recombination	122
DNA repair	177
DNA replication	118
Establishment or maintenance of cell polarity	74
Generation of precursor metabolites and energy	81
Lipid metabolic process	232
Meiotic nuclear division	112
Membrane organization	174
Microtubule cytoskeleton organization	75
Mitochondrial gene expression	167
Mitochondrion organization	146
Mitotic cytokinesis	100
Mitotic sister chromatid segregation	176
mRNA metabolic process	271
Nitrogen cycle metabolic process	16

(continued)

Table 2 (continued)

GO slim term	Number of genes
Nucleobase-containing small molecule metabolic process	191
Nucleocytoplasmic transport	108
Peroxisome organization	22
Polyphosphate metabolic process	2
Protein catabolic process	212
Protein complex assembly	126
Protein folding	84
Protein glycosylation	68
Protein maturation	60
Protein modification by small protein conjugation or removal	98
Protein targeting	103
Regulation of mitotic cell cycle phase transition	165
Regulation of transcription, DNA-templated	415
Ribosome biogenesis	348
Signaling	292
snoRNA metabolic process	33
snRNA metabolic process	19
Sulfur compound metabolic process	109
Telomere organization	45
Transcription, DNA-templated	470
Transmembrane transport	355
tRNA metabolic process	170
Vesicle-mediated transport	329
Vitamin metabolic process	42
Proteins with a biological process annotation not covered by the slim	*27*
Proteins with no slim or biological process annotation	*748*

role is currently unknown in any species (i.e., neither the *S. pombe* protein nor any ortholog has been experimentally characterized in detail).

PomBase also maintains a list of "priority unstudied genes" for genes conserved across taxa, but not yet characterized in any species (http://www.pombase.org/status/priority-unstudied-genes).

6 Ontology Term Views

Each ontology term used in annotations or extensions (GO, Fission Yeast Phenotype Ontology (FYPO), the Sequence Ontology (SO) [18], the chemical ontology ChEBI [19], and the PSI-MOD protein modification ontology [20]) has a term page in PomBase. The top of the term page shows the name, ID, direct links to more general "parent" terms in the ontology, and external links to relevant resources (Fig. 7A). For ontologies used directly in annotations (GO, FYPO, PSI-MOD), genes are associated with the most specific annotated descendant term (Fig. 7B shows a subset of the genes annotated directly to GO:0023052 "signaling" or any of its descendant terms). As on gene pages, the default summary view can be expanded to display annotation extensions, the type of relationship between child and parent terms (e.g., *is_a*, *part_of* or *regulates*), and supporting metadata (Fig. 7C).

7 Publication Pages

Every paper cited in support of PomBase annotations has a publication page that displays citation details, the abstract, and all annotations curated from the publication (Fig. 8). Publication pages are connected from annotation tables and the literature section on gene pages, and from all pages that display the corresponding annotations. The page also acknowledges any community member who has contributed to the annotations derived from the publication.

8 Querying

PomBase offers simple and advanced search tools for querying genes and their annotations. The simple search, available on every page, retrieves individual genes by standard name, systematic ID or an *S. cerevisiae* or human ortholog name.

The advanced search retrieves sets of genes that match criteria specified by an assortment of filters (Fig. 9A). For example, ontology terms can be queried by name or ID to find annotated genes. All genes can be queried by criteria such as name, ID, product description, or chromosomal location. Additional filters are available for features of protein-coding genes. Queries can be combined to narrow down results to genes matching several criteria (Fig. 9B). Queries can be combined using the Boolean operators AND (intersect), NOT (subtract), and OR (union), and saved for reuse (Fig. 9C). For genes matching a query, IDs, names, product

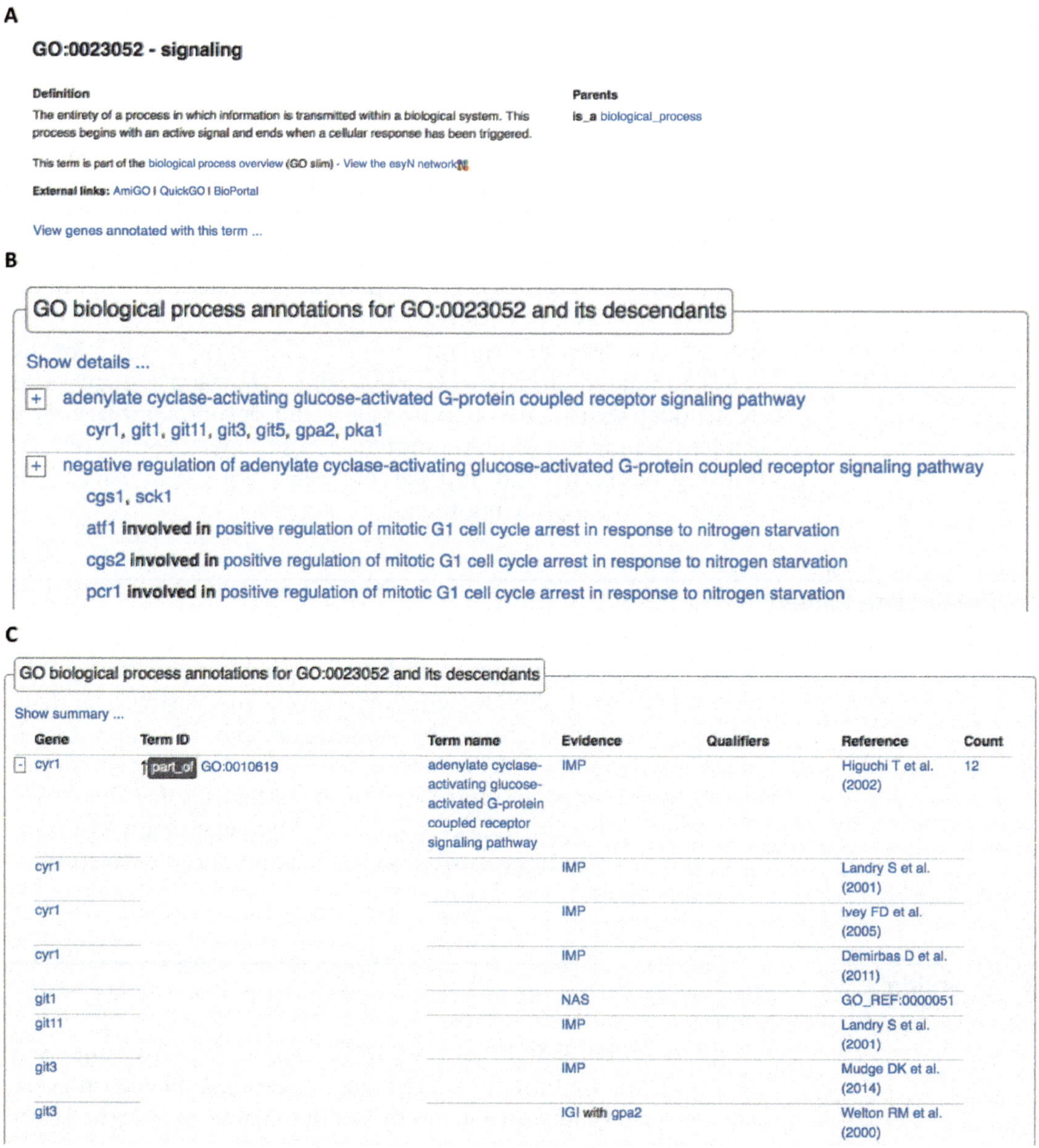

Fig. 7 Ontology term pages. (A) The top of the page shows the term name, ID, and definition, along with immediate parent terms. Links to external resources are provided. (B) The summary view shows genes annotated directly to the term or to any of its child terms, and includes extensions. (C) The detailed view provides additional information such as the relationship of child terms to the parent term, evidence codes and references

The Msd1-Wdr8-Pkl1 complex anchors microtubule minus ends to fission yeast spindle pole bodies.

Contact curators ...

Authors Yukawa M, Ikebe C, Toda T

[✓] Community curation provided by Takashi Toda

Citation J. Cell Biol. 2015 May 25;209(4):549-62

ID PMID:25987607

Links Europe PMC | PubMed

Abstract The minus ends of spindle microtubules are anchored to a microtubule-organizing center. The conserved Msd1/SSX2IP proteins are localized to the spindle pole body (SPB) and the centrosome in fission yeast and humans, respectively, and play a critical role in microtubule anchoring. In this paper, we show that fission yeast Msd1 forms a ternary complex with another conserved protein, Wdr8, and the minus end-directed Pkl1/kinesin-14. Individual deletion mutants displayed the identical spindle-protrusion phenotypes. Msd1 and Wdr8 were delivered by Pkl1 to mitotic SPBs, where Pkl1 was tethered through Msd1-Wdr8. The spindle-anchoring defect imposed by msd1/wdr8/pkl1 deletions was suppressed by a mutation of the plus end-directed Cut7/kinesin-5, which was shown to be mutual. Intriguingly, Pkl1 motor activity was not required for its anchoring role once targeted to the SPB. Therefore, spindle anchoring through Msd1-Wdr8-Pkl1 is crucial for balancing the Cut7/kinesin-5-mediated outward force at the SPB. Our analysis provides mechanistic insight into the spatiotemporal regulation of two opposing kinesins to ensure mitotic spindle bipolarity.

Annotations from this publication:

GO molecular function

Show details ...

+ ATP-dependent microtubule motor activity, minus-end-directed

pkl1 **has substrate** msd1, wdr8 **involved in** protein transport along microtubule to spindle pole body

GO biological process

Show details ...

+ microtubule anchoring at mitotic spindle pole body

msd1, pkl1, wdr8

+ mitotic sister chromatid segregation

msd1, wdr8

+ protein localization to mitotic spindle pole body

msd1 **localizes** pkl1

GO cellular component

Show details ...

+ mitotic spindle

msd1, pkl1, wdr8

+ mitotic spindle pole body

msd1 **during** mitotic M phase

pkl1 **during** mitotic M phase

wdr8 **during** mitotic M phase

Single allele phenotype

Population phenotype

Show details ... Filters: Term No filter

+ sensitive to thiabendazole

msd1Δ, pkl1Δ, wdr8Δ

Cell phenotype

Show details ... Filters: Term No filter

+ abnormal protein localization to microtubule minus-end

msd1Δ **affecting** cut7

wdr8Δ **affecting** cut7

Physical interaction

Gene	Product		Interacting gene	Interacting product	Evidence
pkl1	minus-end directed microtubule motor, kinesin Pkl1	affinity captures	msd1	microtubule-anchoring factor Msd1	Affinity Capture-Western

Fig. 8 Publication pages. The PMID:25987607 page shows publication details and a community curator acknowledgement at the top, and annotations derived from the paper. GO and FYPO annotations have summary and detailed views as on gene and ontology term pages

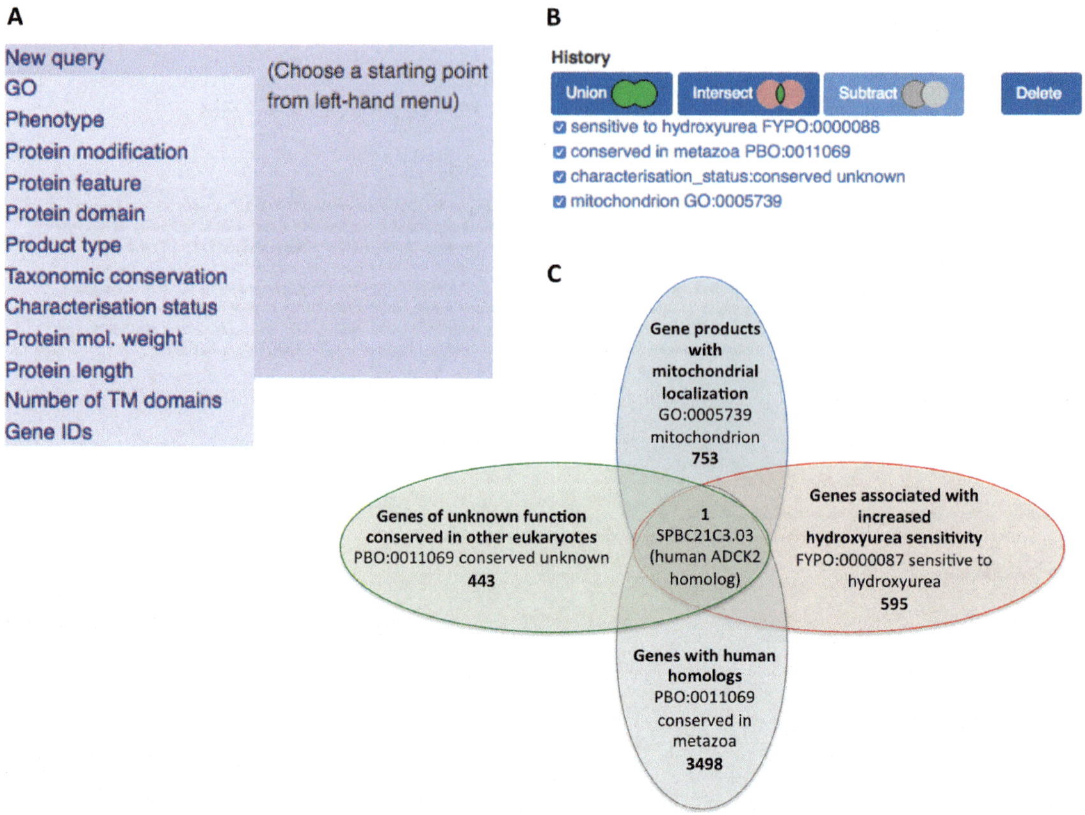

Fig. 9 Query builder filtering. (A) A list of the different filters available to identify genes of interest. (B) The history section can be used to review previously run queries. Queries can be combined using the union, intersect and subtract operators. (C) An example of the results of running and combining queries. 753 genes (August 2017) are annotated to the GO term mitochondrion. Of these, 3498 are conserved in metazoa, 595 genes where any type of allele has been associated with hydroxyuruea sensitivity and 411 genes with the characterization status "conserved unknown," i.e., of unknown function but conserved in other organisms. The union of these four queries identifies one gene

descriptions, and sequences can be downloaded. More flexible download options for query results are slated for addition to the advanced search.

An additional stand-alone motif search tool searches all protein coding sequences to identify genes containing a specified amino acid pattern of interest.

9 Genome Browser and Datasets

The PomBase genome browser supports sequence viewing based on coordinates or feature identifiers. Data tracks are available for sequence-based datasets submitted by the fission yeast community from a variety of high-throughput experiments, including transcriptomic data [21–23], nucleosome positioning [23], replication profiling [24], polyadenylation sites [25, 26], and chromatin binding [27]. (Note: at the time of writing, PomBase is in the process of transitioning from a legacy browser to a JBrowse [28] implementation.)

PomBase also provides a set of static pages describing various aspects of genome-level curation status and links to external community resources. The genome sequence and several annotation datasets (GO, phenotype, and modification data, orthologs, interactions, protein features, etc.) can be downloaded from PomBase's FTP site.

10 Community and Outreach

PomBase makes community engagement a high priority, welcoming data submissions and feedback on the resources we provide.

In addition to using Canto community curation as the primary mechanism for data collection from newly published papers, PomBase invites researchers to submit large-scale datasets for phenotype, expression, and other annotations in spreadsheet-compatible formats as well as datasets suitable to appear on genome browser tracks. The most recent community curation submissions are showcased on the PomBase front page, and PomBase is exploring mechanisms for curation attribution via ORCIDs (https://orcid.org/).

We communicate with fission yeast researchers directly via our 1200-member community mailing list (pombelist) and at workshops and conferences, notably the biennial international *S. pombe* conference. To support PomBase usage, we run a helpdesk and maintain extensive documentation covering PomBase pages, annotation conventions, and Canto features. Advice on data usage and analysis disseminated via the helpdesk becomes incorporated into the extensive FAQ. Documentation and FAQs are available at http://www.pombase.org/help. We run periodic surveys to determine community priorities for new PomBase features and improvements to existing resources, and actively encourage corrections, improvements and suggestions to existing content of PomBase at all times.

Acknowledgments

PomBase is supported by the Wellcome Trust grant 104967/Z/14/Z. We thank the fission yeast community for curation contributions, suggestions, and their constant support of our endeavors. We also thank the biocuration community for effective discussions and the continued endeavor for excellence in data communication.

References

1. McDowall MD, Harris MA, Lock A, Rutherford K, Staines DM, Bahler J, Kersey PJ, Oliver SG, Wood V (2015) PomBase 2015: updates to the fission yeast database. Nucleic Acids Res 43(Database issue):D656–D661. https://doi.org/10.1093/nar/gku1040
2. Wood V, Harris MA, McDowall MD, Rutherford K, Vaughan BW, Staines DM, Aslett M, Lock A, Bahler J, Kersey PJ, Oliver SG (2012) PomBase: a comprehensive online resource for fission yeast. Nucleic Acids Res 40(Database issue):D695–D699. https://doi.org/10.1093/nar/gkr853
3. Kim DU, Hayles J, Kim D, Wood V, Park HO, Won M, Yoo HS, Duhig T, Nam M, Palmer G, Han S, Jeffery L, Baek ST, Lee H, Shim YS, Lee M, Kim L, Heo KS, Noh EJ, Lee AR, Jang YJ, Chung KS, Choi SJ, Park JY, Park Y, Kim HM, Park SK, Park HJ, Kang EJ, Kim HB, Kang HS, Park HM, Kim K, Song K, Song KB, Nurse P, Hoe KL (2010) Analysis of a genome-wide set of gene deletions in the fission yeast *Schizosaccharomyces pombe*. Nat Biotechnol 28(6):617–623. https://doi.org/10.1038/nbt.1628
4. Spirek M, Benko Z, Carnecka M, Rumpf C, Cipak L, Batova M, Marova I, Nam M, Kim DU, Park HO, Hayles J, Hoe KL, Nurse P, Gregan J (2010) S. pombe genome deletion project: an update. Cell Cycle 9(12):2399–2402. https://doi.org/10.4161/cc.9.12.11914
5. Matsuyama A, Arai R, Yashiroda Y, Shirai A, Kamata A, Sekido S, Kobayashi Y, Hashimoto A, Hamamoto M, Hiraoka Y, Horinouchi S, Yoshida M (2006) ORFeome cloning and global analysis of protein localization in the fission yeast Schizosaccharomyces pombe. Nat Biotechnol 24(7):841–847. https://doi.org/10.1038/nbt1222
6. Wood V, Gwilliam R, Rajandream MA, Lyne M, Lyne R, Stewart A, Sgouros J, Peat N, Hayles J, Baker S, Basham D, Bowman S, Brooks K, Brown D, Brown S, Chillingworth T, Churcher C, Collins M, Connor R, Cronin A, Davis P, Feltwell T, Fraser A, Gentles S, Goble A, Hamlin N, Harris D, Hidalgo J, Hodgson G, Holroyd S, Hornsby T, Howarth S, Huckle EJ, Hunt S, Jagels K, James K, Jones L, Jones M, Leather S, McDonald S, McLean J, Mooney P, Moule S, Mungall K, Murphy L, Niblett D, Odell C, Oliver K, O'Neil S, Pearson D, Quail MA, Rabbinowitsch E, Rutherford K, Rutter S, Saunders D, Seeger K, Sharp S, Skelton J, Simmonds M, Squares R, Squares S, Stevens K, Taylor K, Taylor RG, Tivey A, Walsh S, Warren T, Whitehead S, Woodward J, Volckaert G, Aert R, Robben J, Grymonprez B, Weltjens I, Vanstreels E, Rieger M, Schafer M, Muller-Auer S, Gabel C, Fuchs M, Dusterhoft A, Fritzc C, Holzer E, Moestl D, Hilbert H, Borzym K, Langer I, Beck A, Lehrach H, Reinhardt R, Pohl TM, Eger P, Zimmermann W, Wedler H, Wambutt R, Purnelle B, Goffeau A, Cadieu E, Dreano S, Gloux S, Lelaure V, Mottier S, Galibert F, Aves SJ, Xiang Z, Hunt C, Moore K, Hurst SM, Lucas M, Rochet M, Gaillardin C, Tallada VA, Garzon A, Thode G, Daga RR, Cruzado L, Jimenez J, Sanchez M, del Rey F, Benito J, Dominguez A, Revuelta JL, Moreno S, Armstrong J, Forsburg SL, Cerutti L, Lowe T, McCombie WR, Paulsen I, Potashkin J, Shpakovski GV, Ussery D, Barrell BG, Nurse P (2002) The genome sequence of *Schizosaccharomyces pombe*. Nature 415(6874):871–880. https://doi.org/10.1038/nature724
7. Hoffman CS, Wood V, Fantes PA (2015) An ancient yeast for young geneticists: a primer on the *Schizosaccharomyces pombe* model system. Genetics 201(2):403–423. https://doi.org/10.1534/genetics.115.181503
8. Nguyen TT, Chua JK, Seah KS, Koo SH, Yee JY, Yang EG, Lim KK, Pang SY, Yuen A, Zhang L, Ang WH, Dymock B, Lee EJ, Chen ES (2016) Predicting chemotherapeutic drug combinations through gene network profiling. Sci Rep 6:18658. https://doi.org/10.1038/srep18658

9. Rhind N, Russell P (2012) Signaling pathways that regulate cell division. Cold Spring Harb Perspect Biol 4(10). https://doi.org/10.1101/cshperspect.a005942
10. Rosas-Murrieta NH, Rojas-Sánchez G, Reyes-Carmona SR, Martínez-Contreras RD, Martínez-Montiel N, Millán-Pérez-Peña L, Herrera-Camacho IP (2015) Study of cellular processes in higher eukaryotes using the yeast *Schizosaccharomyces pombe* as a model. In: Shah MM (ed) Microbiology in agriculture and human health. https://doi.org/10.5772/60720
11. Rutherford KM, Harris MA, Lock A, Oliver SG, Wood V (2014) Canto: an online tool for community literature curation. Bioinformatics 30(12):1791–1792. https://doi.org/10.1093/bioinformatics/btu103
12. Oliver SG, Lock A, Harris MA, Nurse P, Wood V (2016) Model organism databases: essential resources that need the support of both funders and users. BMC Biol 14:49. https://doi.org/10.1186/s12915-016-0276-z
13. Ashburner M, Ball CA, Blake JA, Botstein D, Butler H, Cherry JM, Davis AP, Dolinski K, Dwight SS, Eppig JT, Harris MA, Hill DP, Issel-Tarver L, Kasarskis A, Lewis S, Matese JC, Richardson JE, Ringwald M, Rubin GM, Sherlock G (2000) Gene ontology: tool for the unification of biology. The gene ontology consortium. Nat Genet 25(1):25–29. https://doi.org/10.1038/75556
14. Consortium GO (2015) Gene ontology consortium: going forward. Nucleic Acids Res 43(Database issue):D1049–D1056. https://doi.org/10.1093/nar/gku1179
15. Huntley RP, Harris MA, Alam-Faruque Y, Blake JA, Carbon S, Dietze H, Dimmer EC, Foulger RE, Hill DP, Khodiyar VK, Lock A, Lomax J, Lovering RC, Mutowo-Meullenet P, Sawford T, Van Auken K, Wood V, Mungall CJ (2014) A method for increasing expressivity of gene ontology annotations using a compositional approach. BMC Bioinformatics 15:155. https://doi.org/10.1186/1471-2105-15-155
16. Harris MA, Lock A, Bahler J, Oliver SG, Wood V (2013) FYPO: the fission yeast phenotype ontology. Bioinformatics 29(13):1671–1678. https://doi.org/10.1093/bioinformatics/btt266
17. Wood V (2005) *Schizosaccharomyces pombe* comparative genomics; from sequence to systems. In: Sunnerhagen P, Piskur J (eds) Topics in current genetics, vol 15. Springer, Berlin, pp 233–285. https://doi.org/10.1007/4735_97
18. Eilbeck K, Lewis SE, Mungall CJ, Yandell M, Stein L, Durbin R, Ashburner M (2005) The sequence ontology: a tool for the unification of genome annotations. Genome Biol 6(5):R44. https://doi.org/10.1186/gb-2005-6-5-r44
19. Hastings J, de Matos P, Dekker A, Ennis M, Harsha B, Kale N, Muthukrishnan V, Owen G, Turner S, Williams M, Steinbeck C (2013) The ChEBI reference database and ontology for biologically relevant chemistry: enhancements for 2013. Nucleic Acids Res 41(Database issue):D456–D463. https://doi.org/10.1093/nar/gks1146
20. Montecchi-Palazzi L, Beavis R, Binz PA, Chalkley RJ, Cottrell J, Creasy D, Shofstahl J, Seymour SL, Garavelli JS (2008) The PSI-MOD community standard for representation of protein modification data. Nat Biotechnol 26(8):864–866. https://doi.org/10.1038/nbt0808-864
21. Marguerat S, Schmidt A, Codlin S, Chen W, Aebersold R, Bahler J (2012) Quantitative analysis of fission yeast transcriptomes and proteomes in proliferating and quiescent cells. Cell 151(3):671–683. https://doi.org/10.1016/j.cell.2012.09.019
22. Rhind N, Chen Z, Yassour M, Thompson DA, Haas BJ, Habib N, Wapinski I, Roy S, Lin MF, Heiman DI, Young SK, Furuya K, Guo Y, Pidoux A, Chen HM, Robbertse B, Goldberg JM, Aoki K, Bayne EH, Berlin AM, Desjardins CA, Dobbs E, Dukaj L, Fan L, FitzGerald MG, French C, Gujja S, Hansen K, Keifenheim D, Levin JZ, Mosher RA, Muller CA, Pfiffner J, Priest M, Russ C, Smialowska A, Swoboda P, Sykes SM, Vaughn M, Vengrova S, Yoder R, Zeng Q, Allshire R, Baulcombe D, Birren BW, Brown W, Ekwall K, Kellis M, Leatherwood J, Levin H, Margalit H, Martienssen R, Nieduszynski CA, Spatafora JW, Friedman N, Dalgaard JZ, Baumann P, Niki H, Regev A, Nusbaum C (2011) Comparative functional genomics of the fission yeasts. Science 332(6032):930–936. https://doi.org/10.1126/science.1203357
23. Soriano I, Quintales L, Antequera F (2013) Clustered regulatory elements at nucleosome-depleted regions punctuate a constant nucleosomal landscape in Schizosaccharomyces pombe. BMC Genomics 14:813. https://doi.org/10.1186/1471-2164-14-813
24. Xu J, Yanagisawa Y, Tsankov AM, Hart C, Aoki K, Kommajosyula N, Steinmann KE, Bochicchio J, Russ C, Regev A, Rando OJ, Nusbaum C, Niki H, Milos P, Weng Z, Rhind N (2012) Genome-wide identification and characterization of replication origins by deep sequencing. Genome Biol 13(4):R27. https://doi.org/10.1186/gb-2012-13-4-r27

25. Mata J (2013) Genome-wide mapping of polyadenylation sites in fission yeast reveals widespread alternative polyadenylation. RNA Biol 10(8):1407–1414. https://doi.org/10.4161/rna.25758
26. Schlackow M, Marguerat S, Proudfoot NJ, Bahler J, Erban R, Gullerova M (2013) Genome-wide analysis of poly(A) site selection in Schizosaccharomyces pombe. RNA 19(12):1617–1631. https://doi.org/10.1261/rna.040675.113
27. Woolcock KJ, Gaidatzis D, Punga T, Buhler M (2011) Dicer associates with chromatin to repress genome activity in *Schizosaccharomyces pombe*. Nat Struct Mol Biol 18(1):94–99. https://doi.org/10.1038/nsmb.1935
28. Skinner ME, Uzilov AV, Stein LD, Mungall CJ, Holmes IH (2009) JBrowse: a next-generation genome browser. Genome Res 19(9):1630–1638. https://doi.org/10.1101/gr.094607.109

Open Access This chapter is licensed under the terms of the Creative Commons Attribution 4.0 International License (http://creativecommons.org/licenses/by/4.0/), which permits use, sharing, adaptation, distribution and reproduction in any medium or format, as long as you give appropriate credit to the original author(s) and the source, provide a link to the Creative Commons license and indicate if changes were made.

The images or other third party material in this chapter are included in the chapter's Creative Commons license, unless indicated otherwise in a credit line to the material. If material is not included in the chapter's Creative Commons license and your intended use is not permitted by statutory regulation or exceeds the permitted use, you will need to obtain permission directly from the copyright holder.

Chapter 5

EuPathDB: The Eukaryotic Pathogen Genomics Database Resource

Susanne Warrenfeltz, Evelina Y. Basenko, Kathryn Crouch, Omar S. Harb, Jessica C. Kissinger, David S. Roos, Achchuthan Shanmugasundram, and Fatima Silva-Franco

Abstract

Fighting infections and developing novel drugs and vaccines requires advanced knowledge of pathogen's biology. Readily accessible genomic, functional genomic, and population data aids biological and translational discovery. The Eukaryotic Pathogen Database Resources (http://eupathdb.org) are data mining resources that support hypothesis driven research by facilitating the discovery of meaningful biological relationships from large volumes of data. The resource encompasses 13 sites that support over 170 species including pathogenic protists, oomycetes, and fungi as well as evolutionarily related nonpathogenic species. EuPathDB integrates preanalyzed data with advanced search capabilities, data visualization, analysis tools and a comprehensive record system in a graphical interface that does not require prior computational skills. This chapter describes guiding concepts common across EuPathDB sites and illustrates the powerful data mining capabilities of some of the available tools and features.

Key words Bioinformatics, Parasite, Pathogen, Genomics, Transcriptomics, Orthology, Fungi, Proteomics, Sequence analysis

1 Introduction

The Eukaryotic Pathogen Database [1, 2] (EuPathDB, http://eupathdb.org) brings the power of bioinformatics to the scientific community by integrating preanalyzed omics data with advanced search capabilities, data visualization and analysis tools that facilitate the discovery of meaningful biological relationships from large volumes of data. EuPathDB provides a sophisticated data mining platform for biologists with no prior computational training to explore omics data in support of their hypothesis driven research.

The resource is organized into 13 sites (Table 1) and supports over 170 eukaryotic parasites, relevant free-living nonparasitic organisms and selected pathogen hosts. EuPathDB resources provide customized sites for accessing genomes and functional

Martin Kollmar (ed.), *Eukaryotic Genomic Databases: Methods and Protocols*, Methods in Molecular Biology, vol. 1757, https://doi.org/10.1007/978-1-4939-7737-6_5, © Springer Science+Business Media, LLC, part of Springer Nature 2018

Table 1
EuPathDB resources

Resource	Address (http://)	Organisms supported
EuPathDB	eupathdb.org	All organisms below
AmoebaDB	amoebadb.org	*Acanthamoeba* (11), *Entamoeba* (5), *Naegleria*
CryptoDB	cryptodb.org	*Chromera*, *Cryptosporidium* (8), *Gregarina*, *Vitrella*
FungiDB	fungidb.org	*Agaricomycetes* (2), *Blastocladiomycetes*, *Chytridiomycetes* (2), *Dothideomycetes*, *Eurotiomycetes* (27), *Leotiomycetes* (2), *Pneumocystidomycetes*, *Pucciniomycetes* (2), *Saccharomycetes* (4), *Schizosaccharomycetes* (3), *Sordariomycetes* (11), *Tremellomycetes* (4), *Ustilaginomycetes* (3), *Zygomycetes* (3) + oomycetes
HostDB	hostdb.org	*Homo sapiens*, *Macaca mulatta*, *Mus musculus*
GiardiaDB	giardiadb.org	*Giardia assemblages* (3), *Spironucleus*
MicrosporidiaDB	microsporidiadb.org	*Anncaliia*, *Edhazardia*, *Encephalitozoon* (4), *Enterocytozoon*, *Mitosporidium*, *Nematocida* (3), *Nosema* (2), *Ordospora*, *Pseudoloma*, *Spraguea*, *Trachipleistophora*, *Vavraia*, *Vittaforma*
PiroplasmaDB	piroplasmadb.org	*Babesia* (3), *Cytauzoon*, *Theileria* (4)
PlasmoDB	plasmodb.org	*Plasmodium* (21)
ToxoDB	toxodb.org	*Cyclospora*, *Eimeria* (8), *Hammondia*, *Neospora*, *Sarcocystis* (2), *Toxoplasma* (18 strains)
TriTrypDB	tritrypdb.org	*Blechomonas*, *Crithidia*, *Endotrypanum*, *Leishmania* (15), *Leptomonas* (2), *Trypanosoma* (8)
TrichDB	trichdb.org	*Trichomonas*
OrthoMCL	orthomcl.org	Proteins from 150 organisms across the tree of life

Number in parentheses indicates the number of organisms from that genus

data based on taxonomic groupings, host data collected during infection, evolutionary relationships based on OrthoMCL clustering, and a portal site that enables queries across all data in EuPathDB. The sites are built on the same web architecture and use the same common vocabulary and organizing logic, which eases the transfer between the search, visualization, and analysis sections and allows users to move between sites without reeducation.

EuPathDB integrates a wide range of data from many sources and repositories (Table 2). The breadth of data broadens the data mining capabilities by providing multiple forms of experimental evidence to search, visualize and analyze. As the data are integrated, they are analyzed with standard workflows, ensuring that data from different sources can be compared. An in-house analysis

Table 2
List of major data types and example techniques

Data type	Example technique	Example source
Genome sequence and annotation	Illumina, 454, PacBio	NCBI GenBank
Orthology profiles	Orthology group assignments via in-house OrthoMCL analysis	In-house analysis
Genome analyses	Splign, Tandem repeat finder, Low complexity finders	In-house analysis
Domain predictions	SignalP, HMMPfam, TMHMM	In-house analysis
Transcriptomics	RNA Sequencing, microarray, ESTs	NCBI SRA, GEO
Proteomics	Mass spec evidence, Quantitative MS evidence	Individual labs, literature
Epigenomics	ChIP-chip, ChIP Seq	NCBI SRA, GEO
Metabolomics	Mass spec evidence, Metabolite	Individual labs, literature
Isolate data	Population resequencing, PopSet sequences	NCBI SRA, PopSet
Host pathogen interactions	Protein array (serum)	Individual labs, literature
Metabolic pathways	N/A	MetaCyc, KEGG, TrypanoCyc, LeishCyc
Compounds	N/A	EBI ChEBI
Phenotypes	CRISPR screens, curation	Broad, AspGD, Individual labs, literature

pipeline also creates orthology profiles across all genomes so that comparisons can be made across organisms.

Data mining in EuPathDB sites can take four general paths. First, record pages compile all available data concerning a feature (gene, SNP, pathway, compound, EST, genomic sequence, etc.) and offer rich data mining opportunities. Second, the search strategy system's unique infrastructure facilitates the exploration of relationships across data sets, data types, and organisms to produce a refined set of features that share biological characteristics of interest.

Third, visualization tools such as the Genome Browser (GBrowse) [3] coupled to EuPathDB's breadth of sequence-based data offer the ability to view different data types in your genomic area of interest. And fourth, tools such as enrichment analyses and a private Galaxy workspace for primary data analyses enhance data mining.

EuPathDB makes it easy to interrogate biological questions relating to issues such as stage-specific expression, gene model integrity or alternative splice variants, and to compile lists of genes that share multiple biological characteristics (e.g., kinases secreted at a particular time, where they may affect host responses). This chapter describes the structure and utility of EuPathDB and illustrates some of the available tools and methods. Since EuPathDB sites are built using the same infrastructure and user interface, the steps described herein can be applied to any EuPathDB site.

2 Using EuPathDB Sites

The exercises below offer example data mining strategies. Because new versions of EuPathDB resources are released about every 2 months and may contain new annotation and functional data, reader results (gene numbers etc.) may vary slightly from that published here.

2.1 Home Pages

EuPathDB home pages are organized using the same layout and provide users with easy access to all the searches, tools, educational material, and helpful database and community information.

2.1.1 Anatomy of EuPathDB Home Pages

1. Visit the EuPathDB home page (http://eupathdb.org) and explore the four sections: the header (Fig. 1A), component link-out section (Fig. 1B), the searches and tools section (Fig. 1C), and the side bar (Fig. 1D).
2. The header (Fig. 1A) is available from all pages and includes a gray menu bar that offers drop-down menus or direct links for accessing most searches, tools, data set information, bulk downloads, and the Galaxy workspace. Above the gray menu bar are two search boxes (Fig. 1A, arrow) for quick access to gene record pages or to a text search that returns genes whose records contain the term(s) of interest. Directly below the search boxes are links to "Login," "Register," or "Contact Us." Although not required to access data records or build search strategies, registering provides access to additional tools and functionality such as the ability to save and share search strategies, to add genes to the "My Basket" and "My Favorites," to add comments on gene records and the Galaxy workspace. The My Strategies page is an important part of the sites, serving as a workspace for creating strategies and viewing search and

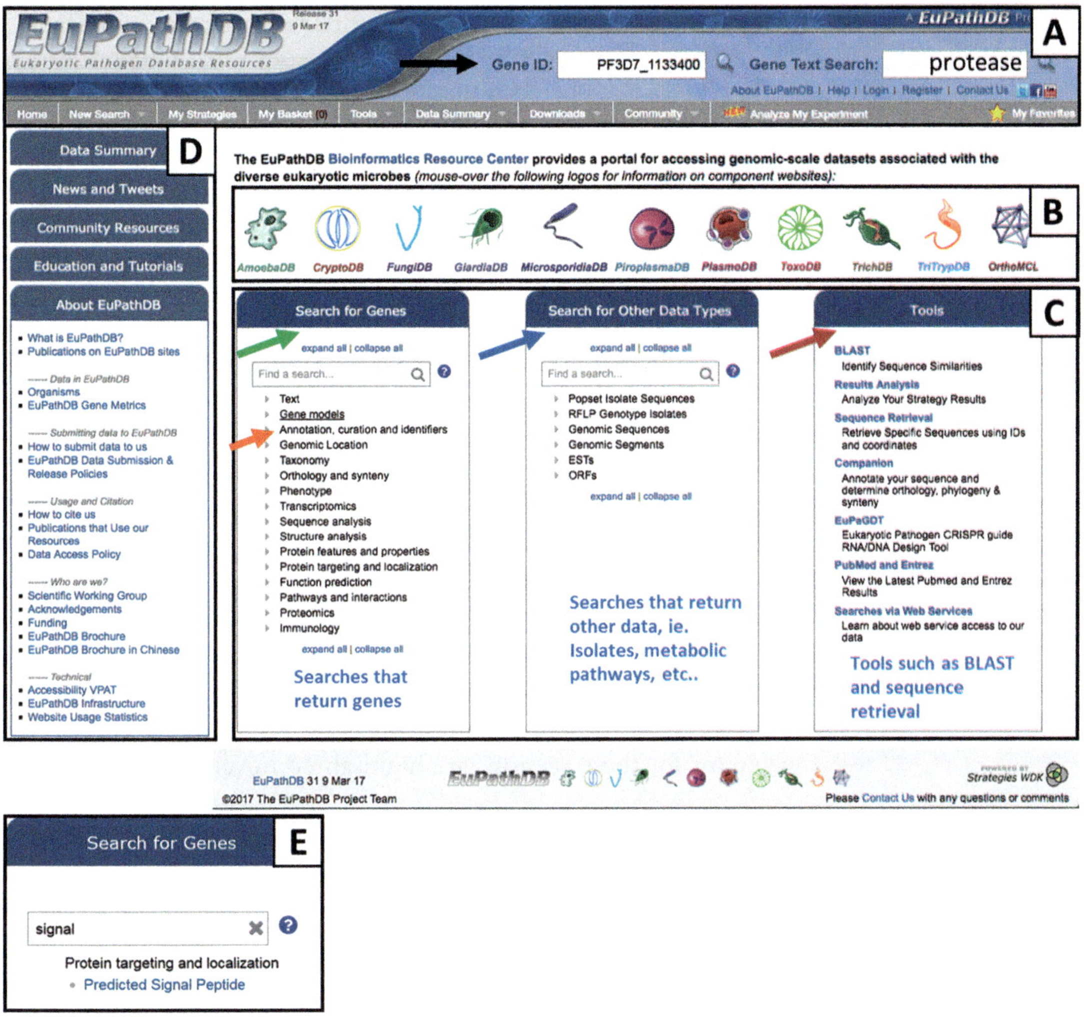

Fig. 1 EuPathDB home page and its main features. (A) The interactive header is visible from any EuPathDB page. The tabs and drop-down menus in the gray menu bar provide access to all EuPathDB searches and tools. (B) The component site link outs section provides direct links to the taxon-specific sites. (C) The core section consisting of three panels: "Search for Genes," "Search for Other Data Types," and "Tools." (D) The side bar contains useful links and information including news releases, community resources and a summary of integrated data. (E) Find a Search Tool. This text search finds available searches within the Search for Genes bubble

strategy results. The "Contact Us" link opens a form for sending questions, comments, and suggestions to our email support line.

3. The component link-out section (Fig. 1B) is available on the EuPathDB home page and at the bottom of all pages of all sites. This section offers direct links to the taxon-specific sites as well as OrthoMCL DB. Click the icons to navigate to the site of your choice.

4. The searches and tools section contains three panels for accessing searches and tools (Fig. 1C). Searches listed under "Search for Genes" (Fig. 1C, green arrow) return only genes while searches that return nongene entities such as SNPs, isolates, or ESTs are available from "Search for Other Data Types" (Fig. 1C, blue arrow). The searches are organized into categories that can be expanded to reveal individual searches. Alternatively, searches can be filtered using the "Find a Search" tool. For example, typing "signal" in the "Find a Search" box of the "Search for Genes" panel filters the searches and reveals the "Predicted Signal Peptide" search within the category "Protein targeting and localization" (Fig. 1E).

   Also available from the home page are tools for BLAST, Results Analysis, Sequence Retrieval, Genome Browser, Companion annotation pipeline [4], and EuPaGDT (Eukaryotic Pathogen CRISPR guide RNA Design Tool) [5] (Fig. 1C, red arrow). The Results Analysis tool enables functional enrichment of output gene lists from the search strategies (*see* Subheading 2.5 for further details).

5. The side bar (Fig. 1D) contains expandable sections for data summary, release notes, Twitter feed, community resources, links to workshop materials, tutorials, and help. Newly added items for these sections are highlighted in yellow.

2.1.2 Gene ID and Gene Text Search Access from Home Pages

There are two ways to access the Gene ID and Gene Text searches: through the search boxes in the header (Fig. 1A, arrow) and through the dedicated search pages categorized in the "Search for Genes" panel on the home page (Fig. 1C, green arrow). Entering a gene ID in the header "Gene ID" box navigates directly to the record page for that gene. Entering a text term or phrase (within quotation marks) in the header "Gene Text Search" box initiates a preconfigured search for genes whose records contain the text term or phrase. The dedicated search pages offer additional options. The Gene ID search page, accessed under the "Annotation, curation and identifiers" category (Fig. 1C, orange arrow), allows a user to search for gene IDs in bulk. A list of gene IDs can be pasted into the text box, uploaded from a file, or converted from a user's basket. The gene text search page can be found in the "Text" category (first category in the list) and allows a user to configure the sections of the gene record that they want to search. For example, the text search can be limited to search only the product description of genes. Both searches support a wild card to perform partial text or ID searches. For example, a text search of the term "phospho*" (the asterisk * is the wild card meaning any character) will return any gene whose record contains any word with the prefix "phospho."

1. *Use the "Gene Text Search" in the EuPathDB home page header to find genes that are likely proteases.* Enter the term "protease"

(without the quotes) in the search box (Fig. 1A) and click on the search icon to the right of the box to initiate a query against all annotated genomes for genes whose records include the term protease. The results (>20,000 genes, results may vary in subsequent database releases) appear in the "My Strategies" section (Fig. 2) and consist of the strategy panel with a graphic representation of the strategy (Fig. 2A), a component website filter which displays the distribution of genes across the taxon-specific sites (Fig. 2B, upper table), an organism table which displays the distribution of genes for the genomes that were queried (ranging from 0 to >400 genes per species) (Fig. 2B, lower table), and the Gene Results consisting of two tabs. The "Gene Results" tab lists gene IDs and associated data for genes returned by the search (Fig. 2C, showing). The "Genome View" tab (Fig. 2C, black arrow) presents a graphic representation of the genomic sequences "painted" with the gene results when there are <10,000 genes in the result. The Analyze Results button (Fig. 2C, blue arrow) opens a tool offering enrichment and other analyses of the gene result.

2. *Explore your result.* The "Gene Results" table contains columns of data associated with the genes that were returned by your search. You can add columns to the table using the "Add Columns" button (Fig. 2C, green arrow.) Look at the product description column. Cathepsin B precursor, GL50803_10217, is returned by the search but does not have the term protease in the product description. In this case, the term protease was found in an InterPro domain and a user comment associated with the gene.

3. *Find several genes using the Gene ID search.* Navigate to the EuPathDB home page by clicking the Home button, the first tab in the header's gray menu bar. Open the Gene ID search page by first clicking on the category "Annotation, curation, identifiers," then clicking on "List of IDs" in the "Search for Genes" panel. On the next page, paste the following list of IDs in the search box and click on the "Get Answer" button:

   TGME49_049180, TA08775, PFD0830w, PCHAS_072830, PBANKA_071930, NCU10053T0, NCLIV_065390, LmxM.06.0860, ECU01_1430, CMU_010300

   The results are displayed as a search strategy including all the genes from the above list in one step.

4. *Notice the gene results.* The filter table contains hits from several different species. Examine the Product description column. These genes are orthologs of dihydrofolate reductase-thymidylate synthase. Try running the same search in PlasmoDB.org (http://plasmodb.org). Since PlasmoDB accesses a reduced taxonomic group of genomes while EuPathDB access all genomes and data, only the *plasmodium*

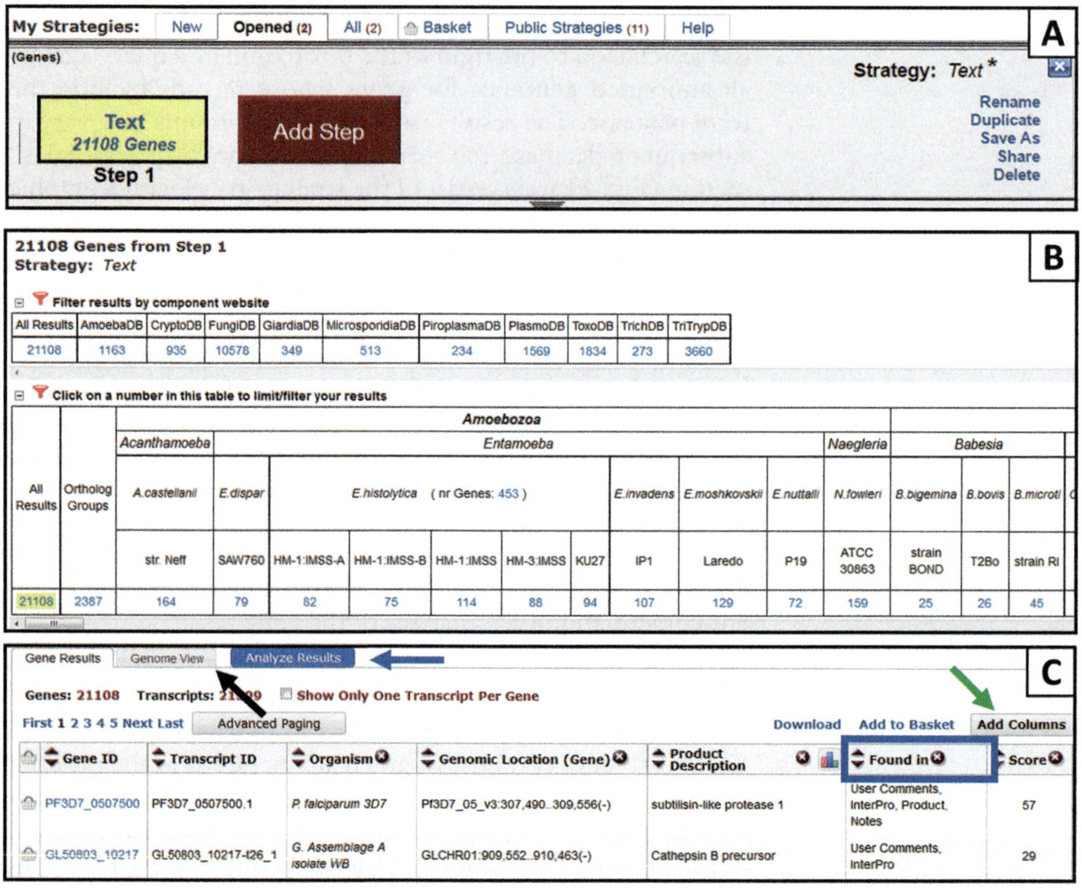

Fig. 2 Result of a text search in EuPathDB. Search results are presented in the My Strategies section and consist of three parts. (A) The strategy panel provides a graphic representation of the search or strategy result. The search result highlighted in yellow is the "active" result and further displayed in the Filter tables (B) and the Gene Result (C). (B) The component site and organism filter tables show the distribution of hits from the result across the taxon-specific sites and the organisms queried, respectively. (C) The result tables currently showing the Gene Result tab which lists all hits for the active search result. The first column, Gene ID, is a link to the record page for that gene

orthologs of dihydrofolate reductase-thymidylate synthase are returned by the PlasmoDB search.

2.2 Exploring Record Pages

Record pages compile all available data for an entity, including genes, SNPs, ESTs, isolates, pathways, compounds, genomic sequences, genomic segments, and ORFs. The following two examples describe the features, navigation, and data content of gene and metabolic pathway record pages.

2.2.1 Gene Record Pages

1. *Anatomy of the gene page*: Visit PlasmoDB (http://plasmodb.org) and enter the gene ID for apical membrane antigen 1, PF3D7_1133400, in the "Gene ID" box in the header.

Designed for easy navigation and access to data of interest, gene record pages contain three major areas, the summary (Fig. 3A, B), the data section (Fig. 3C), and the content navigation (Fig. 3D). The summary provides basic information about the gene (Fig. 3A). The "Shortcuts" that appear in the summary (Fig. 3B) serve two functions: clicking on the magnifying glass icon at the bottom right corner of the thumbnail provides a graphic summary of that data type (Fig. 4A, green arrow); clicking on the image itself, or the title above it, will navigate to that section of the page (Fig. 4A, blue ovals and dashed lines). Several gene page sections contain a "View in genome browser" link which opens the genome browser with the pertinent data tracks open (Fig. 4B, blue arrow) (*see* Subheading 2.4 on data visualization). The "Add to basket" and "Add to favorites" links (Fig. 3A, arrow) will save or bookmark the gene for later use. The "Download Gene" link opens the download tool where the FASTA formatted sequence or all information on the gene page can be downloaded (Fig. 5). The content navigation section on the left side of the gene page (Fig. 3D) serves as a configurable table of contents of all information found in the data section and remains available on the left side of the page as you scroll down. The data section contains all information available for the gene (Fig. 3C). Table 3 describes the data available on the gene page. The data are presented both in graphs and in searchable tables, and sections can be collapsed or expanded using the triangle present in the title of each section (Fig. 3C, black arrow).

2. *Transcriptomics section*: Use the "Contents" navigation menu (Fig. 3D) to navigate to the transcriptomics section of the PF3D7_1133400 gene page by clicking on the section title "Transcriptomics." Alternatively, use the "Search Section Names" tool at the top of the Contents navigation menu to search for the Transcriptomics section. The transcriptomics table (Fig. 6) appears in the data section of the gene page and contains collapsible rows for each data set. Scroll down and click the triangle present in the header of experiments "Polysomal and steady-state asexual stage transcriptomes" (Fig. 6, blue circle) [6] and "Transcriptomes of 7 sexual and asexual life stages" [7]. The rows expand to reveal expression graphs, data tables and coverage plots relative to the data set. Explore the graphs and data tables for these two experiments. At what life cycle stage is the expression highest for Pf3D7_1133400? (answer = schizont stage)

3. *Proteomics section*: To navigate to the proteomics section click on "Proteomics" in the contents navigation or use the "Back to top" arrow to return to the summary section and click on the Proteomics shortcut image. The "Mass Spec.-based

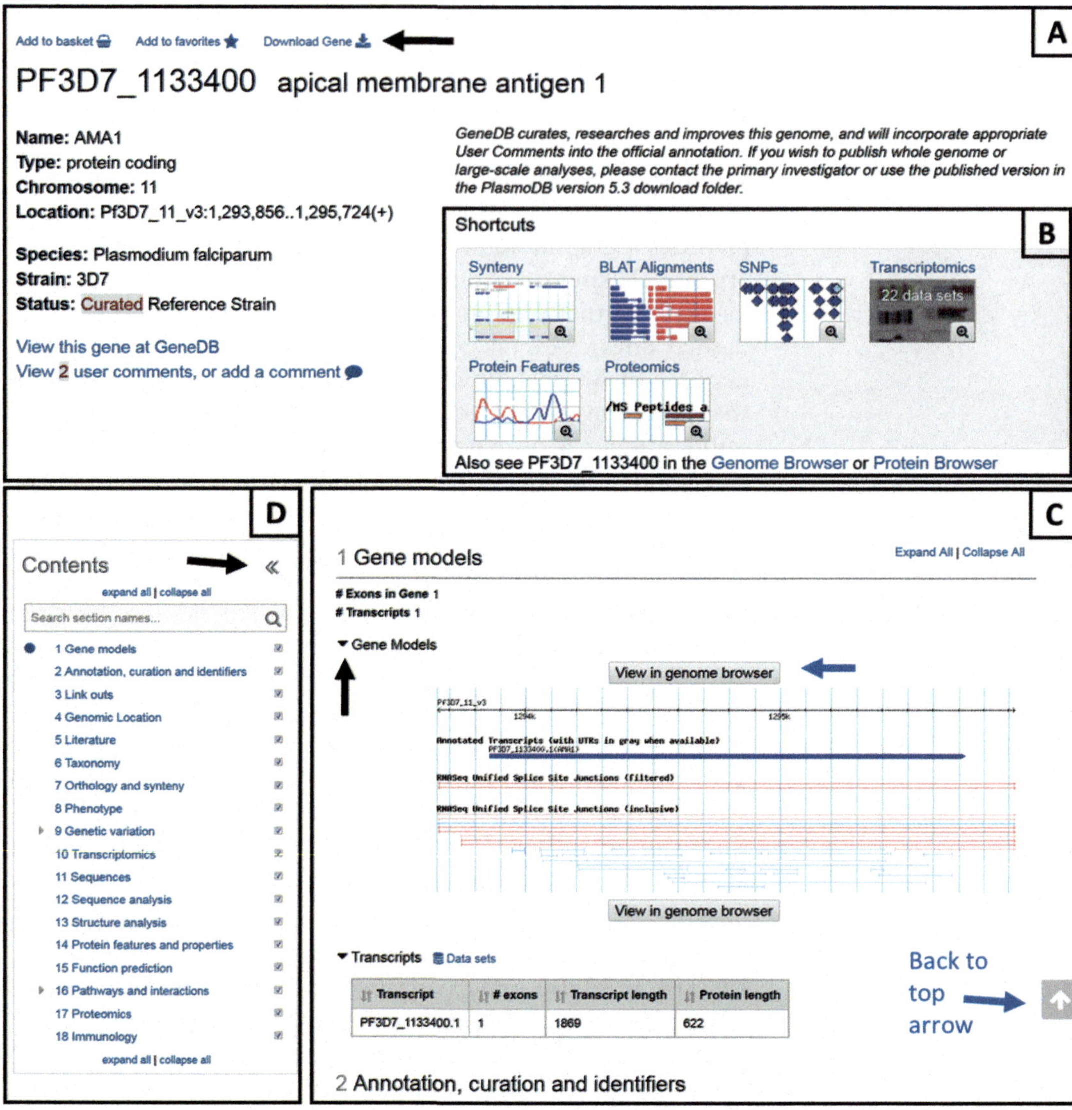

Fig. 3 Gene record page: Main sections. (A) Record pages include an overview section at the top, with basic information including gene ID, product description, or genome location. (B) Shortcuts are available on the right side of the overview, and provide quick navigation links, but also quick views of the images that appear in the data section of the gene record. (C) The data section is displayed below the overview. Organized in consistent, site-wide categories, the data section contains all available information about the gene. (D) The searchable, and collapsible "Contents" menu gives easy access to all the data sections (C). The contents section will remain visible while scrolling the record page and clicking on the double arrow icon will collapse the menu, giving full screen width to the record entry

Expression Evidence Graphic" (Fig. 7A) contains a summary table with a row for each transcript that includes an image of all mapped peptides from each proteomics data set. Hover over the glyphs representing the mapped peptides to obtain details about the peptide (experiment and sample names, sequence, etc.) (Fig. 7B). While there is abundant proteomics data, three

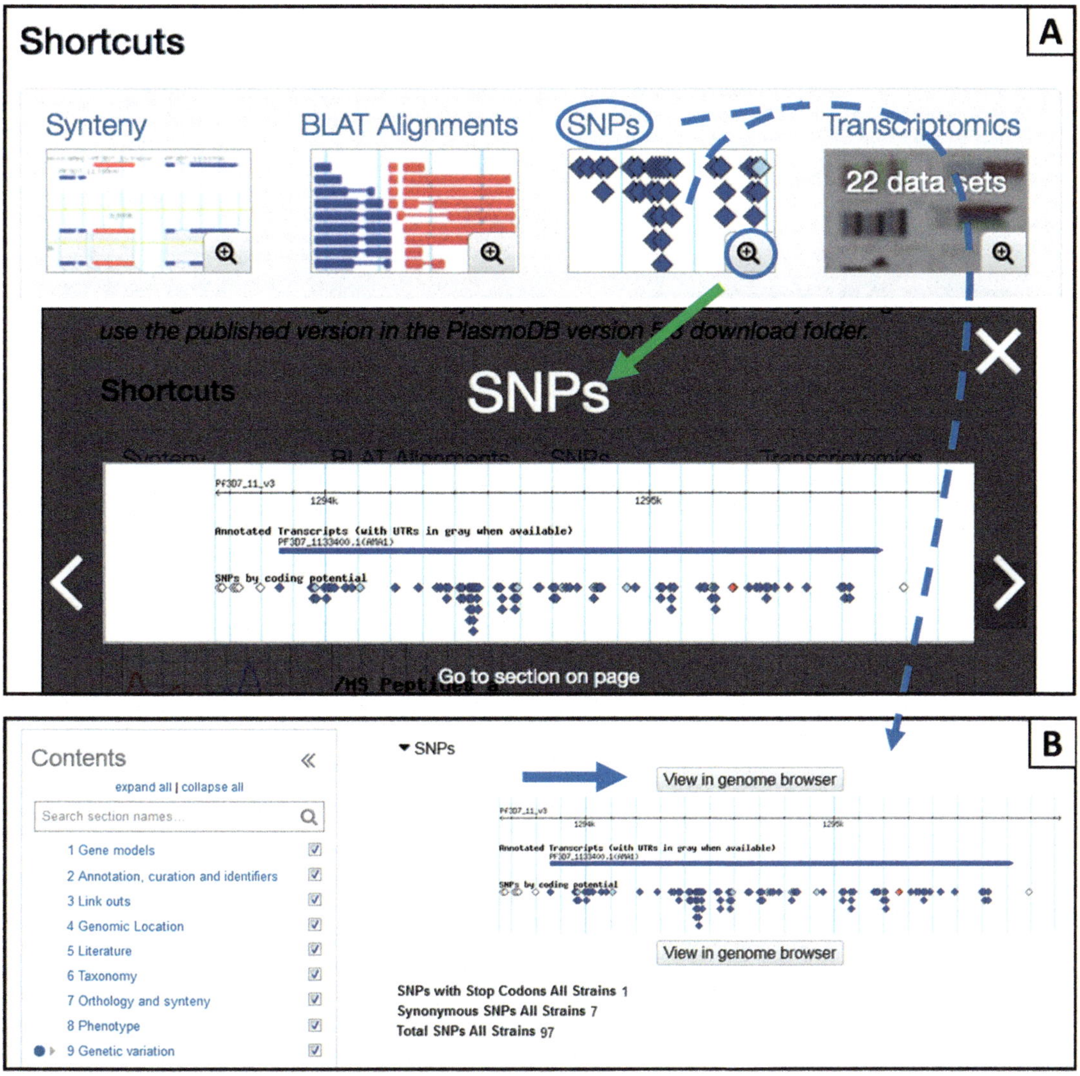

Fig. 4 Gene record page: Shortcuts. (A) Shortcuts can be found at the top of the gene page, on the right side of the overview section. Clicking on the magnifying glass icon (blue circle), will open a graphical display summarizing the data. Clicking on a shortcut image, or on the title above it (blue oval), navigates to the corresponding section of the record page (B)

experiments in particular support expression at the schizont stage—"Schizont Phosphoproteome (3D7)(2012)" [8], "Schizont Phosphoproteome (3D7)(2011)" [9], and "Cytoplasmic and nuclear fractions from rings, trophozoites and schizonts (3D7)" [10]. Each of these has mapped peptides from schizont samples. (Fig. 7A, blue arrows)

4. *Annotation, curation, and identifiers section*: EuPathDB encourages the community to enhance annotations by providing a platform to add comments to the record pages (Fig. 8).

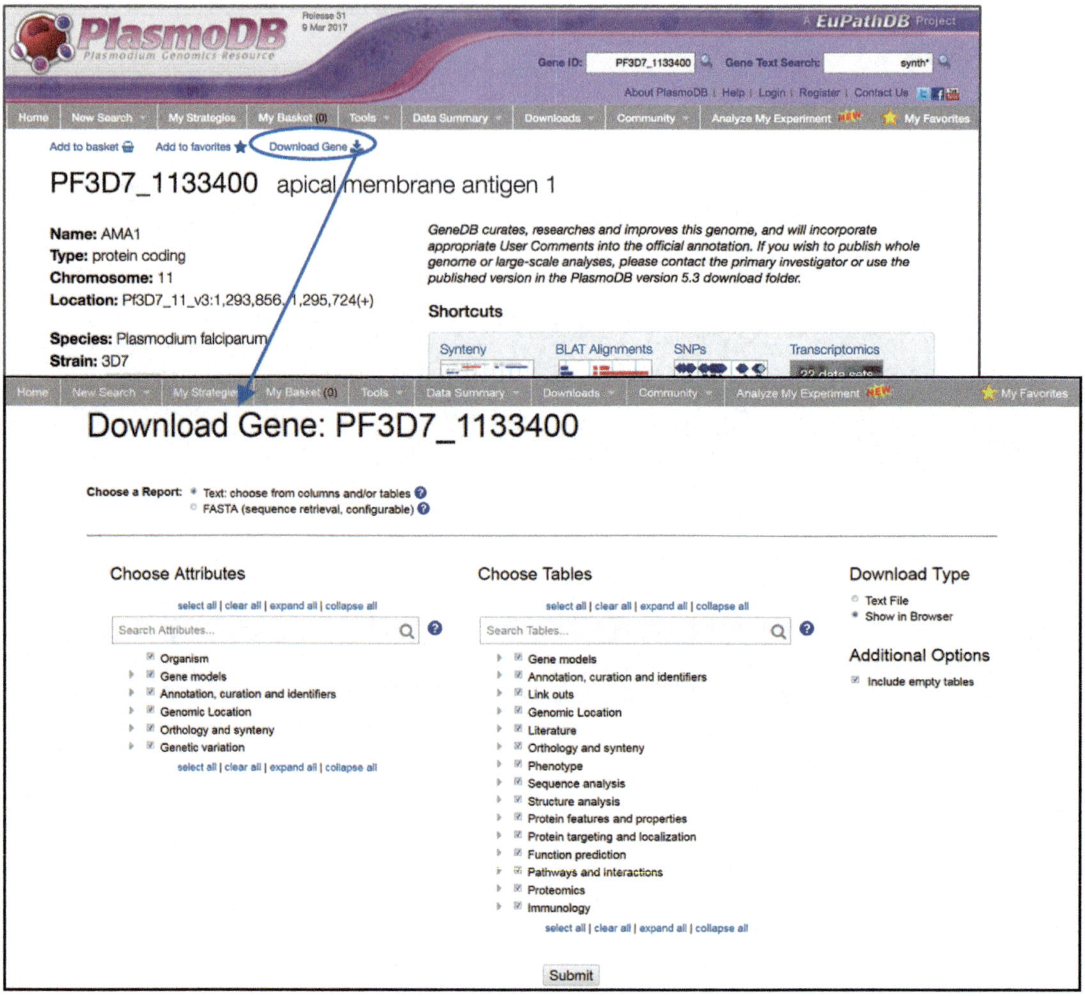

Fig. 5 Gene record page: The "Download Gene" link. Information available in the gene record, including sequences, can be easily exported using the "Download gene" link, located at the top of the overview section. Users can create their own tables, choosing gene attributes of interest

The comment system links knowledge from community experts to gene and other records. Once a user comment is added, it appears immediately on the gene page and becomes searchable through the text search. Some genomes are professionally curated by EuPathDB staff. When appropriate, user comments are integrated into the official annotation for these genomes.

Navigate to this section using the contents navigation menu on the left. This section contains useful information including previous identifiers, gene synonyms, annotation notes and user comments. Note that PF3D7_1133400 has two user comments (Fig. 8A) that are summarized in a table (Fig. 8B). Explore each comment further by clicking on the

Table 3
Gene page sections and content descriptions

Section	Section contents
Gene models	Gene structure, introns, exons, UTRs, alternative transcripts. Includes a gene model graphic and summary of supporting transcriptomic data.
Annotation, curation and identifiers	User comments, notes from curators, community annotation projects, alternative product descriptions, gene names, synonyms and previous identifiers.
Link outs	Links to other databases and resources that serve as alternative or specialized sources for additional information about our gene (ex: Entrez Gene, UniProtKB, PDB, GeneDB, Ensembl…)
Genomic location	Coordinates of the gene at the sequence level. Genome Browser column links to GBrowse centered in the gene of interest.
Literature	Publications containing useful information about the gene. Either automatically retrieved from GenBank records or manually curated.
Taxonomy	Classification of the organism following the NCBI taxonomy.
Orthology and synteny	Ortholog group assignments as predicted by OrthoMCL. This section also contains a synteny graph and a tool for aligning the gene sequence against up to 15 of the genomes included in the database.
Phenotype	Collection of mutant phenotypes, manually curated from publications or inferred from high-throughput phenotyping experiments.
Genetic variation	Alignment tool for exploring differences between isolates. Graphic summary of the SNPs detected in this region, with links to our genomic variation GBrowser tracks.
Transcriptomics	Transcript expression data sets are arranged in a searchable data table with expandable rows. Each data set includes expression data in tabular and graphical format, as well as coverage plots for RNA sequence data sets.
Sequences	Data table containing genomic, mRNA and protein sequences for each transcript.
Sequence analysis	An interactive graphic summary of EST alignments and BLAT hits against the GenBank nonredundant protein sequence database (NRDB).
Structure analysis	3D structure predictions and similar Protein Data Bank (PDB) chains.
Protein features and properties	Protein domains predicted for this gene, displayed both in a graphical representation and in data tables
Function prediction	Complete Gene Ontology annotations plus Enzyme Commission numbers, with links to EC number and GO term descriptions and publications.
Pathways and interactions	Collection of manually curated and computationally predicted metabolic pathways and protein interactions.
Proteomics	Data tables and graphic summaries of proteomic data sets (Mass Spec-based expression evidence and post translation modification data sets).
Immunology	Predicted epitopes from The Immune Epitope Database (IEDB), and host response data sets.

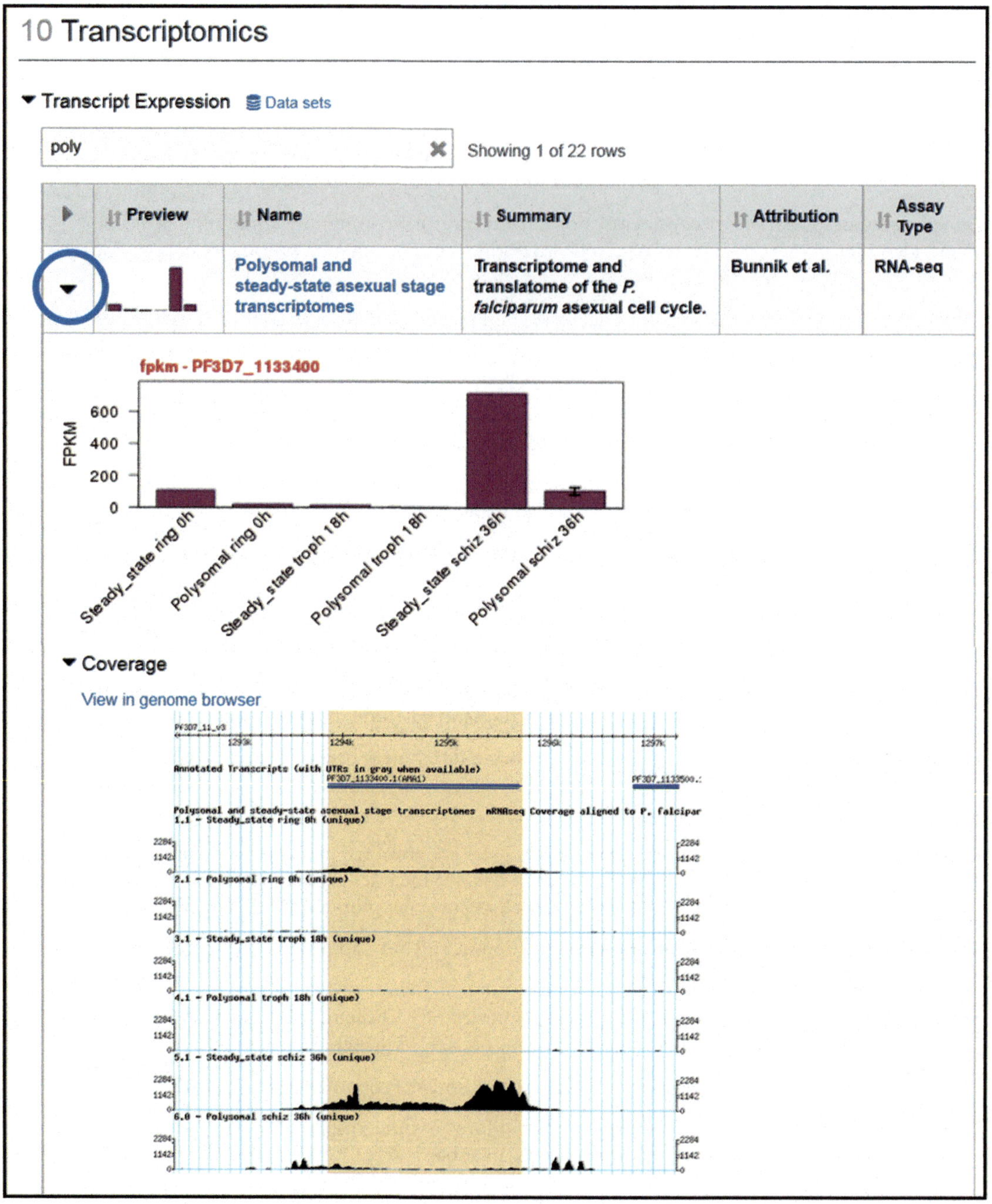

Fig. 6 Transcriptomics table. Transcript expression data sets are organized in searchable data tables, with expandable rows that reveal detailed data. Each data set includes expression data in tabular and graphical format, as well as coverage plots for RNA sequence data sets

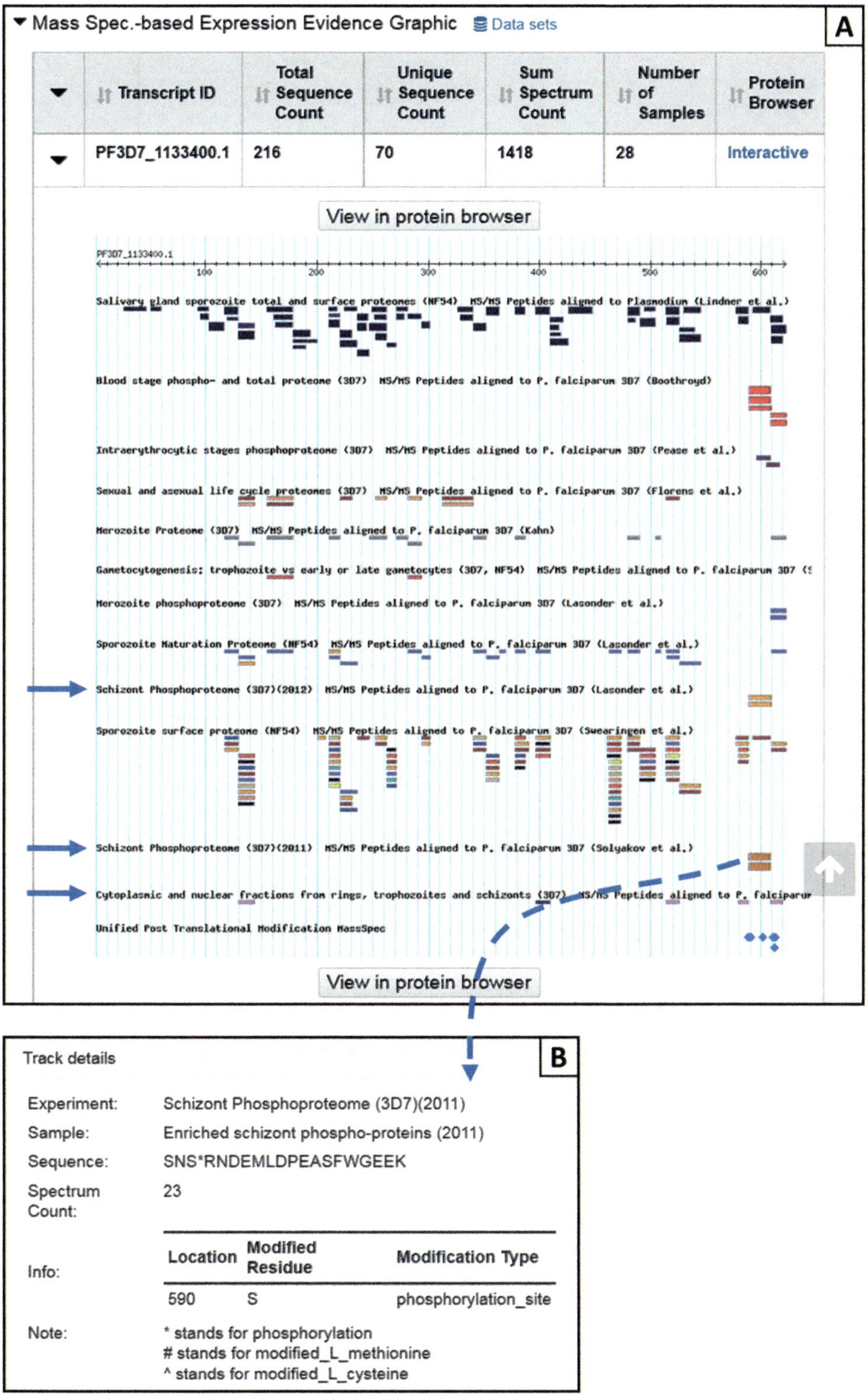

Fig. 7 Proteomics data on gene page. (A) The Mass Spec.-based Expression Evidence Graphic table displays peptides mapped to the gene's protein product. (B) Hover over the glyphs to reveal details concerning the mapped peptides

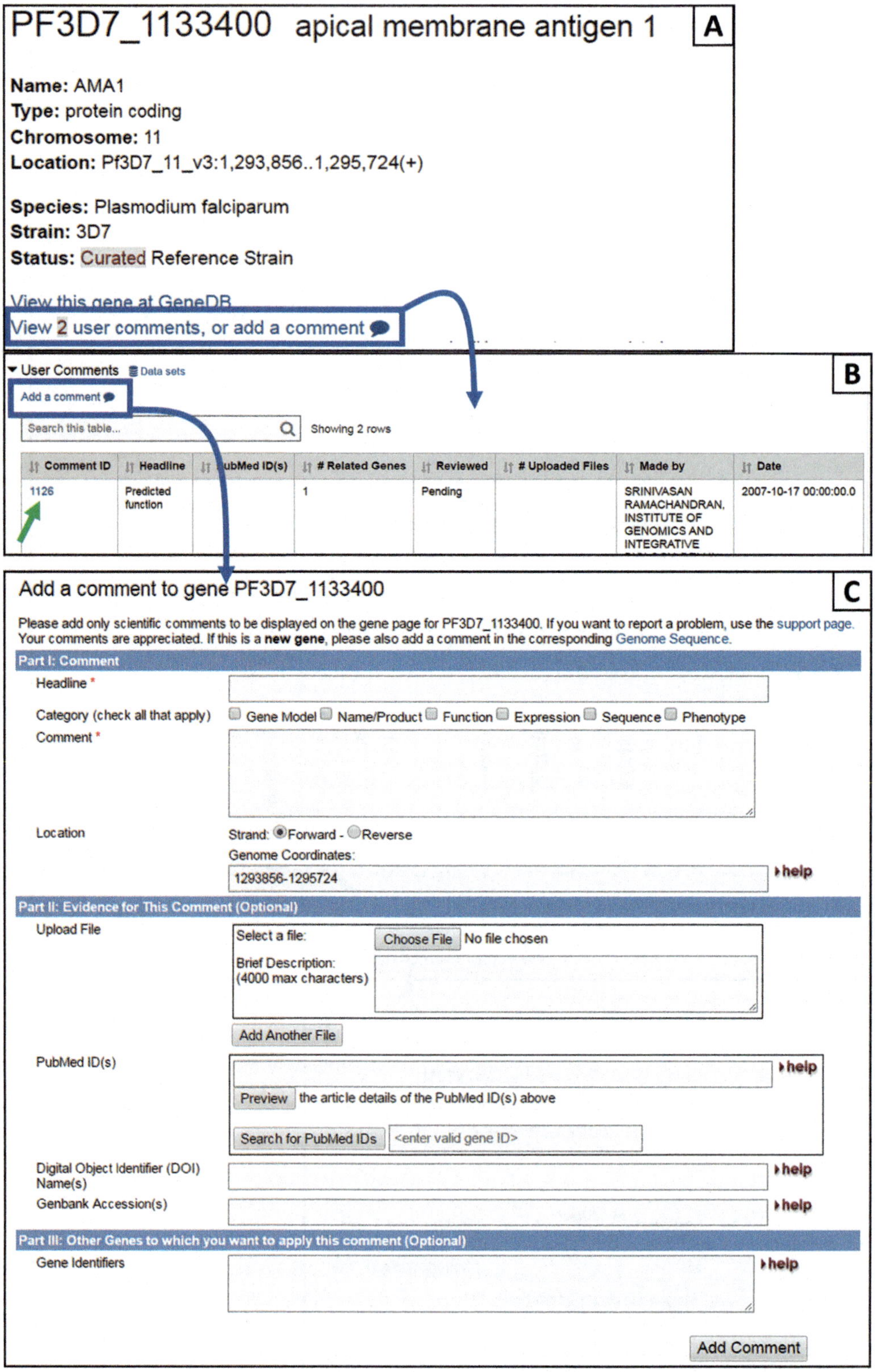

Fig. 8 Submitting user comments. (A) Summary section of PF3D7_1133400 gene record page showing "add a comment" link. (B) User Comments table listing comments and associated information. (C) Form for adding a comment to a gene

comment ID (Fig. 8B, green arrow). To add a new comment, click the "add a comment" link (Fig. 8B, blue box) and complete the form (Fig. 8C). You must be registered and logged in to add a comment. Table 4 gives examples of information to include in a comment.

5. *Orthology and Synteny section*: Explore orthology for the Cyclin-like F box protein 1A in *Trypanosoma brucei* strain TREU 927 (Tb927) in TriTrypDB. Navigate to TriTrypDB (http://tritrypdb.org) and enter Tb927.1.4540 into the Gene ID search box in the header (Fig. 9A, arrow). Use the Contents navigation menu to navigate to the Orthology and Synteny section (Fig. 9B, blue box). The ortholog group ID, OG5_132982 (Fig. 9B, green arrow), to which this gene has been assigned appears as a link to the OrthoMCL database where one can explore the group's features and distribution across a wider range of taxa.

Several interesting things about this gene can be discovered from this table. Based on the gene IDs of the table's entries for Tb927 genes, four paralogs of the Tb927.1.4540 gene can be found on chromosome 1 of Tb927 (i.e., Tb927.1.4560 = organism.chromosome.gene number) (Fig. 9B, green box). The close proximity of these genes (clustered gene numbers in the IDs) suggests that these paralogs may have arisen as a result of tandem duplication. Another paralog is found on chromosome

Table 4
Suggestions for user comment content

Comment type	Example comment
Gene name, including synonym	Purine Phosphoribosyl Transferase, is also known as HPRT, HGPRT, Hypoxanthine Phosphoribosyltransferase, Ppt1, Ppt-1, etc.
Reference	See PMID ##### for functional characterization of this gene. Same reference can be linked to multiple genes, if more than one gene is characterized in the manuscript.
Functional characterization	This "hypothetical protein" has been shown to be a copper transporter based on heterologous expression in *Xenopus* oocytes ... Contact <xxxxx> for further details.
Subcellular localization	GFP tagging demonstrates that this protein localizes to the mitochondrion, as shown in the attached images. See attached image.
Phenotype	Gene knockout has resulted in decreased growth ... Contact <xxxxx> for further details.
Structural information on annotated gene models	The predominant transcript initiation site for this gene has been mapped to ~561 nt upstream of the annotated ATG by 5′RACE and RNAse protection. This conclusion is consistent with available RNA-seq data, but differs from the reference annotation. See attached experimental evidence.

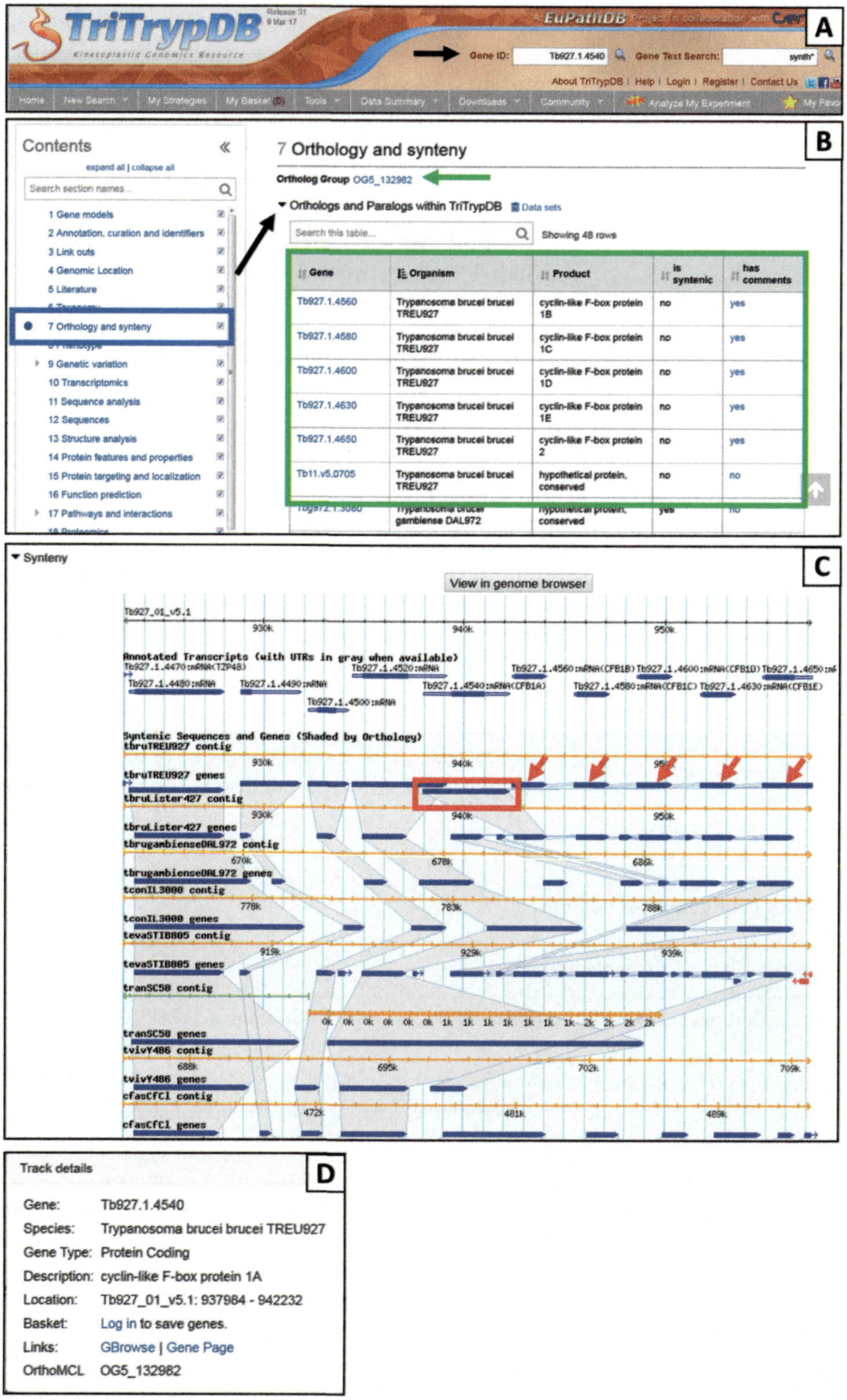

Fig. 9 Orthology and Synteny data on gene pages. (A) Header section of TriTrypDB. Enter the gene ID, Tb927.1.4540 to reach the gene page. (B) Contents navigation panel with section 7 chosen will direct the data section to the Orthology and Synteny section. (C) The gene page Synteny graph showing tracks for ***T. brucei*** TREU927 and ***T. brucei*** Lister 427. (D) Hovering over the glyphs in the Synteny graph reveals details concerning the gene

11 of Tb927 (Tb11.v5.0705) (i.e., Tb11.v5.0705 = organism & chromosome.genome version.gene number) which may have arisen separately.

Close the Orthologs and Paralogs table by clicking the triangle next to the title (Fig. 9B, black arrow). With the table closed, the "Retrieve multiple sequence alignment or multi-FASTA" tool is visible and can be used to conduct a multi-sequence alignment (MSA) using ClustalW across up to 15 organisms from the current database, with outputs in either ClustalW or multi-FASTA format. To use the alignment tool, choose 15 or fewer organisms from the tree in the center of the tool or search for your organisms of choice with the search box. Select an output format and click Submit Query. The results will appear in a separate browser window.

Close the Alignment tool by clicking the triangle next to the title or scroll down to the Synteny graphic that is centered on the Tb927.1.4540. This graph displays output from a Mercator [11] analysis that maps larger regions of orthology across all loaded genomes. The structure of orthologous genomic segments is often conserved, containing similar sets of genes in a similar order. In the graph, synteny is indicated with gray shadowing.

Notice the structure and order of genes in the Synteny graph (Fig. 9C) and hover over the gene glyphs to reveal gene details (Fig. 9D). The parent gene (Tb927.1.4540) (Fig. 9C, red box) is followed downstream by several paralogs (Fig. 9C, red arrows) that we considered while looking at the orthology table. Notice the presence of multiple paralogs in *T. brucei*, *T. evansi*, and *T. congolense*, and the absence of orthologs of the gene in the wider orthologous (syntenic) region of the genome in *Leishmania*, *Endotrypanum*, and *Crithidia*. Interestingly, only some of the genes shown in the table for *T. vivax* and none of the genes shown in the table for *T. rangeli* or *T. grayi* are shown in this graphic. The reason for this is that while these genes are orthologous with Tb927.1.4540, they do not sit on a region of the genome that shares wider orthology (synteny) with the region around this gene in Tb927. This property is also described for each gene in the "Orthologs and Paralogs within TriTrypDB" table with the column headed "Is Syntenic." Links are provided to open this image in the Genome Browser where one can customize the organisms shown, zoom in and out, and add other tracks.

2.2.2 Metabolic Pathway Record Pages

Metabolic pathways from KEGG [12–14], MetaCyc [15], TrypanoCyc [16], or LeishCyc [17, 18] are loaded in EuPathDB sites and mapped to genes that are annotated with appropriate enzyme commission (EC) numbers. Pathway record pages integrate these networks with annotations, gene expression profiles, and orthology data via the Cytoscape [19–21] platform. Metabolic pathways can be retrieved based on several criteria including compounds (substrates or reactants), gene lists, pathway identifiers or names (Fig. 10A).

1. Navigate to the TriTrypDB (http://tritrypdb.org) home page and find the Glycolysis 1 (TrypanoCyc) metabolic pathway: click on the "Metabolic Pathways" category to expand its contents, then select the "Pathway Name/ID" search (Fig. 10A). Begin typing the pathway name (Glycolysis I: GLYCOLYSIS-1 TrypanoCyc) in the "Pathway Name or ID" parameter and then choose the correct pathway from the list that appears (Fig. 10B). Notice that pathway names may appear more than once since they are obtained from multiple sources. Figure 10C–E shows portions of the Glycolysis 1 pathway cycle in *Trypanosoma brucei* as annotated in TrypanoCyc and represented in TriTrypDB.
2. *Explore the pathway.* The organization of the record page is similar to the gene page with a summary at the top, a Contents section for navigation and a data section with tables and images. The interactive pathway image depicts the series of enzymatic reactions as enzyme and compound nodes (Fig. 10C, 1 and 2, respectively) with by-products shown in gray. Small adjustments to the pathway layout can be made by dragging nodes and by-products to new locations. Panning and zooming the view can be achieved with the tools in the top left corner (Fig. 10C, 3), or by clicking and dragging a node or side product to pan or scrolling to zoom. Enzymes that catalyze reactions are displayed in rectangular boxes (Fig. 10C, 1), labeled with an enzyme commission (EC) number if known, and a name or reaction identifier if the EC number is not known. A red outline denotes that at least one gene encoding an enzyme with this EC number is present in the current component database. Compounds are identified using ChEBI identifiers. Where available, the compound structure is shown for primary metabolites (Fig. 10C, 2), whereas side compounds are represented as text. Clicking on any node displays a "Node Details" pop-up (Fig. 10D) which includes links to genes annotated with the EC number for enzymes and a link to the compound record page for compounds.
3. *Annotate the pathway*: Annotate the pathway with expression data that explores differential expression between procyclic (insect) and bloodstream forms of *T. brucei* (Fig. 10E). Choose "Paint Enzymes" and "By Experiment" to pull up a list of all experimental data that can be "painted" on the enzyme nodes. Choose "*T. brucei* brucei TREU927 Bloodstream and Procyclic Form Transcriptomes (Siegel et al.)" [22] and then

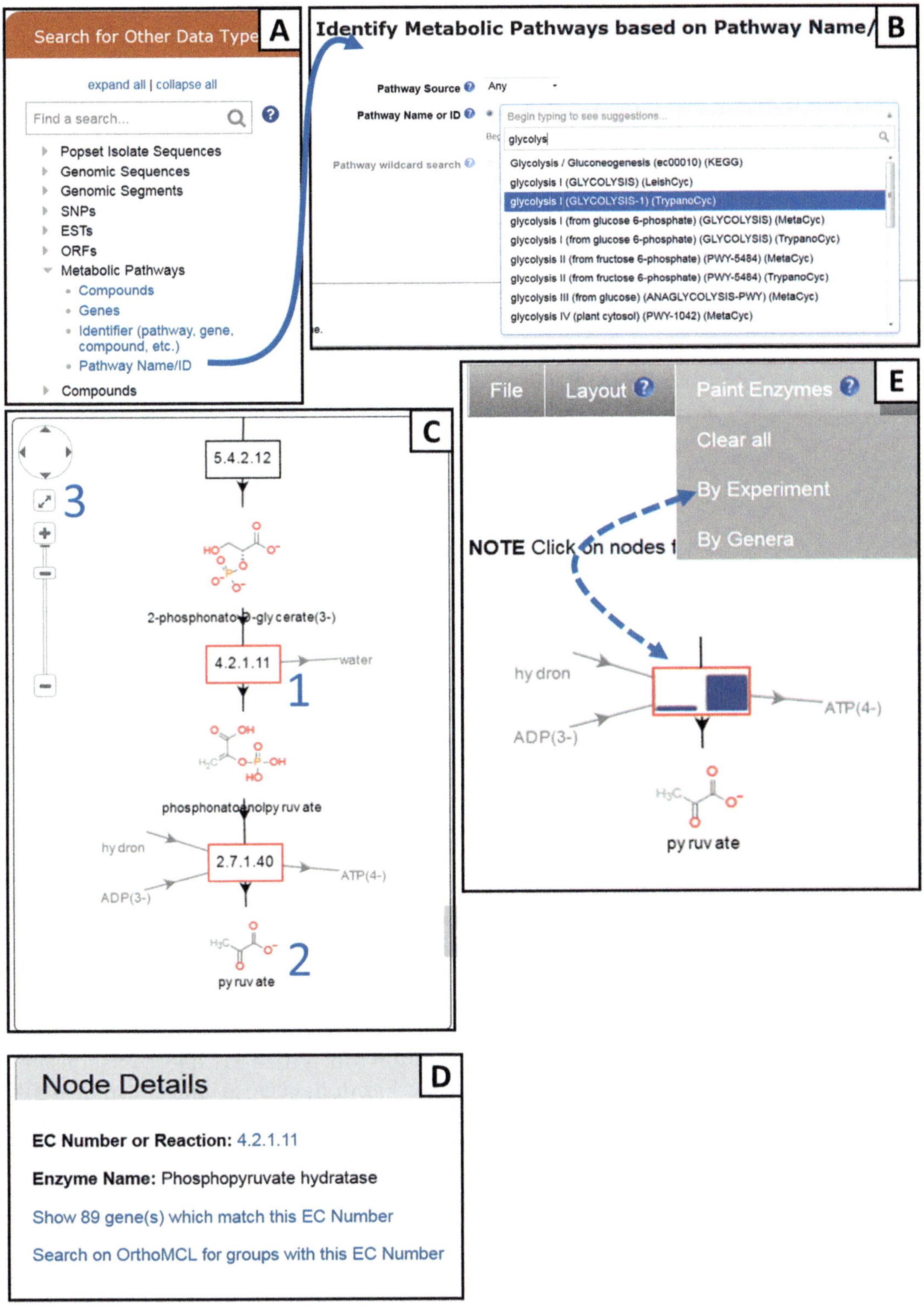

Fig. 10 Metabolic Pathways represented in TriTrypDB. (A) The Search for Other Data Types panel with the Metabolic Pathways category open to reveal the types of searches that return Metabolic Pathway records. (B) The Pathway Name ID search page depicting the "typeahead" function for entering pathway names in the Pathway Name/ID parameter. (C) Partial view of the Glycolycic 1 pathway showing an enzyme node (1), a compound node (2) and the zoom function (3). (D) Node details pop-up that appears when an enzyme or compound node is clicked. (E) Enzyme node painted with expression graph from integrated experimental data

"Paint" to display the expression data in the enzyme nodes. Examine the experimental data in the nodes and notice that several of the enzymes in this pathway are downregulated in procyclic forms compared to bloodstream forms. This can be interpreted as an indication of differential sugar metabolism between the two lifecycle stages, which makes sense given the very different environments they inhabit.

4. Navigate to the 5-aminoimidazole ribonucleotide biosynthesis I (PWY-6121) pathway as above (**step 1** of this section) and explore the tool for annotating the pathway with the distribution of genes across phylogeny. Choose "Paint Enzymes" and "By Genera." From the "Genera Selector" choose Kinetoplastida and Mammalia and then click "Paint." Each enzyme is replaced with a chart showing whether a gene encoding the enzyme is present in three genera from the Kinetoplastida (*Crithidia*, *Leishmania*, *Trypanosoma*), and two genera from the Mammalia (*Homo*, *Mus*, blue). Click on a node to see a larger image of the distribution. Notice that all the enzymes from this pathway are encoded in Mammalia, but none of these enzymes are encoded in any of the represented Kinetoplastida. This pathway is a part of the super pathway involved in de novo purine biosynthesis and this representation agrees with the observation that *Trypanosoma* cannot synthesize purines de novo but instead rely on scavenging from the host. It can be inferred that this is also true of other Kinetoplastida.

2.3 Data Mining with Searches and Strategies

EuPathDB offers over 100 preconfigured searches in a unique and powerful strategy system that allows you to explore relationships across data sets, data types and organisms. Searches query individual data sets that provide evidence for a specific biological property and return a list of records that meet the search criteria and therefore have the biological characteristic defined by the data set. Strategies (Fig. 11A) can be created by adding, subtracting, joining, intersecting, or collocating (Fig. 11B) the results of subsequent searches. The colocation tool is used to explore relationships based on relative genomic location, such as interrogating SNPs located 500 nt upstream of genes. A nesting tool allows you to control the logic when combining search results. Results from any step in a strategy can be analyzed using gene ontology (GO) [23, 24] enrichment, pathway enrichment or genome visualization tools. The following two examples illustrate how to create strategies and leverage orthology in the EuPathDB strategy system.

2.3.1 Strategy Example 1

This example creates a strategy (Fig. 11A) in PlasmoDB (http://PlasmoDB.org) that finds *Plasmodium vivax* proteases that are likely expressed during the gametocyte stage. The strategy employs three searches and uses the Transform by Orthology tool to

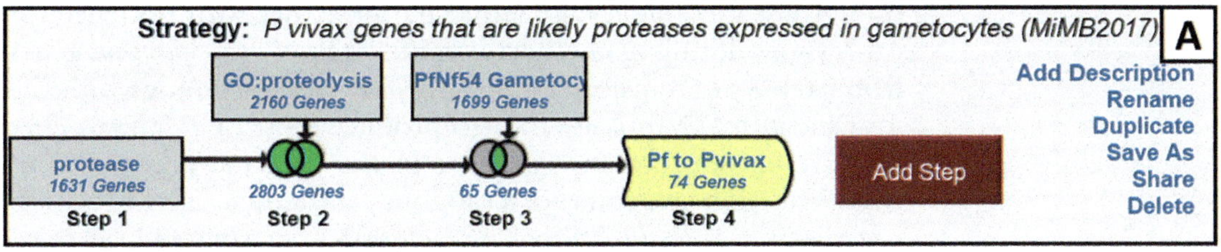

B

Operator	Image	Combined result will contain
1 Intersect 2		IDs that appear in both lists
1 union 2		IDs from list 1 plus IDs from list 2
1 minus 2		IDs unique to 1
2 minus 1		IDs unique to 2
1 relative to 2 (colocation)		IDs of features that are near each other (collocated) in the genome

Fig. 11 Creating strategies by combining search results. (A) PlasmoDB Strategy returning a list of 74 genes that are likely ***P. vivax*** proteases and expressed in gametocytes. The strategy is also available here: http://plasmodb.org/plasmo/im.do?s=2db873c2b03b57bf. Creating this strategy in the current database may produce a different result since genome annotations may be updated with new database releases. (B) Table showing the five options for combining searches into a strategy. When two searches are combined, the two result sets (list of IDs) are merged according to the operator that you specify. If the searches return the same type of genomic feature they can be combined using any of the five operators (i.e., search 1 returns genes, search 2 returns genes). However, searches that return different genomic features (i.e., search 1 returns genes, search 2 returns SNPs) will yield no results when combined with intersect, union or minus operators because there are no IDs in the list of genes (search 1 result) that are present in the list of SNPs (search 2 results). To combine a search that returns genes with a search that returns SNPs, you must use the collocation option (1 relative to 2) to find, for example, genes with SNPs in their upstream regions

convert *P. falciparum* genes into their *P. vivax* orthologs. Steps 1 and 2 return proteases using two different lines of evidence—a text search in step 1 and a GO term search in step 2. These searches are combined with a union to obtain a more comprehensive list of possible proteases. Step 3 returns genes with evidence for expression during the gametocyte stages based on *P. falciparum* RNA sequencing data [25]. Steps 2 and 3 are combined using the intersect operator to produce a list of genes that have both biological properties: these genes are suspected proteases with evidence for expression during gametocyte stages. The *P. falciparum* genes from step 3 are transformed into their *P. vivax* orthologs with the Transform by Orthology tool to produce a list of *P. vivax* genes that are likely proteases expressed in the gametocyte stage. This transformation exploits orthologous clustering of EuPathDB organisms to infer functional characteristics determined in *P. falciparum* to *P. vivax*. The following offers detailed instructions for building the strategy. The completed strategy is also available here: http://plasmodb.org/plasmo/im.do?s=2db873c2b03b57bf.

1. *Find genes that are possible proteases using the text search to query gene records for the term "protease"* (Fig. 12). To reach the search, click on the "Text" category link on the home page "Search for Genes" menu (Fig. 12A). Next click on the "Text (product name, notes, etc.)" link to open the "Text Search: Identify Genes by Text (product name, notes, etc.)" page (Fig. 12B). Each search is loaded with default parameters that can be configured before running the search. The default setting for the "Organism" parameter is set to search all organisms in the database while the default setting for the "Fields" parameter will query every field but "Similar proteins (BLAST hits v. NRDB/PDB)". Type the word "protease" (without the quotes) in the "Text term (use * as wildcard)" box (Fig. 12B, arrow) and click "Get Answer" to initiate the search. The search results (Fig. 12C) are displayed in the "My Strategies" section which consists of a strategy panel with an interactive image of the strategy, followed by an organism filter showing the distribution of hits across the genomes queried, and a result table with the list of genes returned by the search. The first column in the result table is the gene ID and serves as a link to the gene record. Searches and strategies can be saved (Fig. 12C, blue bordered inset) and are given a unique URL that can be used to share the strategy with colleagues.
2. *Expand the list of proteases with a second line of evidence for proteolytic activity.* There may be some proteases that do not have the term "protease" in their record but do have an assigned GO annotation associated with proteolysis. The ontologies are a controlled vocabulary for describing the molecular function, biological process, or subcellular location of a gene product. GO annotations in PlasmoDB were either provided by the sequencing and annotation centers or inferred based on a gene product's similarity to protein domains from the InterPro databases [26].

   To add a GO term step to the search strategy, click on the red "Add Step" button in the strategy panel (Fig. 13A) that opens the "Add Step" pop-up (Fig. 13B). Next, navigate to the GO Term search page, by clicking on "Run a new Search

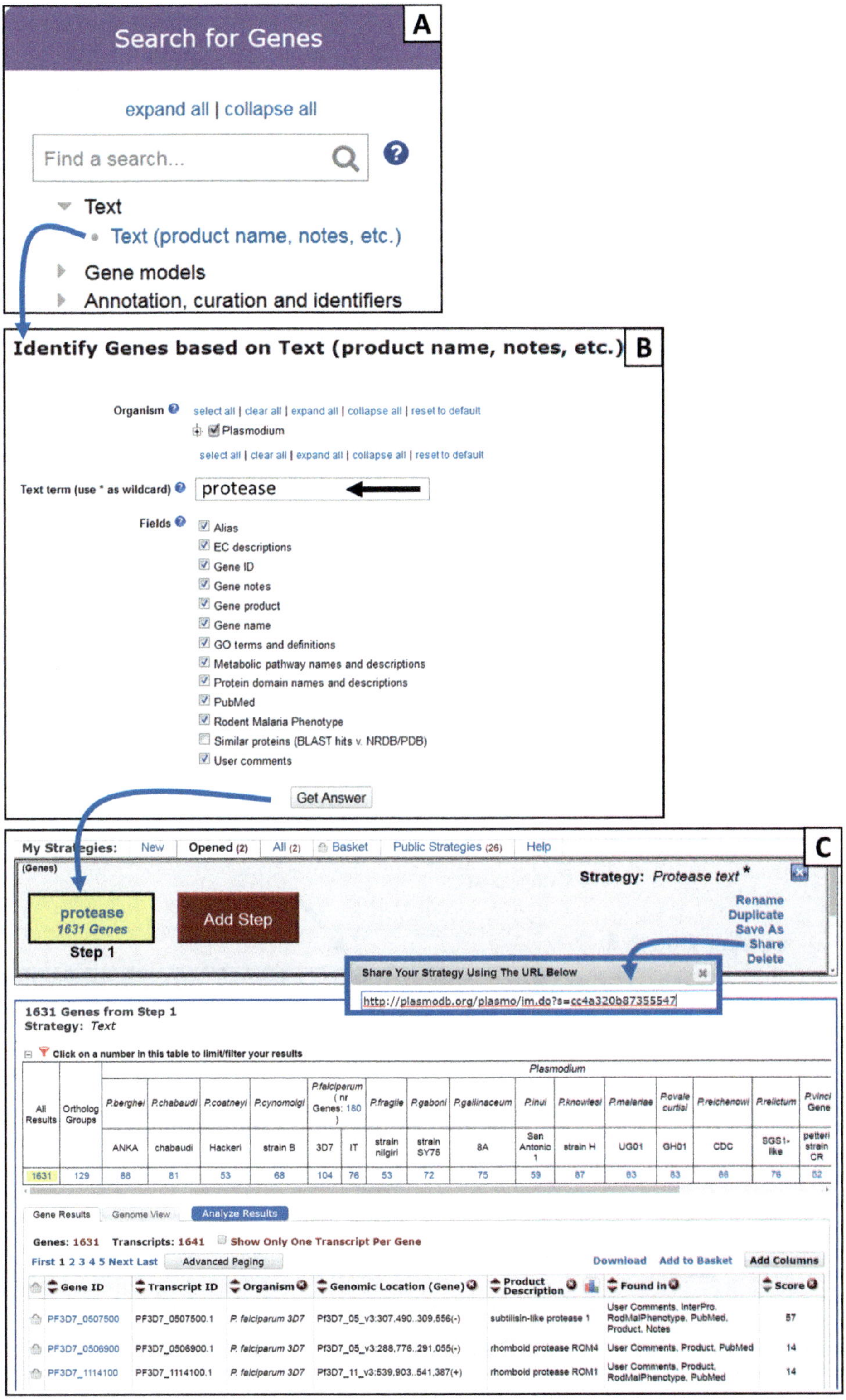

Fig. 12 Text search in PlasmoDB. (A) Home page panel showing access to the Text search page. (B) The Text search page with protease entered for the Text Term parameter. Clicking Get Answer will initiate a search for genes whose records contain the word "protease" in all the Fields chosen. (C) The results of the search as displayed in the "My Strategies" section. The search returned over 1600 genes that are likely proteases

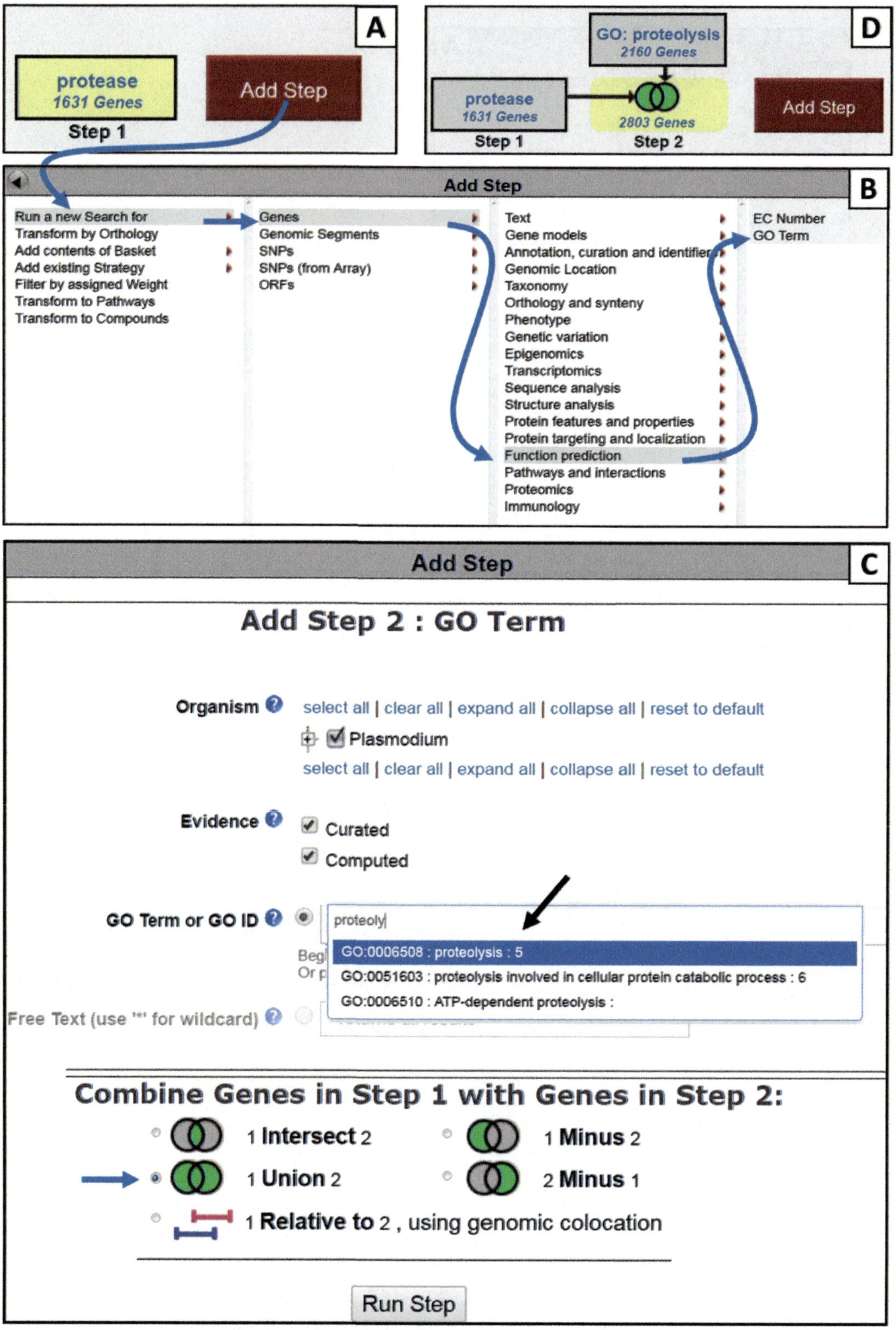

Fig. 13 Creating Step 2 of the PlasmoDB Strategy. (A) The Add Step button for initiating subsequent strategy steps. (B) The Add Step pop-up for choosing the next search in the strategy. All searches are available from this pop-up. (C) The GO Term search depicting the choice of GO Terms using the "GO Term or GO ID" parameter type ahead. (D) The strategy result after running the second search in the strategy—the GO Term search

for," "Genes," "Function Prediction," and "GO Term." Specify the GO Term or GO ID by typing either the GO Term (proteolysis) or ID (GO:0006508) and then choosing the correct term from the list that appears (Fig. 13C, black arrow). Since this is not the first search in the strategy, running this search requires defining how to combine the results of this search with the previous one. Choose to union the two ID lists to add genes discovered in the GO term search that are not already in the list of possible proteases (Fig. 13C, blue arrow). *See* Fig. 11B for more information about combining searches. Click "Run Step" to initiate the search. The resulting strategy (Fig. 13D) contains two steps and returns over 2500 genes whose products are likely to have proteolytic activity based on two lines of evidence, the word protease found in their gene records and/or a GO term assignment of GO:0006508 proteolysis.

3. *Filter by genes highly expressed at the gametocyte stage.* To ensure that our list of proteases is highly expressed at the gametocyte stage, we can intersect the Step 2 results with a search for genes based on transcript expression in the gametocyte stage. Click "Add Step" from the strategy panel and navigate the "Add Step" panel through "Run a new Search for," "Genes," "Transcriptomics," "RNA Seq Evidence." A list of available RNA sequencing data sets and their associated searches will appear (Fig. 14A). Notice that there are no gametocyte RNA seq data sets associated with *P. vivax*, so we will search the *P. falciparum* data set in this step and then transform the results into their *P. vivax* orthologs.

4. Choose the Percentile search (P) for "Female and Male Gametocyte Transcriptomes (Lasonder et al.)" (Fig. 14A, blue borders) to open the search page. The data set contains an RNA sequencing analysis of *P. falciparum* male and female gametocyte samples from a study published in 2016 [25]. To create the Percentile search, EuPathDB obtained the raw sequencing reads, applied a standard RNA seq mapping workflow, ranked expression values from highest to lowest, and then grouped genes into percentile groups in each sample. Running the percentile search using the default "max/min expression percentile" parameters will return the genes whose expression levels are in the top 20% for the samples chosen in the "Samples" parameter.

   Choose both samples, male gametocyte and female gametocyte, from the search page (Fig. 14B, blue arrow). Since the goal is to create a list of genes that are proteases and expressed in gametocytes, choose to intersect the Percentile search with the Step 2 results. Clicking Run Step will initiate the RNA Seq Evidence search, intersect the new results with the Step 2

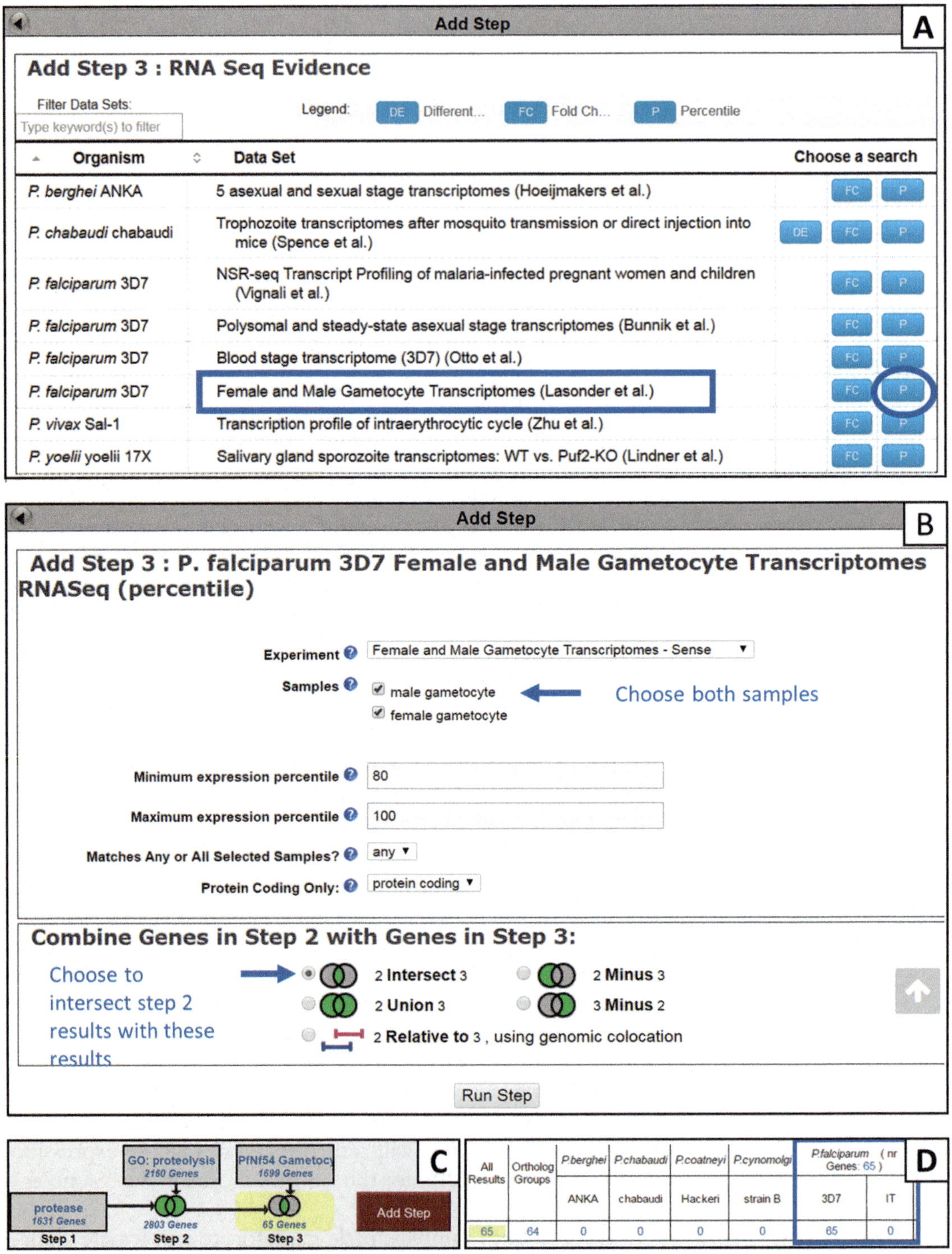

Fig. 14 Creating Step 3 of the PlasmoDB Strategy. (A) The Add step pop-up showing the available searches against RNA sequencing data sets. (B) The search form for the chosen gametocyte RNA sequencing data set. (C) The strategy results after adding Step 3. (D) The organism filter table for Step 3 results. Only ***P. falciparum*** genes are returned in step 3 because the RNA sequencing experiment was performed with ***P. falciparum*** parasites

results and return a Step 3 result that includes genes that are in both result sets. The genes returned in Step 3 result (Fig. 14C) will therefore possess possible proteolytic activity and high gametocyte expression.

Notice the genes in the Step 3 result are only *P. falciparum* genes. This is evident in the organism filter table which shows over 60 genes under *P. falciparum* 3D7 but none in other organisms (Fig. 14D). This is because the RNA sequencing experiment was performed in *P. falciparum*.

5. *Use the "Transform by* Orthology*" tool to transform the P. falciparum gametocyte proteases to their P. vivax orthologs.* Since gametocyte expression data is unavailable for *P. vivax*, this step of the strategy takes advantage of expression data obtained in *P. falciparum* to generate a list of *P. vivax* genes that are likely expressed in *P. vivax*. Click the red "Add Step" button following Step 3 and then choose "Transform by Orthology" in the first column of the pop-up (Fig. 15A). Arrange the "Organism" parameter to include only *P. vivax* Sal1 and leave the "Syntenic Orthologs Only" parameter in the default "No" setting (Fig. 15B). The *P. vivax* Sal1 genes returned by the search will be orthologs of the *P. falciparum* input genes. The resulting four-step strategy returns *P. vivax* genes that are likely proteases expressed in gametocytes (Fig. 15C).
6. *Explore your results*! It is important to perform a critical review of strategy results to determine whether your final gene list truly possess the biological characteristics intended by your search strategy. For example, one quick check can be performed by reviewing the "Product Description" for terms associated with proteolysis.

2.3.2 Example 2: Searching by Orthology and Phyletic Profile

A common use for orthology data in data mining is to find genes that are restricted to particular taxa. For example, vaccine candidates or druggable targets ideally would be proteins that are unique to the pathogen and are not found in the host. Orthology might also be useful for finding genes that are related to a process or organelle. Apicomplexans have a unique four-membraned organelle, the apicoplast, which is thought to have arisen through two endosymbiotic events. This organelle, and the products of the genes that act there, are tempting drug targets, making it important to identify genes that act in the apicoplast.

In this example, we will begin the strategy with a search for *P. falciparum* genes that are likely expressed in the apicoplast. *Plasmodium* harbors a motif that targets proteins to the apicoplast and there is a search in PlasmoDB that returns genes encoding this motif. The search is also available in EuPathDB where the *P. falciparum* results can be transformed to orthologs in other species. In Step 2, the *P. falciparum* apicoplast genes will be transformed to

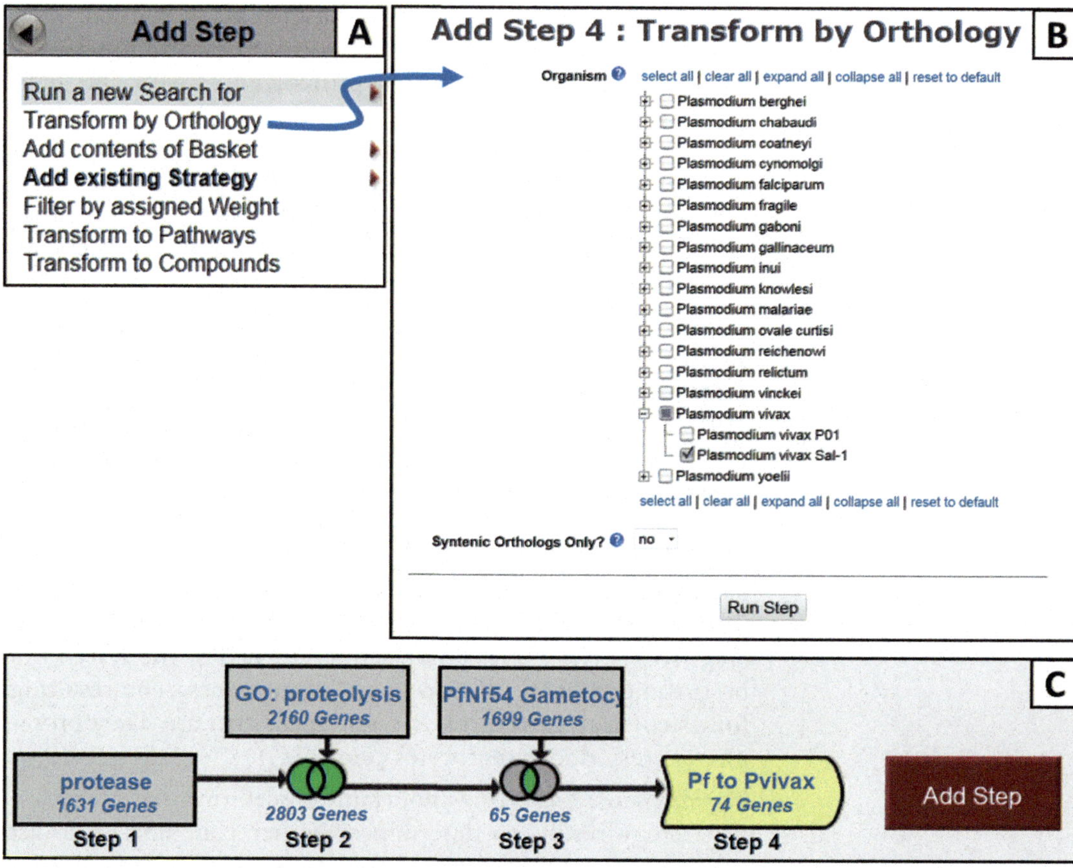

Fig. 15 Transform by Orthology tool. (A) The Add Step pop-up for accessing the tool. (B) The transform by Orthology tool configured to transform to ***P. vivax*** Sal1. (C) The final four step strategy returning 74 ***P. vivax*** genes that likely have protease activity and expressed in gametocytes

their orthologs in *Toxoplasma gondii* strains ME49 and GT1 and the closely related *Neospora caninum*. In Step 3, we will use the Orthology and Phylogenetic Profile search to restrict the list of *T. gondii* and *N. caninum* "apicoplast" genes to those that do not have orthologs in human or *Cryptosporidium*. Since *Cryptosporidium* has lost its apicoplast, any genes in the list of *T. gondii* and *N. caninum* "apicoplast" genes that have orthologs in *Cryptosporidium* are less likely to be apicoplast-specific. We will also remove genes that have orthologs in human since the best parasite druggable targets or vaccine candidates would be genes and proteins that are not present in humans to avoid interactions and side effects. The completed strategy is also available here: http://eupathdb.org/eupathdb/im.do?s=3353bf3401d62d48.

1. Navigate to EuPathDB (http://eupathdb.org) to begin the strategy with a search for *P. falciparum* genes containing a motif that targets proteins to the apicoplast. Note that this search must be carried out in EuPathDB in order to perform

orthology transforms between organisms that are hosted in different component sites. Use the "Search for Genes" panel (Fig. 16A) or the header drop-down menu to view the category "Protein targeting and localization" and open the "P.f. Subcellular Localization" search page (Fig. 16B). Choose "Apicoplast" for the "Localization" parameter and click "Get Answer." The results appear as Step 1 in the strategy panel (Fig. 16C).

2. Transform the *P. falciparum* apicoplast genes to their *T. gondii* ME49, *T. gondii* GT1 and *N. caninum* Liverpool orthologs. Click "Add Step" from the strategy panel and choose "Transform by Orthology." Arrange the "Organism" parameter of the transform tool to include only the three organisms of interest (Fig. 16D). The results of the ortholog transform appear as Step 2 in the strategy (Fig. 16E). While the majority of these genes will act in the apicoplast, some may have additional functions. This gene list can be refined using information from *Cryptosporidium*, a species of apicomplexan that is closely related to *Toxoplasma* and *Neospora* but which has lost its apicoplast. The results can be narrowed to include only genes that are likely to be truly apicoplast-specific by removing genes that have orthologs in *Cryptosporidium.* If the interest is in drug targets, the list can be further refined to exclude genes that have orthologs in vertebrates.

3. Click "Add Step" and navigate the "Add Step" panel through "Run a new search for," "Genes," "Orthology and synteny," and choose "Orthology Phylogenetic Profile" (Fig. 16F). The search opens displaying two parameters. The "Find genes in these organisms" parameter allows selection of the organisms from which genes will be returned. Choose "clear all" and then choose *T. gondii* ME49, *T. gondii* GT1, and *N. caninum* Liverpool from the Apicomplexa category. Use the "Select orthology profile" (Fig. 17A) parameter to define the orthology profile of the genes returned by the search. Arrange green check marks for organisms in which orthologs must be present, and red for organisms in which orthologs cannot be present (red crosses). In this example, all *Cryptosporidium* and all Mammalia should be excluded from the ortholog profile of the genes returned by the search (Fig. 17A) while nothing is required to be included (no green check marks). Then choose to intersect the results of the "Orthology Phylogenetic Profile" with the previous search results (Fig. 17A, arrow). The strategy produces a list of possible apicoplast genes in *T. gondii* ME49, *T. gondii* GT1, and *N. caninum* Liverpool based on *P. falciparum* data from an algorithm that predicts apicoplast targeting based on the presence of a motif.

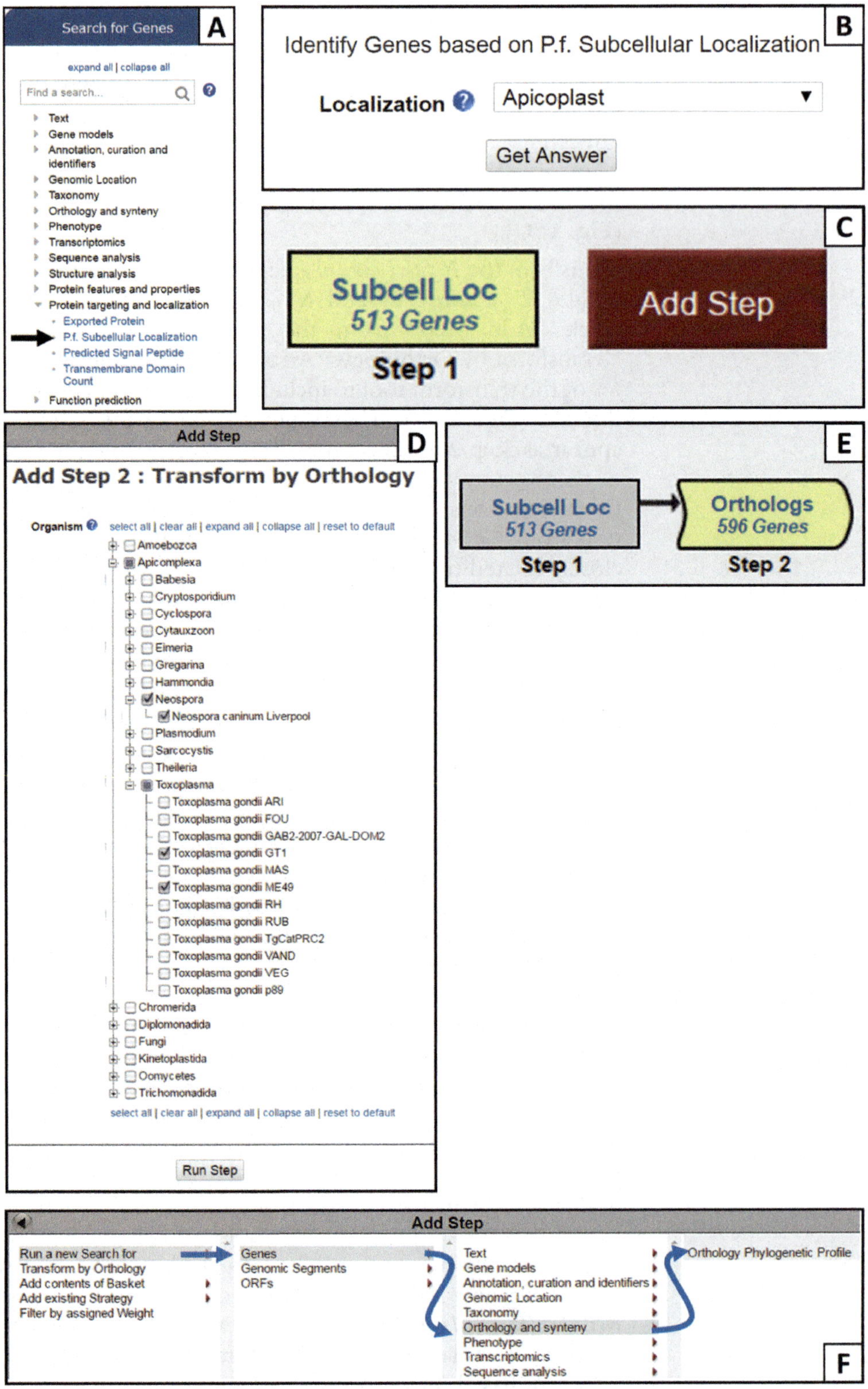

Fig. 16 Find ***T. gondii*** and ***N. caninum*** genes that are predicted to be localized to the apicoplast. (A) The EuPathDB Search for Genes panel with the Protein targeting and localization category opened. The P.f.

2.4 Data Mining with Visualization: Visualization of Genomic Data with GBrowse

GBrowse is a highly configurable tool for visualization of sequence feature data at the genome-wide scale and is embedded into all EuPathDB sites. In this section, we will examine a single gene using GBrowse to visualize data aligned to the genome is the region of the gene—TGME49_200320 hypoxanthine-xanthine-guanine phosphoribosyl transferase, HXGPRT. We will be able to interpret alternative splicing and gene model accuracy.

1. Navigate to ToxoDB (http://toxodb.org) and go to the HXGPRT gene page by entering the gene ID, TGME49_200320, in the Gene ID box in the header (Fig. 18A, blue arrow). Access GBrowse by clicking the "View in genome browser" button from the "Gene models" section (Fig. 18A, green arrow). The initial view on the GBrowse page defaults to the gene region with a track displaying annotated transcripts colored by the direction of transcription as well as tracks for splice site junctions which provide evidence for intron/exon boundaries (Fig. 18B). Hover over the glyphs in the tracks to reveal details. The "Landmark or Region" box shows the coordinates of the displayed region (Fig. 18B, 1). Entering alternative coordinates, a gene ID, or a transcript ID into this box will bring the specified region into view. The "Overview," "Region," and "Details" scales (Fig. 18B, 2) show the entire chromosome or contig, a zoomed view of the chromosome, and the selected region of the chromosome or contig, respectively. Displayed regions are highlighted in yellow along the scale.

   Each track has a toolbar in the track header that can be used to hide (−), remove (x), share (radiowaves), configure (wrench), or access a track description (?) (Fig. 18B, 3). With the configure tool one can change the track dimensions, axes, glyph types and colors, etc. GBrowse layouts can be saved using the "Save Snapshot" utility (Fig. 18B, blue box, login required), or a URL can be generated to share the track using the "File" menu at the top of the page. Downloads of track images can also be obtained from this menu. Finally, personal data tracks can be made or uploaded in the "Custom Tracks" tab.

2. Expand the region to 10 kbp using the drop-down menu in the panning and zooming tool (Fig. 18B, 4). Zooming and relocating can also be achieved through the Landmark tool or by using the mouse to highlight the region of interest in any of these three layers.

Fig. 16 (continued) Subcellular Localization search is accessible here. (B) The P.f. Subcellular Localization search page containing only one parameter. (C) The strategy panel showing the result of Step 1. (D) The Transform by Orthology tool arranged to transform genes from the previous step into ***T. gondii*** ME49, ***T. gondii*** GT1 and ***N. caninum*** Liverpool. (E) The strategy panel after the transformation. (F) The Add Step panel configured to access the Orthology Phylogenetic Profile search

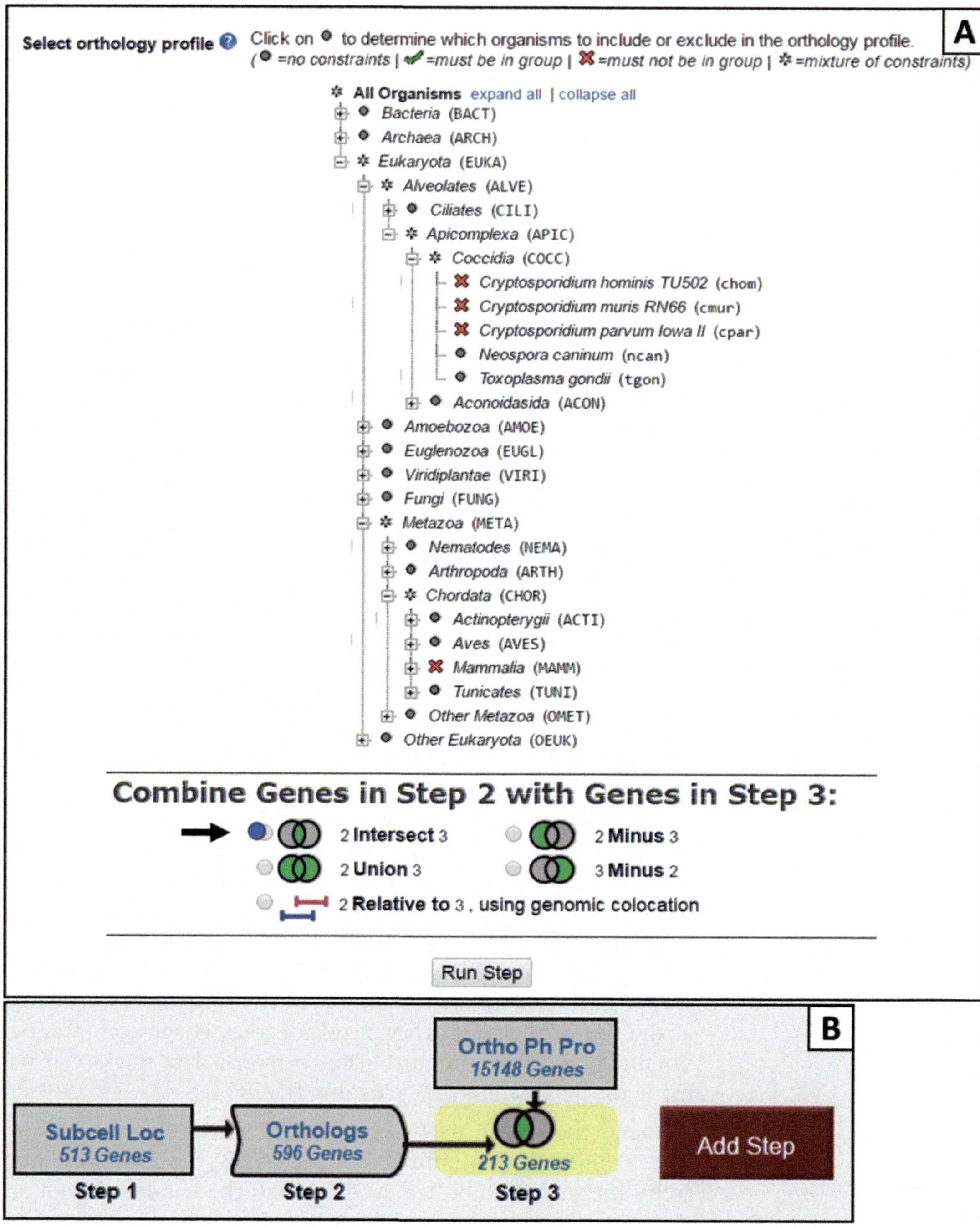

Fig. 17 The Orthology Phylogenetic Profile search. (A) Parameter for defining the orthology-based phylogenetic profile of the genes returned by the search. The phylogenetic profile of a gene is a series of "present" or "absent" calls, reflecting the inclusion of a gene in ortholog groups determined by the OrthoMCL algorithm. As shown, the parameter is configured to return genes that do not have orthologs in ***Cryptosporidium*** or Mammalia. (B) A three-step strategy that returns a refined set of ***T. gondii*** ME49, ***T. gondii*** GT1, and ***N. caninum*** Liv genes that are likely targeted to the apicoplast. The completed strategy is available here: http://eupathdb.org/eupathdb/im.do?s=3353bf3401d62d48

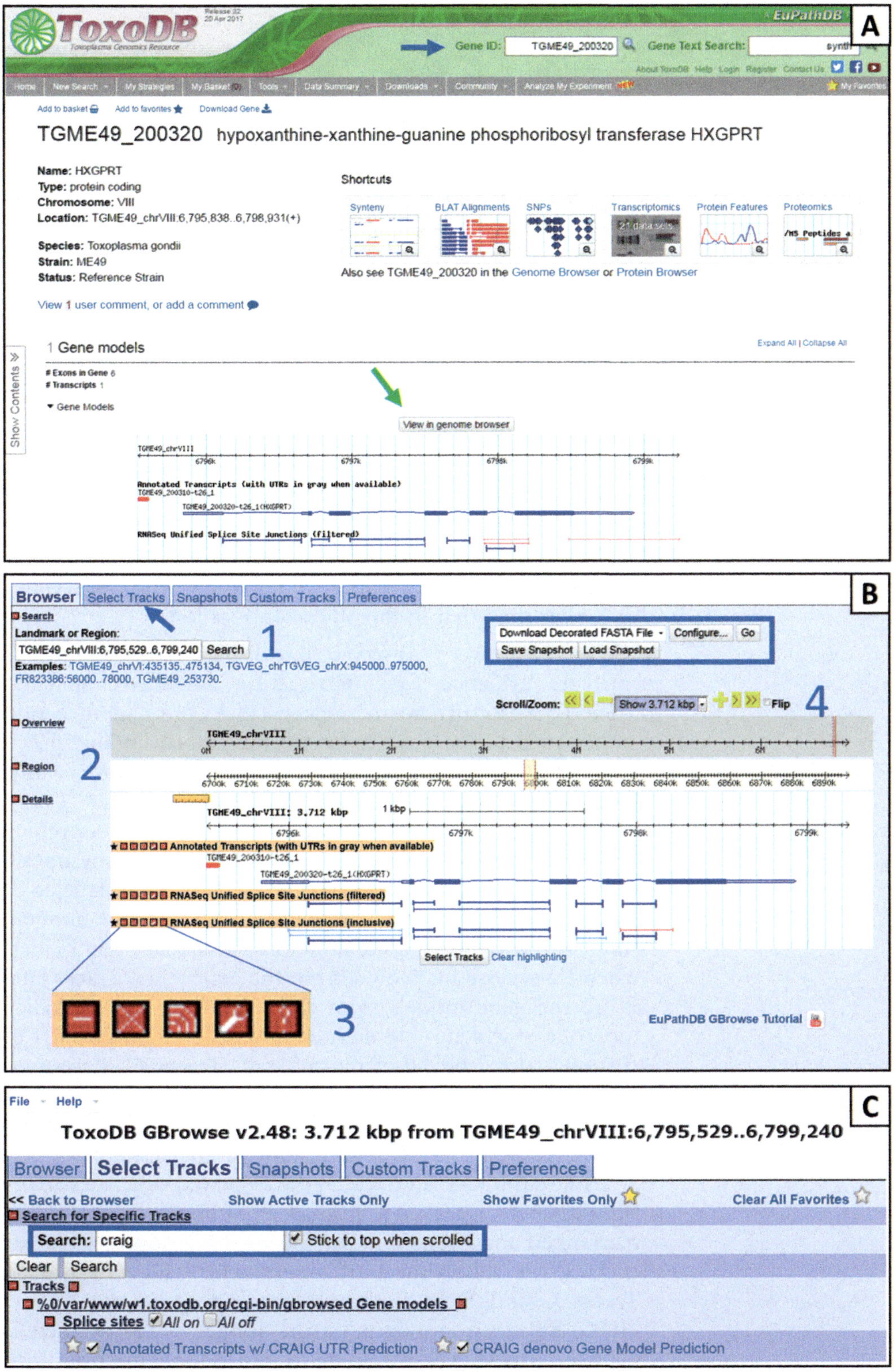

Fig. 18 The Genome Browser main features. (A) The "View in genome browser" link from all gene pages, open the browser in the region of the gene. (B) The browser's main features: the landmark region (1), the Overview, Region and Details scales (2), track controls (3), zoom and scrolling controls (4). (C) The Select Tracks tab for choosing tracks to display in the browser

3. To display additional data tracks, click on the Select Tracks tab (Fig. 18B, arrow). A wide variety of data types are available to display, including gene models, splice site junctions, synteny, sequence variations, epigenetic data sets from ChIP-on-ChIP or ChIPseq, transcriptomics, proteomics, and others. Multiple tracks can be selected, but data from different organisms cannot be displayed at the same time. Tracks are organized by data type according to the same common logic as searches on the home page and a search box (Fig. 18C, blue box) can be used to quickly find tracks of interest. Type "Craig" in the search box and then choose the two tracks labeled "Annotated Transcripts with CRAIG UTR Prediction" and "CRAIG denovo Gene Model Prediction." The tracks are automatically added to the display in the Browser tab. These two tracks are output from the CRAIG algorithm and provide alternative gene models.
4. Return to the Browser tab and compare the gene models between tracks. Note that the 3′ UTR from the CRAIG model is longer than that in the official annotation.
5. Figure 19 shows a GBrowse view displaying 9 data tracks that provide evidence for interrogating alternative splicing in HXGPRT. Return to the Select Tracks tab and turn on the other tracks (Table 5) to create the display in Fig. 19. The GBrowse view is also available at this URL: http://tinyurl.com/m8d4qtp.

   In this display, tracks have been rearranged for convenience. This can be achieved by clicking the title bar of any track and dragging it up or down as required. The tracks labeled A in Fig. 19 show the gene model from the official annotation (upper) and two splice site junction tracks that open by default when we accessed GBrowse from the gene page. Tracks labeled B are the gene models from the CRAIG de novo prediction tool, one of which is highlighted in yellow. Highlighting can be customized in the "Preferences" tab. Tracks C–E show data from a subset of the RNA sequence data sets available in ToxoDB. The *y*-axis represents the number of reads aligned. Note that each of these tracks is only displaying a subset of the available subtracks. In track D, the displayed subtracks are overlaid rather than stacked as in track E. Subtracks can be selected, rearranged and overlaid in the track-specific subtracks menu by clicking on "Showing x of y subtracks" (Fig. 19D, arrow). Tracks C and D both show reads aligned to the 3′ end of HGGPRT corresponding to the longer UTR predicted by CRAIG. Track D shows some evidence for transcription from within first intron, and it can be seen from track E that transcription from exon 3 is lower than other exons, suggesting that

Table 5
Tracks used in the ToxoDB data visualization in GBrowse example

Track title	Category
Annotated Transcripts (with UTRs in gray when available)	Gene Models, Transcripts
RNA Seq Unified Splice Site Junctions (filtered)	Gene Models, Introns
RNA Seq Unified Splice Site Junctions (inclusive)	Gene Models, Introns
Annotated Transcripts w/ CRAIG UTR Prediction	Gene Models, Splice Sites
CRAIG de novo Gene Model Prediction	Gene Models, Splice Sites
Tachyzoite Transcriptome 3 and 4 days post-infection (VEG NcLIV) mRNAseq Coverage aligned to *T. gondii* ME49 (Reid et al.)	Transcriptomics, RNA Seq, *T. gondii* ME49, Linear Scale
Tachyzoite Transcriptome Time Series (ME49) Strand Specific mRNA seq Coverage aligned to *T. gondii* ME49 (Gregory) (linear scale)	Transcriptomics, RNA Seq, *T. gondii* ME49, Linear Scale
Transcriptomes of Cat Enteroepithelial Stages (CZ-H3) Strand Specific mRNAseq Coverage aligned to *T. gondii* ME49 (Hehl Lab)	Transcriptomics, RNA Seq, *T. gondii* ME49, Linear Scale
EST Alignments	Sequence analysis, BLAT and Blast Alignments

exon 3 is sometimes skipped. This agrees with reports of alternative splice forms in this gene. The splice junction track in A shows splice-crossing reads unified from all available RNAsequence data sets. The presence of reads that span exon 2 to exon 4 support the presence of the alternative splice form. Track F shows expressed sequence tag (EST) alignments. These also show evidence of transcripts in which exon 3 is skipped, and additionally lend support to the observed read-through of intron 1.

2.5 Data Analysis

2.5.1 Result Analysis Tool: Enrichment Analysis of a Strategy Result

While EuPathDB's sophisticated search strategy system creates biologically meaningful gene lists, the enrichment analyses aid interpretation by identifying over-represented biologically relevant labels such as GO Terms and metabolic pathways in a result set. EuPathDB offers enrichment analyses for GO Terms, metabolic

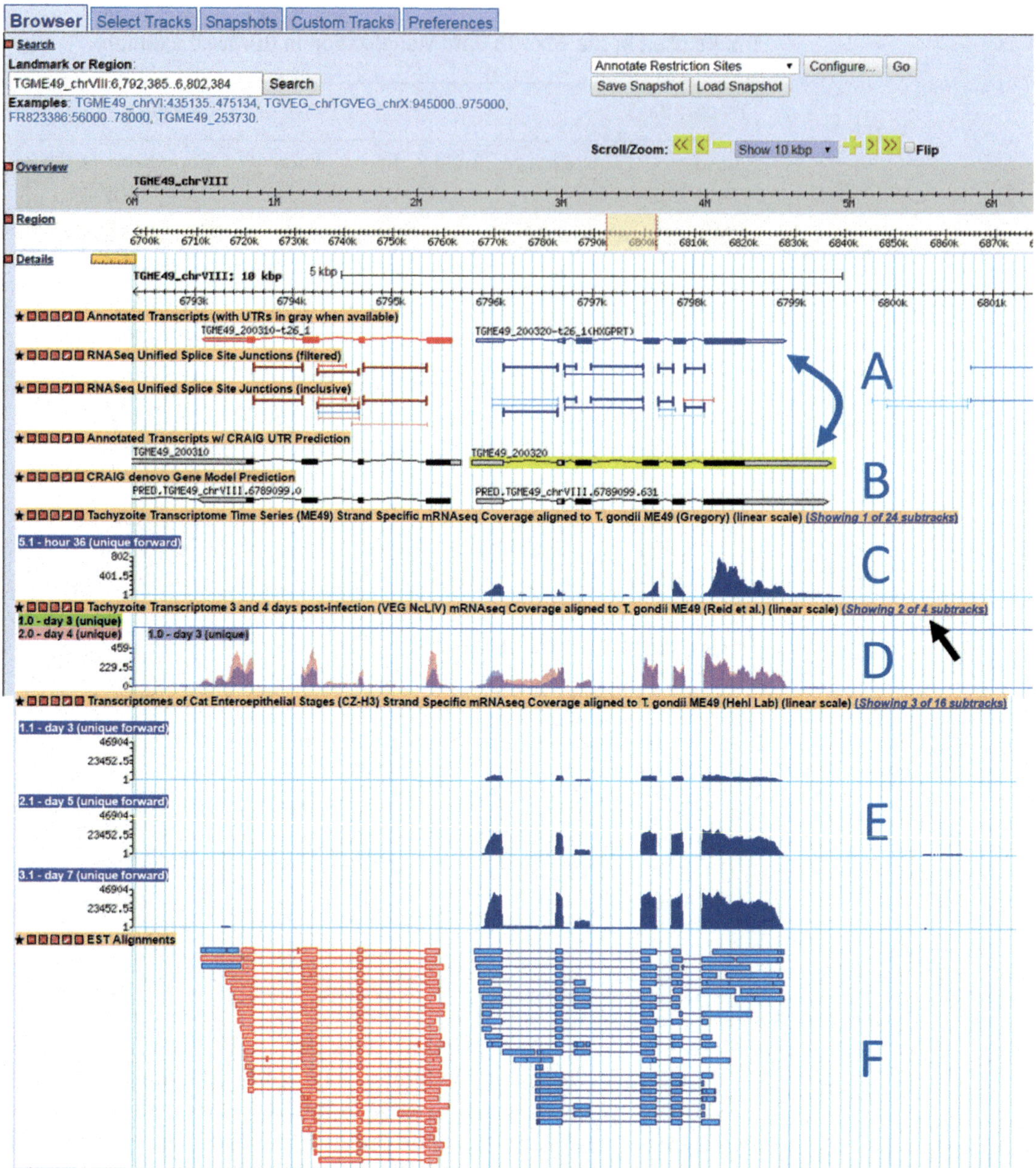

Fig. 19 The Genome Browser for data visualization and mining. (A) TGME49 genome in the region of the HXGPRT gene as displayed in the Genome Browser. Data tracks showing the current gene model and supporting splice junctions (introns) determined from RNA sequencing data. (B) Tracks created from CRAIG gene prediction analysis output. These tracks show an alternative to the official gene model. (C) RNA Sequencing reads from a single tachyzoite sample aligned to the genome. (D) RNA sequencing reads aligned to the genome and displayed with three subtracks overlaid for easy viewing. (E) Three subtracks representing time points of an RNA sequencing experiment measuring transcriptomes of cat enteroepithelial stages. (F) Expressed sequence tag alignments

pathways and words in the gene product description. The enrichment analyses perform a Fisher's Exact Test comparing functional annotations assigned to genes in the result list with the all genes in the genome. In this method, the results of Example 1 will be analyzed for enriched GO Terms and metabolic pathways.

1. *Retrieve the result of the strategy in Example 1: P. vivax* genes that are likely proteases expressed in gametocytes. Access the strategy from your My Strategies section or use this dedicated URL to retrieve our saved strategy: http://plasmodb.org/plasmo/im.do?s=2db873c2b03b57bf. Focus the strategy on the last result by clicking on the Step 4 (orthology) transform box (Fig. 20A, arrow). The active result is highlighted yellow in the strategy panel and its results are displayed in the "Gene Results" tab.
2. *Run a Gene Ontology Enrichment on the Step 4 strategy result.* Click "Analyze Results" (Fig. 20B, arrow) to create a tab for the new analysis. Choose the GO Ontology Enrichment tool (Fig. 20C) to open the tool (Fig. 20D). The Organism parameter reflects the genome from the result list genes, the genome that is "background." The Ontology parameter allows you to choose one of the three GO ontologies to enrich against. GO ontologies are structured, controlled vocabularies that describe gene products in terms of their related biological processes, cellular components, and molecular functions. For statistical reasons, only one ontology may be analyzed at once. If you are interested in more than one, run separate GO enrichment analyses. Choose the "Cellular Component" ontology to look for common cellular location assignments for the result list and click "Submit." The results are displayed as a table of GO IDs and associated data, including *p*-values and adjusted significance parameters (Fig. 20E).
3. Explore your results while paying attention to the significance metrics. There are several enriched GO terms that indicate the product is located in the proteasome complex. Since the strategy finds proteases that are expressed in gametocytes, it follows logically that this set of genes will be enriched in proteasome complex genes. Start a new analysis to find enriched GO terms from the Biological Process ontology and determine if the enriched biological processes are expected based on the strategy.

2.5.2 EuPathDB Galaxy

EuPathDB includes a Galaxy workspace, a web-based bioinformatics analysis platform that houses a large variety of bioinformatics tools to facilitate large-scale data analysis where no programming experience is required [27]. The EuPathDB Galaxy workspace is managed and maintained in partnership with Globus Genomics

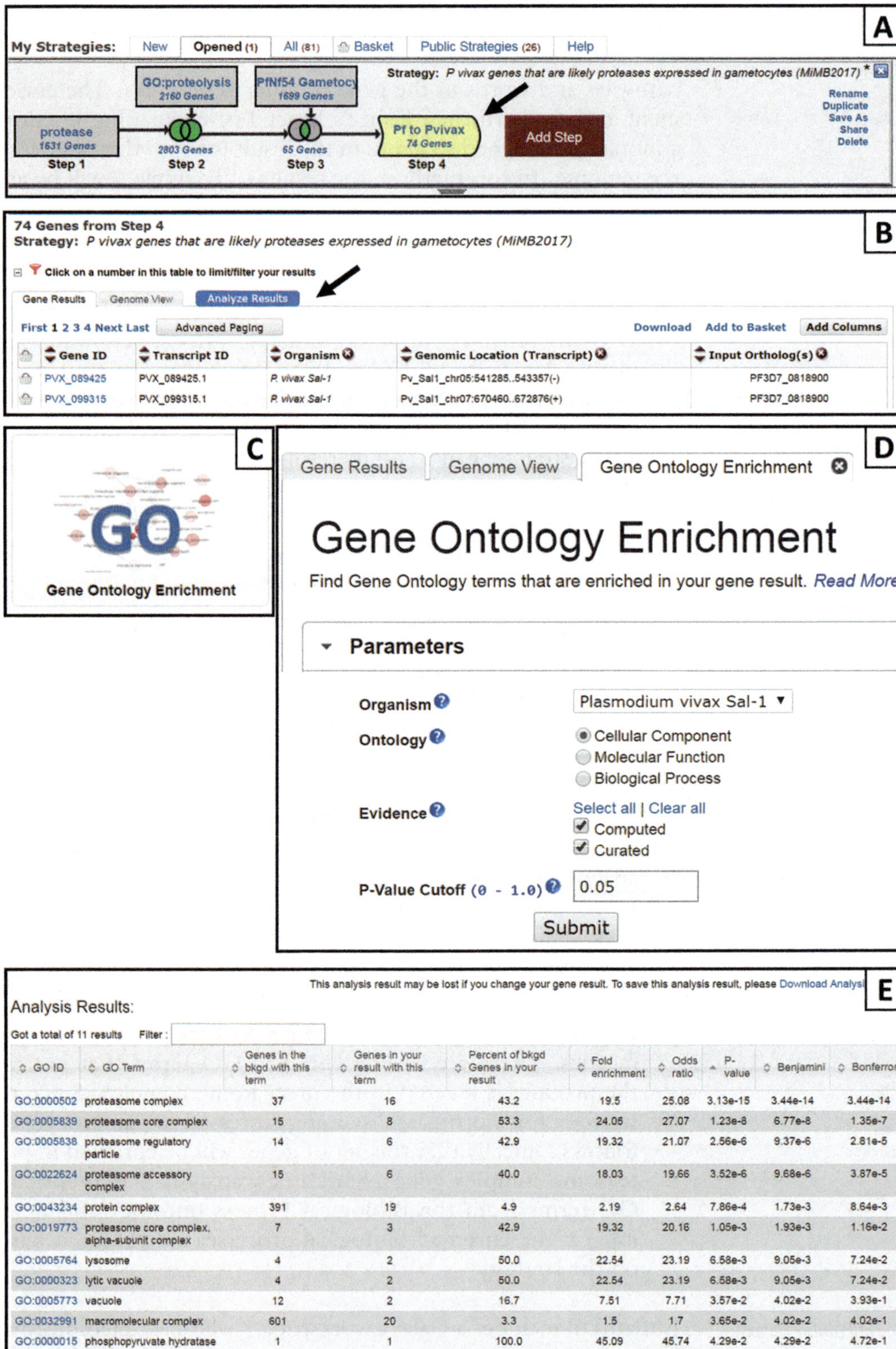

Fig. 20 The Result Analysis Tool. (A) PlasmoDB strategy focused on Step 4. Use this URL to access the strategy http://plasmodb.org/plasmo/im.do?s=2db873c2b03b57bf. (B) The strategy's gene result showing the Analyze Results button. (C) The Gene Ontology Enrichment tool button. (D) The Gene Ontology Enrichment tool showing parameters. (E) Results of a GO enrichment analysis, displaying enriched GO IDs and associated data

(https://www.globus.org/genomics), a cloud-based platform for large-scale sequencing analyses [28]. The EuPathDB Galaxy workspace offers preloaded EuPathDB reference genomes, several RNA seq and SNP calling workflows, private data analysis, data and result sharing with individual EuPathDB Galaxy users or the EuPathDB Galaxy community, and data export.

1. Visit FungiDB (http://fungidb.org) and click on the "Analyze My Experiment" tab located within the main menu in grey (Fig. 21A) to access the Galaxy workspace. To use Galaxy services, one must have a free account with any of the EuPathDB sites and then complete a short Galaxy sign-up process. Once in the Galaxy workspace, a user is directed to the Welcome page (Fig. 21B), which contains a short introduction to the instance and links to several workflows. The left panel (Fig. 21B, 1) offers a number of NGS Applications, microarray, data manipulation, statistical, FASTA, and data management tools. The center panel (Fig. 21B, 2) is controlled by the main Galaxy menu at the top of the page and is linked to the workflow history panel (Fig. 21B, 3) on the right. Its interactive interface allows you to create, run, and save custom workflows, visualize histories and result analysis.
2. Upload a data set from a local computer, the EBI website, or set up an end-point for large data transfers (Fig. 22). The data import tools can be accessed from the left panel under the Get Data option or by clicking on the Upload symbol as shown (Fig. 22A).
3. To begin a preconfigured workflow select the desired type of analysis from the list of workflows in the center panel (Fig. 21B, 2). The workflow will be opened in the center panel with a series of prompts to select filenames and parameter values for each analysis tool. A standard RNA seq workflow includes steps for assessing raw reads quality, trimming and alignment of reads, differential expression calculation, genome mapping, and visualization of results graphically in GBrowse. BigWig files generated during the analysis will be automatically linked to the local EuPathDB GBrowse session, which is not visible to other users (Fig. 21C).
4. Customize a preconfigured workflow by importing a workflow and changing or adding workflow steps within the editor interface (Fig. 22B). Alternatively, new workflows can be created by accessing the Workflow menu at the top of the page and selecting the Create new workflow option (Fig. 21B).

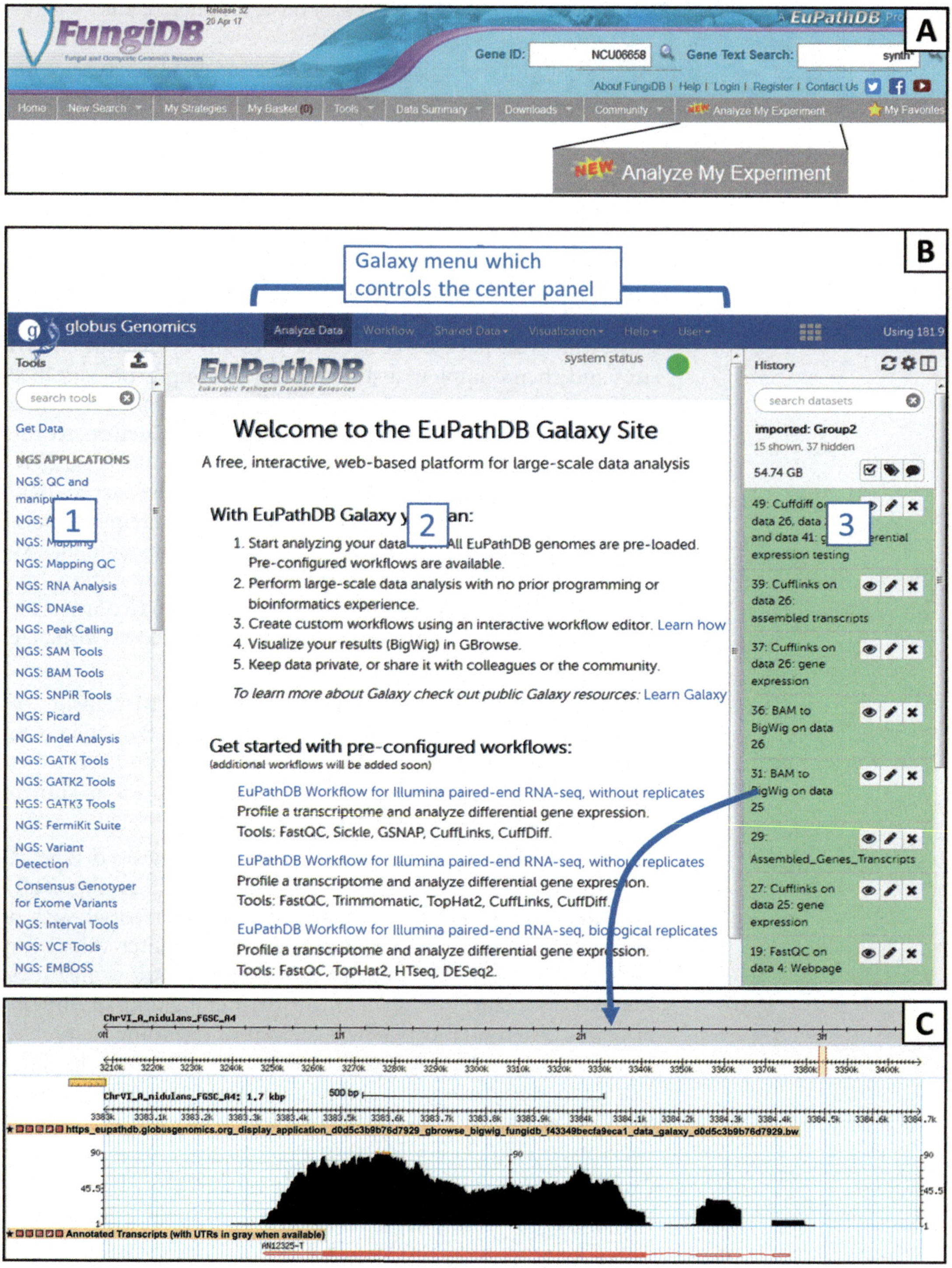

Fig. 21 EuPathDB Galaxy access and main features. (A) Shown in FungiDB. The Galaxy instance can be accessed via the Analyze My Experiment tab, which is conveniently located within the main menu (in grey). (B) From left to right. The workspace has four major components: the left panel (1) lists available large-scale data analysis tools, the center panel (2) which is the main interactive interface and also contains preconfigured workflows for the RNA-seq analysis, and the job history (3) panel on the right. The main panel is controlled via the Galaxy menu at the top. (C) BigWig file displaying RNA seq peaks for a gene in the filamentous fungus ***Aspergillus nidulans***. Files are automatically directed to FungiDB via Display in FungiDB GBrowse links available in the job history panel

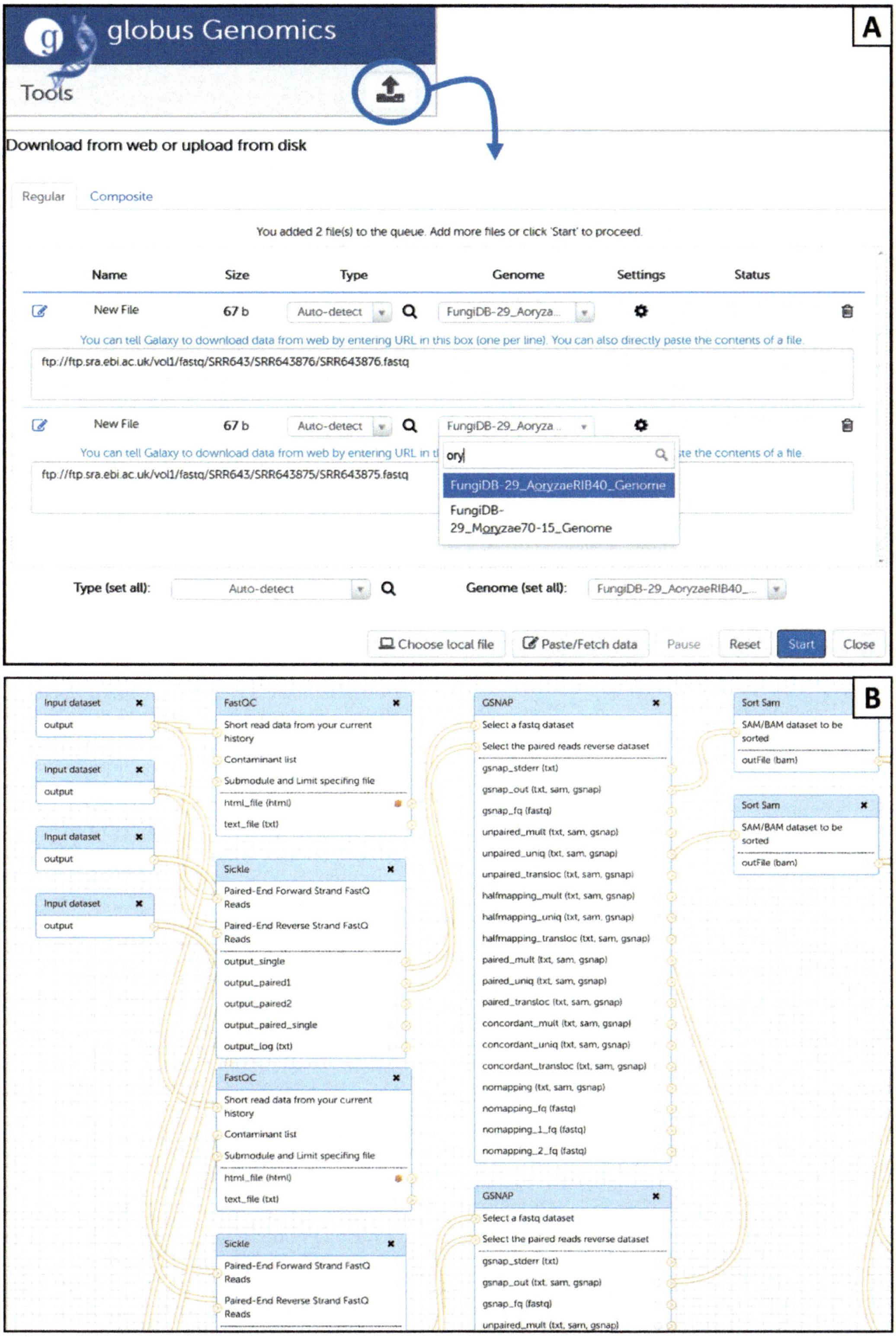

Fig. 22 File Transfer to Galaxy and Workflows. (A) To upload raw read files to Galaxy, the Paste/Fetch data button can be used to specify ftp addresses of the raw reads files at EBI. Genomes can be selected from the Genome drop-down menu. (B) Create workflows in the EuPathDB Galaxy workspace. A portion of the sample RNA seq workflow is shown. This workflow can be modified and saved for later use

Acknowledgments

EuPathDB would like to acknowledge their current funders, the National Institutes of Health (USA), the Wellcome Trust (UK), as well as past funders and The Bill and Melinda Gates Foundation (USA), The Burroughs Wellcome Fund (USA).

References

1. Aurrecoechea C, Barreto A, Basenko EY, Brestelli J, Brunk BP, Cade S, Crouch K, Doherty R, Falke D, Fischer S, Gajria B, Harb OS, Heiges M, Hertz-Fowler C, Hu S, Iodice J, Kissinger JC, Lawrence C, Li W, Pinney DF, Pulman JA, Roos DS, Shanmugasundram A, Silva-Franco F, Steinbiss S, Stoeckert CJ Jr, Spruill D, Wang H, Warrenfeltz S, Zheng J (2017) EuPathDB: the eukaryotic pathogen genomics database resource. Nucleic Acids Res 45(D1):D581–D591. https://doi.org/10.1093/nar/gkw1105
2. Aurrecoechea C, Barreto A, Brestelli J, Brunk BP, Cade S, Doherty R, Fischer S, Gajria B, Gao X, Gingle A, Grant G, Harb OS, Heiges M, Hu S, Iodice J, Kissinger JC, Kraemer ET, Li W, Pinney DF, Pitts B, Roos DS, Srinivasamoorthy G, Stoeckert CJ Jr, Wang H, Warrenfeltz S (2013) EuPathDB: the eukaryotic pathogen database. Nucleic Acids Res 41(Database issue):D684–D691. https://doi.org/10.1093/nar/gks1113
3. Stein LD, Mungall C, Shu S, Caudy M, Mangone M, Day A, Nickerson E, Stajich JE, Harris TW, Arva A, Lewis S (2002) The generic genome browser: a building block for a model organism system database. Genome Res 12(10):1599–1610
4. Steinbiss S, Silva-Franco F, Brunk B, Foth B, Hertz-Fowler C, Berriman M, Otto TD (2016) Companion: a web server for annotation and analysis of parasite genomes. Nucleic Acids Res 44(W1):W29–W34. https://doi.org/10.1093/nar/gkw292
5. Peng D, Tarleton R (2015) EuPaGDT: a web tool tailored to design CRISPR guide RNAs for eukaryotic pathogens. Microb Genom 1(4):e000033. https://doi.org/10.1099/mgen.0.000033
6. Bunnik EM, Chung DW, Hamilton M, Ponts N, Saraf A, Prudhomme J, Florens L, Le Roch KG (2013) Polysome profiling reveals translational control of gene expression in the human malaria parasite Plasmodium falciparum. Genome Biol 14(11):R128. https://doi.org/10.1186/gb-2013-14-11-r128
7. Lopez-Barragan MJ, Lemieux J, Quinones M, Williamson KC, Molina-Cruz A, Cui K, Barillas-Mury C, Zhao K, XZ S (2011) Directional gene expression and antisense transcripts in sexual and asexual stages of Plasmodium falciparum. BMC Genomics 12:587. https://doi.org/10.1186/1471-2164-12-587
8. Lasonder E, Green JL, Camarda G, Talabani H, Holder AA, Langsley G, Alano P (2012) The Plasmodium falciparum schizont phosphoproteome reveals extensive phosphatidylinositol and cAMP-protein kinase A signaling. J Proteome Res 11(11):5323–5337. https://doi.org/10.1021/pr300557m
9. Solyakov L, Halbert J, Alam MM, Semblat JP, Dorin-Semblat D, Reininger L, Bottrill AR, Mistry S, Abdi A, Fennell C, Holland Z, Demarta C, Bouza Y, Sicard A, Nivez MP, Eschenlauer S, Lama T, Thomas DC, Sharma P, Agarwal S, Kern S, Pradel G, Graciotti M, Tobin AB, Doerig C (2011) Global kinomic and phospho-proteomic analyses of the human malaria parasite Plasmodium falciparum. Nat Commun 2:565. https://doi.org/10.1038/ncomms1558
10. Oehring SC, Woodcroft BJ, Moes S, Wetzel J, Dietz O, Pulfer A, Dekiwadia C, Maeser P, Flueck C, Witmer K, Brancucci NM, Niederwieser I, Jenoe P, Ralph SA, Voss TS (2012) Organellar proteomics reveals hundreds of novel nuclear proteins in the malaria parasite Plasmodium falciparum. Genome Biol 13(11):R108. https://doi.org/10.1186/gb-2012-13-11-r108
11. Dewey CN (2007) Aligning multiple whole genomes with Mercator and MAVID. Methods Mol Biol 395:221–236
12. Kanehisa M (2002) The KEGG database. Novartis Found Symp 247:91–101. discussion 101–103, 119–128, 244–152
13. Kanehisa M, Goto S, Hattori M, Aoki-Kinoshita KF, Itoh M, Kawashima S, Katayama T, Araki M, Hirakawa M (2006) From genomics to chemical genomics: new developments in KEGG. Nucleic Acids Res 34(Database issue):D354–D357

14. Kanehisa M, Sato Y, Kawashima M, Furumichi M, Tanabe M (2016) KEGG as a reference resource for gene and protein annotation. Nucleic Acids Res 44(D1):D457–D462. https://doi.org/10.1093/nar/gkv1070
15. Caspi R, Billington R, Ferrer L, Foerster H, Fulcher CA, Keseler IM, Kothari A, Krummenacker M, Latendresse M, Mueller LA, Ong Q, Paley S, Subhraveti P, Weaver DS, Karp PD (2016) The MetaCyc database of metabolic pathways and enzymes and the BioCyc collection of pathway/genome databases. Nucleic Acids Res 44(D1):D471–D480. https://doi.org/10.1093/nar/gkv1164
16. Shameer S, Logan-Klumpler FJ, Vinson F, Cottret L, Merlet B, Achcar F, Boshart M, Berriman M, Breitling R, Bringaud F, Butikofer P, Cattanach AM, Bannerman-Chukualim B, Creek DJ, Crouch K, de Koning HP, Denise H, Ebikeme C, Fairlamb AH, Ferguson MA, Ginger ML, Hertz-Fowler C, Kerkhoven EJ, Maser P, Michels PA, Nayak A, Nes DW, Nolan DP, Olsen C, Silva-Franco F, Smith TK, Taylor MC, Tielens AG, Urbaniak MD, van Hellemond JJ, Vincent IM, Wilkinson SR, Wyllie S, Opperdoes FR, Barrett MP, Jourdan F (2015) TrypanoCyc: a community-led biochemical pathways database for Trypanosoma brucei. Nucleic Acids Res 43(Database issue):D637–D644. https://doi.org/10.1093/nar/gku944
17. Saunders EC, MacRae JI, Naderer T, Ng M, McConville MJ, Likic VA (2012) LeishCyc: a guide to building a metabolic pathway database and visualization of metabolomic data. Methods Mol Biol 881:505–529. https://doi.org/10.1007/978-1-61779-827-6_17
18. Doyle MA, MacRae JI, De Souza DP, Saunders EC, McConville MJ, Likic VA (2009) LeishCyc: a biochemical pathways database for Leishmania major. BMC Syst Biol 3:57. https://doi.org/10.1186/1752-0509-3-57
19. Franz M, Lopes CT, Huck G, Dong Y, Sumer O, Bader GD (2016) Cytoscape.js: a graph theory library for visualisation and analysis. Bioinformatics 32(2):309–311. https://doi.org/10.1093/bioinformatics/btv557
20. Saito R, Smoot ME, Ono K, Ruscheinski J, Wang PL, Lotia S, Pico AR, Bader GD, Ideker T (2012) A travel guide to Cytoscape plugins. Nat Methods 9(11):1069–1076. https://doi.org/10.1038/nmeth.2212
21. Shannon P, Markiel A, Ozier O, Baliga NS, Wang JT, Ramage D, Amin N, Schwikowski B, Ideker T (2003) Cytoscape: a software environment for integrated models of biomolecular interaction networks. Genome Res 13(11):2498–2504. https://doi.org/10.1101/gr.1239303
22. Siegel TN, Hekstra DR, Wang X, Dewell S, Cross GA (2010) Genome-wide analysis of mRNA abundance in two life-cycle stages of Trypanosoma brucei and identification of splicing and polyadenylation sites. Nucleic Acids Res 38(15):4946–4957. https://doi.org/10.1093/nar/gkq237
23. Ashburner M, Ball CA, Blake JA, Botstein D, Butler H, Cherry JM, Davis AP, Dolinski K, Dwight SS, Eppig JT, Harris MA, Hill DP, Issel-Tarver L, Kasarskis A, Lewis S, Matese JC, Richardson JE, Ringwald M, Rubin GM, Sherlock G (2000) Gene ontology: tool for the unification of biology. The Gene Ontology Consortium. Nat Genet 25(1):25–29
24. Gene Ontology C (2015) Gene Ontology Consortium: going forward. Nucleic Acids Res 43(Database issue):D1049–D1056. https://doi.org/10.1093/nar/gku1179
25. Lasonder E, Rijpma SR, van Schaijk BC, Hoeijmakers WA, Kensche PR, Gresnigt MS, Italiaander A, Vos MW, Woestenenk R, Bousema T, Mair GR, Khan SM, Janse CJ, Bartfai R, Sauerwein RW (2016) Integrated transcriptomic and proteomic analyses of P. falciparum gametocytes: molecular insight into sex-specific processes and translational repression. Nucleic Acids Res 44(13):6087–6101. https://doi.org/10.1093/nar/gkw536
26. Finn RD, Attwood TK, Babbitt PC, Bateman A, Bork P, Bridge AJ, Chang HY, Dosztanyi Z, El-Gebali S, Fraser M, Gough J, Haft D, Holliday GL, Huang H, Huang X, Letunic I, Lopez R, Lu S, Marchler-Bauer A, Mi H, Mistry J, Natale DA, Necci M, Nuka G, Orengo CA, Park Y, Pesseat S, Piovesan D, Potter SC, Rawlings ND, Redaschi N, Richardson L, Rivoire C, Sangrador-Vegas A, Sigrist C, Sillitoe I, Smithers B, Squizzato S, Sutton G, Thanki N, Thomas PD, Tosatto SC, Wu CH, Xenarios I, Yeh LS, Young SY, Mitchell AL (2017) InterPro in 2017-beyond protein family and domain annotations. Nucleic Acids Res 45(D1):D190–D199. https://doi.org/10.1093/nar/gkw1107
27. Afgan E, Baker D, van den Beek M, Blankenberg D, Bouvier D, Cech M, Chilton J, Clements D, Coraor N, Eberhard C, Gruning B, Guerler A, Hillman-Jackson J, Von Kuster G, Rasche E, Soranzo N, Turaga N, Taylor J, Nekrutenko A, Goecks J (2016) The Galaxy platform for accessible, reproducible and collaborative biomedical analyses: 2016 update. Nucleic Acids Res 44(W1):W3–W10. https://doi.org/10.1093/nar/gkw343
28. Liu B, Madduri RK, Sotomayor B, Chard K, Lacinski L, Dave UJ, Li J, Liu C, Foster IT (2014) Cloud-based bioinformatics workflow platform for large-scale next-generation sequencing analyses. J Biomed Inform 49:119–133. https://doi.org/10.1016/j.jbi.2014.01.005

Chapter 6

The Ensembl Genome Browser: Strategies for Accessing Eukaryotic Genome Data

Victoria Newman, Benjamin Moore, Helen Sparrow, and Emily Perry

Abstract

The Ensembl Genome Browser provides a wealth of freely available genomic data that can be accessed for many purposes by genetics, genomics, and molecular biology researchers. Herein we present two protocols for exploring different aspects of these data: a phenotype and its associated variants and genes, and a promoter and the epigenetic marks and protein-binding activity associated with it. These workflows illustrate a subset of the data types available through the Ensembl Browser, and can be considered a springboard for further exploration.

Key words Ensembl, Eukaryotic genomes, Phenotypes, Variants, Epigenetic mark

1 Introduction

Genome browsers are resources that integrate data at the genomic level, thereby allowing visualization of related genomic information in one space. These data can include genes, noncoding elements that regulate gene expression, genetic variation and the results of comparative genomics analyses, among other forms of annotation (Fig. 1) [1–4]. Commonly used genome browsers include Ensembl, the UCSC Genome Browser [5] and IGV [6].

The Ensembl project was initially launched in 1999 with the aim of developing methodologies for automatic annotation of (human) genomic sequence with genes and their constituent transcripts [7]. Since that time, the project has broadened substantially in scope; the Ensembl Genome Browser [8], which came online in 2000, now includes reference genomic sequence and annotation for nearly 100 chordate organisms. Ensembl is rapidly incorporating new data, including whole clades of new species' genomes and reference sequence for multiple strains of existing species, such as mouse. In addition, existing annotation is regularly augmented by the inclusion of new data sets. Ensembl's sister site, Ensembl Genomes,

Martin Kollmar (ed.), *Eukaryotic Genomic Databases: Methods and Protocols*, Methods in Molecular Biology, vol. 1757, https://doi.org/10.1007/978-1-4939-7737-6_6, © The Author(s) 2018

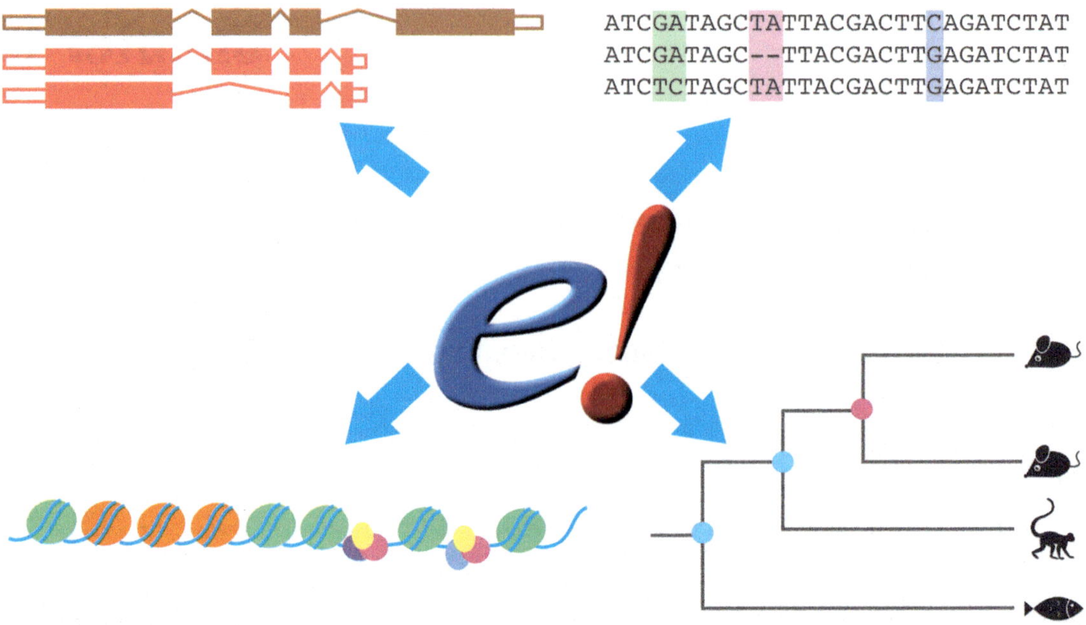

Fig. 1 Ensembl features. Ensembl integrates together gene annotation, genetic variation, gene regulation data, and comparative genomics onto a single genomic platform. Gene annotation is carried out in house, annotating the full intron–exon structure of coding and noncoding transcripts. Short variants, such as SNPs and indels, are pulled into Ensembl from external databases, alongside structural variants and copy-number variants. ChIP-seq and DNase-seq data is used for in-house prediction of regions of open chromatin and regulatory elements such as promoters, enhancers and CTCF binding sites on the genome and their activity in different cell types. Whole genome alignments and gene tree analysis is carried out in house to compare species in Ensembl. These data are presented alongside each other on the genome in the Ensembl browser, and can also be accessed for bulk export through BioMart, programmatically through APIs and as flat-files on the Ensembl FTP site

provides access to nonvertebrate genomes through dedicated portals for Bacteria, Fungi, Plants, Metazoa, and Protists [9, 10].

Ensembl data, annotations, and analyses are updated every 2–3 months, alongside software updates to both the public-facing website and the underlying databases. Prior releases are frozen as archive sites, and from Dec 2013 (Ensembl version 74) will remain accessible via our web interface for at least 5 years following their initial release. A dedicated site is also maintained for the GRCh37 reference human genome assembly, which is annotated with new data on a limited basis (Fig. 2) [11]; partial data from ongoing genome annotation can be accessed via the preview Pre! site.

Data from Ensembl can be accessed at multiple scales. In this chapter, we describe data access through the browser web pages and via BioMart [12], a web-based tool that allows customized retrieval of data from the Ensembl databases. However, data can also be accessed programmatically via our Perl and REST APIs [13, 14] Files containing genome-wide data are available for all species represented in Ensembl via an FTP site [15]; data from all releases

Fig. 2 The Ensembl homepage. The Ensembl homepage provides access to a search function, which can retrieve information associated with, for example, genes, transcripts, proteins, variants, phenotypes, and ontology terms. In addition, links are available to Ensembl's sister site, Ensembl Genomes, as well as to the most-searched genomes and a complete list of annotated genomes. Fully annotated genomes are available on the main Ensembl site, while genomes whose annotation is in process can be browsed on the Ensembl Pre! site. Ensembl maintains web interfaces of archived versions for 5 years. These can be accessed from a link in the lower right-hand corner. Documentation and help pages can be accessed from the homepage, as well as in-house and external tools integrated into the Ensembl web interface. A dedicated page describing data-download strategies is also available and presents links to the point-and-click tool BioMart, which permits bulk download of Ensembl datasets with no requirement for programming expertise, as well as APIs and FTP site

of Ensembl can be retrieved from the FTP site, or from our databases via the Perl APIs, in perpetuity.

Beyond providing access to data related to publicly available genome annotation, Ensembl integrates a number of tools designed to process or analyze your own data. The ID History Converter converts Ensembl IDs from a previous release into their current equivalents, while the Assembly Converter maps genomic coordinates from one version of a genome assembly to another. The Variant Effect Predictor predicts the functional consequences of a set of known and/or novel variants [16]. Sequence alignment using BLAST and BLAT against Ensembl genes, genomes and proteins is also available [17, 18], along with a suite of tools developed as part of the 1000 Genomes Project [19] that can be accessed on the dedicated GRCh37 browser site [11].

In this chapter we describe two workflows showcasing a subset of the data available in the Ensembl browser and indicating possible routes to access them. First, we demonstrate a phenotype-centric search highlighting variation data associated with genes and transcripts. Secondly, we present a gene-centric search illustrating gene and transcript models, and the exploration of regulatory features in the region of a gene. In each case we also indicate strategies for data export via BioMart. Those interested in our annotation methods, in programmatic access to Ensembl data, or in exploring other forms of data and annotation are encouraged to refer to our publications [20].

2 Materials

Computer, Internet connection.

An Internet browser: recent versions of Firefox, Chrome, Safari, and Internet Explorer are supported.

3 Methods

These workflows were written using Ensembl release 88 (March 2017). There may be updates to the data or interfaces if you are using a more recent release.

3.1 WF1: Phenotype-Based Searches and Identification of Associated Genetic Variation

The Ensembl browser can be searched using a variety of terms, including genomic regions, genes, variants, or phenotypes; the following workflow describes a phenotype-based search that highlights data and annotations collated in the Phenotype, Variant, Gene, and Transcript tabs.

Non-melanoma skin cancer—principally basal cell and squamous cell carcinomas—is a relatively common pathology associated with variants in several genes [21].

1. *Getting started*: To explore the phenotype in more detail, type "non-melanoma skin cancer" into the search box on the Ensembl home page, www.ensembl.org, and click the "Go" button. The search autocomplete may retrieve direct links to

suggested results; this will allow you to proceed immediately to **step 2**.

A list of search results will be generated, with "Non-melanoma skin cancer (Human Phenotype)" appearing first. Options on the left-hand side of the page permit restriction by species and/or other categories: click on the different filters individually to apply them to the search results.

2. *Studying loci associated with a phenotype*: Click the "Non-melanoma skin cancer (Human Phenotype)" link to open the Phenotype tab.

   The loci associated with non-melanoma skin cancer are presented in tabular form; their external identifiers, genomic coordinates and associated genes, and the publications in which they were initially described are all listed. Links are provided to further information about the annotation source and relevant publications (in this case, the GWAS catalog [22] *and PubMed* [23]*;* Fig. 3*).*

3. *Studying a variant*: One of the variants associated with non-melanoma skin cancer, rs1805007, falls within the *MC1R* gene. Click the "rs1805007" link to load the Variant tab.

   The Variant tab collates data relating specifically to the variant of interest (A full list of the databases from which Ensembl imports variation data can be found in the documentation [24]*.).*

   An overview of the data is found at the top of the Variant tab (Fig. 4A)*, while a table indicating the phenotypes associated with the variant can be found lower down the page.*

   The most severe consequence linked to rs1805007 is "missense_variant", indicating that the alternative allele at this locus lead to an amino acid substitution. All consequences of the rs1805007 variant can be explored by clicking on the "See all predicted consequences" link. Ensembl uses Sequence Ontology terms to describe variant consequences [25].

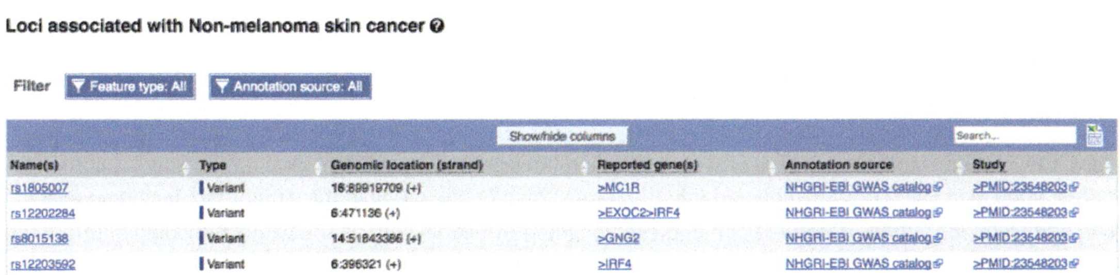
Loci associated with Non-melanoma skin cancer

Filter: Feature type: All | Annotation source: All

Show/hide columns

Name(s)	Type	Genomic location (strand)	Reported gene(s)	Annotation source	Study
rs1805007	Variant	16:89919709 (+)	>MC1R	NHGRI-EBI GWAS catalog	>PMID:23548203
rs12202284	Variant	6:471136 (+)	>EXOC2>IRF4	NHGRI-EBI GWAS catalog	>PMID:23548203
rs8015138	Variant	14:51843386 (+)	>GNG2	NHGRI-EBI GWAS catalog	>PMID:23548203
rs12203592	Variant	6:396321 (+)	>IRF4	NHGRI-EBI GWAS catalog	>PMID:23548203

Fig. 3 The Ensembl phenotype tab. The Ensembl phenotype tab allows you to explore the phenotype ontology associated with a phenotype and any loci (variants, QTLs, or genes) linked to the phenotype. Loci associated with the phenotype shown in a table on the Associated loci page. The buttons above the table allow filtering. Links take you to the database and/or paper where the link between locus and phenotype was made

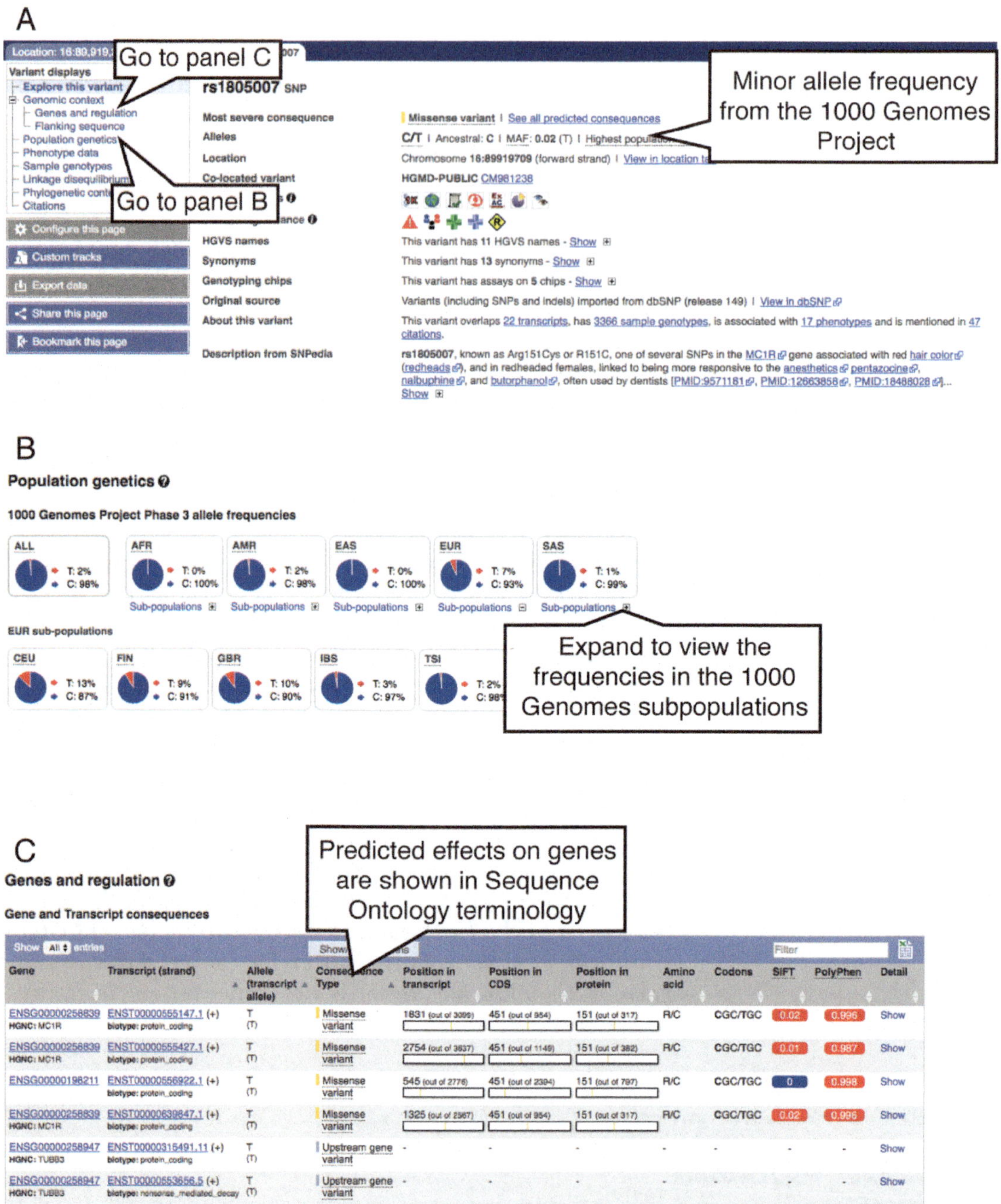

Fig. 4 The Ensembl variation tab. The Ensembl variation tab provides a wealth of information about a particular variant, such as a SNP or indel. (A) A variant summary shown on all pages in the variant tab, including alleles, MAF, and evidence status. The menu at the left-hand side provides links to all the pages providing information on the variant. (B) Pie charts from the Population Genetics page, showing the allele frequencies for the variant in the 1000 Genomes populations. (C) The Genes and Regulation table, listing all genes affected by the variant with details of sequence ontology consequences, position in the gene and protein, and SIFT and Polyphen scores for amino acid changes (where relevant)

Below the consequence, you can see that the reference allele of rs1805007 at the genomic position 16:89919709 is C, and one alternative allele, T, has been observed. Minor allele frequency (MAF) has been calculated for the alternative allele, which was observed in 1000 Genomes Project participants: it was identified in 2% of participants in that study [2, 26].

Navigating to the Variant tab from the Phenotype tab automatically loads a table containing the phenotype data relating to this variant, as mentioned above. Tanning ability, sensitivity to sun, and fair hair and skin color have all been associated with the variant, as has basal cell carcinoma, a form of non-melanoma skin cancer. Collectively, these phenotypes are consistent with the observed linkage between fair complexions and sensitivity to sun exposure.

4. The menu on the left presents additional options. Click "Population genetics" to view allele frequencies in global populations.

   On this page, data from the 1000 Genomes [26], *HapMap* [27], *and NHLBI Exome Sequencing* [28] *Projects and the Exome Aggregation Consortium (ExAC)* [29] *are displayed. The data from the 1000 Genomes Project are shown at the top* (Fig. 4B)*; the pie-charts represent allele frequencies for different superpopulations. Allele frequencies for subpopulations within each superpopulation can be viewed by clicking the "Subpopulations" link beneath the corresponding superpopulation. Allele and genotype frequencies among 1000 Genomes Project participants can also be found in tabular form immediately below the graphical views.*

   The frequency of the T variant allele in 1000 Genomes Project participants is highest among European subgroups, and individuals homozygous for the variant also occur only in these subgroups. This is expected given the phenotypes associated with the variant (Fig. 4B).

5. To explore genes and transcripts with which the variant is associated, click "Genes and regulation" in the menu on the left.

   As we saw previously, the variant lies within the MC1R *gene; the summary table here indicates that it overlaps two independent transcripts of this gene as a missense variant and is a downstream gene variant of a third transcript. Other genes and transcripts affected by the variant, as well as the associated consequences, are also shown* (Fig. 4C).

   In a second table, called "Gene expression correlations," you can find a list of genes whose expression has been found by the GTEx Project to be affected by the variant of interest [30].

   Finally, any regulatory features or motifs in which the variant falls will be listed in two separate tables at the bottom of the page. There are no regulatory features or motifs that overlap the variant rs1805007.

6. *Studying a gene and its transcripts*: Click "ENSG00000258839" in the Genes and regulation table to go directly to the Gene tab, which collates gene-related information, for *MC1R*.

   Navigating to the Gene tab from the Variant tab loads the Variant table, which lists all variants in the Ensembl database that fall within the gene itself or in the region 5 kb upstream or downstream of the gene. The top of the page presents a short overview of MC1R, *including a description of the gene, its genomic location and synonyms, and an option to show a table of all its transcripts. This information can also be found at the top of all subsequent views within the Gene tab. As in other tabs in the Ensembl browser, the menu to the left of the Gene tab presents links to a variety of additional data and annotations* (Fig. 5A).

7. Click "Summary" in the left-hand menu.

   General information about the gene, including a description, synonyms and the genomic location, can be found in this view. A graphical model of the gene's transcripts is shown at the bottom (Fig. 5A).

8. For the complete set of phenotypes associated with *MC1R*, click the "Phenotypes" link in the left-hand menu.

   The three tables list phenotypes associated with the gene, with variants in the gene, and with other species' orthologues of the gene, as predicted by the Ensembl comparative genomics pipeline [3]. *Several phenotypes have been linked to rs1805007, and the* MC1R *gene also plays a role in coat and skin pigmentation in other organisms, suggesting a conserved function.*

9. Click on the "GO: Biological process" link in the left-hand menu.

   The GO, or Gene Ontology, terms related to biological processes which have been associated with the transcripts of the MC1R *gene are displayed in the table* (Fig. 5B) [31, 32]. *Each row of the table contains the GO accession number, a description of the GO term, and the evidence codes, annotation source and stable IDs of transcripts associated with that GO term. Hover over the evidence codes to see their definitions.*

   MC1R*-encoded proteins are involved in signal transduction and the melanin biosynthesis pathway, and are located in the plasma membrane, consistent with a role in pigmentation.*

Fig. 5 (continued) against the genome. The central contig indicates the genome. Positive stranded genes, such as *MC1R* are depicted above the contig. Strand is also indicated by an arrow alongside the transcript name indicating the direction of transcription, and by introns, which are shown pointing upward on positive stranded genes and downward on negative stranded genes. Some transcripts have been removed from this image for size. On all pages in the gene tab, a menu on the left-hand side lists all the pages available for looking at a gene. (B) Three pages are available for looking at the GO terms associated with a gene, conforming to the three categories of terms, Biological process, Molecular function, and Cellular component. These are listed for each gene, including which transcript they are associated with and how they were annotated

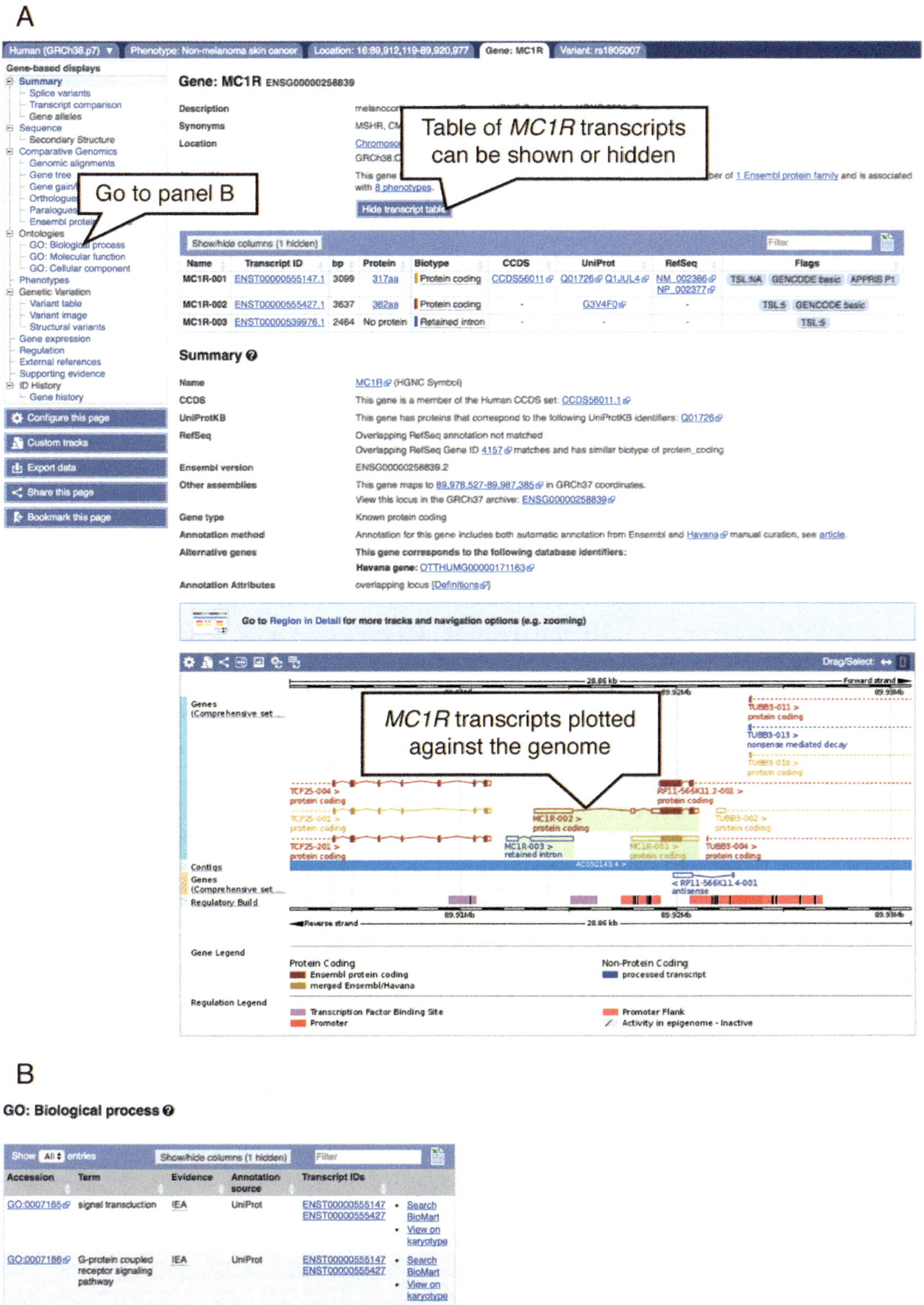

Fig. 5 The Ensembl gene tab. The Ensembl gene tab provides a number of views to look at different aspects of a gene. (A) The gene summary page includes a graphical depiction of the transcripts of the gene, shown

Two further links in the menu at left provide GO term associations regarding the Molecular Function and Cellular Component corresponding to transcripts of the MC1R *gene* (Fig. 5A).

10. Click on the "External references" link in the left-hand menu.

    Links to records in external databases such as EntrezGene [33], *HGNC* [34], *and MIM Gene and MIM Morbid* [35] *can be found on this page.*

11. *Studying a transcript*: Click the "Show transcript table" button in the "Transcripts" section at the top of the page.

    A tabular view of the individual transcripts comprising the gene model can be seen (for more information on the Ensembl gene annotation strategy, see ref. 1*). This table displays information about transcript length and biotype, as well as links to the entries in the CCDS* [36], *UniProt* [37], *and RefSeq* [33] *databases that correspond to particular transcripts.*

    The level of support for a transcript prediction, and its biological relevance, can be inferred from the matching evidence records and associated flags.

12. Click the "ENST00000555147.1" link in the Transcript table; ENST00000555147.1 is the Ensembl stable ID for the *MC1R-001* transcript.

    The MC1R-001 *transcript's biotype is listed as "protein-coding," and the transcript is colored golden in the graphical view. This indicates that it has been independently annotated with identical coordinates by both the Ensembl automated gene annotation and the HAVANA manual gene annotation methods* [1] (Fig. 5A).

    We are now located in the Transcript tab, which is visible in the blue navigation bar at the top of the page, next to the Gene tab. From the left-hand menu of the Transcript tab you can access complete, spliced or translated transcript sequences ("Exons," "cDNA," and "Protein," respectively), as well as graphical and tabular representations of annotated protein domains ("Protein summary" and "Domains & features," respectively). "General identifiers" provides links to related records in external repositories (Fig. 6A).

    You can now click on "Hide Transcript table" in the Gene section at the top of the page to remove the Transcript table from the page view.

13. Click on the "Supporting evidence" link in the left-hand menu.

    This page displays the records used in the annotation in graphical form; all records are hyperlinked to the original data in RefSeq, UniProt, and ENA [38] (Fig. 6B).

14. Click the "Variant table" link in the left-hand menu.

    This table displays the set of variants associated with the MC1R-001 *transcript* (Fig. 7A).

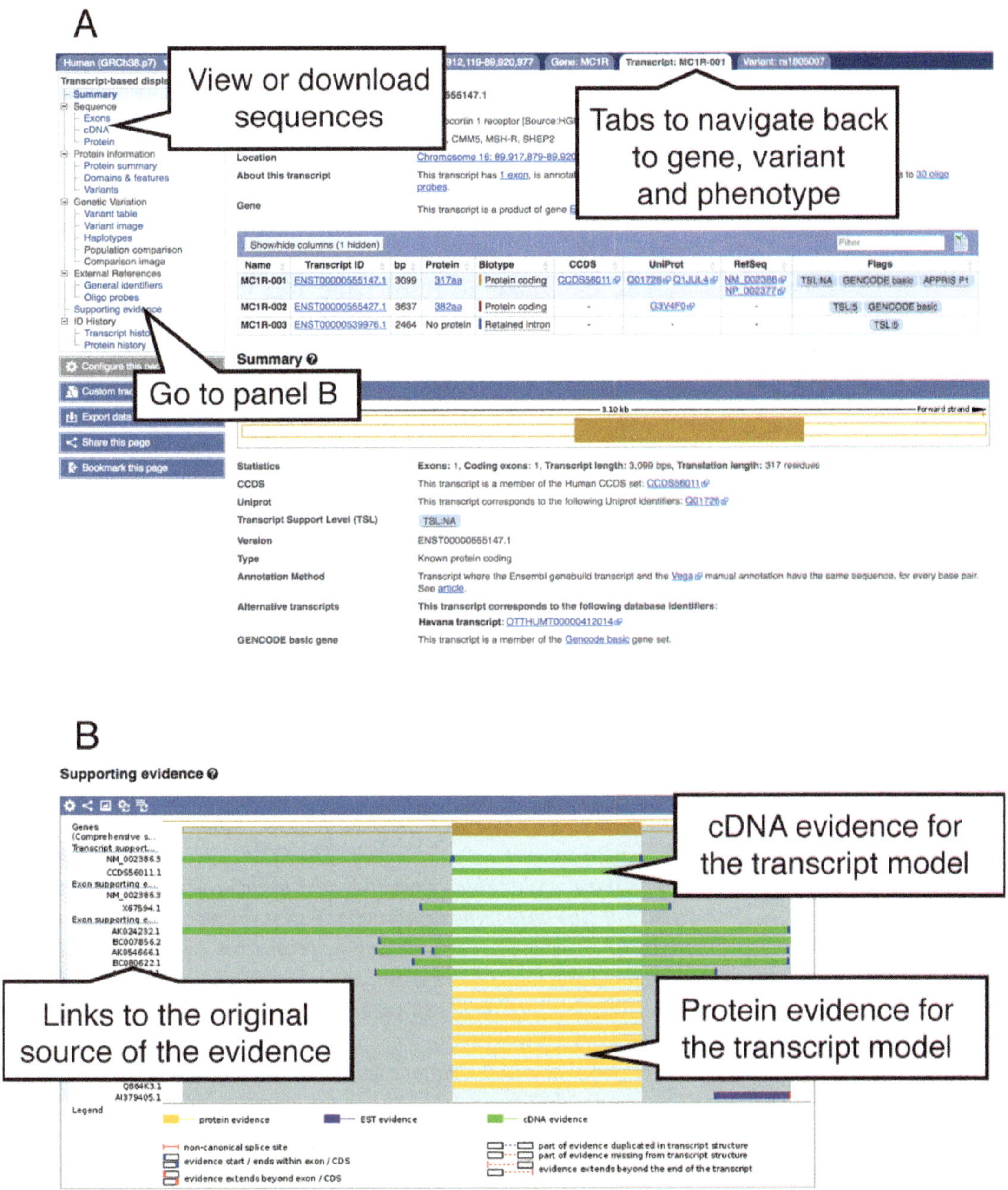

Fig. 6 The Ensembl transcript tab. The transcript tab contains all views for looking at a transcript and its associated protein, where relevant. (A) The left-hand menu on the transcript tab lists all the pages for looking at transcripts and proteins, and differs subtly from the gene tab menu. It has three different sequence views, allowing you to view the exon and intron sequences in a table, an alignment of the cDNA, CDS and peptide sequences, and the protein sequence only. As you open different features, such as genes, transcripts, and variants, tabs appear in the top bar, allowing easy navigation between the different features you've been looking at. (B) The Supporting evidence page shows which cDNA and protein evidence was used to annotate the transcript model

15. Filter the table to view missense variants between amino acid coordinates 150–160.
    (a) Filter the table for missense variants by clicking "Consequence" in the Filter section, then "Turn All Off" and "Missense variant."
    (b) Filter the table to view variants at a specific amino acid coordinate within the translated sequence of the transcript by clicking on "Filter Other Columns," then "AA Coord." Use the sliders to restrict the area for which variants are shown to 150–160.

    You can filter this table in numerous ways, including by consequence, source, and genomic or amino-acid coordinates (Fig. 7B). *For missense variants, there are also options to filter by predicted pathogenicity score, as determined by SIFT* [39] *and/or PolyPhen* [40] *(PolyPhen calculations are available only for human variants). SIFT and PolyPhen pathogenicity predictions have been calculated for rs1805007 and the amino acid substitution is considered deleterious (An additional variant, rs149922657, has been observed at the same position of* MC1R, *but has not been associated with any phenotype.).*

16. Click "Haplotypes" in the left-hand menu.

    This page allows you to view linked variants that tend to be coinherited. As a default, the amino acid identities and coordinates of each haplotype are shown, along with their frequencies in different 1000 Genomes Project populations [26]*; however, clicking "switch to CDS view" at the top of the table will show nucleotide sequences instead* (Fig. 8A). *The fifth haplotype listed in the protein-haplotype table represents our variant of interest. The frequency of this haplotype is, as already seen, higher in the European subgroup. Lower in the table can be found the 151R>H haplotype corresponding to rs149922657, the other variant observed at position 151; this variant was recovered in only two 1000 Genomes Project participants.*

    Clicking on any haplotype will load a table indicating its frequencies in different 1000 Genomes populations in more detail (Fig. 8B), *as well as a sequence view highlighting the nucleotide and amino-acid positions altered, if applicable* (Fig. 8C).

17. *Exporting Ensembl variation data*: Data can be exported from Ensembl at multiple scales. A link to the BioMart tool, which permits the download of customized datasets at intermediate scale, can be found in the navigation bar at the top of all Ensembl pages (Fig. 2) [12]. In the BioMart interface, select the Dataset "Ensembl Variation" (this will also include the release number, which is 88 at the time of writing), then "Human Short Variants (SNPs and indels excluding flagged variants)." To download all variants of ≤50 bp lying within

A

Variant table

Filter: Global MAF: All | SIFT: All | PolyPhen: All | Consequences: All | Filter Other Columns

Show/hide columns

Variant ID	Chr: bp	Alleles	Global MAF	Class	Source	Evidence	Clin. Sig.	Conseq. Type	AA	AA co-ord	SIFT	Poly-Phen
rs780765249	16:89912904	A/C	(-)	SNP	dbSNP	-	-	Upstream gene variant	-	-	-	-
rs780856754	16:89912931	C/T	(-)	SNP	dbSNP	-	-	Upstream gene variant	-	-	-	-
rs141569779	16:89912943	T/G	0.005 (G)	SNP	dbSNP		-	Upstream gene variant	-	-	-	-
rs566980096	16:89912957	G/A	0.000 (A)	SNP	dbSNP		-	Upstream gene variant	-	-	-	-
rs534320022	16:89913001	C/A	0.000 (A)	SNP	dbSNP		-	Upstream gene variant	-	-	-	-
rs747804101	16:89913004	C/A	(-)	SNP	dbSNP	-	-	Upstream gene variant	-	-	-	-
rs552521939	16:89913011	C/T	0.001 (T)	SNP	dbSNP		-	Upstream gene variant	-	-	-	-
rs564872405	16:89913012-89913014	CCA/-	0.001 (-)	deletion	dbSNP		-	Upstream gene variant	-	-	-	-

B

Variant table

Filter: Global MAF: All | SIFT: All | PolyPhen: All | Consequences: Missense variant | AA coord: 150 - 160 | Filter Other Columns

Show/hide columns

Variant ID	Chr: bp	Alleles	Global MAF	Class	Source	Evidence	Clin. Sig.	Conseq. Type	AA	AA co-ord	SIFT	Poly-Phen
rs1805007	16:89919709	C/G/T	0.019 (T)	SNP	dbSNP			Missense variant	R/G	151	0	0.944
rs1805007	16:89919709	C/G/T	0.019 (T)	SNP	dbSNP			Missense variant	R/C	151	0.02	0.982
rs149922657	16:89919710	G/A	0.000 (A)	SNP	dbSNP		-	Missense variant	R/H	151	0.02	0.262
rs1110400	16:89919722	T/C	0.003 (C)	SNP	dbSNP			Missense variant	I/T	155	0	0.864
rs3212365	16:89919724	G/A/C	0.004 (C)	SNP	dbSNP			Missense variant	V/M	156	1	0.529
rs3212365	16:89919724	G/A/C	0.004 (C)	SNP	dbSNP			Missense variant	V/L	156	0.01	0.635
rs201975178	16:89919725	T/C	(-)	SNP	dbSNP		-	Missense variant	V/A	156	0.01	0.816
rs756422682	16:89919727	A/C	(-)	SNP	dbSNP		-	Missense variant	T/P	157	0	0.995

Fig. 7 Table of short variants found within a transcript. The variant table lists all the variants found within a transcript. A similar page can be found in the gene tab listing all the variants in a gene. The table lists the variants, which are links to the variant tab, with their positions, alleles, SO consequences, and predicted protein effects. Buttons above the table allow you to filter to table to only show variants of interest. (A) The unfiltered table for *MC1R-001*. (B) The same table, filtered to only show missense variants between residues 150 and 160. The applied filters are shown above the table and can be easily removed

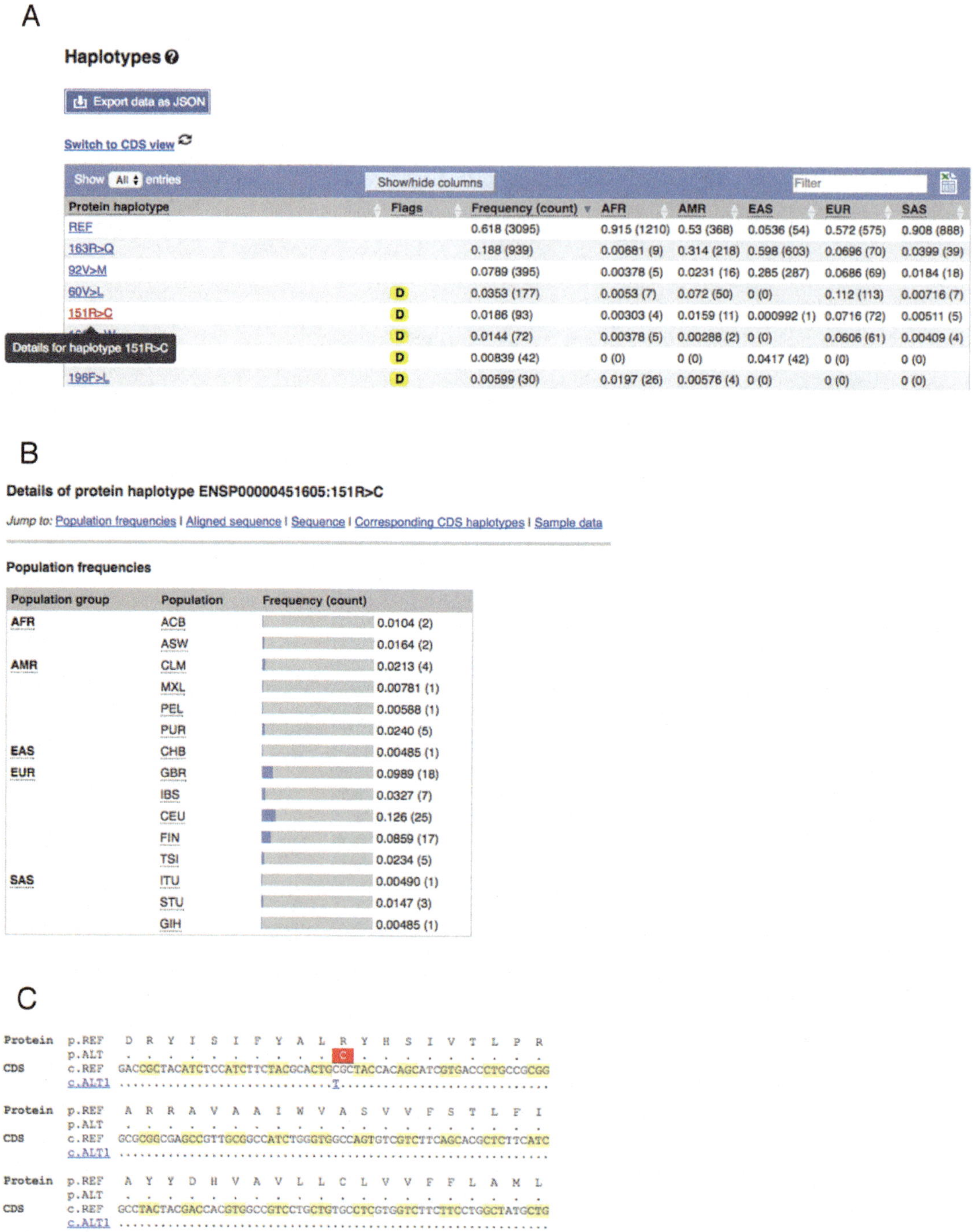

Fig. 8 Representation of protein haplotypes found in 1000 Genomes individuals. For each of the individuals in the 1000 Genomes population, the complete protein and CDS sequences were calculated. Sets of cosegregating variants were defined as protein and transcript haplotypes, their frequencies determined and listed in the Transcript haplotype page. (A) The table lists all the haplotypes found by the amino acid change. Click on the haplotype for more details (shown in panels B and C). By default, the page shows the protein haplotypes, but can be switched to show the CDS haplotypes. (B) The frequency of the selected haplotype across 1000 Genomes subpopulations. (C) An alignment of the reference and haplotype protein and CDS sequences

MC1R, as well as 5 kb upstream and downstream of the gene, filter by "Gene-associated Variant Filters," selecting "Gene stable IDs" and inputting "ENSG00000258839," the stable ID for the *MC1R* gene. You can choose attributes of interest under "Variant" or "Flanking sequences"—for example, the variant name, source, consequence, start and end coordinates, and pathogenicity predictions—which will be listed next to each variant in the output table. Click the "Results" button to view and download the results table (Fig. 9).

3.2 WF2: Gene-Based Searches and Identification of Regulatory Features in a Genomic Region

The following workflow describes a gene-based search and indicates some of the data and annotations collated in the Gene tab and Regulation tab.

The *POU5F1* gene, formerly known as *OCT4*, encodes one of the so-called "Yamanaka factors" implicated in cellular de-differentiation and induction of pluripotency [41, 42]. We can search Ensembl to view the *POU5F1* gene model and associated annotation, including predicted regulatory features.

1. *Getting started*: Type "POU5F1" into the search box on the homepage, www.ensembl.org, or in the upper right corner of any browser page, and click "Go." This will generate a search-results page with "POU5F1 (Human Gene)" as the top hit. Click the title link to navigate directly to the *POU5F1* Gene tab.

   The gene "Summary" containing a graphical representation of the gene model loads by default following navigation from the search results.

2. *Downloading gene sequences*: The sequence of the gene and flanking regions can be downloaded from the Gene tab in two ways.

   (a) To download the sequence in FASTA format for processing in an external tool, simply click the "Export data" button below the left-hand menu.

   This will open a pop-up window that presents customization options.

   (b) To view *POU5F1* sequence in the browser, click "Sequence" in the left-hand menu.

   This opens a display in FASTA format; buttons to download and to BLAST the sequence are shown on this page, and download customization options are similarly available (Fig. 10).

3. *Exploring regulatory features*: Select "Summary" in the left-hand menu. Scroll down to the graphical view of the gene model and locate the Regulatory Build track.

   The Regulatory Build depicts regulatory features that have been annotated based on epigenome-scale data imported from sources such as ENCODE [43], *Roadmap Epigenomics* [44] *and*

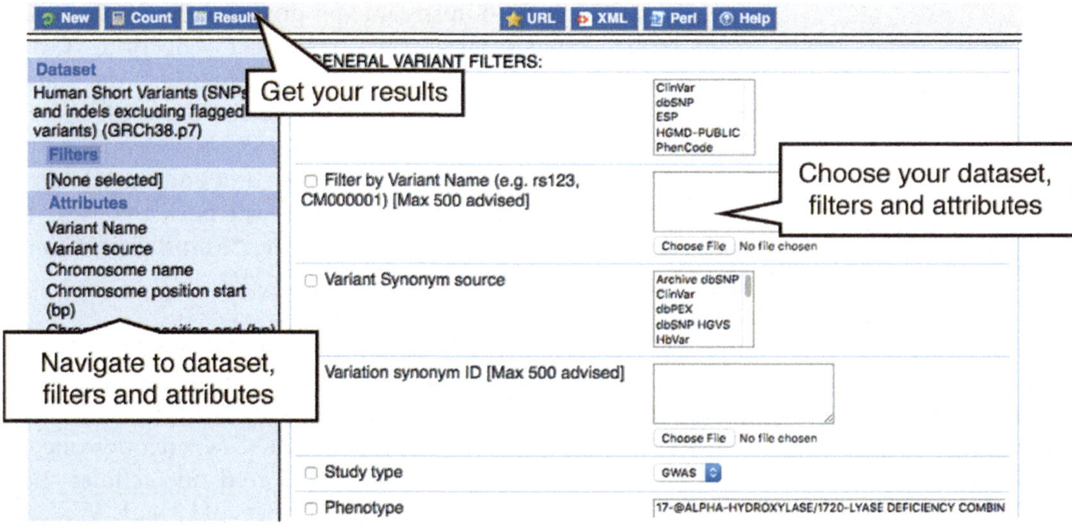

Fig. 9 The BioMart interface. BioMart allows easy export of tables of gene, variant, or regulatory feature data. A video tutorial for BioMart is available at https://www.youtube.com/watch?v=QvGT2G0-hYA&ab_channel=EnsemblHelpdesk

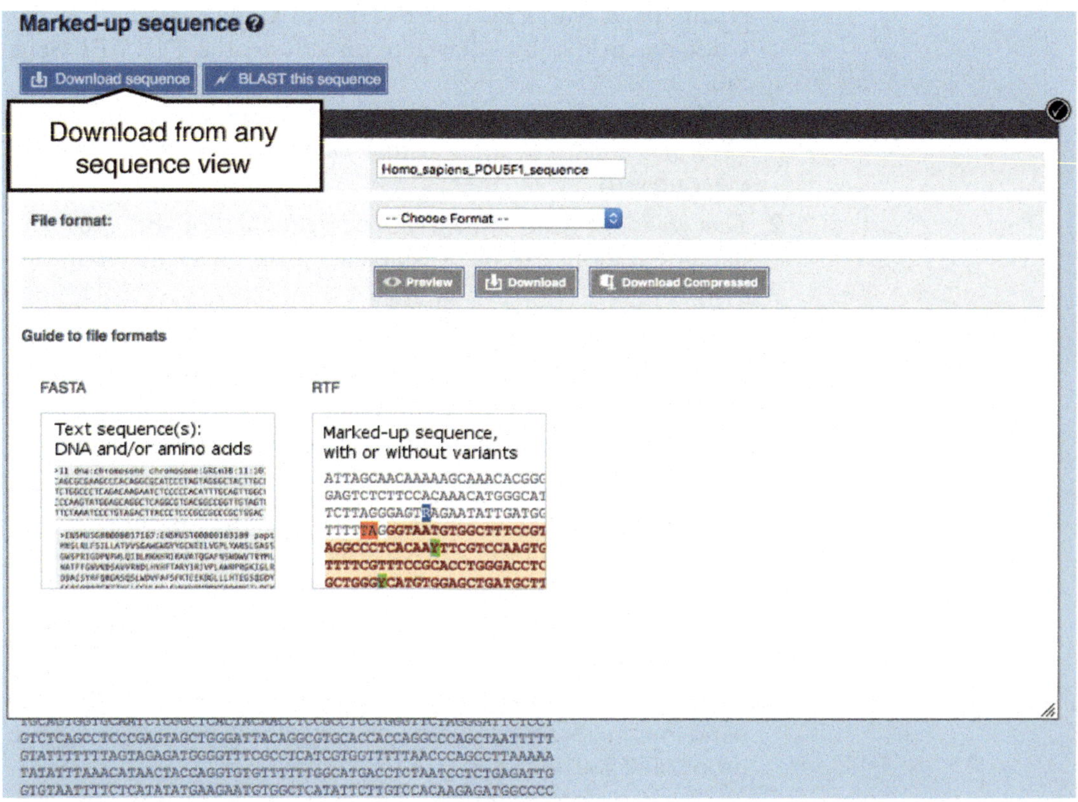

Fig. 10 Exporting gene sequence from Ensembl. All sequence views in Ensembl allow download in either plain FASTA or annotated rich text format (RTF)

Blueprint [45]. *These motifs are color-coded according to the predicted function of the element* (Fig. 11A).

4. Click on the red promoter overlapping the 5′ end of the longest transcript of *POU5F1*, *POU5F1-004*, to open a pop-up box with the stable ID ("ENSR00000195510"), type ("Promoter"), and genomic coordinates of the core element and flanking sequences. Click the stable ID to open the Regulation tab.

   Note: POU5F1 *is transcribed from the reverse strand, and thus the 5' sequences containing the promoter are located to the right of the gene.*

   The Regulation tab displays a graphical representation of the genomic region surrounding the element and a table of the 68 cell types with regulation data currently in Ensembl, organized by activity state. In addition to the Regulatory Build, several tracks are shown by default; these include CRISPR/Cas9 genome-editing sites predicted by the Wellcome Trust Sanger Institute (WTSI) [30], *transcription start sites identified by FANTOM5* [46], *miRNA binding sites imported from Tarbase* [31], *and enhancers identified by VISTA* [29]. *Tracks with no data in the immediate region of the feature are not shown* (Fig. 11B) *(The term "track" refers to a data type that can be plotted against the genome.).*

   Feature activity by cell type can be viewed in graphical form by clicking the "Select cells" button and, in the resulting pop-up, choosing "All on" or selecting individual cell types.

5. To view the element's activation state in individual cell types, click the "Details by cell type" button at the top of the Regulation tab or the link in the left-hand menu. Click the "Select cells" button and then choose "A549" (repressed), "Placenta" (poised), "Pancreas" (inactive), "GM12878" (active). Next, click "Select evidence," then "All on," to load the experimental data available for the cell types of interest.

   You are now viewing data from cell types in which the element is active, inactive, poised, and repressed (Fig. 12). *These activation states are determined on the basis of the histone modifications observed in the region, along with transcription factor and RNA polymerase II or III binding, as well as areas of DNase I hypersensitivity indicating open chromatin* [4].

   Additional tracks can be accessed by clicking the "Configure this page" button, at left, or the cogwheel at the top of the image. These include the evidence underlying the Regulatory Build, as well as comparative genomics analyses and variation data that may provide additional context for the annotated feature.

6. Ensure that both "Peaks" and "Signal" buttons are selected.

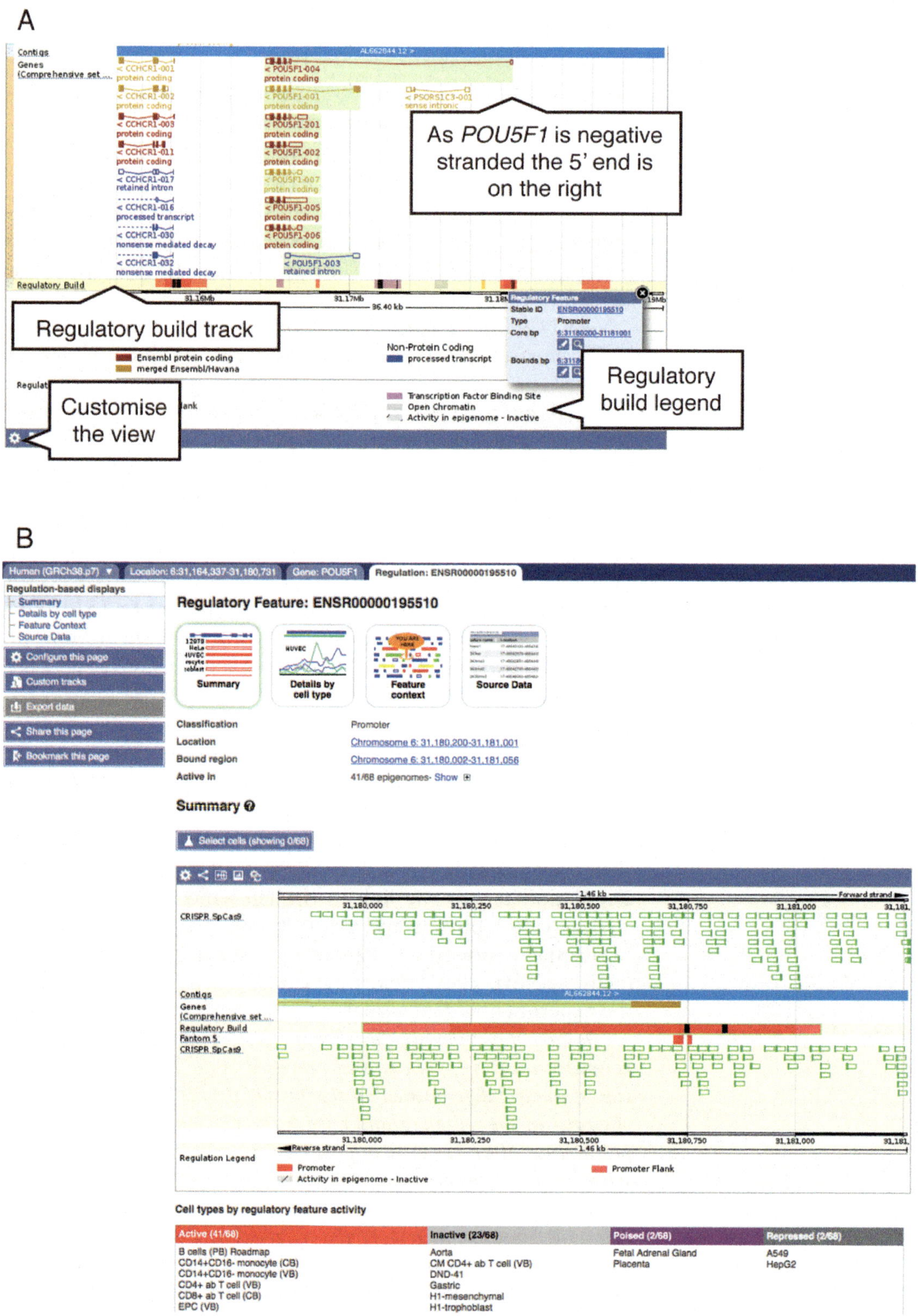

Fig. 11 The Ensembl regulatory build and regulatory features. (A) The regulatory build is shown as a track on the gene image. Clicking on a feature in the track opens a pop-up menu, with a link to the regulatory feature tab. Some transcripts have been removed from this image for size. (B) The summary page of the regulatory feature tab contains a table listing activity in different cell types. The graphic shows the feature in context, along with genes, CRISPR-Cas9 sites and FANTOM5 annotation

This will display a summary of the aligned reads (signal) as well as the peaks for each assay. Annotated features are clickable; for example, clicking on a predicted promoter will indicate any transcription factors known to bind it, along with links to the JASPAR database [47], *where further information on motifs is presented. For other elements, the position of the apex is indicated with black arrowheads* (Fig. 12).

7. To view regulatory features across a larger genomic region, navigate to the Location tab, available to the left of the Gene tab in the navigation bar.

   The Location tab displays three images: a global view of the chromosome of interest, an intermediate-scale view providing an overview of the region flanking the relevant genomic locus (in this case that of POU5F1*), and a final view that presents gene-annotation, comparative genomics and variation tracks by default, along with the Regulatory Build.*

   It is possible to configure the page to view the activity of local regulatory features by cell type, along with the evidence underlying these determinations. As in the Regulation tab, tracks depicting other Ensembl annotations can be added to provide context to the elements shown.

8. Click on the blue "Configure this page" button to add regulatory data tracks for the same cell types: A549, placenta, pancreas, and GM12878.

   This opens a menu listing the many possible tracks available to display on the genome. Categories of tracks are listed on the left. Tracks can be turned on and off by clicking on the box alongside them. To see the activity of regulatory features in different cell types, turn them on within the "Regulatory features" section. In the "Histones & polymerases" and "Open chromatin & TFBS" sections, you will find that tracks are displayed as a matrix, with cell types along the top and evidence to the side (Fig. 13).

9. *Exporting regulatory features with BioMart*: A list of regulatory features, by type, in a genomic region can also be exported via BioMart. Navigate to BioMart, then select "Ensembl Regulation" > "Human Regulatory Features" (it may be necessary to refresh the window by clicking "New" if you have performed a previous query). For features within 5 kb up- and downstream of *POU5F1*, filter for Chromosome 6, Base pair start: 31159337, Base pair end: 31185731. As defaults, "Chromosome Name," "Start (bp)," "End (bp)," and "Feature Type" are selected as Attributes. Add "Regulatory Stable ID" and generate your results.

   Nine features are returned for this genomic region, including the promoter we explored, ENSR00000195510.

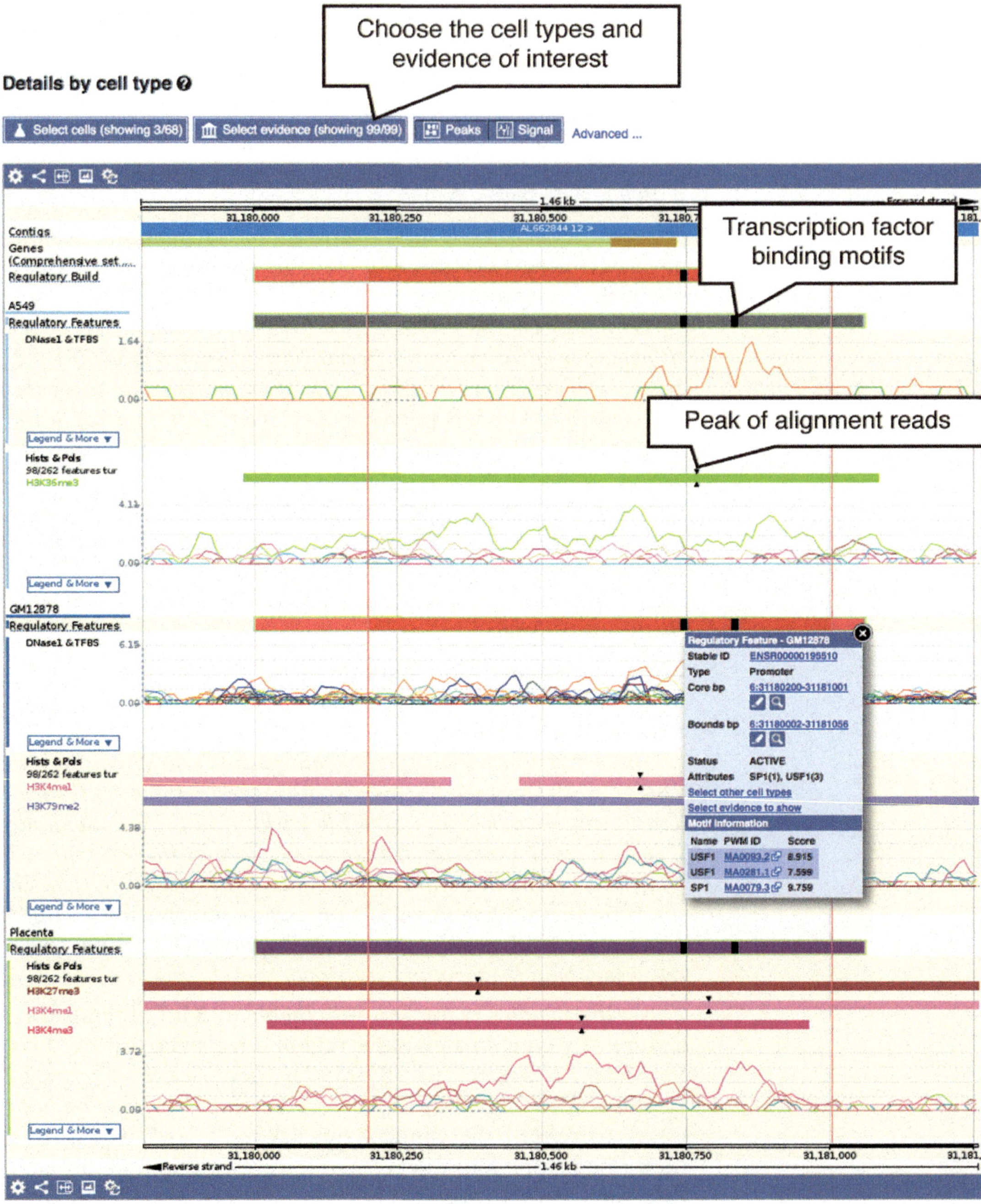

Fig. 12 Evidence for and activity of regulatory features in different cell types. The Details by cell type page in the regulatory feature tab can be manipulated to show the activity of the feature in cell types of interest using the buttons at the top. For each cell type, the feature is shown colour-coded to indicate its activity, with the evidence shown below. The evidence is ChIP-seq and DNase-seq data, and is shown as peaks of significant activity and as signal giving the number of reads. The top of the peak is indicated in the peak bar by pairs of black arrows. Black blocks in the regulatory features indicate the position of transcription factor binding motifs, which are listed in a pop-up when clicked on

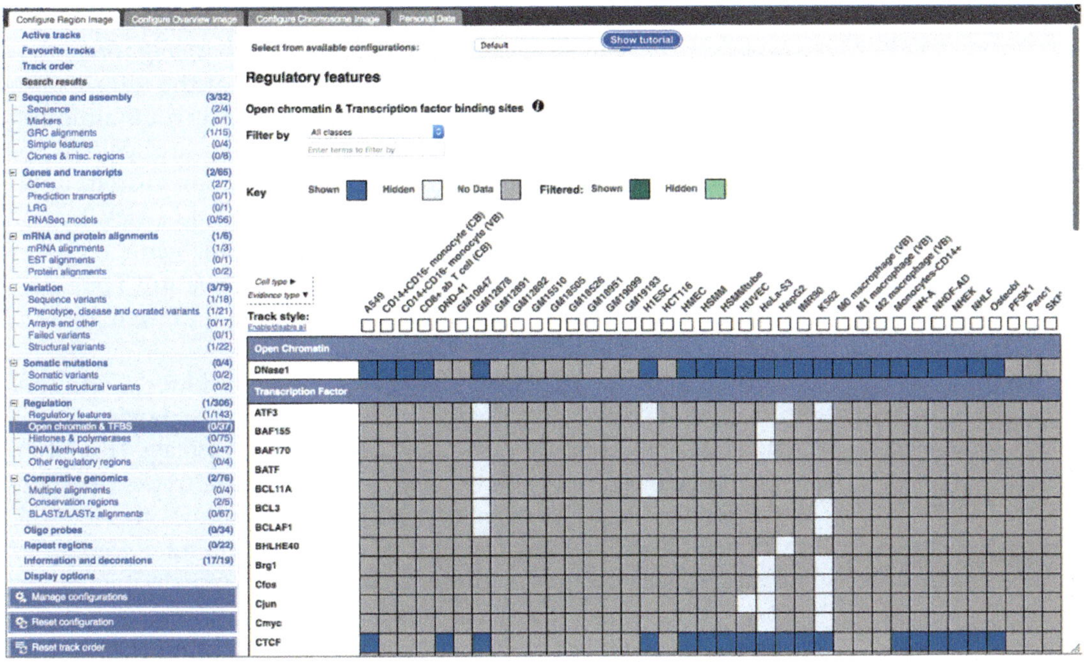

Fig. 13 Adding regulation tracks to a region view. The Region in detail view displays a genomic region and can be customized to show tracks of interest using the Configure this page button. This opens a detailed menu listing all the available tracks, using categories on the left, including regulatory features and evidence. Regulatory evidence can be added using a matrix selector, listing the cell type along the top and type of evidence down the left

4 Discussion

Here, we describe methods to navigate variation and regulation data in the Ensembl browser, focusing on human, although the principle of navigation is relevant to queries in all species.

The typical entry point to a query in the browser is the search function. The Ensembl search is versatile and can retrieve information linked to a variety of inputs—including, but not limited to, genomic locations; gene, transcript, protein and regulatory feature IDs; GO terms; variant IDs; and phenotypes. Unless otherwise specified in the query, search results for human will be returned first; filters displayed on the left-hand side of the results page permit the restriction of results by category (e.g., gene, variant) and by organism.

Selecting a search result will open a tab that collates information on the entity: in the two workflows presented above, we present strategies for accessing the Phenotype, Variant, Gene, Transcript, Location, and Regulation tabs following phenotype- and gene-based searches. As you move from tab to tab in a single query, previously accessed tabs will remain open in the blue navigation bar at the top of the page to facilitate seamless data-retrieval in a minimal number of steps; you can reenter a previous tab simply by clicking on the tab header in the navigation bar (Fig. 6A).

By default, tabs will open with a summary of the information available for each entity (e.g., a transcript or variant), although herein we indicate a few cases where other data are loaded: for example, the Gene Variant table is presented immediately upon navigation from the Variant to the Gene tab. Should a view not be as expected, links to all data and annotations available in a tab can be found in the menu on the left; for Location, Gene and Transcript tabs, these links adhere to a similar framework but present annotations at different scales.

Tabs can be customized by clicking on the blue "Configure this page" button below the left-hand menu, or the cogwheel icons that appear in the upper borders of graphical displays (Fig. 13). Customization allows you to add or remove data tracks that may be useful to interpretation or analysis; for example, to view the evidence underlying an activation-state prediction for a regulatory feature in a cell type of interest (in the Location or Regulation tabs). Other examples of customization include, in the Location tab, the addition of tracks containing the data, imported from external repositories, that were used to annotate transcripts in a genomic region (ENA, UniProt, and RefSeq tracks accessible in Location tab) [11]. Public datasets can also be added from the Track Hub Registry [48], and you can import your own data, in multiple formats [49], for examination in the context of the browser.

Data can be exported directly from the browser by clicking the blue "Export data" buttons found below the left-hand menu in most tabs, or the "Download sequence" buttons above FASTA sequences in the Gene and Transcript tabs. In addition, the BioMart tool described in the workflows presented herein can be used to retrieve custom datasets from our Gene, Variation and Regulation databases, and data can be accessed programmatically from our Perl APIs and REST service. Data from all Ensembl releases can also be downloaded en masse from our FTP site.

Species' sequence data and annotations may be updated several times a year. You should therefore be attentive, when querying Ensembl data, to the current browser version, as annotations are subject to change. Data can, however, still be retrieved directly from archived versions of the browser, as well as via BioMart, while the browser web interface remains online. Following the decommissioning of any browser version, the data remain accessible from our FTP site and APIs, as mentioned above.

A dedicated email helpdesk is available to field any inquiries about Ensembl and we typically reply to messages within two days of receipt. We also hold training workshops upon invitation by research institutes; from 2013 to 2016 we participated in an average of 86 workshops, and trained 2150 students, per year. Our training materials are accessible online [50], along with a number of courses that are available on the Train Online Platform of the European Bioinformatics Institute (EMBL-EBI) [51], and we have published a blogpost outlining the process of hosting your own workshop [52]. Short help videos can be found both on our YouTube channel [53] and, for those who cannot access YouTube, on Youku [54]. We invite the community to contact us via helpdesk@ensembl.org for more information about workshops, with questions regarding the browser, and to suggest features and resources which would assist their work.

Funding/Acknowledgments

Ensembl receives majority funding from the Wellcome Trust (grant number WT108749/Z/15/Z) with additional funding for specific project components from the National Human Genome Research Institute (U41HG007234 and U41HG007823), the Biotechnology and Biological Sciences Research Council (BB/L024225/1, BB/M011615/1, and BB/M020398/1), Open Targets, the Wellcome Trust (WT104947/Z/14/Z, WT200990/Z/16/Z, and WT201535/Z/16/Z) and the European Molecular Biology Laboratory. This project has received funding from the European Union's Horizon 2020 research and innovation programme under grant agreement n° 634143 (MedBioinformatics). This project has received funding from the European Union's Horizon 2020 research and innovation programme under grant agreement n° 733161 (MultipleMS).

References

1. Aken BL, Ayling S, Barrell D et al (2016) The Ensembl gene annotation system. Database (Oxford) 2016. https://doi.org/10.1093/database/baw093
2. Chen Y, Cunningham F, Rios D et al (2010) Ensembl variation resources. BMC Genomics 11:293. https://doi.org/10.1186/1471-2164-11-293
3. Herrero J, Muffato M, Beal K et al (2016) Ensembl comparative genomics resources. Database (Oxford) 2016. https://doi.org/10.1093/database/baw053
4. Zerbino DR, Johnson N, Juetteman T et al (2016) Ensembl regulation resources. Database (Oxford) 2016. https://doi.org/10.1093/database/bav119
5. Kent WJ, Sugnet CW, Furey TS et al (2002) The human genome browser at UCSC. Genome Res 12(6):996–1006. https://doi.org/10.1101/gr.229102. Article published online before print in May 2002
6. Robinson JT, Thorvaldsdottir H, Winckler W et al (2011) Integrative genomics viewer. Nat Biotechnol 29(1):24–26. https://doi.org/10.1038/nbt.1754
7. Hubbard T, Barker D, Birney E et al (2002) The Ensembl genome database project. Nucleic Acids Res 30(1):38–41
8. The Ensembl Browser. http://www.ensembl.org
9. Kersey PJ, Allen JE, Armean I et al (2016) Ensembl Genomes 2016: more genomes, more complexity. Nucleic Acids Res 44(D1):D574–D580. https://doi.org/10.1093/nar/gkv1209
10. The Ensembl Genomes Browser. http://www.ensemblgenomes.org
11. Aken BL, Achuthan P, Akanni W et al (2017) Ensembl 2017. Nucleic Acids Res 45(D1):D635–D642. https://doi.org/10.1093/nar/gkw1104

12. Kinsella RJ, Kahari A, Haider S et al (2011) Ensembl BioMarts: a hub for data retrieval across taxonomic space. Database (Oxford) 2011:bar030. https://doi.org/10.1093/database/bar030
13. Ruffier M, Kahari A, Komorowska M et al (2017) Ensembl core software resources: storage and programmatic access for DNA sequence and genome annotation. Database (Oxford) 2017(1). https://doi.org/10.1093/database/bax020
14. Yates A, Beal K, Keenan S et al (2015) The Ensembl REST API: Ensembl data for any language. Bioinformatics 31(1):143–145. https://doi.org/10.1093/bioinformatics/btu613
15. The Ensembl FTP site. ftp://ftp.ensembl.org
16. McLaren W, Gil L, Hunt SE et al (2016) The Ensembl variant effect predictor. Genome Biol 17(1):122. https://doi.org/10.1186/s13059-016-0974-4
17. Kent WJ (2002) BLAT–the BLAST-like alignment tool. Genome Res 12(4):656–664. https://doi.org/10.1101/gr.229202. Article published online before March 2002
18. Altschul SF, Gish W, Miller W et al (1990) Basic local alignment search tool. J Mol Biol 215(3):403–410. https://doi.org/10.1016/S0022-2836(05)80360-2
19. Clarke L, Zheng-Bradley X, Smith R et al (2012) The 1000 genomes project: data management and community access. Nat Methods 9(5):459–462. https://doi.org/10.1038/nmeth.1974
20. Ensembl Publications. http://www.ensembl.org/info/about/publications.html
21. Zhang M, Song F, Liang L et al (2013) Genome-wide association studies identify several new loci associated with pigmentation traits and skin cancer risk in European Americans. Hum Mol Genet 22(14):2948–2959. https://doi.org/10.1093/hmg/ddt142
22. The GWAS catalog. https://www.ebi.ac.uk/gwas/
23. Europe PMC. https://europepmc.org/
24. Sources of Ensembl variation data. http://www.ensembl.org/info/genome/variation/sources_documentation.html
25. Eilbeck K, Lewis SE, Mungall CJ et al (2005) The sequence ontology: a tool for the unification of genome annotations. Genome Biol 6(5):R44. https://doi.org/10.1186/gb-2005-6-5-r44
26. Genomes Project Consortium, Auton A, Brooks LD et al (2015) A global reference for human genetic variation. Nature 526(7571):68–74. https://doi.org/10.1038/nature15393
27. Goldstein DB, Cavalleri GL (2005) Genomics: understanding human diversity. Nature 437(7063):1241–1242. https://doi.org/10.1038/4371241a
28. Exome Variant Server. NHLBI GO Exome Sequencing Project (ESP). http://evs.gs.washington.edu/EVS/
29. Visel A, Minovitsky S, Dubchak I et al (2007) VISTA enhancer browser–a database of tissue-specific human enhancers. Nucleic Acids Res 35(Database issue):D88–D92. https://doi.org/10.1093/nar/gkl822
30. Hodgkins A, Farne A, Perera S et al (2015) WGE: a CRISPR database for genome engineering. Bioinformatics 31(18):3078–3080. https://doi.org/10.1093/bioinformatics/btv308
31. Vlachos IS, Paraskevopoulou MD, Karagkouni D et al (2015) DIANA-TarBase v7.0: indexing more than half a million experimentally supported miRNA:mRNA interactions. Nucleic Acids Res 43(Database issue):D153–D159. https://doi.org/10.1093/nar/gku1215
32. Gene Ontology Consortium (2015) Gene ontology consortium: going forward. Nucleic Acids Res 43(Database issue):D1049–D1056. https://doi.org/10.1093/nar/gku1179
33. O'Leary NA, Wright MW, Brister JR et al (2016) Reference sequence (RefSeq) database at NCBI: current status, taxonomic expansion, and functional annotation. Nucleic Acids Res 44(D1):D733–D745. https://doi.org/10.1093/nar/gkv1189
34. HGNC database of human gene names. http://www.genenames.org/
35. Online Mendelian Inheritance in Man. McKusick-Nathans Institute of Genetic Medicine, Johns Hopkins University (Baltimore, MD). https://www.omim.org/
36. Pruitt KD, Harrow J, Harte RA et al (2009) The consensus coding sequence (CCDS) project: identifying a common protein-coding gene set for the human and mouse genomes. Genome Res 19(7):1316–1323. https://doi.org/10.1101/gr.080531.108
37. The UniProt Consortium (2017) UniProt: the universal protein knowledgebase. Nucleic Acids Res 45(D1):D158–D169. https://doi.org/10.1093/nar/gkw1099
38. Toribio AL, Alako B, Amid C et al (2017) European nucleotide archive in 2016. Nucleic Acids Res 45(D1):D32–D36. https://doi.org/10.1093/nar/gkw1106

39. Ng PC, Henikoff S (2003) SIFT: predicting amino acid changes that affect protein function. Nucleic Acids Res 31(13):3812–3814
40. Adzhubei IA, Schmidt S, Peshkin L et al (2010) A method and server for predicting damaging missense mutations. Nat Methods 7(4):248–249. https://doi.org/10.1038/nmeth0410-248
41. Takahashi K, Yamanaka S (2006) Induction of pluripotent stem cells from mouse embryonic and adult fibroblast cultures by defined factors. Cell 126(4):663–676. https://doi.org/10.1016/j.cell.2006.07.024
42. Okita K, Ichisaka T, Yamanaka S (2007) Generation of germline-competent induced pluripotent stem cells. Nature 448(7151):313–317. https://doi.org/10.1038/nature05934
43. ENCODE Project Consortium (2012) An integrated encyclopedia of DNA elements in the human genome. Nature 489(7414):57–74. https://doi.org/10.1038/nature11247
44. Roadmap epigenomics Consortium, Kundaje A, Meuleman W et al (2015) Integrative analysis of 111 reference human epigenomes. Nature 518(7539):317–330. https://doi.org/10.1038/nature14248
45. Fernandez JM, de la Torre V, Richardson D et al (2016) The BLUEPRINT data analysis portal. Cell Syst 3(5):491–495.e495. https://doi.org/10.1016/j.cels.2016.10.021
46. Fantom Consortium, Forrest AR, Kawaji H et al (2014) A promoter-level mammalian expression atlas. Nature 507(7493):462–470. https://doi.org/10.1038/nature13182
47. Bryne JC, Valen E, Tang MH et al (2008) JASPAR, the open access database of transcription factor-binding profiles: new content and tools in the 2008 update. Nucleic Acids Res 36(Database issue):D102–D106. https://doi.org/10.1093/nar/gkm955
48. The Track Hub Registry. https://trackhubregistry.org
49. Data formats compatible with Ensembl. http://www.ensembl.org/info/website/upload/index.html - formats
50. The Ensembl Training Site. http://training.ensembl.org
51. EMBL-EBI's Train Online Platform. https://www.ebi.ac.uk/training/online/
52. Hosting an Ensembl Workshop. http://www.ensembl.info/blog/2017/01/05/so-you-want-to-run-an-ensembl-workshop/
53. The Ensembl Helpdesk YouTube channel. https://www.youtube.com/user/EnsemblHelpdesk
54. The Ensembl Helpdesk Youku channel. http://i.youku.com/i/UMzM1NjkzMTI0?spm=a2h0j.8191423.subscription_wrap.DD~A

Open Access This chapter is licensed under the terms of the Creative Commons Attribution 4.0 International License (http://creativecommons.org/licenses/by/4.0/), which permits use, sharing, adaptation, distribution and reproduction in any medium or format, as long as you give appropriate credit to the original author(s) and the source, provide a link to the Creative Commons License and indicate if changes were made.

The images or other third party material in this chapter are included in the chapter's Creative Commons License, unless indicated otherwise in a credit line to the material. If material is not included in the chapter's Creative Commons License and your intended use is not permitted by statutory regulation or exceeds the permitted use, you will need to obtain permission directly from the copyright holder.

Chapter 7

Mouse Genome Informatics (MGI) Is the International Resource for Information on the Laboratory Mouse

MeiYee Law and David R. Shaw

Abstract

Mouse Genome Informatics (MGI, http://www.informatics.jax.org/) web resources provide free access to meticulously curated information about the laboratory mouse. MGI's primary goal is to help researchers investigate the genetic foundations of human diseases by translating information from mouse phenotypes and disease models studies to human systems. MGI provides comprehensive phenotypes for over 50,000 mutant alleles in mice and provides experimental model descriptions for over 1500 human diseases. Curated data from scientific publications are integrated with those from high-throughput phenotyping and gene expression centers. Data are standardized using defined, hierarchical vocabularies such as the Mammalian Phenotype (MP) Ontology, Mouse Developmental Anatomy and the Gene Ontologies (GO). This chapter introduces you to Gene and Allele Detail pages and provides step-by-step instructions for simple searches and those that take advantage of the breadth of MGI data integration.

Key words MGI, MGD, GXD, MouseMine, Mouse, Alleles, Phenotypes, Genotypes, Disease models, Gene expression, Functional annotations, Strains, IMSR, MTB

1 Introduction

MGI made its first appearance on the World Wide Web in June 1994 as the Mouse Genome Database (MGD), and it has grown to comprise these interacting databases: MGD [1], Gene Expression Database (GXD) [2], Mouse Tumor Biology Database (MTB) [3], and the International Mouse Strain Resource (IMSR) [4]. These databases work in concert to make MGI the essential resource for research on the laboratory mouse.

The major goal of MGI is to help researchers uncover the genetic foundations of human diseases by utilizing information from mutant mouse phenotypes and disease models, and from mouse gene function and expression data. MGI has curated data from over 230,000 publications, and also downloads, curates, and integrates data from external sources, such as high-throughput phenotyping and expression projects, and for sequences, SNPs,

Martin Kollmar (ed.), *Eukaryotic Genomic Databases: Methods and Protocols*, Methods in Molecular Biology, vol. 1757, https://doi.org/10.1007/978-1-4939-7737-6_7, © Springer Science+Business Media, LLC, part of Springer Nature 2018

homology and genome locations. MGI then makes its curated data available for downloading by researchers and databases.

Researchers, as well as other research groups' databases, rely on MGI as the official source of mouse gene, allele and strain nomenclature. Expert curation makes MGI the authoritative source for mouse functional and phenotype annotations, and mouse sequence-to-gene associations. Searches of MGI are aided by annotations to defined, hierarchical vocabularies: the Mammalian Phenotype (MP) Ontology [5], the Gene Ontologies (GO) [6], the Human Phenotype Ontology (HPO) [7], and Mouse Developmental Anatomy [8]. Mouse models of human diseases are annotated to the Disease Ontology (DO) [9]. Data include over 300,000 MP annotations, 300,000 GO annotations and 78,000 expression assays, including data from over 230,000 publications.

GXD compiles and integrates gene expression information for mouse embryonic development, focusing on classical types of expression data (Subheading 7). Researchers can search information about the expression profiles of transcripts and proteins in different mouse strains and mutants and quickly find out when and where a gene is expressed, and what genes are expressed in specific tissues and at specific developmental stages.

For researchers studying cancer, the Mouse Tumor Biology Database (MTB) (Subheading 11) specializes in providing information on the mouse as the model system for human cancers. Alleles affecting tumorigenesis are also described in MGI, and for stocks contributed to public repositories, MGI allele detail pages link to the International Mouse Strain Resource (IMSR) (Subheading 2.3, **step 6**), a database of mouse strains, stocks, and mutant ES cell lines available in worldwide repositories. You can also search the IMSR directly for inbred, congenic, mutant, and genetically engineered stocks.

The integration of these resources, precise curation to standardized nomenclatures and vocabularies, and query tools that permit quick and refined searches, enable researchers to collect and analyze data relevant to their research. These tools allow researchers to find disease models, associate phenotypic alleles with their genes, uncover strain background effects, and find available knockout mice and mice with recombinase-carrying and conditional alleles.

In addition to data-specific search tools, MGI also prepares dozens of weekly database reports and provides several tools for querying with lists of genes, alleles and IDs, and for generating tab-delimited files of search results. An easy-to-use Batch Query supports queries using gene symbols and a variety of IDs. A more powerful tool, MouseMine [10] (Subheading 10), offers iterative querying, built-in enrichment analysis, and API support. A Batch Search for gene expression data returns a dynamic tissue-by-gene matrix view that facilitates a comparison of expression patterns between genes.

Gene Detail pages (Subheading 2.2) provide a synopsis of a gene's functions, where it is expressed and the phenotypes of mutant alleles on specific genetic backgrounds, along with ways to access subsets of data, such as homologs, sequences, and references.

2 Homepage and Quick Search

Prominently featured on the MGI Home Page (Fig. 1) is the Quick Search field, followed by links for search and analysis tools organized by topic. Individual tools can also be reached by using the drop-down menus in the dark blue banner at the top of the page.

For simple keyword searches, the Quick Search will produce relevant results for most users. Keywords may be disease, phenotype, anatomy or function terms, protein motifs, or part of a gene or allele name. The Quick Search field also searches most IDs maintained within the MGI database. These include not only MGI IDs, but sequence, Ensembl, VEGA, dbSNP, and vocabulary IDs, such as GO, Mammalian Phenotype (MP), Disease Ontology (DO), and OMIM. PubMed IDs are also supported and are a quick way to find approved nomenclature and curated data associated with a publication. The Quick Search supports the asterisk (*) as a wildcard character.

2.1 Quick Search Example

Variant alleles of the human *APOE* gene are implicated in hypercholesterolemia, atherosclerosis and Alzheimer's disease. In order to find information on the mouse *Apoe* gene:

1. Go to the Mouse Genome Informatics (MGI) (http://www.informatics.jax.org/) web site, enter *apoe* in the Quick Search field at the top of the web page and click the Quick Search button (Note that MGI searches are case-insensitive.).
2. Your Quick Search results are sorted into three main sections. The first one, Genome Features, displays nomenclature matches of current official, former and unofficial (synonyms) symbols and names. The second section, Vocabulary Terms, displays matching GO, MP, OMIM, and protein domain terms. The third section displays matching accession IDs. In the rare case that these results fall short of your expectations, a final section offers the option to perform a Google search of MGI. This Google search also searches reference abstracts and curator notes and may find obscure terms and phrases.
3. Click on the *Apoe* gene symbol to see its Gene Detail page (Figs. 2 and 3), the page with links to all MGI data associated with this mouse gene.

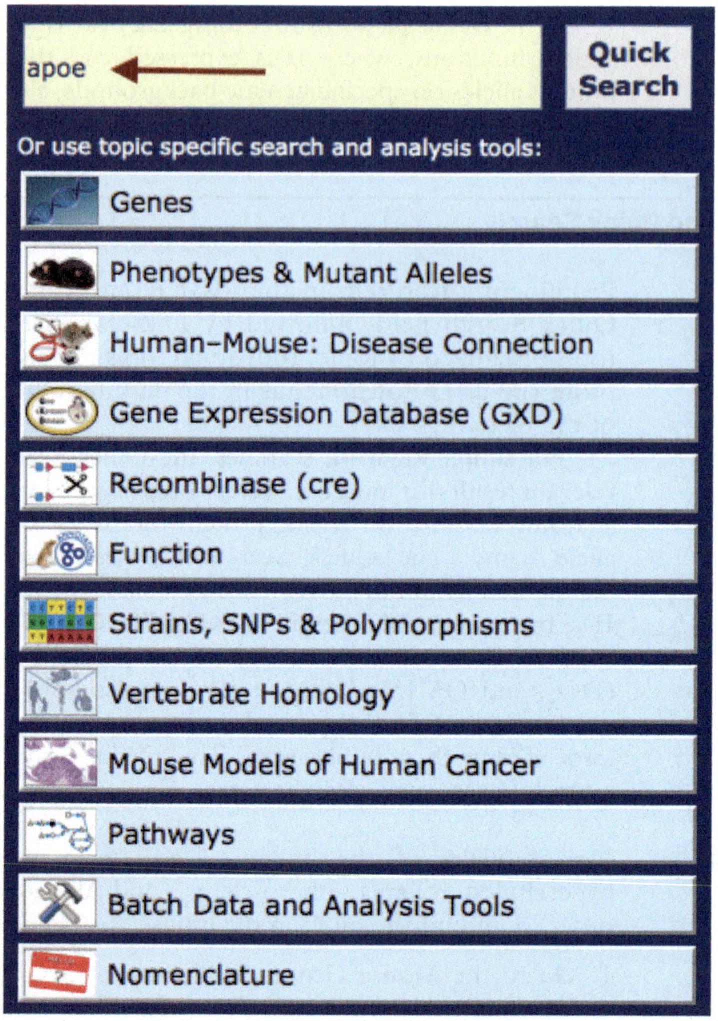

Fig. 1 A portion of the MGI Home Page showing the Quick Search field with a search for the *Apoe* gene. Below the Quick Search field are links to search tools and resources arranged by topic

2.2 Gene Detail Page

1. The top two sections of the Gene Detail page (Fig. 2A) provide the official marker symbol and name, and genome location. Use the more/less buttons to expand or collapse sections. Expand the **Location & Maps** section to access genome browsers. The MGI Mouse Genome Browser uses the common open-sourced Jbrowse.
2. Expand the **Homology** section (Fig. 2B) to access mapping data and sequences for human, chimpanzee, rhesus macaque, rat, dog, cattle, chicken, western clawed frog, and zebrafish homologs.

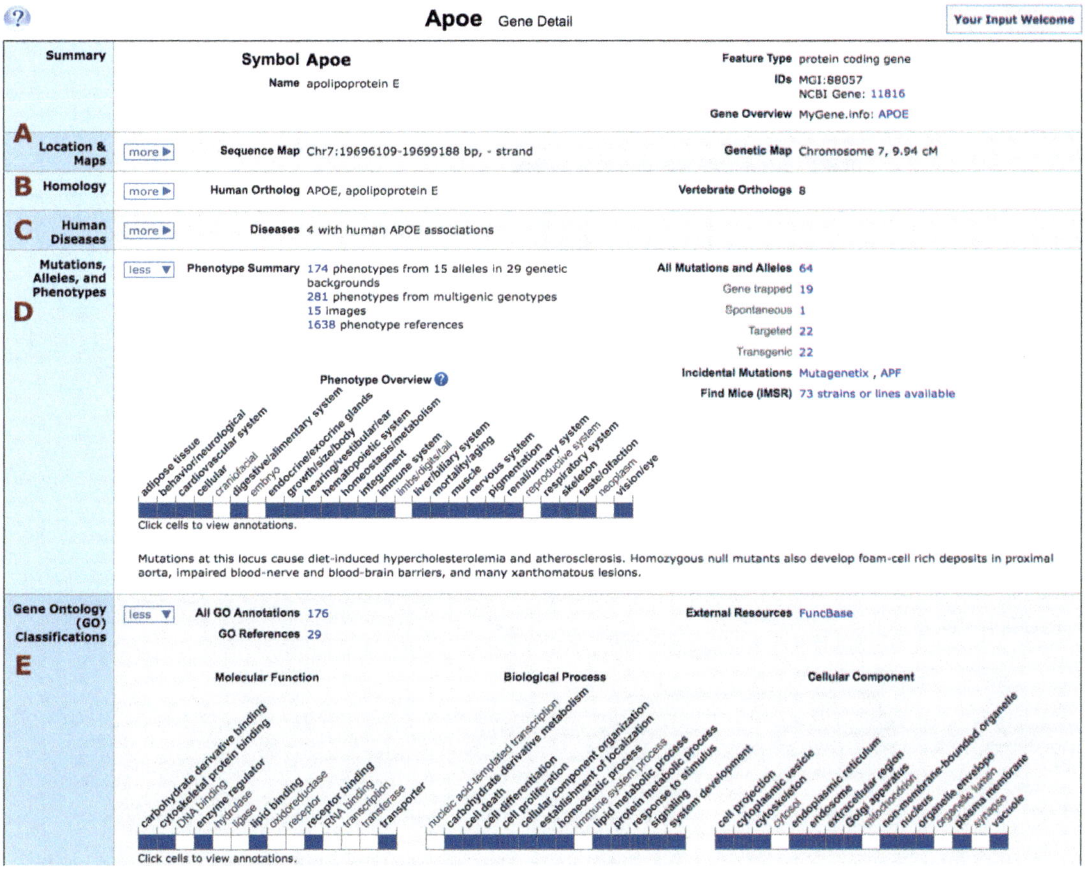

Fig. 2 The top portion of the *Apoe* Gene Detail page, the page with links to all information on the mouse gene. Use the more/less buttons to expand and contract sections. All sections link to more detailed information. See text for details

3. Expand the **Human Diseases** section (Fig. 2C) to see a table of human diseases associated with the human *APOE* gene as well as diseases modeled using alleles of the mouse *Apoe* gene.
4. The next three sections use blue-celled grids to summarize the tissues where the gene is expressed, its functions and the phenotypes of mutant alleles. These cells link to specific supporting data.
5. In the **Mutations, Alleles and Phenotypes** section (Fig. 2D), a Phenotype Overview displays high-level terms from the Mammalian Phenotype (MP) Ontology and the blue cells mark those with phenotypic annotations to genotypes with alleles of the gene. This section also provides links for several ways to view summaries of *Apoe* phenotype data and alleles.
6. The **Gene Ontology (GO) Classifications** section (Fig. 2E) displays selected defined vocabulary terms for the Molecular Functions, Biological Processes and Cellular Components of

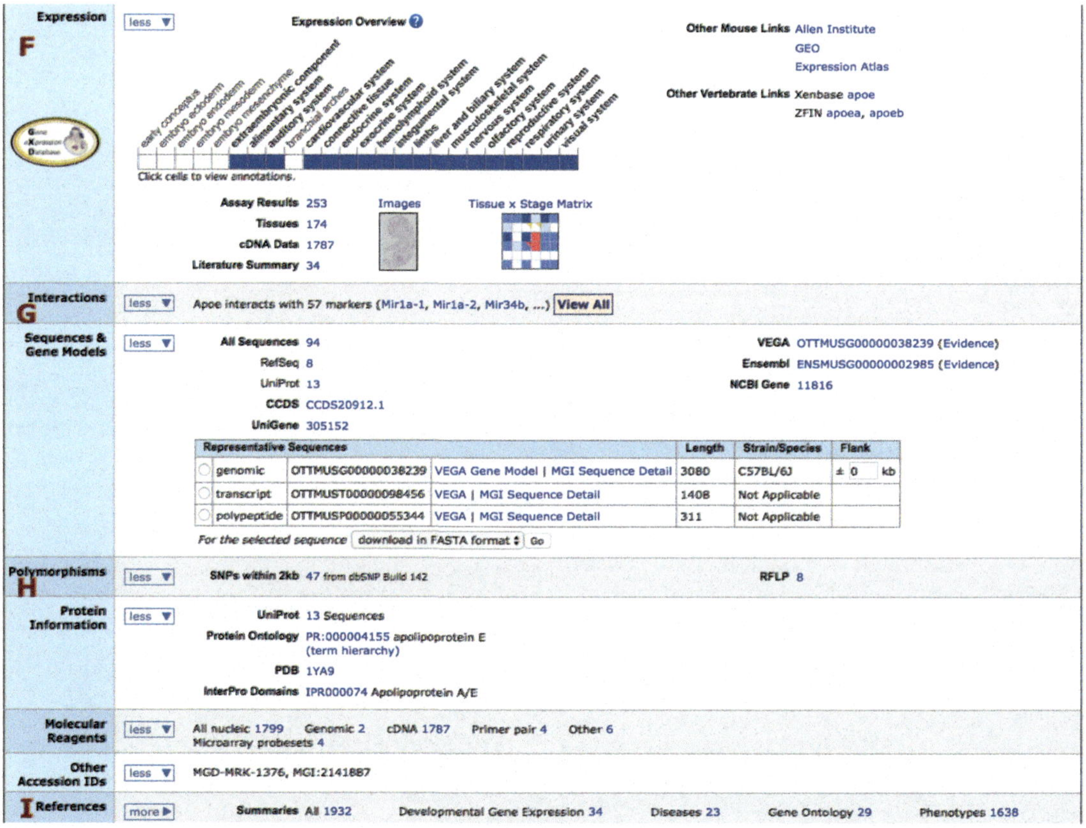

Fig. 3 The bottom portion of the *Apoe* Gene Detail page. Use the more/less buttons to expand and contract sections. All sections link to more detailed information. See text for details

the gene's products. Follow the link to all GO classifications annotated to this gene and then click on a term to view its place in the vocabulary hierarchy and to find all genes annotated to that term.

7. The **Expression** section (Fig. 3F) includes a link to all the expression images for the gene, as well as a link to a dynamic tissue-by-developmental stage matrix that provides an overview of the spatiotemporal expression patterns of the gene. Use the blue triangles in the matrix to navigate down the Mouse Developmental Anatomy hierarchy and view annotations to lower-level terms.

8. The **Interactions** section (Fig. 3G) indicates the number of microRNAs with which the gene is known or suspected of interacting, along with links to up to three of those markers' detail pages as well as a link to View All the validated and predicted interacting markers.

9. The **Polymorphisms** section (Fig. 3H) links to a table of SNPs obtained from NCBI's dbSNP resource, mapped to within 2 kb of the gene, for up to 88 mouse strains.
10. **References** (Fig. 3I) curated to the mouse gene with links to reference details, abstracts, PubMed and Journals, and links to curated data associated with each publication.

2.3 Allele Detail Page

In the **Mutations, Alleles and Phenotypes** section of the *Apoe* Gene Detail page (Fig. 2D), click on the **All Mutations and Alleles** link to view a summary of the known mouse alleles of this gene.

Click on the Allele Symbol for a targeted allele, $Apoe^{tm3(APOE*4)}_{Mae}$, to view its Allele Detail page (Fig. 4).

1. At the top of the page is a "Your Input Welcome" button (Fig. 4A). Use this to report omissions and errors, and to ask questions. A detailed help document is marked by a blue question mark on the top left corner of the page.
2. The **Nomenclature** section (Fig. 4B) lists the official allele symbol, unique MGI identifier, the gene symbol, linked to the Gene Detail page, and phenotype images.
3. The **Mutation origin** and **Mutation description** sections (Fig. 4C) describe the background of the mutation and generation method. In this allele, a DNA fragment containing exons 2–4 of the human *APOE* gene (the APOE4 isoform) replaced an equivalent portion of the mouse *Apoe* gene. The J: number links to the source reference, typically a publication in PubMed.
4. The table in the **Phenotypes** section (Fig. 4D) shows high-level mammalian phenotype terms associated with their genotypes. Use the "*show* or *hide* all annotated terms" links to expand (and contract again) the table to see lower level terms, linked to definitions (*see* also Subheading 3, Vocabulary Browsers). Only terms with data are displayed in the table. Click on a genotype button, such as hm2 (for the second homozygous genotype described) to see all annotations for that genotype as well as references and curatorial notes.
5. The **Disease models** table (Fig. 4E) displays the human Disease Ontology (DO) terms by genotype (listed in Genotype/ Background table in the Phenotype section). The disease terms link to the MGI Disease Ontology Browser.
6. The **Find Mice (IMSR)** section (Fig. 4F) offers information on where to obtain mutant stocks. If stocks are available carrying this allele, or carrying any mutation of this gene, links are provided to the International Mouse Strain Resource (IMSR, http://www.findmice.org) [4], a database of mouse stocks

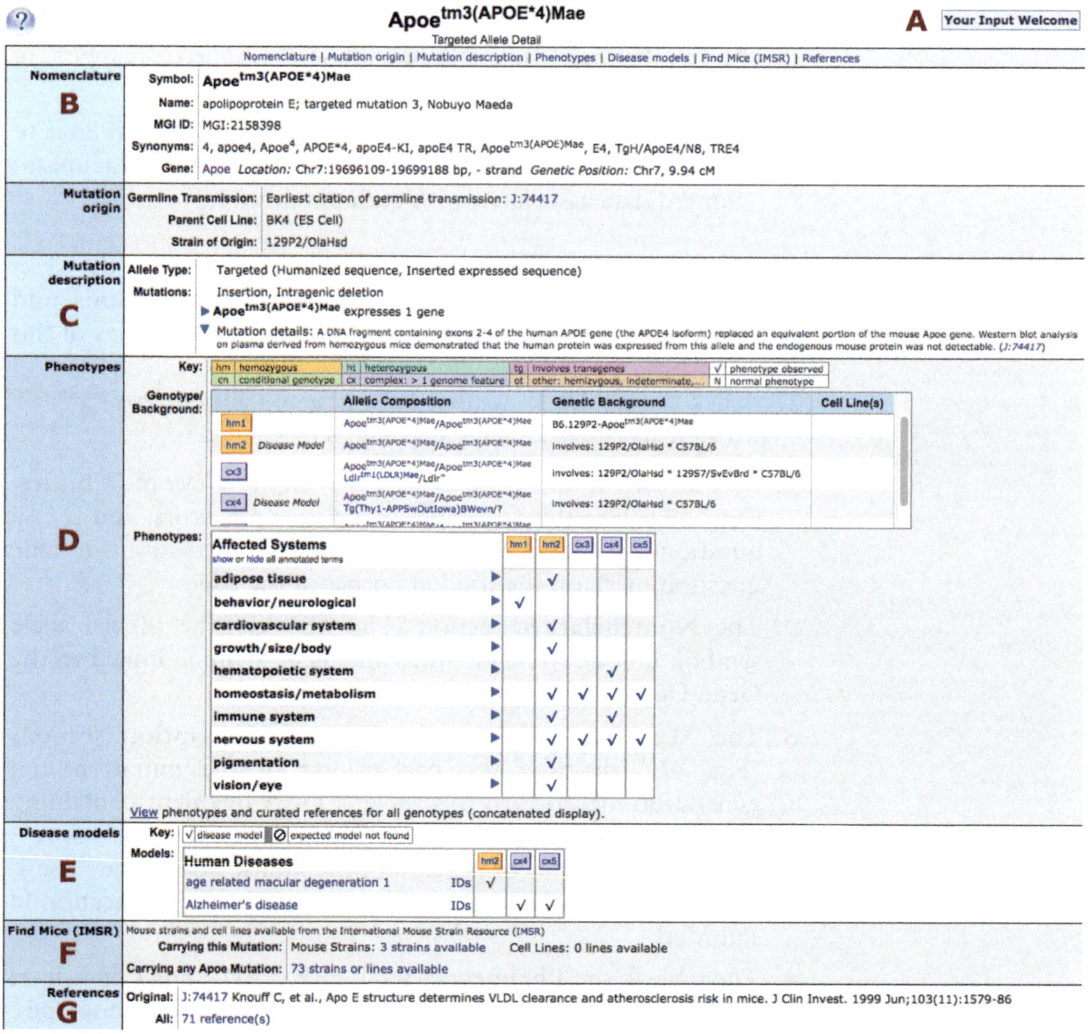

Fig. 4 The *Apoe*$^{tm3(APOE*4)Mae}$ Allele Detail page has information on official nomenclature, a description of the mutation and detailed phenotypes associated with genotypes containing the allele. In the Phenotypes section of the page, click on the "View phenotypes for all genotypes (concatenated display)" link to view curated details about this allele. See text for details

available worldwide through public repositories. The IMSR does not house any stocks but provides links for ordering stocks and contacting the appropriate repositories.

7. The **References** section (Fig. 4G) links to all references associated with the allele. Each reference is assigned with a traceable "J number" that can be further filtered by author, journal, reference type, and published date. Links to PubMed and Journals are also provided.

3 Vocabulary Browsers

MGI uses defined, structured vocabularies or ontologies for gene function, mouse phenotypes, and mouse anatomy. Their use provides researchers with a foundation for consistent querying and computational analysis. Data include over 78,000 expression assays, annotated to the Mouse Developmental Anatomy, and over 300,000 MP annotations and 300,000 GO annotations.

3.1 Mammalian Phenotype (MP) Browser Example

To access a browser, go to the drop-down Search menu beneath the MGI logo on the upper left hand side of an MGI web page and select one of the Vocabularies options. For this example, choose the Mammalian Phenotype (MP) Browser (Fig. 5).

1. You can search or browse the ontology. In this example, browse by clicking on nervous system phenotype in the Phenotype Tree View.
2. Click on abnormal nervous system morphology. (A small black triangle preceding a term means the term has subterms. You can use the triangles to expand and collapse branches of the tree.)
3. Click on abnormal neuron morphology and then on abnormal motor neuron morphology.
4. Click on the number of genotypes.

This displays a table of all the genotypes (combination of alleles and strain background) annotated to abnormal motor neuron morphology or one of its subterms, and the reference ID for the source of the annotation, linked to reference details.

The Gene Ontology (GO) Browser works in a very similar manner. The Disease Ontology Browser and Anatomy Browsers have different, but intuitive interfaces. We are working to provide all the vocabulary browsers with very similar interfaces.

4 Genes and Markers Query Form

The Quick Search and Vocabulary Browsers offer simple keyword searches. However, MGI data integration allows users to perform much more specific searches. We have developed a number of search tools geared toward building refined queries focused on returning specific types of data, such as genes, alleles, or expression assay results. Near the top of these forms, and on many other web pages, a large Question Mark icon links to detailed user help documentation that explains how to use forms and interpret result pages. The next example describes a simple Genes and Markers Query Form search.

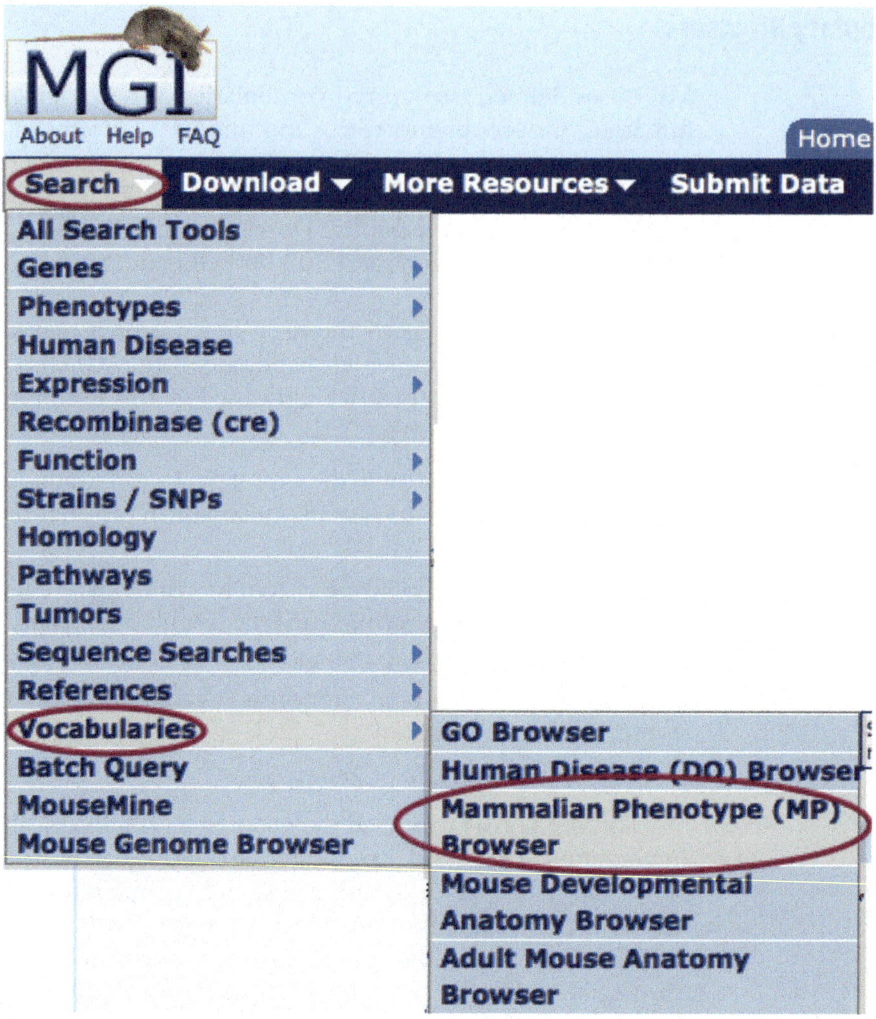

Fig. 5 Use the drop-down Search Menu to access individual search forms. One way on how to access the Mammalian Phenotype (MP) Browser is shown

4.1 Genes and Markers Query Form Example

You are interested in a quantitative trait locus (QTL) defined by the coordinates, chr2:109680000–131040000. How can you see all the genes mapped in this region?

1. Go to Search beneath the MGI logo on the upper left hand side of an MGI web page and select Genes → Genes & Markers Query.
2. In the Feature type section, check "protein coding gene." (Click Show to see all the types and the counts for each.)
3. In the Genome location section of the page, select Chromosome 2 and in the Genome Coordinates field, enter *109680000–131040000.*

4. Click the Search button.

Now that you have all the candidate genes in this region, you can search for stronger candidates by adding phenotype or GO terms to your search parameters. For this example:

1. Use the "Click to modify search" button above the results. Also notice that you have four Export options.
2. In the Gene Ontology (GO) classifications field, enter: *ion transport.*
3. In the Phenotypes/Disease field at the bottom of the form, enter: *seizures.*
4. Click the Search button (Fig. 6 shows the form filled out for the second, modified, search).

This returns all protein coding genes within this region on mouse Chromosome 2 that have a GO annotation to ion transport and also have alleles associated with seizures.

5 Phenotypes, Alleles & Disease Models Search

Identifying mouse models for desired phenotypes can be a challenge without any prior knowledge of genes that contribute to the phenotypes. Using this flexible search tool, we can search for mouse models with any phenotype term, with or without additional constraints to the search criteria. Though you can search with any keyword, using ontology terms may yield more accurate search results. Links to browse supported ontologies can be found on the search form. Here we describe a simple query using this search form.

5.1 Phenotypes, Alleles & Disease Models Search Example

To query for phenotype data and return alleles, go the Search menu beneath the MGI logo and select: Phenotypes → Phenotypes, Alleles & Diseases Query.

1. In the Phenotype/Disease field, enter: *tremors AND MP:0001405.*
2. Select *Targeted* under Generation Method in the Categories section.
3. Under Allele Attributes, select *Conditional ready* and *No functional change*. Click on the *Generation Method* and *Allele Attributes* headings for term definitions.
4. Click the Search button.

The search results returns all conditional ready alleles curated to genotypes exhibiting tremors and impaired coordination phenotypes (the MP term with the ID, MP:0001405). Allele

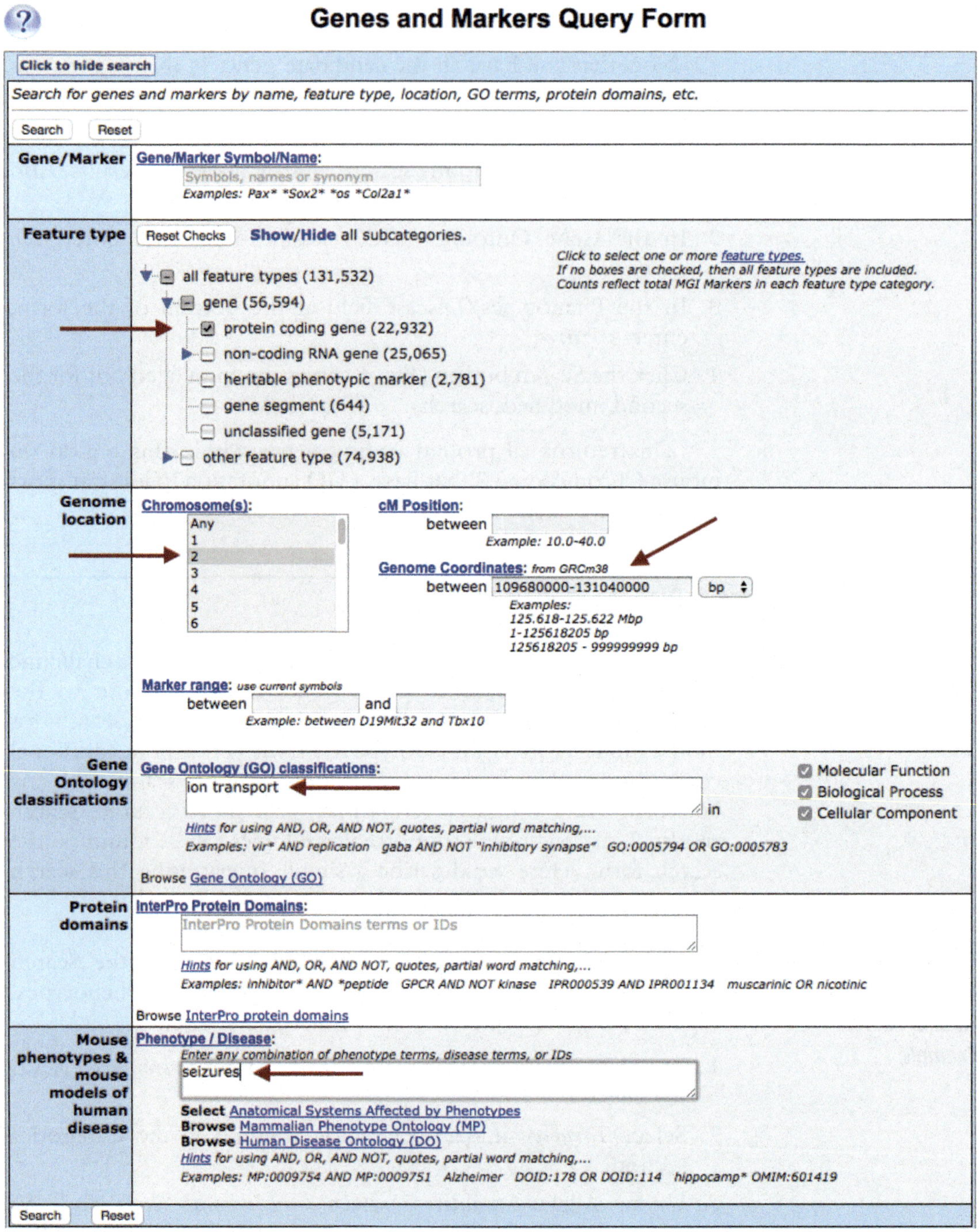

Fig. 6 The Genes and Markers Query Form filled out for a search for protein coding genes involved in ion transport and mapped to mouse Chromosome 2 between build GRCm38 coordinates, 109680000–131040000, and associated with seizure alleles

symbols in your results link to Allele Detail pages (*see* Subheading 2.3). You can also export your results as tab-delimited text or spreadsheet files.

6 Recombinase (cre) Activity

Use the Recombinases tab at the top of most MGI web pages to search for cre and other recombinase carrying alleles. A simple search form permits searches by the anatomical structure where recombinase activity was assayed and/or the driver or promoter. Search results show the systems where recombinase activity was detected and not detected and link to allele details and any available stocks known to the IMSR. Your search results offer the ability to filter by driver, inducer, and for the anatomical systems in which the recombinase was detected or not detected.

7 The Gene Expression Database (GXD)

As an integral component of MGI, the Gene Expression Database (GXD) [2] collects and integrates expression data from RNA in situ hybridization, immunohistochemistry, RT-PCR, Northern blot and Western blots experiments, with a particular focus on endogenous gene expression during mouse development. The time and space of gene expression is annotated in standardized ways using the hierarchical Mouse Developmental Anatomy ontology. You can use the Mouse Developmental Anatomy Browser to search or browse the anatomy and look up expression data for specific anatomical structures.

7.1 Gene Expression Data Query

In this example we will use the Gene Expression Data Query (which enables expression searches using many different parameters) to find expression assays that detected expression in the endocrine system for genes involved in lipid homeostasis.

1. Select the Gene Expression Data Query on the GXD Home Page (http://www.informatics.jax.org/expression.shtml), or use the Search Menu beneath the MGI logo on the upper left hand side of any MGI page to select: Expression → Gene Expression Data Query. Three tabbed query forms are provided. The Standard Search, and two more specialized forms, a Differential Expression Search to find genes expressed in some anatomical structures or developmental stages but not others, and a Batch Search for searching with lists of gene symbols or IDs.
2. Use the Standard Search tab and in the Genes section of the page, use the right hand field and start typing: *lipid homeostasis* and select the term from the list. Note that the number of genes annotated to the term is shown.

3. In the Anatomical structure or stage section, Find assay results where expression is "detected in."
4. In the Anatomical Structures field, start typing: *endocrine system* and select the term, *endocrine system, TS17–28*, from the list. GXD annotates expression data to standardized developmental, Theiler, stages (TS) [11]. TS17–28 indicates that the endocrine system exists during those stages.
5. To reduce the number of results for this example, limit the search to Assay type: Immunohistochemistry. To do this quickly, click the check mark for "Find expression data in any assay type" to uncheck all assay types, and then select Immunohistochemistry.
6. Click the Search button.

The query form filled out as above is shown in Fig. 7.

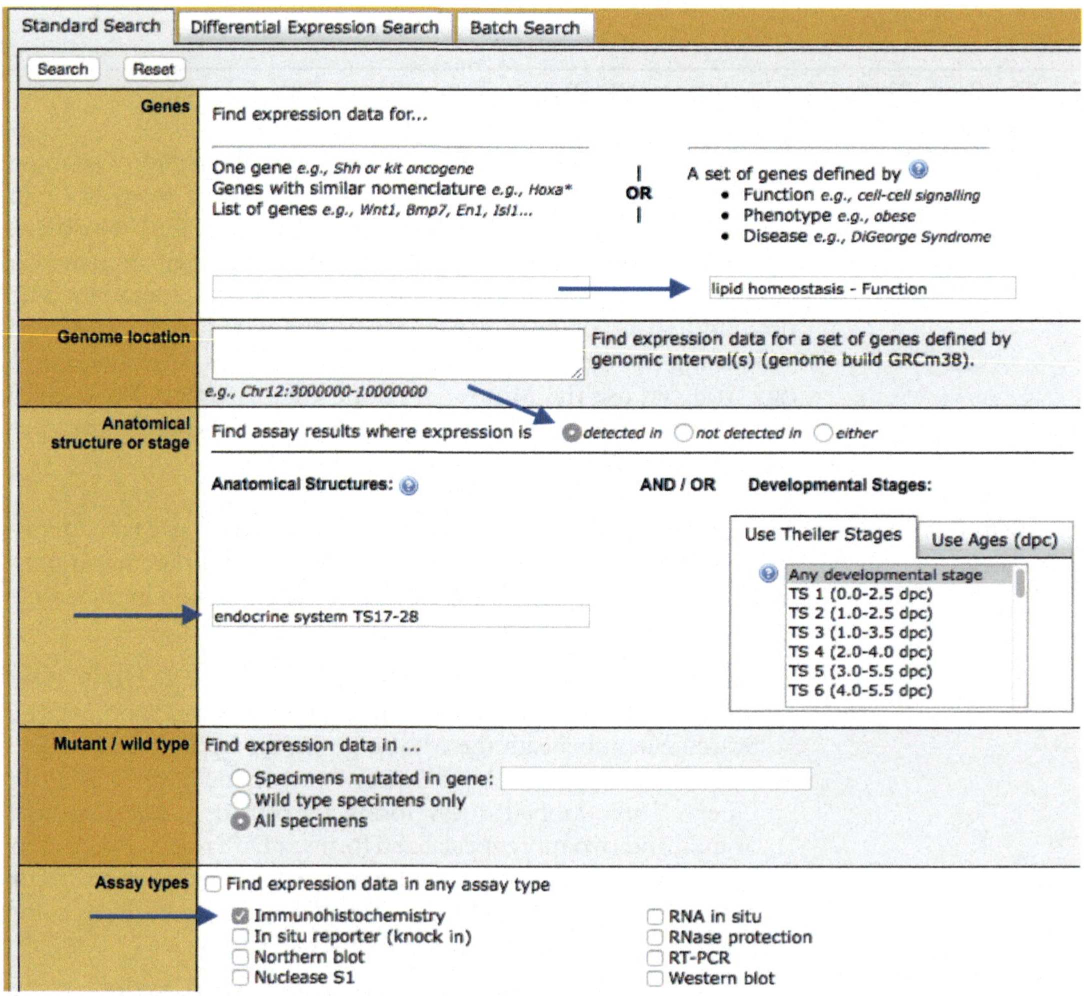

Fig. 7 The Gene Expression Data Query form filled out for a search for immunohistochemistry assays of genes involved in lipid homeostasis, that detected expression in the endocrine system

Tabbed sections offer different ways to view your search results. The first three tabs, Genes, Assays and Assay Results, provide progressively detailed views of your search results. The Images tab shows all the expression images that meet the search criteria. The two matrices tabs offer interactive overviews of temporal and spatial expression patterns and the ability to navigate to more detail. The Assay Results tab is returned by default and lists the assay results that meet the search criteria and links to each assay's details, reference and the appropriate gene detail page. To view the assay details, click on the *data* link in the Result Detail column.

If you want to alter your query:

Above your results, click: "Click to modify search." (The banner title changes to Click to hide search.) The window that appears above your results contains your original search criteria.

8 Data Submission

MGI encourages electronic data submissions. Click on Submit Data in the dark blue banner on MGI web pages to access simple forms for contributing nomenclature, phenotype, recombinase specificity and expression data. Data will be integrated in MGI and properly referenced. Researchers can also submit prepublication data to MGI to obtain official nomenclature for genes and alleles. If desired, data will be kept private until publication.

9 Human–Mouse: Disease Connection (HMDC)

Genetically defined mouse models of human diseases exhibit a range of phenotypes analogous to their human counterparts. Data from these models can be leveraged by researchers interested in finding candidate genes for complex diseases and examining the affects of genetic background on phenotypes. The Human–Mouse: Disease Connection (HMDC) permits enquiries into mouse genetic and comparative data from the human or mouse perspective. Clinical researchers can explore the mouse genome for models of human phenotypes and diseases, and find potential candidate models where mice display a range of phenotypes similar to humans. Mouse geneticists can easily explore relationships between mouse phenotypes and the human diseases that are associated with mouse genotypes.

MGI annotates mouse models of human diseases to the Mammalian Phenotype (MP) Ontology and the Human Disease Ontology (DO). Online Mendelian Inheritance in Man (OMIM) makes gene associations for human diseases and the human phenotypes associated with those genes are from the Human Phenotype

Ontology (HPO). These data are integrated into MGI and our Human–Mouse: Disease Connection (HMDC) tool enables researchers to build queries starting from a human or mouse aspect.

9.1 Human–Mouse: Disease Connection Search

This example shows how you can find human phenotypes and diseases associated with genes in two regions of human chromosomes, as well as those for their mouse homologs.

1. Click on Human–Mouse: Disease Connection on the MGI Home Page or the Human Disease tab at the top of most other MGI web pages.
2. Toggle "Please select a field" to "Genome Location."
3. Select the species and build: Human (GRCh38).
4. Enter the region: *19:43340000–44900000* (This format is also supported: Chr19:43340000–44900000)
5. Click the Add button.
6. Toggle the second "Please select a field" to "Genome Location."
7. Select the species and build: Human (GRCh38).
8. Enter the region: *21:25000000–26970000.*
9. Toggle the default Boolean from AND to OR.
10. Click the Search button.

    The search finds human genes mapped to those regions and returns them, their mouse homologs, and phenotypes and diseases associated with all the returned genes (Fig. 8).
11. Scroll out to the right to see diseases associated with these genes. Blue squares indicate mouse data and tan is for human data.
12. Click on the blue square at the intersection of *PLAUR* and immune system to see the mouse *Plaur* genotypes with immune system abnormalities.

10 Batch Queries and MouseMine

MGI offers several tools for querying with lists of genes, alleles and IDs and most web searches provide tab-delimited data export options. Weekly database reports can be downloaded from an FTP site (http://www.informatics.jax.org/downloads/reports/index.html). Data from these files can be used to generate custom data sets by inputting them in a simple to use Batch Query (http://www.informatics.jax.org/batch), or a more powerful tool, MouseMine (http://www.mousemine.org/). MouseMine [10], built on the InterMine software framework, offers iterative

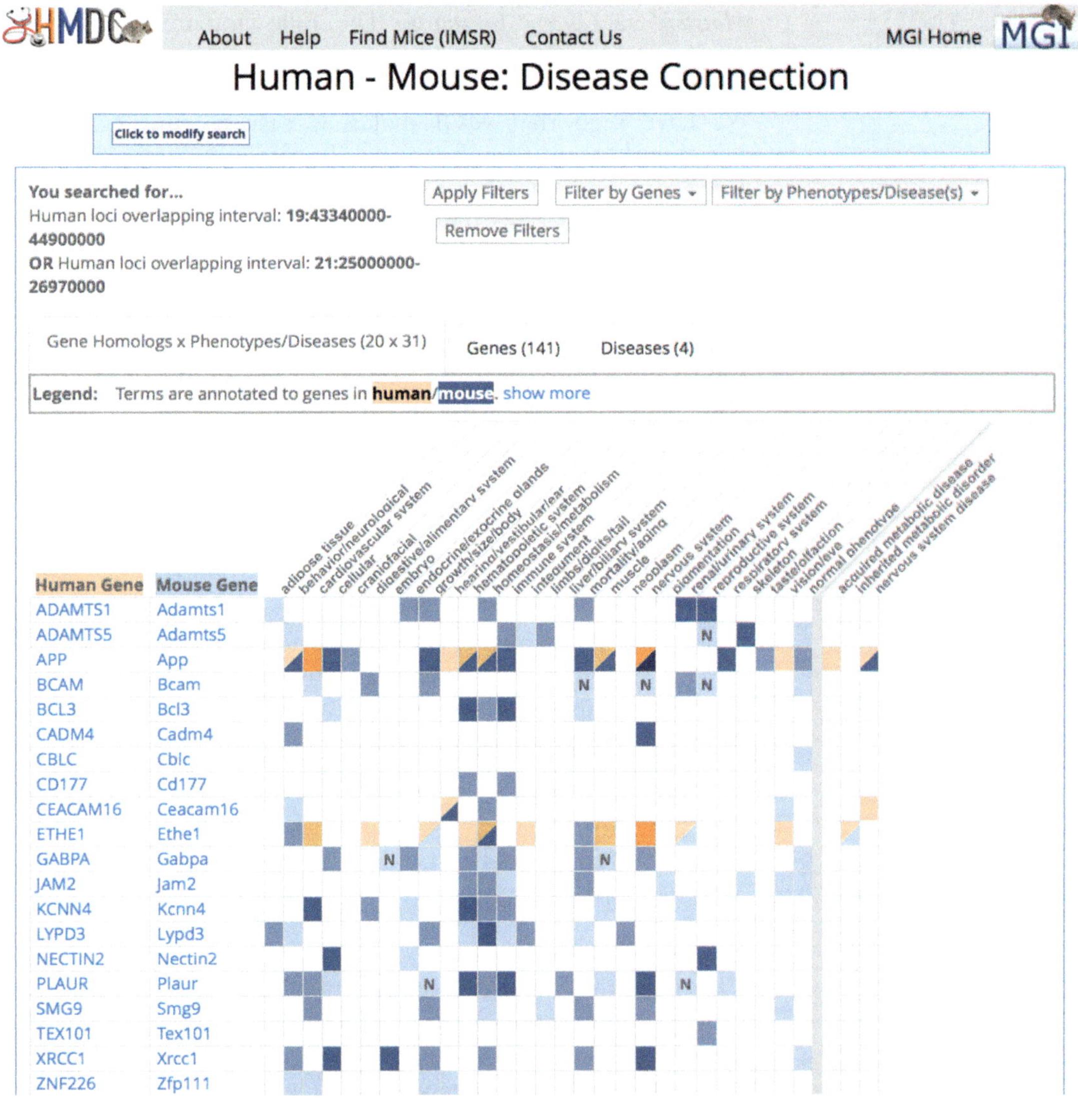

Fig. 8 The default view of the results of a Human–Mouse: Disease Connection Search for regions of two human chromosomes. The results show the human genes in those regions, their mouse orthologs, high-level phenotype terms and diseases associated with variants of those genes. Blue colors show mouse data and tan colors are human data. The colored cells link to more details. A bold **N** in a phenotype column indicates that no abnormal phenotypes of that type were observed in any allele of the gene. The results can be filtered by genes and by phenotypes/diseases

querying, built-in enrichment analysis, and API support. Both the Batch Query and MouseMine return gene expression results but when searching for gene expression data, the Gene Expression Database's Batch Query is preferable because it returns the above shown multitabbed data summaries, including image summaries and the interactive tissue-by-gene matrix view as part of its multitabbed data summary.

MouseMine Query Template: The following example shows how to use a MouseMine template and add additional information to its results.

We have seen that MGI makes it easy to find genotypes associated with particular phenotypes but its web forms currently do not allow you to export genotype details in tab-delimited formats. For example, if you used the Mammalian Phenotype (MP) Browser to find genotypes associated with emphysema (MP:0001958) and abnormal lung elastance (MP:0011043), you can use MouseMine to get a text file of those genotypes.

1. Go to Search beneath the MGI logo on the upper left hand side of an MGI web page and select MouseMine.
2. Click on the Templates tab near the top of the page.
3. Click on Mammalian phenotypes (MP terms) → Mouse genes and models.
4. Enter *MP:0001958, MP:0011043* in the LOOKUP field and click the Show Results button.
5. Toggle the 25 results so you can see all the results.

   This returns a list of genes and transgenes, and genotypes associated with these phenotypes terms and their subterms. Now you can modify the query:
6. Click on the Manage Columns button and then click on Add a Column.
7. Click on MGI Type the first time it is listed (under Subject).
8. Click Add 1 new column.
9. Click Apply changes.
10. This adds the gene feature types to the table. Now you can remove columns you don't want to see by clicking the X in the column heading and you can filter by clicking the last icon in a column heading, the graph icon (Fig. 9).
11. Click the Export button to select your formatting options.

11 Mouse Tumor Biology Database (MTB)

Early in the twentieth century, many of the first inbred strains of mice generated by William Ernest Castle and Clarence Cook Little were used to demonstrate the role of genetics in cancer. The laboratory mouse has continued to develop as model system of human cancers and MTB (http://tumor.informatics.jax.org) [3] endeavors to support this system by providing current information on spontaneous and induced tumors in genetically defined mice. Data include digital histopathologic images annotated to tumor subtypes. MTB also is the source of data for Patient Derived Xenograft

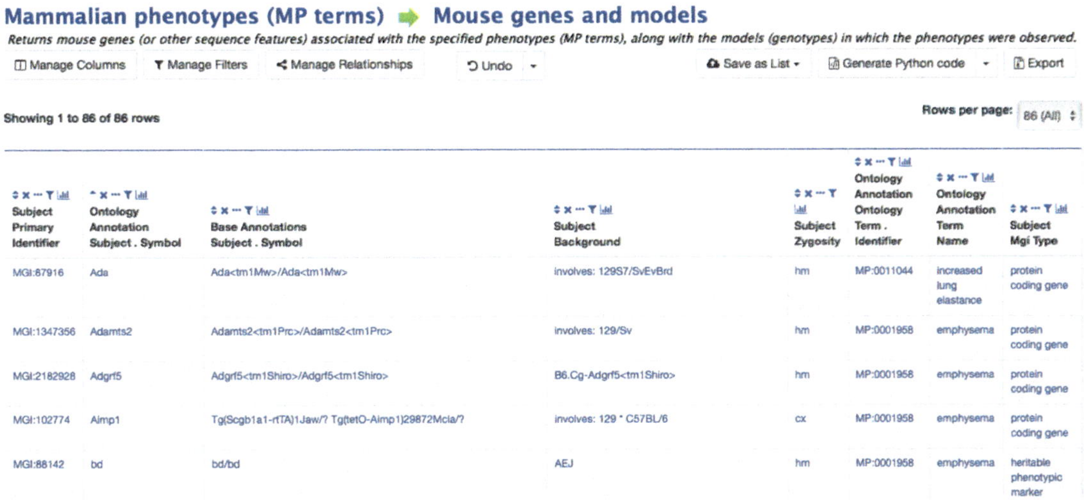

Fig. 9 MouseMine results of a search for Mammalian Phenotype (MP) terms. Columns can be easily removed or filtered and additional columns of data can be added. Your final results can be exported and you can save lists and your search code

(PDX) models. PDX models for cancer research are created by the implantation of human cells and tumor tissue into immune compromised NOD-SCID-Gamma2 mouse hosts.

11.1 PDX Model Search

This example shows how you can find SNPs and Variants for a given tumor.

1. On the MGI Home Page, click on Tumors in the drop-down Search menu beneath the logo, or click on Mouse Models of Human Cancer in the main part of the page.
2. In the Additional Resources on the left hand side of the page, click on *PDX Model Search.*
3. In the Search by primary cancer site section, select Breast and click the Search button.
4. In your results, click on Model ID: *TM00089.*
5. Click to expand the Variant Summary section (Fig. 10).
6. You can also expand the Gene Expression section or see Copy Number Variants by expanding the Gene CNV section. Further down the page are histology images, a drug response curve, and circus plots.

12 Summary

MGI is updated with new data weekly and regularly adds new data sets and refines its user interface and search tools.

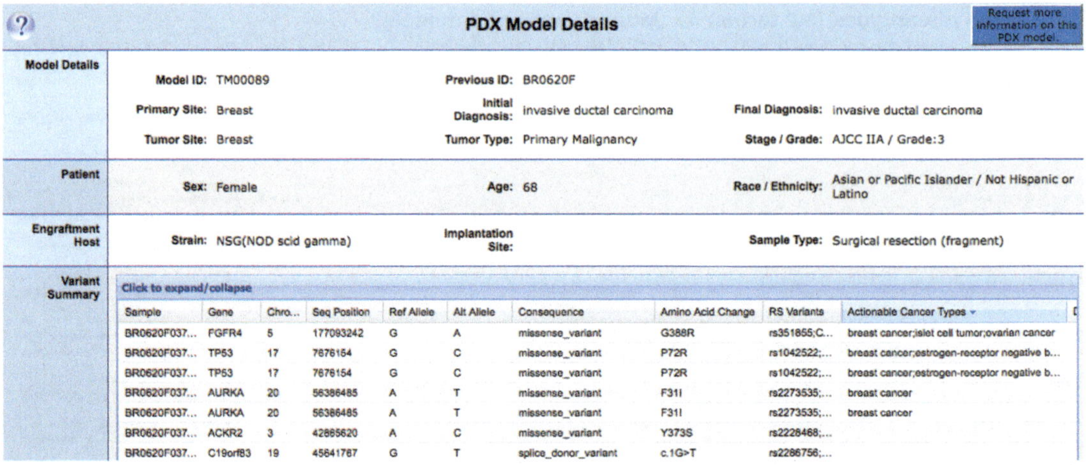

Fig. 10 MTB PDX Model Details, showing information on the engrafted human tumor and a portion of the variant details. Cropped out of the image are RNA Seq data, Gene Copy Number Variant values, histology images, circos plots, and a drug response curve

You can reach MGI's dedicated User Support group with questions and comments at mgi-help@jax.org. We provide email technical support, hands-on training and encourage community annotations, collaborations and suggestions. User Support moderates two email mouse community bulletin boards: (1) mgi-list, a forum for topics in mouse genetics, as well as MGI and community news; (2) mgi-technical-list, a forum for technical information about accessing MGI data, using APIs, and linking to MGI web pages. You can subscribe to these lists here: http://www.informatics.jax.org/mgihome/lists/lists.shtml.

We are engaged in the collaborative efforts of the Alliance of Genome Resources (AGR, http://www.alliancegenome.org) to develop shared tools and compatible interfaces for biomedical web resources for the major model organisms.

Acknowledgments

These MGI projects are supported by NIH/NHGRI grants: HG000330 and HG002273, NIH/NICHD grant: HD062499, and NIH/NCI CA089713.

References

1. Blake JA, Eppig JT, Kadin JA, Richardson JE, Smith CL, Bult CJ, Mouse Genome Database Group (2017) Mouse Genome Database (MGD)-2017: community knowledge resource for the laboratory mouse. Nucleic Acids Res 45:D723–D729
2. Finger JH, Smith CM, Hayamizu TF, McCright IJ et al (2017) The mouse Gene Expression Database (GXD): 2017 update. Nucleic Acids Res 45:D730–D736
3. Bult CJ, Krupke DM, Begley DA et al (2015) Mouse Tumor Biology (MTB): a database of

mouse models for human cancer. Nucleic Acids Res 43:D818–D824

4. Eppig JT, Motenko H, Richardson JE et al (2015) The International Mouse Strain Resource (IMSR): cataloging worldwide mouse and ES cell line resources. Mamm Genome 26:448–455
5. Smith C, Eppig JT (2015) Expanding the mammalian phenotype ontology to support automated exchange of high throughput mouse phenotyping data generated by large-scale mouse knockout screens. J Biomed Semantics 6:11
6. Drabkin HJ, Christie KR, Dolan ME et al (2015) Application of comparative biology in GO functional annotation: the mouse model. Mamm Genome 26:574–583
7. Köhler S, Vasilevsky NA, Engelstad M et al (2017) The Human Phenotype Ontology in 2017. Nucleic Acids Res 45:D865–D876
8. Hayamizu TF, Baldock RA, Ringwald M (2015) Mouse anatomy ontologies: enhancements and tools for exploring and integrating biomedical data. Mamm Genome 26:422–430
9. Kibbe WA, Arze C, Felix V et al (2014) Disease Ontology 2015 update: an expanded and updated database of human diseases for linking biomedical knowledge through disease data. Nucleic Acids Res 43:D1071–D1078
10. Motenko H, Neuhauser SB, O'Keefe M et al (2015) MouseMine: a new data warehouse for MGI. Mamm Genome 26:325–330
11. Theiler K (1989) The house mouse: atlas of embryonic development. Springer, New York

Chapter 8

A Primer for the Rat Genome Database (RGD)

Stanley J. F. Laulederkind, G. Thomas Hayman, Shur-Jen Wang, Jennifer R. Smith, Victoria Petri, Matthew J. Hoffman, Jeff De Pons, Marek A. Tutaj, Omid Ghiasvand, Monika Tutaj, Jyothi Thota, Melinda R. Dwinell, and Mary Shimoyama

Abstract

The laboratory rat, *Rattus norvegicus*, is an important model of human health and disease, and experimental findings in the rat have relevance to human physiology and disease. The Rat Genome Database (RGD, http://rgd.mcw.edu) is a model organism database that provides access to a wide variety of curated rat data including disease associations, phenotypes, pathways, molecular functions, biological processes and cellular components for genes, quantitative trait loci, and strains. We present an overview of the database followed by specific examples that can be used to gain experience in employing RGD to explore the wealth of functional data available for the rat.

Key words Rat, Database, Quantitative trait locus, Ontology, Genomics, Gene

1 Introduction

The Rat Genome Database (RGD) provides the scientific community with a public source for a variety of information related to the laboratory rat (http://rgd.mcw.edu) [1]. RGD incorporates manually curated data and information obtained through electronic resources into a comprehensive and dynamic database containing information on genes, strains, quantitative trait loci (QTLs), simple sequence length polymorphisms (SSLPs), sequences, maps, cell lines, and orthologs, all with supporting references. RGD also provides a collection of visualization and analysis applications to assist researchers in effectively utilizing the information available in

---

Dedication We wish to dedicate this chapter to the memory of our colleague and longtime RGD curator Dr. Victoria Petri, who recently passed away. Her legacy lives on through the RGD Pathway Ontology and Pathway Diagrams that she created.

Martin Kollmar (ed.), *Eukaryotic Genomic Databases: Methods and Protocols*, Methods in Molecular Biology, vol. 1757, https://doi.org/10.1007/978-1-4939-7737-6_8, © Springer Science+Business Media, LLC, part of Springer Nature 2018

the database. This integration of manually curated data with electronically imported data obtained from major public data repositories (e.g., NCBI, UniProt), combined with diverse analysis tools, makes RGD a uniquely valuable resource to the scientific community.

This chapter focuses on the disease, phenotypic, genomic, and functional information that is available in the database and the RGD tools available to analyze that information. An overview of the RGD home page will be presented, followed by sections on search functions, report pages, data portals, and tools for analysis and visualization.

2 Navigating the RGD Home Page

The RGD home page at http://rgd.mcw.edu/ provides entry points to all the basic resource categories in RGD (Fig. 1). Resource categories are arranged on tabs at the top of the page and are divided into the seven major areas of the RGD web site: Data, Analysis and Visualization, Diseases, Phenotypes & Models, Genetic Models, Pathways, and Community (Fig. 1A). Each tab provides quick access to the corresponding section of the RGD web site. Expanded data and tool links are also available in the center of the home page (Fig. 1B). Note that the Keyword Search text box (Fig. 1E) is located in the top right corner of most RGD pages.

Other features of the home page include:

1. Below and left of center is a section containing links to RGD video tutorials (Fig. 1C), which provide general introductions to various analysis tools and sections of the RGD web site.
2. The bottom left portion of the home page has a list of upcoming conferences which may be of interest to RGD users (Fig. 1D). Each line in the list is a link to the home page of that particular conference.
3. Below and right of center is the RGD home page revolving banner (Fig. 1F), which announces new information and features at RGD. Each banner item is linked to information about, or an example of, what is described in the item. Below the banner is a chronological list of news items that appear or have previously appeared on the revolving banner (Fig. 1G). All of the listed items are hyperlinked to the corresponding web page.
4. At the top of the home page and most RGD pages, a link to the help pages can be found as the first choice on the left side of the menu bar (Fig. 1). The home page for RGD help lists many links to individual help pages for data pages, tool pages, portal pages, and a glossary of terms used throughout the site.

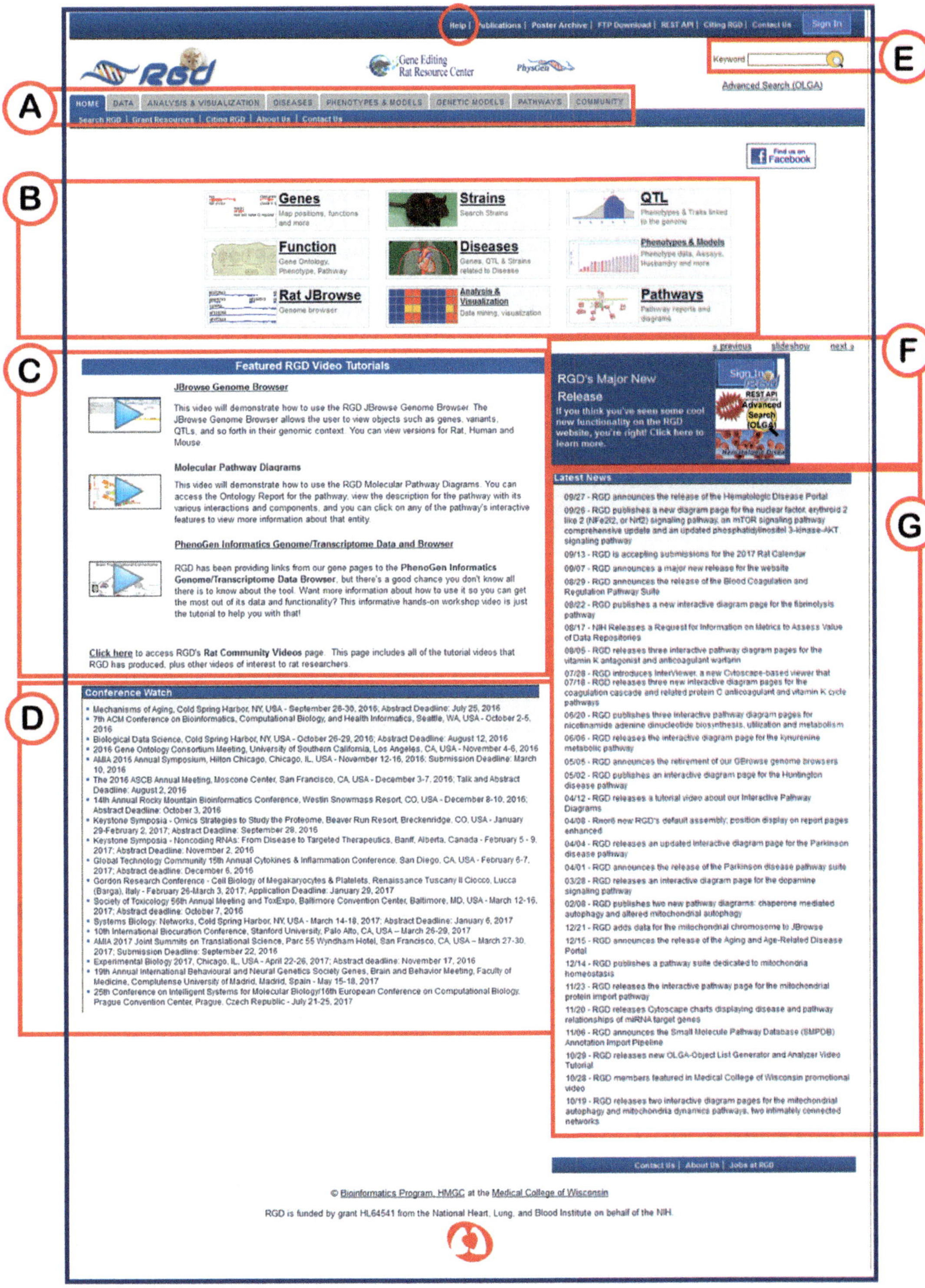

Fig. 1 The RGD home page. (A and B) Tabs and links to all of the data and analysis tools at RGD are found here. (C) Links to RGD video tutorials. (D) Links to websites of scientific conferences relevant to RGD users. (E) Keyword search text box. (F) Rotating news banner. (G) Chronological list of news items. Red circle at top of page indicates link to help page

2.1 Using the RGD Search Functions

The RGD web site has a number of ways to search for data, depending on the scope of the specific information desired. The keyword search text box, available in the upper right-hand corner of most RGD web pages (Fig. 1E), provides a fast way to get at specific data, such as gene or QTL information, when a name, keyword, or accession number is known. It searches across most object types (genes, QTLs, strains, homologs, SSLPs, ESTs, and references) and many data types. It also searches the many controlled vocabularies used at RGD, including the gene ontology (biological process, molecular function, and cellular component), mammalian phenotype ontology, human phenotype ontology, pathway ontology, and disease vocabulary. Also, one can search specific object types and ontologies directly.

2.1.1 Performing an Object-Specific Search

To perform an object-specific search the Data tab under the RGD logo of any RGD web page is selected to access the "Data" page (Fig. 2).

1. A data type is selected by clicking the data name or adjacent icon. For example if "GENES" is selected, a new page is returned for gene-specific searches (Fig. 3A). On all object search pages and the ontology search home page, there are "example searches" above the main keyword search box. These examples are hyperlinked words which are provided as sample searches.

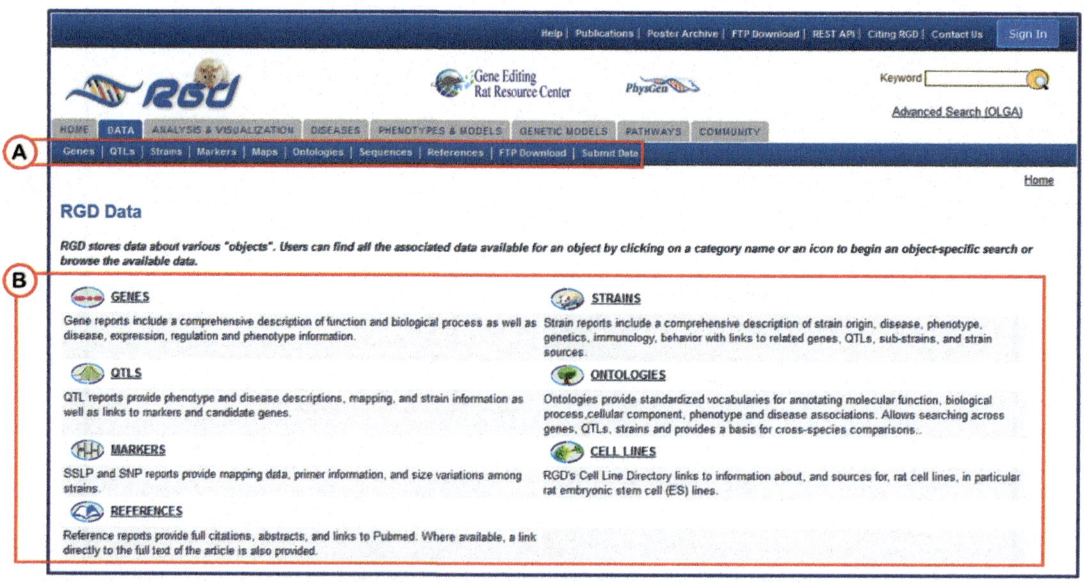

Fig. 2 The RGD Data page. Specific data types are selected by clicking a link on the menu bar (A) under the tabs or by clicking an icon/term from the middle of the page (B)

Fig. 3 Object-Specific Search Pages. (A) Gene Search with free text search box (1) and limiting options underneath the text box. (B) QTL search with multiple text box options for searching. (C) Strain Search with keyword search box

2. A "Keyword" search box on the left side of the gene search page is used with a selection of options to limit the results (Fig. 3A1).
3. On the gene results page the entries are listed by species, with rat as the default display (Fig. 4A). If it is desirable to view gene lists in other species, the "Mouse," "Human," "Chinchilla," "Bonobo," "Dog," "Squirrel," or "All" tab at the top of the gene list may be selected.
4. The results can be sorted alphabetically or numerically on any column. To sort by chromosome, select "Chr" in the first "Sort By" drop-down menu and "Descending" or "Ascending" in the second drop-down menu on the upper right side of the results page (Fig. 4A1).
5. To download the results to an Excel or other type of file, click the "CSV" or "TAB" link at the top of the results list (Fig. 4A2).
6. To print the results or to view the results in RGD's GViewer tool, click the "Printer" or "Genome Viewer" link, respectively, also at the top of the results list.
7. Clicking "Other Analysis Tools" opens a pop-up window with multiple options to analyze the gene list in various tools at RGD.
8. To display a certain gene report page, select that gene record by clicking the symbol, which is hyperlinked to the gene report page.

The QTL and strain searches (Fig. 3B, C) and search results pages (Fig. 4B, C) work much the same way as the gene search/results pages. While rat, mouse, and human QTLs are available at RGD, the strain data availability is restricted to rat.

2.1.2 Using the Ontology Search

The ontology/vocabulary browser can be accessed via the "Function" icon on the RGD home page, and from the menu bar via the "Ontologies" link in the data list of the Data page (http://rgd.mcw.edu/wg/data-menu) (Fig. 2). Any of the 20 ontologies/vocabularies can be browsed from the top node by clicking the appropriate name in the list on the Ontologies page.

1. Clicking on an ontology/vocabulary opens the browser with three horizontal frames listing parent terms, selected term with siblings, and child terms (Fig. 5A, B).
2. When each term is clicked, it moves to the center column with its siblings and the adjacent columns refresh to show parent terms to the left and child terms to the right (Fig. 5B–D). This allows easy navigation of the ontology in both directions, with three levels of terms being visible at all times. An "A" icon to

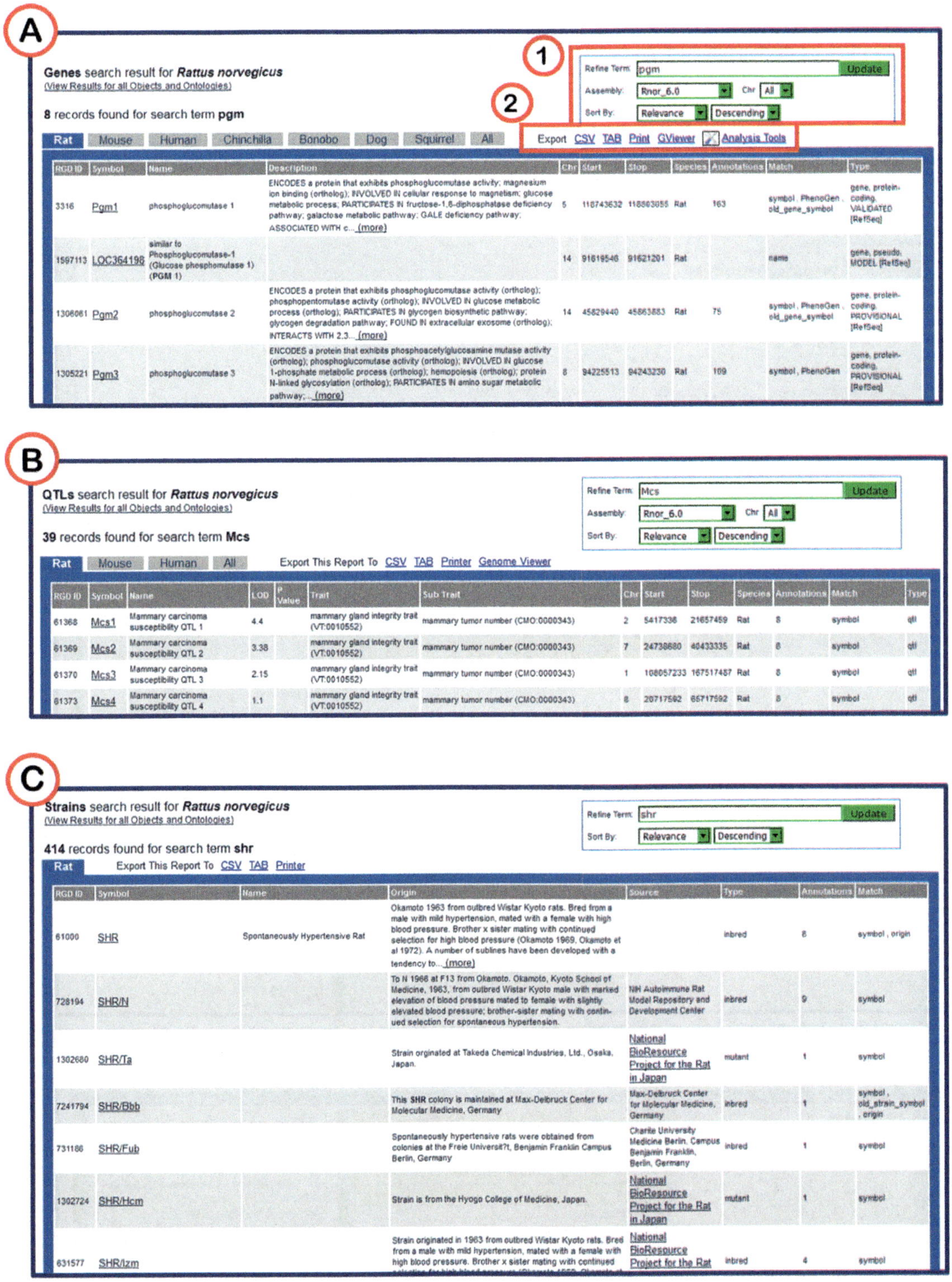

Fig. 4 Object-Specific Results Pages. (A) Genes search results. (1) Search editing and sorting options. (2) Data export options. (B) QTL results page. (C) Strains result page

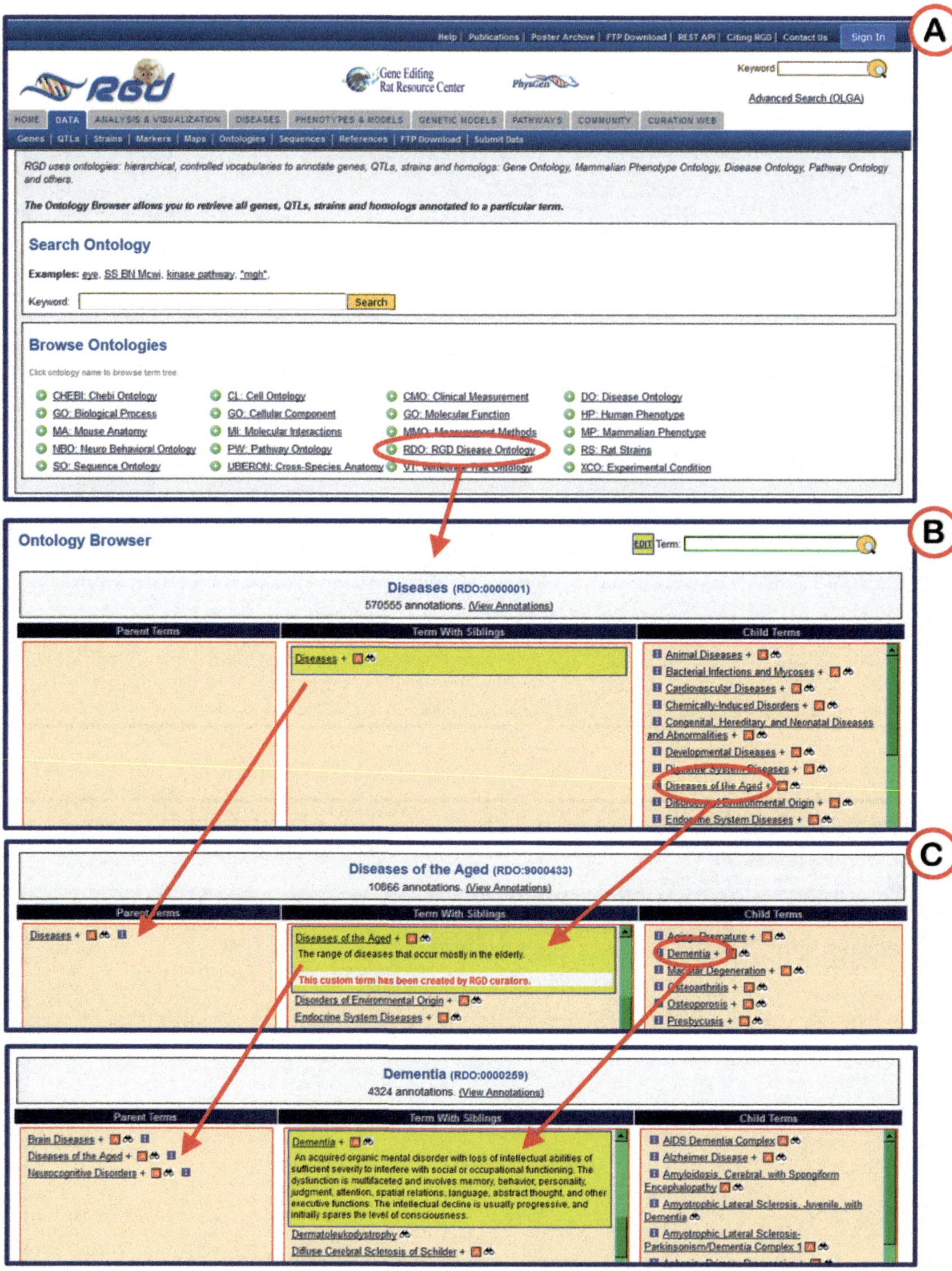

Fig. 5 Ontology Term Browser. The RGD Disease Ontology is selected (indicated by red oval) in (A), followed by selection of terms "Diseases of the Aged" in (B) and "Dementia" in (C). Red arrows show movement of terms from one column to another when a term is selected

the right of a term means that gene annotations for that term exist in RGD, and clicking on the "A" icon takes the user to the ontology report page for that specific term.

Another option to find ontology/vocabulary terms is by using the keyword search in the middle of the ontology search page (Fig. 6A).

1. Any searched entry will lead to a results page with a box showing an "Ontology Name" column which lists all ontologies with terms which at least partially match the query term and a "Terms" column which lists the number of matching terms in each ontology (Fig. 6B).
2. Clicking any ontology in the list returns a list of all the terms in that ontology which contain the query term (Fig. 6C). An underlined term means that annotations exist in RGD for that term. Additional columns in the table include accession ID number, a count of annotations to the term and its children, and links ("browse tree" and "branch" icons) to the ontology browser.
3. The term results list provides access to either the term browser (Fig. 5) or the ontology report page with all annotations listed for that term (Fig. 6D).

Ontology term report pages feature definitions, synonyms, and annotations across all data types made to the specific term and its child terms.

1. A visual display of annotated objects is presented in a genome viewer (GViewer) (Fig. 6D) which illustrates the genomic location of each gene or other data object. The approximate location of genes, QTLs, and congenic segments are shown in an ideogrammic view of chromosomes.
2. The species tabs underneath the GViewer access separate lists of annotated rat, mouse, human, or other species genes (Fig. 6D1). Above the species tabs is a check box to display genes annotated to the ontology term or both the term and its child terms. There are also two drop-down menus for sorting the annotation gene list by any of its columns, a download button, and a check box for an expanded view. The Symbols, JBrowse links, Evidence source, and Reference IDs in the gene list are all hyperlinked to relevant information at RGD and at external databases.
3. The bottom of the page shows a conventional, vertical view and a graph representation (Fig. 7) of the branch of the ontology where the selected term resides. Clicking on a branch icon next to the annotation count for any of the listed terms accesses the main RGD ontology term browser.

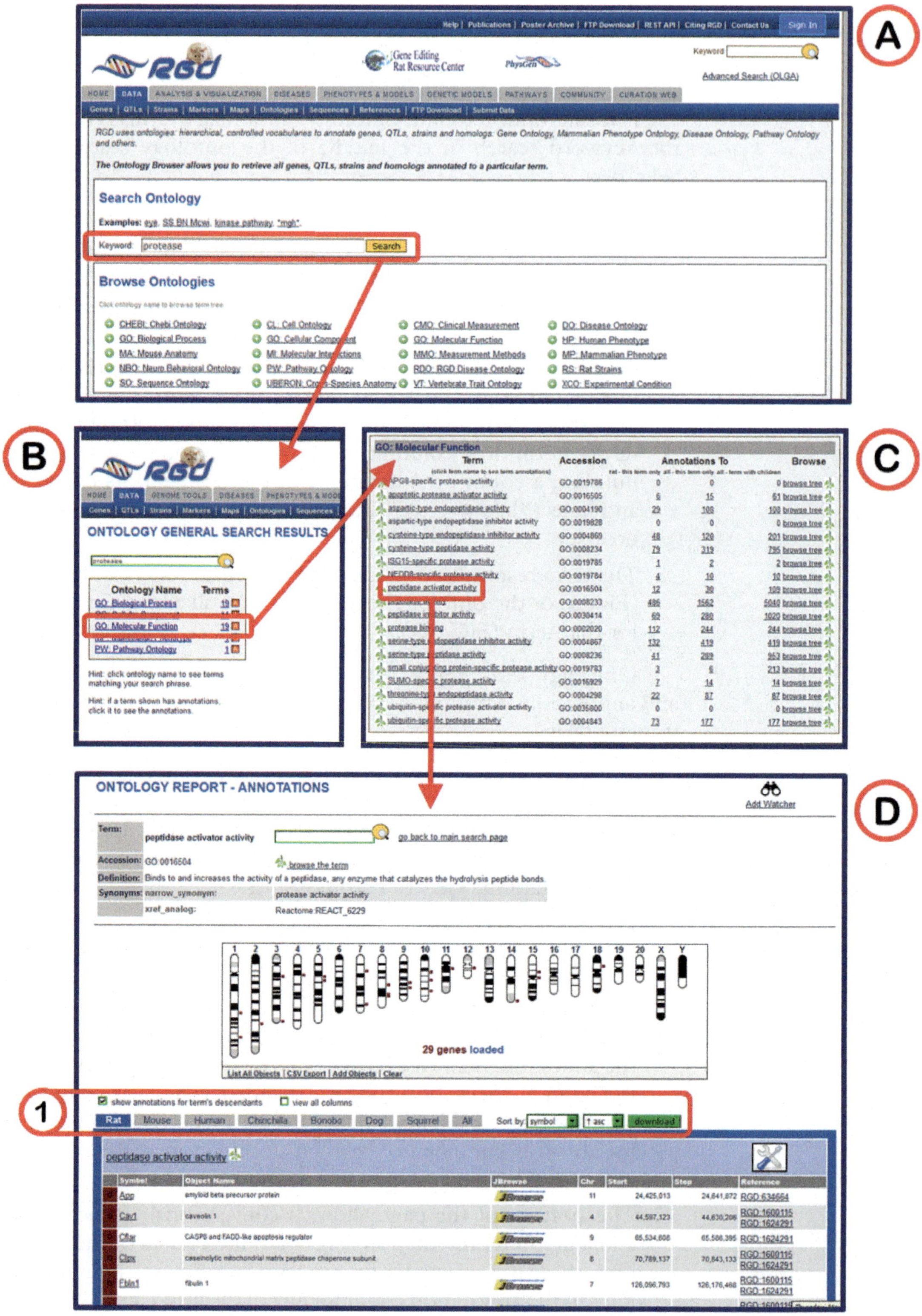

Fig. 6 Ontology General Search. (A) Ontology home page with keyword search indicated (red box). (B) General search results across all listed ontologies/vocabularies for "protease." (C) Detailed list of Molecular Function terms matching the searched word. Clicking on any term in the list links to the ontology report page for that term.

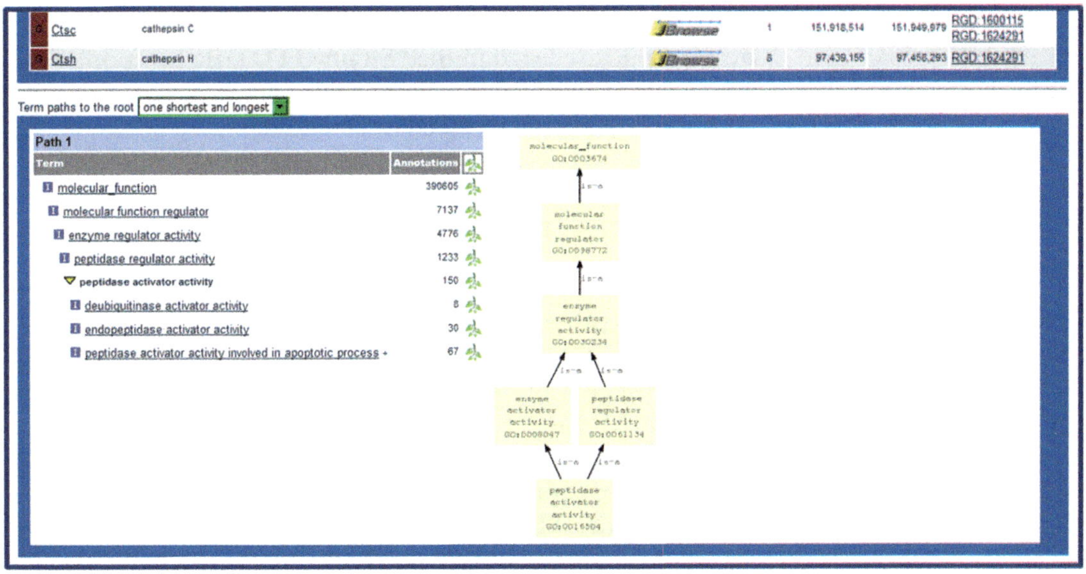

Fig. 7 Bottom of Ontology Report Page. Underneath the annotated object list are text and graph representations of the ontology branch(s) containing the selected term

2.2 Data-Specific Report Pages

RGD data report pages compile all of the curated data for the specific object: gene, QTL, variant, or strain. For example, click on one of the annotated genes in the rat species tab (Fig. 6D).

1. RGD report pages contain many of the same elements regardless of the data type. These include official nomenclature for the object, annotations in the form of both ontology terms and free-text notes, and links to related information in other databases. In addition, reports for genomic and genetic data types include information on mapping and a link to various genome browsers to permit viewing of the object in its genomic context (Fig. 8A).
2. Some of the data elements are reciprocally linked to information of other data types. A link on a gene report page, for instance, will lead to a QTL report page which will, in turn, link back to the gene (Fig. 9).
3. Typically, a QTL will be defined by the cross of two strains. The source strain(s) of a QTL are listed with hyperlinked symbols in the general information section of a QTL report page. Those links lead to the report pages for the respective strains

Fig. 6 (continued) (D) The ontology report page for "peptidase activator activity." The top of the page defines the ontology term, lists synonyms, and provides external database links, followed by a GViewer display of all the genes annotated to the ontology term. Below the GViewer display is a list of all genes annotated to the term and its child terms. (1) Species-specific tabs and sorting controls for the list are indicated in red box

(Fig. 10). Reciprocally, the QTLs that are derived from a particular strain are listed in the "Strain QTL Data" section of the strain report page. Those listed QTLs are linked back to the appropriate QTL report page.

3 Data Portals in RGD

There are particular types of data at RGD that are presented in their own sections of the web site called "portals." Disease information is currently divided into 11 different "portals," separated by disease category. Phenotype data is accessible through the "Phenotypes & Models" portal, which includes quantitative PhenoMiner data, strain medical records, and more. Finally, the pathway portal contains both molecular pathway and physiological pathway diagrams. Whereas the physiological pathways are limited to a few interactive diagrams, the pathway portal currently has 200 interactive molecular pathway diagrams across five nodes (classic metabolic pathway, signaling pathway, regulatory pathway, disease pathway, and drug pathway) of the Pathway Ontology. Related pathway diagrams are organized in "suites" and "suite networks."

Access to the portals is provided by the tabs at the top of most RGD web pages and by the icons in the middle of the RGD home page. In addition disease portals may be accessed by icons found in the "RGD Disease Portals" section of the "Annotation" portion of gene, QTL, and strain report pages. Also, pathway diagrams may be accessed directly from the appropriate term entry in the RGD ontology term browser and from the ontology report page of the appropriate pathway term.

3.1 Disease Portals

The RGD Disease Portals home page (Fig. 11) has icons that link to the individual disease portals. RGD maintains a growing list of disease portals, each designed to be an entry point for researchers to access consolidated data and tools related to a particular category of disease.

1. On a portals main page disease terms can be selected by the use of two drop-down menus (Fig. 12A) and the count of objects annotated to the selected disease(s) are summarized in a matrix below the drop-down menus (Fig. 12B).
2. Directly below the matrix summary is an ideogrammic view of chromosomes (an instance of GViewer) of all the annotated objects in their approximate chromosomal locations. The chromosome view is interchangeable between rat, mouse, and human, with an option to view one species synteny while viewing the primary species.

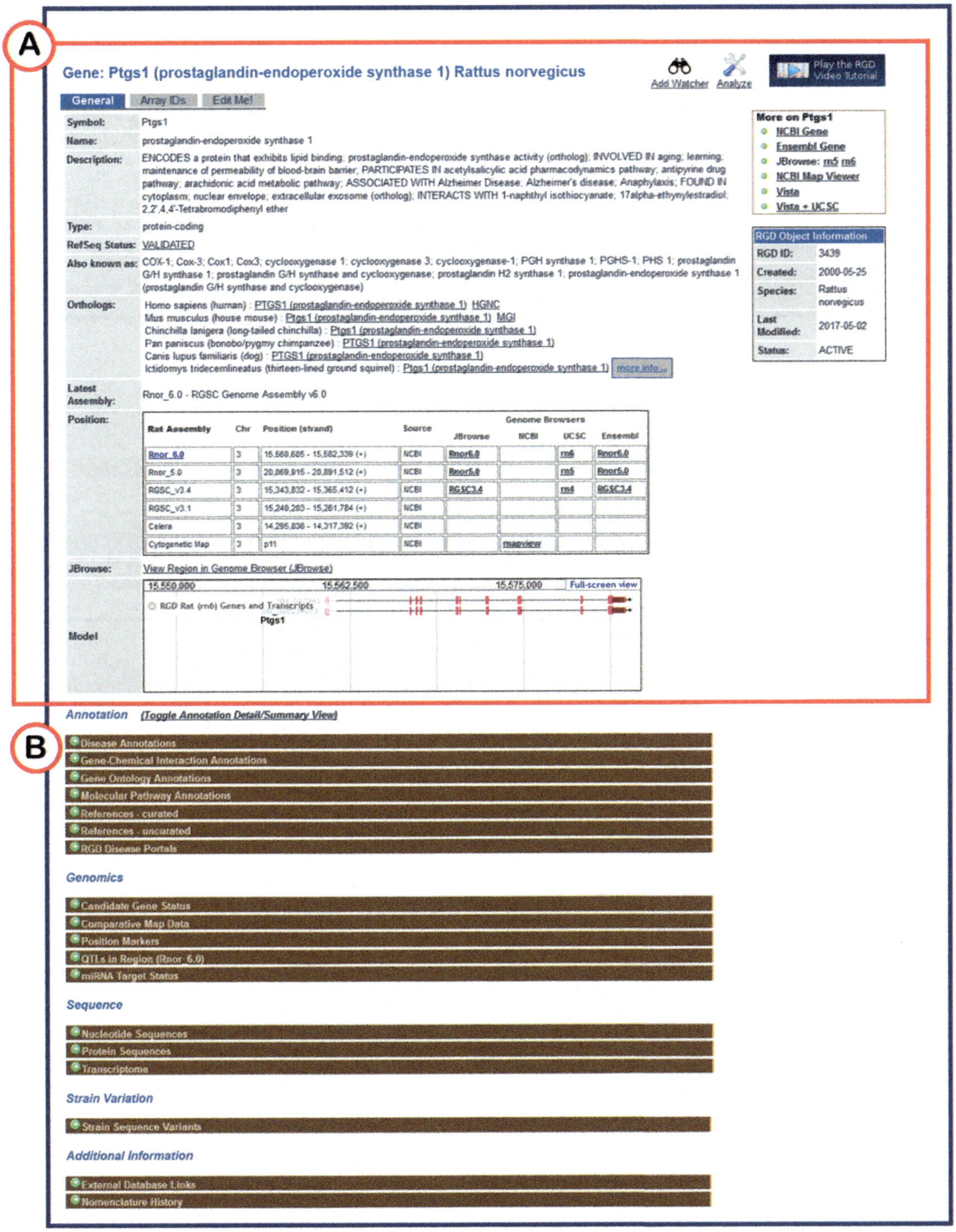

Fig. 8 Gene Report Page for Rat Ptgs1. (A) The top half of the page contains general information, ortholog assignments, genomic positions, JBrowse model, and links to external sites. (B) The bottom half of the page has annotations in various categories, genomic information, sequence information, and more, all in expandable, labeled bars

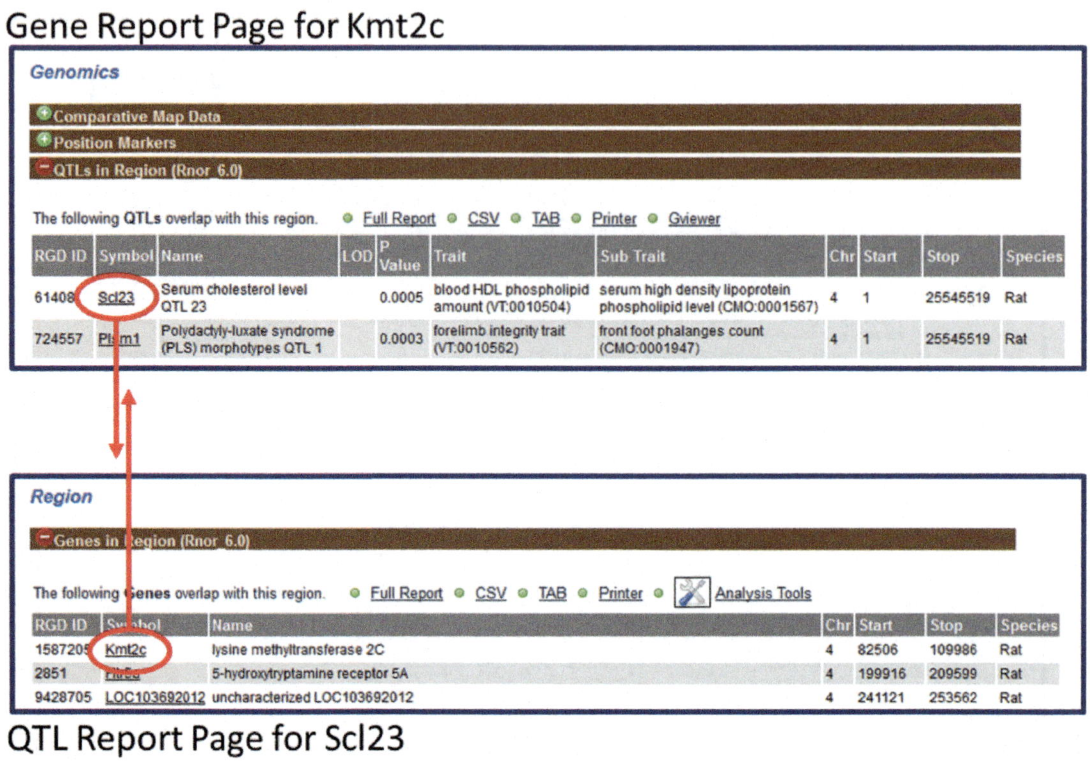

Fig. 9 Reciprocal Links between Gene and QTL Report Pages. The rat gene Kmt2c is located within the QTL Scl23. The Scl23 link in the "QTLs in Region" section connects (downward arrow) to the report page for Scl23. Reciprocally, the Kmt2c link in the "Genes in Region" section of the Scl23 QTL report page connects (upward arrow) to the report page for Kmt2c

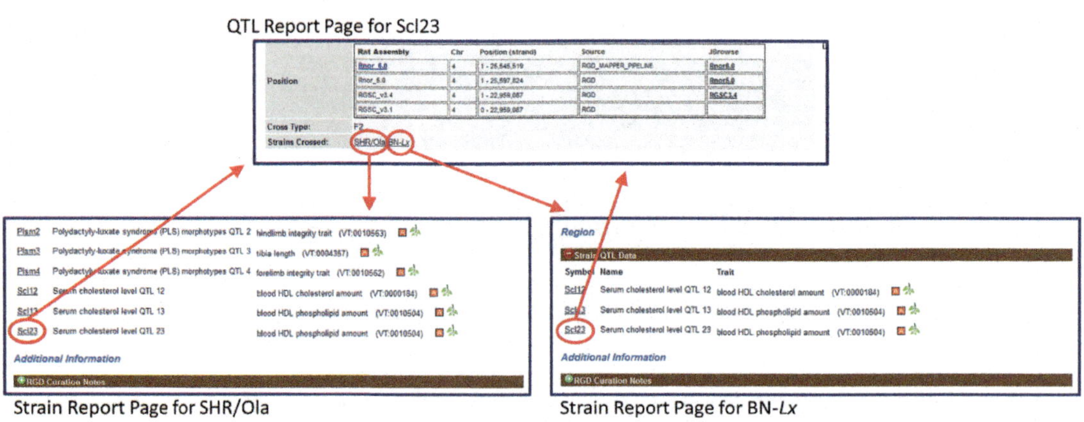

Fig. 10 Reciprocal Links between QTL and Strain Report Pages. The rat QTL Scl23 was generated by crossing rat strains SHR/Ola and BN-*Lx* (marked by red ovals). The symbols listed as "Strains-Crossed" on the QTL page link to the strain pages (downward arrows). The QTL symbol "Scl23" is found on both strain pages in the "Strain QTL Data" sections (red ovals). The QTL symbols link to the QTL report page (upward arrows)

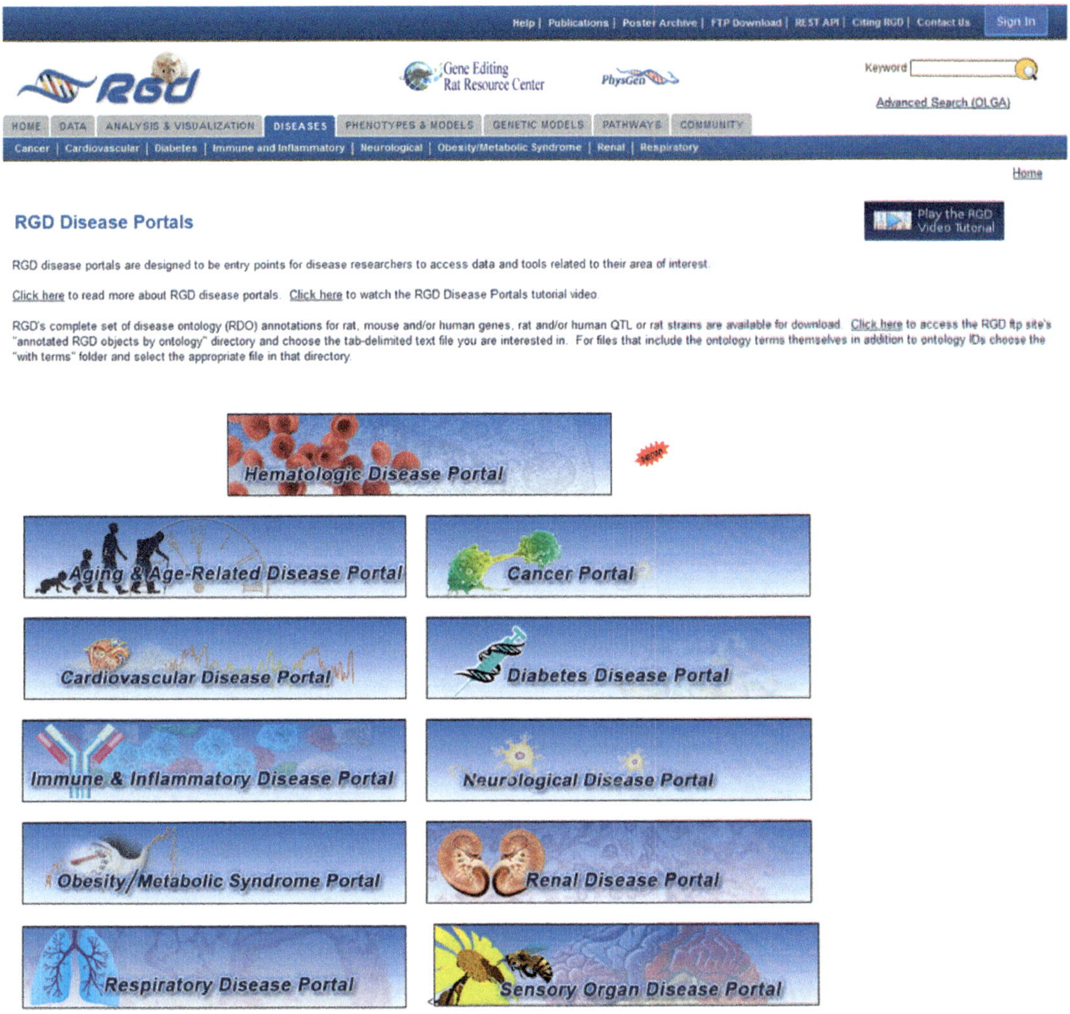

Fig. 11 Disease Portals Home Page. Icons for all current disease portals are shown here. All icons link to their respective portals

3. Below the GViewer are scrollable lists of the annotated objects: genes, QTLs, and strains.
4. Finally, at the bottom of the page are enrichment charts for GO terms annotated to all the rat genes in the disease list.

In addition to disease-related genes, QTLs and strains, RGD's disease portals contain sections for Phenotypes, Biological Processes and Pathways related to the disease category covered by that portal. The data for those categories are accessed and presented in the same way as the disease data. These other data categories, along with links to tools and related information, are accessed via the menu tabs along the top of each portal page (Fig. 12).

Fig. 12 Hematologic Disease Portal Home Page. (A) Drop-down menus for selection of disease category and specific disease. (B) Numerical summary of results for the selected disease category/disease. (C) GViewer display of results with approximate positions of all genes, QTLs, and strains. (D) Lists of genes, QTLs, and strains annotated to selected disease category/disease. (E) Enrichment charts showing Gene Ontology annotations for all selected disease-annotated genes

3.2 Phenotypes & Models Portal

The Phenotypes & Models Portal contains data related to rat strains and phenotypes, as well as essential information for conducting physiological research, identifying disease models, and community forums for gathering feedback from the scientific community. The various sections of the portal have icons on the portal home page linking to the respective data or tools (Fig. 13).

Fig. 13 Phenotypes & Models Home Page

1. "Meet Joe Rat" has general information about the phylogenetics of the laboratory rat, laboratory techniques using rat, strain availability from vendors, and data submission.
2. The "Phenotypes" section is another way to access data from the Program for Genomics Applications (PGA) project (a large scale phenotyping project which has collected data for consomic, ENU mutant, and knock-out rat strains) (http://pga.mcw.edu). In this case static tables of data are available by multiple links based on type of physiological data.
3. "Strain Medical Records" contains the same data as "Phenotypes," but with a strain-specific view.
4. "PhenoMiner" is a link to the quantitative phenotype tool to be described later.
5. The "Strains & Models" section is a hub to access any of the rat strain information at RGD, including strain search, strain commercial availability, disease models, animal husbandry, and links to outside sources of information.

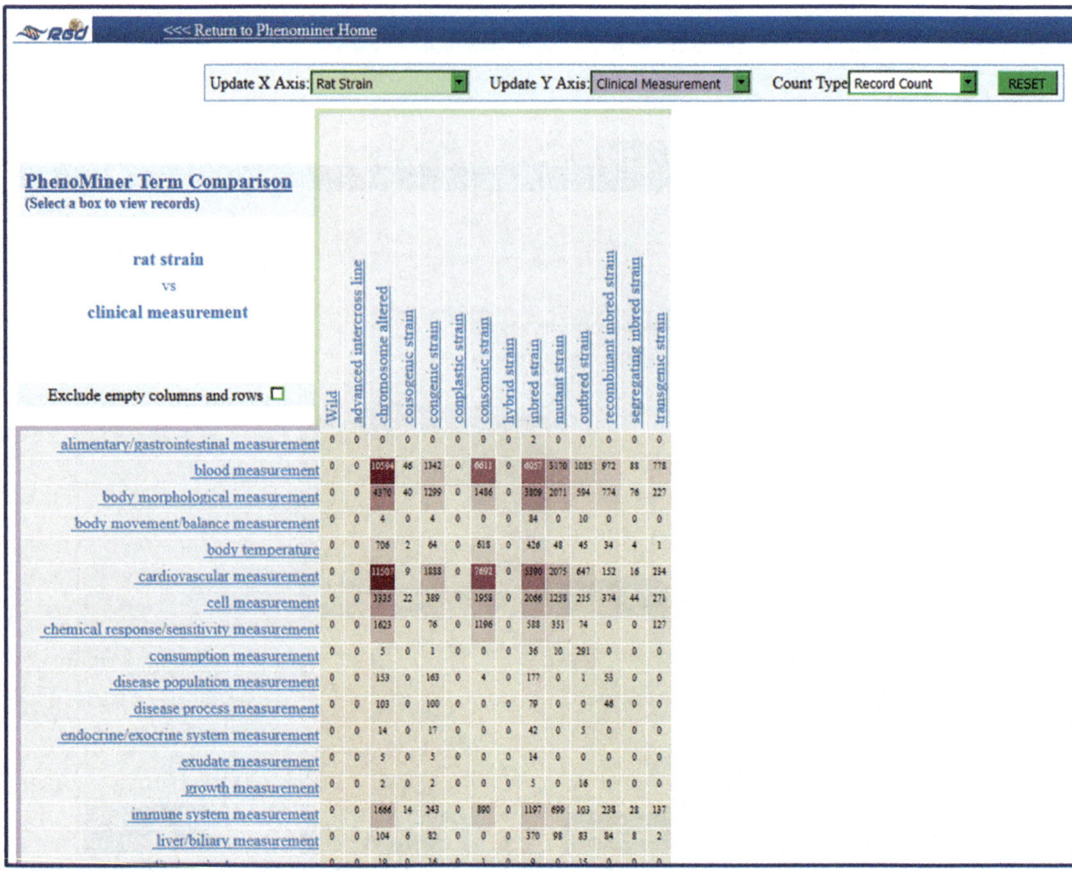

Fig. 14 PhenoMiner Term Comparisons Default Page

6. The final icon on the Phenotypes & Models home page is "PhenoMiner Term Comparisons". This accesses a tool similar to the heat map in the Gene Annotator (GA; *see* Subheading 4.5) tool, with the default view being rat strain on one axis and clinical measurement on the other axis (Fig. 14). The heat map is interactive with the axes changeable via drop-down menu and by clicking on terms on the axes to access child terms. Clicking any numbered square in the heat map accesses that data in PhenoMiner.

3.3 Pathway Portal

The home page of the pathway portal provides access to the list of molecular pathway diagrams via two separate icons ("Individual Diagram Pages" and "Pathway Suites and Suite Networks") (Fig. 15) and to the list of physiological pathway diagrams (Fig. 15C) via its own icon. Both diagram list pages are also accessible by the links at the top of the page under the "Pathways" tab. The molecular pathway diagrams (Fig. 16) are designed with Elsevier's Pathway Studio software (http://support.pathwaystudio.com/) and feature

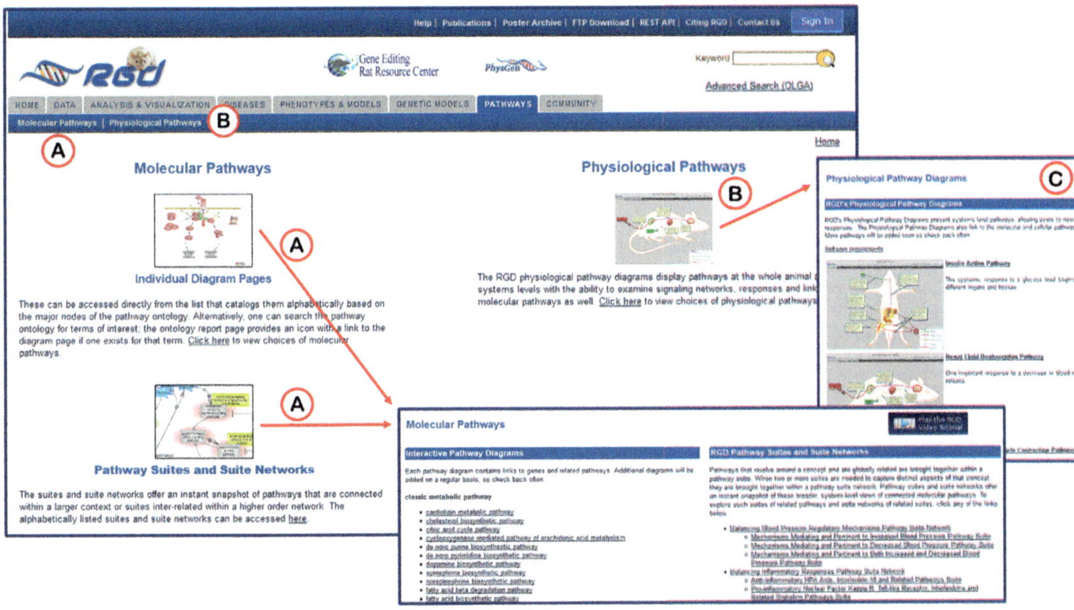

Fig. 15 Pathway Portal Home Page. (A) Links to the page containing lists of all interactive pathway diagrams and pathway suites/networks. (B) Links to the page containing the list of physiological pathway diagrams. (C) Both the icons and underscored names of the pathways are links to the individual diagram pages. Because the physiological diagrams are made with RGD software, the user side software requirements can be found by using the link above the pathway icons

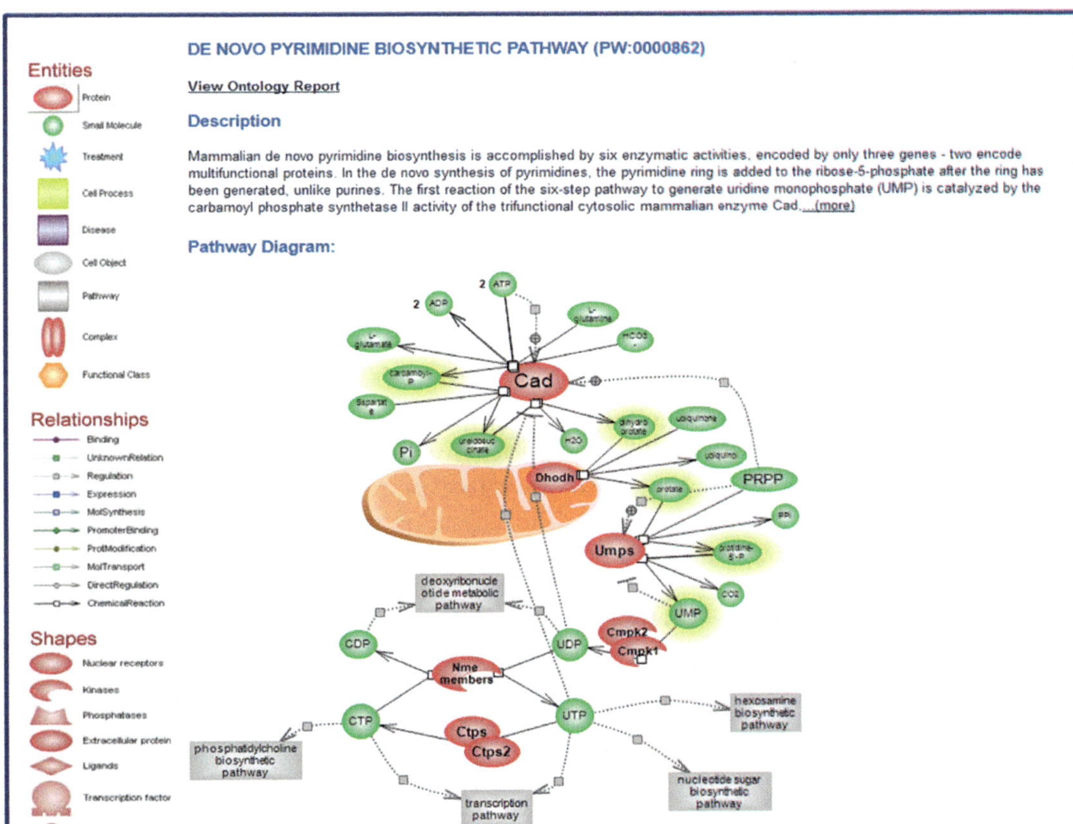

Fig. 16 De Novo Pyrimidine Biosynthetic Pathway Diagram. The diagram is accompanied by a text description above it and a key to the left of it

hyperlinks from most of the objects in the diagram to RGD pages representing the respective term, gene, chemical, or associated secondary pathway. Beneath the diagrams on the molecular pathway pages are several lists:

1. "Genes in Pathway": genes with annotations to the title term of the diagram and to child terms of the title pathway from the Pathway Ontology [2]. This is the same type of list found on the ontology term report page for the title term (*see* Subheading 2.1.2, Figs. 6D and 17A).
2. "Additional Elements in Pathway" is a list found on some pathway diagram pages. This list may include small molecules, gene groups, other pathways, etc. (Fig. 17B).
3. "Pathway Gene Annotations": A list of disease terms associated with the genes involved in the pathway. This list toggles between disease term to genes and gene to disease terms (Fig. 17C).
4. A list of additional pathways with which the genes are involved. This list toggles between pathway term to genes and gene to pathway terms (Fig. 17D). The terms and gene symbols of all three lists link to the appropriate report pages in RGD.
5. Some diagram pages also have a list of phenotype terms associated with the genes involved in the pathway. This list toggles between phenotype term to genes and gene to phenotype terms (Fig. 17E).
6. Below the gene lists there is a reference list of publications associated with the diagrammed pathway.
7. Below the references is an ontology graph that shows the diagrammed term and all its ancestor terms up to the root term.
8. Finally, at the bottom of the page is a link to download the pathway diagram into a user's instance of Pathway Studio.

The physiological pathway diagrams present systems level pathways, allowing users to view the interactions at the whole animal level or drill down to explore the complex networks of signals and responses. The Physiological Pathway Diagrams also link to the molecular and cellular pathways giving users a glimpse into how these pathways work together to build the physiological systems (Fig. 18).

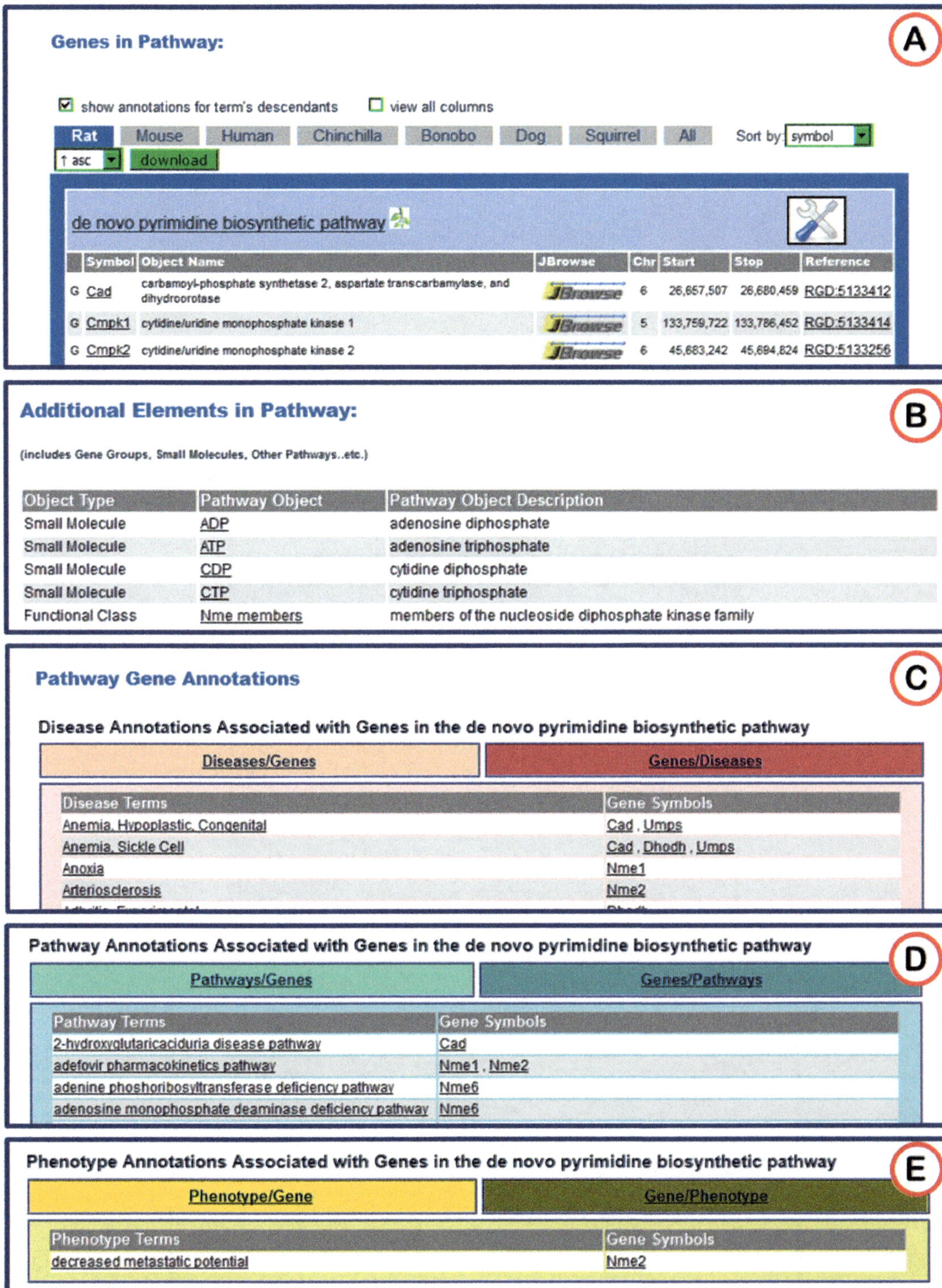

Fig. 17 Gene to Pathway and Gene to Disease Lists. A number of gene/term lists are found on pathway diagram pages below the diagram. (A) A list of genes annotated to "de novo pyrimidine biosynthetic pathway" and its child terms. The list includes links to RGD gene report pages, JBrowse, and reference pages. (B) A list of additional elements in pathway. (C) A list of disease terms/genes that can be toggled by the title bar to genes/disease terms. All the disease terms link to ontology report pages and the gene symbols link to gene report pages. (D) A list of additional pathways associated with genes annotated to the diagrammed pathway. (E) A list of phenotypes associated with the genes annotated to the diagrammed pathway

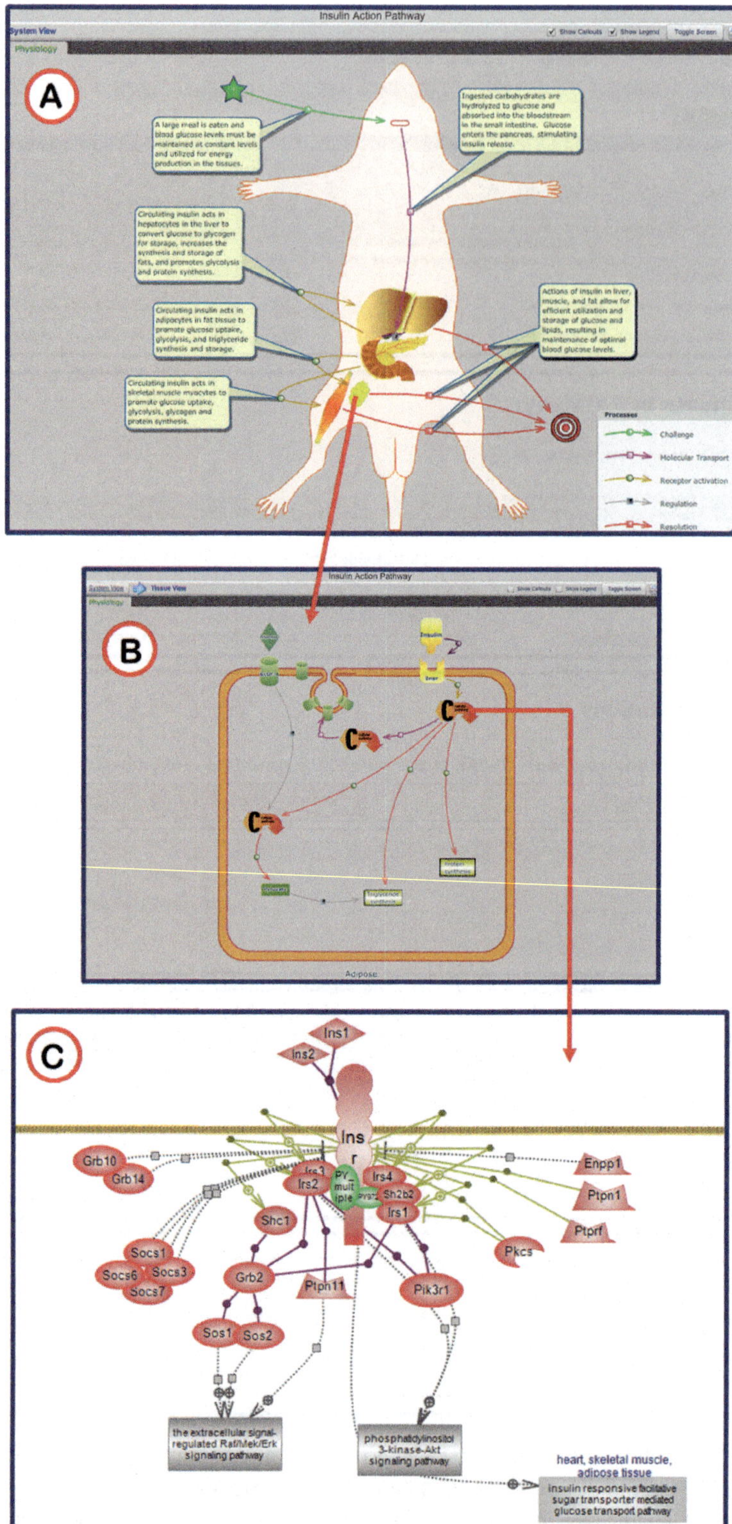

Fig. 18 Insulin Action Pathway Diagram. (A) The system view gives the overall view of insulin action. The callout pop-ups for the separate processes can be toggled on and off by way of the check box in the upper right. In off mode individual callouts will show if the cursor hovers over a circle or square in the center of a process arrow.

4 Data Analysis and Visualization Tools

4.1 Overview

Some of the data analysis tools at RGD are database-specific instances of freely available software. These include JBrowse, RatMine, and InterViewer (Cytoscape). The remaining analysis/visualization tools described in this section were developed at RGD: Gene Annotator, GViewer (Genome Viewer), OLGA (Object List Generator & Analyzer), PhenoMiner, and Variant Visualizer. They all provide different views or different types of analysis of the data in RGD. All of the tools may be accessed by the "Analysis & Visualization" icon in the middle of the RGD home page or the "Analysis & Visualization" tab near the top of most RGD pages.

4.2 InterViewer

InterViewer, RGD's Cytoscape-based (http://www.cytoscape.org/) [3] protein–protein interaction viewer, takes one or more gene symbols, RGD gene IDs, or UniProtKB protein IDs for rat, mouse, human, and/or dog (Fig. 19) and displays pairwise protein interactions for them, with information about the types of interactions and links to the associated genes in RGD and the originating interaction records at IMEX [4, 5].

1. A basic InterViewer results display is shown in Fig. 19. The results page features an interactive graphic display (linked to interactive thumbnail display), a list of interactions, detail/control options, and a legend for the graphic display.
2. Clicking on any circle in the display generates a detail box in the details/control frame, which gives information about the protein and provides a link to the UniProt page for that protein. Simultaneously a pop-up appears with the UniProt link and a link to the corresponding gene page at RGD.
3. The interaction edges between circles can also be clicked. Again a detail box appears with information about the specific interaction. A link to the PubMed source(s) of information is included.
4. Options to print the graphic display are available via links in the upper right hand of the page. A download link for the interactions list is available at the upper right side of the list.

Fig. 18 (continued) Organs may be identified by hovering the cursor over the organ image. (B) Detailed cellular pathways are visualized by clicking on the organ images (red arrow). As in the organismal view, hovering over parts of the diagram with reveal pop-up labels. (C) Individual molecular pathways are accessed by clicking any of the "Cellular pathway" icons in the cellular diagram (red arrow)

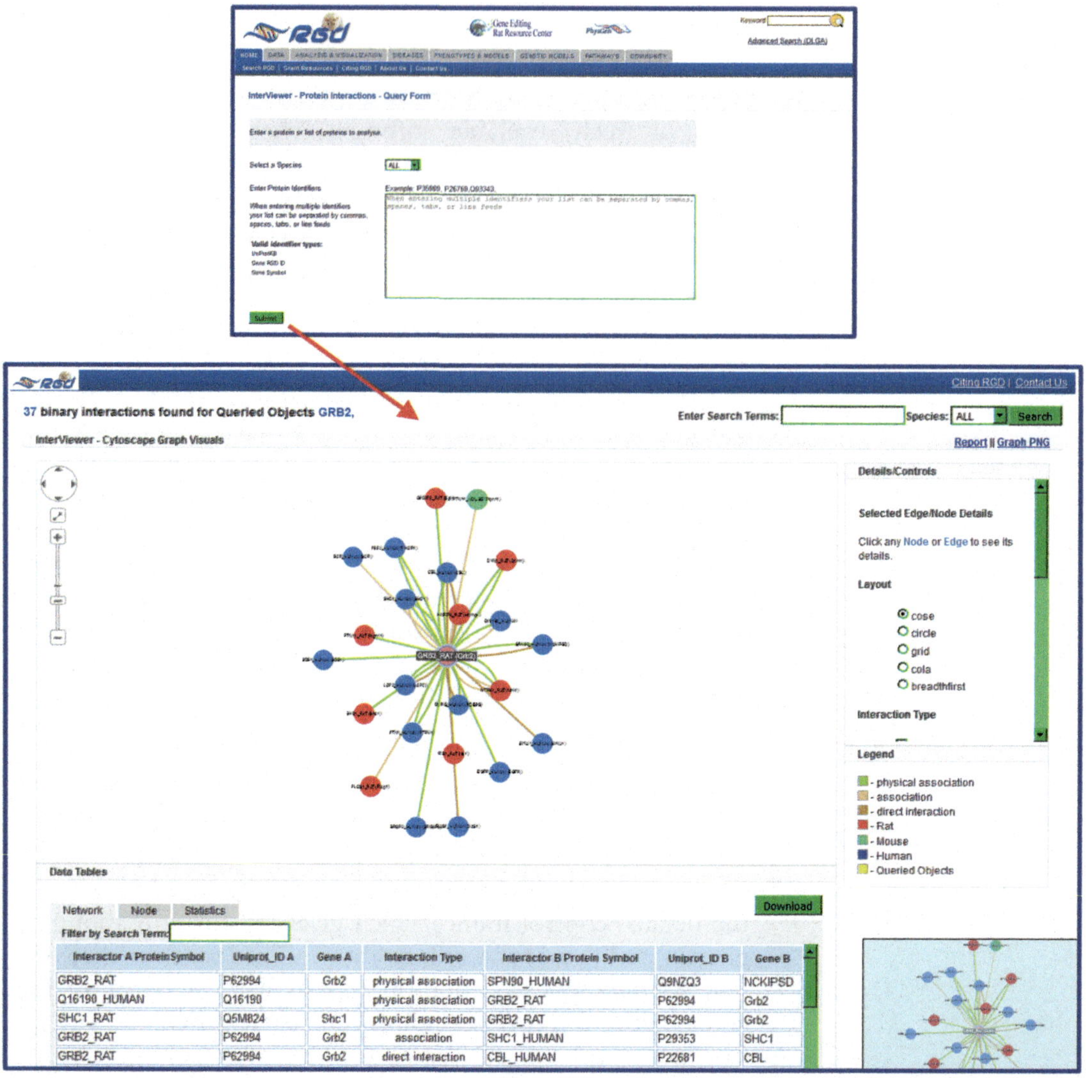

Fig. 19 Interviewer Search/Results. The target protein (rat Grb2) that initiated the search is shown in the center of the graphic display. Individual proteins are indicated by color-coded circles (red—rat, green—mouse, blue—human). The types of interactions are designated by color-coded lines between the circles

4.3 JBrowse

The JBrowse genome browser [6–8] from the Generic Model Organism Database project (http://www.gmod.org) is an interactive tool which allows researchers to visualize a variety of genetic and phenotypic data types in their genomic context. Virtually all of the data within the Rat Genome Database have been associated with the genome sequence in one way or another. As fundamental datasets such as genes, quantitative trait loci, microsatellite and SNP markers, and sequence resources such as ESTs, are aligned with the genome sequence, they bring with them phenotypic and

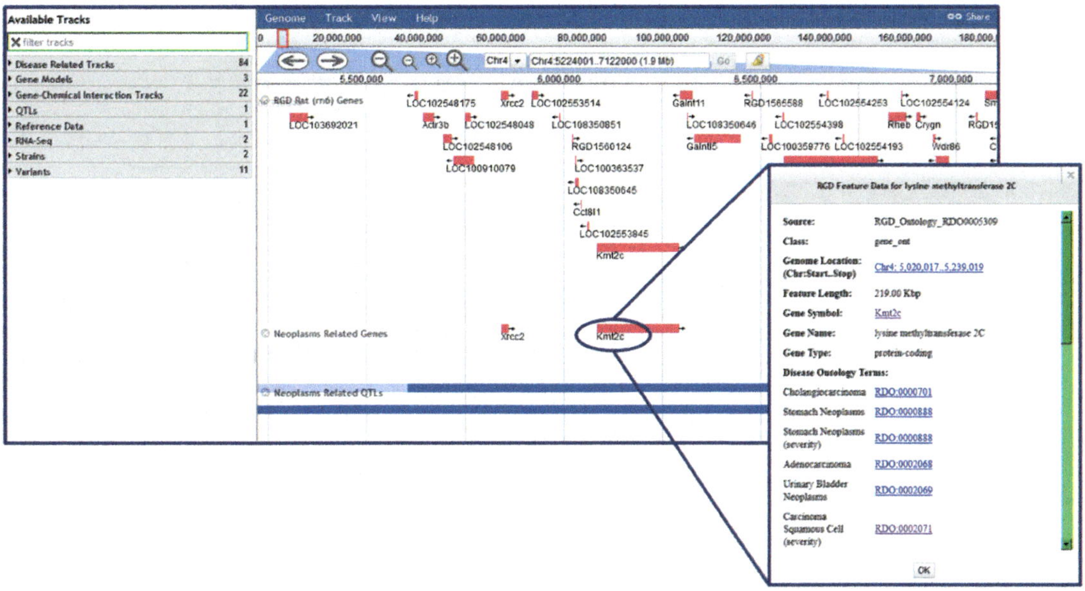

Fig. 20 JBrowse Display. Display shows RGD genes on chromosome 4 and neoplasm-related genes and QTLs. Pop-up shows details for the Kmt2c gene

other information. This information includes gene-chemical interaction data, genetic associations with disease, RNA-Seq data, synteny views of rat, mouse, and human genomes, and many types of variant/mutation data. Any or all of these can be accessed via the JBrowse genome browser and their relationship to the genomic sequence explored.

Any data object in JBrowse can be clicked to access a pop-up window containing relevant information for that object (Fig. 20). Included in that information are links to the RGD report page for that object, the annotation report page, an RGD data analysis page, or external database page with relevant information.

To allow comparison across species, RGD provides instances of JBrowse for mouse, human, and other species in addition to the rat version, which includes assemblies Rnor 6.0, Rnor 5.0, and Rnor 3.4. Multiple assemblies are provided for mouse (GRCmv37, GRCmv38) and human (GRChv36.3, GRChv37, and GRChv38.7) to make it easy to look at data that was generated for a specific genome assembly. The rat JBrowse v3.4 includes synteny tracks for rat v5.0, mouse v37, human v36, and human v37 (Fig. 21). The corresponding mouse and human browsers contain the reciprocal synteny tracks.

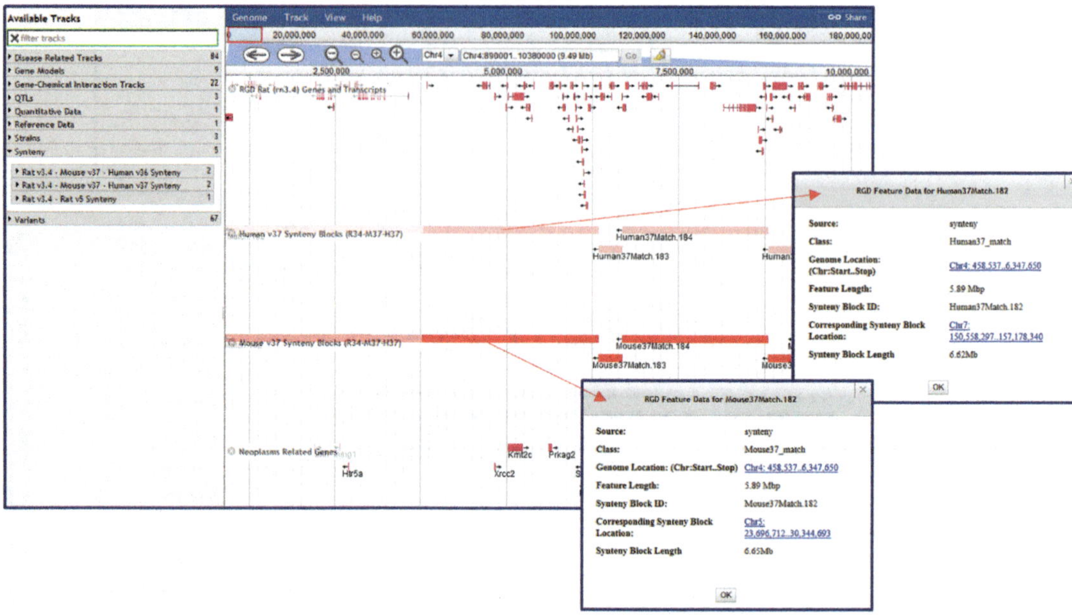

Fig. 21 JBrowse Synteny Display. Display shows RGD genes on chromosome 4 and neoplasm-related genes aligned with human and mouse synteny tracks. Pop-up windows (indicated by red arrows) show details of human and mouse synteny blocks that match the rat chromosome segment containing the Kmt2c gene, among other neoplasm-related genes

4.4 RatMine

RatMine (Fig. 22) integrates data on function, disease, phenotype, variation, and comparative genomics from RGD, UniProtKB (http://www.uniprot.org/), Ensembl (http://www.ensembl.org), NCBI (https://www.ncbi.nlm.nih.gov/), PubMed (https://www.ncbi.nlm.nih.gov/pubmed) and KEGG (http://www.genome.jp/kegg/) to form a web-based data warehousing, mining and analysis tool tailored to the needs of rat researchers. Datasets derived from querying this data or from uploading researchers' own data can be saved, manipulated and/or downloaded for use in other applications.

RatMine also has interaction datasets imported from BioGrid (Biological General Repository for Interaction Datasets) (https://thebiogrid.org) [9] and IntAct (http://www.ebi.ac.uk/intact) [5]. The BioGRID database manually curates the biomedical literature for genetic, protein, and chemical interaction data for major model organisms and humans. IntAct is a molecular interaction database that provides data derived from literature curation or direct user submissions to IntAct.

A key component of RatMine and of InterMine instances in general is the "MyMine" feature (Fig. 22A). Logging in as a

Fig. 22 RatMine Home Page. (A) The tab for MyMine, a personalizing feature of RatMine, a feature allowing saving of queries and datasets within RatMine. (B) RatMine data is accessible by API

specific user allows one to keep object lists (genes, etc.), user-created queries, and a history of activity. An API (application program interface) allows queries to run in RatMine from various web-based programs (Perl, Python, Ruby, or Java) (Fig. 22B).

4.5 Gene Annotator

The Gene Annotator (GA) takes a list of gene symbols, RGD IDs, GenBank accession numbers, Ensembl identifiers, and/or a chromosomal region, and retrieves annotation data from RGD. The tool will retrieve annotations from most ontologies used at RGD for genes and their orthologs, as well as links to additional information at other databases. The entry page (Fig. 23A) is very similar to the InterViewer entry page.

1. The first GA page after a search is an annotation/external link/species selection page where everything is selected by default (Fig. 23B).

Fig. 23 The Gene Annotator (GA) Tool. (A) The start page of the GA tool with large text box for entry of gene lists or object identifiers. (B) The results selection page where category of ontology annotations, external links, and orthologs may be chosen. (C) Results page listing annotations and external links for the rat, human, and mouse A2M genes. (D) Annotation distribution shows what percentage of the gene list is annotated to lists of terms in various ontologies/vocabularies. (E) The Comparison Heat Map shows an interactive matrix comparing numbers of annotated genes between terms of the disease vocabulary and the pathway ontology

2. Clicking the submit button returns a page with all annotations for the first gene (and selected orthologs) in the list. The lists include links to RGD gene pages, ontology term pages, annotation pages, and external data pages (Fig. 23C).
3. A list of links at the top of the page allows the user to pick a particular type of analysis to view (Annotation Distribution or Comparison Heat Map) or to send the gene list to another tool by selecting the "All Analysis Tools" link.
4. On the "Annotation Distribution" page (Fig. 23D) there are enrichment lists of terms by category, which rank the terms according to how many of the searched genes are annotated to those particular terms. Each entry in the list can be opened to see which genes and which specific terms are in the annotations. Subsets of annotations can be displayed by selecting at least two of the check boxes which appear to the right of every term in the lists.
5. The "Comparison Heat Map" (Fig. 23E) compares the number of genes annotated to selected terms of two different ontologies by displaying how many genes are annotated to specific terms in one ontology while also being annotated to the selection of terms in the other ontology. The default view shows annotations to disease terms versus annotations to pathway terms. The drop-down menus on the left side of the page allow the ontology to be changed on either axis of the heat map. Also, clicking on any of the hyperlinked terms on either axis of the heat map will reset that respective axis to include just child terms of the selected term. By clicking on any nonzero numbered square in the heat map, a list is returned of all genes annotated to both terms which cross at that spot in the matrix.
6. Finally, by selecting "All Analysis Tools," an "Analyze Gene List" pop-up appears (Fig. 24) with icons of various RGD analysis tools that link to those respective tools. Clicking a tool sends the gene list from the GA tool to the selected tool.
7. To download annotations retrieved by the tool at any stage in viewing results, the "This Gene" or "All Genes" links are available at the upper right corner of the results pages.

4.6 GViewer

The Genome Viewer (GViewer) provides users with a complete genome view of genes, QTLs, and congenic strains annotated to a molecular function, biological process, cellular component, phenotype, disease, or pathway. The tool will search for matching terms from the Gene Ontology, Mammalian Phenotype Ontology, RGD Disease Ontology or Pathway Ontology. The search page for GViewer (Fig. 25) features an autocomplete text box accompanied by an ontology selection section where any or all available

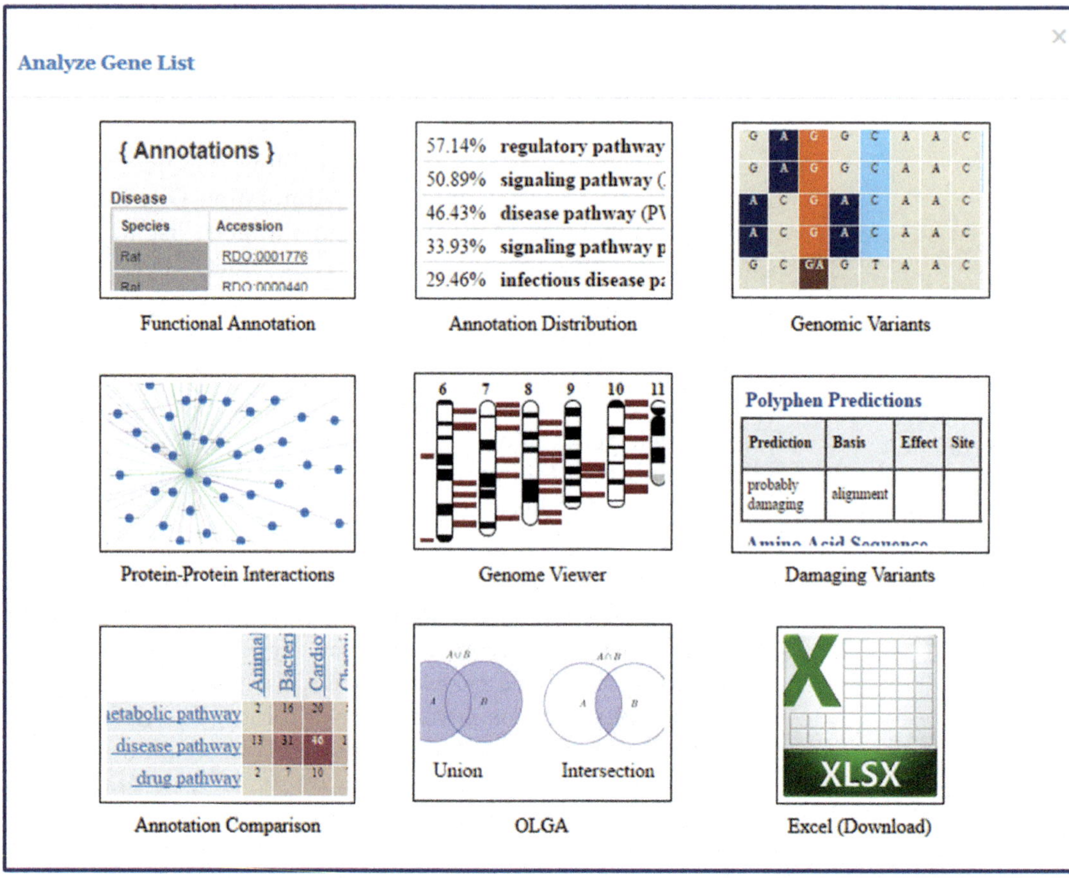

Fig. 24 "Analyze Gene List" Pop-up. A pop-up window with a set of icon links to various tools to which gene lists may be transferred. This pop-up is accessible in the GA tool, as well as in other RGD tools, on gene report pages, and on ontology term report pages

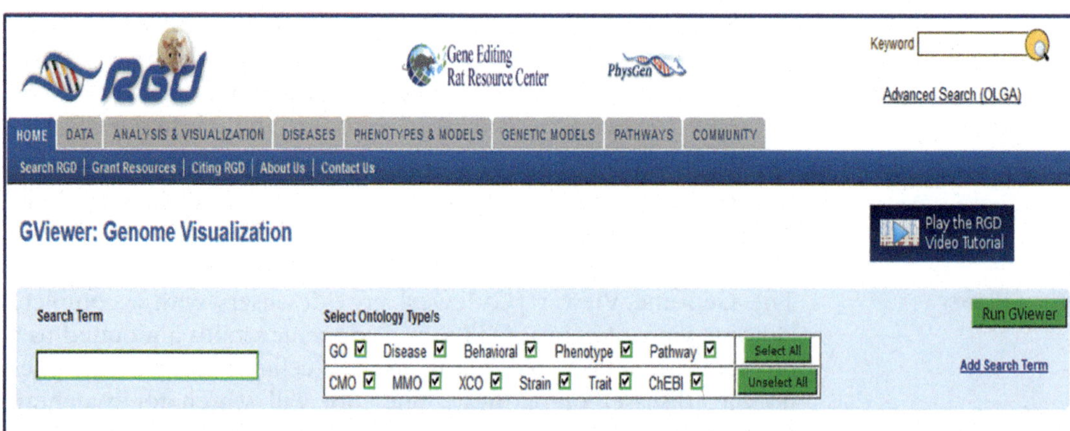

Fig. 25 GViewer Search Page. The GViewer search page has one text box and optional additional text boxes ("Add Search Term" link) with multiple ontology check boxes to restrict the search. A link to a tutorial video is available on the right side of the page

ontologies may be chosen. Complex searches may be made by clicking "Add Search Term" on the right side of the page. Each succeeding text box is accompanied by a drop-down menu with the Boolean-type choices of OR, AND, NOT. The returned results include all annotations to the chosen term and its child terms.

The main feature of the GViewer results page is the ideogrammic view of chromosomes with genes, QTLs, and congenic strains marked by color-coded bars in their approximate positions according to genome coordinates (Fig. 26). This GViewer graphic is used on every ontology term report page to visualize annotated objects. Below the chromosome view in the GViewer tool is a list of returned objects (alphabetical under type of object) and the searched ontology term(s), all of which link to the appropriate RGD report page. Clicking on any chromosome opens the zoom pane (Fig. 27A) that is displayed beneath the chromosome view. The zoom pane features a horizontal view of the chosen locus with a closer view featuring labeled genes, QTLs, and congenic strains. The zoom pane can be scrolled and zoomed.

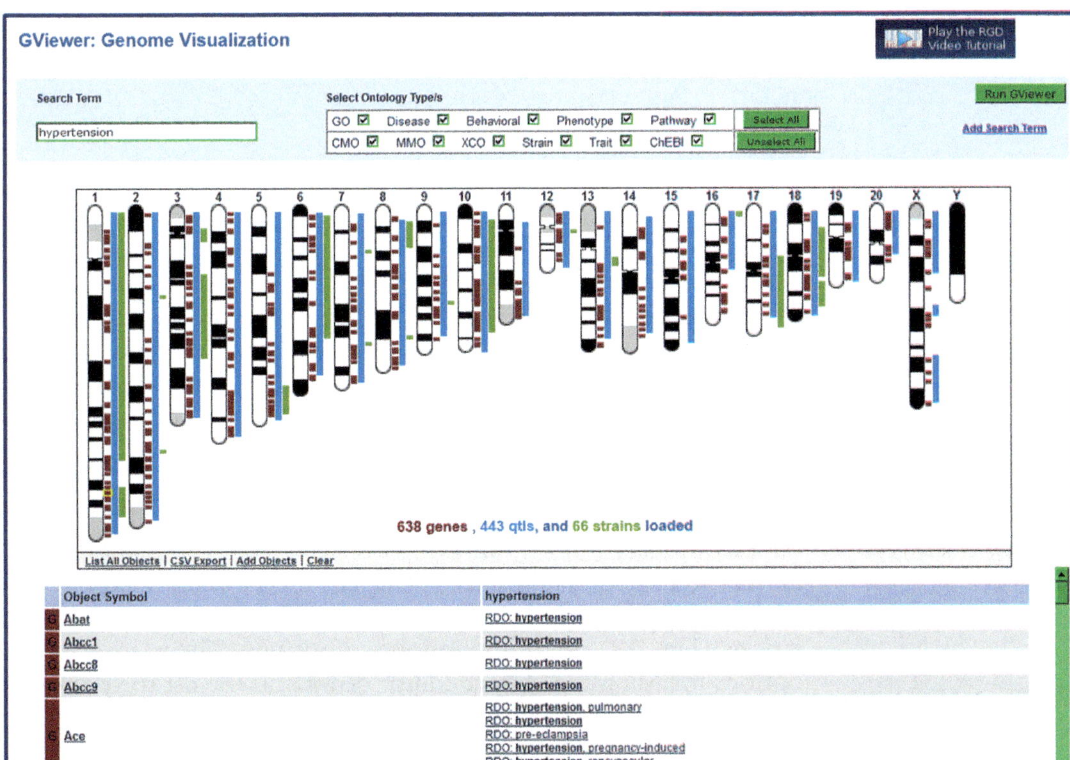

Fig. 26 GViewer Genome Visualization. The ideogrammic view of chromosomes with the location of returned genes marked in brown, QTLs in blue, and strains in green. A list of returned objects matched with their exact ontology term hit is shown below the genome display

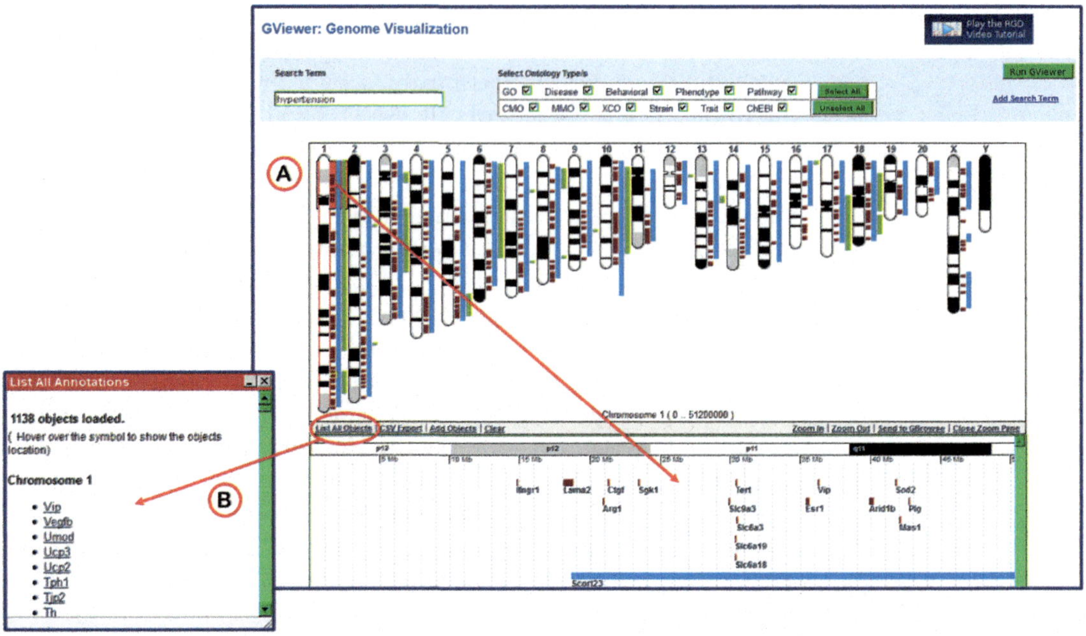

Fig. 27 GViewer Genome Visualization Functionality. (A) Clicking on the end of chromosome 1 opens the zoom pane to show a closer view of genes and QTL in the locus. (B) Clicking on "List All Objects" opens a pop-up window with a chromosome by chromosome list of all objects returned by the GViewer search

Alternate views of the data and other options are available on the bottom menu of the chromosome display. "List all objects" displays a pop-up window (Fig. 27B) to show a chromosome by chromosome list of annotated objects. All of those symbols in the list link to a close-up view of the object in the zoom pane. Other options on the chromosome view bottom menu are links for data download (CSV export) and an "Add Object" function which allows the user to add any specific gene, QTL, or congenic strain that isn't already displayed.

4.7 OLGA: Object List Generator & Analyzer

OLGA, the Object List Generator & Analyzer tool, is a list builder for rat, mouse, human, and other species genes, rat, mouse, or human QTLs, or rat strains, using any of a variety of querying options. Find objects in RGD using any of RGD's functional annotations (Fig. 28), or using genomic positions. There are various options for using the generated list: download the list to your own computer for analysis or loading into other tools, send the list to RGD's GA Tool for functional analyses, InterViewer for interaction analysis, GViewer for genomic visualization, or for rat and human, send the list to the Variant Visualizer to find sequence variants.

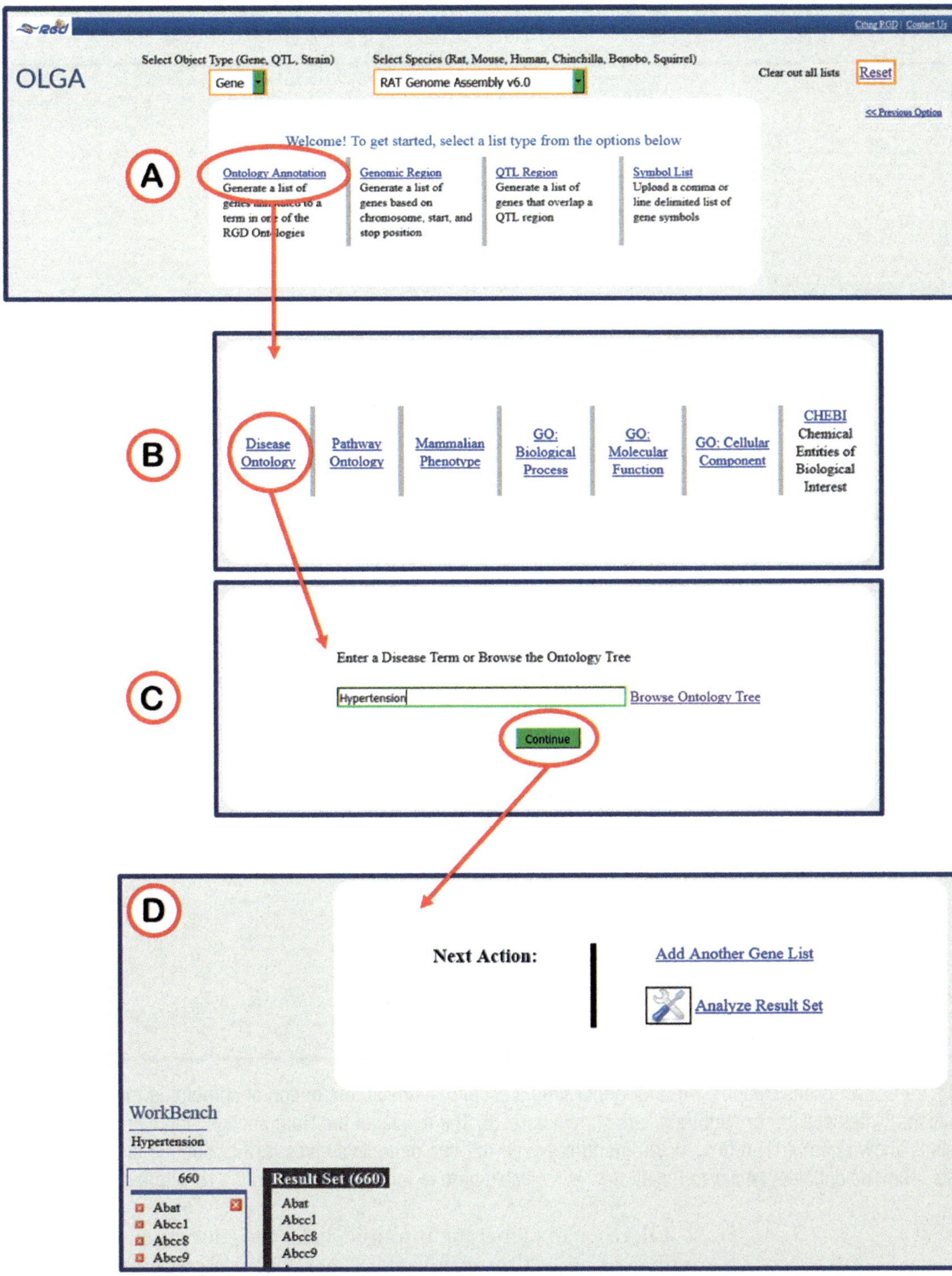

Fig. 28 Object List Generator & Analyzer (OLGA) tool. (A) OLGA selection screen for object, species/assembly, ontology, and genomic region. (B) Ontology/vocabulary selection screen. (C) Autocomplete text entry box. (D) Results page with links to "Add Another Gene List" (which adds another A, B, and C to the search process) and the "Analyze Result Set" pop-up window

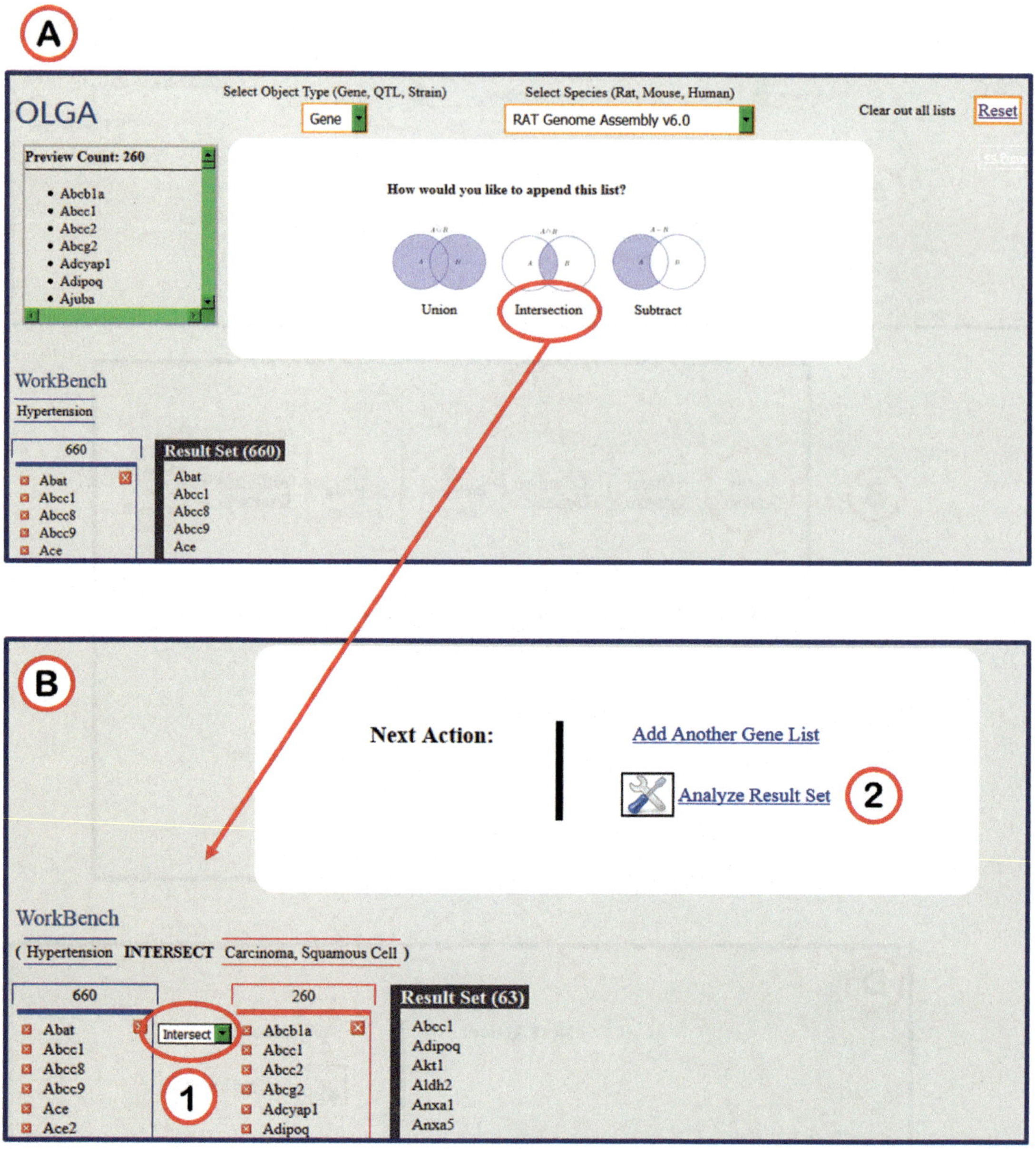

Fig. 29 OLGA Results Options. (A) After generating a second gene list, the option of combining the two lists by "Union," "Intersection," or "Subtract" are shown here. (B) The results of the "Intersection" function on two gene lists is shown here. (B1) A drop-down menu between the two gene lists gives further options to compare the lists. (B2) The option of "Analyze Result Set" is available here to further manipulate the resultant set of genes

OLGA can generate multiple lists and integrate them in a number of ways. OLGA gives the user a choice of union, intersection, or subtraction for combining a second list with a previously generated list (Fig. 29A). A drop-down menu between results lists (Fig. 29B1) gives the user an easy way to switch to another integration option. After two or more lists have been integrated, the result list may be transferred to another tool via the "Analyze Result Set" link (Figs. 24 and 29B2).

4.8 PhenoMiner

The purpose of the PhenoMiner tool is to integrate phenotypic data from different rat strains, collected by a variety of measurement methods under various experimental conditions. The data in PhenoMiner is comprised of results from the PhysGen Program for Genomic Applications (http://pga.mcw.edu) [10], the National BioResource Project—Rat (http://www.anim.med.kyoto-u.ac.jp/nbr/) (a large scale phenotyping project which has collected data for inbred rat strains) [11], and manual annotation from the rat physiological literature.

The PhenoMiner start page (Fig. 30A) features a choice of rat strains, clinical measurements, measurement methods, or experimental conditions to begin a search for quantitative phenotypic data.

1. Selecting one of the four parameters returns a search box and a list of specific options (Fig. 30B). For example the "rat strains" selection involves a simple search and/or drilling through the rat strain nomenclature in a vocabulary tree.
2. After a selection is made, a tally of results is made (Fig. 31A1) and the other three categories remain as options (Fig. 31A2). The selection of terms from the clinical measurement ontology, experimental conditions ontology, and the measurement methods ontology are all set up similarly to the strain selection. Each consecutive selection limits the remaining selections based on what strains were measured for what parameter, by which method, and under what condition.
3. A running tally of results obtained at each step of the query building process is provided. "Generate Report" may be clicked at any point after at least one category has been selected (Fig. 31A3).
4. All the data available for the selected parameters is displayed or made accessible on the PhenoMiner report page (Fig. 31B). The top of the report page has a series of drop-down menus and check boxes that correspond to the item choices made on the previous pages and to the data displayed in the bar graph in the center of the page. The presentation of the data can be manipulated by altering the choices in the drop-down menus or the check boxes.
5. The data is also presented in table form below the bar graph (Fig. 32). The table is sortable on any column. There are also options of download, expanded form, and help at the top of the table.

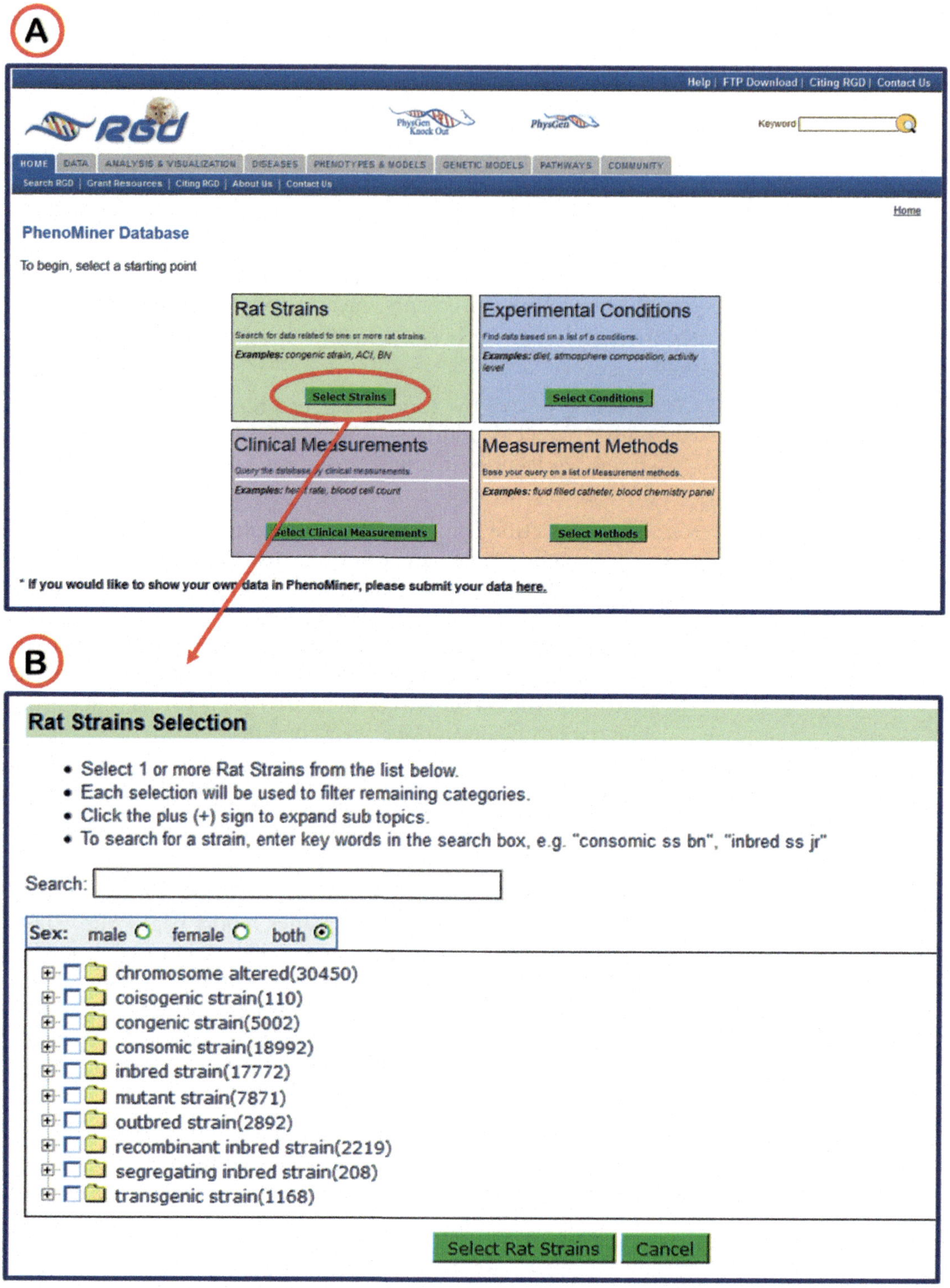

Fig. 30 PhenoMiner Start Page. Selection of "Rat Strains" (A) opens a new window (red arrow) with a text search box and a browsable vocabulary tree (B)

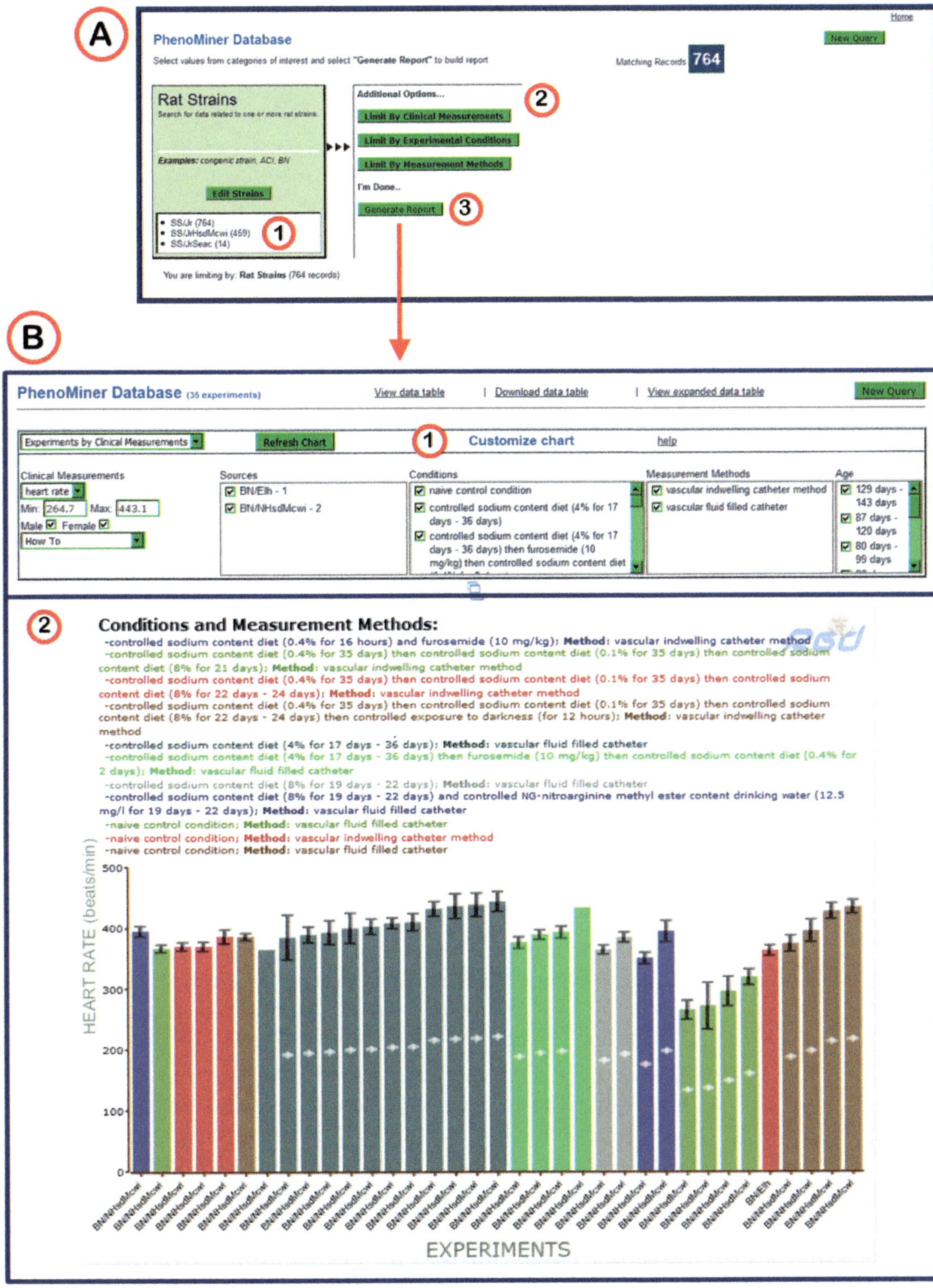

Fig. 31 PhenoMiner Selection Process and Report Page. (A) Intermediate selection page with scoreboard (1) of selected strains, more limiting options (2), and a "Generate Report" link (3). (B) PhenoMiner Report Page with customizable result options (1), featuring a color-coded bar graph (2) with all selected strains, methods, and conditions

Options: View chart | Download data table | View expanded data table | Which do I want

# of Animals (Click to Sort)	Clinical Measurement (Click to Sort)	Strain (Click to Sort)	Sex (Click to Sort)	Value (Click to Sort)	Units (Click to Sort)	Condition 1 (Click to Sort)	Condition 2 (Click to Sort)	Condition 3 (Click to Sort)	Condition 4 (Click to Sort)
21	heart rate	BN/Elh	male	362.0	beats/min	naive control condition	-	-	-
8	heart rate	BN/NHsdMcwi	male	408.0	beats/min	controlled sodium content diet (4% for 17 days - 36 days)	-	-	-
8	heart rate	BN/NHsdMcwi	female	409.4	beats/min	controlled sodium content diet (4% for 17 days - 36 days)	-	-	-
1	heart rate	BN/NHsdMcwi	male	432.8	beats/min	controlled sodium content diet (4% for 17 days - 36 days)	then furosemide (10 mg/kg)	then controlled sodium content diet (0.4% for 2 days)	-
3	heart rate	BN/NHsdMcwi	female	388.8	beats/min	controlled sodium content diet (4% for 17 days - 36 days)	then furosemide (10 mg/kg)	then controlled sodium content diet (0.4% for 2 days)	-
4	heart rate	BN/NHsdMcwi	male	384.8	beats/min	controlled sodium content diet (4% for 17 days - 36 days)	-	-	-
8	heart rate	BN/NHsdMcwi	female	431.1	beats/min	controlled sodium content diet (4% for 17 days - 36 days)	-	-	-
5	heart rate	BN/NHsdMcwi	male	375.5	beats/min	controlled sodium content diet (4% for 17 days - 36 days)	then furosemide (10 mg/kg)	then controlled sodium content diet (0.4% for 2 days)	-
4	heart rate	BN/NHsdMcwi	female	392.7	beats/min	controlled sodium content diet (4% for 17 days - 36 days)	then furosemide (10 mg/kg)	then controlled sodium content diet (0.4% for 2 days)	-
4	heart rate	BN/NHsdMcwi	male	399.8	beats/min	controlled sodium content diet (4% for 17 days - 36 days)	-	-	-
9	heart rate	BN/NHsdMcwi	female	443.1	beats/min	controlled sodium content diet (4% for 17	-	-	-

Fig. 32 PhenoMiner Results Table. This is located below the column chart on the PhenoMiner Report Page. Options for download and table expansion are shown at the top of the table. The table is sortable by clicking the title of any column and all of the strain symbols link to the respective strain report pages

4.9 Variant Visualizer

Variant Visualizer is a viewing and analysis tool for rat strain-specific sequence polymorphisms and human ClinVar variants. Select rat strains or human assembly of interest, define one or more genomic regions and, if desired, set parameters for the type(s) of desired variants and the tool will return all of the single nucleotide variants (SNVs) which match the input criteria, including information on read depth, zygosity, conservation score and more.

1. Beginning on the main page of the tool (Fig. 33), the user selects a genome assembly, followed by selection pages to limit output by strain, gene, genomic position, or function of gene product (link to OLGA).
2. The user then chooses what kind of variants to see in the graphic display of results. This includes choices of type of variant, genomic feature, potential effect of variant upon protein product, and statistics involved in the variant calls (Fig. 33D).
3. The graphic display of results shows a horizontal view of DNA sequence of strain/assembly compared to reference sequence (Fig. 34). Variants are identified by chromosome coordinate and base designation. It is easy to compare many rat strains simultaneously since the variants are stacked up on each other based on chromosome coordinate in a scrollable display frame. For details of any variant, clicking on the base will open a pop-up window with details (Fig. 34C).

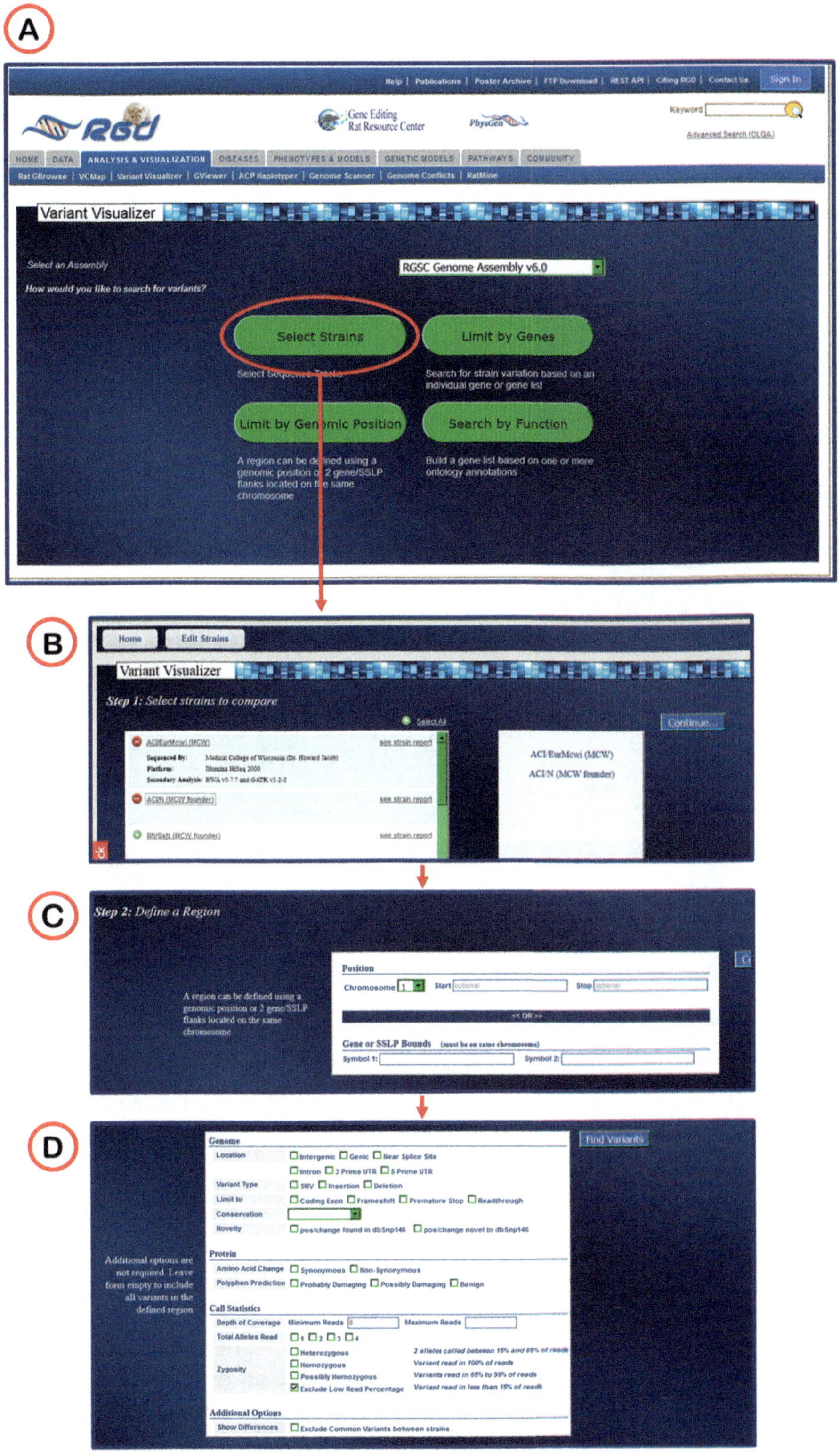

Fig. 33 Variant Visualizer Selection Screens. (A) The Variant Visualizer home page with drop-down menu for assembly selection and selection buttons for strain (red oval), gene(s), genomic position, and function. (B) Strain selection page listing all strains annotated to the selected assembly. The first two strains in the list have been selected and appear in the box on the right. (C) Region/Genomic position selection page with drop-down menu for chromosome and text boxes for start/stop coordinates. (D) Selection page for filtering on variant type, variant location, variant at protein level, and call statistics for variants

Fig. 34 Variant Visualizer Results Display. (A) The reference sequence of selected region. (B) Selected Strains. (C) Pop-up window with details of variant (red oval) selected from result display. (D) Options of editing prior selections of strain, gene, annotation, or coordinates

4. From the sequence display page optional views and links to further analysis are available from a pop-up selection box accessed by a link in the upper right corner of the display page (Fig. 35). The options include an overview plot of the data, a distribution graph of the data, a link to the GA tool (Gene Annotator—*see* Subheading 4.5) for functional analysis of the selected region, a download link for the data, and help documentation.

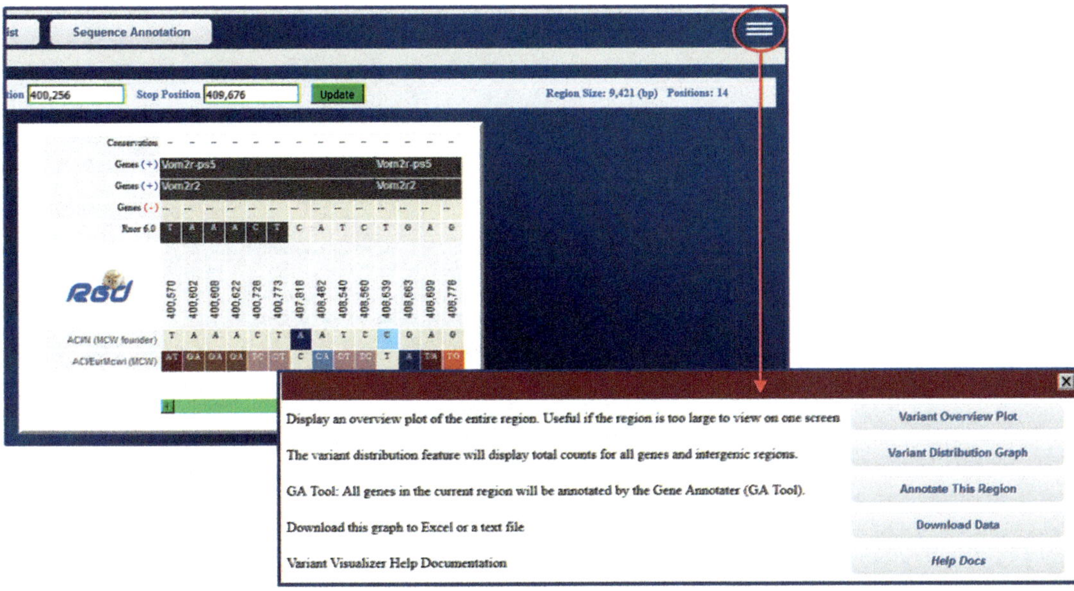

Fig. 35 Further Options on Variant Visualizer Results Page

4.10 VCMap

The Virtual Comparative Map (VCMap) tool was originally developed at RGD to explore the syntenic relationships between rat, mouse, and human genomes. A newer version of VCMap has been developed by a collaboration of Iowa State University, University of Iowa, Medical College of Wisconsin and RGD. The tool was updated by Bio::Neos (http://bioneos.com/). The current VCMap expands both the versatility and utility of this valuable tool by incorporating pig, chicken, cow, and horse genomes. Access to the tool is available through Chrome, Firefox, and Internet Explorer web browsers through a Java applet. To run the applet an exception in the Java security settings should be added for http://www.animalgenome.org, the site which hosts VCMap.

The link on the RGD tools page connects to the VCMap hosted at http://www.animalgenome.org/ (Fig. 36). The Start button launches the Java applet followed by a selection window with a choice of species, map choice (genomic, radiation hybrid, and genetic), map version (RGSC_v3.4 for rat), and chromosome (Fig. 36B). After the anchor species is chosen, additional species may be successively chosen for comparison (Fig. 37A) by using the "load" function under the "Map" menu. The chromosomes are displayed vertically, with the coordinates of the anchor chromosome increasing from top to bottom. The chromosome sections of the other species are arranged such that orthologs and syntenic regions line up, as much as possible, along the horizontal axis. Using the zoom slider on the left side of the display window and

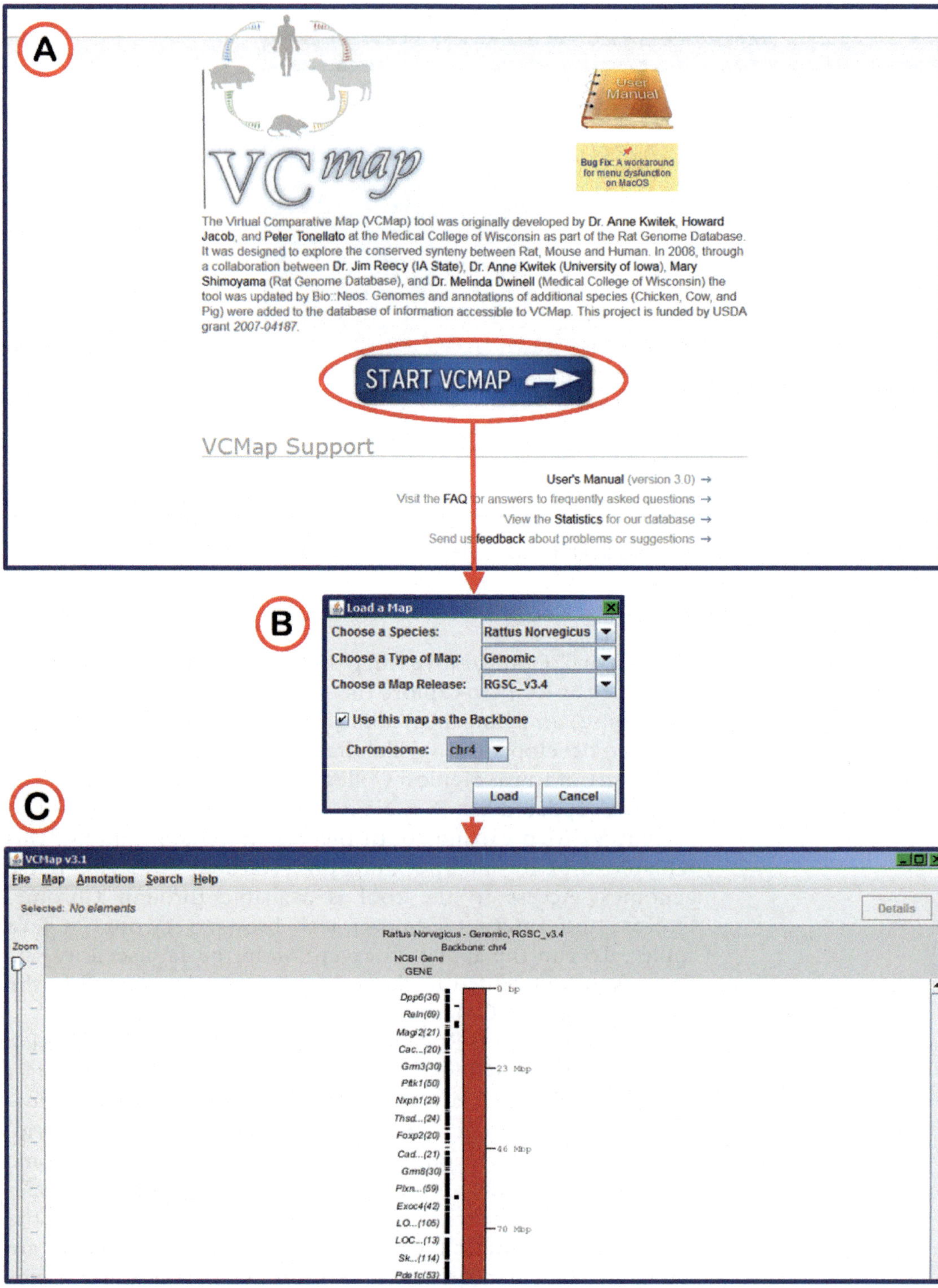

Fig. 36 VCMap Start Page and Map Selection. (A) VCmap start page. (B) Pop-up map selection window for choosing species, type of map, map release, and chromosome. (C) Result display for rat chromosome 4

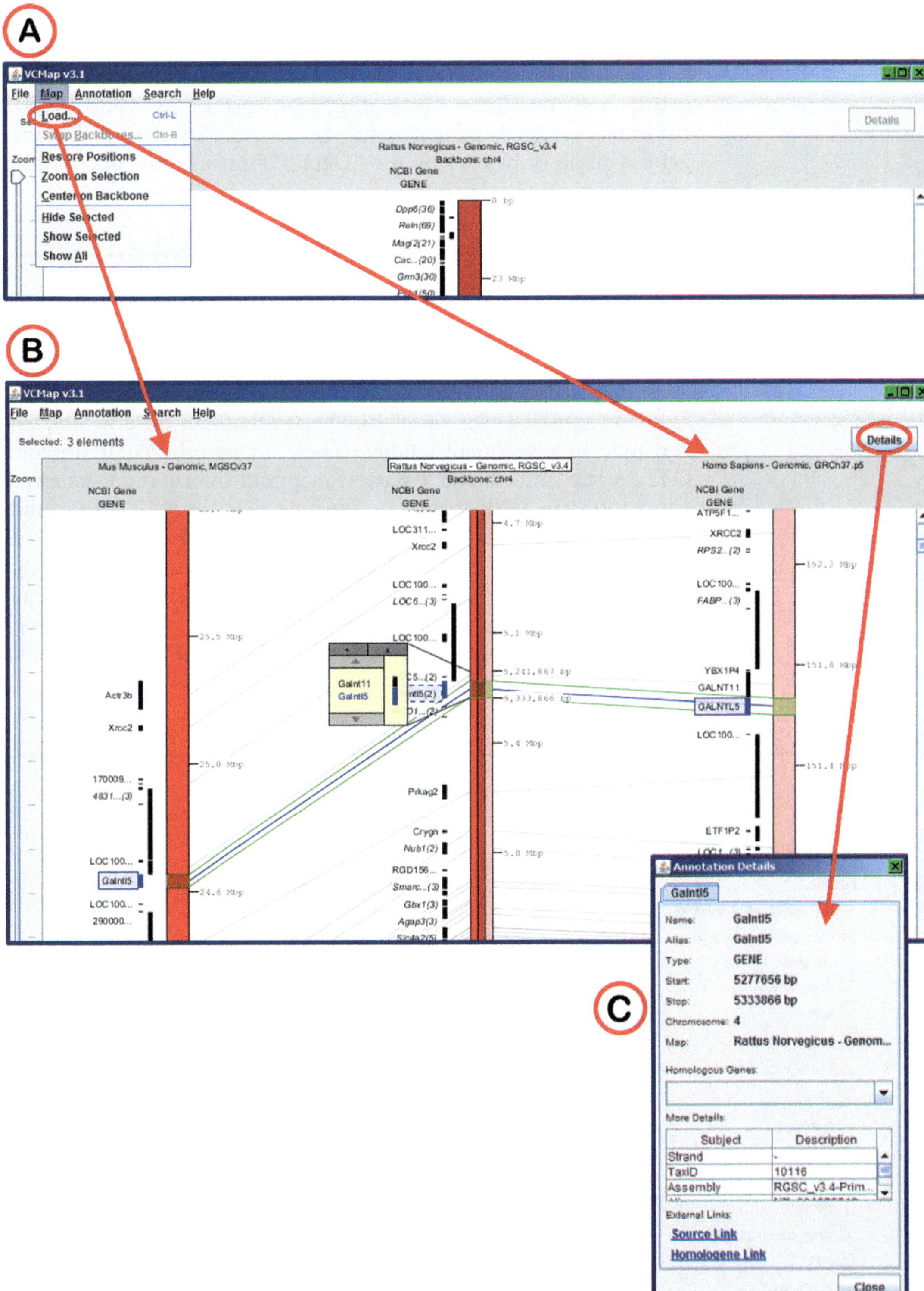

Fig. 37 Using VCMap for Interspecies Comparisons. (A) Choosing mouse and human maps for comparison to rat. (B) Display of mouse and human homologous chromosomes as compared to rat. The Galntl5 gene orthologs have been selected and are highlighted in blue across the three species. (C) Annotation details pop-up window for rat Galntl5

the scroll bar on the right side of the display, specific genes may be viewed. By clicking on a specific gene symbol in the display, highlighting appears on that symbol, the ortholog symbols on the other maps, and on lines connecting the orthologs (Fig. 37B). Information details of genes may be accessed by double-clicking gene symbols or by clicking the "Details" button at the upper right when a gene symbol is highlighted.

5 Automated Access to RGD Data

RGD has facilitated bulk download and other automated access to curated and other data. Data is accessible by both FTP download (Fig. 38) and REST API (Fig. 39). The RGD FTP site maintains regularly updated files of all RGD data that can be downloaded and used in subsequent studies. These include the curated gene, QTL, strain and marker datasets, mapping information, genome annotation (in GFF format), sequence files for RGD data, and RGD-developed ontologies/vocabularies. The FTP site can be reached by clicking the "FTP Download" link found in the menu bar on the upper right of most RGD web pages. This link will lead to the FTP site (ftp://rgd.mcw.edu/pub/), where one can browse the files available for download. The "REST API" link may be found adjacent to the FTP link.

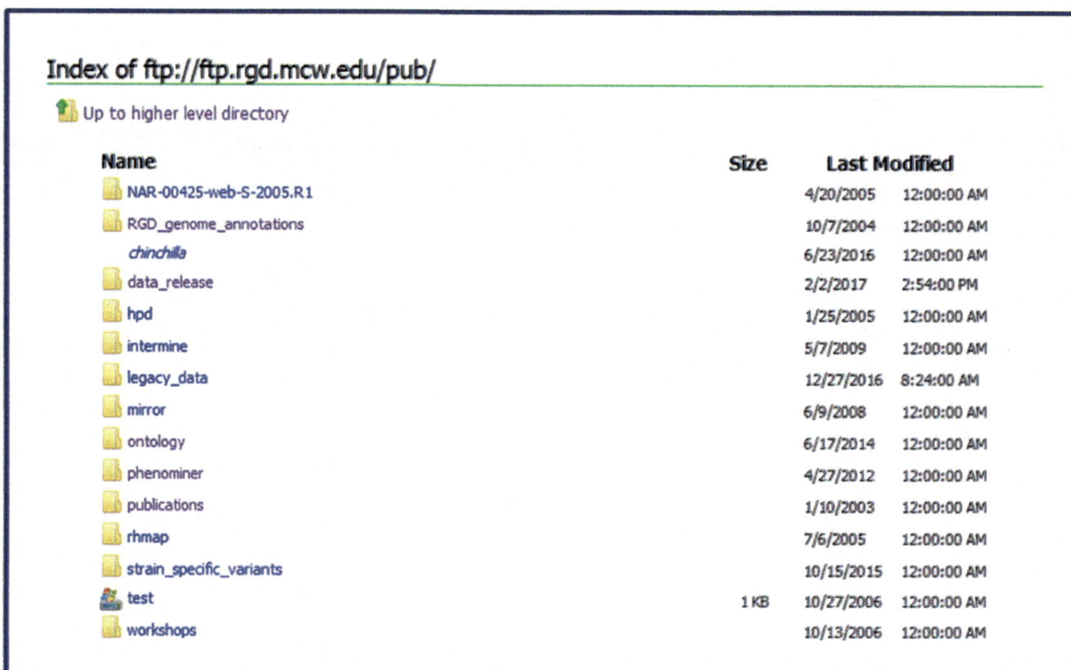

Fig. 38 FTP Download Page

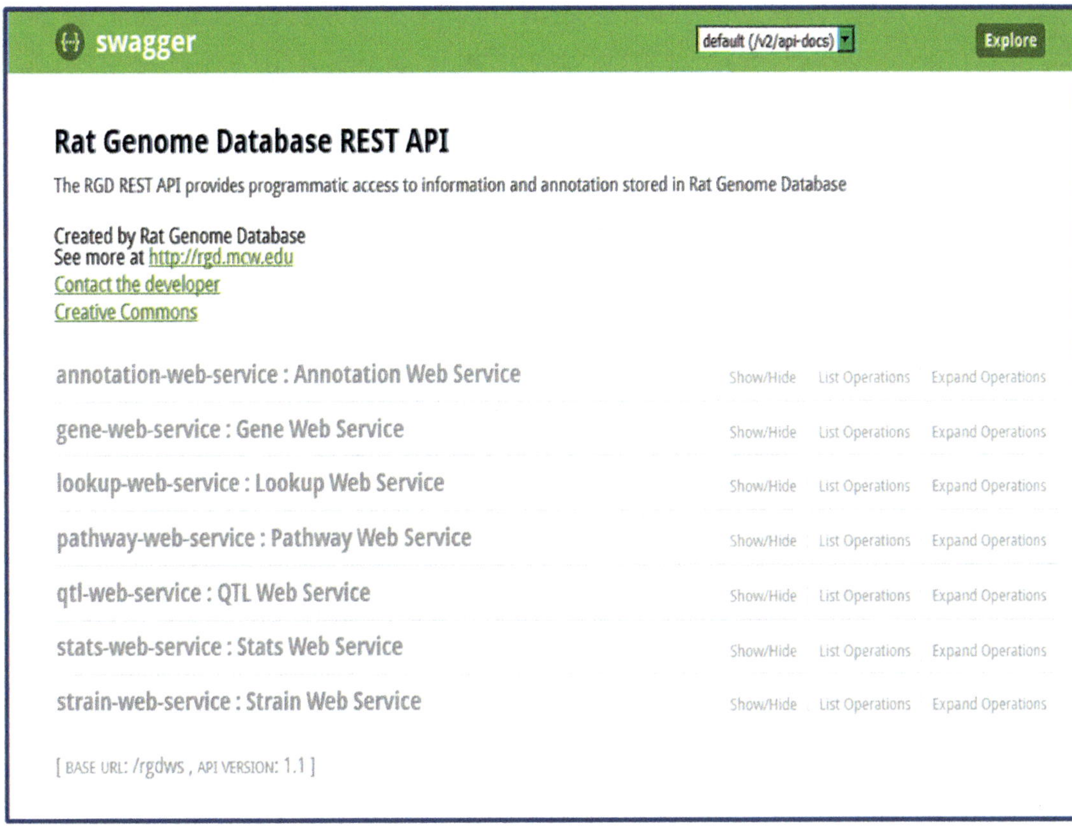

Fig. 39 REST API page

6 Summary

The Rat Genome Database was established in 1999 as a resource to support the emerging genomic data for the rat. This role has continued to expand with continuing work on the rat reference genome sequence (current assembly is Rnor_6.0—RGSC Genome Assembly v6.0), strain-specific DNA sequencing [12], expanded SNP discovery, and large-scale phenotyping projects such as the PhysGen project (http://pga.mcw.edu) and NBRP [11], all needing to be integrated with existing and newly published research data. As the amount of data has grown, so has the challenge of mining relevant information and defining its meaning in the broader context of biomedical science. With this in mind, much effort has gone into the development and incorporation of biomedical ontologies such as the Gene Ontology [13], the Mammalian Phenotype Ontology [14], the Pathway Ontology [2], and others [15]. These are incorporated into the search and analysis tools, greatly facilitating the discovery of information and interpretation of its meaning.

Many researchers using the rat as a model system are ultimately studying a specific phenotype or disease with the goal of applying this knowledge to humans. To meet this need, RGD has developed "disease portals" that present RGD data and tools from the perspective of a particular disease. The disease portals allow researchers to visit a single page that is focused on a single disease area like cardiovascular, neurological, or respiratory disease. These disease categories are being expanded in an ongoing process of targeted curation to create more portals devoted to particular disease areas that will cater directly to researchers working in those areas. The rest of RGD is accessible via these portals, but researchers will find the items of greatest interest first, reducing the challenge of finding the data and interpreting its meaning. Similarly, the Phenotypes & Models portal and the Pathways Portal focus on specific areas of research, which allows easier access to targeted searches for relevant data.

In addition to the portal style of data organization, the access to different software tools at RGD is an important part of the database. Ranging from annotation-based analysis to sequence-based analysis, the options are extensive to manipulate both RGD data and user-uploaded data. Further analysis may be done with downloaded data via the FTP site or the REST API.

Acknowledgments

RGD is supported by the National Heart, Lung, and Blood Institute on behalf of the National Institutes of Health [HL64541].

References

1. Shimoyama M, De Pons J, Hayman GT, Laulederkind SJ, Liu W, Nigam R, Petri V, Smith JR, Tutaj M, Wang SJ, Worthey E, Dwinell M, Jacob H (2015) The Rat Genome Database 2015: genomic, phenotypic and environmental variations and disease. Nucleic Acids Res 43(Database issue):D743–D750. https://doi.org/10.1093/nar/gku1026
2. Petri V, Jayaraman P, Tutaj M, Hayman GT, Smith JR, De Pons J, Laulederkind SJ, Lowry TF, Nigam R, Wang SJ, Shimoyama M, Dwinell MR, Munzenmaier DH, Worthey EA, Jacob HJ (2014) The pathway ontology - updates and applications. J Biomed Semantics 5(1):7. https://doi.org/10.1186/2041-1480-5-7
3. Shannon P, Markiel A, Ozier O, Baliga NS, Wang JT, Ramage D, Amin N, Schwikowski B, Ideker T (2003) Cytoscape: a software environment for integrated models of biomolecular interaction networks. Genome Res 13(11):2498–2504. https://doi.org/10.1101/gr.1239303
4. Orchard S, Kerrien S, Abbani S, Aranda B, Bhate J, Bidwell S, Bridge A, Briganti L, Brinkman FS, Cesareni G, Chatr-Aryamontri A, Chautard E, Chen C, Dumousseau M, Goll J, Hancock RE, Hannick LI, Jurisica I, Khadake J, Lynn DJ, Mahadevan U, Perfetto L, Raghunath A, Ricard-Blum S, Roechert B, Salwinski L, Stumpflen V, Tyers M, Uetz P, Xenarios I, Hermjakob H (2012) Protein interaction data curation: the International Molecular Exchange (IMEx) consortium. Nat Methods 9(4):345–350. https://doi.org/10.1038/nmeth.1931

5. Orchard S, Ammari M, Aranda B, Breuza L, Briganti L, Broackes-Carter F, Campbell NH, Chavali G, Chen C, del-Toro N, Duesbury M, Dumousseau M, Galeota E, Hinz U, Iannuccelli M, Jagannathan S, Jimenez R, Khadake J, Lagreid A, Licata L, Lovering RC, Meldal B, Melidoni AN, Milagros M, Peluso D, Perfetto L, Porras P, Raghunath A, Ricard-Blum S, Roechert B, Stutz A, Tognolli M, van Roey K, Cesareni G, Hermjakob H (2014) The MIntAct project—IntAct as a common curation platform for 11 molecular interaction databases. Nucleic Acids Res 42(Database issue):D358–D363. https://doi.org/10.1093/nar/gkt1115
6. Skinner ME, Holmes IH (2010) Setting up the JBrowse genome browser. Curr Protoc Bioinformatics 32:9.13.1–9.13.13. https://doi.org/10.1002/0471250953.bi0913s32
7. Westesson O, Skinner M, Holmes I (2013) Visualizing next-generation sequencing data with JBrowse. Brief Bioinform 14(2):172–177. https://doi.org/10.1093/bib/bbr078
8. Buels R, Yao E, Diesh CM, Hayes RD, Munoz-Torres M, Helt G, Goodstein DM, Elsik CG, Lewis SE, Stein L, Holmes IH (2016) JBrowse: a dynamic web platform for genome visualization and analysis. Genome Biol 17:66. https://doi.org/10.1186/s13059-016-0924-1
9. Chatr-Aryamontri A, Oughtred R, Boucher L, Rust J, Chang C, Kolas NK, O'Donnell L, Oster S, Theesfeld C, Sellam A, Stark C, Breitkreutz BJ, Dolinski K, Tyers M (2017) The BioGRID interaction database: 2017 update. Nucleic Acids Res 45(D1):D369–d379. https://doi.org/10.1093/nar/gkw1102
10. Dwinell MR (2010) Online tools for understanding rat physiology. Brief Bioinform 11(4):431–439. https://doi.org/10.1093/bib/bbp069
11. Serikawa T, Mashimo T, Takizawa A, Okajima R, Maedomari N, Kumafuji K, Tagami F, Neoda Y, Otsuki M, Nakanishi S, Yamasaki K, Voigt B, Kuramoto T (2009) National BioResource Project-Rat and related activities. Exp Anim 58(4):333–341
12. Hermsen R, de Ligt J, Spee W, Blokzijl F, Schafer S, Adami E, Boymans S, Flink S, van Boxtel R, van der Weide RH, Aitman T, Hubner N, Simonis M, Tabakoff B, Guryev V, Cuppen E (2015) Genomic landscape of rat strain and substrain variation. BMC Genomics 16:357. https://doi.org/10.1186/s12864-015-1594-1
13. Ashburner M, Ball CA, Blake JA, Botstein D, Butler H, Cherry JM, Davis AP, Dolinski K, Dwight SS, Eppig JT, Harris MA, Hill DP, Issel-Tarver L, Kasarskis A, Lewis S, Matese JC, Richardson JE, Ringwald M, Rubin GM, Sherlock G (2000) Gene ontology: tool for the unification of biology. The Gene Ontology Consortium. Nat Genet 25(1):25–29. https://doi.org/10.1038/75556
14. Smith CL, Goldsmith CA, Eppig JT (2005) The Mammalian Phenotype Ontology as a tool for annotating, analyzing and comparing phenotypic information. Genome Biol 6(1):R7. https://doi.org/10.1186/gb-2004-6-1-r7
15. Laulederkind SJ, Tutaj M, Shimoyama M, Hayman GT, Lowry TF, Nigam R, Petri V, Smith JR, Wang SJ, de Pons J, Dwinell MR, Jacob HJ (2012) Ontology searching and browsing at the Rat Genome Database. Database (Oxford) 2012:bas016. https://doi.org/10.1093/database/bas016

Chapter 9

Bovine Genome Database: Tools for Mining the *Bos taurus* Genome

Darren E. Hagen, Deepak R. Unni, Aditi Tayal, Gregory W. Burns, and Christine G. Elsik

Abstract

The Bovine Genome Database (BGD; http://bovinegenome.org) is a web-accessible resource that supports bovine genomics research by providing genome annotation and data mining tools. BovineMine is a tool within BGD that integrates BGD data, including the genome, genes, precomputed gene expression levels and variant consequences, with external data sources that include quantitative trait loci (QTL), orthologues, Gene Ontology, gene interactions, and pathways. BovineMine enables researchers without programming skills to create custom integrated datasets for use in downstream analyses. This chapter describes how to enhance a bovine genomics project using the Bovine Genome Database, with data mining examples demonstrating BovineMine.

Key words Bovine, Cattle, *Bos taurus*, Genome database, Genome annotation, Orthology, Pathway, Gene expression, Single nucleotide polymorphism, Data mining, InterMine, BovineMine

1 Introduction

The goal of the Bovine Genome Database (http://bovinegenome.org) is to support the efforts of bovine genomics researchers by providing easy access to the bovine reference genome assembly and annotations of genome features via a graphical genome browser, sequence database searching, data mining, and bulk data download [1]. Data provided by BGD includes annotation data gathered from external sources as well as computed data to improve genome annotation. BGD provides tools for manual annotation, viewing multiple genome assemblies, and exploring a tissue-specific gene expression atlas for both RefSeq and Ensembl genes. In addition, BGD hosts a data mining warehouse, BovineMine, which integrates tissue-specific gene expression levels, QTL and external sources of functional annotation, such as pathways and interactions, for fine-scale data mining and generation of custom data sets. BGD includes both the Ensembl

Martin Kollmar (ed.), *Eukaryotic Genomic Databases: Methods and Protocols*, Methods in Molecular Biology, vol. 1757, https://doi.org/10.1007/978-1-4939-7737-6_9, © Springer Science+Business Media, LLC, part of Springer Nature 2018

and RefSeq gene sets, and provides database cross reference tools to seamlessly convert from one to the other.

2 Methods

2.1 Website Navigation

The navigation bar of Bovine Genome Database provides access to all of the web-based informatics tools, which include data mining (BovineMine), BLAST searching, genome browsing (JBrowse), manual gene annotation (Apollo), various quick lookup tools and data download. Since the bovine reference genome has been revised several times, an "Assembly History" page provides information about previous and current reference genome assemblies to mitigate confusion.

2.2 BGD Lookup Tools

The Lookup Tools tab in the BGD navigation bar provides access to quick lookup tools to perform chromosome or gene identifier conversion or select RNA-seq JBrowse tracks that provide expression information for specific genes (Fig. 1A). These tools are actually very simple search interfaces that submit queries to BovineMine (described below) for quick lookups of single identifiers. BovineMine is recommended for more complex queries or searching a list of IDs. The Lookup Tools will change depending on needs associated with specific reference genome assemblies. For example, the current Annotation Assembly Tool, which provides locations of genes (Ensembl, RefSeq, OGSv2) on two bovine assemblies (UMD3.1 and Btau_4.6.1) will be updated upon the release of a new bovine reference genome assembly.

2.3 JBrowse, Apollo and the Faceted Track Selector

Genome browsing is provided using JBrowse [2] as implemented by Apollo [3]. From a user-perspective, the only differences between BGD JBrowse and BGD Apollo are the gene editing functions and the user annotation pane that are available only when logged into Apollo. The evidence tracks are identical across the browsers. All BGD users can access JBrowse, while only users registered for annotation can access Apollo. The remainder of this section will focus on BGD specific data and browser features. BGD JBrowse includes over 300 "tracks," so BGD implements a faceted track selector (Fig 1B, C) that is available by clicking the "Select Tracks" tab in the upper left corner of the browser (Fig. 1D). The track selector organizes tracks into the following data types: Gene Prediction, LiftOver Gene Prediction, Genome Assembly Gaps, Alternate Assembly Alignment, Protein Homolog Alignment, Microarray Probe Alignment, Variation, QTL Region, Repeat, Combined Cufflinks, and several different RNA-seq data types.

The Gene Prediction tracks are organized according to source database (e.g., RefSeq or Ensembl) and the gene type assigned by the source database (e.g., protein-coding, several noncoding gene types, pseudogenes, gene models with errors due to assembly

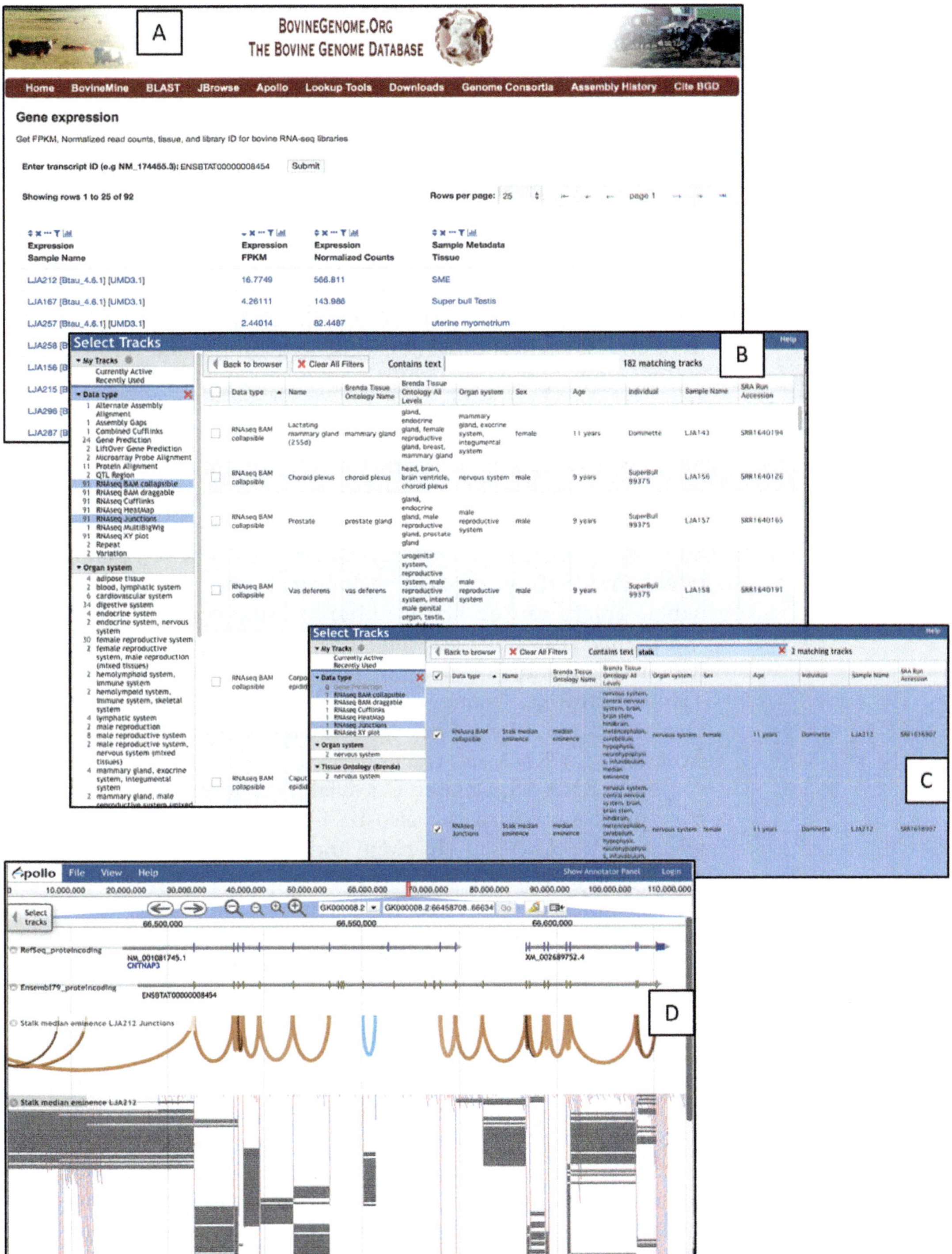

Fig. 1 (A) To assist users in selecting JBrowse tracks to visualize expression of a specific gene, BGD includes a quick lookup tool to retrieve precomputed expression levels based on RNA-seq The tool is available via the Lookup Tool pull-down menu of the BGD navigation bar. The lookup tool pulls data from BovineMine, where additional RNA-seq metadata can be retrieved. (B, C) Once a tissue or sample of interest is identified, the JBrowse faceted track selector facilitates choosing RNA-seq tracks for viewing. RNA-seq tracks are categorized according to both Organ System and Brenda Tissue Ontology, and are searchable based on metadata such as sample name, SRA identifier, and tissue. (D) Each RNA-seq dataset is available in six different visualizations depending on the user's objective. This figure shows RNA-seq Junctions and Collapsed BAM tracks

issues). The RefSeq tracks also include gene models based on the ab initio Gnomon pipeline, and a track called "ambiguous genes," which are gene loci that code both protein-coding and noncoding transcripts. If a particular RefSeq gene or transcript is not found in the expected track (e.g., RefSeq protein coding), you should view the "ambiguous genes" and "frameshift genes" tracks. LiftOver Gene Prediction tracks include genes that were predicted based on a previous or alternate assembly version, and are actually alignment tracks rather than gene models. Genes in the LiftOver tracks may have errors, such as missing exons, frameshifts, and internal stop codons. Any gene in a LiftOver track may exist at multiple chromosome coordinates.

In addition to data type categories, RNA-seq tracks are organized according to both Organ System, from the Mouse Anatomy Ontology [4] and the Uberon Anatomy Ontology [5], and Brenda Tissue Ontology [6]. After highlighting one or more data types, organ systems or tissues in the left panel, only tracks under those categories appear in the table in the right panel. Tracks can be further filtered using a text search in the "Contains Text" box for any information found in the table, which includes data type and track name for all tracks, Brenda Tissue Ontology Name, Brenda Tissue Ontology All Levels, Organ System, Sex, Age, Individual, and SRA Experiment Accession for RNA-seq tracks.

Each RNA-seq dataset is available in several different track configurations, which can also be filtered with the faceted track selector. The RNA-seq HeatMap and XY-Plot tracks are useful for zoomed-out views of the chromosome, to see areas with high expression levels. The other RNA-seq tracks are more suitable for zoomed-in views used to compare transcript isoforms with RNA-seq alignments. The Cufflinks tracks contain assembled contigs, and show up as either histograms depicting feature density when zoomed out, or gene models depicting exon/intron structure when zoomed in sufficiently. The Junctions tracks show arcs connecting parts of spliced reads; these are useful for quickly visualizing discrepancies in exon/intron structure or split/merge gene model disagreements when viewed along with a gene prediction tracks (Fig. 1D). The BAM tracks require a sufficient zoom-in level. The Collapsed BAM track is the less compute intensive of the two BAM tracks, and shows individual read alignments and spans within spliced read alignments (Fig. 1D). The Draggable BAM tracks are the most compute intensive, and require the highest zoom level. This track allows annotators to drag single reads to the Apollo annotation panel, but does not provide advantages over the Collapsible BAM track when simply viewing in JBrowse.

Once suitable tracks are identified, they are selected by clicking the corresponding box on the left side of the table. Clicking "Back to Browser" above the left side of the table hides the track selector. Within the browser, tracks can be rearranged by dragging the track label.

2.4 BLAST

The BGD BLAST search interface leverages the SequenceServer platform [7]. When the search database is a genome assembly, BLAST hits are linked to JBrowse based on match coordinates. When the search database is coding sequence, transcript or peptide, BLAST hits are linked to a JBrowse location based on the hit identifier. SequenceServer also provides downloadable tab-delimited or BLAST XML reports and graphical overviews of the matches.

2.5 BovineMine

BovineMine is a powerful data mining tool that allows scientists with limited programming skills to exhaustively explore the bovine genome and associated biological data from a variety of external sources. BovineMine, which is accessible from the BGD navigation bar, employs the InterMine data warehousing system [8] to integrate the data and allow you to generate customized data sets. The data currently available in BovineMine include reference genome assemblies, genes, proteins, protein families and domains, orthologs and homologs, biochemical pathways, interactions, Gene Ontology annotations, cattle quantitative trait loci (QTL), variation, and publications (Table 1). BGD-specific data, including computed variant effects and RNA-seq-based gene expression data, permits you to mine tissue specific gene expression levels in the context of genomic variation data. In addition to the bovine reference genome assembly, BovineMine includes the reference genome assemblies and gene sets of sheep and goat to allow researchers of nonbovine ruminants to leverage the extensive amount of available bovine genomics data.

2.5.1 BovineMine Navigation Bar and Homepage

The BovineMine navigation bar is available on all BovineMine pages (Fig. 2). Tabs in the navigation bar direct you to the relevant entry points for different analysis types, and also include information about the data sources (Data Sources) and a diagram depicting data integration (Data Model). The Help tab opens a new browser window with a written tutorial and short videos demonstrating BovineMine features.

The Quick Search and Quick List tools on the BovineMine homepage (Fig. 2) allow you to quickly search the database for a single keyword (Quick Search, more in Subheading 2.5.3) or a list of identifiers (Quick List, more in Subheading 2.5.5). Halfway down the home page is a set of tabs that provide access to pre-defined template queries organized into categories (GENE, EXPRESSION, PROTEINS, FUNCTION, HOMOLOGY, INTERACTIONS,VARIATION, ALIAS AND DBXREF, ENTIRE GENE SET). The ENTIRE GENE SET category lists

Table 1
BovineMine data sources

Data Source	Reference
AnimalQTLdb	[10]
BioCyc	[18]
BioGRID	[19]
Bovine genome assembly UMD3.1.1	[20]
Bovine HapMap	[21]
Bovine official gene set	[22]
dbSNP	[23]
dbVar	[24]
EnsemblCompara	[25]
Ensembl Genes	[26]
Goat genome assembly ARS1	[27]
Gene Ontology	[28]
HomoloGene	[29]
IntAct	[30]
InterPro	[31]
KEGG	[32]
OrthoDB	[33]
PubMed	[34]
Reactome	[35]
RefSeq	[36]
Sequence Read Archive	[37]
Sheep genome assembly OAR_v3.1	[38]
SNPchiMp	[39]
TreeFam	[40]
UniProt	[41]
UniProt-GOA	[42]

templates for querying entire gene sets, such as retrieving all microRNA for a given organism. The ALIAS AND DBXREF category provides templates that convert identifiers between gene sets, allowing you to relate gene identifiers between the Ensembl and RefSeq gene sets. Upon the release of an updated bovine reference genome, template queries will be added to relate gene

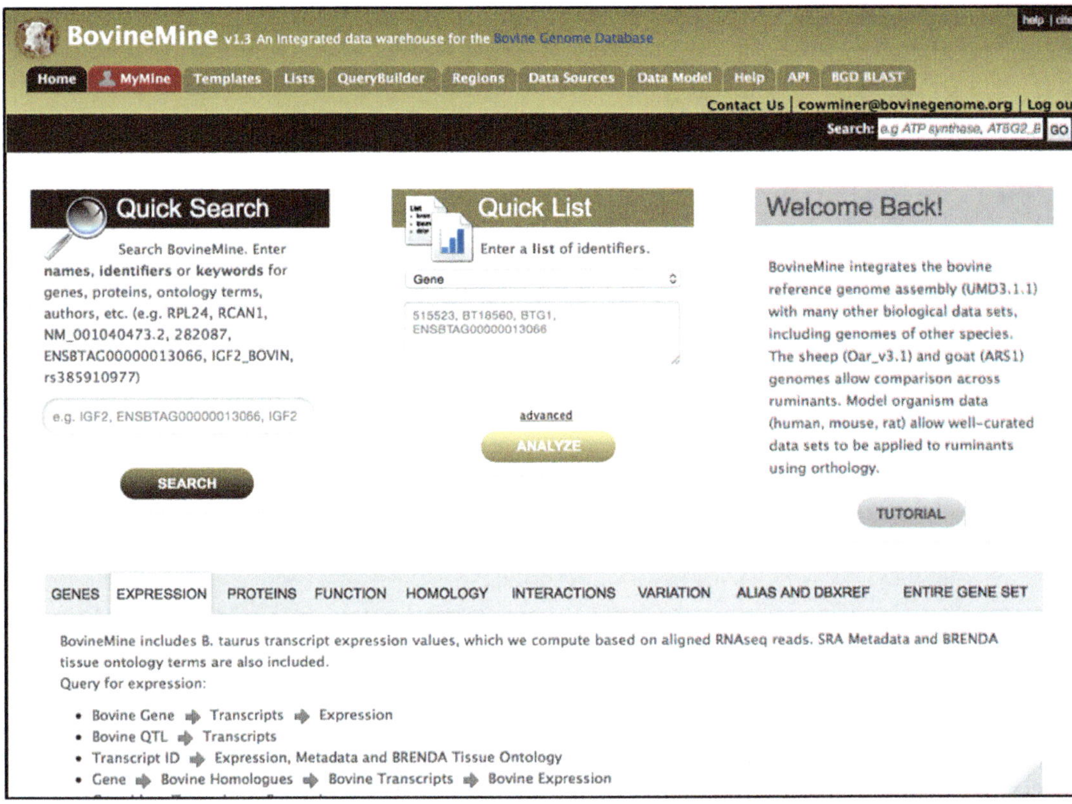

Fig. 2 The BovineMine homepage allows you to initiate analysis by using a query form for "Quick Search" or "Quick List" or by using predefined template queries organized into major data categories. Alternatively, you could select a tab in the BovineMine navigation bar for the more advanced search options available with the QueryBuilder, List Tool, and Regions search

identifiers between the old and new gene sets. Subheading 2.5.7 provides further details regarding template queries.

2.5.2 MyMine

Although you may use BovineMine anonymously, creating a MyMine account provides several advantages. While logged in to MyMine, your query history is automatically saved and you can save lists, queries and template queries for use during later sessions. Your work is maintained even in the case of accidental server disconnection. You can also organize your lists with tags and share them with other MyMine users by selecting the "Share with users" link. You can register or log in by either choosing the "Log In" link in the header or by choosing the red MyMine tab in the navigation bar. Registration requires only an email address to be entered in the "Username" area and setting a password.

2.5.3 Quick Search

You can input keywords for text searching in two areas, the Quick Search box on the BovineMine homepage and the keyword search box below the navigation bar on all BovineMine pages. Since Quick

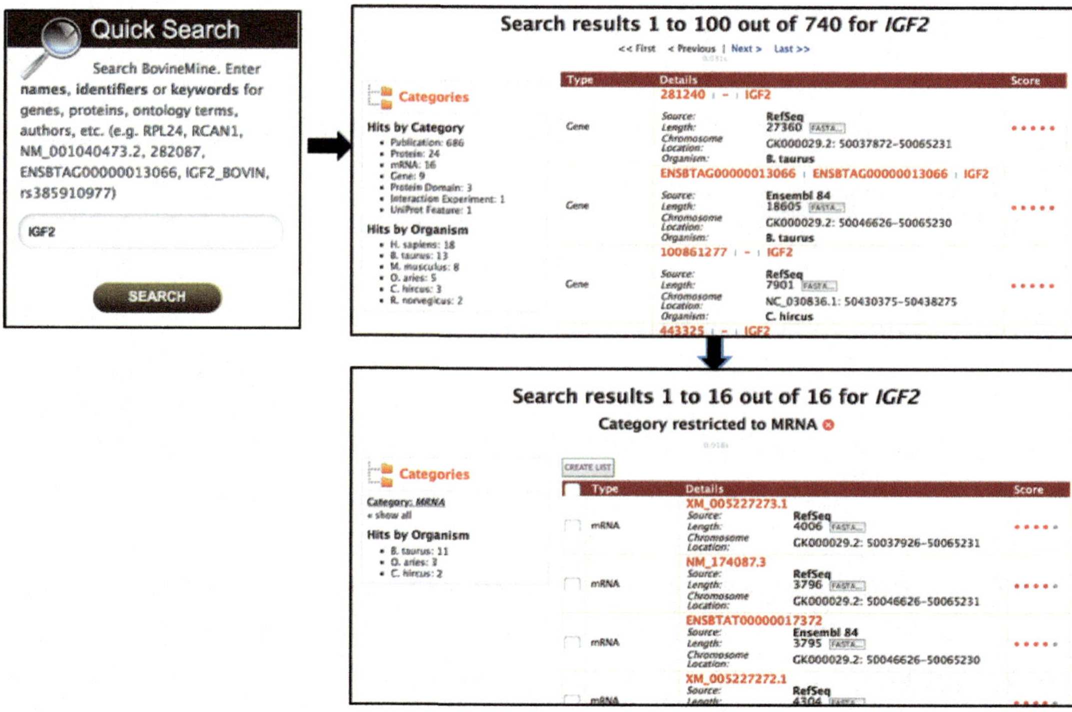

Fig. 3 The quick search box employs a full text search and identifies all records containing the search term. A search summary page allows you to filter by feature class and/or species. Filtered summaries can be further saved as a list by using the "Create List" button. Clicking the orange name of a record leads to a report page (not shown)

Search is a full text search, you can enter any kind of identifier or text string, including the wildcard '*'. For example, after entering the gene symbol IGF2 into the Quick Search box and clicking "Search," you will see a page summarizing the query, and over 700 results for which IGF2 is included in the text (Fig. 3). You can filter results by clicking a data category or organism name within the box on the left of the page (Fig. 3). Details are provided about each result, including a score indicating the similarity of your query to the result. Clicking an individual result provides the report page for that object.

2.5.4 Report Page

BovineMine generates a report page with detailed information customized by data class (e.g., genes, transcripts, proteins, variants) for each data object. Every report page is a collection of tables, each of which can be sorted, filtered and exported. The gene report page is subdivided into sections including Summary, Transcripts, Protein, Function, Homology, Interactions, Publication, and Other. The Summary information includes the gene identifier, gene symbol, description, chromosomal location, strand information, and other identifiers. The Transcript section lists transcript identifiers, each of which is linked to a transcript report page that includes a table of expression values based on alignment of Illumina RNA-seq data from tissues of L1 Dominette 01449, as well as SRA metadata. The Transcript section of the Gene report includes graphical views of the transcripts; clicking a transcript graphic opens BGD JBrowse and allows you to view the transcripts in genomic context, along with other JBrowse tracks (Subheading 2.3). The Transcript section also allows you to download fasta-formatted sequences. The Proteins section lists protein identifiers that link to Protein report pages with information such as protein family, GO annotations, InterPro domains, and curated notes from UniProt. The Function section of the Gene report provides GO annotations with evidence codes. The Homology section lists mammalian homologs.

2.5.5 List Tool

The List Tool allows you to create unique datasets based on lists of identifiers. As opposed to the Quick Search function, which returns any report page containing the provided keyword, the List Tool performs a database lookup to return objects based on their primary identifiers, which include gene IDs or symbols, transcript IDs, and protein IDs. Two list functions are available, the Quick List tool found on the BovineMine homepage and the full List Tool, available via the Lists tab in the BovineMine navigation bar. The Quick List tool is a streamlined version of the List Tool, and can be used by manually entering gene or protein identifiers separated by commas, spaces, tabs or line breaks.

The advanced-function List Tool allows you to choose from among a larger array of data types, select a species, and enter a list either manually or by choosing a file to upload (Fig. 4A). Clicking Lists in the navigation bar will lead you to an Upload page or the list View page. You can toggle between the two pages using "Upload" or "View" in the black bar below the BovineMine navigation bar. List analysis begins by clicking the "Create List" button using the main List Tool Upload page or "Analyze" using Quick List. BovineMine then performs a lookup and returns an intermediate page (Fig. 4B) prompting you to name the list and possibly to provide further input to select from among duplicate identifiers from different data sets. Once you click "Save a list of X Genes" (where X is a number), your list is saved for the duration of your session if you are not logged in, or permanently if you are logged in to MyMine. The resulting page provides your list in a table form with other relevant information (Fig. 4C). The "Save as List" drop-down menu allows you to resave the list if you forgot to name it before, or to select different columns in the table to save. Your saved lists can be used in BovineMine queries (Subheadings 2.5.7 and 2.5.8). The list analysis page also contains enrichment widgets that operate on gene lists. Enrichment widgets are described further in Subheading 2.5.6.

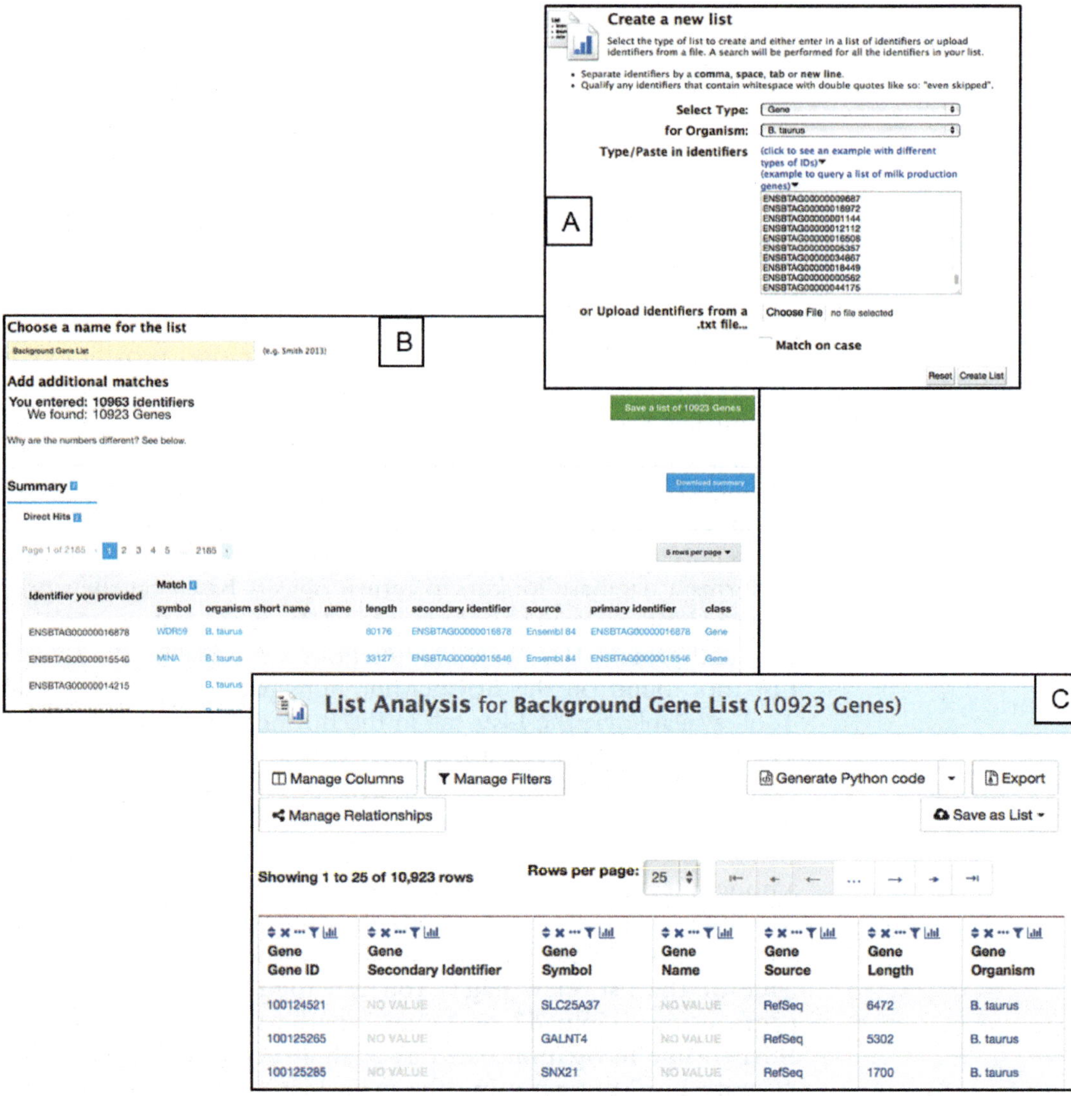

Fig. 4 (A) The List Tool Upload page takes lists of identifiers as input. The correct data type must be selected. (B) After a database lookup is performed, a page listing matching identifiers is provided with options to name and save the list. (C) List analysis results in a table, which, like all tables in BovineMine, may be altered using icons within the column headers and using the Manage Columns, Manage Filters, and Manage Relationships buttons. This figure shows the creation of the DE Gene List in example Subheading 2.5.6.

The List View page shows user-saved lists highlighted in purple and premade BovineMine lists with a white background. The View page also includes set operation functions. Finding the difference, union or intersection between two or more saved lists leads to the creation of a new list (Fig. 5).

An advantage of the List Tool is that it enables you to remove redundancy in a list, or determine the source of an identifier. Identifiers submitted to the list function may include any combina-

Fig. 5 The List View page shows user-made lists highlighted in purple. Once a list intersection is performed, a new list appears. This figure shows the intersection performed in the first example in Subheading 2.5.11.

tion of IDs or symbols, and will return results as long as they are found in BovineMine and are of the selected data type. For example, 515523 (RefSeq gene id), BTG1 (gene symbol), and ENSBTAG00000006858 (Ensembl gene id) each return a result when the "Gene" data type is selected. Thus, the List Tool allows you to perform meta-analyses of published studies without requiring you to have a priori knowledge of the data source. Entering the gene IDs and symbols given above, choosing "Gene" and "B. taurus" from the respective pull-down menus, and clicking "Create List" returns results for these genes in cattle. A green button allowing you to "Save a list of 2 Genes" suggests the results need manual inspection as we submitted a list of three genes. In this instance, the gene symbol BTG1 produced multiple records, one each for the RefSeq and Ensembl gene IDs associated with BTG1, so BTG1 is not included in the count of two genes. You are asked to choose which record to include in the list by clicking the "Add" button to the right of the record or the "Add All" button in the upper right corner of the "Duplicates" section. After making a choice, RefSeq gene in this example, you have the option to name the list before finally saving the list by clicking "Save a List of 3 Genes." This takes you to the List Analysis Page, which includes a table listing the three genes.

2.5.6 Enrichment Widgets

The BovineMine List tool analyzes each list of genes for enrichment of gene ontology, pathways, and publications, and presents the data using widgets (Fig. 6). Options to select test correction, *p*-value and background population are given for each widget. For gene enrichment analysis, the default background data set is all genes in the organism that have annotations of the type being calculated. Since BovineMine may include several gene sets for a single organism, (ex: RefSeq, Ensembl, OGS), it is recommended that you change the background population to a gene set appropriate for your subject gene list. Premade gene lists are provided for each gene set in BovineMine, and can be selected within an enrichment widget by clicking the "Change" button (Fig. 6). The List Tool also makes it easy to create more refined background gene lists for specific questions. For example, you can create lists of all expressed genes to use as the background to test for enrichment in differentially expressed genes.

Analysis of Differentially Expressed Genes Using List Enrichment Widgets

This example demonstrates use of the BovineMine List tool for enrichment analysis in a differential gene expression study. We will use a dataset of differentially expressed genes between chorionic tissue extracted 34 days after artificial insemination (AI) and extra embryonic tissue 18 days post AI from the study reported by Biase et al. [9]. Download the file, Dataset_S01, using this link: http://www.pnas.org/content/suppl/2016/12/08/1520945114.DCSupplemental/pnas.1520945114.sd01.xlsx

1. The first step is to create a gene list that will serve as a background gene list in enrichment analysis. In this case, we will use all expressed genes. Click the List tab. Choose "Gene" from the "Select Type" pull-down menu and "B. taurus" from the "Organism" menu.
2. Select all Gene IDs from column B of the spreadsheet and paste them into the List Tool text box. You might have noticed that the gene IDs included IDs from both Ensembl and NCBI. There is no need to separate them. It appears that Ensembl IDs were the main source of IDs in this study, and RefSeq IDs were used when an Ensembl ID did not exist for a particular gene. Click "Create List."

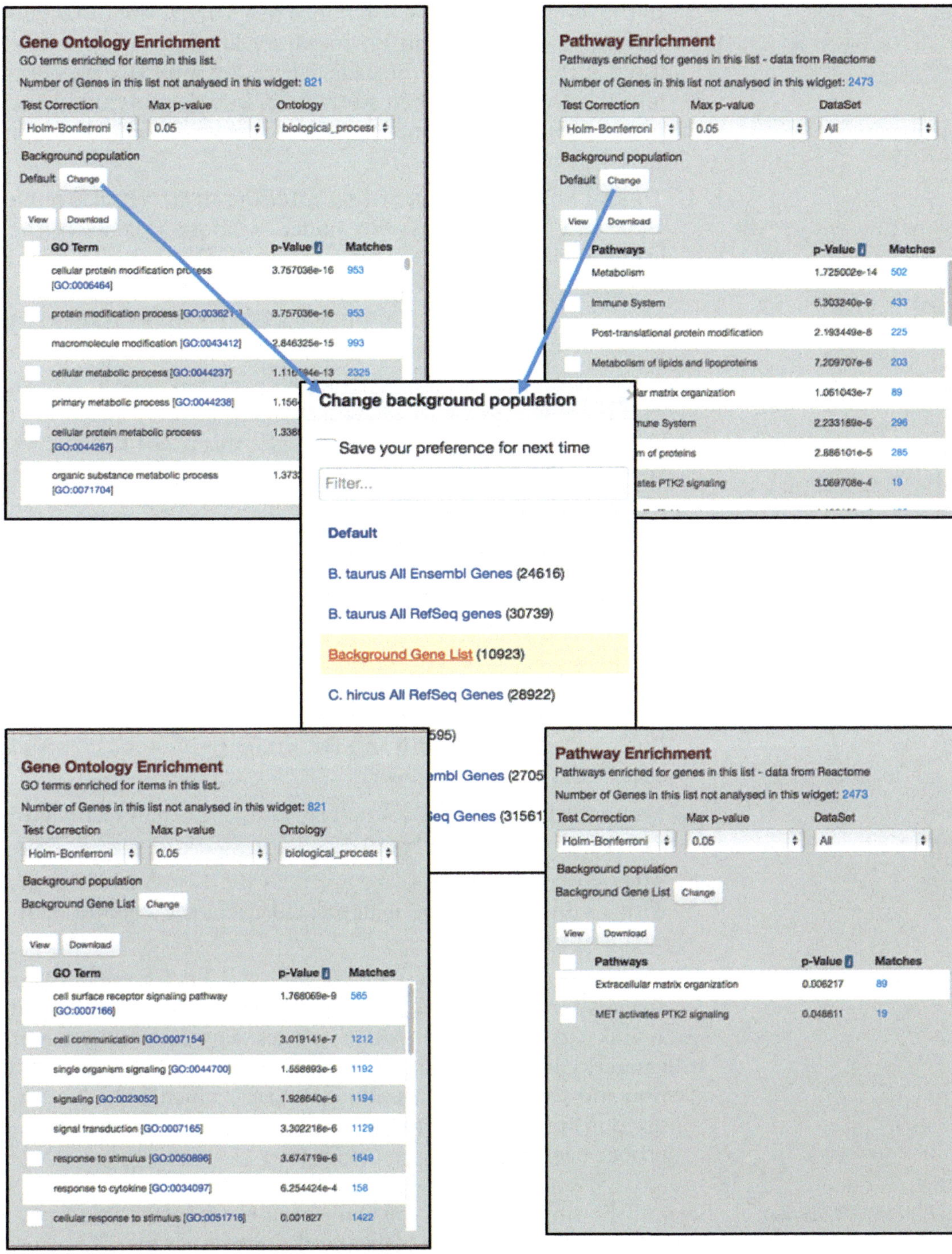

Fig. 6 Using the List Tool with a list of gene identifiers generates enrichment widgets that include options to select test correction, p-value, and background population. It is recommended that BovineMine users always select an alternative background population, since the default background population is all genes annotated with the appropriate datatype; these genes may be from more than one gene set (e.g. both Ensembl and RefSeq). Alternative gene lists, including premade gene set lists and all user-saved lists, are available by clicking "Change" under to "Background Population." This figure, showing the analysis from example Subheading 2.5.6, demonstrates the difference in enrichment results after changing the background population from the default to the list of all expressed genes identified in the study [9]

3. In the resulting page you will notice that 10,923 out of 10,963 identifiers were found in BovineMine. The missing genes are listed at the bottom of the page. These Ensembl IDs are missing because the published study used an older Ensembl gene set, and some of the genes have been removed from the newer Ensembl gene set.
4. To save your entered gene list, enter the name "Background Gene List" into the text box under "Choose a name for the list" and click the green "Save a list of 10,923 Genes" button. An advantage of the BovineMine enrichment tool is that you were able to create the appropriate background gene list even though your data contains identifiers from different gene sets.
5. After saving the list, a new table of genes will appear. Notice that an enrichment analysis has automatically been performed and appears below the table. However, this is not the analysis you care about. This analysis compared your background list of all expressed genes to the default background list of all bovine genes in BovineMine. It is not recommended that you ever use the default background list in BovineMine, because BovineMine contains multiple gene sets, causing the default background list to be redundant.
6. The next step is to create the test gene list and perform an enrichment analysis with the List Tool. Click the List Tool tab and in the text box, paste only the gene IDs found in column B from Dataset_S01 with an FDR <0.05 (rows 4–4617). And click "Create List" (Fig. 4A).
7. Name the new list "DE Gene List" and click the green "Save a list of 4595 Genes" (Fig. 4B).
8. After saving the list you are taken to the List Analysis page, with a table that can be manipulated using the column management tools (Fig. 4C).
9. Further down the page you will find widgets for enrichment analysis for publications, gene ontology, and pathways as well as links to orthologs in other species. For each enrichment widget, click the "Change" button for Background population and select "Background Gene List," which is the list you saved in **step 4**. The enrichment analysis will be rerun with the proper background data set (Fig. 6).

2.5.7 Template Queries

Beyond the simple search, you can query BovineMine using predefined template queries, available by clicking either the Templates tab in the navigation bar or one of the Template category tabs in the middle of the BovineMine home page. Clicking a template query name takes you to an interface that may be prepopulated with examples and usually requires your input. For example, choosing the "Bovine Gene → Transcripts → Expression" template under the EXPRESSION category produces a web form with up to five fields for your input (Fig. 7), with the Tissue field being optional. In most cases you would modify the input identifier and enter a cutoff in one of the expression level fields. You can click "Show Results" button to run the query or click "Edit Query" to go to the QueryBuilder (described in Subheading 2.5.8) to modify the query by adding or removing search constraints or selecting additional information to be included in the output.

Template Query Example

Suppose you are interested in the gene Ago2, a key player of RNAi-dependent gene silencing. You would like to know in which tissues Ago2 is expressed, in order to make an informed decision about which tissues to focus on. Since BovineMine includes tissue levels computed on RNA-seq data of over 90 tissues from the reference genome individual (L1 Dominette 01449) and her sire and calf, you could use a simple template query to identify tissues with Ago2 expression.

1. Click the "Templates" tab in the toolbar at the top of the BovineMine webpage and select the template, "Bovine Gene → Transcripts → Expression" (Fig. 7).
2. In the Gene field, you could enter "Ago2," the NCBI gene ID "404130" or the ENSEMBL gene ID "ENSBTAG00000001579." Note that if you enter a gene symbol, you may retrieve the gene from more than one gene set. For this example, we will use the NCBI gene ID "404130."
3. Leave the tissue filter off so the results table will include information for all tissues.
4. Select ">=" from the pull-down menu and enter "10" in the FPKM section.
5. Click "Show Results".

This example returns a table with four rows, each showing a tissue for which Ago2 is expressed with an FPKM of ten or greater (Fig. 7). These results would allow you to focus on muscular tissue, choroid plexus, or testis for future expression studies. Furthermore, the isoform column indicates that the NCBI gene includes only one transcript isoform (NM_205794.1). In cases where the gene codes for multiple isoforms, each row would provide the expression data for an individual transcript isoform in that tissue.

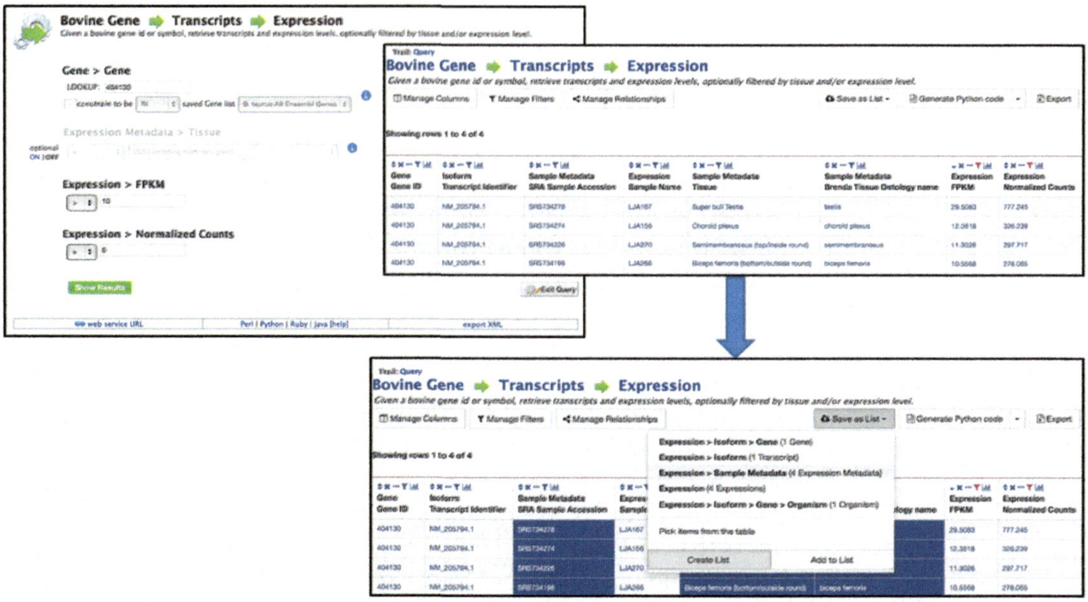

Fig. 7 Predefined template queries are provided for commonly used queries. The results are provided in tables, which can be altered using the icons in the column headers or the column management buttons above the table. Furthermore, you can save lists of different identifiers within the table by clicking "Save as List" and choosing the column of identifiers you wish to save. This figure shows the example provided in Subheading 2.5.7

2.5.8 QueryBuilder for Custom Queries

If there is no template query to fit your needs, you can use the QueryBuilder to construct your own query. Although the QueryBuilder is not intuitive and requires experimentation to gain familiarity, mastering QueryBuilder to create an integrated data set is likely to be far more efficient than learning a scripting language such as Perl or Python.

Before initiating query construction, you may wish to view the BovineMine data network diagram available by clicking the Data Model tab in the BovineMine navigation bar to visualize connections between data sources. In addition, the Identifier Relationship Table available on the Data Model page, and in Table 2 of this chapter, indicates which gene and protein identifiers can be used to retrieve different data sets. For example, Table 2 shows that to retrieve KEGG pathways for bovine genes, you need to input bovine RefSeq identifiers. If your dataset has Ensembl identifiers, you can use the "Gene ID → Database Cross Reference ID" template query to convert identifiers prior to constructing your query.

After choosing the "QueryBuilder" tab in the BovineMine navigation bar, start building a query using the pull-down menu

Table 2
Primary gene or protein identifiers used in data sets for different species at BovineMine

Data class	Species / Data set	*Bos taurus*	*Capra hircus*	*Ovis aries*	*Sus scrofa*	*Equus caballus*	*Canis familiaris*	*Mus musculus*	*Rattus norvegicus*	*Homo sapiens*
Gene	Ensembl	E		E						
Gene	RefSeq	R	R	R						
Protein	UniProt	U	U	U				U	U	U
Gene	BioCyc	E						E		E
Gene	BioGRID	R						R	R	R
Gene	Ensembl Compara	E			E	E	E	E	E	E
Gene	Gene symbols	E, R	R	E, R				E, R	E, R	E, R
Gene	GO annotations	E, R	R	E, R				R	R	R
Gene	Homologene	R					R	R	R	R
Gene	IntAct	E		E				E	E	E
Gene	KEGG	R			R	R	R	R	R	R
Gene	OrthoDB	E	R	E	E	E	E	E	E	E
Gene	Publications	E, R	R	R				R	R	R
Gene	TreeFam	E						E	E	E
Gene	Reactome	E, R								
Gene	UniProt	E, R	R	E, R				R	R	R

(continued)

Table 2
(continued)

Data class	Species Data set	*Bos taurus*	*Capra hircus*	*Ovis aries*	*Sus scrofa*	*Equus caballus*	*Canis familiaris*	*Mus musculus*	*Rattus norvegicus*	*Homo sapiens*
Transcript	Expression	E, R								
Transcript	dbSNP variant consequences	E, R								
Protein	InterPro	U	U	U				U	U	U
Protein	Reactome	U						U	U	U

Information in this table is useful when building a query that involves a dataset that uses a specific source of gene identifier. The first three rows of the table indicate which primary gene and protein data sets are available for each species. The second part of the table indicates which of the first three data sets are related to other data sets in BovineMine, and which type of identifier is used for those relationships. The Data Class column in the second part of the table indicates the data class that has a direct relationship to the data set listed in the second column

E = Ensembl, *R* = RefSeq, *U* = UniProt

under "Select a Data Type to Begin a Query" (Fig. 8A). Once you select a data type, the Model Browser opens and reveals the hierarchical structure of the BovineMine data model, beginning with the data class you selected (Fig. 8B). Stemming from the main data class are its attributes, subclasses and references to other data classes. A "+" symbol next to a data class allows you to open that part of the tree to reveal attributes and subclasses. Query construction begins by selecting either SHOW or CONSTRAIN to the right of a class or attribute. Choosing CONSTRAIN generates a pop-up box that allows you to enter an identifier for which the class will be searched (Fig. 8C). When you choose CONSTRAIN at the class level, any attribute of the selected class can be used as a constraint. For example, if you select CONSTRAIN next to "GENE," either a gene ID or gene symbol, both listed as GENE attributes, could be used to constrain the search. If you are logged into MyMine and have already saved a list, the pop-up box provides the additional option of selecting the list to constrain multiple searches, as long as the list is composed of the appropriate data type. For example, if you choose to construct a query by selecting the "Gene" data class, the list provided as a constraint must be a list of gene identifiers, not transcript or protein identifiers. Choosing SHOW next to an attribute adds the attribute as an output column. The "Fields selected for output" section below the Model Browser shows blue boxes that signify output columns. You can drag the boxes to modify column order, add descriptions to columns, and indicate whether a column should be sorted using the A-> Z function. You can run the query by clicking "Show results" or you can download the XML to share the query with others. While logged into your MyMine account you can save the query as a template for reuse or additional editing. The query output table can be manipulated or filtered by using the Column Manager tools (described in Subheading 2.5.10).

QueryBuilder Example: Genes underlying QTL for Residual Feed Intake

BovineMine includes curated bovine QTL data from AnimalQTLdb [10], and allows you to identify QTL for a trait of interest along with genes or other genomic features underlying the QTL regions. In this example, we will retrieve RefSeq genes underlying QTL for the trait "Residual Feed Intake" (Fig. 8).

1. Click the "QueryBuilder" tab in the BovineMine navigation bar.
2. In the box titled "Select a Data Type to Begin a Query," use the scrolling menu and select the class "QTL" (Fig. 8A). This action highlights "QTL" and enables the "Select" button. Now click "Select."
3. You are presented with a view of the Model Browser and Query Overview (Fig. 8B). The Model Browser, rooted at the QTL data class, shows the attributes of QTL and data classes that are

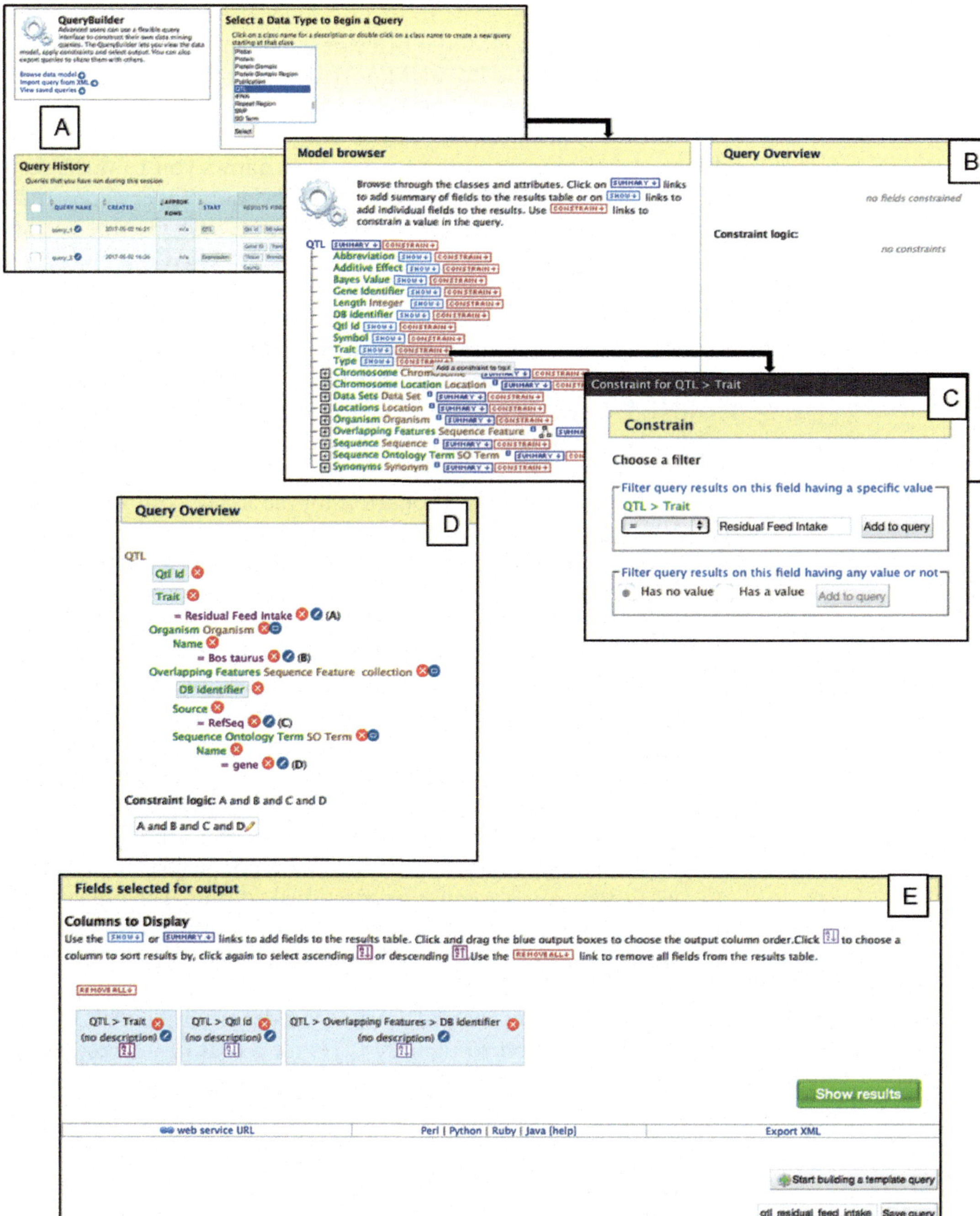

Fig. 8 These figures show the steps in the QueryBuilder example in Subheading 2.5.8. The QueryBuilder allows you to develop custom queries incorporating multiple data sources. (A) From the QueryBuilder page, you can browse the data model, import queries in XML format, or view saved queries. The QueryBuilder entry page also reports your most recent query history. Query creation begins by selecting a data class and clicking select. (B) The model browser allows you to create query constraints. (C) When you click CONSTRAIN a pop-up menu allows you to enter information. Here the QTL output is constrained to the Trait "Residual Feed Intake." (D) The final Query Overview from the Subheading 2.5.8 example shows that the query will output three columns (QTL

connected to QTL. Since you would like to see data for a specific trait of interest, click CONSTRAIN next to "Trait." A pop-up box appears and you can now enter the trait "Residual Feed Intake" (not case sensitive) and select "=" from the pull-down menu (Fig. 8C). Click "Add to query." If you are unsure of the trait nomenclature, you can use wildcards in the constraint, for example "*feed*", and later filter for the correct trait in your output table. In order to filter, you would need to show the trait in the output. After you have clicked "Add to Query," notice the constraint is shown in the Query Overview. The constraint can be edited by clicking the blue pencil symbol, or removed by clicking the red X.

4. Although you constrained the query for a specific trait, you should include the trait in the output to ensure you did not make an error in entering the constraint, and to keep pertinent information together in your analysis to refer back to in the future. To do so, go back to the Model Browser, and click SHOW next to "Trait" under "QTL." Notice in the Query Overview the word "Trait" is now shown in a blue box to signify it will be included in the output.
5. We would like the output table to show the AnimalQTLdb QTL identifier. To do so, click SHOW next to the attribute "Qtl Id" under QTL in the Model Browser.
6. The next step is to constrain the organism. The Model Browser shows that the "Organism" class is connected to QTL. Click the "+" next to "Organism" to expand its list of attributes. Click CONSTRAIN next to the "Name" attribute. You will again see a pop-up box with drop-down menu options. Select "=" and "Bos taurus" from their respective menus. Click "Add to query." Notice that several more lines related to Organism have been added in the Query Overview, and these are indented to indicate that the Organism is an attribute of QTL. The blue box symbol next to the word "Organism" indicates that your query is joining data collections to each other (the QTL collection and the Organism collection). Clicking this symbol allows you to indicate whether a relationship is required or optional, with the default being required.

Fig. 8 (continued) ID, QTL Trait, and Sequence Feature DB identifier) and the query has four constraints: (QTL Trait = Residual Feed Intake, Organism Name = Bos taurus, Overlapping Features Source = RefSeq, Sequence Ontology Term Name = Gene). (E) The output columns and their order in the final results table are illustrated by blue boxes that may be rearranged by dragging and dropping. In addition to running, a custom query can be named and saved, exported for sharing and developed into template query

7. The next few steps are to output identifiers of genes within the QTL regions.
8. Scroll down in the Model Browser, and expand the attributes of the "Overlapping Features" class. Click SHOW next to "DB identifier." This will create a column in the output table listing identifiers of any genome features (e.g., genes, transcripts, exons) with chromosomal coordinates that overlap the QTL region coordinates.
9. Next you will limit the list of feature identifiers to the RefSeq gene set. Click CONSTRAIN next to the "Source" attribute under Overlapping Features. Within the pop-up box, select "=" and "RefSeq" from the pull-down menus. Add this filter to the query by clicking "Add to query."
10. The next step is to further limit the list of identifiers to only genes using a Sequence Ontology term. The Sequence Ontology [11] is a standard controlled vocabulary used to describe components of genomes and genomic data. Click the "+" to expand "Sequence Ontology Term" under "Overlapping Features," making sure that you stay within the correct subtree by following the vertical line that descends from "Overlapping Features." If you are not careful, you could erroneously select the "Sequence Ontology Term" collection that descends directly from QTL. Click CONSTRAIN next to "Name" under the correct "Sequence Ontology Term" collection. From the pull-down menu of the pop-up box, select "gene" and "Add to query."

At this point, query construction is complete. Looking at the Query Overview, notice that your query will output three columns: QTL ID, QTL Trait and Sequence Feature DB identifier (Fig. 8D). Your query has four constraints: QTL Trait = Residual Feed Intake, Organism Name = Bos taurus, Overlapping Features Source = RefSeq, Sequence Ontology Term Name = Gene. Notice that the query joins three data collections: QTL, Organism and Overlapping Features. Also notice, the title of each data collection includes one or more words shown in brown font. If you click a word shown in brown, the Model Browser window will automatically adjust to show the corresponding data collection in the tree (Fig. 9). This is convenient way to navigate to the correct region of the tree to select more attributes for a data collection.

Below the Model Browser the "Columns to Display" section shows the columns to be included in the output (Fig. 8E). The output column order is not necessarily the same as shown in the Query Overview, but the order in which you added the column during query construction. You can drag the boxes to reorder the columns. With symbols in the blue boxes, you can eliminate columns, indicate that the output should be sorted based on a column, and add column descriptions.

Fig. 9 After adding each query component, the view in the Model Browser resets to the top of the tree. The Query Overview provides a trick that allows you to easily jump back to a region of the Model Browser tree to select more outputs or constraints. (A) After the Sequence Feature DB Identifier was added to the query, the Model Browser view was reset. In the Query Overview, notice the words "Sequence Feature" in brown font. (B) Clicking the word "Sequence Feature" in the Query Overview allows you to navigate back to the associated data collection in the Model Browser to select additional Sequence Feature attributes or constraints

You have several options once query construction is complete. You can click "Show results" to run the query (Fig. 8E). You can select "Export XML" and copy the XML code to share the query with other users. If you export the XML, you can use the back button of your web browser to go back to the query to run it. If you are logged into MyMine, you can save the query by entering a name in the text box and clicking "Save query." The saved query can be retrieved in the "Queries" section under MyMine. Finally, you can click the "Start building a template query" which will open a very similar page with your constructed query and additional options that allow you to name and describe the template.

Clicking "Save template" or "Save and Run" will promote the query to a finalized template which can be now found under the "Template" tab of the BovineMine navigation bar.

In the output of this example you will see the three columns, including one labeled "Overlapping Features DB identifier." These identifiers are RefSeq Gene IDs, because you constrained the overlapping features to be genes from the RefSeq gene set. Above the table you will see the number of rows of output. If you would like to save the genes for further analysis, you can click "Save as List" and then click "QTL → Overlapping Features." In the pop-up box, you can rename the list and then click "Create List." When you saved the list of overlapping features, you might have noticed that the number of features you saved is smaller than the number of rows in your table as some of the genes were found under more than one QTL. Keep in mind, that Animal QTLdb curates QTL from the literature, and assigns unique identifiers to independently published QTL, even if they occur in the same region.

Once the list is saved, you can use it with any template query that takes gene identifiers as input. Examples are "Gene → Gene Ontology" and "Gene → Pathway". You can see the list via the Lists tab in the navigation bar. If necessary, click "View" in the black bar under the BovineMine navigation bar to toggle to the List View page. If you click on the name of the list, you will be presented with the List analysis page. For this example, you will notice the output columns are labeled as "sequence features" rather than genes, so gene enrichment analysis is not performed when you view the list. In order to run gene enrichment on the identifier list, use a template query, such as "Gene → Chromosomal Location," that is sure to output all the genes, using this sequence feature list as the identifier constraint. Running a gene-based template query will output the identifiers as genes rather than "sequence features," as you can see in the output column headers. When you save the gene IDs again, you will notice the default list name starts with "Gene" rather than "Sequence Feature," and the resulting list can be used with the List Tool for a gene list analysis, including enrichment.

2.5.9 Regions Search

The "Regions" tab on the BovineMine navigation bar takes you to a web form that allows you to search for genomic features based on submitted chromosomal locations (chromosome IDs and coordinates). The Regions search is particularly useful for bovine researchers who identify genomic variants through genome wide association studies and desire to know what genomic features exist within a specified range of the variant. Conversely, if you are interested in identifying all SNPs within a specified distance of a gene, the gene coordinates can be uploaded and the range adjusted to a desired distance. The results page provides all the features that overlap each region queried by the Regions search. The overlapped features can be exported in csv, tab delimited, gff3, or fasta file formats. Additionally, a new list can be created with the overlapped features by choosing a feature type for the "Create list by feature type" option. The second example in Subheading 2.5.11 utilizes the Regions search tool.

2.5.10 Column Management

The output tables of each query, list, or region search may be altered using the available tools. The top of each column header has a series of icons that may be used to sort the contents of the column, delete the column, hide the column from view, filter based on column contents, or summarize the results of the column. These functions can also be achieved using the "Manage Columns" or "Manage Filters" buttons near the top of the page. The Manage Columns button allows you to remove entire columns, add columns, and prioritize sort order. After choosing the Manage Columns button, a pop-up window appears with a list of the columns currently available in the output table. Columns can be removed by clicking the red button to the right of a given column. Clicking the green "Add columns" button in the upper right portion of the pop-up window reveals a hierarchical data model browser similar to the one found in QueryBuilder. You can add new columns by selecting additional fields. The green button will now state "Add X new columns," with X being the number of new fields chosen. The changes are finalized by clicking the blue "Apply Changes" button at the bottom. Columns can be filtered based on content by choosing the filter icon after the new column is added to the output table. All active filters can be edited or removed by clicking "Manage Filters." The second example in Subheading 2.5.11 utilizes column management tools.

2.5.11 Two Step-by-Step Examples of Meta-Analysis Using BovineMine

The Lists and Regions Search Tools are particularly useful for meta-analyses. An example is the meta-analysis of SNPs associated with fertility traits reported by Ortega et al. [12], in which SNPs common across different studies were identified. The comparison was relatively straightforward because the different studies used the same SNP assay. Sometimes the overlap of exact SNP identifiers or coordinates between studies does not occur due to differences in assays between studies. Another approach is to ask whether an SNP identified in one study is located within the vicinity of an SNP identified in another study. We will provide two examples of SNP meta-analyses. In the first example, we will identify identical SNPs across studies. In the second example, we will identify SNPs from one study located within 20 kb of SNPs from another study.

GWAS Meta-Analysis Example 1

Say you have identified a set of SNPs using the Bovine Illumina HD array, and would like to see how these SNPs align with the SNPs associated with the trait "Number of Services per Conception" (NSC) from Ortega et al. [12]. A challenge is that you would like

Table 3
Identifiers for step 1c of the GWAS Meta-analysis in the first example described in Subheading 2.5.11

ARS-BFGL-NGS-34049
ARS-BFGL-NGS-67989
ARS-USMARC-670
BFGL-NGS-116469
BovineHD0100038154
BovineHD0300002060
BovineHD0500025143
BovineHD0600006358
BovineHD0700015348
BovineHD0700015365
BovineHD0700028738
BovineHD1300008936
BovineHD1400004569
BovineHD1800015205
BovineHD2500000403
Hapmap41181-BTA-120938

to compare identifiers from the Bovine Illumina HD array with dbSNP rs identifiers from the publication.

1. The first step is to create a list using your study SNP IDs.
   (a) Click the Lists tab in the BovineMine navigation bar.
   (b) Choose "SNP" from the "Select Type" pull-down menu and "B. taurus" from the "for Organism" menu.
   (c) The list of identifiers in Table 3, from the Bovine Illumina HD chip, is a simulated dataset created for this example. Paste these IDs into the Lists Tool text box. After pasting identifiers from a PDF file, you should ensure that each row is on its own line and there are no extra whitespaces. Sometimes you may find that converting a PDF to a Word document may help to format information from tables.
   (d) Click "Create List."
   (e) In the resulting webpage, save the list with the name "Study SNP List" in the text box and click the green "Save a list of 16 SNPs" button. You are taken to the

page showing the list you created. Notice the SNP chip identifiers have been replaced with dbSNP rs IDs.

2. The next step is to create a list of SNPs associated with the trait "Number of Services Per Conception" from the Ortega et al. [12] publication.
   (a) If possible, access the web page for that publication (http://www.sciencedirect.com/science/article/pii/S0022030217301819). Scroll down to Table 3 in the article, and download the table as a comma-separated value (csv) file. If you change the file extension to .txt, you will be able to open it in Excel (See the next step if you cannot access the publication).
   (b) Paste the SNP identifiers from column 1 of the Ortega et al. [12] Table 3 into the text box. The SNP IDs are provided in Table 4 of this chapter in case you cannot access that publication.
   (c) Click "Create List."
   (d) After database lookup, the result indicates that one of the SNP IDs is found in more than one BovineMine dataset.

Table 4
Identifiers for step 2b in the GWAS Meta-analysis in the first example described in Subheading 2.5.11

rs133674837
rs110217852
rs137601357
rs109621328
rs133747802
rs109443582
rs109137982
rs41893756
rs109262355
rs109830880
rs110828053
rs111015912
rs134264563
rs109813896
rs109629628
rs110660625

Select either of those, as the allele does not matter, and click "Add" so it will be included in your final list. Provide a name for the list such as "SNPs for NSC" and click "Save a list of 16 SNPs."

3. The last step of this example is to perform a List intersection. If you have navigated away from the previous page, click "Lists" in the BovineMine navigation bar to get back to the List tool. If you are presented with the Upload page rather than a page showing your lists, click "View" in the black bar below the BovineMine navigation bar. You will see both "Study SNP List" and "SNPs for NSC" (Fig. 5). Any list that you have created is highlighted in purple, with "MY" in the left corner. Check the box next to each list. Click "Intersect" in the "Action" bar above the list, and enter a new name, such as "Study_SNP_vs_SNPs_for_NSC" (Fig. 5). The intersection shows that three of your study SNPs were also identified by the Ortega et al. [12] study. Click the name of the list to see the SNPs.

GWAS Meta-Analysis Example 2

Looking for exact matching SNPs between studies might have been too stringent since your population and assay were different from those used in the Ortega et al. [12] study. You would now like to see whether any of your SNPs are located within 20 kb of the SNPs identified in the previous study.

1. Since you have already created SNP lists, the next step is to create a list of regions for the "SNPs for NSC" identifiers to use in the Regions Search tool.
   (a) If you are not already viewing your lists, click "Lists" in the BovineMine navigation bar, and if necessary, click "View" in the black bar below the navigation bar.
   (b) Click the name of the list "SNPs for NSC."
   (c) Click "Manage Columns" above the table (Fig. 10A).
   (d) In the pop-up menu, remove the unneeded Reference Allele and Alternate Allele columns by clicking the red circles (Fig. 10B).
   (e) Click the green "+Add a Column" box (Fig. 10B).

Fig. 10 (continued) Columns menu allows you to remove existing columns and add new columns. (C) If you choose to add new columns, a hierarchical data model tree is provided, allowing you to select attributes as new output columns. The attributes you select are highlighted in blue. (D) After selecting attributes and clicking "Apply Changes," you return to the main Manage Columns menu, where you can reorder the columns using up and down arrows. (E) Clicking "Apply Changes" in the Manage Columns menu generates the new Table. (F) Clicking "Export" opens the Download menu. In this example, we wished to export only the SNP coordinates, not the SNP IDs, so we use the toggle next to "SNP > SNP id" to remove it from the list of columns to be output in the tsv file

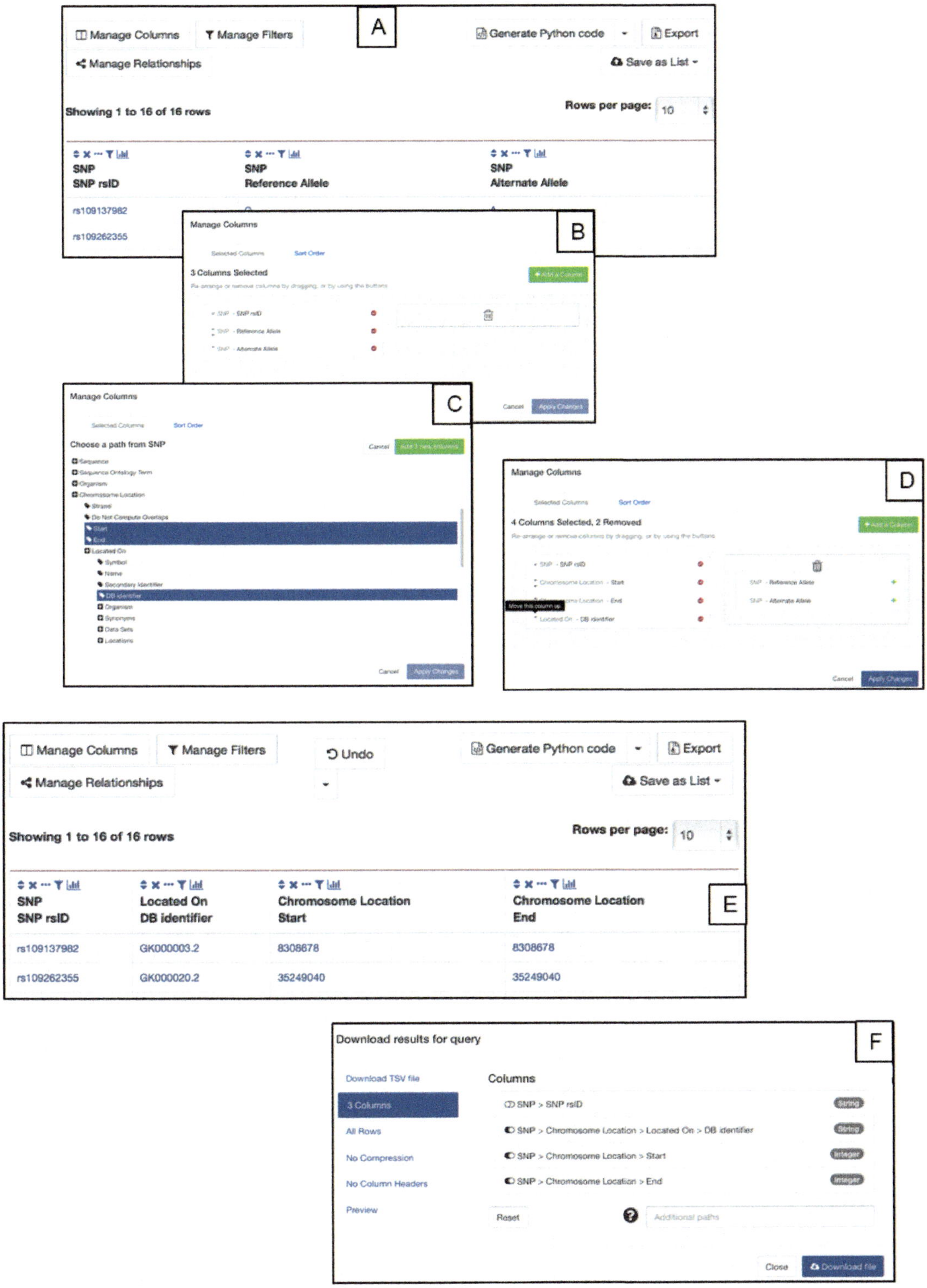

Fig. 10 This figure shows column management steps used in the second example described in Subheading 2.5.11. (A) Above every table are the column management buttons "Manage Columns," "Manage Filters," and "Manage Relationships." Tables can also be modified using the icons in the column headers. (B) The Manage

(f) Click the "+" sign next to "Chromosome Location." Then click "Start" and "End". Although the SNP is a single coordinate, the Region Search tool requires a Start and End coordinate (Fig. 10C).

(g) To retrieve the chromosome ID, click the "+" sign next to "Located On," and then click "DB identifier" (Fig. 10C).

(h) Click the green "Add 3 Columns" box (Fig. 10C).

(i) Notice columns have been added to the column list. However, the order needs to be modified to correctly format the location information for the Regions Search. Use the up arrow to move "Located On >> DB identifier" above "Chromosome Location >> Start". Then click the blue "Apply Changes" box (Fig. 10D).

(j) The next step is to export the coordinates so they can be uploaded to the regions search. Click the "Export" button above the table (Fig. 10E).

(k) Enter a file name. Then mouse over "All Columns" to change that default setting so that only selected columns are exported. Click the toggle next to "SNP > SNP rsID" so that it will not be included in the output, since the Regions Search tool does not accept identifiers (Fig. 10F).

(l) Click "Download File" and save the file on your computer.

2. The next steps are to perform a Regions Search for all known SNPs within 20 kb of your coordinates. Click the "Regions" tab in the BovineMine navigation bar.

   (a) In the Regions Search menu (Fig. 11) select "B. taurus" from the "1. Select Organism" pull-down menu.

   (b) Click the square next to "2. Select Feature Types" to uncheck all options and then click the box next to "SNP" as the chosen feature option.

   (c) Upload the file you just saved.

   (d) Type "20 kb" into the text box of "4. Extend your regions at both sides."

   (e) Click "Search" to run the Regions Search tool.

3. After the Region Search is successfully run you are presented with an output page listing each of the regions and the numbers of SNPs identified within the regions (Fig. 11). At this point you could download any results for any region separately, with a choice of formats, or you could create an SNP list for each region separately. For this example, we want to save a list of all the SNPs, so use the "Create List by feature type" button above the output after selecting "SNP" in the pull-down menu and click "Go." This action creates a new list that is visible on the List View page. You are not given the opportunity

Fig. 11 The Regions search upload menu takes input of genomic coordinates in the text box, or uploaded as a file. Examples of accepted coordinate formats are provided above the text box. To extend the search region, either enter the distance in the text box or use the slider. The output of a Regions search is a page that lists each region with the numbers of features found. You can download data for each region in various formats, or you can generate a list of features found in all regions using "Create List by Feature Type." This figure shows the Regions search performed in the second example described in Subheading 2.5.11

to provide a name for the list; it is assigned a default name that starts with "all_regions_SNP_list". If you are logged into MyMine, you can change the name by clicking the MyMine tab, selecting Lists in the red tool bar, and clicking the pencil symbol next to the list name to edit it.

4. The next step is to perform a List intersection to identify SNPs from your study that are also found in the list of all SNPs that are within 20 kb of SNPs identified by Ortega et al. [12]. Go to the List View page by clicking "View" in the black bar

below the BovineMine navigation bar. Check the box next to each list. Click "Intersect" in the "Action" bar above the list, and enter a new name, such as "Study_SNP_vs_SNPs_within 20 kb_SNPS_for_NSC_intersection". The intersection shows that nine of your study SNPs are located within 20 kb of the SNPs identified by the Ortega et al. [12] study.

The two examples above provide us with very limited information. You can use the lists in a variety of ways to acquire more information. For example, you can use column management functions to determine whether these SNPs are located within genes.

1. If you are not already on a List Analysis page showing your list, click on the List name on the List View page.
2. Click "Manage Columns" above the List Analysis table.
3. Click the green "+Add a Column" button.
4. Add the following columns:
   (a) Scroll down, open "Overlapping Features" and select both "DB identifier" and "Source".
   (b) Scroll down further, open "Sequence Ontology Term" and select "Name," being sure that you selected the "Sequence Ontology Term" within "Overlapping Features."
5. Click the green "Add 3 new columns" button.
6. Click "Apply changes."
7. You can see the new columns added to the output table (Fig. 12). You can see in the Sequence Ontology Term column that there are several feature types. To get only genes, click the histogram icon in the column header, check "gene," and select "Restrict table to matching rows" in the blue "Filter" pull-down menu (Fig. 12).
8. Now only rows with genes are shown. The "Overlapping Features Source" column indicates the genes are from three gene sets.
9. To limit the output to RefSeq genes filter the column using the histogram icon in the Overlapping Features Source column, similar to **step 7**.
10. Save this list of overlapping features using the "Save as List" pull-down menu, naming the list "RefSeq Genes Overlapping SNPs." You will notice that you will save "7 Sequence Features" even though you have 8 rows in the table, due to one gene (515321) overlapping two SNPs.
11. Since the list consists of gene IDs, you can use it in any template query that accepts gene IDs as input. Go to the BovineMine home page and locate the "Gene → Pathways"

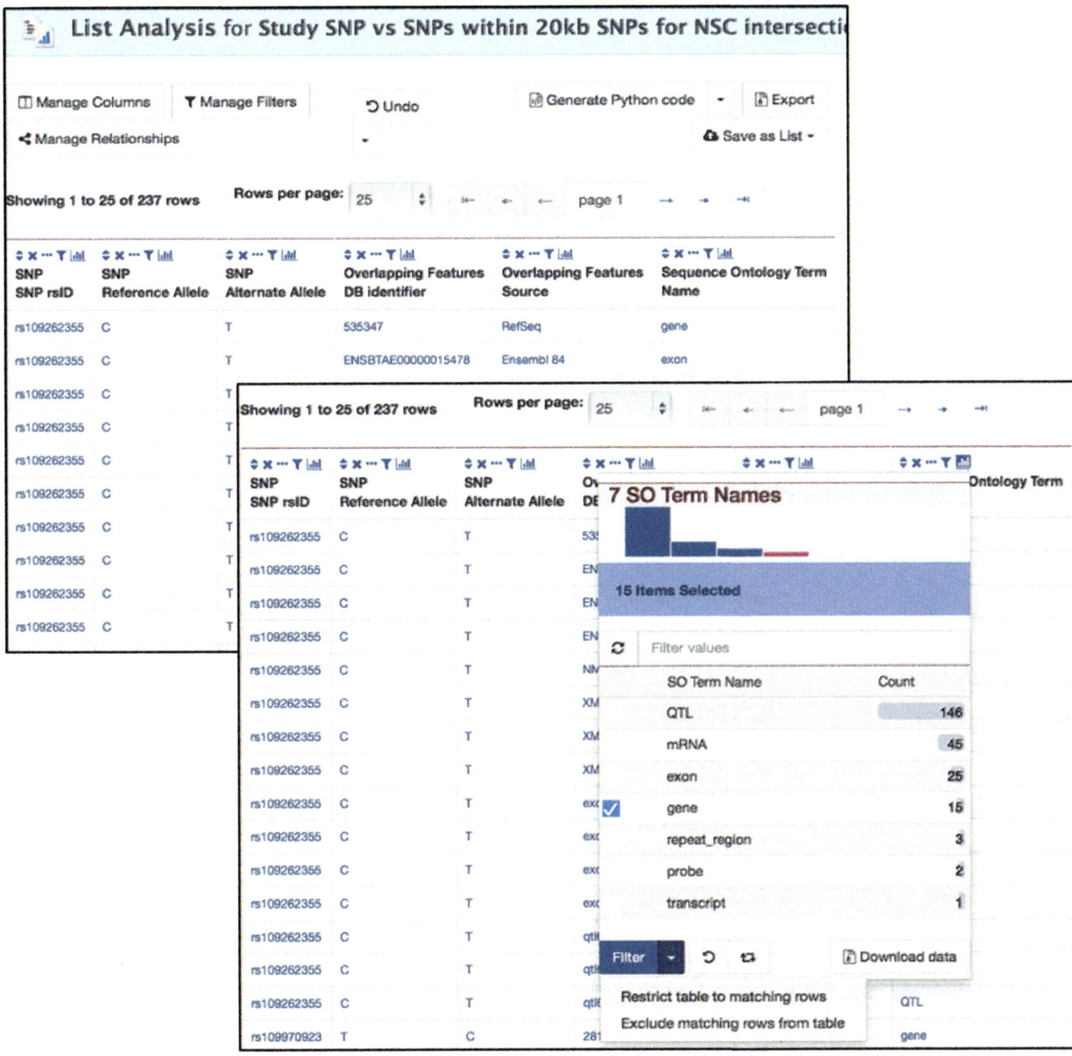

Fig. 12 This figure shows the table generated in **step 7** of the second example described in Subheading 2.5.11. The histogram icon in each column header produces a column summary. Clicking this icon in the Sequence Ontology Term Name column shows all the feature types overlapping the SNPs. Checking the box next to "gene," then selecting "Restrict table to matching rows" in the Filter pull-down menu modifies the table so that only gene features are listed. Following this step, a list of gene IDs can be generated

template query under the Function template category. Click the template name, and check the box next to "constrain to be," make sure that "IN" is selected, and choose the list "RefSeq Genes Overlapping SNPs" in the pull-down menu. (Fig. 13). The output shows that your genes are annotated with both Reactome and KEGG pathways.

The previous examples showed you how to compare SNP lists from different studies using a combination of the List Tool and Regions Search, identify genes overlapping SNPs, and retrieve

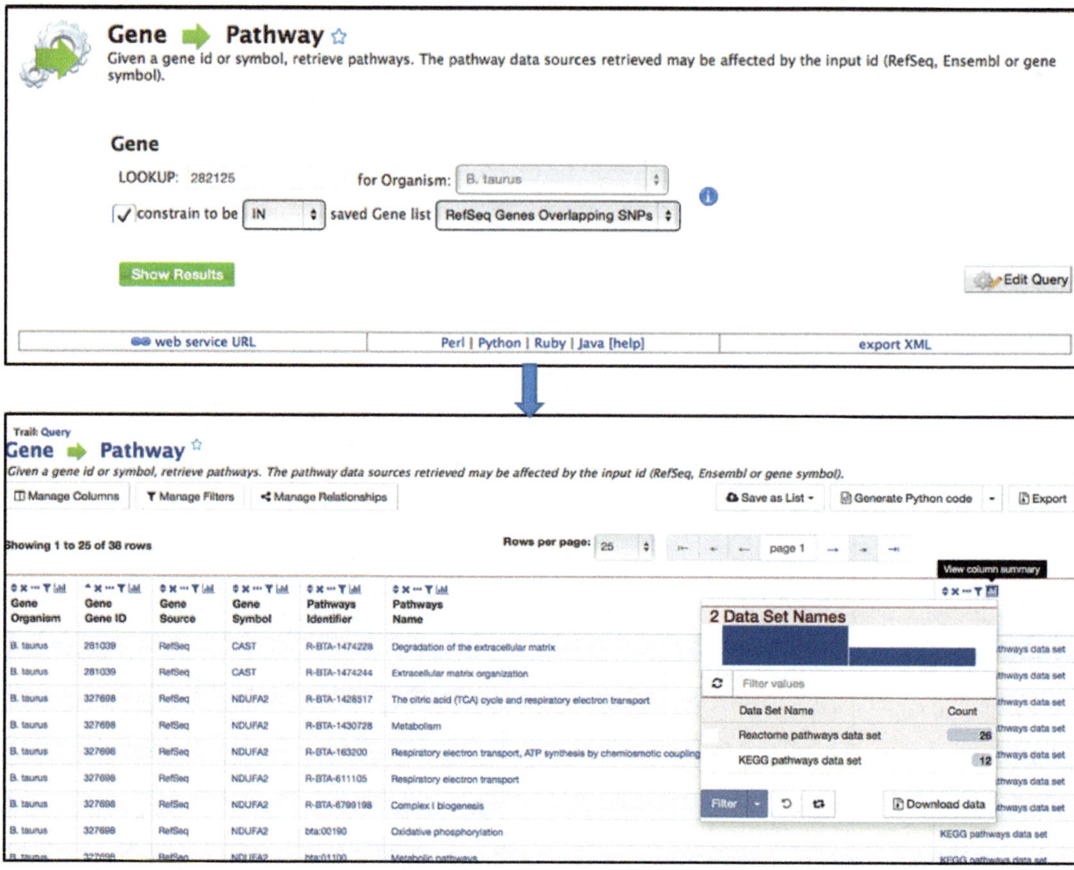

Fig. 13 Template queries that take a single identifier as input can also be used with lists of identifiers of the appropriate data type. When lists are available, the option to "constrain to be" is automatically provided. A pull-down menu with a default "IN" can be changed to "NOT IN." All available lists of the appropriate data type, including user-made lists, are provided for selection in the list pull-down menu. This figure shows the result of using the "Genes Overlapping SNPs" list from the second example described in Subheading 2.5.11 in the Gene → Pathway template query

pathway information for the genes. BovineMine provides many alternatives at each of the steps. For example, rather than filtering the features overlapping SNPs for genes, you could have filtered for QTL, saved a list of QTL IDs, and then used a template query to determine the traits. Another use of BovineMine related to GWAS is to identify genes in the vicinity of SNPs, along with GO terms or pathway information, in order to perform SNP-based Gene Set Enrichment Analysis (GSEA-SNP) (e.g., [13]). One could create a gene annotation data set for genes within a specified distance to all SNPs on an SNP chip by first using the List Tool to create an SNP list, followed by saving a list of SNP coordinates to be loaded into Regions search box, and then performing the Regions search with "Gene" selected as the feature type. In the absence of SNP IDs, as may be the case in a genome resequencing project, the Regions search rather than the List Tool could be the starting point. After the gene list is saved, it can be used with the template queries to gather information about the genes.

3 Notes

3.1 BovineMine Release for This Chapter

The most current release of BovineMine can be accessed from the main BGD navigation bar, or with the following URL: http://bovinegenome.org/bovinemine/. This chapter is based on BovineMine release 1.3, which will always be available here: http://bovinegenome.org/bovinemine-release-1.3/. You can perform the examples anonymously, but it is advisable to login to your MyMine account so that you can save your work.

3.2 Computing Expression Values and Variant Effects

We trim Illumina RNA-seq reads for adaptors using Fastq-MCF (https://code.google.com/p/ea-utils/wiki/FastqMcf); trim for quality using DynamicTrim [14] and align reads to the bovine UMD3.1 genome assembly using TopHat2 [15]. We determine FPKM (Fragments Per Kilobase of transcript per Million mapped reads) and normalized read counts for each expression dataset for transcripts in the Ensembl and RefSeq gene sets using cuffquant and cuffnorm, which are part the Cufflinks package [16]. We use the Ensembl Variant Effect Predictor [17] to predict variant effects for the bovine gene sets.

Acknowledgments

The authors would like to thank Colin M. Diesh for his contributions to the development of BovineMine, BGD BLAST, and BGD JBrowse/Apollo. This work was supported by funding from the European Union's Seventh Framework Programme for research, technological development, and demonstration under grant agreement no. 613689.

References

1. Elsik CG, Unni DR, Diesh CM, Tayal A, Emery ML, Nguyen HN, Hagen DE (2016) Bovine Genome Database: new tools for gleaning function from the Bos Taurus genome. Nucleic Acids Res 44(Database issue):D834–D839. https://doi.org/10.1093/nar/gkv1077
2. Buels R, Yao E, Diesh CM, Hayes RD, Munoz-Torres M, Helt G, Goodstein DM, Elsik CG, Lewis SE, Stein L, Holmes IH (2016) JBrowse: a dynamic web platform for genome visualization and analysis. Genome Biol 17:66. https://doi.org/10.1186/s13059-016-0924-1
3. Lee E, Helt GA, Reese JT, Munoz-Torres MC, Childers CP, Buels RM, Stein L, Holmes IH, Elsik CG, Lewis SE (2013) Web Apollo: a web-based genomic annotation editing platform. Genome Biol 14(8):R93. https://doi.org/10.1186/gb-2013-14-8-r93

4. Hayamizu TF, Baldock RA, Ringwald M (2015) Mouse Anatomy Ontologies: enhancements and tools for exploring and integrating biomedical data. Mamm Genome 26(9–10):422–430. https://doi.org/10.1007/s00335-015-9584-9
5. Mungall CJ, Torniai C, Gkoutos GV, Lewis SE, Haendel MA (2012) Uberon, an integrative multi-species anatomy ontology. Genome Biol 13(1):R5. https://doi.org/10.1186/gb-2012-13-1-r5
6. Gremse M, Chang A, Schomburg I, Grote A, Scheer M, Ebeling C, Schomburg D (2011) The BRENDA Tissue Ontology (BTO): the first all-integrating ontology of all organisms for enzyme sources. Nucleic Acids Res 39(Database):D507–D513. https://doi.org/10.1093/nar/gkq968
7. Priyam A, Woodcroft BJ, Rai V, Munagala A, Moghul I, Ter F, Gibbins MA, Moon H, Leonard G, Rumpf W, Wurm Y (2015) SequenceServer: a modern graphical user interface for custom BLAST databases. bioRxiv. https://doi.org/10.1101/033142
8. Smith RN, Aleksic J, Butano D, Carr A, Contrino S, Hu F, Lyne M, Lyne R, Kalderimis A, Rutherford K, Stepan R, Sullivan J, Wakeling M, Watkins X, Micklem G (2012) InterMine: a flexible data warehouse system for the integration and analysis of heterogeneous biological data. Bioinformatics 28(23):3163–3165. https://doi.org/10.1093/bioinformatics/bts577
9. Biase FH, Rabel C, Guillomot M, Hue I, Andropolis K, Olmstead CA, Oliveira R, Wallace R, Le Bourhis D, Richard C, Campion E, Chaulot-Talmon A, Giraud-Delville C, Taghouti G, Jammes H, Renard JP, Sandra O, Lewin HA (2016) Massive dysregulation of genes involved in cell signaling and placental development in cloned cattle conceptus and maternal endometrium. Proc Natl Acad Sci U S A 113(51):14492–14501. https://doi.org/10.1073/pnas.1520945114
10. ZL H, Park CA, XL W, Reecy JM (2013) AnimalQTLdb: an improved database tool for livestock animal QTL/association data dissemination in the post-genome era. Nucleic Acids Res 41(Database issue):D871–D879. https://doi.org/10.1093/nar/gks1150
11. Eilbeck K, Lewis SE, Mungall CJ, Yandell M, Stein L, Durbin R, Ashburner M (2005) The Sequence Ontology: a tool for the unification of genome annotations. Genome Biol 6(5):R44. https://doi.org/10.1186/gb-2005-6-5-r44
12. Ortega MS, Denicol AC, Cole JB, Null DJ, Taylor JF, Schnabel RD, Hansen PJ (2017) Association of single nucleotide polymorphisms in candidate genes previously related to genetic variation in fertility with phenotypic measurements of reproductive function in Holstein cows. J Dairy Sci 100(5):3725–3734. https://doi.org/10.3168/jds.2016-12260
13. Neibergs HL, Settles ML, Whitlock RH, Taylor JF (2010) GSEA-SNP identifies genes associated with Johne's disease in cattle. Mamm Genome 21(7–8):419–425. https://doi.org/10.1007/s00335-010-9278-2
14. Cox MP, Peterson DA, Biggs PJ (2010) SolexaQA: at-a-glance quality assessment of Illumina second-generation sequencing data. BMC Bioinformatics 11:485. https://doi.org/10.1186/1471-2105-11-485
15. Kim D, Pertea G, Trapnell C, Pimentel H, Kelley R, Salzberg SL (2013) TopHat2: accurate alignment of transcriptomes in the presence of insertions, deletions and gene fusions. Genome Biol 14(4):R36. https://doi.org/10.1186/gb-2013-14-4-r36
16. Trapnell C, Williams BA, Pertea G, Mortazavi A, Kwan G, van Baren MJ, Salzberg SL, Wold BJ, Pachter L (2010) Transcript assembly and quantification by RNA-Seq reveals unannotated transcripts and isoform switching during cell differentiation. Nat Biotechnol 28(5):511–515. https://doi.org/10.1038/nbt.1621
17. McLaren W, Pritchard B, Rios D, Chen Y, Flicek P, Cunningham F (2010) Deriving the consequences of genomic variants with the Ensembl API and SNP effect predictor. Bioinformatics 26(16):2069–2070. https://doi.org/10.1093/bioinformatics/btq330
18. Caspi R, Billington R, Ferrer L, Foerster H, Fulcher CA, Keseler IM, Kothari A, Krummenacker M, Latendresse M, Mueller LA, Ong Q, Paley S, Subhraveti P, Weaver DS, Karp PD (2016) The MetaCyc database of metabolic pathways and enzymes and the BioCyc collection of pathway/genome databases. Nucleic Acids Res 44(D1):D471–D480. https://doi.org/10.1093/nar/gkv1164
19. Chatr-Aryamontri A, Oughtred R, Boucher L, Rust J, Chang C, Kolas NK, O'Donnell L, Oster S, Theesfeld C, Sellam A, Stark C, Breitkreutz BJ, Dolinski K, Tyers M (2017) The BioGRID interaction database: 2017 update. Nucleic Acids Res 45(D1):D369–D379. https://doi.org/10.1093/nar/gkw1102
20. Zimin AV, Delcher AL, Florea L, Kelley DR, Schatz MC, Puiu D, Hanrahan F, Pertea G, Van Tassell CP, Sonstegard TS, Marcais G, Roberts M, Subramanian P, Yorke JA, Salzberg SL (2009) A whole-genome assembly of the domestic cow, Bos Taurus. Genome Biol 10(4):R42. https://doi.org/10.1186/gb-2009-10-4-r42

21. Bovine HapMap Consortium, Gibbs RA, Taylor JF, Van Tassell CP, Barendse W, Eversole KA, Gill CA, Green RD, Hamernik DL, Kappes SM, Lien S, Matukumalli LK, McEwan JC, Nazareth LV, Schnabel RD, Weinstock GM, Wheeler DA, Ajmone-Marsan P, Boettcher PJ, Caetano AR, Garcia JF, Hanotte O, Mariani P, Skow LC, Sonstegard TS, Williams JL, Diallo B, Hailemariam L, Martinez ML, Morris CA, Silva LO, Spelman RJ, Mulatu W, Zhao K, Abbey CA, Agaba M, Araujo FR, Bunch RJ, Burton J, Gorni C, Olivier H, Harrison BE, Luff B, Machado MA, Mwakaya J, Plastow G, Sim W, Smith T, Thomas MB, Valentini A, Williams P, Womack J, Woolliams JA, Liu Y, Qin X, Worley KC, Gao C, Jiang H, Moore SS, Ren Y, Song XZ, Bustamante CD, Hernandez RD, Muzny DM, Patil S, San Lucas A, Fu Q, Kent MP, Vega R, Matukumalli A, McWilliam S, Sclep G, Bryc K, Choi J, Gao H, Grefenstette JJ, Murdoch B, Stella A, Villa-Angulo R, Wright M, Aerts J, Jann O, Negrini R, Goddard ME, Hayes BJ, Bradley DG, Barbosa da Silva M, Lau LP, Liu GE, Lynn DJ, Panzitta F, Dodds KG (2009) Genome-wide survey of SNP variation uncovers the genetic structure of cattle breeds. Science 324(5926):528–532. https://doi.org/10.1126/science.1167936
22. Bovine Genome Sequencing and Analysis Consortium, Elsik CG, Tellam RL, Worley KC, Gibbs RA, Muzny DM, Weinstock GM, Adelson DL, Eichler EE, Elnitski L, Guigo R, Hamernik DL, Kappes SM, Lewin HA, Lynn DJ, Nicholas FW, Reymond A, Rijnkels M, Skow LC, Zdobnov EM, Schook L, Womack J, Alioto T, Antonarakis SE, Astashyn A, Chapple CE, Chen HC, Chrast J, Camara F, Ermolaeva O, Henrichsen CN, Hlavina W, Kapustin Y, Kiryutin B, Kitts P, Kokocinski F, Landrum M, Maglott D, Pruitt K, Sapojnikov V, Searle SM, Solovyev V, Souvorov A, Ucla C, Wyss C, Anzola JM, Gerlach D, Elhaik E, Graur D, Reese JT, Edgar RC, McEwan JC, Payne GM, Raison JM, Junier T, Kriventseva EV, Eyras E, Plass M, Donthu R, Larkin DM, Reecy J, Yang MQ, Chen L, Cheng Z, Chitko-McKown CG, Liu GE, Matukumalli LK, Song J, Zhu B, Bradley DG, Brinkman FS, Lau LP, Whiteside MD, Walker A, Wheeler TT, Casey T, German JB, Lemay DG, Maqbool NJ, Molenaar AJ, Seo S, Stothard P, Baldwin CL, Baxter R, Brinkmeyer-Langford CL, Brown WC, Childers CP, Connelley T, Ellis SA, Fritz K, Glass EJ, Herzig CT, Iivanainen A, Lahmers KK, Bennett AK, Dickens CM, Gilbert JG, Hagen DE, Salih H, Aerts J, Caetano AR, Dalrymple B, Garcia JF, Gill CA, Hiendleder SG, Memili E, Spurlock D, Williams JL, Alexander L, Brownstein MJ, Guan L, Holt RA, Jones SJ, Marra MA, Moore R, Moore SS, Roberts A, Taniguchi M, Waterman RC, Chacko J, Chandrabose MM, Cree A, Dao MD, Dinh HH, Gabisi RA, Hines S, Hume J, Jhangiani SN, Joshi V, Kovar CL, Lewis LR, Liu YS, Lopez J, Morgan MB, Nguyen NB, Okwuonu GO, Ruiz SJ, Santibanez J, Wright RA, Buhay C, Ding Y, Dugan-Rocha S, Herdandez J, Holder M, Sabo A, Egan A, Goodell J, Wilczek-Boney K, Fowler GR, Hitchens ME, Lozado RJ, Moen C, Steffen D, Warren JT, Zhang J, Chiu R, Schein JE, Durbin KJ, Havlak P, Jiang H, Liu Y, Qin X, Ren Y, Shen Y, Song H, Bell SN, Davis C, Johnson AJ, Lee S, Nazareth LV, Patel BM, Pu LL, Vattathil S, Williams RL Jr, Curry S, Hamilton C, Sodergren E, Wheeler DA, Barris W, Bennett GL, Eggen A, Green RD, Harhay GP, Hobbs M, Jann O, Keele JW, Kent MP, Lien S, McKay SD, McWilliam S, Ratnakumar A, Schnabel RD, Smith T, Snelling WM, Sonstegard TS, Stone RT, Sugimoto Y, Takasuga A, Taylor JF, Van Tassell CP, Macneil MD, Abatepaulo AR, Abbey CA, Ahola V, Almeida IG, Amadio AF, Anatriello E, Bahadue SM, Biase FH, Boldt CR, Carroll JA, Carvalho WA, Cervelatti EP, Chacko E, Chapin JE, Cheng Y, Choi J, Colley AJ, de Campos TA, De Donato M, Santos IK, de Oliveira CJ, Deobald H, Devinoy E, Donohue KE, Dovc P, Eberlein A, Fitzsimmons CJ, Franzin AM, Garcia GR, Genini S, Gladney CJ, Grant JR, Greaser ML, Green JA, Hadsell DL, Hakimov HA, Halgren R, Harrow JL, Hart EA, Hastings N, Hernandez M, Hu ZL, Ingham A, Iso-Touru T, Jamis C, Jensen K, Kapetis D, Kerr T, Khalil SS, Khatib H, Kolbehdari D, Kumar CG, Kumar D, Leach R, Lee JC, Li C, Logan KM, Malinverni R, Marques E, Martin WF, Martins NF, Maruyama SR, Mazza R, McLean KL, Medrano JF, Moreno BT, More DD, Muntean CT, Nandakumar HP, Nogueira MF, Olsaker I, Pant SD, Panzitta F, Pastor RC, Poli MA, Poslusny N, Rachagani S, Ranganathan S, Razpet A, Riggs PK, Rincon G, Rodriguez-Osorio N, Rodriguez-Zas SL, Romero NE, Rosenwald A, Sando L, Schmutz SM, Shen L, Sherman L, Southey BR, Lutzow YS, Sweedler JV, Tammen I, Telugu BP, Urbanski JM, Utsunomiya YT, Verschoor CP, Waardenberg AJ, Wang Z, Ward R, Weikard R, Welsh TH Jr, White SN, Wilming LG, Wunderlich KR, Yang J, Zhao FQ (2009) The genome sequence of taurine cattle: a window to ruminant biology and evolution. Science 324(5926):522–528. https://doi.org/10.1126/science.1169588
23. Sherry ST, Ward MH, Kholodov M, Baker J, Phan L, Smigielski EM, Sirotkin K (2001) dbSNP: the NCBI database of genetic variation. Nucleic Acids Res 29(1):308–311

24. Lappalainen I, Lopez J, Skipper L, Hefferon T, Spalding JD, Garner J, Chen C, Maguire M, Corbett M, Zhou G, Paschall J, Ananiev V, Flicek P, Church DM (2013) DbVar and DGVa: public archives for genomic structural variation. Nucleic Acids Res 41(Database issue):D936–D941. https://doi.org/10.1093/nar/gks1213
25. Vilella AJ, Severin J, Ureta-Vidal A, Heng L, Durbin R, Birney E (2009) EnsemblCompara GeneTrees: complete, duplication-aware phylogenetic trees in vertebrates. Genome Res 19(2):327–335. https://doi.org/10.1101/gr.073585.107
26. Aken BL, Achuthan P, Akanni W, Amode MR, Bernsdorff F, Bhai J, Billis K, Carvalho-Silva D, Cummins C, Clapham P, Gil L, Giron CG, Gordon L, Hourlier T, Hunt SE, Janacek SH, Juettemann T, Keenan S, Laird MR, Lavidas I, Maurel T, McLaren W, Moore B, Murphy DN, Nag R, Newman V, Nuhn M, Ong CK, Parker A, Patricio M, Riat HS, Sheppard D, Sparrow H, Taylor K, Thormann A, Vullo A, Walts B, Wilder SP, Zadissa A, Kostadima M, Martin FJ, Muffato M, Perry E, Ruffier M, Staines DM, Trevanion SJ, Cunningham F, Yates A, Zerbino DR, Flicek P (2017) Ensembl 2017. Nucleic Acids Res 45(D1):D635–D642. https://doi.org/10.1093/nar/gkw1104
27. Bickhart DM, Rosen BD, Koren S, Sayre BL, Hastie AR, Chan S, Lee J, Lam ET, Liachko I, Sullivan ST, Burton JN, Huson HJ, Nystrom JC, Kelley CM, Hutchison JL, Zhou Y, Sun J, Crisa A, Ponce de Leon FA, Schwartz JC, Hammond JA, Waldbieser GC, Schroeder SG, Liu GE, Dunham MJ, Shendure J, Sonstegard TS, Phillippy AM, Van Tassell CP, Smith TP (2017) Single-molecule sequencing and chromatin conformation capture enable de novo reference assembly of the domestic goat genome. Nat Genet 49(4):643–650. https://doi.org/10.1038/ng.3802
28. Gene Ontology Consortium (2015) Gene Ontology Consortium: going forward. Nucleic Acids Res 43(Database issue):D1049–D1056. https://doi.org/10.1093/nar/gku1179
29. Resource Coordinators NCBI (2016) Database resources of the National Center for Biotechnology Information. Nucleic Acids Res 44(D1):D7–19. https://doi.org/10.1093/nar/gkv1290
30. Kerrien S, Aranda B, Breuza L, Bridge A, Broackes-Carter F, Chen C, Duesbury M, Dumousseau M, Feuermann M, Hinz U, Jandrasits C, Jimenez RC, Khadake J, Mahadevan U, Masson P, Pedruzzi I, Pfeiffenberger E, Porras P, Raghunath A, Roechert B, Orchard S, Hermjakob H (2012) The IntAct molecular interaction database in 2012. Nucleic Acids Res 40(Database issue):D841–D846. https://doi.org/10.1093/nar/gkr1088
31. Finn RD, Attwood TK, Babbitt PC, Bateman A, Bork P, Bridge AJ, Chang HY, Dosztanyi Z, El-Gebali S, Fraser M, Gough J, Haft D, Holliday GL, Huang H, Huang X, Letunic I, Lopez R, Lu S, Marchler-Bauer A, Mi H, Mistry J, Natale DA, Necci M, Nuka G, Orengo CA, Park Y, Pesseat S, Piovesan D, Potter SC, Rawlings ND, Redaschi N, Richardson L, Rivoire C, Sangrador-Vegas A, Sigrist C, Sillitoe I, Smithers B, Squizzato S, Sutton G, Thanki N, Thomas PD, Tosatto SC, Wu CH, Xenarios I, Yeh LS, Young SY, Mitchell AL (2017) InterPro in 2017-beyond protein family and domain annotations. Nucleic Acids Res 45(D1):D190–D199. https://doi.org/10.1093/nar/gkw1107
32. Kanehisa M, Furumichi M, Tanabe M, Sato Y, Morishima K (2017) KEGG: new perspectives on genomes, pathways, diseases and drugs. Nucleic Acids Res 45(Database issue):D353–D361. https://doi.org/10.1093/nar/gkw1092
33. Zdobnov EM, Tegenfeldt F, Kuznetsov D, Waterhouse RM, Simao FA, Ioannidis P, Seppey M, Loetscher A, Kriventseva EV (2017) OrthoDB v9.1: cataloging evolutionary and functional annotations for animal, fungal, plant, archaeal, bacterial and viral orthologs. Nucleic Acids Res 45(D1):D744–D749. https://doi.org/10.1093/nar/gkw1119
34. NCBI Resource Coordinators (2017) Database resources of the National Center for Biotechnology Information. Nucleic Acids Res 45(D1):D12–D17. https://doi.org/10.1093/nar/gkw1071
35. Fabregat A, Sidiropoulos K, Garapati P, Gillespie M, Hausmann K, Haw R, Jassal B, Jupe S, Korninger F, McKay S, Matthews L, May B, Milacic M, Rothfels K, Shamovsky V, Webber M, Weiser J, Williams M, Wu G, Stein L, Hermjakob H, D'Eustachio P (2016) The Reactome Pathway Knowledgebase. Nucleic Acids Res 44(D1):D481–D487. https://doi.org/10.1093/nar/gkv1351
36. O'Leary NA, Wright MW, Brister JR, Ciufo S, Haddad D, McVeigh R, Rajput B, Robbertse B, Smith-White B, Ako-Adjei D, Astashyn A, Badretdin A, Bao Y, Blinkova O, Brover V, Chetvernin V, Choi J, Cox E, Ermolaeva O, Farrell CM, Goldfarb T, Gupta T, Haft D, Hatcher E, Hlavina W, Joardar VS, Kodali VK, Li W, Maglott D, Masterson P, McGarvey KM, Murphy MR, O'Neill K, Pujar S, Rangwala SH, Rausch D, Riddick LD, Schoch C, Shkeda

A, Storz SS, Sun H, Thibaud-Nissen F, Tolstoy I, Tully RE, Vatsan AR, Wallin C, Webb D, Wu W, Landrum MJ, Kimchi A, Tatusova T, DiCuccio M, Kitts P, Murphy TD, Pruitt KD (2016) Reference sequence (RefSeq) database at NCBI: current status, taxonomic expansion, and functional annotation. Nucleic Acids Res 44(D1):D733–D745. https://doi.org/10.1093/nar/gkv1189

37. Kodama Y, Shumway M, Leinonen R, International Nucleotide Sequence Database Consortium (2012) The Sequence Read Archive: explosive growth of sequencing data. Nucleic Acids Res 40(Database issue):D54–D56. https://doi.org/10.1093/nar/gkr854

38. Jiang Y, Xie M, Chen W, Talbot R, Maddox JF, Faraut T, Wu C, Muzny DM, Li Y, Zhang W, Stanton JA, Brauning R, Barris WC, Hourlier T, Aken BL, Searle SM, Adelson DL, Bian C, Cam GR, Chen Y, Cheng S, DeSilva U, Dixen K, Dong Y, Fan G, Franklin IR, Fu S, Fuentes-Utrilla P, Guan R, Highland MA, Holder ME, Huang G, Ingham AB, Jhangiani SN, Kalra D, Kovar CL, Lee SL, Liu W, Liu X, Lu C, Lv T, Mathew T, McWilliam S, Menzies M, Pan S, Robelin D, Servin B, Townley D, Wang W, Wei B, White SN, Yang X, Ye C, Yue Y, Zeng P, Zhou Q, Hansen JB, Kristiansen K, Gibbs RA, Flicek P, Warkup CC, Jones HE, Oddy VH, Nicholas FW, McEwan JC, Kijas JW, Wang J, Worley KC, Archibald AL, Cockett N, Xu X, Wang W, Dalrymple BP (2014) The sheep genome illuminates biology of the rumen and lipid metabolism. Science 344(6188):1168–1173. https://doi.org/10.1126/science.1252806

39. Nicolazzi EL, Caprera A, Nazzicari N, Cozzi P, Strozzi F, Lawley C, Pirani A, Soans C, Brew F, Jorjani H, Evans G, Simpson B, Tosser-Klopp G, Brauning R, Williams JL, Stella A (2015) SNPchiMp v.3: integrating and standardizing single nucleotide polymorphism data for livestock species. BMC Genomics 16:283. https://doi.org/10.1186/s12864-015-1497-1

40. Schreiber F, Patricio M, Muffato M, Pignatelli M, Bateman A (2014) TreeFam v9: a new website, more species and orthology-on-the-fly. Nucleic Acids Res 42(Database issue):D922–D925. https://doi.org/10.1093/nar/gkt1055

41. UniProt Consortium (2015) UniProt: a hub for protein information. Nucleic Acids Res 43(Database issue):D204–D212. https://doi.org/10.1093/nar/gku989

42. Huntley RP, Sawford T, Mutowo-Meullenet P, Shypitsyna A, Bonilla C, Martin MJ, O'Donovan C (2015) The GOA database: gene ontology annotation updates for 2015. Nucleic Acids Res 43(Database issue):D1057–D1063. https://doi.org/10.1093/nar/gku1113

Chapter 10

Navigating Xenbase: An Integrated *Xenopus* Genomics and Gene Expression Database

Christina James-Zorn, Virgilio Ponferrada, Malcolm E. Fisher, Kevin Burns, Joshua Fortriede, Erik Segerdell, Kamran Karimi, Vaneet Lotay, Dong Zhuo Wang, Stanley Chu, Troy Pells, Ying Wang, Peter D. Vize, and Aaron Zorn

Abstract

Xenbase is the *Xenopus* model organism database (www.xenbase.org), a web-accessible resource that integrates the diverse genomic and biological data for *Xenopus* research. It hosts a variety of content including current and archived genomes for both *X. laevis* and *X. tropicalis*, bioinformatic tools for comparative genetic analyses including BLAST and GBrowse, annotated *Xenopus* literature, and catalogs of reagents including antibodies, ORFeome clones, morpholinos, and transgenic lines. Xenbase compiles gene-specific pages which include manually curated gene expression images, functional information including gene ontology (GO), disease associations, and links to other major data sources such as NCBI:Entrez, UniProtKB, and Ensembl. We also maintain the *Xenopus* Anatomy Ontology (XAO) which describes anatomy throughout embryonic development. This chapter provides a full description of the many features of Xenbase, and offers a guide on how to use various tools to perform a variety of common tasks such as identifying nucleic acid or protein sequences, finding gene expression patterns for specific genes, stages or tissues, identifying literature on a specific gene or tissue, locating useful reagents and downloading our extensive content, including *Xenopus* gene-Human gene disease mapping files.

Key words *Xenopus*, Genome database, Polyploid genome, Gene expression analysis, Anatomy ontology, BLAST, GBrowse, Textpresso

1 Introduction

Modern cell and developmental biologists have relied on the large externally developing embryos of amphibians, particularly in the African clawed frogs of the genus *Xenopus*, since the late 1950s. Early cloning experiments in *Xenopus* demonstrated that differentiated cells contained the full complement of nuclear material, the principle of genomic equivalence [1, 2], and this finding revolutionized the understanding of cell differentiation, and thus paved the way, decades later, to induce pluripotent stem cells which in

Martin Kollmar (ed.), *Eukaryotic Genomic Databases: Methods and Protocols*, Methods in Molecular Biology, vol. 1757,
https://doi.org/10.1007/978-1-4939-7737-6_10, © Springer Science+Business Media, LLC, part of Springer Nature 2018

turn has revolutionized regenerative biomedical research. While *Xenopus* has been an outstanding system to make fundamental discoveries such as these, it has also played a major role in understanding pathological processes and elucidating the function of an increasing number of human disease genes (reviewed in [3]). Importantly, as the major nonmammalian tetrapod model in biomedical research, *Xenopus* research bridges the gap between the mammalian models and the more evolutionarily distant vertebrates such as teleosts [3].

Today, genomic data is at the core of all modern experimental design and interpretation. Xenbase is the *Xenopus* Model Organism Database (MOD), launched in 2005 (*see* [4]), and now running in a virtual environment [5], whose mission is to integrate and widely disseminate key molecular, cell, developmental, and bioinformatic data about *Xenopus*. We aim to accelerate discovery and to support the use of *Xenopus* for modeling human disease. To this end, Xenbase content is integrated with other MODs (MGI, Zfin, Geisha, WormBase; *see* Table 1 for a full list of abbreviations and website links used) and human disease databases (OMIM, Decipher, MalaCards, Gene Cards, HGNC). Our system associates *Xenopus* genes through "Gene Pages" to the orthologous human genes, and reciprocal data exchanges with numerous external databases and knowledgebases (e.g., NCBI, Entrez Gene, UniProtKB, and Ensembl). Thus, Xenbase not only supports *Xenopus* researchers but also makes *Xenopus* data broadly available to researchers in diverse fields, from cell and developmental biology, to environmental toxicology and human disease research.

The DNA sequencing revolution of the 2000s quickly focused on model organisms, and the first amphibian species to be sequenced was the diploid Western clawed frog *Xenopus tropicalis* [6]. The larger *Xenopus* species, the African clawed frog, *X. laevis*, which is widely used as the nonmammalian tetrapod model in biomedical research, posed a more intractable problem to sequence because it is an allotetraploid ($2n = 36$). *X. laevis* likely arose via the interspecific hybridization of two diploid progenitors with $2n = 18$, followed by subsequent genome doubling which restored meiotic pairing and disomic inheritance [7]. The sequencing, genome assembly, and annotation of *X. laevis* was, not surprisingly, very complicated [8] and took several years to complete [7]. Simultaneous integration of the two *X. laevis* homologs (referred to as "L" and "S" for long and short chromosomes, respectively [9]) into the Xenbase genome module was finalized in 2016. As a result, Xenbase currently provides cell and developmental biologists the most up-to-date genomic information based on both frog species, and this data is displayed on our genome browser and on

Table 1
Glossary of abbreviations for online resources, databases, and tools referred to in text, and/or linked to from Xenbase Gene Pages, with website address

Resource	Description	Website address
Allen Brain Atlas	A comprehensive database with a suite of tools to view neurobiology in humans, mouse and nonhuman primates.	www.brain-map.org
CRB	Center for Xenopus Biological Resources, based in France.	xenopus.univ-rennes1.fr
Decipher	Mapping database to comparison Human clinical phenotypic and genomic data.	decipher.sanger.ac.uk
DRYAD	A curated data repository for scientific and medical literature.	datadryad.org
Ensembl	A genome browser for comparative vertebrate genomics.	www.ensembl.org/index.html
Eurexpress	A Transcriptome Atlas Database for the Mouse Embryo.	www.eurexpress.org/
EXRC	European Xenopus Resource Center based in UK.	xenopusresource.org
FlyBase	A Database for *Drosophila* Genes and Genomes.	flybase.org
GBrowse	An interactive tool used by most MODs to manipulate and display genomes.	
Geisha	A Chicken Embryo Gene Expression Database.	geisha.arizona.edu/geisha/index.jsp
GeneCards	The Human Gene Database with integrated genomic, transcriptomic, proteomic, genetic, clinical and functional information.	www.genecards.org
Genomicus	Genomes in Evolution. A genome browser to display genes/genomes across taxa, through time and in predicted ancestral species.	www.genomicus.biologie.ens.fr/genomicus-88.01/cgi-bin/search.pl
GitHub	Online version control depository, where open source software and code, like the Xenopus Anatomy Ontology, is available.	github.com/
GO	The Gene Ontology, from the GO Consortium.	www.geneontology.org
HGNC	Human Gene Nomenclature Committee.	www.genenames.org
iHOP	Information Hyperlinked Over Protein.	www.ihop-net.org/UniPub/iHOP/
IMPC	International Mouse Phenotyping Consortium.	www.mousephenotype.org

(continued)

Table 1
(continued)

Resource	Description	Website address
JBrowse	JBrowse is a new genome browser which will replace GBrowse on Xenbase, (over ~2 years phase-out period) because GBrowse is no longer supported or being developed.	jbrowse.org
JGI-*Xenopus*	Joint Genome Institute, *Xenopus* genome project.	jgi.doe.gov/xenopus-frog-genome-project-on-cbc/
JGI-Metazome	Genome database that organizes the proteomes of metazoans into gene families in evolutionary context.	metazome.jgi.doe.gov/pz/portal.html
JGI/KOG	Functional protein annotations from fungal genomics resource at Joint Genome Institute.	genome.jgi.doe.gov/help/kogbrowser.jsf
KEGG	The Kyoto Encyclopedia of Genes and Genomes.	www.kegg.jp/kegg
MalaCards	Human Disease Database with clinical and genetic annotations.	www.malacards.org
MGI	Mouse Genomic Informatics.	www.informatics.jax.org
miRBase	A searchable database of published miRNA sequences and annotations.	www.mirbase.org/index.shtml
NBRP	National BioResource project, based in Japan.	www.nbrp.jp/report/reportProject.jsp?project=xenopus
NCBI	National Center for Biotechnology Information. Hosts a extensive range of biomedical and genomic databases and analysis tools to support advances in science and human health.	www.ncbi.nlm.nih.gov
NCBI/BLAST	The Basic Local Alignment Search Tool (BLAST) finds regions of local similarity between sequences.	blast.ncbi.nlm.nih.gov/Blast.cgi
NCBI/Entrez Gene	A portal to gene-specific content based on NCBI's RefSeq project, model organism databases and others.	www.ncbi.nlm.nih.gov/gene
NCBI/GEO	Gene Expression Omnibus, functional genomics data repository at NCBI.	www.ncbi.nlm.nih.gov/geo
NCBI/HomoloGene	A tool to construct putative homology groups from gene sequences.	www.ncbi.nlm.nih.gov/homologene
NCBI/SRA	Sequence Read Archive, stores raw sequence data from next-generation sequencing projects.	trace.ncbi.nlm.nih.gov/Traces/sra/sra.cgi
NXR	National Xenopus Resource based in USA.	www.mbl.edu/xenopus

(continued)

Table 1 (continued)

Resource	Description	Website address
OBO Foundry	*Open Biomedical Ontologies.*	www.obofoundry.org
OMIM	Online Mendelian Inheritance in Man, An Online Catalog of Human Genes and Genetic Disorders.	omim.org
Panther	Protein Annotation Through Evolutionary Relationship, a large-scale gene function analysis tool.	pantherdb.org
RRID	Research Resource Identifiers, which are persistent and unique identifiers we use to reference research resources, such as antibodies and transgenic *Xenopus* lines.	scicrunch.org/resources
The Human Protein Atlas	Database of protein coding genes, their expression and localization at tissue and cellular levels.	www.proteinatlas.org
TrEMBL	A computer-annotated supplement of SwissProt that contains all the translations of EMBL nucleotide sequence entries not yet integrated in SwissProt.	www.uniprot.org/uniprot
Uberon	Integrated multispecies anatomy ontology, available on GitHub.	uberon.github.io
UniProtKB/ Swiss-Prot	A protein sequence and function database.	www.uniprot.org
WormBase	A database for genetics, genomics and biology of *C. elegans* and related nematodes.	www.wormbase.org
XenMARK	Heatmap-based *Xenopus* gene expression image annotation tool.	genomics.crick.ac.uk/apps/ XenMARK
XenMine	Multitool analysis resource for published Xenopus genomic data.	www.xenmine.org
XGNC	Xenopus Gene Nomenclature Committee, the scientific group charged with gene nomenclature review and approval, coordinated by Xenbase.	
Zfin	The Zebrafish Information Network.	zfin.org

Gene Pages, with both the *X. tropicalis* and corresponding *X. laevis* L and S genes. Combined with an extensive catalog of curated literature, that covers over 48,000 published *Xenopus* articles, and a vast catalog of manually curated, tissue-specific gene expression images (66,000+), Xenbase is the go-to site for the most-up-to-date genomic *Xenopus* data. In addition, Xenbase hosts a vast amount of technical and reference material on *Xenopus* development, anatomy (including the extensive Xenopus Anatomy Ontology (XAO) [10]), and husbandry. Xenbase also provides an online hub for researchers, as we host personal profiles and laboratory descriptions, list conferences, workshops, a jobs board, discussion forum, and an array of links to other resources.

This chapter aims to give a practical guide on how to access the major features of Xenbase in a step-by-step manner, first covering how to navigate the home page, the extensive data on Gene Pages, then how to use the Quick Search Menu. We continue with a discussion of how to utilize Xenbase to its full potential- the remaining topics are presented in the order as they appear of the drop-down menus (except Gene Pages), going from left to right. We discuss how to find markers for a specific organ system, download large NextGen Sequence (NGS) data, use genomic tools (like BLAST and GBrowse), find guidelines on gene and transgenic nomenclature, and locate *Xenopus* specific protocols or reagents.

2 Navigating the Xenbase Home Page

The home page (http://www.xenbase.org) combines the horizontal navigation bars that are common to all Xenbase pages with additional information in subject based "tiles" and an additional vertical navigation bar. The tiled lay-out covers the same areas that are accessible via drop-down menus in the header. Many search functions are also available in a quick search bar (aka the mini-bar), in the top right corner of the home page. Centrally placed on the Xenbase home page is a rotating image carousel, where we spotlight the latest high impact *Xenopus* research publications, and which serves as a community notice board covering, for example, conferences and workshops, awards and journal special issues. These are reiterated in the "Announcements" column on the right-hand side of the home page. This side column also gives links to static content on the website, including an introduction to *Xenopus* as a model organism, links to various features and data on Xenbase, the *Xenopus* Stock Centers, and other databases and external resources useful to *Xenopus* researchers.

3 Genes and Gene Pages

Xenbase is fundamentally a "gene-centric" database. The Genes module is a catalog of genes in the diploid *X. tropicalis* and polyploid *X. laevis*—all three genes (one *X. tropicalis* gene, and two *X. laevis* genes) are represented on a single "umbrella" Gene Page, which details all information about the *Xenopus* gene and its products. Each Gene Page carries a stable Xenbase Gene Page ID (e.g., XB-GENEPAGE-483057 is the *bmp4* Gene Page), and each gene has its own stable Xenbase gene identification number. Here we describe the information on a Gene Page, and the how to find a specific Gene Page.

3.1 How to Find a Specific Gene Page

1. Select "Gene Search" under the "Genes" menu to find specific Gene Pages or gene families. The default is to "search all," but to scale down or speed up results, choose one of the more specific search options which include a partial or full gene name (e.g., "bone" or "bone morphogenetic protein 4"), gene symbol (e.g., *bmp4*) or synonym (e.g., bmp-4), orthologs (if any with different symbols/names), or gene function (e.g., "morphogenetic protein" which will return all *bmp* genes as well as related gene families). The menu will autofill with the matched text highlighted in yellow.
2. Alternatively, enter an NCBI accession number, Entrez gene ID, Unigene ID, OMIM ID, GO ID, or GO term. Also, a Xenbase accession number such as a "Gene Page ID" can be entered (e.g., XB-GENEPAGE-483057) to find Gene Page(s).
3. Checkboxes permit you to filter results to include only "manually curated Gene Pages" or "Gene Pages with expression images."
4. Gene Pages can also be browsed alphabetically.
5. The "Advanced search" offers additional filters: to text-match specific letter combination (e.g., "*rsp*") or parts of names (e.g., "receptor").

3.2 Gene Pages

The most utilized, useful data and salient features for each gene are presented on the "Summary" tab on the Gene Page, under the following headings (as an example, enter "bmp4" into the quick search bar in the top right corner of the Xenbase hompage):

1. *Summary*: Official gene symbol and full name, synonyms, gene function, protein function, a list of cocited interactants (and a thumbnail of an interactive graphical display of interactants), and associated OMIM diseases are all detailed on the top of the Gene Page. Images that summarize the gene expression throughout a range of embryonic stages are shown to the right. Click the + link (the Xenbase symbol that additional text is available) to see all OMIM associations (if present), and click the link to "Nomenclature history" to open the Wiki tab, where changes to gene names and gene symbols are recorded. *Xenopus* gene names and symbols are identical to human gene names, whenever possible, and orthology to human genes is usually assigned by synteny. Gene names for *X. laevis* homeologs are appended with "L homeolog" or "S homeolog" to distinguish the sub-genome with which they are associated.

2. *Xenbase Gene ID*: Xenbase IDs are allocated to each species/sub-genome specific gene. The chromosome location is indicated when known, and scaffold positions are given in cases where the location has not been fully determined (e.g., due to incomplete or in-progress genome annotations).
3. *Molecules* section lists and links-out to NCBI/Entrez Gene IDs, nucleotide, and protein data at Swiss-Prot and/or TrEMBL. mRNA RefSeq data has BLAST functionality (click on the rocket icon) and sequence files in FASTA format (click magnifying glass icon to pop up sequence file), can be viewed for any listed sequence. Complete data for Nucleotides and Proteins associated with the gene are listed on relevant "tabs" at the top of the Gene Page.
4. *Genomic* data is illustrated by gene model snapshots from the genome browser JBrowse. Clicking on these options will open the full view of the gene in JBrowse. The default display is the most current genome with an annotated model for the gene displayed, and earlier versions and GBrowse view can be selected from drop-down menus under each gene model snapshot.
5. *Expression* section links out to Ensembl and UniGene entries, and RNA-Seq profiles illustrating temporal and tissue expression patterns.
6. *Data Mining* section allows researchers to access a specific gene's entry on XenMine, a comprehensive toolbox for NGS data analysis that is part of the Intermine project, and is hosted by Stanford University.
7. *Phenotype* section currently links to the morpholino screen data produced by the Smith Lab at the Gurdon Institute at the University of Cambridge. Full phenotype curation is a major priority for Xenbase in the coming year, and phenotype annotations will be posted in this section on Gene Pages.
8. *Orthology* section provides direct links to the orthologous genes recorded in human (OMIM:gene, HGNC, and GeneCards) and the relevant other model organism databases: mouse (MGI), zebrafish (Zfin), chicken (GEISHA), fruit fly (FlyBase) and worm (WormBase).
9. *Publications* lists the first article to mention the gene, and the most recent article. Click on the journal reference in parenthesis to see the Article Page in Xenbase, or click the "View All Papers" link to go to the complete list (which can also be access via the Gene Literature tab). A camera icon indicates the paper has images displayed.
10. *Functional Ontologies* section provides gene-specific links to GO terms (sourced from UniProt), gene information at PANTHER [11] (a Gene Ontology consortium project),

KEGG orthology entry for this gene, and KOG classification of the gene (sourced from JGI).

11. *Reagents* section provides links to reagents and resources tailored to the relevant gene. A link is provided to our list of design tools for CRISPR/Cas constructs that includes information on which *Xenopus* genome builds are compatible with the various tools. Links are provided to several sources for sequence clones including the EXRC and GE Dharmacon, and to our own catalog of antibodies, morpholinos and ORF clones used in *Xenopus* research involving the specific gene (*see* Subheading 9 for more reagent details). We also provide links to the details of Affymetrix array probe-sets for *Xenopus* (note these require an Affymetrix log-in to access).

3.3 Expression Tab: Viewing Gene Expression Data and Images

Xenbase displays 66,000+ in situ hybridization and immunohistochemistry images that are posted on the Gene Page "Expression" tab, under two main headings: "Community Submitted" (mostly unpublished images from large scale screens) and "Literature Images" (from journal articles). Curators manually annotate the observed gene expression in these images using terms from the Xenopus Anatomy Ontology, the XAO, to generate a gene expression annotation table for each curated image. Out of the 15,878 Gene Pages currently in Xenbase (v4.7, January 2018), c. 24% (3775 genes) have gene expression images, mostly from in situ hybridization (a camera icon indicates that images are posted for that gene). Additionally, about 95% of genes have expression data from RNA-Seq and EST Transcriptome profiles and/or developmental stage profiles determined by microarray analyses (*see* Fig. 1B, C).

Gene expression data is organized on the "Expression" tab, under the following headings:

1. *Anatomy terms*: XAO terms compiled from manual curation by Xenbase, and NBCI cDNA libraries. Use the [+/−] toggle to expand or hide terms.
2. *Anatomy stages* in which gene expression has been recorded, often unfertilized egg to adult frog stage.
3. *RNA-Seq and EST Transcriptome profiles*. We link out to:
   (a) Gurdon Institute EST database (e.g., *X. tropicalis bmp4*)
   (b) Unigene EST Profiles, with heat map of tissue-specific expression (e.g., *X. tropicalis bmp4*)
   (c) RNA-Seq and microarray profiles, publications available with temporal expression data including Owens et al. [12] and Sessions et al. [7]. Profiles from Yanai et al. [13] have been retired.

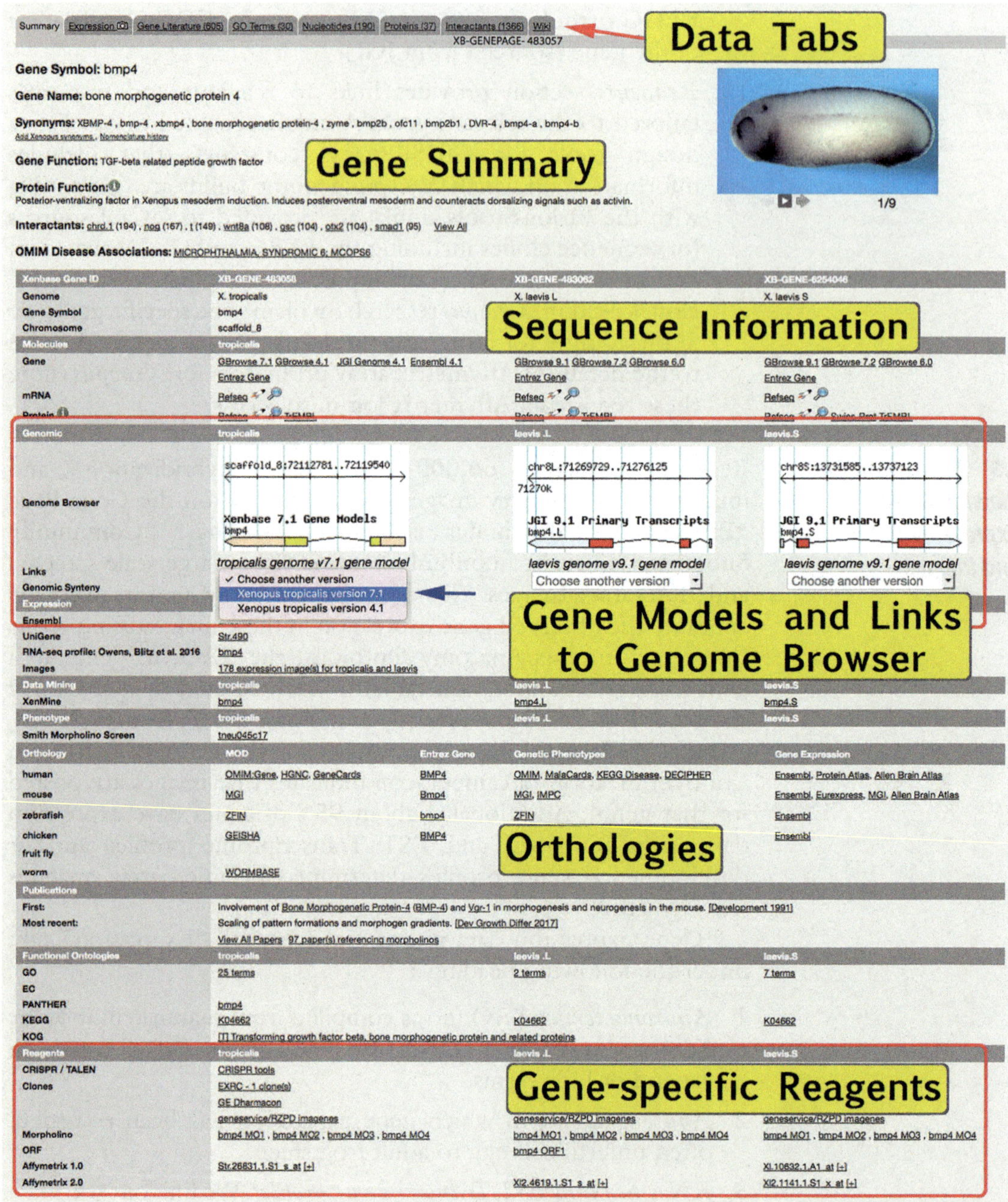

Fig. 1 Xenbase is built around the "Gene Page," where a file-like tab system provides comprehensive coverage of data about each gene. This example is the "Summary" tab for 'bone morphogenetic protein 4' (*bmp4*) with the Xenbase gene page ID 'XB-GENEPAGE-483057'. The salient features of gene (official name, synonyms, gene and protein function, cocited interactants, and human disease associations) are all shown in the upper summary panel, along with a developmental expression series (where available). Sequence information and JBrowse snapshots of the gene models are shown for *X. tropicalis* and *X. laevis* L and *X. laevis* S homeologs (upper red box). To view a different gene model, select from "choose another version" (blue arrow). The rest of the Gene Page provides links to more data covering orthology, first and most recent publications, and functional ontology, with curated gene-specific reagents (e.g., MOs, primary antibodies, and ORFeome clones) in the lower panels (lower red box). Frequently accessed tabs include the "Expression" tab (details in B–D below with a camera icon that indicates presence of images, "Gene Literature," and "GO terms" from UniProtKB,

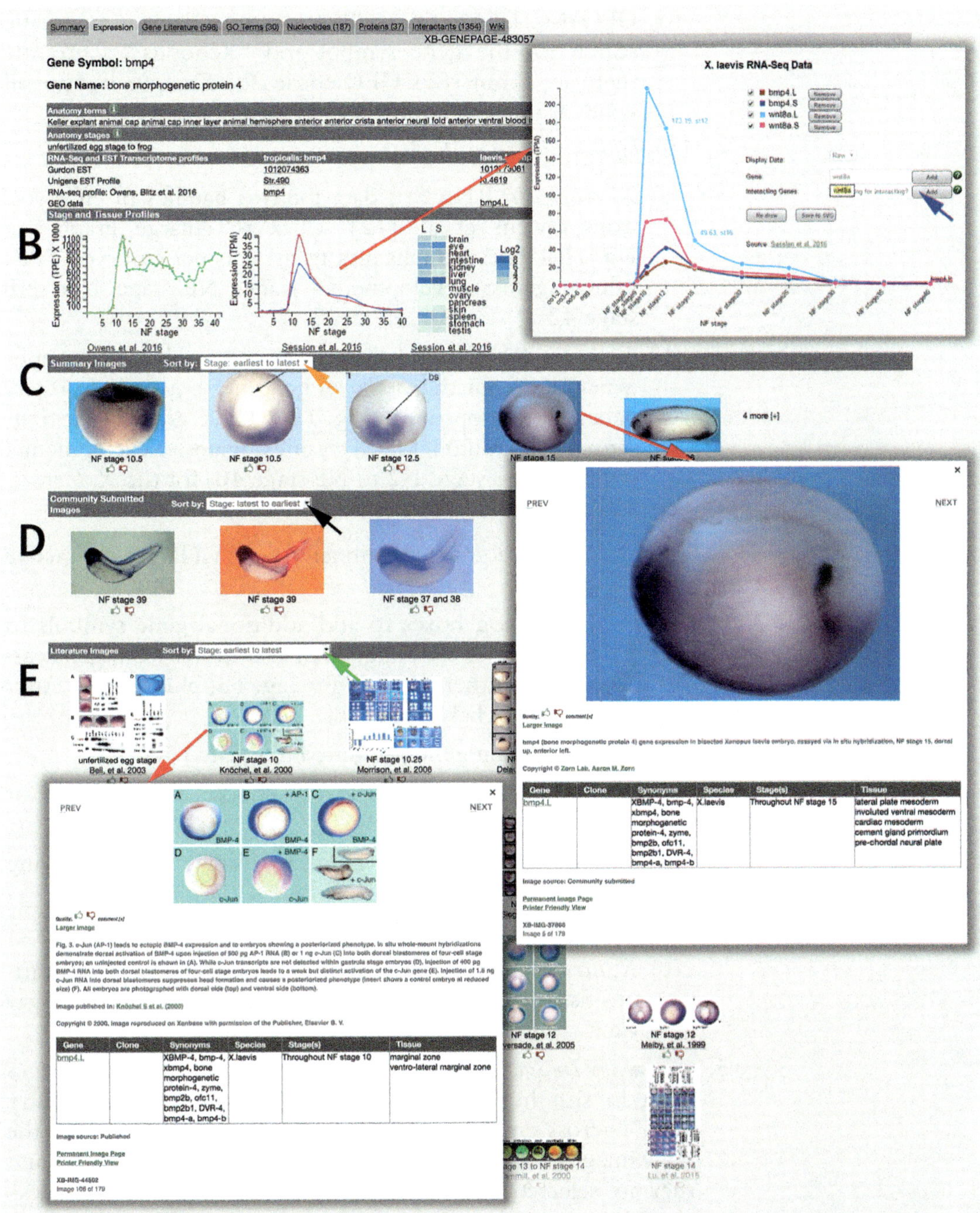

Fig. 1 (continued) where number in parenthesis indicates number of citations or terms respectively. (B) The Expression tab of a Gene Page displays gene expression data in several useful formats. Interactive graphs plot *X. laevis* L and S homeolog expression from RNA-Seq data [7] with ability to add more genes (blue arrow) to the graph via dialog boxes (red arrow, click to pop-up). Heat-maps from adult tissues compare *X. laevis* L and S homeologs, data from [7]. (C) Summary images are selected to represent gene expression over a range of embryonic stages and can be sorted by stage (orange arrow). (D) Community submitted images from large scale screens, which generally use ISH and IHC, can also be sorted by stage (black arrow). (E) Literature images from published articles can be sorted by stage or publication date (green arrow), and include link to the Xenbase Article Page. Click on the image to pop up a larger image (red arrow), along with caption and annotation table

(d) GEO data: links to this NCBI resource and runs an automatic search for the gene symbol and "Xenopus." Currently there are about 185k GEO entries for *Xenopus*, but not all genes are represented.

4. *Developmental Stage Profiles*:

(a) *X. tropicalis* RNA-Seq data for two batches of embryos, from Owens et al. [12]. Click to enlarge graph (*see* Fig. 1B). These graphs plot transcripts per embryo (TPE) values against developmental stages NF stage 1 to NF stage 42.

(b) *X. laevis* RNA-Seq data is displayed in dynamic graphs generated from the *X. laevis* genome sequencing project data [7] (*see* pop-out in Fig. 1B). These graphs plot transcripts per million (TPM) values against developmental stage (oocyte stage 1-2 to NF stage 40) for the *X. laevis* L and *X. laevis* S homeologs.

- Click the graph thumbnail to open a larger interactive graph.
- Use dialog boxes to add additional gene symbols to plot: type ahead suggests gene symbols from gene catalog, and there is no limit (Fig. 1B, blue arrow). After selecting click "Add."
- Interacting genes is limited to cocited genes.
- Use "Display data" box to choose either "Raw" or "log2" transformation.
- Mouse over a data point to display the underlying value and the stage.
- Click "save to svg" button to download graph.

(c) *X. laevis* L versus *X. laevis* S homeolog expression in various tissues illustrated via a heatmap. Click to open a larger view.

5. *Summary Images.* A curated selection of gene expression images from in situ hybridization (ISH) or immunohistochemistry (IHC) across embryonic developmental stages. These are the same images that appear in Summary section of Gene Page, and they are selected from either "Community Submitted images" or "Literature Images" by Xenbase curators (Fig. 1C). Click image to enlarge and view annotation table.

6. *Community Submitted Images* come mostly from large scale screens, and are generally ISH. Laboratory of origin holds the copyright to these images. Double click the image to enlarge it and view the annotation table.

7. *Literature images* display the curated figures from research papers where we have redisplay permission or which are open

access. Figures are often multipaneled, and gene expression annotation table is viewed by double clicking on the figure. These images may be protected by copyright; if so, this is indicated.

Notes/Troubleshooting on viewing Expression on Gene Page:

- Some genes are very well studied with hundreds of images posted. Click the [+] to toggle between more and less data [−].
- Use "Sort By" to organize by developmental stage: "earliest to latest" or "latest to earliest."
- Literature images are also sortable by earliest or latest publication data.
- Use thumbs up or thumbs down tool to vote for high quality images
- Xenbase welcomes high quality images via community submission to populate poorly studied genes! Submit new gene expression images via the "Contact Us" (email: xenbase@ucalgary.ca) in the footer of every Xenbase page.
- As there is strong conservation in gene expression in the vast majority of the expressed orthologs and in situ probes designed for one species generally work equally well in the alternate *Xenopus* species [14], gene expression tables are largely accepted as applicable to both species, although there are exceptions.
- Species (*X. tropicalis* or *X. laevis*) is indicated in the image caption for community submitted and large scale screen data.

3.4 Other Gene Page Tabs

At the top of each Gene Page, a series of file-like "tabs" collate additional gene-specific data as follows:

1. *Gene Literature* lists all articles that refer to the gene in its data or text.
2. *GO Terms* provide a quick overview of the cellular role of a gene and can also be used for analysis of high-throughput proteomics data. GO terms are presented under the three categories—Molecular Function, Biological Process, Cellular Component (sourced from UniProt). Click on the GO term for a full definition or the information button for evidence metadata.
3. *Nucleotides* tab provides links to all gene models and mRNA data from JGI, Ensembl, NCBI, Unigene clusters, mRNA and ESTs for the gene. The rocket icon will autofill a BLAST request, and the magnifying glass icon will provide a pop-up of the sequence in FASTA format. Click on Clone name or Accession number for more details.
4. *Proteins* tab links to all protein model data from JGI, NCBI, Ensembl and protein sequence from specific accessions in NCBI

Protein, RefSeq and Swiss-Prot/UniProKB. The rocket icon will auto fill a BLAST request and the magnifying glass icon will provide a pop-up of the sequence in FASTA format.

5. *Interactants*: An interactive graph illustrates the genes cocited with the gene of interest, which is placed in the center of the graph.
   - Drag the nodes to move them, and set them in place.
   - Double click to release node position.
   - Number of cocitations are marked on the edges of the graph.
   - Click on the gene symbol to go to the corresponding Gene Page.
   - Graph is downloadable in two formats: use buttons "save to svg" or "save to png."

Cocited genes are then listed in ranked descending order in two columns, with links to Gene Pages and to literature (e.g., 1358 genes have been cocited with *bmp4*, the top hit being *chrd.1* (chordin, gene 1) in 190 articles; status June 2017). Finally, links are also provided to IHOP (Information Hyperlinked over Proteins) for both *X. tropicalis* and *X. laevis*. In the near future, interactants will include data on physically interacting proteins from human networks, and also genes in coexpression or coregulated networks.

6. *Wiki*: Nomenclature changes are recorded on the Wiki tab, which can be also accessed by clicking the "Nomenclature History" link. In addition, the Wiki is used to record any information about a gene that is not recorded elsewhere on Xenbase, such as synteny analysis methods, reagent or protocol notes. Registered users can add to Wiki content.

3.5 Notes/Troubleshooting Genes Module

- Genes can also be searched using the Quick Search Menu (*see* Subheading 4 below).
- Can't find a gene? If you cannot find a Gene Page for a gene of interest, try our Search Help page for hints. *Xenopus* genes are following human gene nomenclature, so searching by an old name may not work. We store old or "legacy" gene names as synonyms. If your search fails, it may mean Xenbase does not have that gene name or symbol in the database. Try the human, mouse, chicken, or zebrafish gene symbol. If this fails also, the ultimate gene finder requires you to BLAST the Xenbase genome database as detailed in Subheading 5.
- Gene nomenclature issues? Xenbase is the clearing house for *Xenopus* gene nomenclature. Gene Nomenclature Guidelines are posted under the Genes menu. As gene nomenclature is

updated constantly by the HGNC, many gene names and symbols completely change over time. Although gene symbol synonyms are a powerful tool to track down the new name for a gene, they also can be misleading, especially when the same gene symbol has been used/reused in different species/model organisms. NCBI databases record a more comprehensive list of legacy synonyms and symbols than Xenbase, as we try to concentrate on just those symbols used/referred to in *Xenopus* literature. Note that our gene search *does not* search Wiki entries, which is where gene nomenclature changes are recorded, however the "Search with Google" in the Quick Search menu does search the Wiki (and everywhere else). Suggest adding a gene name, missing synonyms, or report errors or omissions by contacting Xenbase (xenbase@ucalgary.ca).

- Why is not there an L or S model for this gene? Not all *X. laevis* genes have both *X. laevis* homeologs. After the hybridization event that created *X. laevis*, there was a genome reduction that resulted in loss of some homeologous genes with a higher proportion of S genes being removed than L genes [7]. It is also possible that the homeologous locus is still being assembled fully, or both gene models exist, but only one has been properly annotated.
- Where did the A and B genes go? With the discovery that *X. laevis* contains two independently interacting legacy genomes that can be distinguished from each other, "A" and "B" genes were migrated to the more informative L and S nomenclature.

4 Quick Search Menu

The quickest way to get to the most popular and well-used content on Xenbase is to use the **Quick Search Menu** (aka the mini-bar) in the top right hand corner of the home page and every Xenbase page (in red box, Fig. 2). Select the search topic from the dropdown options, and enter a term to search as follows:

1. *Genes*: Enter a partial or full gene symbol (e.g., "fgf") or partial gene name (e.g., "fibroblast") to return all "fgf" family genes as well as "fgfr" genes, genes with "FGF" in the name, function or synonyms. Get precise, single gene return by entering an exact gene name or symbol (e.g., "fgf3," or 'fibroblast growth factor 3') (*see* Fig. 3A).
2. *Xenbase with Google*: Search Xenbase for *any* text, e.g., a partial article title or phrase, gene symbol, clone ID, or author with the Google search option to pull *every* match in the Xenbase database, including Wiki entries. Searching for "fgf3," for example, returns the Gene Page record, in situ data, expres-

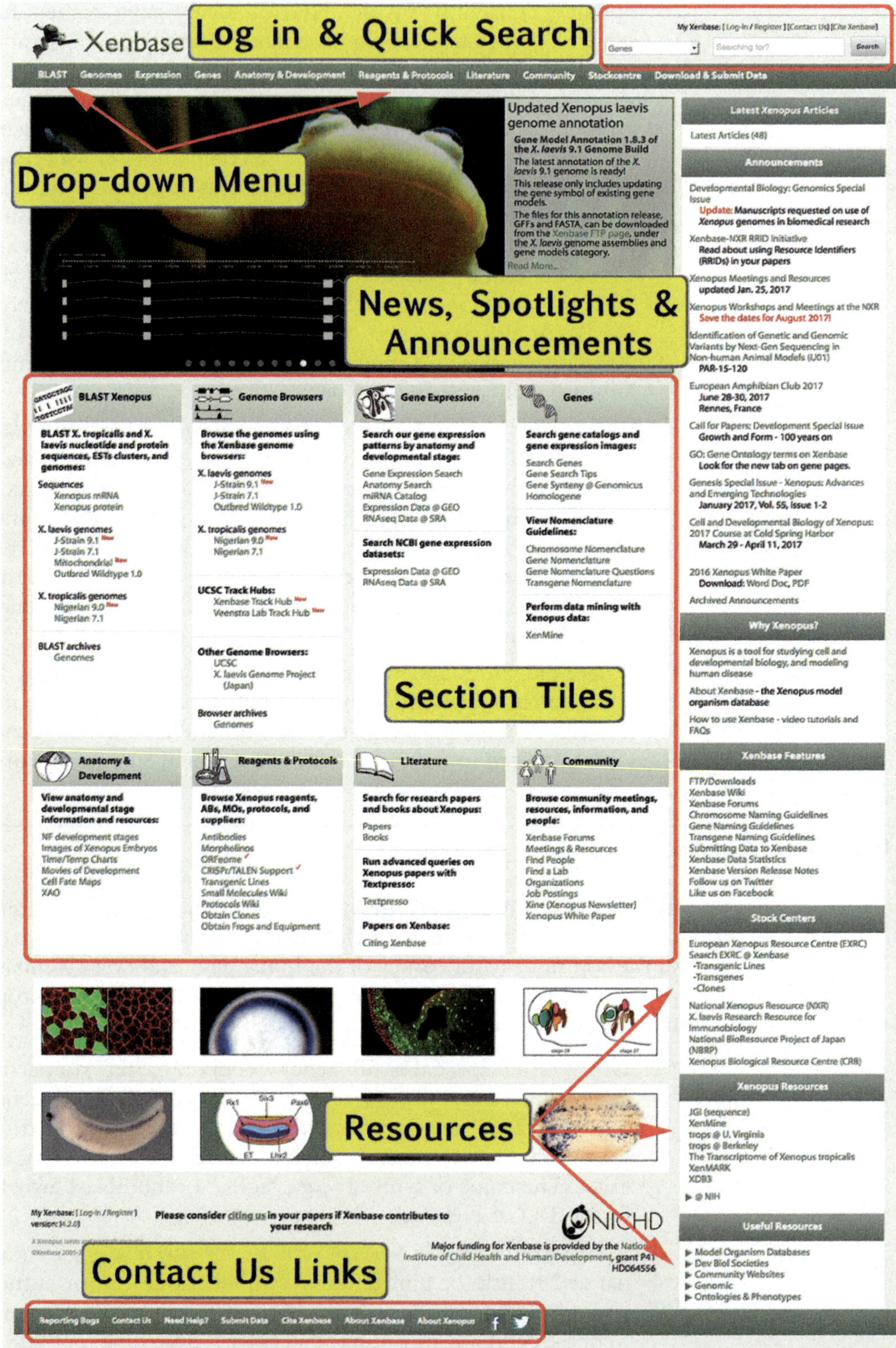

Fig. 2 The Xenbase home page (http://www.xenbase.org) features a rotating image carousel to spotlight new articles and announce relevant news to the *Xenopus* community. Log-in and the Quick Search minibar are in the upper right corner. The drop-down menu bar spans the top of the web page, and reiterates the links in the

sion profiles, literature for that gene, the anatomy term expression page (for which it has gene expression curations), a list of potential gene regulatory network interactants and cocited genes, as well as all ORFeome clones and plasmids mapped to this gene. We control which pages google indexes, so if you note something missing from these search results please let us know and staff will ensure that the missing content is included in future crawls.

3. *Anatomy Items*: Enter an anatomical term (e.g., "heart," *see* Fig. 3B) to find all "Xenopus Anatomy Ontology" (XAO) terms [10] used in gene expression annotations. Select a term as it autofills from the XAO, text matches are highlighted in yellow, in addition to showing all elements that are "part of" the term (e.g., "cardiac mesoderm"). Selecting any option from drop down will take you to the specific XAO term page.
4. *People*: Find any of the 1900+ researchers with Xenbase profiles. Enter any part of a person's first or last name and it autofills a list, highlighting in yellow the text match. Hit search to display all results.
5. *Labs*: Find any of the over 270 *Xenopus* research labs with Xenbase profiles. Laboratories are generally named with group leader's last name (e.g., Smith Lab).
6. *Organizations*: Find contact information for stock centers and other organizations that supply reagents, frogs and husbandry equipment, as well as publishers of key life science journals and scientific societies (e.g., NXR, *see* Fig. 3C).
7. *Paper Authors*: Enter a surname to search all authors of all 48,000+ *Xenopus* research papers in the literature module. Enter any part of an author's last name, and autofill options will highlight matched text in yellow (*see* Fig. 3D). Select a specific author or hit search to display all results. This search will also find letter combinations, e.g., "vg" will find all instances in both the first and last names and as an author's initials.
8. *Paper Title*: Enter the entire paper title to find a specific paper, or a partial title or any word or phrase from the title of a published article to run a quick literature search for *Xenopus* specific articles on a topic (e.g., "left–right" to return all papers on "left–right patterning," "left–right asymmetry," and "left–right axis determination"; Fig. 3E).

Fig. 2 (continued) subject tiles below. Additional links to *Xenopus* resources are in the side column, and social media and contact Xenbase links are in the footer. This layout visually describes the database architecture and is designed to accommodate different workflows and preferences

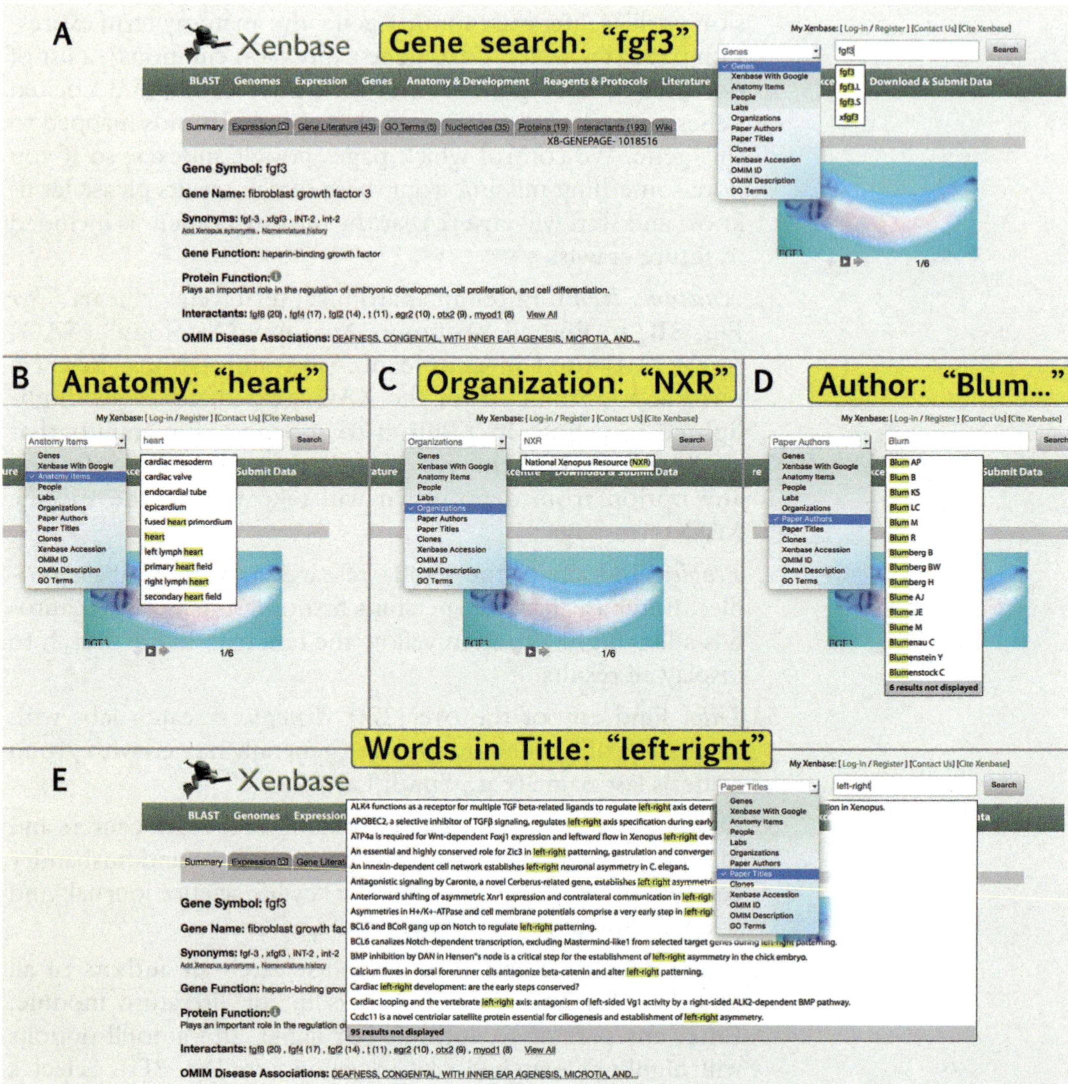

Fig. 3 Quick Search Menu is located at the upper right corner on every Xenbase page. Options available from drop downs include: (A) Select "Genes" to search for partial of full gene name or symbol (e.g., "*fgf*"); (B) "Anatomy Items" link an XAO term page search (e.g., "heart"); (C) "Organization" quickly finds contact details for stock centers of suppliers (e.g., NXR); (D) "Paper Authors" will text-match partial and full names to *Xenopus* literature (e.g., "Blum"); (E) "Paper Titles" searches full or partial article titles, and effectively searches for keywords

9. *Clones*: Search for data from over one million clone entries in the Xenbase database. Enter either the gene symbol to which the clone/plasmid specifies (e.g., "fgf3") or an existing clone ID number (e.g., IMAGE:7029804 or xl301j22).

10. *Xenbase Accession*: This search finds specific data using the unique Xenbase identifiers with our numbering and cataloguing systems. After working with Xenbase data, researchers may

record a specific Xenbase accession number to easily return to this specific database page. The following are examples of valid Xenbase Accession numbers:

- XB-GENE-484294 (Gene Page)
- XB-ART-53013 (Article Page)
- XB-PERS-3515 (Person/Researcher Page)
- XB-LAB-702 (Lab Page)
- XB-ANTIBODY-14574796 (Antibody Page)
- XB-MORPHOLINO-17249870 (Morpholino Page)

11. *OMIM ID*: Enter an OMIM ID number for any disease from the Online Mendelian Inheritance in Man (OMIM) database to find associated *Xenopus*-Human disease model data. For example, enter "219700," the OMIM ID for "Cystic Fibrosis," to return two Gene Pages associated with this disease, *cftr* and *tgfb1*. This is a quick way to find the *Xenopus* literature from cell biology to phenotypic models that are applicable to, or associated with, a specific human disease.
12. *OMIM Description*: Enter a term from the OMIM disease name or description (e.g., "diabetes") to return all homologous *Xenopus* Gene Pages to discover all *Xenopus* literature and associated genomic data associated with that specific human disease or family of diseases. This is a fast way to find the known *Xenopus* gene expression data and associated literature from cell biology to phenotypic models, that is applicable to, or associated with, a range of related or similar human diseases and syndromes.
13. *GO Terms*: Gene Ontology (GO) terms cover three areas: molecular function, biological process and cellular component. Enter a full or partial GO term (e.g., "axial") and the drop-down menu autofills and text matches highlight in yellow. Mouse down to select the specific term of interest and hit search. Single returns will direct to the gene page, and multiple returns will be shown in a table. Click the gene symbol to go to that Gene Page, where this GO term, and all others annotated for the gene, are listed under the GO Terms tab. Approximately 8400+ GO terms are currently associated with *X. laevis* genes (both L and S) and 7400+ GO terms are currently associated with *X. tropicalis* genes. As Xenbase further develops this feature, reciprocal data exchange with the GO Consortium will update and add more GO terms to *Xenopus* genes. Xenbase curators will also manually add GO annotations extracted from the published literature to Gene Pages and Articles Pages.

Notes/Troubleshooting the Quick Search Menu

- Note that there is no wildcard (*) search in the quick search menu.
  - If you get no results, check for typing errors (remove all spaces before or after the text, check for erroneous spelling, symbols, or Greek letters that did not copy correctly, or extra punctuation marks), as this is an "exact" text match algorithm, so only perfect matches will be returned, then search again.
  - For GO term searches, users must select a specific GO term from the drop-down menu to return results.
  - For a comprehensive text match search of the entire paper, not just the title, use Textpresso; *see* Subheading 6 below.
  - Google is continually adding more content from Xenbase to their search engine, however, the Google search may not include *all* content from Xenbase.
  - If you cannot find what you are looking for, try choosing one of the specific search areas from the menu, and ensure that you are searching the right item from the appropriate menu option.

4.1 Accessing Xenbase Features from the Main Menus and Home Page Tiles

The following sections cover how to access and use the database features of Xenbase via the Main Navigation Menu, remembering that these options are reiterated on the Home Page Tiles. All topics discussed can be accessed via both options, and we discuss them here in order of the main menu, from left to right, excluding the Gene Search, which is covered above in Subheading 3.

5 BLAST Menu

BLAST (Basic Local Alignment Search Tool) is a tool that finds regions of similarity between two nucleotide or protein sequences [15].

Use BLAST to:

- Identify sequence fragments.
- Calculate sequence conservation across taxa.
- Identify orthologs across taxa.
- Check for target versus off-target sites for a PCR primer or morpholino (MO).

The main BLAST menu offers options to align the query sequence against *Xenopus* mRNA, *Xenopus* proteins, various genome versions and the mitochondrial genomes for three *Xenopus* species (*X. laevis*, *X. borealis*, and *X. victorianus*).

5.1 How to Use Xenbase BLAST

1. Choose from the alignment program query options (e.g., blastn: DNA query to DNA database or blastp: Protein query to protein database) (red arrow, Fig. 4A).

2. Choose the target database or genome build to which you want to compare/align your sequence (e.g., *Xenopus laevis* and *tropicalis* mRNA or *X. laevis* J-strain 9.1) (black arrow, Fig. 4B). Xenbase BLAST allows users to compare nucleotide or protein sequences to the latest (and legacy) *X. laevis* and *X. tropicalis* genome builds, mRNA and protein sequences, with these options available in the second drop-down menu labeled "Database."
3. Enter (i.e., type or copy and paste) a single query sequence into the data box in GenBank/FASTA format (green arrow, Fig. 4A). Alternatively, upload a query sequence file using the "choose file" dialog box (orange arrow, Fig. 4A).
4. For almost all *Xenopus*-to-*Xenopus* comparisons, the default "Options" settings will result in a high scoring, statistically significant alignment, although more advanced users can choose a custom set of options.
5. Click the "Submit Job" button to compute the sequence alignment.
6. BLAST results are displayed graphically in three sections:

   (a) A color key for alignment scores grades the alignment matches from red (highest scores) to black (lowest scores) (Fig. 4B.1). Click on the color-coded bars to skip directly to the high scoring pair alignments described below (*see* Fig. 4B.3).

   (b) An "Overview of Results" table has five columns: "Hit ID" (i.e., accession ID number of hit or scaffold number), "Hit Description" (name of the sequence hit), Gene Page (i.e., link via gene symbol), HSP "Score" and *E*-value. Click "Hit ID" (black arrow, Fig. 4B.2) to open the match on GBrowse. Click "Hit Description" (Fig. 4B.2, white arrow) to skip to the High-scoring Segment Pair (HSP) alignments, with computed percent identity, and links to chromosome locations and GBrowse.

   (c) Click the gene symbol (green arrow, Fig. 4B.2) to go to the Xenbase "Gene Page."

5.2 How to Use BLAST to Inform Design of Xenopus-Specific Primer or Morpholino

1. Select "blastn - DNA-to DNA query."
2. Select the database "*Xenopus laevis* and *tropicalis* mRNA."
3. Enter the sequence into the query sequence box (e.g., CTCACTGGACATCCAGGTCTGAG, a potential *scl4a1* PCR primer sequence).
4. Click "Submit Job." Results are displayed in the same formats as shown in Fig. 4B in the above example, indicating that 24/24 bases match *X. laevis scl4a1.L* homeolog, and 23/24 bases match *X. laevis scl4a1.S* homeolog.

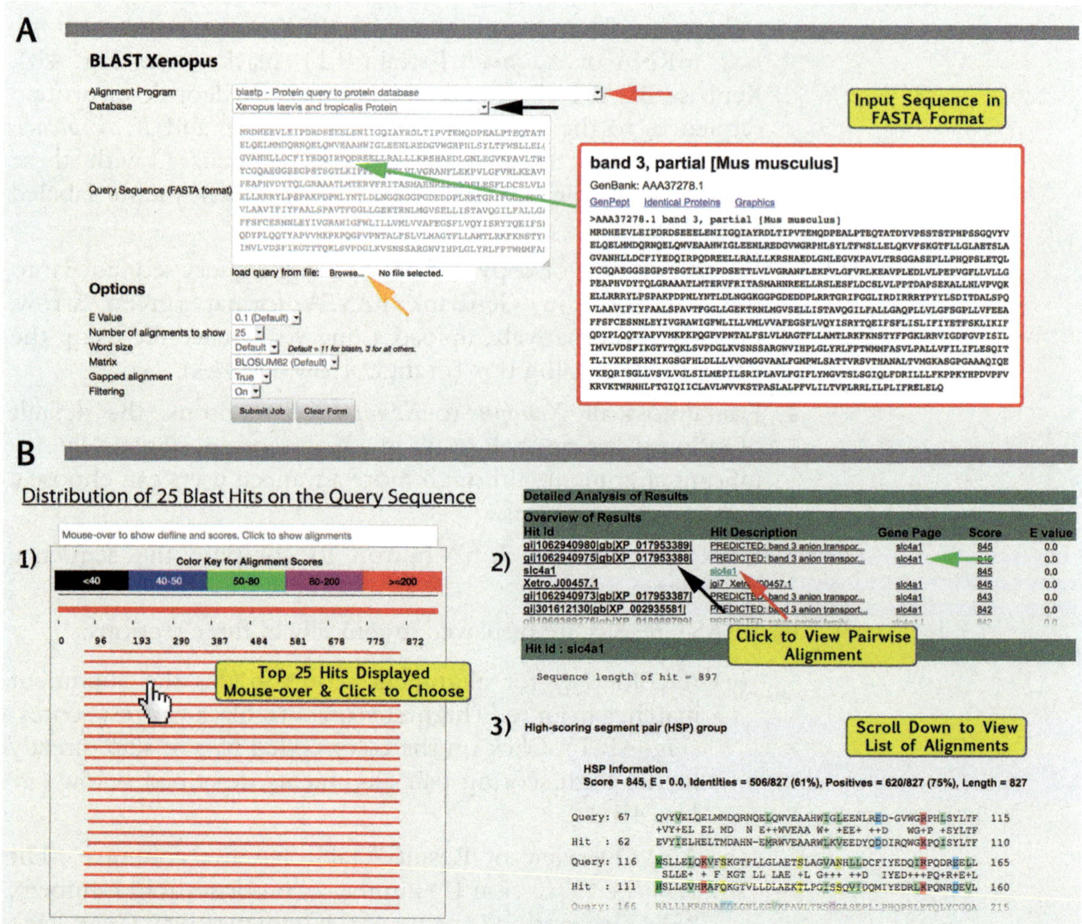

Fig. 4 Using BLAST on Xenbase. (1) Access BLAST from drop-down menu or tile; (2) Choose Alignment program and (3) the database to which your search will be aligned; (4). Paste the query sequence into the box, in FASTA format; (5) Set and adjust options and (6) click "Submit Job" button. (A) In this example to assess evolutionary conservation of the protein Slc4a1 between mouse and frog, we used "blastp" (protein query-to-protein database) and entered the amino acid sequence for mouse Slc4a1 (Gene ID: 20533; protein_id=AAA37278.1) in FASTA format. We selected "*X. laevis* and *X. tropicalis* proteins" from database options. (B) Results of BLAST for mouse Slc4a1 vs. *Xenopus* are displayed sequentially in three formats: (1) Distribution of the top 25 hits on the query sequence with red indicating alignment scores >200; (2) Table with "Overview of Results" showing high scoring segment pair alignments (with alignment score). (3) Click hit ID or scroll down page to view pairwise alignments and identity calculated as a percentage

5.3 Troubleshooting BLAST

- BLAST searches are usually almost instant, but occasionally can take some time to complete. Very long sequences (e.g., a scaffold), sequences with repeats, or sequences with low complexity increase the chance of a BLAST run being slow, or even timing out. In these cases, try entering a smaller sequence, or change the *E*-value to get a more sensitive alignment.

- If a BLAST query results in no alignments, check that the correct database and BLAST program has been selected, increase the *E*-value, or rerun the same BLAST. A warning message will be displayed if an incorrect database is selected for the selected alignment program.
- Mitochondrial genomes form a distinct data unit in BLAST and therefore must be selected from the option in the main menu. No mitochondrial annotations are currently available.
- If BLAST times out after ~30 s, it can be due to heavy use of the service. Try again during an "off peak" time slot and if problems persist, please contact Xenbase. This typically only occurs with very large jobs or complex tblastn or tblastx runs. Once again, feel free to email us if this occurs.
- There is no fully annotated genome available for *X. victorianus*, only mtDNA.

6 Genomes Menu

6.1 Download *Xenopus* Genomes

The Xenbase Data Downloads page provides access to genome assemblies, gene models, sequences, and database reports. Most files are in a tab-delimited format. Use the toggle [+] to see all files. Click the [readme] link to view information on the files, including the header row for these files. To download a file, click on the corresponding FASTA link. More files are located at our FTP File Browser.

6.2 GBrowse

GBrowse is an open source, browser based, interactive genome visualization software that allows gene models to be viewed within the genome next to RNA-seq and ChIP-Seq data. Xenbase GBrowse can be accessed from the menu bar, via BLAST against a genome, or clicking a snapshot on a Gene Page, or a snapshot morpholino page. Xenbase hosts the most recent and several legacy genome assemblies for both *X. tropicalis* and *X. laevis*. The main view of GBrowse on Xenbase shows all selected tracks for the chosen genome. Tracks can include gene model annotations, RNA-Seq alignments, ChIP-Seq alignments, and morpholino alignments. Tracks are binned into categories, such as gene models, tissue RNA-Seq, stage RNA-Seq, and methylation ChIP-Seqs. To customize which tracks are displayed, click the "Select Tracks" tab, and use checkboxes.

6.2.1 How to Use GBrowse

1. Open GBrowse via the Genomes menu, or home page tile, by selecting a genome model version (e.g., *X. laevis 9.1 (J-Strain) on GBrowse*).
2. Use the "Landmark or Region" dialog box (black arrow, Fig. 5A) to search for a scaffold position (e.g., chr9_10S:3,571,719..3,581,718).

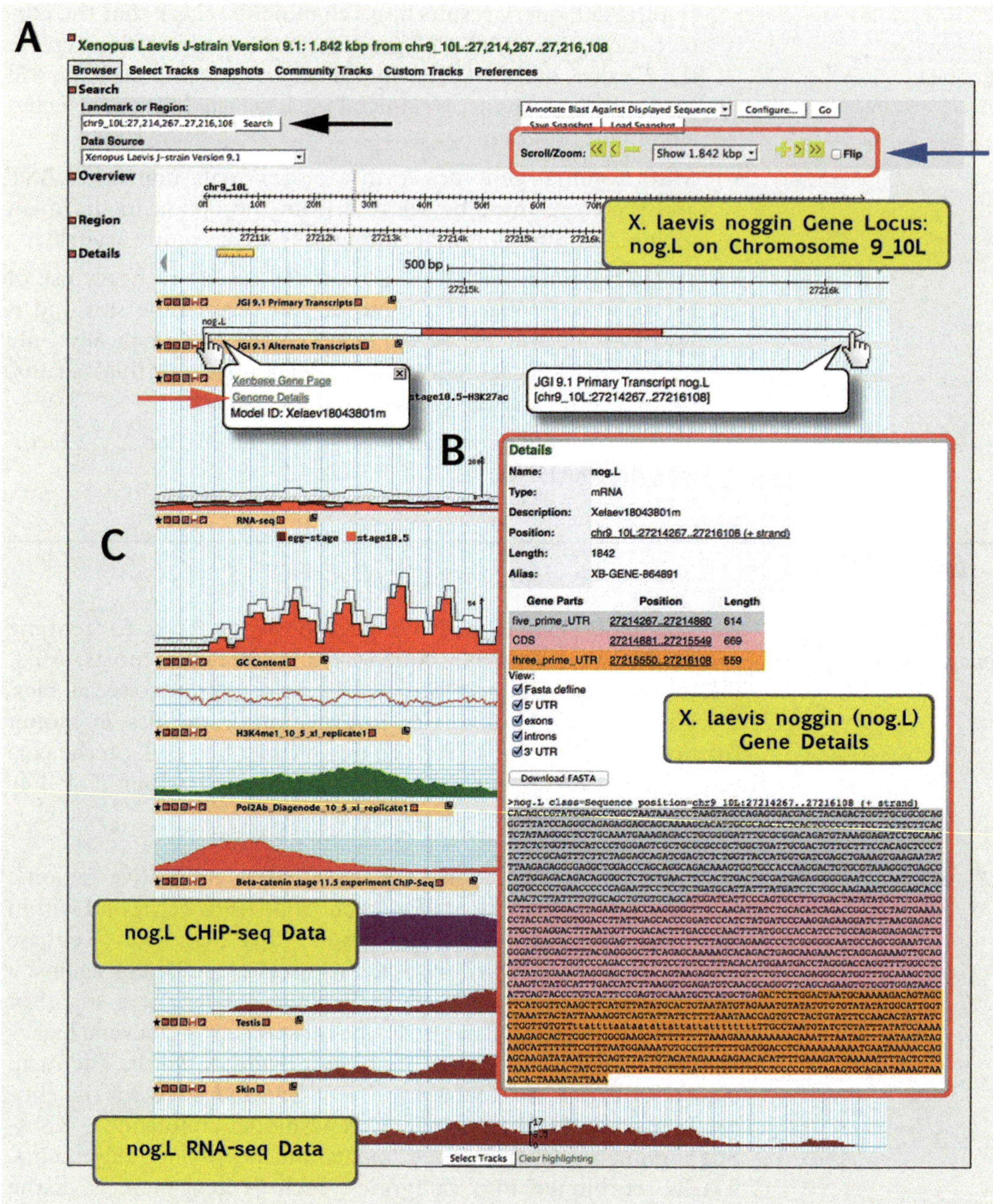

Fig. 5 Using GBrowse, in conjunction with BLAST, to visualize alignments against gene models and Next-Gen sequence data. (A) In this example, we BLAST *X. tropicalis nog* mRNA against *X. laevis* 9.1 genome. GBrowse gene model details are shown for *X. laevis nog.L*. Moving the cursor over the gene model (hand cursor) generates a pop-up with links to "Xenbase Gene Page" and "Genome Details." Click on Genome Details (red arrow) to access metadata (pop up in B) about the given gene model, including the specific *Xenopus* gene (L or S) to which the model is associated. (B) The Gene Details provides a quick overview of the structure of the model, and size of exons, introns, and 5′ and 3′ UTRs. The interactive sequence section allows the UTRs and introns to be toggled on and off for easy copying and can be edited to include/exclude exons. (C) Next-Gen RNA-Seq and ChIP-Seq datasets are shown below the gene models

3. If the scaffold position is unknown, enter a gene symbol (e.g., *pax3*), to identify which chromosome the gene is on (e.g., chr5L or chr5S) then click to choose a region to view (e.g., pax3.L, chr5L:123,000,255..123,046,707) from the results table.
4. Scroll/Zoom tools allow you to move left and right along a GBrowse view (blue arrow, Fig. 5A). Alternatively, use the drop-down menu to select options from 100 bp to 2 Mbp to zoom in and out. This is very helpful when identifying surrounding gene models.
5. Click on specific track to access additional information.
6. Hover/mouse over a track to popup its precise scaffold position.
7. Click on a gene model to give a pop-up box that provides a link to the Xenbase Gene Page, as well as gene model details for the given transcript. Click the gene model "Details" (red arrow, Fig. 5A) to show metadata for the model (e.g., nog.L in box, Fig. 5B), including the type, position, and length of each exon, and an interactive FASTA display, which allows the sequence to be copied for further use.
8. Click and hold/drag a track to rearrange track position.
9. On each track, a series of buttons on the far left side allow users to save a track as a favorite [star], show or hide a track [−], turn off at rack [×], share [radio], save [disc icon], or configure [tool icon] tracks. The [?] button gives more information including an option to download the data for the track.
10. Use check boxes on the "Select Tracks" tab to customize the data displayed (e.g., include or exclude BAC and Fosmid end data or Methylation ChIP-Seq data) and whether to show the RNA-Seq and ChIP-Seq data stacked in Topoview.
11. Additional tabs along the top of GBrowse allow users to save "Snapshots," view "Community Tracks" and upload "Custom Tracks."
12. Changes to the color scheme and grid width can be set in "Preferences" tab.

6.2.2 Troubleshooting GBrowse

- Some gene models (e.g., *pax1.S*) in GBrowse may give "Xelaevis" model IDs and not link to Gene Pages. The frequency of these legacy mappings will decrease with ongoing improvements to the gene model annotations.
- Occasionally tracks within GBrowse will not display, and will show a rendering error. Changing the zoom level will usually fix this problem.
- If the gene search does not work, a gene symbol synonym may be being searched, rather than the official gene symbol. Refer to Xenbase Gene Pages for the official gene symbol.

- If a gene model is still not being found, it is best to BLAST the sequence against the genome (as shown in Fig. 5), and then follow the BLAST links to try to identify the correct model.
- JBrowse [16], a newer genome browser with increased functionality, has just been launched on Xenbase (under "Genomes" menu, X. laevis v9.2 on JBrowse is now at the top of the list). We will continue to support both genome browsers for ~2 years, as GBrowse is phased out, and new genomic data will only be added to JBrowse.

6.3 Xenbase UCSC Track Hub

Track hubs are web-accessible directories of genomic data that can be viewed on an external genome browser, and are helpful tools for quickly visualizing large genome-wide data sets, including numerous custom tracks [17]. Xenbase hosts a University of California, Santa Cruz (UCSC) Track Hub that can be loaded into a UCSC instance. A link to the track hub is accessible from the Xenbase home page under the Genomes menu. The UCSC Track Hub includes gene models for *X. tropicalis* v7.1, v8.0, and v9.0, and *X. laevis* v9.1 genome builds, plus a large number of RNA-Seq and ChIP-Seq tracks. Xenbase is currently processing additional NGS datasets including them in the track hub (*see* RNA-Seq (red) and ChIP-Seq (orange) tracks in Fig. 5C).

6.4 Other Genome Assemblies

Xenbase also links out to other genome resources from the Genomes menu, including the Japanese National Institute of Genetics *X. laevis* genome project.

7 Expression Menu

Gene expression patterns can be searched via two routes on Xenbase, both under the Expression Menu. The first method is the "Expression Search" and the second method is the "Anatomy Search." These two options take you to two *different* types of gene expression search, and can help answer different questions—a more detailed explanation follows.

7.1 Method 1: Expression Search

The "Search Gene Expression" interface offers numerous user-defined criteria to include/exclude data types, the goal of which is to filter a large catalog of images to answer specific or general questions. As such it can be used to find all examples of gene expression in a specific gene, tissue, or combinations of these, as well as a range of additional criteria. Put simply, there are three tiers of filters that can be selected. First, choose from variables such as species or embryonic stage (details in Subheading 7.1.1). Secondly, choose the anatomy terms to search (details in Subheading 7.1.2), and third, add optional filters based on experimenter (i.e., laboratory or researcher) or database associations (details in Subheading

7.1.3). An example of a three-tiered query might be for "genes expressed in the 'pronephric duct' in tadpole stages (NF stage 28 to NF stage 35 and 36) from the Lienkamp laboratory screen," or "all genes expressed in the foregut progenitor tissues, but not the heart, in gastrula stage embryos." Equally a query might focus on the unknown tissues, "where is *shh* expressed besides the notochord"?

7.1.1 Gene Expression Search Options

The top section of the "Search Gene Expression" interface (Fig. 6A, B) presents a set of options to effectively "filter" the annotated expression database. Fields include gene symbol, clone name or sequence, species, and/or developmental stage. If a gene symbol is entered, the search effectively "filters" all gene expression image data for that gene (which are also shown in the Expression tab of the Gene Page), using a user-defined set of criteria (e.g., all *shh* expression in *X. tropicalis* at NF stage 28), thus omitting nonapplicable and/or redundant and/or semiautomated curations, which are common to large scale screens. Note that not all fields need to be entered, and combinations are acceptable. The options include:

1. Enter a gene symbol in the top entry box (e.g., *shh*, Fig. 6A red arrow), a clone or Affymetrix ID (*Optional*).
2. Select the "Search Synonyms" box (black arrow Fig. 6A) so that legacy names will also be searched (i.e., *xshh* and *vhh-1* are legacy gene symbols for the gene now called "sonic hedgehog" with the gene symbol "*shh*") (*Optional*).
3. Specify either *X. tropicalis* or *X. laevis* from the menu box, or the default "*Xenopus*" returns all data (*Optional*).
4. In the expandable box, paste your sequence in FASTA format or simply provide a GenBank accession identifier (e.g., mRNA accession BC166395 for *X. tropicalis shh*; Fig. 6B). Set the *E*-value in the box to the right (the default is 0.1) (*Optional*).
5. Limit by developmental stage via drop-down menus to select a start and end embryonic stage range. Options include specific stages (e.g., NF stage 10.5) or general terms (e.g., blastula). Click the + and – buttons to add and remove stage(s). Use the "All Stages" and "Any Stages" radio buttons to select the Boolean operator for your search criteria (AND or OR, respectively) (*Optional*).
6. Continue to next section to choose additional filters, or scroll to the bottom of the page and hit "Search."

7.1.2 Specifying Anatomy (XAO) Terms to Include and/or Exclude Organs/Tissues

The central section of the Search Gene Expression interface will set up a query of the database for expression patterns in specific organ(s), tissue(s), or cell types by using terms from the XAO. Queries can be submitted as follows:

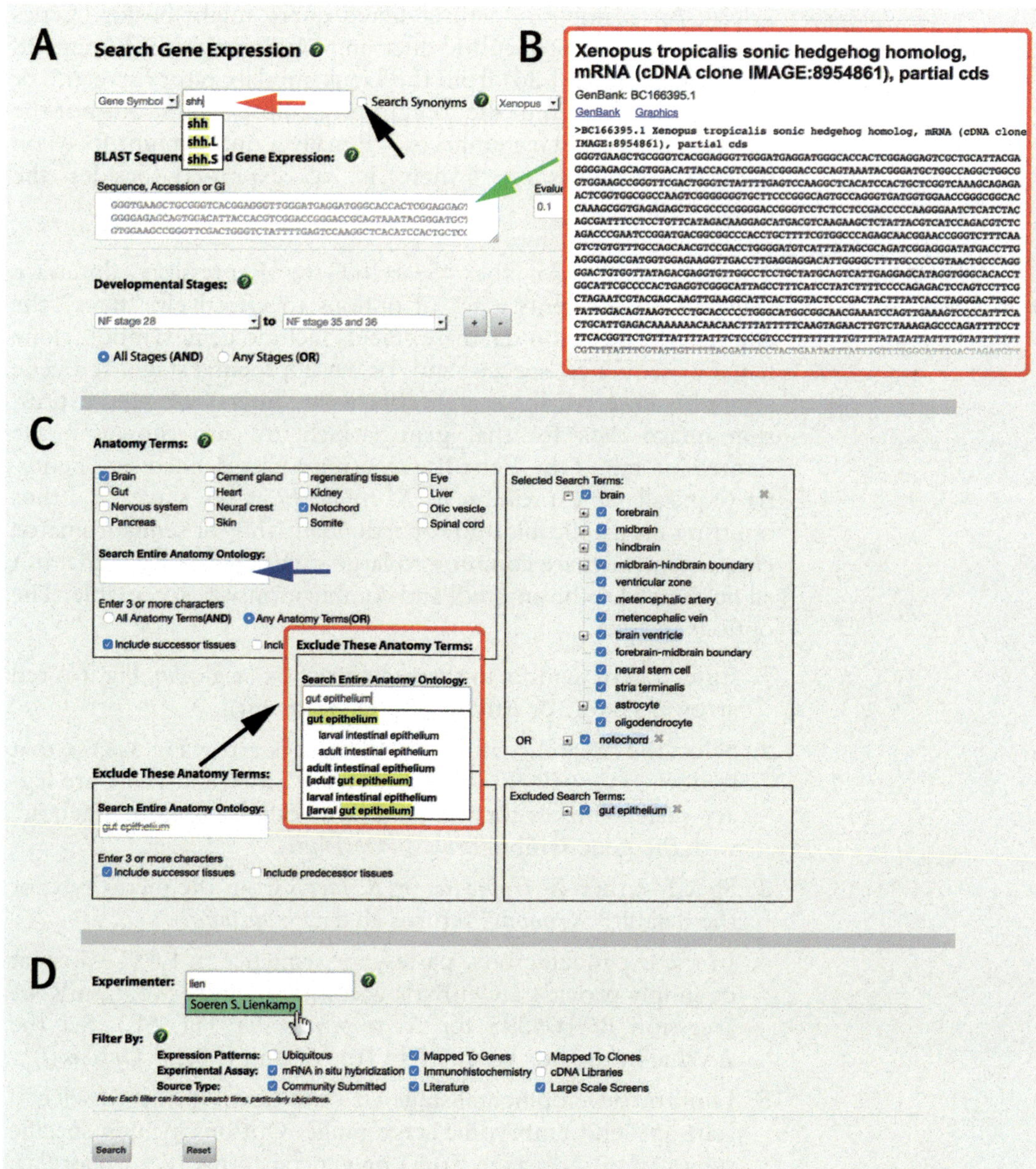

Fig. 6 Search Gene Expression via "Expression" menu. (A) Gene expression can be searched by entering a gene symbol (red arrow), or by entering a sequence in FASTA format into the dialog box (green arrow, *see* (B). Additional filters can include search synonyms (checkbox, black arrow) and choosing which species (*X. laevis* or *X. tropicalis*, the default is for both). (C) Anatomical terms (organs, tissues, or cell types) can be included (top box) and/or excluded (black arrow). Select XAO terms either by marking check boxes, or manually entering terms. Selected anatomy terms move right to the "Selected Search Terms" box. Toggle between child terms using + and – buttons, mark or unmark checkboxes to select/deselect terms (selected terms have tick in a small blue box). Choose "All Anatomy terms" (AND) or "Any Anatomy terms" (OR) functionality via radio buttons. Choose to "Include predecessor tissues" and/or "Include successor tissues" via checkboxes as needed. (D) The bottom fields include additional filter options including "Experimenter," which autosuggests (highlighted in green) from the list of paper authors and *Xenopus* community members

- A "free standing" XAO query (e.g., all records of expression in the "brain").
- A "combined" XAO query (e.g., all records of expression in "brain" AND/OR "notochord").
- An "include-exclude" XAO query (e.g., all records that include "brain" but exclude "notochord" (e.g., "brain" NOT "notochord").
- Any XAO query (option 1, or 2, or 3 above) in conjunction with 1 or more options chosen in the top section as described above (e.g., all *shh* expression in *X. tropicalis* at NF stages 28–34, expressed in "brain" AND/OR "notochord," but NOT in "liver diverticulum").

Select XAO terms as follows:

1. Choose from a set of 16 common anatomy terms, available as checkboxes. In the example in Fig. 6C, "brain" and "notochord" have been checked, and they automatically move to the "Selected Search Terms" box to the right.
2. Enter anatomy term(s) (using three or more characters) using the "Search Entire Anatomy Ontology" suggestion box (Fig. 6C, blue arrow). All terms that match your text will autofill in bold below, with matched text highlighted in yellow. Synonyms of an XAO term appear in square brackets. Mouse down to select a term from this menu to add it to the list of search terms (Fig. 6C).
3. The XAO sub-parts of a term can be expanded by clicking the [+] icon (e.g., "brain" has parts including "hindbrain," "midbrain," and "forebrain"). By default, all subparts are checked, but users can exclude any from the search by unchecking them.
4. To *exclude* an anatomy term(s) enter your "excluded" term(s) in the section marked "Exclude These Anatomy Terms" in the same manner as those in the included terms box above (e.g., "gut epithelium" is excluded in Fig. 6C).
5. Choose to "Include predecessor tissues" and/or "Include successor tissues" via checkboxes as needed. This option applies to embryonic anlage terms, such as "anterior neural tube," which *develops_into* successor tissues, "brain" that *has_parts*, "hindbrain," "midbrain," and "forebrain".
6. Continue to next section to choose additional filters, or scroll to the bottom of the page and hit "Search."

7.1.3 Specifying "Experimenter" and Using "Filter By" Options

Specifying "Experimenter" and using "Filter By" options are the third tier of filters for a gene expression query (Fig. 6D). This is an excellent way to find all, or a subset of, the images from large data sets that Xenbase hosts. These large data sets include an angiogenesis screen (Patient Lab, *see* [18]), retinal marker screen (Perron Lab and Pollet Lab, *see* [19]), pronephric marker screens (Brandli Lab, *see* [20]; Lienkamp lab, *see* [21]); MO-synphenotype screen (Smith Lab, *see* [22]), XenMARK images [23]; or ISH images for clones supplied by the European *Xenopus* Resource Centre (EXRC), among others. To find images from published literature and community submissions:

1. Enter a full, or partial, first or last name of a researcher or author in the "Experimenter" field. Use cursor to select a name from the autofilled options.
2. Additional advanced filtering options to either reduce or increase the number of results can be selected via checkboxes as follows:
   (a) *Expression patterns*: Ubiquitous (i.e., annotated with five or more tissues); Mapped to Genes, or Mapped to Clones.
   (b) *Experimental Assay type*: ISH, IHC or cDNA libraries.
   (c) *Source types*: Community submitted, Literature, or "Large Scale Screens" (e.g., defined from cDNA libraries).
3. Scroll to the bottom of the page and hit "Search."

7.1.4 Navigating the Gene Expression Search Results

Here we show two examples of gene expression queries and guide the user through features of the results tables. The first example, shown in Fig. 7A, is gene expression for the gene "sonic hedgehog" (*shh*), which is an early marker of notochord, but is later expressed in the foregut. Here we combine a gene symbol and an XAO term in a query. We checked the upper level XAO term "gut," which is a synonym for "alimentary system," to include all parts of the gut from embryo to adult frog stages without stage restriction. We choose AND, but deselected predecessor and successor tissue. Figure 7A shows a subset of the returns, with the source (e.g., citations from literature or community submitted data) to left, then species, a thumbnail of the data image, NF stages, and the XAO terms annotated (and thus matched) to the image to the right. In a second example Gene expression query, the Experimenter "Lienkamp" returns all annotated images from 52 genes in a screen for pronephric markers submitted and published from this

Fig. 7 (continued) and the XAO terms annotated (and thus matched) to that image to the right. Click on the image to enlarge it and view annotation table. Click on the source to open the Article or Lab page. (B) Gene expression query output for XAO term "pronephric kidney" plus the Experimenter "Lienkamp," returns all annotated images from a screen for pronephric markers, plus any images from publications with this author (not shown). Images can be filtered for more specific terms using the "Modify Search" button (black arrow). (C) Adding the anatomy term "pronephric duct" filters the results (shown in B) to a smaller, annotated set of images from the "Lienkamp" laboratory

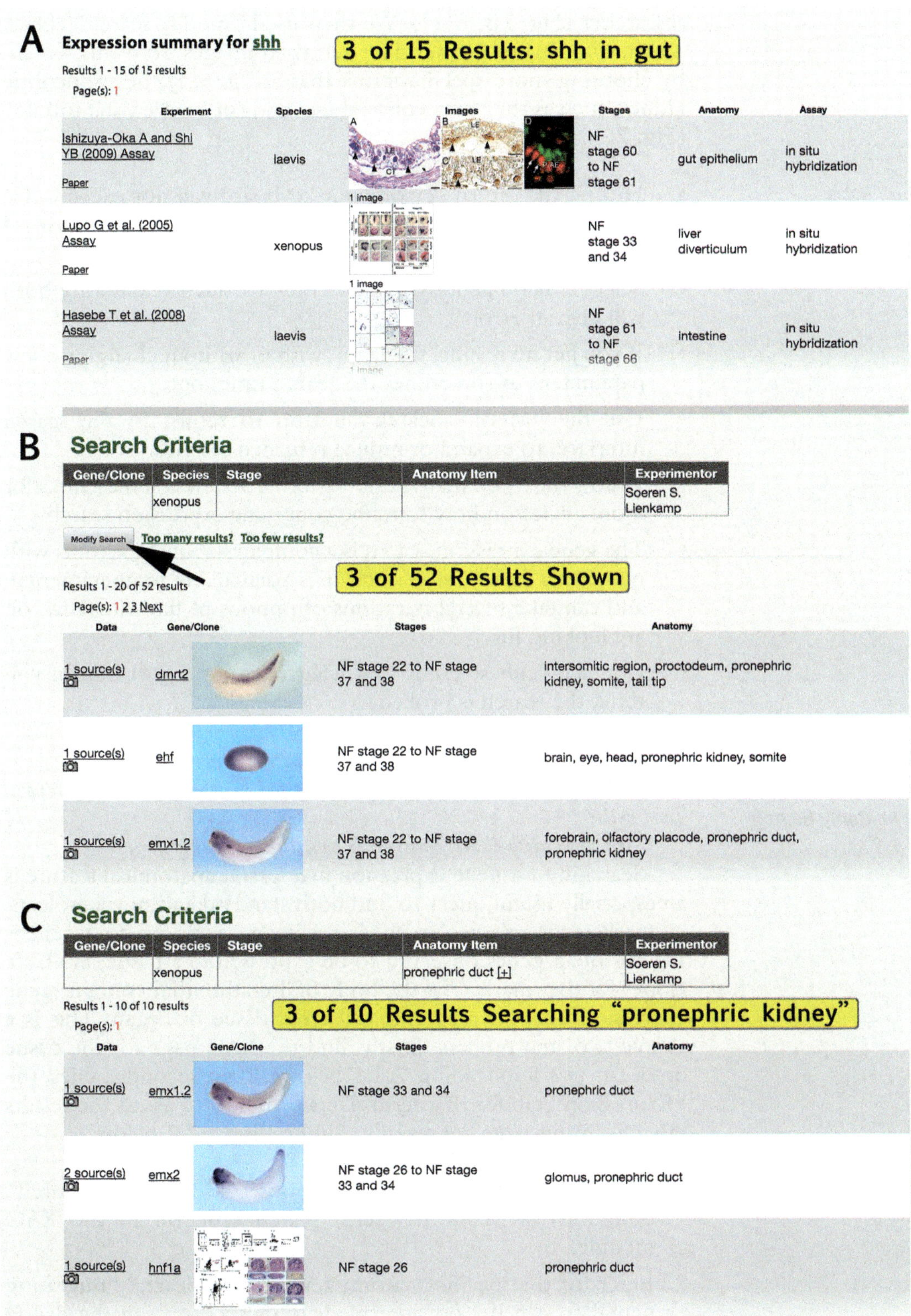

Fig. 7 Gene expression search results. (A) Gene expression query output for the gene "sonic hedgehog" (*shh*) and the XAO term "alimentary system." A subset of the returns, with the experimental source (e.g., citations from literature or community submitted data) to left, then species, a thumbnail of the data image, NF stages,

researcher (Fig. 7B; [21]). We then used "modify search" (black arrow, Fig. 7B) to further filter returned images from this screen, by choosing more specific terms that are *part_of* the pronephric kidney, such as the "pronephric duct" and/or "early distal tubule" (Fig. 7C).

7.1.5 Notes/ Troubleshooting the Gene Expression Search

- Hitting the return key on your keyboard will **not** execute this search—always click the "Search" button—bottom left-hand corner of the screen.
- Select either a gene symbol *or* enter a sequence: entering both will give an error.
- If you get no results, try again, with or without changing a few parameters, as sometimes the search times out.
- Use the "Modify Search" button to return to the search interface, to expand or reduce returned results.
- Follow the "Too many results?" or "Too few results?" links for more advice on how to refine your gene expression search.
- The gene expression search is a complex set of algorithms with numerous variables: as such it is particularly temperamental, and can take several iterations of options to find the data you are looking for.
- Contact Xenbase (xenbase@ucalgary.ca) to report bugs if you think the search is broken.

7.2 Method 2: Gene Expression via Anatomy Search

What are the best markers for cardiac mesoderm?

Which genes are known to be expressed on the migrating neural crest cells?

Are there any clones/plasmids available for this gene?

Searching for gene expression in a *specific* anatomical feature is an especially useful query to find both standard and novel molecular markers for a tissue/organ via the image catalog; a comprehensive list of all genes observed to be expressed in a tissue; available clones for that marker; or the body of literature that contains gene expression data for a specific cell type, tissue or organ. This is a simple two-step process. Firstly, find the XAO page for the tissue or organ (*see* Subheading 7.2.1 below), then secondly, click the "Expression" tab for that term. Details of how to assess the results table from this page are given in Subheading 7.2.2 below.

7.2.1 Finding the XAO Term Page

1. Under the "Expression" menu, choose the "Anatomy Search" option to arrive at the term search function in the XAO module.
2. Enter for the specific anatomy term (e.g., "heart," "migrating neural crest cell," or "intermediate mesoderm") in the dialog box. The matched text will autofill: matches are highlighted in yellow, synonyms are in square brackets. Menu options also

include XAO ID numbers (e.g., heart, XAO ID: 0000064) or anatomy page number (e.g., heart, XB-ANAT-63).

3. Hit Search (or Browse All).
4. Use the + and – buttons in the navigable view of the entire XAO, located to the right of the page (*Optional*).
5. Use Nieuwkoop & Faber (NF) stage restrictions from drop-down menus, using either broad categories (e.g., "early tailbud stage" to "tadpole stage") or precise stages (e.g., NF stage 20 to NF stage 28) to focus results (*Optional*).
6. Multiple matches (e.g., "heart," "primary heart field," "left lymph heart") are displayed in a table. Click term name to go to the XAO term page (Fig. 8A). Single results go directly to the XAO term which has an "Expression" tab, just like a Gene Page.
7. Click "Expression" tab to display gene expression for this specific XAO term.

7.2.2 Assessing Results Table on "Expression" Tab of an XAO Term

The genes annotated as having expression in the XAO term appear in a table, with gene symbols to the left, and associated data types for that gene organized in four columns: Images, Clones, Papers (i.e., articles/literature), then a combined Total count of records (*see* Fig. 8B). Each column can be sorted in descending order by clicking the column tile. All table entries are underlined indicating they are live links to further data.

1. Click the *View All* link to open the entire list of genes matched to the XAO term.
2. Click "Images" column header to find the top marker genes.
3. Click the gene symbol (e.g., *nkx2-5* or *hand2*) to open that Gene Page (hand cursors, Fig. 8B).
4. Click the number of images available for a gene to see annotated gene expression images matching the XAO term.
5. Click the number of clones to execute a query for clones for that gene.
6. Click the number of papers to see a full literature list associated with the XAO term and the gene of interest.

Here we use the XAO module to explore gene expression in "heart" (Fig. 8A, XAO page XB-ANAT-63, for "heart," XAO I:0000064). After selecting the "Expression" tab, the top 100 results for "genes expressed in heart" are shown from 5300+ records on the first page of the results. Genes expressed in "heart" are ranked in descending order by total count of data records), and here have been reordered by "Images" (black arrow, Fig. 8B): *nkx2-5* has 69 images, *tnni3 has* 56 images, *hand1* has 28 images, etc. Further down the column, more, but less well-studied, genes with heart expression can be assessed (e.g., *hand2*, 6 images) (blue arrow, Fig. 8B).

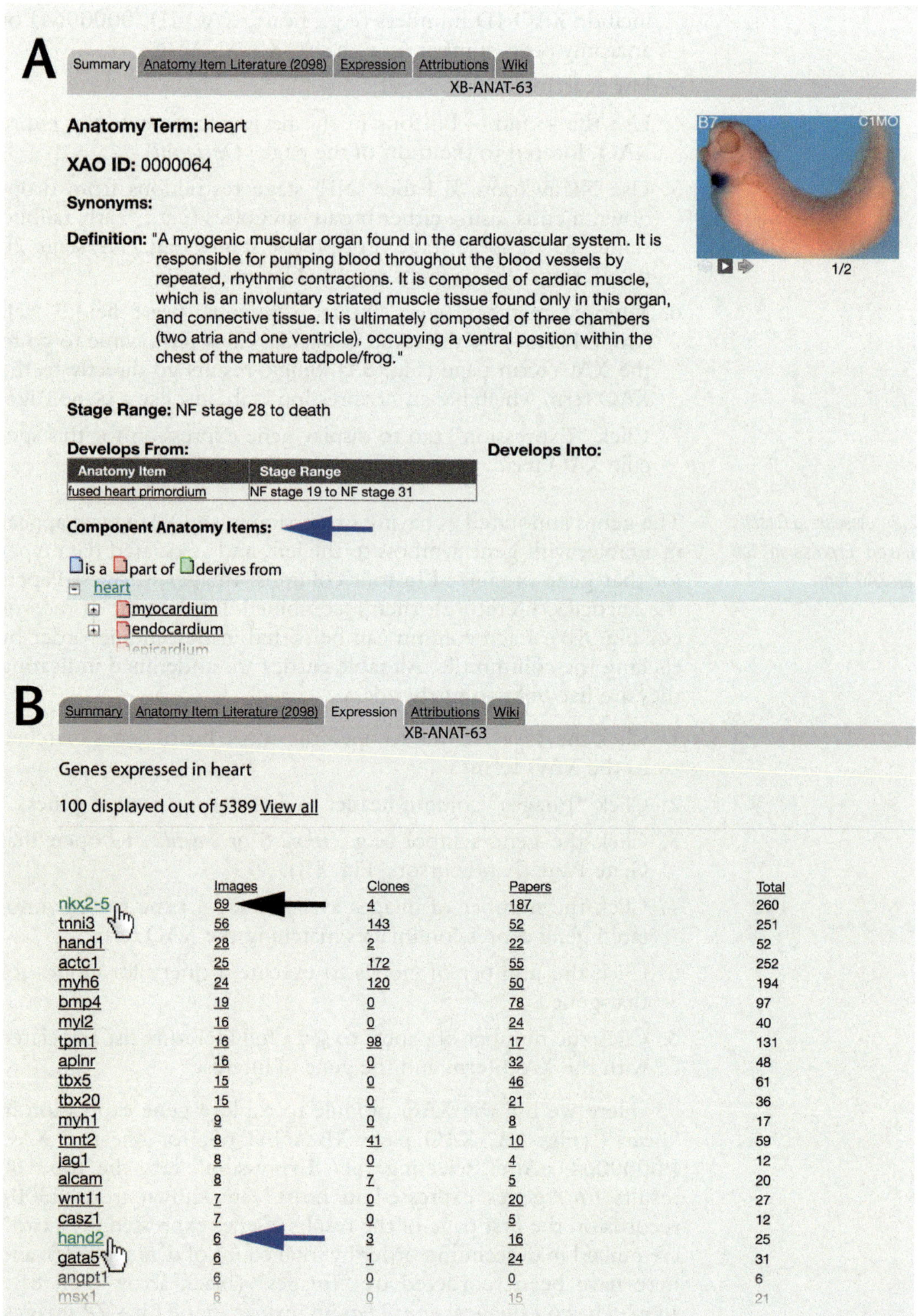

Fig. 8 Using the Anatomy Search to explore gene expression, and the XAO. Here we use the XAO module to ask "Which genes are expressed in the heart?" (A) Each XAO term page gives the term definition, NF stage restrictions for its use and relationships to other terms (see "Component Anatomy Items," blue arrow). (B) From the

7.2.3 Notes/Troubleshooting the Anatomy Search for Gene Expression

- Additional routes get to this feature: from the home page "Gene expression" tile/"Anatomy Search" link, or from the "Anatomy and Development" tile, choose "XAO," then the "Search Anatomy" tab.
- Adult tissue terms (e.g., "bladder") and many cell types (e.g., "cementoblast") have few gene expression annotations.
- "Attributions" on this page is an attribution to the definition of the XAO term.
- A Wiki is provided to record notes not recorded elsewhere on Xenbase.
- If no data is available for a particular class of data (i.e., no clones) clicking on the zero will execute a query for clones, but will show no results.
- For higher level ontology terms, such as "heart," data returned using this search includes matches for predecessor and successor tissues (e.g., "cardiac mesoderm" and "endocardial tube"). Use the Expression search (described above, Subheading 7.1) to exclude these results.

There are three additional expression data sets under the "Expression" Menu. These are:

7.3 miRNA Catalog

MicroRNAs (miRNAs) are small, noncoding RNAs that play a role in regulating gene expression [24, 25]. The data in the miRNA Catalog contains miRNA in situ expression in *Xenopus* embryos that was submitted by courtesy of the Wheeler Laboratory [24] and XenMARK [25]. The miRNAs have been correlated with *Xenopus* records in miRBase to provide more information. Click the miRNA links (e.g., xtr-miR-133a) to view more information about the miRNA, including in situ images.

7.4 Expression Data at GEO

Select this menu option to run a preset search for *Xenopus* NGS data sets through the NCBI Gene Expression Omnibus (GEO) database.

7.5 RNA-Seq Data at the NCBI SRA

Select this menu option to run a preset search for *Xenopus* sequence data through the NCBI Sequence Read Archive (SRA) database.

Fig. 8 (continued) XAO term page for heart click the "Expression" tab. The first few results (of top 100) are shown, resorted by clicking "images" to be ranked in descending order by number of images (e.g., *nkx2-5* 69 images, *tnni3*, 56 images, and *hand1*, 28 images etc.) with data from clones, papers and total columns on the left. Lower down the column lesser known genes with heart expression are shown (e.g., *hand2*, 6 images). Mouse over (hand cursor) to select

8 Anatomy and Development Menu

The Anatomy and Development section of Xenbase covers a wide range of reference material used by researchers and students. The following headings can be selected via either drop-down menu or home page tile. A brief description of the content available under each subject follows.

8.1 Organ Atlas

The organ systems in *Xenopus* are illustrated here with a variety of imaging methodologies, including confocal microscopy. Currently, the organ atlas covers only heart and pronephric kidney development, and is undergoing a significant expansion to cover more organ systems in the future (e.g., cranial cartilages from the XenHead project [26], the nervous system and musclular skeletal system).

8.2 NF Developmental Stages

A complete *Xenopus laevis* stage series (NF stage 1–NF stage 66) [http://www.xenbase.org/anatomy/alldev.do] based on Nieuwkoop and Faber [27] illustrations are shown. A new developmental stage series, the Zahn drawings, and complementary bright field photographs, all of which are open access, posted here on Xenbase [26]. The Zahn drawings can be downloaded and used in the laboratory setting to illustrate gene expression domains, phenotypes, and other changing patterns during normal and abnormal development, and can be reused under the creative commons license under which they will be published. Examples of the new images, which include anterior, dorsal, and ventral views, perspectives not included in Nieuwkoop and Faber [27], are shown in Fig. 9.

8.3 Images of Xenopus Embryos

These image files are in the Wiki, and are generally whole-mount microscopy, illustrating each developmental stage as a researcher would see the live embryo.

8.4 Development Stage/Temperature Charts

The rate of *Xenopus* development is influenced by temperature, and although *X. laevis* and *X. tropicalis* embryos develop at similar rates, *X. tropicalis* tolerate a narrower range of temperatures [14]. The charts provide a standard reference with which to plan experiments and were supplied by the Khokha Laboratory, Yale University.

8.5 Movies of Xenopus Development

High quality movies of the developing *Xenopus* embryos are provided as educational resources, covering key developmental processes including cleavage, gastrulation and neurulation, and the synchronous development of *Xenopus laevis* embryos during early embryogenesis.

8.6 Cell Fate Maps

Cell fate is illustrated with mouse-over animations in forward direction (blastomere-to-tissue) from NF stage 5 (16-cell) to NF stage 10.5 (beginning of gastrulation) (Fig. 10A), and reverse direction (tissue-to-blastomere) (Fig. 10B), based on the classic studies by Moody [28, 29], and Bauer et al. [30]. To use these dynamic fate maps, simply move the cursor over the blastomere to highlight which cells in later stage embryos are derived from the 16-cell and 32-cell stage blastomeres. The 16-cell blastomeres and their descendants appear in orange (upper panels), while 32-cell descendants will appear in blue (lower panels). Due to the two-dimensional nature of the illustrations, and that NF stage 8 and NF stage 10.5 embryos are shown as sections, only some derivatives of the blastomeres of the NF stage 8 (32-cell) embryo show up in blue on later stage figures. In the reverse fate maps (Fig. 10B), move the cursor over an anatomy term to highlight blastomeres that make major contribution (in red), a minor contribution (in green) or rarely contribute (in orange) cells to the adult tissue.

8.7 The Xenopus Anatomy Ontology

The *X*enopus *A*natomy *O*ntology, aka the XAO, is a comprehensive set of anatomical terms that describe the entire course of development and organogenesis in *Xenopus* from unfertilized egg to the adult frog [31]. The XAO forms the backbone of our gene expression curation and is updated frequently in response to the latest research and community input. The goal of the XAO is to describe all anatomical structures in a formal language hierarchy, with each term being defined and related to other terms. XAO terms have "is_a", "part_of", "develops_from", and "develops_into" relationships, as well as specific developmental timing boundaries, using NF stages [27], such that each term has "starts_during" and "ends_during" stage relationships (*see* [31]). Cross referencing the XAO to mouse and human phenotype ontologies will ensure the interoperability of Xenbase phenotype annotation (new feature to be launched on Xenbase), with human disease phenotype.

8.7.1 Downloading the XAO

The latest XAO (v5 released January 2017) is available for download from Xenbase, in either OWL or OBO formats, and from the Open Biomedical Ontologies site (OBO Foundry).

8.7.2 Requesting New XAO Terms

Request new XAO terms via GitHub (a log-in is required). New term requests require a definition and additional supporting information about relationships to other terms, developmental timing, cross-references to other ontologies (e.g., Uberon or ZFA), and literature reference(s). Submissions suggesting many new XAO terms, can be made as a file attachment through the GitHub portal.

8.7.3 Illustrating the XAO

Xenbase is currently working to illustrate XAO terms and developmental stages with exemplary figures from anatomical dissections,

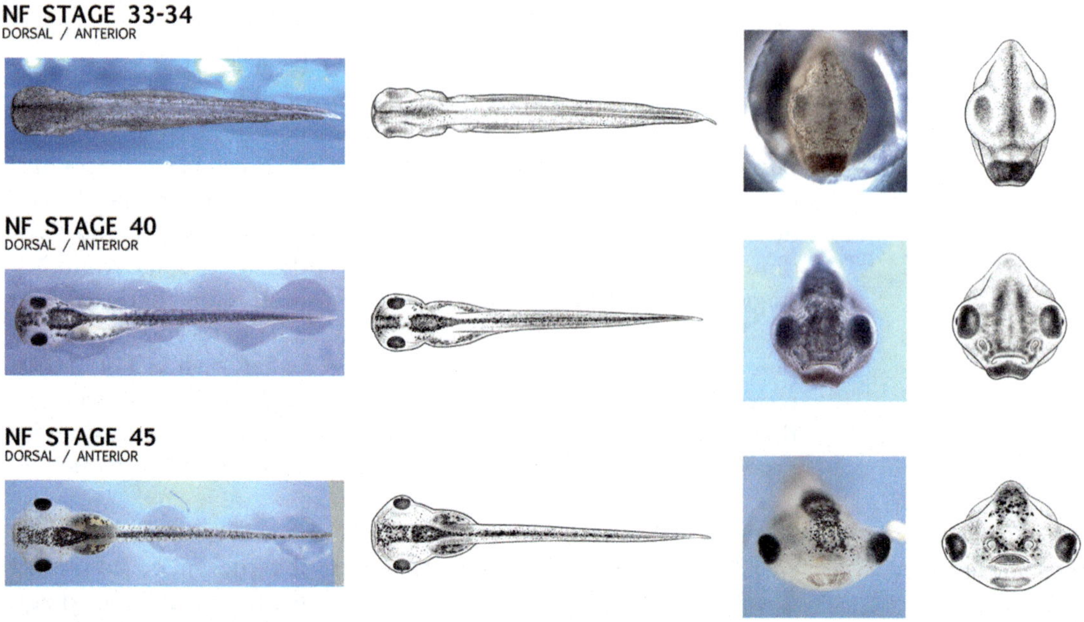

Fig. 9 New developmental series illustrations: Zahn series. The newly published, open access, Zahn drawings will be posted on Xenbase under the Anatomy and Development menu. This developmental stage series for *Xenopus*, based on multiple individuals, includes views that have not been previously published (e.g., dorsal and anterior as show here) as well as ventral views (not shown). The drawings call attention to morphological changes during critical stages of organogenesis, with a focus on changes in the shape and size of the head as seen in the images here demonstrating changes through NF stages 33 and 34, NF stage 40 and NF stage 45. The image series also includes bright field photographs (left) to compare with drawings (right). Images reproduced here are open access, and appear in Zahn, Levin and Spencer Adams, (2017) Development. 14:January 1, 2017; 144 (15): 2708–2713

histology, whole mount microscopy, textbook figures and/or with marker gene expression. Images will appear on the XAO term page. Figure 8A illustrates the XAO page for "heart," showing its definition, relationships, and place in the hierarchy of other ontology terms and a key marker gene as an in situ hybridization. We will post a variety of images including gene expression, dissections and histology to illustrate XAO terms. Note that ontology phrases can be read in both forward and reverse order in the hierarchy, for example "endocardium" is *part_of* "heart," while "heart" *has_parts* "endocardium" is also true. Contact Xenbase (xenbase@ucalgary.ca) to suggest or submit images to illustrate the XAO term pages and/or for the anatomy atlas.

8.8 Notes/Troubleshooting

Anomalies in the time temperature charts have been reported by some researchers and this table is currently under revision. Contact Xenbase (xenbase@ucalgary.ca) with any questions.

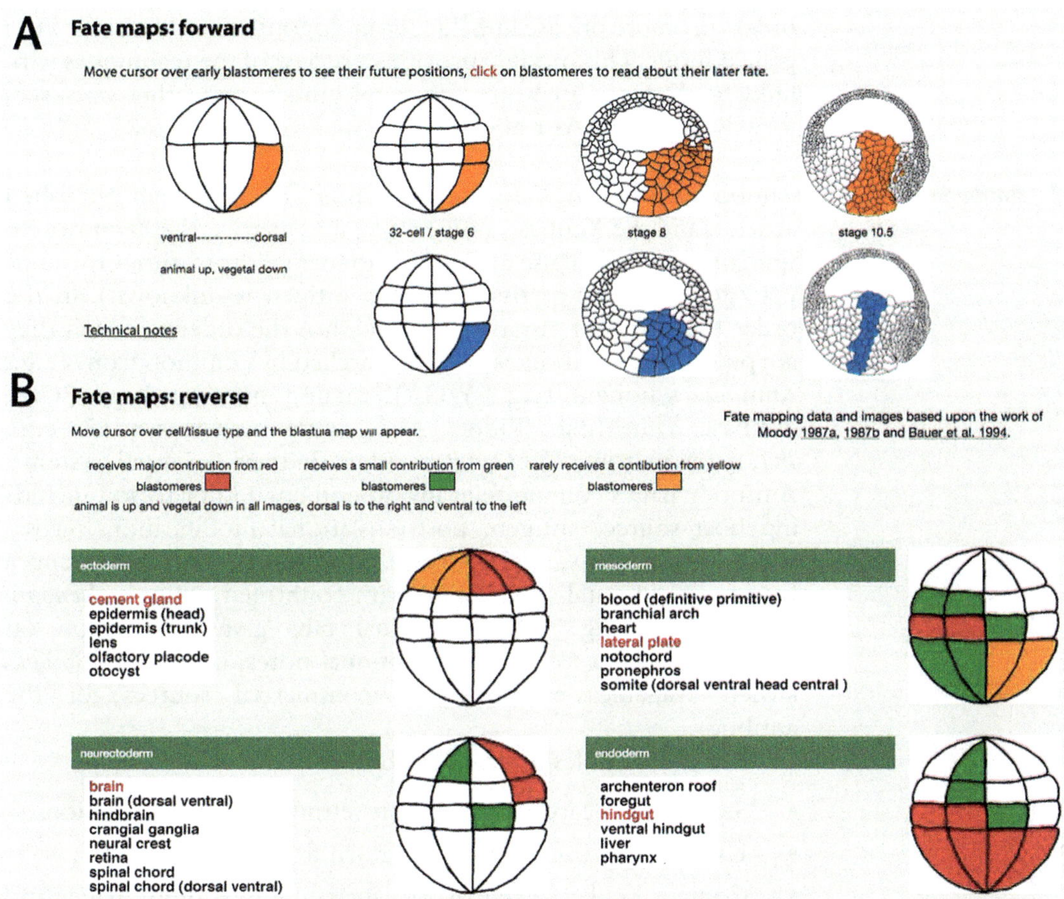

Fig. 10 Dynamic cell fate maps by Xenbase, based on classic studies by Moody [28, 29] and Bauer et al. [30]. (A) Cell fate in a forward direction, from blastomere to tissue. To use these animations, move the cursor over a blastomere, and the cells in later developmental stages are highlighted for NF stage 5 (16-cell) in orange, and NF stage 6 (32-cell) embryos in blue. Click on any blastomere to see its derivatives. (B) Cell fate in the reverse direction (i.e., tissue from blastomere). To use this tool, mouse over an anatomy term (e.g., "cement gland"), from a primary germ layer category (e.g., ectoderm, neurectoderm, mesoderm, or endoderm) to highlight the blastomeres that contribute to these tissues. Color coding indicates the degree of contribution from the NF stage 6 (32-cell) embryo: major (red), minor (green), or rarely incorporates cells (orange)

9 Reagents and Protocols Menu

Access the following modules and catalogs under the "Reagents & Protocols" menu of the main website banner or the tile on the home page.

9.1 CRISPr and TALEN Support

New genome editing technologies work well in *Xenopus* [32, 33]. CRISPr/Cas and TALEN/ZFN editing technologies function by inducing site-specific DNA strand breaks anywhere in the *Xenopus* genome. Mutations are induced by inefficient, error-prone nonhomologous end joining (NHEJ). In addition, site-specific DNA breaks promote precise knockin, homologous recombination (HR) gene editing. This module provides a review of the techniques with links to *Xenopus* literature, protocol guides, and other resources for CRISPr and TALENS.

9.2 Antibodies

Antibodies used in *Xenopus* research are curated from published articles, and the Xenbase antibody catalog has over 1200 entries (at time of press). Xenbase antibodies are named from either the antigen/gene symbol or tissue (where antigen is unknown), in the order they exit our curation pipeline, not the order in which they are published. Antibodies may be searched by common name (e.g., Xlim-1), synonym (e.g., WGA), catalog number (e.g., 3G8), Xenbase name (e.g., Kidney Ab2), antigen gene symbol (e.g., *lhx1*), or anatomy/XAO terms (e.g., kidney or visual system). Antibody pages contain relevant experimental information including host source, antigen, posttranslational modifications, cross-species interactivity, RRID (Research Resource Identifiers), and a list of experimental applications with confirmed utility in *Xenopus* research (*see* Fig. 11). Additional tabs give information on "Attributions," a "Wiki" for additional notes, images in *Xenopus* (when available) and links to commercial sources for the antibody.

Notes/Troubleshooting Antibodies

- Start with "Search All" (default setting) to cover all options.
- Use browser back button to return from the Wiki pages.
- Contact Xenbase (xenbase@ucalgary.ca) to submit images and additional usage data for validated antibodies.
- Text box does not autofill from XAO terms or gene symbols.
- GO terms may also be matched to Antibody entries.

9.3 Morpholinos

Morpholinos (MOs) are chemically modified oligonucleotides used to reduce the expression of a gene of interest [34, 35]. MOs knockdown gene expression by inhibiting mRNA translation, blocking RNA splicing, or inhibiting miRNA activity and maturation [34]. MOs have been shown to be effective in both *X. laevis* and *X. tropicalis* [36] and are widely used in experimental *Xenopus* embryology. Xenbase has manually curated 2400+ published *Xenopus*-specific MOs. How to use the MO search interface, and an example MO entry (e.g., sox2 MO1) is illustrated in Fig. 12. Search for published MOs through the interface, via the MO name or target gene symbol (e.g., *sox2*), a MO sequence, or use the "Alphabetic Search" (Fig. 12A). Note that the search for "sox2" also returns MOs for "sox21" (blue arrows, Fig. 12A). Each MO is assigned a unique name based on the targeted mRNA/gene (Fig. 12B), and we record any synonyms, the 5′ to 3′ sequence of the oligonucleotide, and whether it is designed to be splice-blocking or translation-blocking. We BLAST the MO sequence to identify on-target and off-target hits (Fig. 12C), and display the

Fig. 11 Xenbase Antibody catalog is accessed under the Reagents & Protocols menu/tile. (A) To find antibody entries, use the "Search All" (default) by entering antigen gene symbol, catalog number, or common name (red arrow). Select an antibody from the results table (e.g., Sox2 Ab1, black arrow) to open the antibody entry. (B) Each Antibody page includes Xenbase name, common name, source and catalog number, an image illustrating reactivity plus details such as tissue-specific expression (XAO terms). (C) Properties including validated activity and citation (including RRID numbers) are recorded when available (blue arrow). (D) Immunogen details and, (E) "Reported Usage" (e.g., western blot, immunofluorescence; orange arrow) are recorded. (F) Publications using the antibody are listed as "First" and "Most recent," with a "View All Papers" option, which is reiterated on the "Attributions" tab

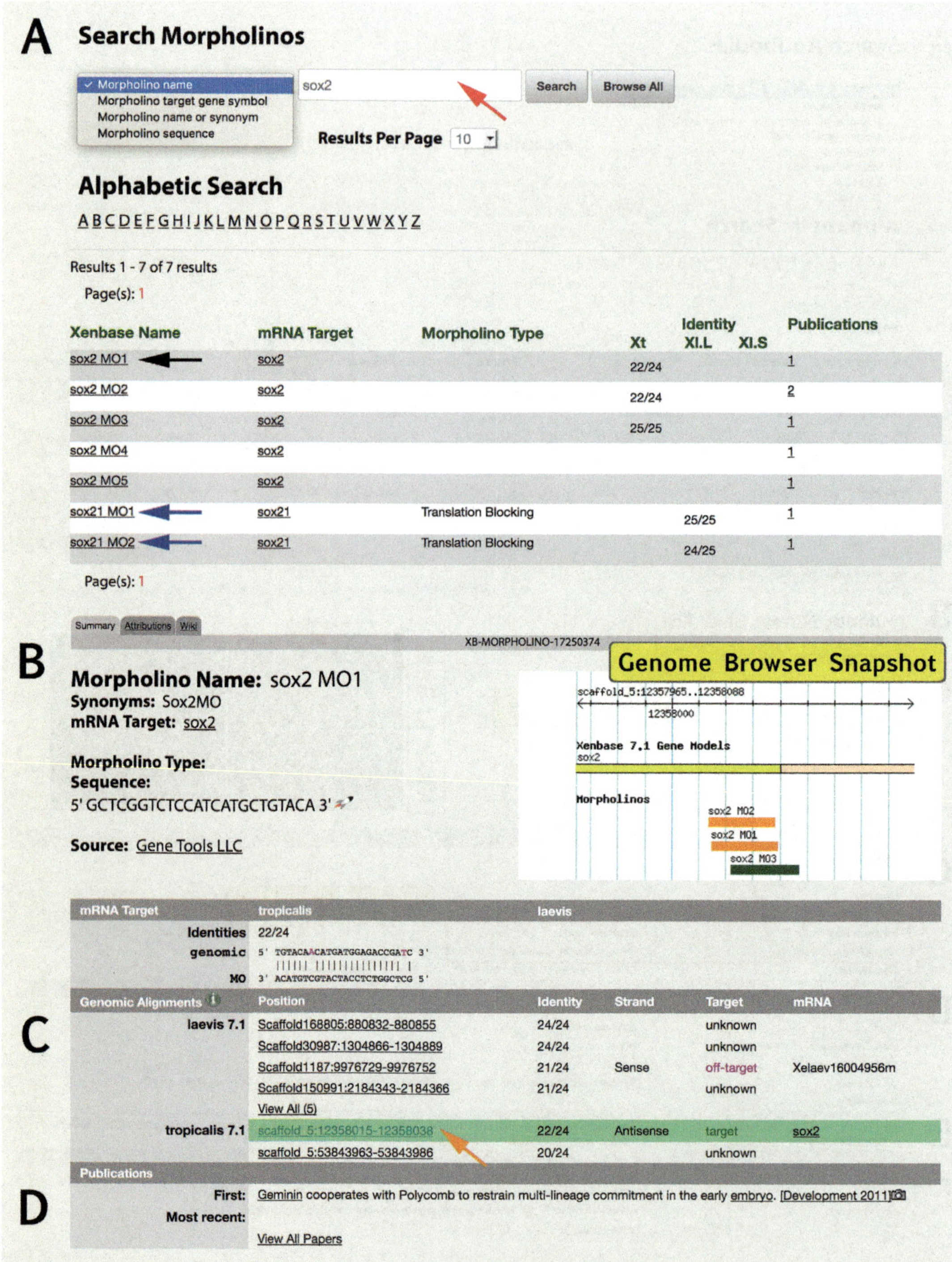

Fig. 12 Xenbase morpholino catalog is accessed via the "Search Morpholino" interface, under the Reagents & Protocols menu/tile. (A) Enter a gene symbol or MO sequence (red arrow) and click the Search button. Select the MO (black arrow), from the results table to open the MO page. (B) Each MO page includes all recorded details and a GBrowse snapshot showing the aligned position on the MO to the target mRNA. (C) Genomic alignments illustrate on-target (green) and off-target (pink) hits for *X. tropicalis*, *X. laevis* L and *X. laevis* S. Select a scaffold (highlighted in green) to view in GBrowse (orange arrow). (D) Publications using the same MO are listed, with a "View All Papers" option, which is reiterated on the 'Attributions' tab

MO's scaffold position in a GBrowse snapshot of each MO page (Fig. 12B), as well as give scaffold positions orange arrow, Fig. 12C). All curated MOs are mapped to the genome and are displayed in the full genome view on GBrowse. Note that MOs are listed on Gene Pages under "Reagents," and are also displayed on associated Article Page(s), the latter being accessible under "publications" section (Fig. 12D).

Our catalog of MOs can be searched under the Reagents and Protocols menu via the Search Morpholinos using the steps:

1. Choose search options from drop-down menu and enter text in the search field (red arrow, Fig. 12A). Options include:
   - Target gene/mRNA symbol (e.g., *sox2*).
   - MO name (e.g., sox2 MO2).
   - MO synonym, as used in a publication(s) (e.g., MO-sox2).
   - MO sequence (e.g., AGCTCGGTCTCCATCATGCT GTAC).
2. Hit Search button.
3. A single search hit will go straight to that MO page (e.g., sox2 MO2: XB-MORPHOLINO-17250375). Multiple search hits (such as resulting from a search for "sox2") will be displayed in a table (Fig. 12A).
4. Click the Xenbase MO name from left hand column to examine details on the MO page (black arrow, Fig. 12A).
5. Click the GBrowse snapshot (Fig. 12B) or the specific scaffold position (orange arrow, Fig. 12C) to view MOs in the full genome browser tool,
6. Use browser back button to return to the MO page or search results table.

Notes/Troubleshooting MOs

- An "Alphabetic Search" is also enabled on the MO search interface.
- MOs can also be found using the Quick Search Menu, using the Xenbase accession number option (e.g., XB-MOR PHOLINO-17249151).
- Nucleotide searches must be exact matches to find a specific MO, and exclude the 5′ prefix and 3′ suffix.
- "Browse All" will provide an alphanumeric list of the entire MO catalog of 2400+ entries, so using at least a gene symbol or synonym will help finding a specific record.
- Click on the GBrowse snapshot to view positional information or search for off-target interactions.
- Xenbase curates MOs (and numbers them in serial sequence) in the order in which they exit our curation pipeline, not in order of publication.

- Gene symbol search uses text-matching, so that a search for MOs to the gene symbol "*apln*" will return MOs for *apln*, *aplnr*, and *hapln3*.
- Phenotypes generated using a specific MO will be posted on each MO Page as well as on the Article Page, when Phenotypes are launched on Xenbase.

9.4 ORFeome

The *Xenopus* ORFeome project generated a comprehensive set of 8600+ full-length, end-sequence validated, high quality open reading frame clones in the Gateway cloning system, suitable for recombinant protein expression [37]. ORF sequences represent 7800+ unique genes, including 2724 genes with human ortholog disease association (e.g., optn ORF1, associated with glaucoma (OMIM #137760)). In total the ORFeome clones represent approximately 40% of the nonredundant *X. laevis* genome [37]. ORFeome reagents allow high-throughput in vivo functional-genomic screening of frog genes in a manner previously not feasible. Details of each ORF clone are given on individual "ORF Page" including gene symbol, gene name, Entrez ID, 5′ and 3′ sequence, predicted translation, confidence estimates, and links to suppliers.

Notes/Troubleshooting ORFeome clones

- Xenbase ORF Page IDs in the form "XB-ORF-#" (e.g., XB-ORF-17287509), can be used on the ORF search page or from the quick search menu using Xenbase Accession option.
- Xenbase does not supply ORFeome clones. Researchers must contact the supplier(s) directly.

9.5 Small Molecules Wiki

Useful small molecules are manually catalogued from published literature in a Wiki format. Entries are listed under the following headings:

1. *Alphabetic list* with literature references.
2. *Small Molecules affecting Pathways*: e.g., Retinoic Acid: Citral; or Hedgehog: cyclopamine.
3. *Small Molecules affecting Biological Functions and Processes*: e.g., angiogenesis: suramin, apoptosis: cyclohexaminde, or neurotransmission: diazepam.
4. *Drugs by class*: e.g., kinases: KT5720, a specific, cell-permeable inhibitor of protein kinase A (PKA).

Notes/Troubleshooting Small Molecules Wiki

- Registered users can add to this Wiki.
- Minimal information required includes a description, genes/pathways/functions affected, source/supplier, reference(s), as well as a structural diagram (such as those available in PubChem).

- Click the Textpresso link at the bottom of a Wiki entry, to run a search for the term in *Xenopus* articles.

9.6 Protocols Wiki

Entries are listed under the following headings:

1. *Books for Xenopus Research and Protocols*
2. *Online Resources*
   (a) *Journal of Visualized Experiments* (JOVE) video demonstrations, showing protocols and techniques (e.g., host transfer methods of oocyte fertilization; dissections of retinal tissue; electroporation; live-cell imaging for quantitative analysis; and patch clamp and perfusion techniques).
   (b) *Cold Spring Harbor* (CSH) *Xenopus* Protocols.
3. *General Research Protocols* covering Animal Husbandry, Lab solution recipes and reagents, Generating Embryos, Transgenesis, in situ Hybridization, Immunohistochemistry, ChIP-Seq protocols, Histology, Embryo Staining Protocols, Immunohistochemistry and Protein Protocols, Nucleic Acid Protocols, Oocyte Transfer Technique, *Xenopus* Oocyte and Egg Extracts, and *Xenopus* Tissue Culture.

Notes/Troubleshooting Protocols Wiki

- Reference books, text-books and chapters without PubMed IDs cannot be added to the literature module in Xenbase.
- Both JOVE and CSH *Xenopus* Protocols require institutional licensing to access.
- Registered Xenbase users can add to the Protocols Wiki.
- Protocols are submitted to Xenbase by specific laboratories, as indicated, and researchers should contact the lab directly (via Xenbase Laboratory profile) to troubleshoot the specific protocol. Contact Xenbase if there are errors or updates in the protocol as published.

9.7 Search Clones

The clone catalog in Xenbase houses data on over one million plasmids developed from both *X. laevis* and *X. tropicalis.* Access the clone search interface via Reagents and Protocols menu, and choose from search options: search all, gene symbol (e.g., *aldh1a2.L*), NCBI/GenBank accession number, clone name (e.g., IMAGE:3421129), source tissue (e.g., cornea, gonad, or oocyte), clone page ID (e.g., XB-CLONE-242125). Each *Clone Page* gives the clone name, the gene to which it is mapped, and source species, as well as sequence data, Unigene accession number, source/external databases, a description of tissue or embryonic stage from which it was generated, the vector details, and a vector map.

Notes/Troubleshooting Clones

- Wildcard * can be used to search clones.
  - Use the checkbox to filter clones only available from the EXRC.
  - Alphabetic search is for clone name, not gene symbol (these often do not match).

9.8 Clone Libraries

Several cDNA libraries were used to generate the *Xenopus* ESTs, many of which are from *the Xenopus Gene Collection* Library. The Library name and tissue used are listed here for *X. tropicalis* and *X. laevis*.

9.9 Vectors

Select this menu option to see a list of standard vectors used in *Xenopus* research. Click the vector name (e.g., pCS107) to view the plasmid and phagemid Vector Page, with map and supplier details. Use the + toggle to see the sequence if available.

9.10 Obtain Frogs

This table gives contact details of Stock Centers, commercial suppliers of frogs, oocytes, wet lab and aquarium equipment.

10 Literature Menu

The Xenbase literature module houses an extensive catalog of 49,000+ *Xenopus* research papers that are available from NCBI's PubMed service, 90% of which are searchable by Textpresso, and a books catalog. Articles are automatically uploaded each week by text-matching "*Xenopus*" or "*Silurana*" in the title, abstract or keywords of newly released papers. Latest *Xenopus* research articles added to Xenbase are listed at the top of the Announcements column on the home page. Each article is represented on an "Article Page" (details below in Subheading 10.1), a recent example of which is shown in Fig. 13. Articles can be searched via the quick search menu (discussed above, Subheading 4), and via the "Literature" menu/tile link, by entering an author name, partial or full paper title, or using a Xenbase accession number (e.g., XB-ART-45000).

10.1 Article Pages

An Article Page header includes the Xenbase accession number (sequentially numbered as they are added to the database), and each article page carries the following information:

1. *Reference*: Journal, Title and Authors appear at the top of each entry, followed by a full Abstract, PubMed ID and PMC ID. PubMed/PMC links redirect to the article record at those resources.
2. *Article* link goes directly to journal website. If the article link is missing, click the PubMed Link to access the journal website. This may require a subscription.
3. *Grant support* gives details of sources of funding (when available), with a link to the funding website.

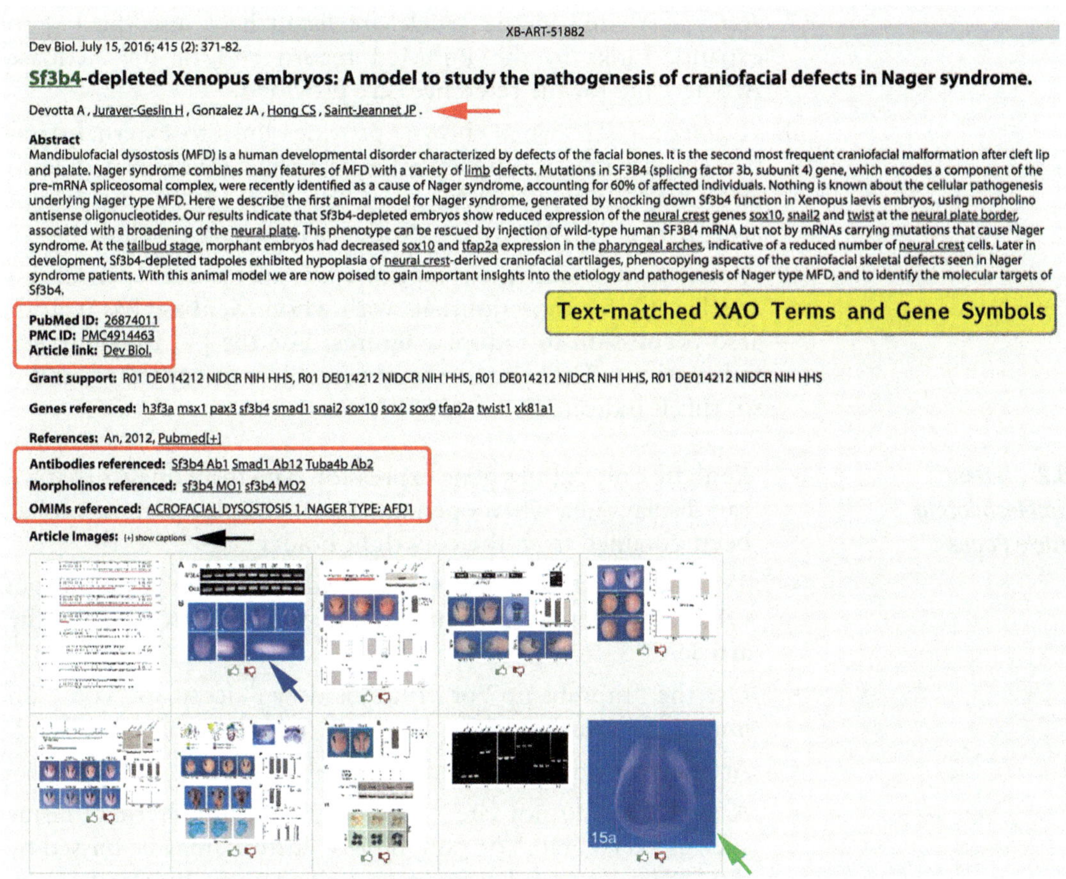

Fig. 13 Publications in the Xenbase Literature module are represented on an "Article Page." We assign a database accession number (e.g., XB-ART-51882) and pull the full abstract and associated data from PubMed. Authors with Xenbase profiles have their name underlined (red arrow), indicating a link to their personal profile page. The abstract is automatically text-matched for gene names, gene symbols, and XAO terms (underlined). Direct links are provided for PubMed, PMC, and the Journal entries for the article (upper red box), enabling quick access to the full article and PDFs. Genes referenced in the article are either mentioned in abstract or added manually by curators, as are antibodies and morpholinos (lower red box). Matched and curated terms are underlined and linked to Gene Page(s), XAO, AB, and MO pages, respectively. Figures from the article (if available) are shown as thumbprints, and references cited in the paper follow. Use the [+] to expand to show captions (black arrow) or the full reference list. Double click the image to open the larger figure and the annotation table. Additionally, links to OMIM diseases and GO terms reverenced in the research, and to raw or supplementary data (e.g., NCBI/GEO and DRYAD) are shown when available. Images with new gene expression can be selected from figures and be placed as 'summary images' on a Gene Page (green arrow)

4. *Genes referenced* in the article come from both text matching and manual curation.
5. *Antibodies referenced* lists the curated primary antibodies used in the research.
6. *Morpholinos referenced* lists the curated morpholinos used in the research.

7. *References* cited in the article are listed here, use the [+] to expand. Links to the PubMed record and/or the Xenbase Article Page for the references are provided.
8. *Resources URL* (when shown) provides links to external databases where raw or supplementary data files are deposited, such as NCBI/GEO or DRYAD (e.g., XB-ART-42630 with microarray data at NCBI/GEO). Not all article have this link.
9. *Article Images* and their captions are posted for open access articles and for those journals with whom Xenbase has negotiated permission to redisplay figures. Use the [+] toggle to see full captions. These images may be subject to copyright, and if so, this is indicated.

10.2 Notes/Troubleshooting Article Pages

- Xenbase only curates gene expression data from images that we can display (i.e., when open access or redisplay permission has been obtained from the copyright holder).
- Two new data links will soon be added to the Article Page: ORFeome clones and curated Phenotypes described in the article.
- Use the "thumbs up" or "thumbs down" icons to "vote" on image/data quality.
- Articles can only be uploaded via PubMed ID.
- Articles that do not cite "Xenopus," and use alternate terms such as "vertebrate" or "amphibian" instead, may be missed by the loader, yet can be manually uploaded via PubMed ID by Xenbase staff and registered users. To do this, log in, click the "Literature/papers" menu heading, then the "Click here to manually add an article by PubMed ID" link.
- Books and book chapters are entered separately under the Literature/Books menu and can be searched via Title, Author, ISBN number, publisher, or Xenbase accession (e.g., XB-BOOK-202).

10.3 Textpresso

Advanced querying of the full text of most *Xenopus* papers can be achieved by using a Xenbase-specific instance of Textpresso, an information extraction and processing software package for scientific literature developed by the California Institute of Technology, and used under license on Xenbase. Simple keyword searches can be entered in a free text field, and/or up to five category terms can be chosen from a preset list which includes higher level GO terms, experimental techniques and relational terms. The most advanced and flexible option for a Textpresso search is the "Query Language" which is available toward the top of the page, in the menu under the Textpresso logo. This tool allows the user to assemble sets of commands to build complex queries, and allows users not only to specify a variety of query fields, categories, and keywords, but also to set specific thresholds for the number of instances of those keywords and combine previously defined searches using Boolean operators (i.e., AND, OR, NOT).

Notes/Troubleshooting Textpresso

1. The user guide for Textpresso can be found here on Xenbase: http://www.xenbase.org/cgi-bin/textpresso/xenopus/user_guide
2. The "Advanced Search" allows users to specify which sections of the paper should be searched. Select "on" next to the "Advanced search options" text.
3. "Body" option is used for papers that could not be properly sectioned into the other elements. "Scope" option allows a choice between "sentence," "document," and "field." "Sort by" option offers a variety of criteria including year of publication, number of citations, title, author, or score.
4. Option available to exclude supplementary data and/or abstracts.
5. Filter by "Author," "Journal," "Year," or "Doc ID" (i.e., PubMed ID).
6. Too many results? Exclude terms using filter option by entering "+" or "−" signs in front of terms. Enter the field type (e.g., author or title) in square brackets and phrases in double quotes (e.g., "+Patel-Zheng[author]"). Click on the "Filter!" button to activate.
7. Too few results? Broaden the scope by allowing the search to include synonyms for your keywords or by increasing the sections of the literature searched, including "unsectioned" manuscripts.

11 Community Support

Xenbase has several features designed to facilitate and encourage communication within the *Xenopus* research community. This includes an extensive catalog of individual *Xenopus* researchers, research labs, institutes such as Universities, governmental and nongovernmental organizations, funding bodies, publishers, and companies. Xenbase also provides announcements for upcoming meetings relevant to *Xenopus* research, job postings, the "Xine" community newsletter, Xenbase forums and links to a variety of relevant professional societies.

11.1 Find Researcher, Lab, and Organization Profiles

Register with Xenbase to make a personal and/or lab profile page, with contact details and a description of your research and link papers, which are shown on the "Publications" tab. Click the "Register" link in the top right corner of the site and follow prompts to record your name and contact information, and log-in details. Lab Pages are associated with individual profiles of the Principle Investigator as well as Lab members (e.g., postdocs, graduate students). Find a People,

Lab, or Organization profile (e.g., a *Xenopus* supplier or stock center), by clicking the Community Menu/Tile to the search interface. Search via researcher name, institution or research interests.

11.2 Job Postings

A jobs board is available for registered users to post open positions in a laboratory or institution, and covers all levels of research from entry-level lab technician and student scholarships to professorships and division directors.

11.3 Xine

Xine is a *Xenopus* community email group hosted at the National Xenopus Resource (NXR). Xine's goal is to inform researchers about important developments and to disseminate information of wide interest. Xenbase archived Xine releases and supplements from 2001–2011. Xine depends on contributions from members of the *Xenopus* community for its content. Not a member? Join here: https://lists.mbl.edu/mailman/listinfo/xenopus. If you encounter problems, contact the Xine editor (xenopus@mbl.edu).

11.4 Xenbase Forums

Xenbase forums, an avenue for researchers to take part in Xenopus-related discussion, was relaunched in May 2017, and users must register to post items.

11.5 Meetings and Resources

We provide details to upcoming conferences, workshops, and technical courses relevant to the *Xenopus* research community. These are updated biannually.

11.6 Xenopus White Papers

The *Xenopus* Community White Papers, which provide a succinct, authoritative report and literature review of recent *Xenopus* research, are posted on Xenbase to maximize their distribution. White Papers provide recommendations on how continued, focused funding from the National Institutes of Health (NIH) can maximize the impact of biomedical research using the *Xenopus* system (*see* [3]), and therefore they serve as a very useful resource. Researchers are encouraged to reference the most recent *Xenopus* Community White Papers in their NIH grant proposals.

11.7 International Xenopus Board

We host information and history about the International *Xenopus* Board (IXB). The IXB was incorporated in 2015 as a tax-exempt, nonprofit organization with a remit to organize a biennial International *Xenopus* Conference, organize an annual Resources and Emerging Technologies meeting, represent and promote communication among *Xenopus* researchers, and promote the development and use of *Xenopus* resources.

Xenbase provides a list of the current members of the IXB and posts documents containing a report on the creation of the IXB, minutes from previous Resources and Emerging Technologies meetings, the by-laws and certificate of incorporation of the IXB.

11.8 Notes/Troubleshooting Community Support Pages

Contact Xenbase (xenbase@ucalgary.ca) to post a meeting or workshop, or if you find omissions, errors, or bugs in these pages.

12 Stock Center Support

The international *Xenopus* community has established several *Xenopus* stock centers for obtaining reagents and frogs for biomedical and immunological research. A major goal of Xenbase is to support the Stock Centers by curating and cataloguing the frogs (lines and strains) and other reagents they supply. *Xenopus* lines and strains are given Research Resource Identifiers (RRID) numbers, a measure which helps disambiguate which lines were used in the research, promoting reproducibility, rigor, and transparency. The catalog is searchable through the "Lines and Strains" option on the "Stock Center" menu, or the "Transgenic lines" link in the "Reagents & Protocols" tile of the Xenbase home page. There are currently five Stock Centers worldwide:

1. *NXR*: The National *Xenopus* Resource is the US frog stock center and training center for advanced technologies (Contact: xenopus@mbl.edu).
2. *EXRC*: The European *Xenopus* Resource Centre is the stock center in the UK (Contact: EXRC@xenopusresource.org).
3. *CRB*: Centre de Ressources Biologiques *Xénopes* (Biological Resource Center *Xenopus*) is the stock center in France (Contact: crb-xenopes@univ-rennes1.fr).
4. *NBRP*: The National BioResource Project, is the stock center in Japan (Contact: oakkashi@hiroshima-u.ac.jp).
5. *URMC*: The *X. laevis* Research Resource for Immunology is based at the University Rochester, NY (Contact: jacques_robert@urmc.rochester.edu).

12.1 Xenopus Lines

Standardized nomenclature is critical to make research accessible to the broader scientific community and to ensure consistency and provenance, and researchers are encouraged to follow the working Transgenic Nomenclature Guidelines which are posted on Xenbase under the Genes tile on the home page. Xenbase names transgenic (Tg) and mutant lines according to these guidelines, which were developed in consultation with the *Xenopus* stock centers, following best practices used by all other model organisms (*see* review in [38]). We only curate published, stable lines, and those available at stock centers using these criteria. Common names or Stock Center shorthand names are recorded as synonyms. For example the Tg line called line "511" or line "275," has been named officially as *Xla.Tg(actc1:GFP)*^{Amaya}. It is a *X. laevis* (*Xla*) line with a transgene (*Tg*) which has the promoter for the X. *laevis* cardiac actin gene (*actc1*), driving expression of a green fluorescent protein (*GFP*); the line comes from the Amaya lab (^{Amaya}). Each Tg line page (e.g., XB-LINE-935) has an "Attributions" tab which details associated papers, and the "Transgene" tab gives details of the transgenic constructs used to produce the line.

12.2 Xenopus Strains

Wild type strains of *Xenopus* are named with a species code (*Xla* or *Xtr*), geographic origin or common names, and supplier/source (e.g., *Xla.J-strain*^{EXRC} or *Xtr.Nigerian*^{NBRP}).

12.3 Transgenes

We curate transgenic constructs from published research articles. "Transgenes" are searchable via the "Stock Center" menu. While some of the transgenes are the basis for Tg lines, many have been used for transient transgenesis or are expression constructs. Transgenes are named following the mutant and Tg line nomenclature guidelines, with two exceptions: we omit species prefix and the originating lab code.

12.4 Notes/ Troubleshooting

- Consult nomenclature guidelines for Genes, Chromosomes and Tg/Mutant lines when naming your transgenic constructs and frogs.
  - Gene nomenclature: http://www.xenbase.org/gene/static/geneNomenclature.jsp
  - Chromosome nomenclature: http://www.xenbase.org/gene/static/chromosomeNomenclature.jsp
  - Transgenic and Mutant Line nomenclature: http://www.xenbase.org/gene/static/tgNomenclature.jsp
- Contact Xenbase (xenbase@ucalgary.ca) if you experience omissions, errors or bugs in these pages, or if you need advice with nomenclature.

13 Downloading and Submit Data Menu

Xenbase hosts an assortment of files of general utility on our FTP server. Users can find regularly updated reports from Xenbase database content relating to gene expression, sequence information, accession numbers and genomic location, and other specific mappings files (e.g., *Xenopus* gene-human disease gene mappings), by navigating through the expandable folders on the FTP page. Additionally, users are encouraged to submit their own data to be shared with the broader research community. We can accept gene expression images (see template page for minimum metadata required), protocols, development movies and cell-fate maps, and welcome discussions on hosting other types of data. Click "Submit your data" option from the menu to open a data submission form for uploading the files.

14 New Features and Future Developments on Xenbase

- Phenotypes including anatomical, gene function, and gene expression as phenotype.
- JBrowse will replace GBrowse as genome viewer tool, with a 2-year phase-out period.
- RNA-Seq and ChIP-Seq data will be shown in stacked views.
- Illustrated XAO terms and expanded Organ Atlas, including images from XenHead [26] and EctoMap projects.
- Many NGS and genome feature tracks added to genome browser, older NGS content will be mapped to latest genome builds.
- Expanded curation of mutant and transgenic lines available from *Xenopus* Stock Centers.
- "How to use Xenbase" videos will be expanded to cover more topics.
- Educational pages on *Xenopus* in biomedical research will be expanded.
- OMIM disease, GO terms and References cited will be posted on Article Pages.
- Protein–protein interaction data and enhanced gene network support.
- Revamped gene expression interface.
- RRID numbers for all antibodies.

Acknowledgements

Funding for Xenbase is provided by the Eunice and Kennedy Shriver National Institute of Child Health and Human Development, grant P41 HD064556 (Zorn and Vize, Joint-PIs). We thank James Coulombe (NIH/NICHD), the Xenbase EAB members, and *Xenopus* researchers around the world for continued support and feedback that help us set priorities for development and new features on Xenbase.

References

1. Gurdon JB (1960) The developmental capacity of nuclei taken from differentiating endoderm cells of Xenopus laevis. J Embryol Exp Morphol 8:505–526
2. Gurdon JB, Elsdale TR, Fischberg M (1958) Sexually mature individuals of Xenopus laevis from the transplantation of single somatic nuclei. Nature 182(4627):64–65
3. Sater AK, Moody SA (2017) Using Xenopus to understand human disease and developmental disorders. Genesis 55(1-2). https://doi.org/10.1002/dvg.22997
4. Bowes JB, Snyder KA, Segerdell E, Gibb R, Jarabek C, Noumen E, Pollet N, Vize PD (2008) Xenbase: a Xenopus biology and genomics resource. Nucleic Acids Res 36(Database issue):D761–D767. https://doi.org/10.1093/nar/gkm826
5. Karimi K, Vize PD (2014) The Virtual Xenbase: transitioning an online bioinformatics resource to a private cloud. Database (Oxford) 2014:bau108. https://doi.org/10.1093/database/bau108
6. Hellsten U, Harland RM, Gilchrist MJ, Hendrix D, Jurka J, Kapitonov V, Ovcharenko I, Putnam NH, Shu S, Taher L, Blitz IL, Blumberg B, Dichmann DS, Dubchak I, Amaya E, Detter JC, Fletcher R, Gerhard DS, Goodstein D, Graves T, Grigoriev IV, Grimwood J, Kawashima T, Lindquist E, Lucas SM, Mead PE, Mitros T, Ogino H, Ohta Y, Poliakov AV, Pollet N, Robert J, Salamov A, Sater AK, Schmutz J, Terry A, Vize PD, Warren WC, Wells D, Wills A, Wilson RK, Zimmerman LB, Zorn AM, Grainger R, Grammer T, Khokha MK, Richardson PM, Rokhsar DS (2010) The genome of the Western clawed frog Xenopus tropicalis. Science 328(5978):633–636. https://doi.org/10.1126/science.1183670
7. Session AM, Uno Y, Kwon T, Chapman JA, Toyoda A, Takahashi S, Fukui A, Hikosaka A, Suzuki A, Kondo M, van Heeringen SJ, Quigley I, Heinz S, Ogino H, Ochi H, Hellsten U, Lyons JB, Simakov O, Putnam N, Stites J, Kuroki Y, Tanaka T, Michiue T, Watanabe M, Bogdanovic O, Lister R, Georgiou G, Paranjpe SS, van Kruijsbergen I, Shu S, Carlson J, Kinoshita T, Ohta Y, Mawaribuchi S, Jenkins J, Grimwood J, Schmutz J, Mitros T, Mozaffari SV, Suzuki Y, Haramoto Y, Yamamoto TS, Takagi C, Heald R, Miller K, Haudenschild C, Kitzman J, Nakayama T, Izutsu Y, Robert J, Fortriede J, Burns K, Lotay V, Karimi K, Yasuoka Y, Dichmann DS, Flajnik MF, Houston DW, Shendure J, DuPasquier L, Vize PD, Zorn AM, Ito M, Marcotte EM, Wallingford JB, Ito Y, Asashima M, Ueno N, Matsuda Y, Veenstra GJ, Fujiyama A, Harland RM, Taira M, Rokhsar DS (2016) Genome evolution in the allotetraploid frog Xenopus laevis. Nature 538(7625):336–343. https://doi.org/10.1038/nature19840
8. Vize PD, Liu Y, Karimi K (2015) Database and informatic challenges in representing both diploid and tetraploid Xenopus species in Xenbase. Cytogenet Genome Res 145(3-4):278–282. https://doi.org/10.1159/000430427
9. Matsuda Y, Uno Y, Kondo M, Gilchrist MJ, Zorn AM, Rokhsar DS, Schmid M, Taira M (2015) A new nomenclature of Xenopus laevis chromosomes based on the phylogenetic relationship to Silurana/Xenopus tropicalis. Cytogenet Genome Res 145(3-4):187–191. https://doi.org/10.1159/000381292
10. Segerdell E, Bowes JB, Pollet N, Vize PD (2008) An ontology for Xenopus anatomy and development. BMC Dev Biol 8:92. https://doi.org/10.1186/1471-213X-8-92
11. Mi H, Muruganujan A, Casagrande JT, Thomas PD (2013) Large-scale gene function analysis with the PANTHER classification system. Nat Protoc 8(8):1551–1566. https://doi.org/10.1038/nprot.2013.092
12. Owens ND, Blitz IL, Lane MA, Patrushev I, Overton JD, Gilchrist MJ, Cho KW, Khokha MK (2016) Measuring absolute RNA copy numbers at high temporal resolution reveals transcriptome kinetics in development. Cell Rep 14(3):632–647. https://doi.org/10.1016/j.celrep.2015.12.050
13. Yanai I, Peshkin L, Jorgensen P, Kirschner MW (2011) Mapping gene expression in two Xenopus species: evolutionary constraints and developmental flexibility. Dev Cell 20(4):483–496. PubMed ID: 21497761
14. Khokha MK, Chung C, Bustamante EL, Gaw LW, Trott KA, Yeh J, Lim N, Lin JC, Taverner N, Amaya E, Papalopulu N, Smith JC, Zorn AM, Harland RM, Grammer TC (2002) Techniques and probes for the study of Xenopus tropicalis development. Dev Dyn 225(4):499–510. https://doi.org/10.1002/dvdy.10184
15. Altschul SF, Gish W, Miller W, Myers EW, Lipman DJ (1990) Basic local alignment search tool. J Mol Biol 215(3):403–410. https://doi.org/10.1016/S0022-2836(05)80360-2
16. Skinner ME, Uzilov AV, Stein LD, Mungall CJ, Holmes IH (2009) JBrowse: a next-generation genome browser. Genome Res 19(9):1630–1638. https://doi.org/10.1101/gr.094607.109
17. Raney BJ, Dreszer TR, Barber GP, Clawson H, Fujita PA, Wang T, Nguyen N, Paten B, Zweig AS, Karolchik D, Kent WJ (2014) Track data hubs

enable visualization of user-defined genome-wide annotations on the UCSC Genome Browser. Bioinformatics 30(7):1003–1005. https://doi.org/10.1093/bioinformatics/btt637

18. Ciau-Uitz A, Pinheiro P, Kirmizitas A, Zuo J, Patient R (2013) VEGFA-dependent and -independent pathways synergise to drive Scl expression and initiate programming of the blood stem cell lineage in Xenopus. Development 140(12):2632–2642. https://doi.org/10.1242/dev.090829
19. Parain K, Mazurier N, Bronchain O, Borday C, Cabochette P, Chesneau A, Colozza G, El Yakoubi W, Hamdache J, Locker M, Gilchrist MJ, Pollet N, Perron M (2012) A large scale screen for neural stem cell markers in Xenopus retina. Dev Neurobiol 72(4):491–506. https://doi.org/10.1002/dneu.20973
20. Raciti D, Reggiani L, Geffers L, Jiang Q, Bacchion F, Subrizi AE, Clements D, Tindal C, Davidson DR, Kaissling B, Brandli AW (2008) Organization of the pronephric kidney revealed by large-scale gene expression mapping. Genome Biol 9(5):R84. https://doi.org/10.1186/gb-2008-9-5-r84
21. Kaminski MM, Tosic J, Kresbach C, Engel H, Klockenbusch J, Muller AL, Pichler R, Grahammer F, Kretz O, Huber TB, Walz G, Arnold SJ, Lienkamp SS (2016) Direct reprogramming of fibroblasts into renal tubular epithelial cells by defined transcription factors. Nat Cell Biol 18(12):1269–1280. https://doi.org/10.1038/ncb3437
22. Rana AA, Collart C, Gilchrist MJ, Smith JC (2006) Defining synphenotype groups in Xenopus tropicalis by use of antisense morpholino oligonucleotides. PLoS Genet 2(11):e193. https://doi.org/10.1371/journal.pgen.0020193
23. Gilchrist MJ, Pollet N (2012) Databases of gene expression in Xenopus development. Methods Mol Biol 917:319–345. https://doi.org/10.1007/978-1-61779-992-1_19
24. Ahmed A, Ward NJ, Moxon S, Lopez-Gomollon S, Viaut C, Tomlinson ML, Patrushev I, Gilchrist MJ, Dalmay T, Dotlic D, Munsterberg AE, Wheeler GN (2015) A database of microRNA expression patterns in Xenopus laevis. PLoS One 10(10):e0138313. https://doi.org/10.1371/journal.pone.0138313
25. Armisen J, Gilchrist MJ, Wilczynska A, Standart N, Miska EA (2009) Abundant and dynamically expressed miRNAs, piRNAs, and other small RNAs in the vertebrate Xenopus tropicalis. Genome Res 19(10):1766–1775. https://doi.org/10.1101/gr.093054.109
26. Zahn N, Levin M, Adams DS (2017) The Zahn drawings: new illustrations of Xenopus embryo and tadpole stages for studies of craniofacial development. Development 144(15):2708–2713. PubMed ID: 28765211
27. Nieuwkoop PD, Faber J (1994) Normal table of Xenopus laevis (Daudin): a systematical and chronological survey of the development from the fertilized egg till the end of metamorphosis. Garland Pub, New York, NY
28. Moody SA (1987) Fates of the blastomeres of the 16-cell stage Xenopus embryo. Dev Biol 119(2):560–578
29. Moody SA (1987) Fates of the blastomeres of the 32-cell-stage Xenopus embryo. Dev Biol 122(2):300–319
30. Bauer DV, Huang S, Moody SA (1994) The cleavage stage origin of Spemann's Organizer: analysis of the movements of blastomere clones before and during gastrulation in Xenopus. Development 120(5):1179–1189
31. Segerdell E, Ponferrada VG, James-Zorn C, Burns KA, Fortriede JD, Dahdul WM, Vize PD, Zorn AM (2013) Enhanced XAO: the ontology of Xenopus anatomy and development underpins more accurate annotation of gene expression and queries on Xenbase. J Biomed Semantics 4(1):31. https://doi.org/10.1186/2041-1480-4-31
32. Blitz IL, Biesinger J, Xie X, Cho KW (2013) Biallelic genome modification in F(0) Xenopus tropicalis embryos using the CRISPR/Cas system. Genesis 51(12):827–834. https://doi.org/10.1002/dvg.22719
33. Bhattacharya D, Marfo CA, Li D, Lane M, Khokha MK (2015) CRISPR/Cas9: an inexpensive, efficient loss of function tool to screen human disease genes in Xenopus. Dev Biol 408(2):196–204. https://doi.org/10.1016/j.ydbio.2015.11.003
34. Nasevicius A, Ekker SC (2000) Effective targeted gene 'knockdown' in zebrafish. Nat Genet 26(2):216–220. https://doi.org/10.1038/79951
35. Heasman J, Kofron M, Wylie C (2000) Beta-catenin signaling activity dissected in the early Xenopus embryo: a novel antisense approach. Dev Biol 222(1):124–134. https://doi.org/10.1006/dbio.2000.9720
36. Nutt SL, Bronchain OJ, Hartley KO, Amaya E (2001) Comparison of morpholino based translational inhibition during the development of Xenopus laevis and Xenopus tropicalis. Genesis 30(3):110–113
37. Grant IM, Balcha D, Hao T, Shen Y, Trivedi P, Patrushev I, Fortriede JD, Karpinka JB, Liu L, Zorn AM, Stukenberg PT, Hill DE, Gilchrist MJ (2015) The Xenopus ORFeome: a resource that enables functional genomics. Dev Biol 408(2):345–357. https://doi.org/10.1016/j.ydbio.2015.09.004
38. Knowlton MN, Smith CL (2017) Naming CRISPR alleles: endonuclease-mediated mutation nomenclature across species. Mamm Genome 28:367. https://doi.org/10.1007/s00335-017-9698-3

Chapter 11

Using ZFIN: Data Types, Organization, and Retrieval

Ceri E. Van Slyke, Yvonne M. Bradford, Douglas G. Howe, David S. Fashena, Sridhar Ramachandran, Leyla Ruzicka, and ZFIN Staff*

Abstract

The Zebrafish Model Organism Database (ZFIN; zfin.org) was established in 1994 as the primary genetic and genomic resource for the zebrafish research community. Some of the earliest records in ZFIN were for people and laboratories. Since that time, services and data types provided by ZFIN have grown considerably. Today, ZFIN provides the official nomenclature for zebrafish genes, mutants, and transgenics and curates many data types including gene expression, phenotypes, Gene Ontology, models of human disease, orthology, knockdown reagents, transgenic constructs, and antibodies. Ontologies are used throughout ZFIN to structure these expertly curated data. An integrated genome browser provides genomic context for genes, transgenics, mutants, and knockdown reagents. ZFIN also supports a community wiki where the research community can post new antibody records and research protocols. Data in ZFIN are accessible via web pages, download files, and the ZebrafishMine (zebrafishmine.org), an installation of the InterMine data warehousing software. Searching for data at ZFIN utilizes both parameterized search forms and a single box search for searching or browsing data quickly. This chapter aims to describe the primary ZFIN data and services, and provide insight into how to use and interpret ZFIN searches, data, and web pages.

Key words ZFIN, Zebrafish Information Network, *Danio rerio*, Database, Genetics, Genomics

1 Introduction

The Zebrafish Information Network (ZFIN; http://zfin.org) is the community resource for genetic and genomic data about the zebrafish (*Danio rerio*). ZFIN was established by Monte Westerfield in 1994 following the first open international zebrafish research meeting, held at Cold Spring Harbor. By 1996 ZFIN curators were already playing a role in establishing nomenclature guidelines for genes, mutants, and transgenic lines in zebrafish, as well as curating mutants from the published literature [1]. By 2000, cross

*The members of the ZFIN Staff are listed in the Acknowledgments.

Martin Kollmar (ed.), *Eukaryotic Genomic Databases: Methods and Protocols*, Methods in Molecular Biology, vol. 1757, https://doi.org/10.1007/978-1-4939-7737-6_11, © Springer Science+Business Media, LLC, part of Springer Nature 2018

linking and integration of ZFIN with other genomic data sets including GenBank, Vega, UniProt, and Ensembl had begun. In 2003 additional data types including orthology, Gene Ontology, Morpholinos, and more complex genotypes were added to the system. In subsequent years, support was added for more complex curation of gene expression and phenotype data [2–4] as well as records for transgenic constructs, antibodies, TALENs, CRISPRs, the zebrafish anatomy ontology (ZFA) [5], the zebrafish experimental conditions ontology (ZECO) [36], a zebrafish data mine (ZebrafishMine), and also zebrafish genetic and chemical models of human disease. ZFIN staff now provide expert curation of these data types from the published literature and directly submitted data sets [6]. Over the past 23 years ZFIN has grown into a mature knowledgebase, well integrated into the larger ecosystem of genetic and genomic data service providers that are critical for researchers today. With this increased complexity come new challenges in navigating the zfin.org web site and obtaining bulk downloads of data of interest. In this chapter, all the major pages in ZFIN and methods for searching the database are described, including advanced searching tips and how to get the most out of ZFIN data.

2 Data Pages

As a genomic and genetic database, ZFIN web pages convey information about genomic sequences, genes, alleles, and gene function via direct curation and inferred from expression and phenotype. ZFIN contains only data that have been curated from papers or submitted by researchers, and therefore does not contain all possible data about zebrafish. ZFIN web pages follow a consistent layout with a ZFIN identifier, a header that starts with the name of the object followed by the object's most important characteristics. The rest of the page is dedicated to providing annotated data and relevant information that are specific to the data object. All pages end with a Citation link which lists all papers that have referenced the particular data object. ZFIN data pages are highly interconnected allowing for easy navigation between various data objects and associated annotations. The Gene page is one of the most highly connected pages providing access to many other data pages (Fig. 1).

2.1 Gene and Pseudogene

Genes are the archetypal DNA element curated by ZFIN. Because of its core position in our database, the Gene page can include significantly more data than other DNA element pages, including links to the DNA elements that are products of, are located within, or that contain, the gene. The Gene page at ZFIN is the place where all data pertaining to a specific gene are brought together to give the most complete picture possible of a gene's function.

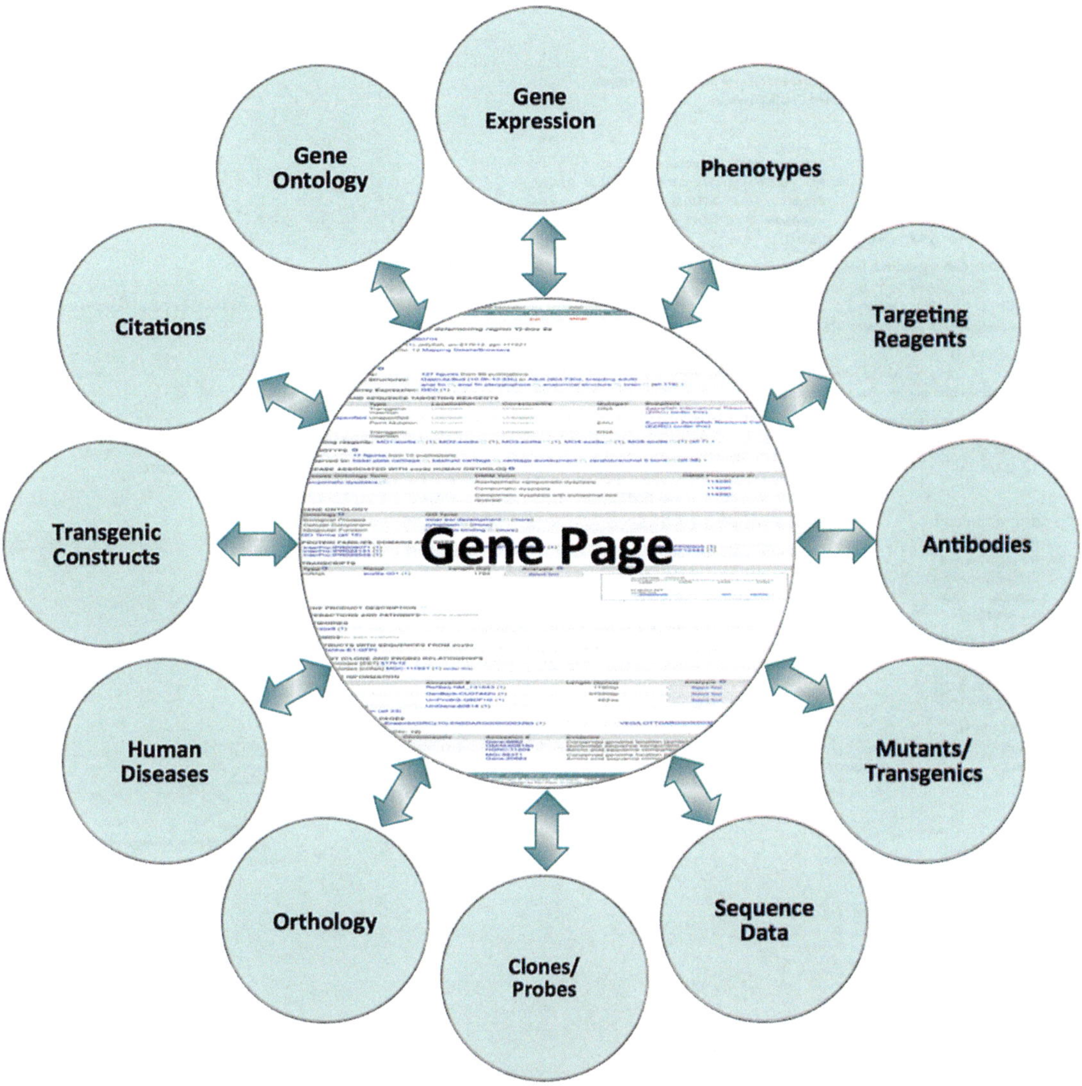

Fig. 1 ZFIN is an extensively crosslinked gene-centric resource. Many data pages in ZFIN share reciprocal hyperlinks between related records. The gene record and corresponding gene page provide the central hub around which much of the data and zfin.org website are organized, respectively. Gene web pages (center) share reciprocal hyperlinks with pages for 12 major related data types (peripheral). Many of these data type pages are also linked to pages other than the gene page. For clarity, those links are not represented

Here we will discuss the data available on the Gene page roughly in the order they are displayed (Fig. 2). All data sections discussed below will appear on all Gene pages, however because all information is not known for every gene, sections without information display "No data available." When there is an abundance of data a small sample is shown, usually 3–5 lines or links, followed by a link to the entire set of data that indicates how much more information is available.

Fig. 2 Gene page for *cxcr4b*

The page begins with the official gene name and symbol (Fig. 2A), which should be used in publications. Any names previously used to refer to the gene as well as old locus names and systematic names are located in the "Previous Names" section. A nomenclature history link is included at the bottom of the basic gene information section. Gene and Pseudogene nomenclature follows the guidelines outlined at: https://wiki.zfin.org/display/general/ZFIN+Zebrafish+Nomenclature+Guidelines. Following the names is the Location section where the chromosomal location is displayed. For a more comprehensive view of the information and to see any historic mapping data, the adjacent "Mapping Details/Browsers" link can be used to see all the available mapping information for the gene.

The next sections of the Gene page collate all the curated data about a gene's function. These include where in the fish the gene is expressed, known alleles and knockdown reagents of the gene, as well as phenotypes and associated diseases caused by disruption of the gene. The "Expression" section on the Gene page (Fig. 2B) focuses primarily on the expression of the gene in a wild-type background. A link to all curated gene expression, both in wild-type (WT) and mutant fish, as well as expression under a variety of different experimental conditions, is followed by the links to data generated in large scale WT in situ screens [7]. The "Wild-type Stages, Structures" portion of the Gene Expression section lists the stage range during which the gene expression has been curated, followed by a list of anatomical structures where expression of the gene in WT Fish with standard experimental conditions has been curated. The last link in the expression section takes you to the Gene Expression Omnibus (GEO) [8] page for microarray expression data for the gene.

The "Mutations and Sequence Targeting Reagents" section (Fig. 2C) lists all known alleles of the gene, as well as providing a list of the published sequence targeting reagents. Summary information about individual alleles is provided in a tabular format, including a link to the Feature page for each allele that provides more details about each mutation (*see* Subheading 2.3), as well as links to international resource centers that supply that mutant. The table is expandable to accommodate large numbers of alleles. The list of sequence targeting reagents links to pages for individual Morpholinos, CRISPRs and TALENs (*see* Subheading 2.6).

The "Phenotype" and "Disease" sections provide information about phenotypes and diseases caused by gene disruption. The "Phenotype" section has a "Data" area that supplies links to figures for which phenotype annotations have been made to Fish that contain alleles of the gene, or utilize gene-specific sequence targeting reagents or both (Fig. 2D). For more information about Fish please *see* Subheading 2.5. In addition, the "Observed In" portion of the "Phenotype" section lists anatomical structures and biological processes that have an abnormal phenotype caused solely by disrupting the gene. The "Disease" section on the Gene page provides information about the diseases associated with the human ortholog of the zebrafish gene (Fig. 2E). The table lists the *Online Mendelian Inheritance in Man* (OMIM) disease name as well as links to the OMIM page, and the corresponding Disease Ontology term (DO) [9, 10], with links to the DO term page where more information can be found about the disease. The information in the table is produced by computational mappings of ZFIN curated orthology, data available in the genemap and mim2gene files from OMIM (https://omim.org/downloads/), and cross references to OMIM found in the DO [2, 11].

The next two sections are dedicated to providing information about gene product functions and protein products. The "Gene Ontology" section provides a brief tabular summary of the molecular functions and biological processes the gene is involved in as well as the cellular components in which the gene product has been observed. Clicking on the link below the Gene Ontology table opens a new page that displays all of the Gene Ontology (GO) annotations for the gene (Fig. 2F), including both those added by ZFIN curators and collaborating organizations, and those determined computationally based on protein sequence, gene family functions, or assignment of keywords at UniProt [12, 13]. The GO annotation uses evidence codes from the Evidence and Conclusion Ontology (ECO) which can be used to determine if an annotation is curated from experimental data or generated by an automated method [14, 15]. While the GO section is about gene product function, the "Protein Families, Domains and Sites" section provides links to additional resources where further information about protein products of the gene can be obtained. This section contains links to InterPro [16] for information about protein families and functional domains, PROSITE [17] for protein domains, families and functional sites, Pfam [18] for information about protein families, and occasionally the Enzyme commission number from ExPASy [19] (Fig. 2G).

Several sections of the Gene page exist to provide quick access to information about gene products and reagents. These sections typically contain a list of links to other ZFIN pages, providing at a glance the quantity of information available, as well as a link to the detailed information. These include the "Transcripts," "Constructs with Sequences From," "Antibodies," and "Gene Product Description" sections. The "Transcript" section lists the transcripts of the gene with the type of transcript, the name which links to the more detailed transcript page, the length, a button that populates BLAST searches at ZFIN, NCBI, Ensembl, UCSC, and Vega, and a GBrowse image showing all the transcripts of the gene in a genomic context (Fig. 2H). The transcript records are added to ZFIN and linked to ZFIN gene records during a data exchange with Ensembl, where the transcript records are generated. The "Constructs With Sequences From" section lists constructs that contain either promoter or coding sequence of the gene, linking to the ZFIN construct pages where more detailed information can be found (Fig. 2I). The "Antibodies" section lists antibodies that recognize the gene product, with links to the Antibody page for more detailed information (*see* Subheading 2.7) (Fig. 2I). The "Gene Product Description" section opens a pop-up containing protein analysis provided by UniProt when available (Fig. 2I).

Several sections provide links to external resources. The "Other Gene Pages" section has links to the corresponding NCBI Gene page, and the Ensembl and Vega Gene pages (Fig. 2L). These links are established via collaboration with the external resources to provide the most accurate and up-to-date information about the sequence and related genomic information. Other sections that contain only external links are the "Plasmids" section (Fig. 2I), which links to Addgene [20] when they have plasmids that contain sequence from the gene, and the "Interactions and Pathways" section (Fig. 2I), which links to SignaFish (http://signafish.org) [21] when they have information about protein function or gene interactions involved in signaling pathways.

The following sections display sequence related data from several different perspectives. The "Segment (Clone and Probe) Relationships" section displays the relationships between genes and various DNA sequences (Fig. 2J). These sequences are usually bacterial artificial chromosome constructs such as, BACs, PACs or Fosmids produced during the Sanger zebrafish genome sequencing project [22], or are ESTs, cDNAs or sequence-tagged sites (STSs) from recombination based mapping efforts, expression screens, or the Zebrafish Gene Collection. These markers are displayed with the relationship to the gene. A marker, such as a gene, has the relationship "*contained in*" when it is a smaller part of a marker representing a larger piece of DNA (such as a BAC) and the relationship "*encodes*" where the marker, such as an EST, is a smaller part of the coding sequence of the whole gene. The links in this section go to the corresponding ZFIN marker pages which have data similar to the Gene page about the various clones, omitting information that is not relevant to the clone type.

The "Sequence Information" section (Fig. 2K) gives a preview of nucleotide and protein sequences associated with the gene by displaying the following sequences when available: a Reference Sequence or predicted sequence from NCBI, a genomic sequence from NCBI, a polypeptide sequence from UniProt, and a sequence cluster from UniGene. If there is no associated sequence of a particular type, then the type is not displayed on the Gene page. The link underneath the preview section shows how many sequences of all types are associated with the gene. Following the link leads the user to the sequence detail page where all the sequences associated with the gene are displayed. RNAs are listed first, starting with the longest sequence and ending with the shortest, followed by Genomic sequences. Peptide sequences are listed next, with UniProt sequences first, followed by Reference Protein sequences, and GenPept sequences. Each peptide category starts with the longest sequences and ends with the shortest. The sequence cluster information is displayed at the bottom of the table. Following the Sequence Information table is a list of all associated markers and for each of those markers the associated nucleotide and peptide sequence.

The last data section on the Gene page is the "Orthology" section which lists human, mouse, and fly orthologs of the zebrafish gene. It should be noted that human and mouse orthology data are curated from papers and generated by ZFIN curators as they do sequence analysis, whereas fly orthology is added by curators only when reported in the literature [23]. For each ortholog, the species and gene name are displayed, along with the location and related accession numbers. The evidence for orthology, using codes from ECO, follows with citations for each line of evidence to the right. Due to the teleost genome duplication a zebrafish gene can have multiple orthologs in human and mouse, whereas vertebrate gene family expansion can lead to multiple mammalian orthologs for the same zebrafish gene. The "Orthology" section was recently updated to enable display of 1:many orthologous relationships; for an example see *rdh12* https://zfin.org/ZDB-GENE-040718-9. Orthology information can be downloaded using the "Download Curated Orthology" link underneath the data table. The teleost genome duplication and the challenging task of assembling the zebrafish genome have precluded reliable computational analysis of zebrafish orthology. In the future, as orthology algorithms are improved, computed orthology data may be added to this section.

2.2 Other DNA Elements

ZFIN has pages for a variety of DNA elements that range from STS to BAC. The DNA element pages follow the pattern of all ZFIN data pages. In addition the DNA element pages include a section that contains information about the library construction where available. Most of the pages have a section that displays the GenBank sequence corresponding to the DNA element. When the DNA element has a relationship with another element, a "Marker Relationships" section is displayed.

Most DNA elements were added to ZFIN before the genome sequence was completed and were used for genetic mapping purposes. cDNAs and ESTs used as probes in large scale in situ screens for expression are other sources of DNA elements. The data pages for the DNA elements used in these screens have the corresponding expression images linked directly to the cDNA or EST page, as well as providing the information displayed on the gene record that encodes that sequence.

Because transgenic lines have become an important research tool, we add DNA elements that are essential for making the constructs to ZFIN. These records, referred to as Engineered Regions or Engineered Foreign Genes, use a display page similar to other DNA elements, with a name section at the top, followed by other data sections. Engineered regions are small functional pieces of DNA used in transgenic constructs, such as a nuclear localization sequence or a protein binding site. Engineered regions have a note that describes the effect or function of the motif. Engineered foreign genes are genes from other organisms such as Green Fluorescent protein. These engineered regions and engineered foreign genes are linked to the constructs that contain them.

2.3 Genomic Feature

The Genomic Feature page provides detailed information about mutant alleles and transgenic insertions. The allele number, i.e., laboratory line designation, is listed as the genomic feature and is comprised of the two or three letter laboratory line designation assigned to the institution where the originating laboratory resides and a unique number. Lab line designations can be obtained by contacting the nomenclature coordinator at nomenclature@zfin.org. Current laboratory line designations can be viewed at http://zfin.org/action/feature/line-designations. The header of the Genomic Feature page lists known synonyms, the affected genes, and transgenic constructs used to create transgenic features (Fig. 3A). For transgenic features where the construct has not

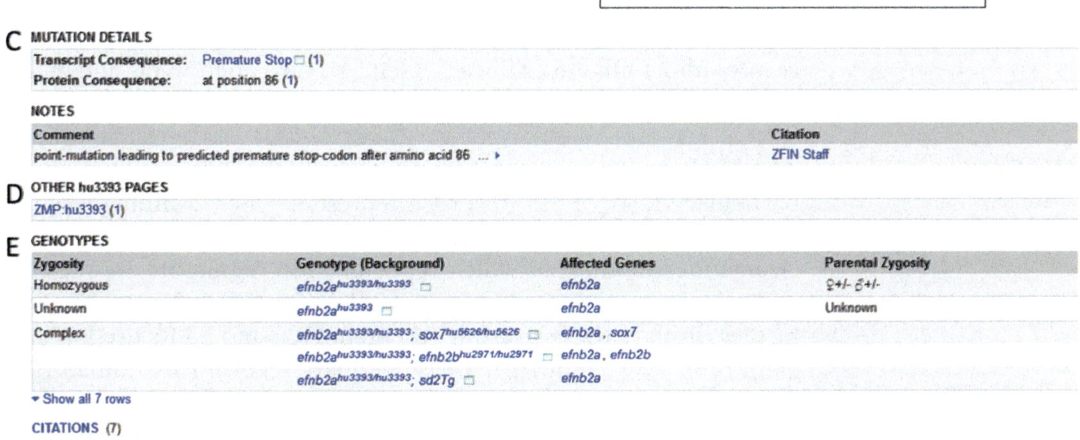

Fig. 3 Genomic feature page for hu3393

inserted into a gene locus, a note is displayed 'This feature is representative of one or more unknown insertion sites' to reflect that the information about the insertion is unknown. Also included in the header is information about the type of mutation, the protocol used to induce the mutation, laboratory of origin, genomic location, associated sequences, and links to current sources. A GBrowse image of the surrounding genomic context is displayed for features where genomic build and coordinates were submitted (Fig. 3B). In addition, the "Mutation Details" section reports information about the mutation including nucleotide substitutions for point mutations, the number of base pairs inserted or deleted for insertions, deletions or indels, the transcript consequences that result from the mutation such as frameshift or premature STOP, and protein information including the amino acid change, position of amino acid change, and resulting changes to the polypeptide as reported in the primary literature (Fig. 3C). Also listed in the mutation details section are notes specific to the genomic feature. The genomic feature page links to external web pages in the "Other Pages" section (Fig. 3D). These links often link to the original source of the feature data, and can provide additional information about the genomic feature. The "Genotypes" table lists all of the genotypes that the genomic feature is associated with, and is ordered with simple homozygous genotypes listed first, then heterozygous genotypes, and finally complex genotypes (Fig. 3E). The genotype table provides information about the affected gene and zygosity of the associated allele, in addition to links to the genotype page.

2.4 Genotype

The Genotype page provides the genotype name, previous names, the zygosity of the genomic features, the background, affected genes and current sources, if any, in the header (Fig. 4A). The "Genotype Composition" table provides a snapshot of the genomic features involved in the genotype, including associated constructs, laboratory of origin, zygosity and parental zygosity if known (Fig. 4B). The "Fish Utilizing Genotype" table lists all the Fish that include the genotype, linking to individual Fish pages as well as providing links to affected Gene pages, phenotype and gene expression results (Fig. 4C).

2.5 Fish

The Fish page provides information about the data annotated to a particular Fish. Fish records represent the combination of "Genotype + Sequence Targeting Reagent (STR)," providing a clear understanding of which reporters and regulators are present and which genes have been mutated or knocked down [2]. The header displays the Fish name, the genotype and STRs used, if any (Fig. 5A). The "Human Disease Model" section lists human diseases that have been modeled using the Fish, along with links to the associated publication (Fig. 5B). The "Gene Expression" table

A **Genotype:** ***vu119Tg***
Previous Name: Tg(b-actin:mgfp)^{vu119} (1)
Background: Unspecified
Affected Gene:
Current Source: No data available

Note:
The last 20 amino acids of c-Ha-Ras is fused to eGFP and targets it to the plasma membrane; this construct is under the control of the medaka beta-actin promoter. (1)

B **GENOTYPE COMPOSITION**

Genomic Feature	Construct	Lab of Origin	Zygosity	Parental Zygosity
vu119Tg	*Tg(Ola.Actb:Hsa.HRAS-EGFP)*	Lila Solnica-Krezel Lab - Vanderbilt University	unknown	Unknown

C **FISH UTILIZING** ***vu119Tg***

Fish	Affected Genes	Phenotype	Gene Expression
vu119Tg		2 figures from Roxo-Rosa *et al.*, 2015	14 figures from 11 publications
vu119Tg + MO1-itga7	*itga7*	Fig. 5 from Postel *et al.*, 2008	Fig. 5 from Postel *et al.*, 2008
vu119Tg + MO1-ube2ib	*ube2ib*	Fig. 6 from Nowak *et al.*, 2006	Fig. 6 from Nowak *et al.*, 2006
vu119Tg + MO2-ilk	*ilk*	2 figures from Postel *et al.*, 2008	2 figures from Postel *et al.*, 2008
vu119Tg + MO2-itga7	*itga7*	Fig. S5 from Postel *et al.*, 2008	Fig. S5 from Postel *et al.*, 2008
vu119Tg + MO3-pkd2 + MO9-pkd2	*pkd2*	3 figures from Roxo-Rosa *et al.*, 2015	2 figures from Roxo-Rosa *et al.*, 2015
vu119Tg + MO4-cftr	*cftr*	Fig. 2 from Roxo-Rosa *et al.*, 2015	Fig. 2 from Roxo-Rosa *et al.*, 2015
vu119Tg + MO2-lmx1bb + MO3-lmx1ba	*lmx1ba*, *lmx1bb*	Fig. 3 from McMahon *et al.*, 2009	Fig. 3 from McMahon *et al.*, 2009
vu119Tg + MO3-pkd2 + MO4-cftr + MO9-pkd2	*cftr*, *pkd2*	Fig. 2 from Roxo-Rosa *et al.*, 2015	Fig. 2 from Roxo-Rosa *et al.*, 2015
vu119Tg + MO2-lmx1bb + MO3-lmx1ba + MO4-tp53	*lmx1ba*, *lmx1bb*, *tp53*	Fig. 3 from McMahon *et al.*, 2009	Fig. 3 from McMahon *et al.*, 2009
vu119Tg + MO9-pkd2			

CITATIONS (37)

Fig. 4 Genotype page for *vu119Tg*

provides information about the genes annotated as having expression in the Fish, along with structures where the expression was noted, and links to the associated figures and publications (Fig. 5C). The "Phenotype" table lists the phenotypes observed in the Fish, the experimental conditions where the phenotypes were observed, and links to the figures and publications (Fig. 5D).

2.6 Sequence Targeting Reagents

The use of injectable reagents to knock down specific gene expression was one of the techniques that propelled the growth of zebrafish as a model organism. There are currently three main types of reagents used in zebrafish to target specific genes: Morpholinos (MO) [24], CRISPRs [25] and TALENs [26]. These sequence targeting reagents (STR) have dedicated ZFIN pages summarizing curated data associated with the STR. The page starts with the name and targeted gene (Fig. 6A). All STRs are given systematic names based on the type of reagent and the targeted gene, with sequential numbers assigned as new STRs are added to the database. For example, the first MO added to ZFIN that targets the gene *pax2a* is named MO1-*pax2a*. The same pattern is used for both CRISPRs (CRISPR#-[gene symbol]), and TALENs (TALEN#-[gene symbol]). Previous names are also recorded.

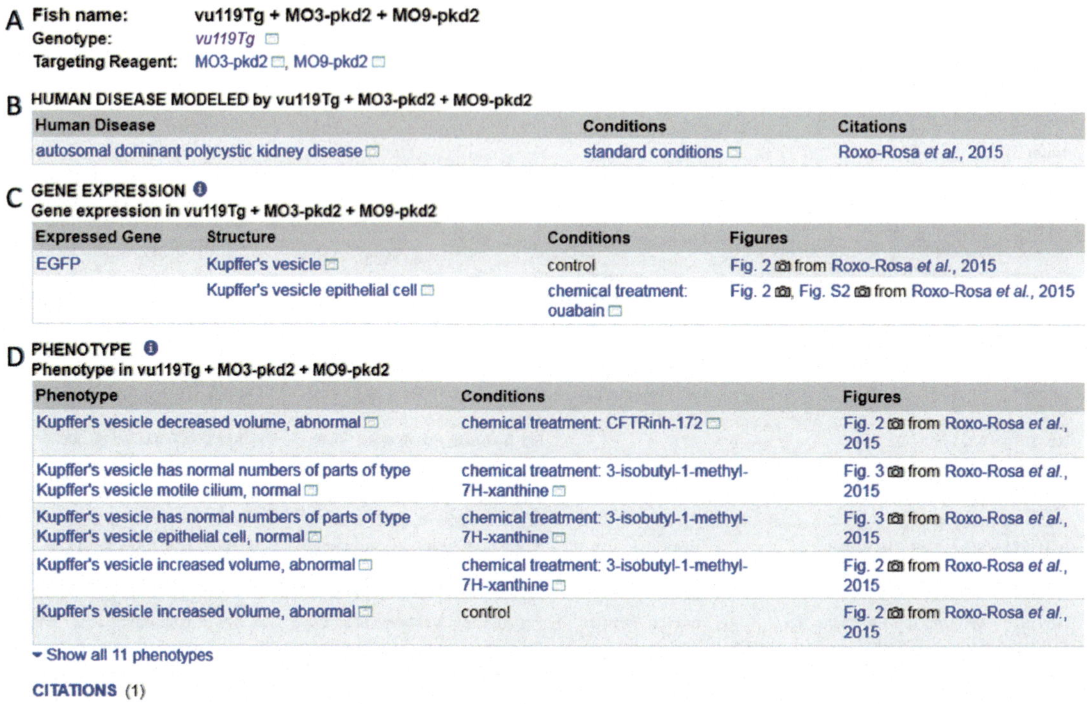

A

Fish name: vu119Tg + MO3-pkd2 + MO9-pkd2
Genotype: *vu119Tg*
Targeting Reagent: MO3-pkd2, MO9-pkd2

B

HUMAN DISEASE MODELED by vu119Tg + MO3-pkd2 + MO9-pkd2

Human Disease	Conditions	Citations
autosomal dominant polycystic kidney disease	standard conditions	Roxo-Rosa *et al.*, 2015

C

GENE EXPRESSION
Gene expression in vu119Tg + MO3-pkd2 + MO9-pkd2

Expressed Gene	Structure	Conditions	Figures
EGFP	Kupffer's vesicle	control	Fig. 2 from Roxo-Rosa *et al.*, 2015
	Kupffer's vesicle epithelial cell	chemical treatment: ouabain	Fig. 2, Fig. S2 from Roxo-Rosa *et al.*, 2015

D

PHENOTYPE
Phenotype in vu119Tg + MO3-pkd2 + MO9-pkd2

Phenotype	Conditions	Figures
Kupffer's vesicle decreased volume, abnormal	chemical treatment: CFTRinh-172	Fig. 2 from Roxo-Rosa *et al.*, 2015
Kupffer's vesicle has normal numbers of parts of type Kupffer's vesicle motile cilium, normal	chemical treatment: 3-isobutyl-1-methyl-7H-xanthine	Fig. 3 from Roxo-Rosa *et al.*, 2015
Kupffer's vesicle has normal numbers of parts of type Kupffer's vesicle epithelial cell, normal	chemical treatment: 3-isobutyl-1-methyl-7H-xanthine	Fig. 3 from Roxo-Rosa *et al.*, 2015
Kupffer's vesicle increased volume, abnormal	chemical treatment: 3-isobutyl-1-methyl-7H-xanthine	Fig. 2 from Roxo-Rosa *et al.*, 2015
Kupffer's vesicle increased volume, abnormal	control	Fig. 2 from Roxo-Rosa *et al.*, 2015

Show all 11 phenotypes

CITATIONS (1)

Fig. 5 Fish page for *vu119Tg*+ MO3-pkd2 + MO9-pkd2

Sequence targeting reagents are distinguised by their sequences (Fig. 6B). The pair of CRISPR targets, the target of the TALEN, or the sequence of the MO are used to distinguish the reagents and are recorded on each STR page. ZFIN curators endeavor to validate all published target sequences and to contact authors when the published sequences do not match the current genome build. There are several reasons a STR sequence may not match the current build sequence. When a sequence reporting error is identified by communicating with authors, the sequence is corrected in the record in ZFIN. Authors may or may not update the publication itself. Because builds change over time it is possible that the reference sequence has changed so that the reagent does not target in the expected manner or that a reagent will not work in a different strain because of sequence differences. ZFIN encourages researchers always to check the sequence of reagents and their desired target before ordering or using a published reagent. To facilitate this, a "Select Sequence Analysis Tool" button is provided immediately adjacent to the STR sequence. The location of the STR on the current genomic build is provided in a genome browser image in the "Target Location" section. Alignment of the STR to the genome is generated by ZFIN using the short read aligner Bowtie (Fig. 6C) [27]. It should be noted that if the STR sequence does not align to the genome a GBrowse image is not displayed and conversely if the STR aligns to multiple genomic locations the locations can be

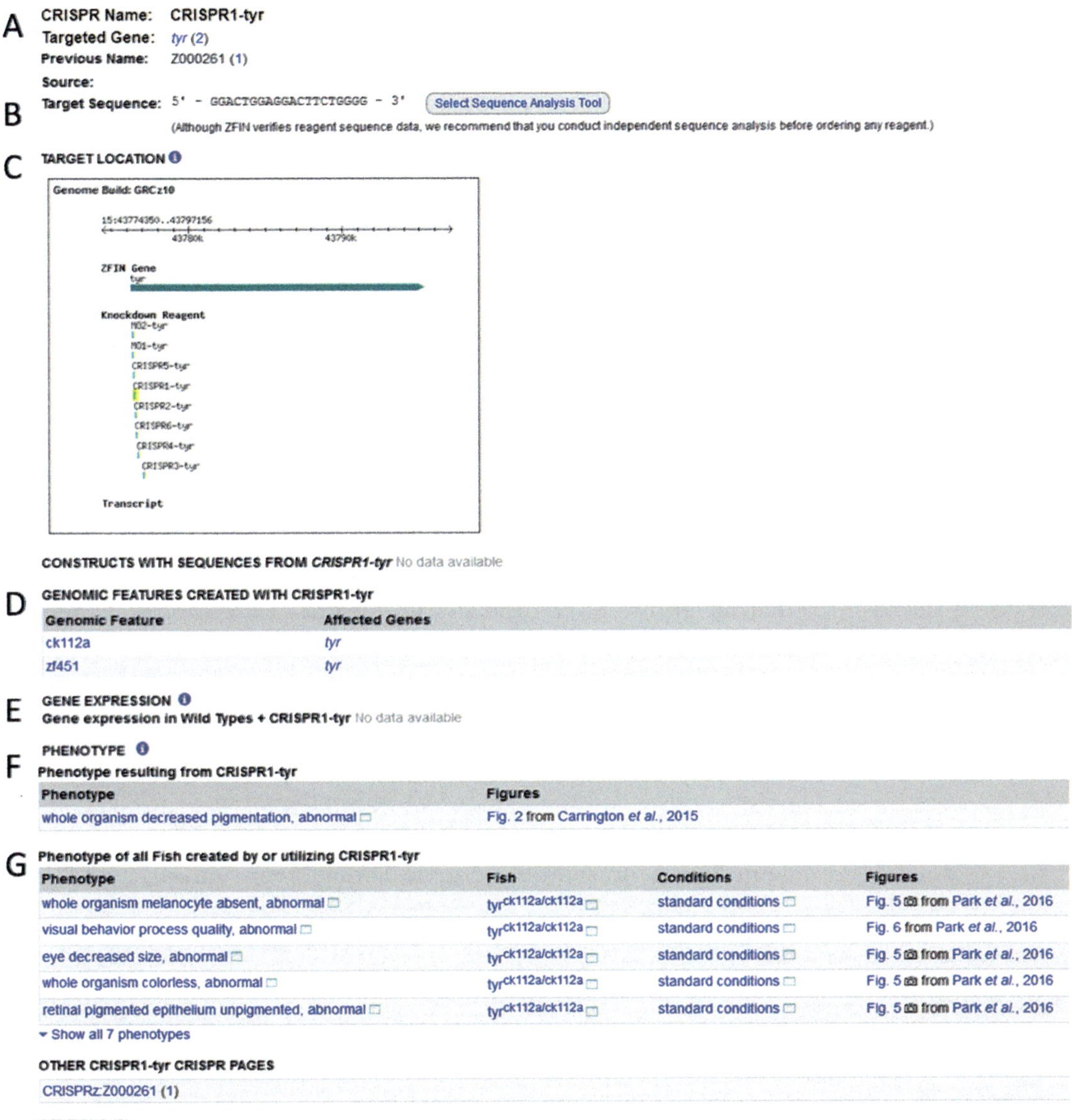

Fig. 6 Sequence targeting reagent page for CRISPR1-tyr

viewed by clicking on the "View All Target Locations" link (for example see http://zfin.org/ZDB-MRPHLNO-130117-2). Clicking on the GBrowse image will redirect the user to the full genome browser centered on the genomic location of the STR. One of the primary uses of CRISPRs and TALENs is to generate germ-line mutations. The "Genomic Features Created With" section lists all features created using the STR and lists the affected genes (Fig. 6D).

Expression and phenotype curated for experiments where the STRs were used is displayed in sections near the bottom of the page. The "Gene Expression" section displays expression only recorded in Fish where only the STR of interest is used in a WT background and standard or generic control experimental conditions (Fig. 6E). The "Phenotype" section includes two phenotype data tables. The "Phenotype resulting from STR" Table summarizes phenotype data observed in Fish composed of the single STR in a WT background with generic control or standard experimental conditions (Fig. 6F). These phenotypes are attributable to the STR treatment itself. The second table, "Phenotype of all Fish created by or utilizing STR," summarizes all phenotypes in which the STR was used, including those involving more complex genotypes or experimental treatments (Fig. 6G).

2.7 Antibody

The ZFIN database contains only those antibodies that recognize zebrafish tissue or gene products, as described in published literature. This is important because many widely used antibodies do not cross-react with zebrafish. A ZFIN antibody record can be associated with one or more genes, or with no gene. A gene–antibody connection is made only when supported by published evidence that an antibody recognizes the product of a specific zebrafish gene. Antibody pages have links to corresponding Gene pages and vice-versa.

The antibody page begins with a header section that contains the name, synonyms and essential information about the antibody such as the host organism, antigen gene, sources and a link to the ZFIN antibody wiki (*see* Subheading 5.5 and Fig. 7A). ZFIN antibody names have the general form "Ab#-name", where name is a gene symbol or gene family stem symbol, and # is a

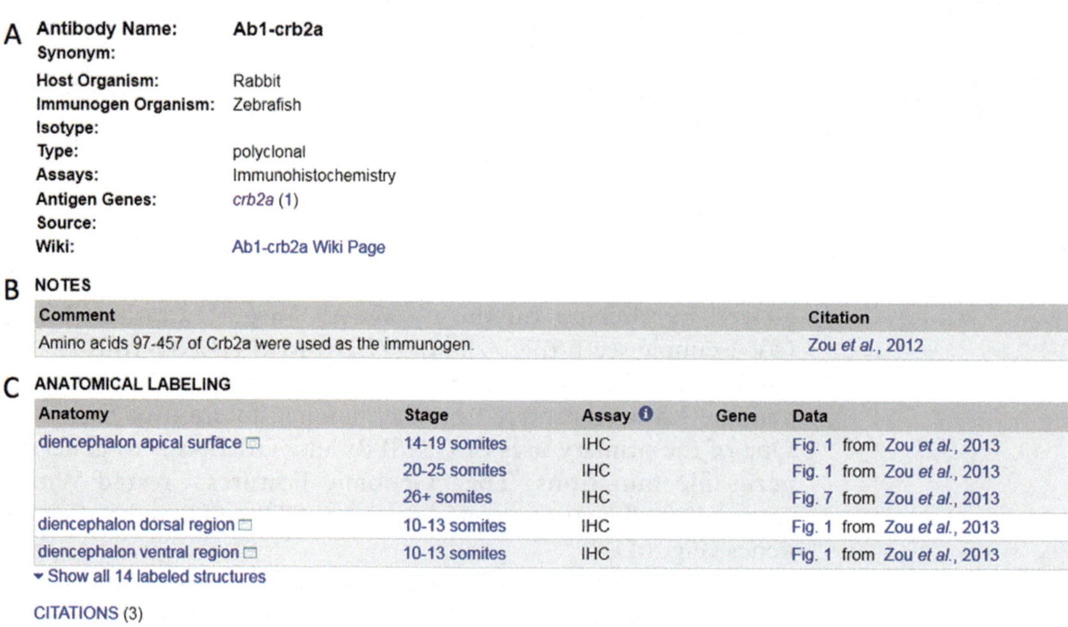

A

Antibody Name:	**Ab1-crb2a**
Synonym:	
Host Organism:	Rabbit
Immunogen Organism:	Zebrafish
Isotype:	
Type:	polyclonal
Assays:	Immunohistochemistry
Antigen Genes:	*crb2a* (1)
Source:	
Wiki:	Ab1-crb2a Wiki Page

B NOTES

Comment	Citation
Amino acids 97-457 of Crb2a were used as the immunogen.	Zou *et al.*, 2012

C ANATOMICAL LABELING

Anatomy	Stage	Assay	Gene	Data
diencephalon apical surface	14-19 somites	IHC		Fig. 1 from Zou *et al.*, 2013
	20-25 somites	IHC		Fig. 1 from Zou *et al.*, 2013
	26+ somites	IHC		Fig. 7 from Zou *et al.*, 2013
diencephalon dorsal region	10-13 somites	IHC		Fig. 1 from Zou *et al.*, 2013
diencephalon ventral region	10-13 somites	IHC		Fig. 1 from Zou *et al.*, 2013

▾ Show all 14 labeled structures

CITATIONS (3)

Fig. 7 Antibody page for Ab1-crb2a

number assigned sequentially as antibodies are added to ZFIN. Whenever possible, we use official gene symbols or gene family stem symbols. For example, Ab1-crb2a labels the product of the *crb2a* gene, whereas Ab2-crb is a less-specific label of multiple members of the *crb* gene family. Sometimes the antibody name is not composed of a gene or gene family symbol but rather the commonly used manufacturer name. To facilitate searches multiple synonyms are added to the header, including clone IDs and supplier product numbers. Reporting the supplier product number, or ZDB-ATB ID in publications greatly enhances the ability of fellow researchers and ZFIN curators to identify the precise antibody that was used. The notes section contains detailed information about the immunogen sequence, when provided by the author, along with the citation of the paper (Fig. 7B) The "Anatomical Labeling" section provides information about which anatomical structures are labeled, which developmental stages and assay types the antibody has been utilized with, along with the accompanying citations (Fig. 7C).

2.8 Construct

The Construct page provides information about constructs that are used to generate transgenic insertions. The header lists the construct name and synonyms (Fig. 8A). A construct map is displayed, if available, providing a visual representation of the promoters, coding sequence and regulatory elements contained in the construct (Fig. 8B). The "Construct Components" section provides information and links to ZFIN pages for promoter, coding sequences and engineered regions included in the construct (Fig. 8C). The "Genomic Features That Utilize Construct" section lists the genomic features that have been created using the construct, with affected genes if known (Fig. 8D). The "Transgenics That Utilize Construct" section lists all of the Fish that have genomic features that utilized the construct, with links to the corresponding Fish page, affected genes, curated phenotype, and gene expression (Fig. 8E). Links to GenBank are provided in the "Sequence Information" section if a sequence accession is provided for the construct (Fig. 8F), and links to external resources for the construct can be found in the "Other Construct Pages" section (Fig. 8G).

Construct nomenclature is guided by the nomenclature committee and is outlined in the nomenclature guidelines https://wiki.zfin.org/display/general/ZFIN+Zebrafish+Nomenclature+Guidelines#ZFINZebrafishNomenclatureGuidelines-4.3. Construct names begin with Tg, Gt, or Et identifying the construct as a general transgenic, gene trap, or enhancer trap construct, respectively. The salient features of the transgene are represented within parentheses in the construct name such that enhancers or promoters are listed to the left of the colon and cod-

A **Gene Trap Construct Name:** *Gt(GBT-P9)*
Sequence Ontology ID :
Synonyms: *GBT-P9, P9* (1), *pGBT-P9, gene-breaking transposon P9* (1), *Gt(Cca.Actb:GFP)*

B
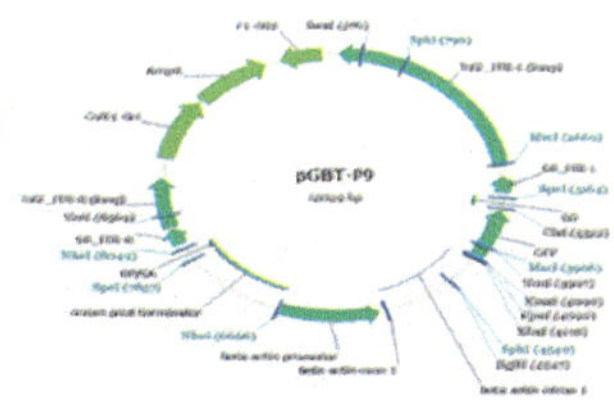

C **CONSTRUCT COMPONENTS**

Coding Sequences:	GFP (1)
Contains:	T2K (1)

D **GENOMIC FEATURES THAT UTILIZE *Gt(GBT-P9)***

Genomic Feature	Affected Genes
mn30Gt	*cct8*
xu015Gt	*mtor*

E **TRANSGENICS THAT UTILIZE *Gt(GBT-P9)***

Fish	Affected Genes	Phenotype	Gene Expression
$cct8^{mn30Gt/+}$(AB)	*cct8*		
$cct8^{mn30Gt/+}$	*cct8*	Fig. 2 from Petzold *et al.*, 2009	Fig. 3 from Petzold *et al.*, 2009
$cct8^{mn30Gt/mn30Gt}$	*cct8*		Fig. 3 from Petzold *et al.*, 2009
$fbxw7^{vu56/vu56}$; $mtor^{xu015Gt/xu015Gt}$	*fbxw7*, *mtor*	Fig. 3 from Kearns *et al.*, 2015	Fig. 3 from Kearns *et al.*, 2015
$mtor^{xu015Gt/+}$	*mtor*	text only from Ding *et al.*, 2011	Fig. 3 from Ding *et al.*, 2011

▾ Show all 7 transgenic lines

F **SEQUENCE INFORMATION** No data available

G **OTHER *Gt(GBT-P9)* GENE TRAP CONSTRUCT PAGES**

zfishbook:P9 (1)

CITATIONS (7)

Fig. 8 Construct page for Gene Trap *Gt(GBT-P9)*

ing sequences are placed to the right of the colon. Some constructs have only promoters or coding sequence. In those cases a colon is not included in the name. Construct names do not contain the names of the engineered regions found in the construct.

3 Data Source Pages

The majority of data at ZFIN comes from published research publications. ZFIN has records for more than 25,000 research publications. Currently more than 225 records for research publications are added to the ZFIN database every month. All publications are associated with genes and markers, mutant or transgenic alleles, antibodies, MOs, CRISPRs and TALENs, a process referred to as "Indexing," but only publications that contain prioritized information are fully curated. The curation policy is to curate prioritized publications fully, curating all data types that currently are supported by the ZFIN database. With the ever increasing number of new zebrafish papers being published, curation of older literature for newer datatypes is not usually done. As the database infrastructure expands to cover more new data types, only subsequently published data of those types are added. Thus data curation from publications is completed at the time of curation, but previously curated publications will typically not have the newer data types curated. Expression curation began in 2005 and Phenotype curation began in 2007 so articles published prior to those years will not have those data curated. Older expression and phenotype data are available in ZFIN but were usually added via a data load of directly submitted data. Currently ZFIN prioritizes curation of publications that include: new disease models, new genes, new mutants and their phenotype, and expression for genes that do not have any recorded expression. When curation of a paper publication is completed, an email is sent to the corresponding author and any other authors linked to the paper so the authors can review the curated information and notify ZFIN of any mistakes before they become widely disseminated. The information about data contained in research publications is displayed on two types of pages, the publication page, which contains bibliographic data about a publication in addition to links to data contained in the publication, and the figure page where the data curated from that particular figure are displayed.

3.1 Publication

Publication records are displayed on the publication page. The Publication page is populated with publication records loaded directly from PubMed via a script the day after the publishers create records at PubMed. The records are created with the information provided by the publisher which generally includes publication title, authors, keywords, and abstract. Each publication receives a ZFIN identifier (ZDB-PUB-#). This unique identifier, displayed at the top of the Publication page (Fig. 9A), serves as the unique identifier within ZFIN and is used in all the download files that reference publications. The ZFIN identifier is used as the primary publication identifier because a significant number of publication records in ZFIN are historic or are from journals not included in PubMed and thus do not have a PubMed ID. The Publication page is where all information, both publisher provided and ZFIN curated data, from a paper is aggregated. When the publisher changes the status of the publication from "epub ahead of print" to published, ZFIN attempts to obtain a copy of the paper. If the publication is not open access and the University of Oregon library does not have a subscription, we contact the corresponding authors directly. Once an electronic copy is obtained, the publication can be indexed and subsequently curated.

The publication title is the first information displayed on the publication page, followed by the author list and publication year (Fig. 9B). If the publication has the status "epub ahead of print"

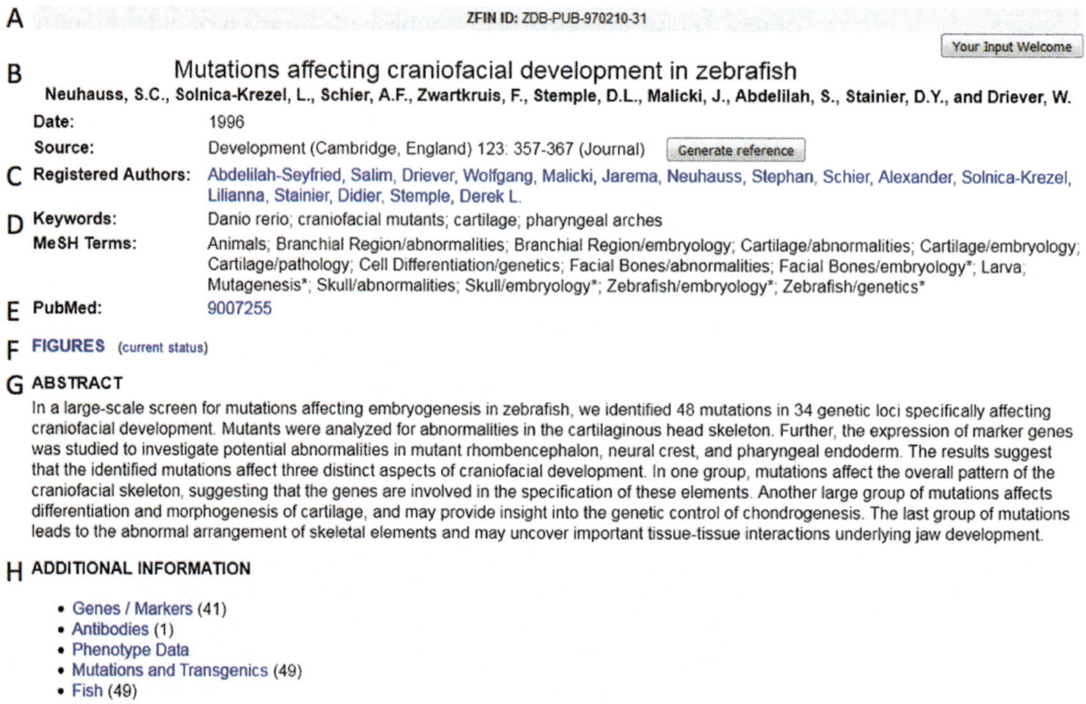

A ZFIN ID: ZDB-PUB-970210-31

Your Input Welcome

B Mutations affecting craniofacial development in zebrafish

Neuhauss, S.C., Solnica-Krezel, L., Schier, A.F., Zwartkruis, F., Stemple, D.L., Malicki, J., Abdelilah, S., Stainier, D.Y., and Driever, W.

Date: 1996

Source: Development (Cambridge, England) 123: 357-367 (Journal) Generate reference

C **Registered Authors:** Abdelilah-Seyfried, Salim, Driever, Wolfgang, Malicki, Jarema, Neuhauss, Stephan, Schier, Alexander, Solnica-Krezel, Lilianna, Stainier, Didier, Stemple, Derek L.

D **Keywords:** Danio rerio; craniofacial mutants; cartilage; pharyngeal arches

MeSH Terms: Animals; Branchial Region/abnormalities; Branchial Region/embryology; Cartilage/abnormalities; Cartilage/embryology; Cartilage/pathology; Cell Differentiation/genetics; Facial Bones/abnormalities; Facial Bones/embryology*; Larva; Mutagenesis*; Skull/abnormalities; Skull/embryology*; Zebrafish/embryology*; Zebrafish/genetics*

E **PubMed:** 9007255

F **FIGURES** (current status)

G **ABSTRACT**

In a large-scale screen for mutations affecting embryogenesis in zebrafish, we identified 48 mutations in 34 genetic loci specifically affecting craniofacial development. Mutants were analyzed for abnormalities in the cartilaginous head skeleton. Further, the expression of marker genes was studied to investigate potential abnormalities in mutant rhombencephalon, neural crest, and pharyngeal endoderm. The results suggest that the identified mutations affect three distinct aspects of craniofacial development. In one group, mutations affect the overall pattern of the craniofacial skeleton, suggesting that the genes are involved in the specification of these elements. Another large group of mutations affects differentiation and morphogenesis of cartilage, and may provide insight into the genetic control of chondrogenesis. The last group of mutations leads to the abnormal arrangement of skeletal elements and may uncover important tissue-tissue interactions underlying jaw development.

H **ADDITIONAL INFORMATION**

- Genes / Markers (41)
- Antibodies (1)
- Phenotype Data
- Mutations and Transgenics (49)
- Fish (49)

Fig. 9 Example of a ZFIN publication page for ZDB-PUB-970210-31

the year that the record was added to PubMed is displayed. This is followed by the journal name. The next data field is a list of registered authors (Fig. 9C). Registered authors are researchers who have a personal record in ZFIN (Subheading 5.1). A link is made between published papers and the authors who have records. Please bring it to our attention if we have missed an attribution or misattributed a publication.

The list of keywords provided by the journal is displayed followed by the MeSH terms added by PubMed (Fig. 9D). MeSH terms are frequently added significantly after publication so this field may be empty for newer publications. ZFIN loads MeSH term updates weekly so when the papers have been indexed by MeSH, the terms will be added to the publication record at ZFIN. The PubMed identifier (PMID) is also displayed and serves as a link to PubMed, this is followed by the link to the journal (Fig. 9E). A link labeled "Figures" will be displayed if there are any data attributed to figures in the paper or if ZFIN has permission from the publisher to display the figure or if the paper was published as Open Access (Fig. 9F). The abstract is displayed as it was submitted to PubMed (Fig. 9G).

Underneath the abstract is the "Additional Information" section (Fig. 9H). This section provides links to any data curated from the publication. Because ZFIN is a genetic database, papers that discuss gene function are given highest priority for curation. The list of possible topics is in Table 1. In the case where there is only

Table 1
Data types curated from publications

Link name	Destination
Genes/Markers	List of Gene or Marker pages
Morpholino	List of Morpholino pages
CRISPR	List of CRISPER pages
TALEN	List of TALEN pages
Clones and Probes	List of cDNA or EST pages
Engineered Foreign Genes	List of EFG pages
Antibodies	List of Antibody pages
Expression and Phenotype Data	All the figures in the paper that have either expression or phenotype data or both
Mapping Details	Table of mapped markers and locations
Mutations and Transgenics	Table of Alleles or Transgenic insertions with construct
Fish	List of Fish
Orthology	Table of orthology from the paper
Human Disease/Zebrafish Model Data	Table of Disease Models from the paper

one item in the list of data, users are taken directly to the data page instead of the list view.

As new data types are curated new topic are added to the list of curated topics in the "Additional Information" section. Other data types may no longer be curated when newer research methods have superseded older methods. An example is new genetic mapping data, which are no longer curated since the genome sequence was finished, and serves as the preferred source for genomic location. Thus the Mapping Data link to historic mapping data remains for publications already curated for genetic mapping but we do not curate mapping data for newer publications.

3.2 Figure

Most publications include figures that illustrate the reported data. These figures can be accessed by clicking the "Figures" link located just above the abstract on the Publication page (Fig. 9F). When ZFIN has copyright permission, the images and captions associated with the figures are displayed on the figure page with accompanying curated data for gene expression and phenotypes. The best way to ensure your publication record has images displayed at ZFIN is to publish in journals that support open access.

Each figure on the page has a summary of the expression and phenotype data, with links to all ZFIN data used in annotations for that figure. Detailed annotations for each figure can be found by opening the "Expression/Labeling details" and "Phenotype details" links located below each figure with associated summary data. For a detailed discussion of expression and phenotype curation please *see* [2, 4, 28]. Briefly gene expression is recorded as a gene or engineered foreign gene in some Fish, under specific experimental conditions at a particular developmental stage, that is expressed in some anatomical structure. Lack of expression in a structure is recorded, if deemed noteworthy by the authors. Antibody expression follows the same pattern except that an antibody rather than a gene is associated with a labeling pattern; antibodies targeting a specific gene product are shown in the section "Gene Expression Details". Expression in an anatomical structure means expression was observed somewhere within the structure. If the structure is not specifically identified or in the case of an RTPCR assay performed on whole embryo tissue, the annotation is associated with "whole organism". Consequently, annotations to "whole organism" can refer to ubiquitous expression in the organism or expression in some unspecified part of the whole organism. Phenotype is recorded as a Fish under specified conditions at a developmental stage having an entity such as anatomical structure, biological process, cellular component, with some quality such as size, morphology, or color, that is normal, abnormal, ameliorated or exacerbated. Ameliorated or exacerbated are based on the author's statements about the interactions of multiple mutations, or environmental factors and mutations, and are based on comparisons within the paper.

4 Ontology Term Pages

ZFIN uses several ontologies to structure the data that are curated. For example, without ontologies it would not be possible to find an expression pattern in the cardiac ventricle when a user searches for genes expressed in the heart. The relationship between the cardiac ventricle and the heart, (part_of), is encoded in the ZFA. All anatomical structures referred to in data at ZFIN are taken from the ZFA. Several additional ontologies are used to structure other domain specific annotations include gene function (Gene Ontology) and human diseases (Disease Ontology). Each term in these ontologies, and other ontologies in use at ZFIN, has its own term page in ZFIN.

4.1 Zebrafish Anatomy Ontology

The Zebrafish Anatomy (ZFA) term page displays information about each anatomy term including synonyms, a definition, developmental stages when the structure is present, relationships to

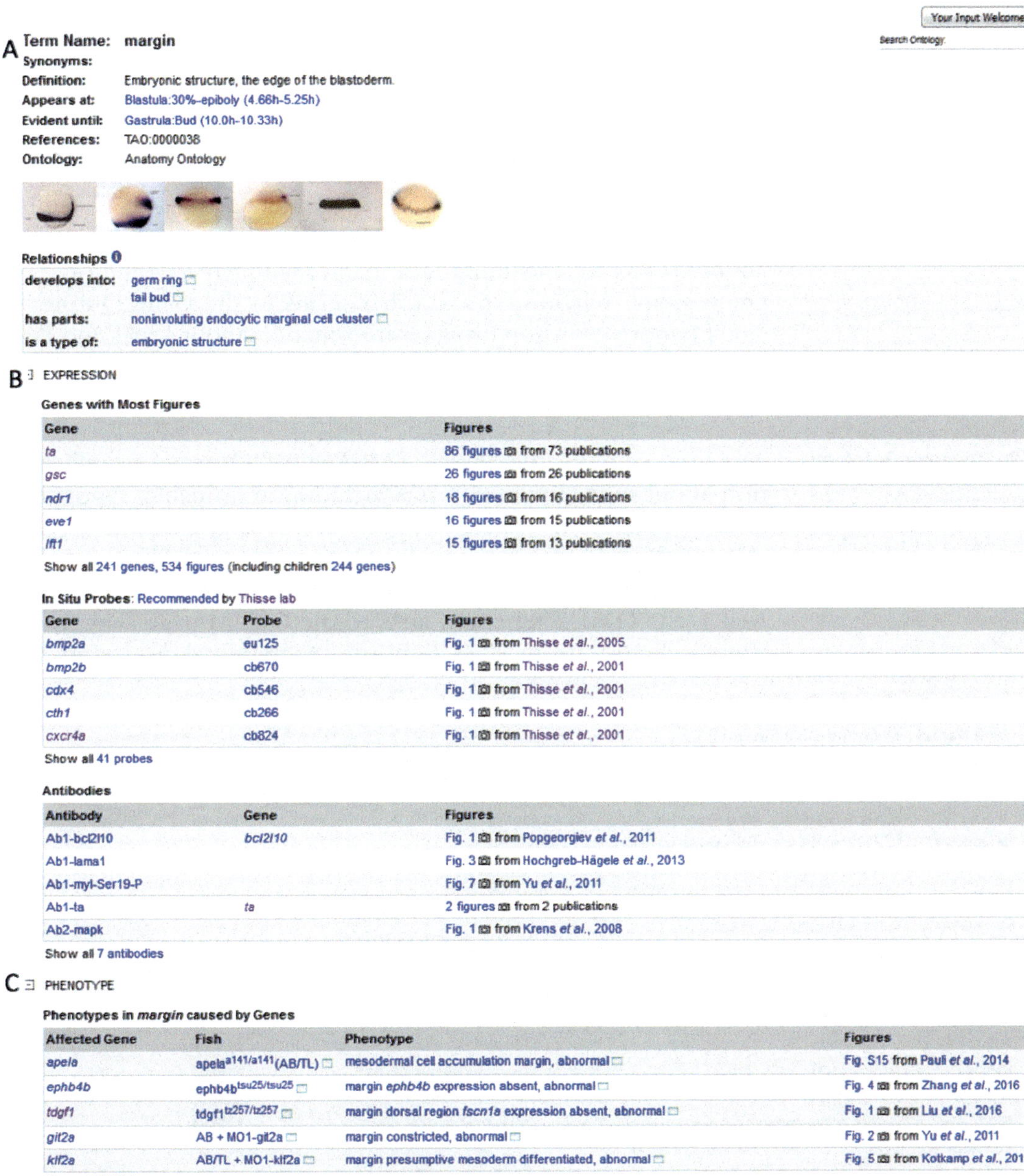

Fig. 10 Zebrafish Anatomy Ontology term page for margin

other anatomy terms that reflect what the structure develops from or into, and which system the structure is part of (Fig. 10A). The "Expression" section, when expanded, provides tables with information about genes or antibodies that are expressed in the anatomical structure with links to figures and publications that have described the expression (Fig. 10B). At the bottom of each table is a link to see all the expression in the structure and all the expression in any substructure, a more specific part of the structure. The "Phenotype" section displays phenotypes in which the particular anatomical structure is affected (Fig. 10C). Expansion of the Phenotype section displays a table that lists the affected gene, Fish, phenotype statements and links to figures and associated publications. Below the phenotype table a link is provided to show all phenotype annotations for the structure, including substructures.

4.2 Gene Ontology

The Gene Ontology (GO) [29]) term page provides information about GO terms, including term name, synonyms, definition, references, and relationships to other terms in the Gene Ontology. The Phenotype section provides information about genes and Fish that have a phenotype annotation utilizing the GO term, with links to associated figures and publications.

4.3 Disease Ontology

The term page for the Disease Ontology (DO) provides information about human disease terms including term name, synonyms, disease definition, references, and relationships in the header (Fig. 11A). The "Genes Involved" table lists the Human genes associated with a disease, the corresponding zebrafish ortholog, as well as the OMIM term and links to the OMIM page (Fig. 11B). The "Zebrafish Models" section lists experimental disease models

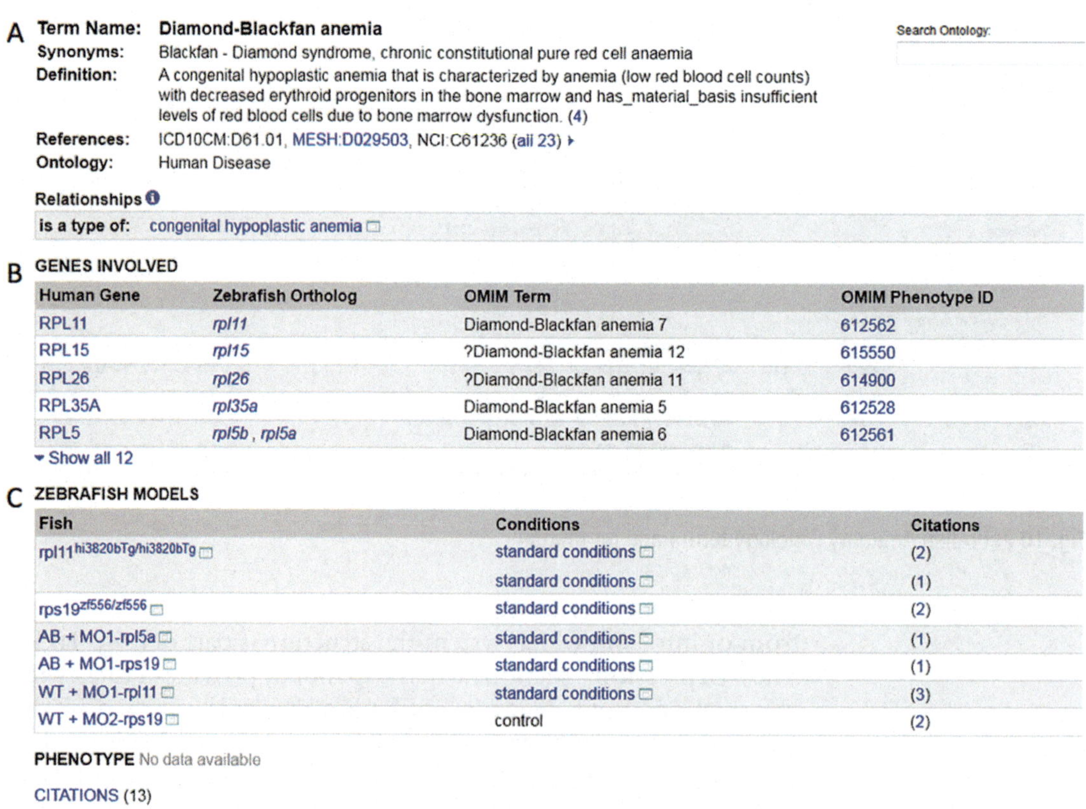

Fig. 11 Disease Ontology term page for Diamond-Blackfan anemia

created by genetic manipulation and/or experimental conditions with a link to the publication describing the model (Fig. 11C).

5 Community Pages and Resources

Records for people, laboratories, and companies are some of the oldest in the ZFIN database, with the earliest being created in 1996. These pages provide information about the scientists in the zebrafish research community, and the companies that provide reagents utilized by the community. ZFIN also supports a community wiki where new protocols or new antibody records can be created by community members. Community members can comment on protocols and antibody records, sharing their experience and insight.

5.1 Person Records

Person records serve several functions: they provide a list of all of someone's zebrafish publications in one place; they allow researchers to find and contact each other; and they allow communication about community events and important announcements to go out, by email, to the entire community. There are more than 8000 person records in ZFIN. To create a Person record, researchers should contact zfinadmn@zfin.org and provide their name, address, email address, and optionally ORCiD and the laboratory with which they are associated. An email is then sent with login information and they are added to the mailing list. When logged into ZFIN users can edit their record, update address and email, update the password, add research interests and nonzebrafish publications, and opt out of the email list if desired. If a user chooses not to use their login to update their records, new information can be added by emailing zfinadmn@zfin.org. ZFIN encourages the addition of an ORCiD to person records and publications to help with disambiguation of authorship. Once a person record is created it is the responsibility of the researcher to update their record.

5.2 Lab Records

Lab Records have basic information about the laboratory in a similar manner to Person records. Lab records are created following the same procedure as Person records and the request can be made simultaneously. One difference in the process of creating a laboratory record is that a laboratory member who has a ZFIN Person record must be specified as the primary contact. This person should be someone who can deal with request for reagents and requests to update information pertaining to the laboratory since laboratories are responsible for updating the information on their own record. A statement of research interest and a laboratory photo can be added to the Lab record. People with ZFIN records can be linked to the laboratory and given a role in the laboratory such as "Graduate Student" and "Primary Investigator". All publications linked to any member of the laboratory will appear automatically on the Lab page even if the papers were not written while in the current laboratory.

Lab records at ZFIN are necessary if a laboratory is creating and publishing new mutant or transgenic lines. The lines generated by a laboratory will appear on their laboratory record with a unique identifier consisting of an institution line designation and a number. Each institution that produces fish lines should acquire an institutional line designation by contacting the zebrafish nomenclature coordinator at nomenclature@zfin.org. Historically, line designations were assigned to laboratories, but the proliferation of laboratories producing fish has resulted in a change to assigning the designations on an institutional basis. All laboratories producing fish lines at the same institution should seek to coordinate and share a single institutional line designation. Currently there are more than 1100 laboratories with records in ZFIN.

5.3 Company Records

Records for companies that supply antibodies used in publications are created so that the reagent can be linked to a source (*see* Subheading 2.7 for more details). Companies that cater to the zebrafish research community are also welcome to request a record for their company. Currently there are more than 200 Company records at ZFIN. If a company also creates mutant lines a laboratory record is set up for the company so that the lines they create can be associated with the Company's laboratory record. ZFIN does not endorse any companies.

5.4 Protocol Wiki

ZFIN hosts the zebrafish research community wiki (https://wiki.zfin.org), offering one section for protocols and another for antibodies. The protocols section is where zebrafish researchers share experimental protocols and tips with the rest of the research community. Protocols are organized into sections corresponding to the chapters of *The Zebrafish Book, 5th edition* [30], covering a range of topics from general care and breeding of zebrafish to molecular and behavioural methods. Anyone in the research community can get an account on the wiki by visiting the wiki home page and clicking the "Sign up" link at the top right corner. Once logged in, members can post new protocols into the community protocol wiki. The protocol wiki currently includes 299 protocols (196 from the Zebrafish Book, 31 added by the community). Feel free to add new protocols to the appropriate section or add comments on any existing protocol.

5.5 Antibody Wiki

In addition to antibodies in the ZFIN database, ZFIN hosts an antibody wiki to encourage community interaction and input. Every ZFIN antibody has a corresponding page on the wiki. In addition, registered users can create community-submitted antibody wiki pages. Users can add usage notes and protocols, both to

ZFIN-generated antibody wiki pages and to community-submitted antibody wiki pages. There are currently 1170 community-submitted antibodies and 2630 ZFIN-generated antibodies.

6 Data Retrieval Pages

ZFIN strives to provide multiple points of access for all data and to meet the data access needs of all our users. Parametrized search forms are provided to support searches for data that meet specific criteria, returning only records that fill all the criteria entered and if there are no data, no results are returned. For users who prefer to explore ZFIN data, we provide a single box search interface. Single box search supports searching, but is intended to facilitate data exploration by always returning some results even if the results are not exactly the same as the terms used to search. When users want to query data programmatically, ZFIN provides two options: ZebrafishMine and data download files. ZebrafishMine provides a query builder interface that interacts with a data warehouse populated with a subset of ZFIN data [23], whearas the Download files provide raw data that users can download and manipulate with their favorite tools. All of these data access points are accessible from the ZFIN homepage.

6.1 Homepage

The ZFIN homepage links to various resources needed for conducting research using zebrafish. At the top of the page is the header that includes tabs labeled “Research”, “General Information”, “ZIRC” (Fig. 12A), links to the download files, a link to the ZFIN login page, an RSS feed button to receive a feed from wiki.zfin.org, and a Twitter “follow” button; follow us at @ zfinmod (Fig. 12B). The Research tab has links to parameterized search functions for ZFIN data pages. The General information tab has links to searches for community information as well as links to the wiki, newsgroup pages and other general information topics. The ZIRC tab gives quick access to the Zebrafish International Resource Center web pages and search forms. The header appears at the top of every page at ZFIN, giving easy access to search forms and other information. Directly under the header of the home page is a search box that is the gateway to the single box search (Fig. 12C). Entering anything, or nothing in the box and hitting the “Go” button will take you to the single box search interface (*see* Subheading 6.2 for more information about using this search).

Content on the home page is displayed in two columns. The left column lists links to ZFIN parameterized searches, displayed in bold text (Fig. 12D). The parametrized searches are distinct for each data type, so the search you use will determine the type of information you can retrieve and with what information you can search (Table 2). The results pages for the parameterized searches retain the filled in search form at the bottom of the results section,

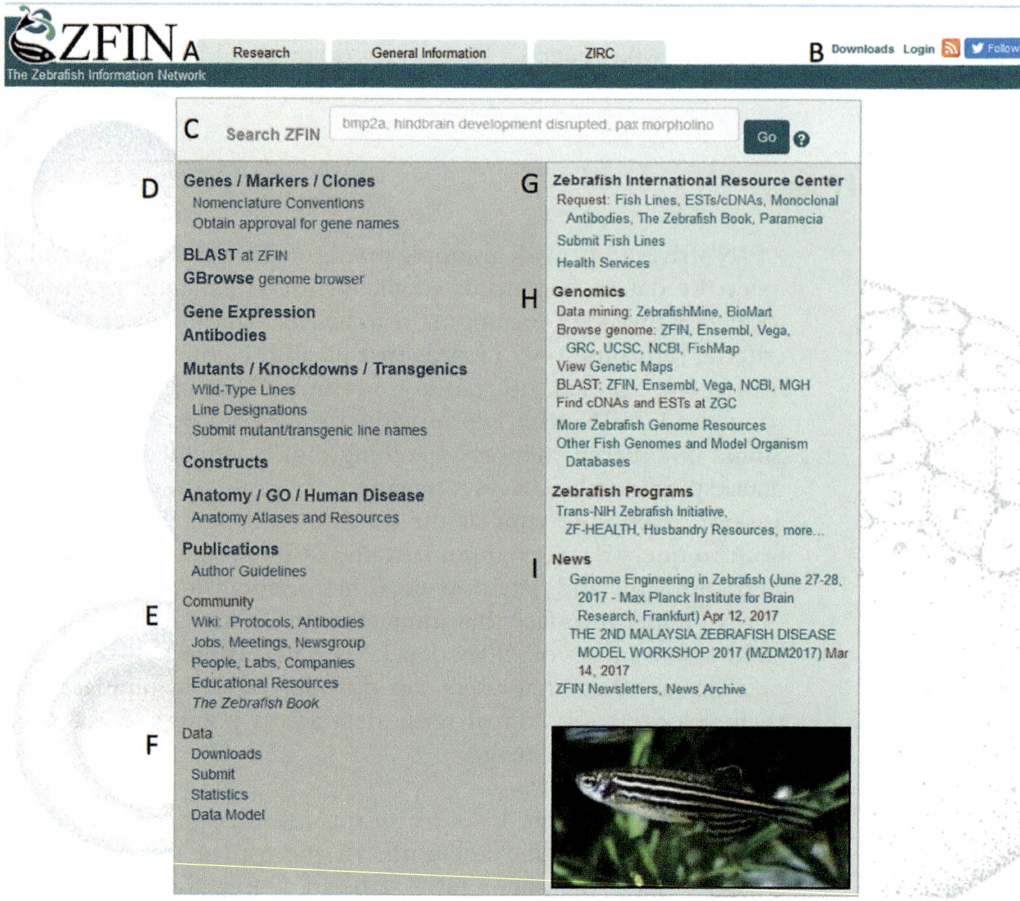

Fig. 12 ZFIN home page

which support variations and refinements of the current search. Below the individual parameterized search links on the home page are links to informational pages relevant to the search link with which they are associated. The left column also has a "Community" section which provides links to the ZFIN wiki for protocols and antibodies, community information links to job postings and newsgroups, and search links for Person, Laboratory and Company records (Fig. 12E). If a user is logged in they will see a link to edit their personal record and/or Lab Record. The community section also provides an Educational Resources link that leads to a page of external links for primary, secondary, and post-secondary educators who want to use zebrafish in the classroom. The final link in the community section is to an online edition of *The Zebrafish Book*, 4th edition [31]. The current version (v5) can be ordered using the "The Zebrafish Book" link in the right column on the home page. The "Data" section is the final section of the left column containing links to the download files (Subheading 6.4), an article on how to submit data directly to ZFIN, statistics about the data in ZFIN, and the ZFIN Data Model (Fig. 12F).

Table 2
Parameterized searches at ZFIN

Search name	Sections of Chapter where data discussed	Description of function
Genes/Markers/ Clones	2.1, 2.2, 2.6, 2.8	Searches all gene, pseudogene, DNA elements, STRs, Engineered Regions and Engineered Foreign Genes.
BLAST	NA	BLAST search with data sets specific to ZFIN data.
GBrowse	6.5	Link to ZFIN GBrowse.
Gene Expression	2.1, 3.2	Allows searching by Gene of interest, submitting Author, Anatomy structure where expression occurs and to limit data to WT, Tg or expression with images either from the literature or directly submitted to ZFIN.
Antibodies	2.7	Allows specification of the antigen gene, the structure it labels, or other specific attributes of the antibody.
Mutants/ Knockdowns/ Transgenics	2.3, 2.4, 2.5	Allows searching by gene name or line number, the structure or process with a mutant phenotype along with the mutation type.
Anatomy/GO/ Human Disease	4	Search for terms from ontologies. There is an autocomplete function to help narrow the search. The search form also provides a link for requesting terms that are missing from the anatomy ontology.
Publications	3.1	Publications can be searched by author, title, journal or keywords or ZDB-PUB or date or publication type.
People	5.1	Search for individuals by name or address as well as by keywords in their biography and research interest.
Labs	5.2	Search for Laboratories by name or address as well as by keywords in their biography and research interest.
Companies	5.3	Search for Companies by name or address or by Products.

The right column of the Home Page primarily contains links to external resources for zebrafish researchers. The "Zebrafish International Resource Center" section provides links to ZIRC web pages and resources (Fig. 12G). The "Genomics" section includes links to the data mining resources, ZebrafishMine and BioMart; genome browsers at ZFIN, Ensembl, UCSC, GRC, Vega, NCBI, and FishMap; and various BLAST servers (Fig. 12H). At the bottom of the "Genomics" section are links to more comprehensive lists of resources, the "More Zebrafish Genome Resources" link and "Other Fish Genomes and Model Organism Databases". The "News" section is at the bottom of the right column providing links to user submitted news posts and meeting announcements (Fig. 12I).

6.2 Single Box Search

The single box search is a widely used data exploration paradigm, providing rapid refinement of search results through the use of filters, also called facets. ZFIN uses the single box search as the main search interface, with single-box text field entry points found at the top of the ZFIN home page and in the upper right of every ZFIN data page. Searches typically begin with text entered into the search box, which has an autosuggest feature to provide a quick lookup of terms, notably genes, alleles, transgenics, and terms from GO, ZFA, and DO which are extensively used in ZFIN data annotations.

Clicking “return” in the search box, with or without entered text, leads to the search results page. Data categories are shown on the left side of this page. A category is a high-level type of result, Gene/Transcript, Expression, Phenotype, Fish, Antibody, etc. The number of records in each category that match the entered search string is also shown. Each category has an associated set of attributes, called facets. Some facets are specific to a single category, whereas other facets can apply to multiple categories.

After selecting a category, the left side of the search page is occupied by the “facet panel”. Search results can then be filtered by selecting values of interest from the facet panel. Each facet provides a list of valid values on which the result set can be filtered. Each of those values is labeled with a number indicating how many of the current results would remain after selecting that particular facet value. Selecting a facet value adds it to a list near the top of the search page, commonly called “the breadbox.” Clicking an item in the breadbox removes it from the search. Every time the user selects a facet value, the results and facets are immediately updated. Each of the newly updated facet values will be true for at least one of the newly updated results, ensuring that selecting a facet will never yield zero results.

Facets often have many values. For clarity, only the four values with the most associated data are shown in the facet panel. When there are more than four, a “Show All” link is displayed beneath the fourth facet value in the facet panel. Clicking “Show All” will bring up a window with all the facet values for that facet, along with a search box. Hovering the mouse pointer over a facet value will display (+) and (−) signs immediately to the left of the value. Clicking (+) will select that value, just like clicking it. Clicking (−) does the opposite—it will exclude any results that match that facet value.

To demonstrate how single box search works two different methods are described below for finding genes that are expressed in Purkinje cells.

Method 1: Go to the faceted search single box entry, click “return” in the search box with no search string entered. Select the Gene/Transcript category. The facets change to those relevant for this category. Select “Gene” from the Type facet in the facet panel to limit results to just gene records. In the Expression/Anatomy facet of the facet panel, use the Show All link to find and select Purkinje cell. The result shows all the genes that have curated expression in Purkinje cells (Fig. 13A).

Method 2: Type "Purkinje cell" in the search box and click "return". Results include any record that matches the string "Purkinje cell". Select the "Expression" category from the left panel. The facets change to those relevant for the Expression category. Limit the results to expression of zebrafish genes (vs. GFP etc.) by clicking "Any Zebrafish Gene" from the facet panel. Select show all in the "Expressed in Anatomy" facet and search for and select "Purkinje cell" to limit the expression results to genes expressed in the Purkinje cell not genes that contain "Purkinje

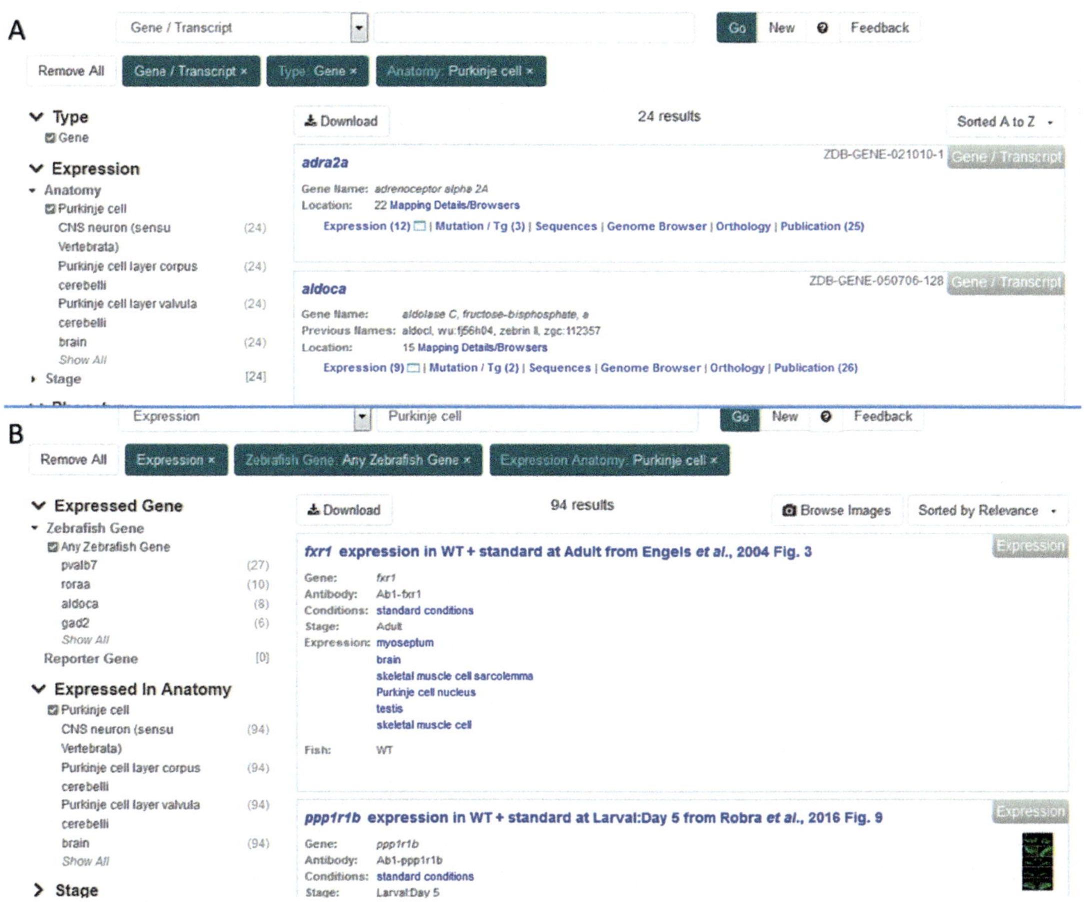

Fig. 13 Two examples of Single Box Search using two different search strategies to converge on the same data. (A) Selecting the Gene/Transcript category, then specific facet values for Type = Gene and Expression Anatomy = "Purkinje cell" produces a list of 24 genes that have curated gene expression in Purkinje cells. (B) Searching for expression in Purkinje cells using the text search. A search for Purkinje cell in the "Expression" category, then limit results to zebrafish genes and expression in Purkinje cells. Both methods produce genes and expression data for Purkinje cells

cell" in the name. The figures with gene expression in Purkinje cells are displayed in the results panel (Fig. 13B).

Single box searching also supports more advanced options including Boolean operators and wild card searches. For example, searching in the expression category for "heart AND retina" will return expression results that include both heart and retina or their parts. It will not return results that match only heart or only retina or their parts. Asterisk (*) is the wildcard character. At times using a wildcard character can be very helpful to find a larger dataset. For example a search for terms related to the heart by using the search string "cardi" returns terms containing "cardiac," "cardio," or "cardium," etc., as expected. Changing the search string to "*cardi" returns all those results plus additional terms which include words like endocardium, pericardium, and myocardium, which were not found by the original "cardi" search.

6.3 ZebrafishMine

ZebrafishMine (http://zebrafishmine.org), is a data warehouse built on the InterMine platform that allows customizable search and download options, and provides web services [32, 33]. ZebrafishMine includes periodic updates of data from ZFIN and the Panther homology database (http://pantherdb.org/).

ZebrafishMine's predefined searches, called "templates," support in-depth exploration of a variety of data types. Popular templates can be found in the middle section of the ZebrafishMine home page, and a complete list of templates can be retrieved by clicking on the "Templates" tab. Users can edit preexisting templates, or create their own.

ZebrafishMine supports the creation and manipulation of "lists", which contain information about one type of object (such as genes). A few lists are displayed on the ZebrafishMine home page, and a complete set of lists can be accessed by clicking on the "Lists" tab and selecting "View." Users can create their own lists, and use any list in template searches. A list can be combined with, or subtracted from another list.

How to use templates and lists: To demonstrate how to work with templates and lists, the following instructions describe how to use an existing template to create a list of zebrafish mutants that have been used to model human disease, and then find which subset of the mutants are available from the ZIRC.

1. Click on the "Templates" tab at the top of the home page (http://zebrafishmine.org).
2. Find the template "Human Disease → Zebrafish Model (mutants)".
3. Alternatively, filter the template list by typing "disease" in the "Filter" box (Fig. 14A).

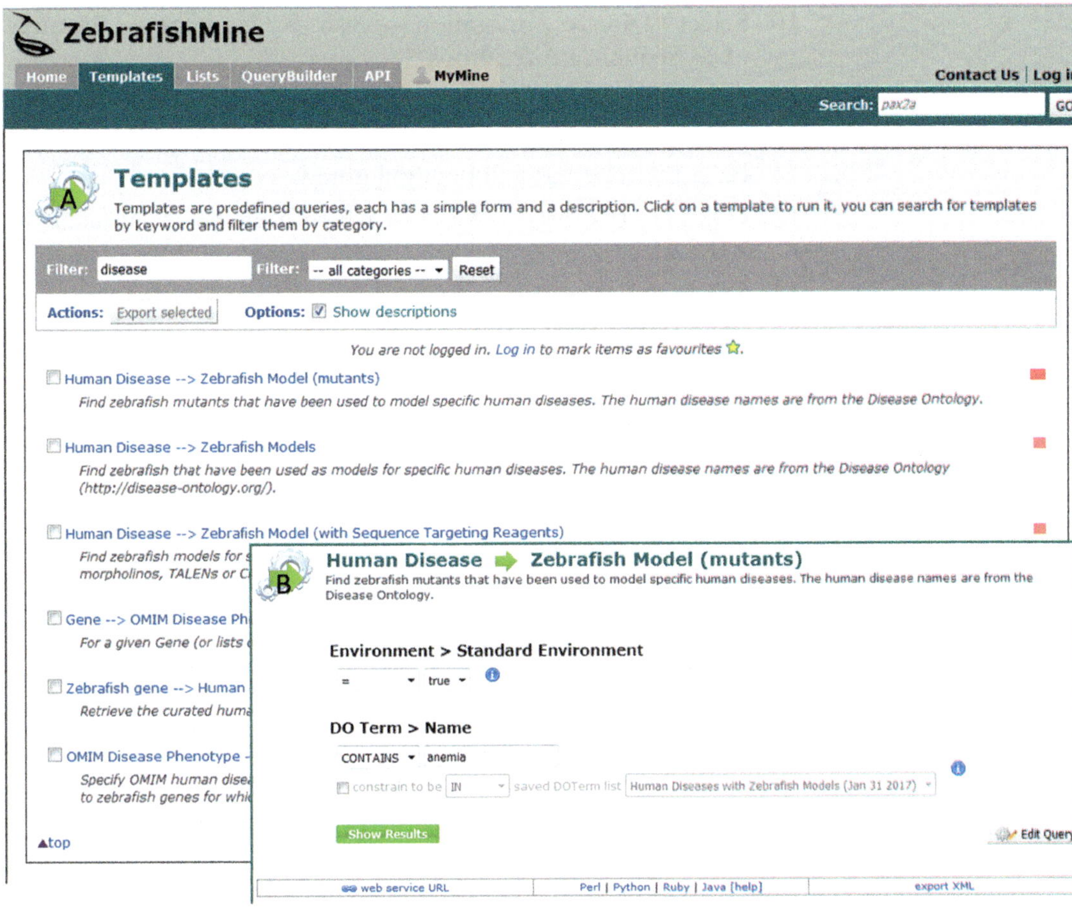

Fig. 14 ZebrafishMine Templates—Human Disease. (A) Templates are preset queries against the ZebrafishMine data warehouse. Filtering templates for "disease" returns 5 templates. Selecting "Human Disease → Zebrafish Model (mutants)"brings up the template. (B) "Human Disease → Zebrafish Model (mutants)" allows queries by Disease Ontology terms and returns disease models that consist of mutant fish in standard environmental conditions. The query can be edited using the "Edit Query" button on the bottom right that takes users to the QueryBuilder interface. The bottom bar contains links to the web service URL for the specific query as well as links to download the query in a variety of scripting languages or in XML format

4. Click on the template to see the details. Use this template to find zebrafish mutants that have been used to model specific human diseases (Fig. 14B).
5. Change the default search value to a wildcard (*).
6. The default value in the search box is "anemia". To search for all diseases, replace "anemia" with a wildcard (*).
7. Click the "Show Results" button to run this search.
8. Create a list from the search results.
9. Click on the "Save as List" button on the top right.

10. Select "Disease Annotation → Fish → Genotype → Features (123 Sequence Alterations)".
11. In the subsequent pop-up, name your list. Click "Create List".
12. Click on the "Lists" tab, and select "View". The new list will be at the top, and will be highlighted.
13. Compare two lists to find out which mutants are available at ZIRC.
14. Select the checkbox next to the new list, and the checkbox next to the "Mutants and transgenics available at ZIRC" list.
15. From the "Actions" menu, select "Intersect".
16. In the pop-up, name the new list, and save.
17. The new list contains zebrafish mutants that have been used to model human diseases, and which are available at ZIRC.

Downloading search results: Search results can be downloaded by clicking the "Export" button in the top right corner of any search results page. A pop-up window will appear allowing customization of the exported file. Clicking on ".tsv", will allow you to change the file format option; opening a new window that allows selection of a download format such as: tsv, csv, JSON, XML. Using the menu on the right of the window allows specification of the columns and rows to be downloaded, file compression, and a destination for the file.

Editing or building templates: Users can modify existing templates or build custom templates in ZebrafishMine's custom "Query Builder." To modify a preexisting template in Query Builder, click on the "Edit Query" button on the template page. Detailed instructions on using Query Builder can be found here:

http://flymine.readthedocs.org/en/latest/query-builder/Documentationquerybuilder.html#querybuilder.

Saving templates and lists: Users can create an account on ZebrafishMine to save lists, queries, and templates. To create an account, click on "Log in" at the top right hand corner of the home page. On the Log in page, click on "Create account now" and follow the instructions. Saved items can be found by clicking the MyMine tab when logged in.

6.4 Downloads

ZFIN makes data available for programmatic consumption via download files at https://zfin.org/downloads. The download files are the most complex data presentation at ZFIN. These files are accessible from the Downloads link found in the header and from the Data section of the home page. Downloadable files are organized by categories that represent the data that can be found within the associated files (Table 3). Data within each file are accessible by clicking on the file title, which redirects the user to a web based

Table 3
Download files list by category and file name

Category	File name
Anatomical ontologies	Zebrafish Anatomy Term
	Zebrafish Anatomy Term Relationships
	Zebrafish Anatomy Term Synonyms
	Zebrafish Development Stage Terms
	Zebrafish Stage Series
Antibody data	Antibody
	Expressions in wild-type
Fish data	Experiment Details
	Fish Components
	Genotype Backgrounds
	Genotype Features
	Genotype Features (removed or displaced genes)
	Innocuous/Phenotypic Construct Details
Gene expression	Expression data for wildtype fish
	Expression Environment Description
	Expression Experiment-Figure
	Zebrafish Gene Expression by Stage and Anatomy Term
	ZFIN Antibody Expression Assay Records
	ZFIN Genes with Expression Assay Records
Genetic marker relationships	Genetic Marker Relationships
Genetic markers	Genetic Marker
	Previous ZFIN IDs (includes former and current ZFIN IDs for merged data)
	SNP Data
Genomic feature data	All Genomic Features
	Construct Components
	Genomic Features and their affected genes
	Transgenic Insertions
Human disease	Human Disease/Models
Image data	Image-Figure translations
Knockdown reagent data	CRISPR
	Morpholino

(continued)

Table 3 (continued)

Category	File name
	TALEN
Mapping data	Mapping Data from the 6 Zebrafish Mapping Panels
Orthology data	Drosophila and Zebrafish Orthology
	Human and Zebrafish Orthology
	Mouse and Zebrafish Orthology
	Phenotypic Zebrafish genes with Human Orthology
Phenotype data	Antibody Labeling Phenotypes
	Fish with Phenotypes
	Fish-Figure relation
	Gene Expression Phenotypes
	Gene Expression Phenotypes
	Phenotype Environment Description
	Phenotype for Zebrafish genes with Human Orthology
	Phenotype of Zebrafish Genes
Previous names	Previous Names
Publications	Fish-Publication relation
	Gene-Publication relation
	Genotype-Publication relation
	ZFIN Bibliography
	ZFIN Marker IDs/UniProtKB IDs to ZFIN Pub IDs and PubMed IDs
	ZFIN Publication IDs: ZFIN IDs to PubMed IDs
Sequence coordinates	Assembly: Ensembl
	Complete Assembly Clones: Ensembl
	Transgenic Insertion: ENSEMBL
	ZFIN Genes with Antibodies: Ensembl
	ZFIN Genes with Expression: Ensembl
	ZFIN Genes with Phenotype: Ensembl
	ZFIN Genes: Ensembl
	ZFIN Knockdown Reagents: Ensembl
Sequence data	Transcripts

(continued)

Table 3
(continued)

Category	File name
	ZFIN Gene IDs indirectly and directly associated with Sequence Accessions via cDNA and EST
	ZFIN Gene IDs indirectly associated with Sequence Accessions via cDNA and EST
	ZFIN Marker associations to Ensembl IDs
	ZFIN Marker associations to GenBank sequence data
	ZFIN Marker associations to GenPept protein data
	ZFIN Marker associations to InterPro protein data
	ZFIN Marker associations to NCBI Gene data
	ZFIN Marker associations to RefSeq sequence data
	ZFIN Marker associations to Sanger Vega data
	ZFIN Marker associations to UniGene sequence data
	ZFIN Marker associations to UniProt protein data
Wildtype lines	Wildtype Lines
ZFIN Identifiers	ZFIN Identifiers

text representation or by downloading the data by clicking the file type button. The Downloads page also provides information about individual file column headers, the file size and the number of records in the file. An archive of download files is available at https://zfin.org/downloads/archive. The files have been archived daily since March 28, 2012.

There are a variety of files available for download, ranging from simple files in which all necessary data are contained within the file to complex data files that require joining with additional files to provide a more complete data representation. The simple files are usually a direct report of all ZFIN records for a particular data type, like Genetic Markers. The complex files provide all annotated data, including data collected under a wide range of environments where the environments can cause changes to the observed gene expression or phenotype. To obtain a complete set of data requires joining many individual files using unique identifiers. The unique identifiers are either ZFIN object identifiers that begin with "ZDB-" followed by the data type and a number, or ontology identifiers that have the ontology prefix and a colon followed by a number. Many of the downloads files are equivalent to a single database table so the Data Model, linked from the home page, is useful for understanding how the files should be linked together. The ontologies referred to in the files can be found at http://www.obofoundry.org/ [34].

The simple files for gene expression and phenotype data are "Expression Data for Wildtype Fish" and "Phenotype of Zebrafish Genes," respectively. The file "Expression Data for Wildtype Fish" is the compilation of all gene expression data for individual genes in WT fish under standard conditions and corresponds to the data found in "Wild-type Stages, Structures" in the Expression section on the Gene page. The data displayed in the All Expression Data link in the expression section of the Gene page are represented by the download file "ZFIN Genes with Expression Assay Records". This file is a complex download file containing all curated gene expression in any WT, mutant or transgenic line and in any environment. To demonstrate the difference between the simple and the more complex download files a concept map of the other files necessary to obtain a complete data representation for gene expression is shown in Fig. 15. The only file needed when using the file "Expression Data for Wildtype Fish" is the ZFIN Bibliography and that file is needed only if you wish to see in which publication the expression was reported. The only external resource that is referred

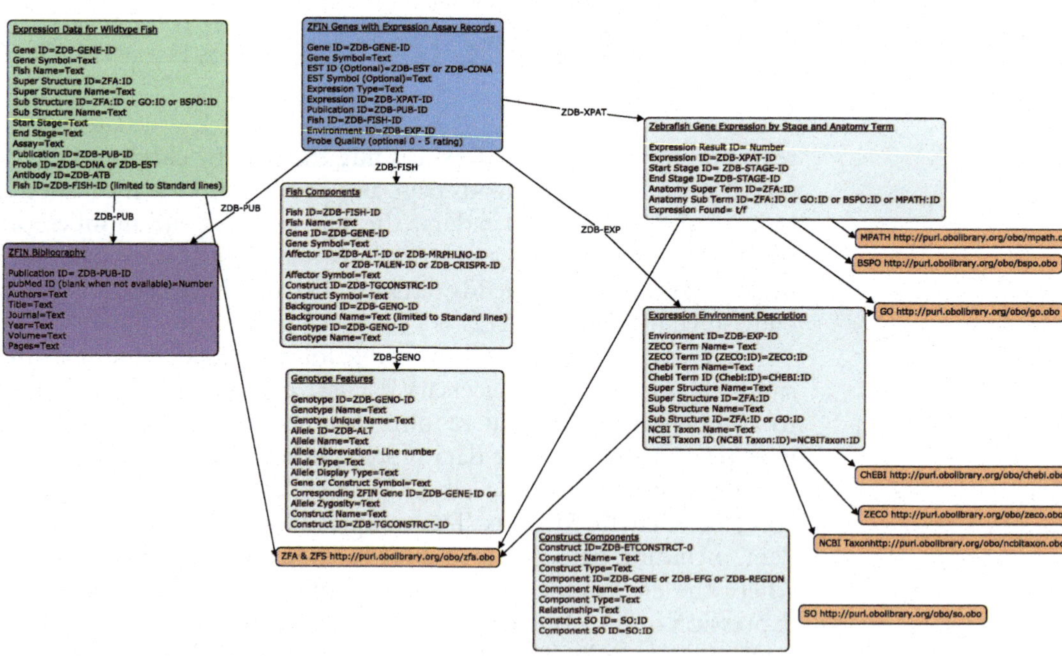

Fig. 15 File dependencies "Expression data for Wildtype Fish" compared with "ZFIN Genes with Expression Assay Records" using a concept map. "Expression data for Wildtype Fish" refers to only one ontology, the Zebrafish Anatomy Ontology, and one additional download file "ZFIN Bibliography". In contrast "ZFIN Genes with Expression Assay Records" references five files in addition to "ZFIN Bibliography" The set of files refer to eight separate ontologies to provide the complete picture of the Fish, environmental conditions or treatments present during the expression measurements and the structures where the expression occurs

to is the ZFA, which has more detailed term information and definitions of the terms. The file "ZFIN Genes with Expression Assay Records" has concepts that are broken out into three separate files to define Fish, environment, structures and stages. To obtain a full picture of what components are part of a Fish requires two additional files. The five files refer to eight different ontology files. These files and the "ZFIN Bibliography" file are used to link the expression to a publication.

The extensive use of ontologies for data annotation, especially the various anatomy ontologies, leads to several caveats when interpreting data. The first is that expression or phenotype in a structure listed means only that the expression or phenotype was observed somewhere in the structure. For expression annotations listing whole organism as the structure, this simply means there was expression somewhere in the fish. This annotation usually means the primary literature didn't specify in which structures expression was observed, or they used an assay such as RTPCR where the entire embryo was utilized. The second similar caveat is that when a stage range is provided for expression or phenotype and a structure is named, that expression or phenotype in that structure was observed sometime during the time period. So if an author says expression was observed in the optic cup, during the embryonic stage, which ranges from fertilization to hatching, it can be inferred from the data that expression was observed sometime in the overlap between 0 and 72 hpf, the embryonic stages, and 19–30 hpf, the stages where the optic cup exists. The third caveat is that there is an implied relationship between substructure and superstructure as well as between entities in the phenotype file. The last thing to be aware of when using complex data files is that when an abnormal phenotype is unexpectedly absent or expected expression is *not* observed, then those data are curated as such and identifiable by a flag in the file.

6.5 GBrowse

The zebrafish genome sequence and associated annotations can be viewed at ZFIN using GBrowse, a genome browser developed by the Generic Model Organism Database (GMOD) [35]. The GBrowse graphical interface is interactive and customizable, allowing the exploration of the zebrafish genome. Detailed information from manual and automated annotations provide evidence for over 26,000 protein-coding genes [22]. These annotations are made available as tracks in the GBrowse interface. The genome browser can be accessed through the GBrowse genome browser link on the home page. Alternatively, GBrowse can be launched directly from the detail pages for genes, transcripts, STRs, and Mutants/Transgenic pages by clicking on genome browser images.

Upon launching GBrowse, multiple tracks are displayed (Fig. 16). This selection of tracks can be customized to either include or exclude tracks of interest using buttons located to the left of the track name. In addition, users can reorder the tracks

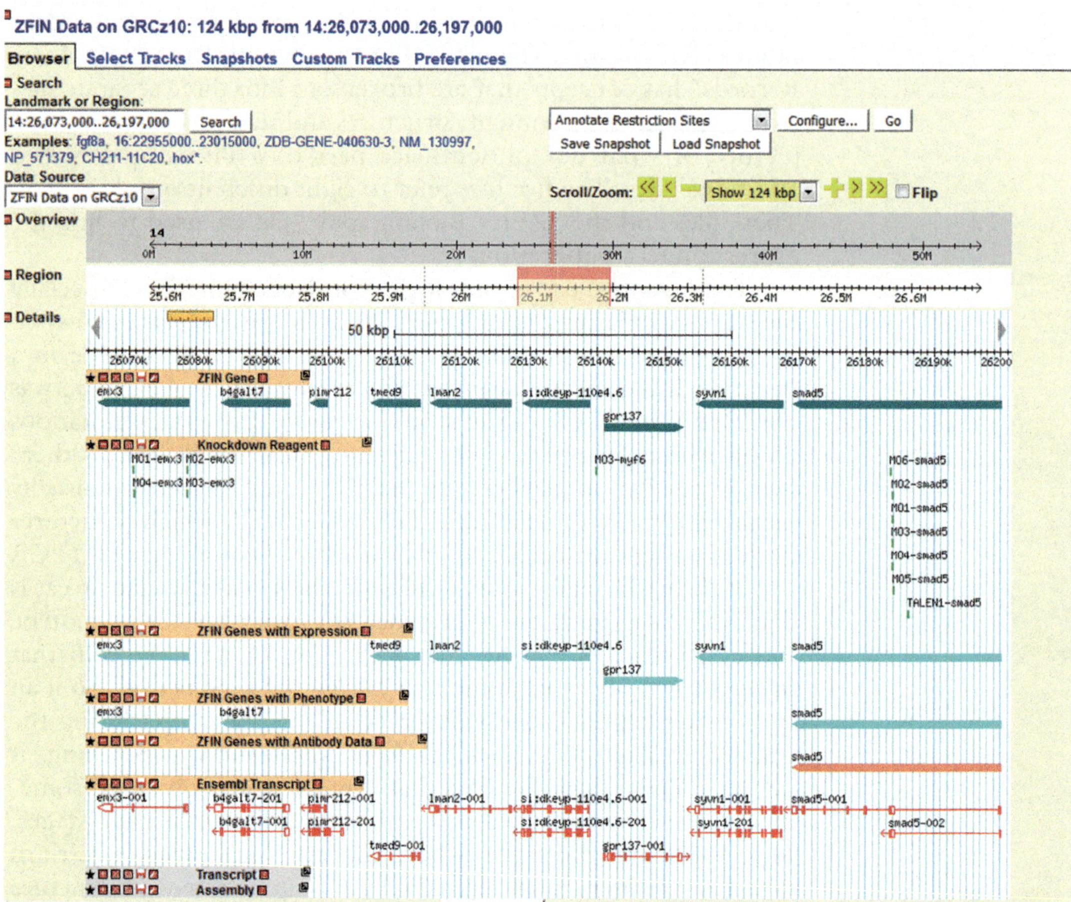

Fig. 16 ZFIN GBrowse integration. Information about gene expression, phenotype, knockdown reagents (STRs) along with Ensembl transcripts is provided via GBrowse. The Knockdown Reagent track is generated by ZFIN using bowtie analysis and is a unique resource provided by ZFIN. On the left side of every track name there is a set of icons that allows users to customize their browsing experience; hiding, expanding, removing the track from the display, sharing or saving a track of interest, and configuring the track display, color, number of features

according to their preference by a simple "click and drag" operation. Gene and transcript tracks are provided based on ZFIN, Vega and Ensembl annotations. Other customized tracks include ZFIN Genes with Expression, and ZFIN Genes with Phenotype. Tracks for ZFIN Genes with Antibodies and Knockdown Reagents present users with prospective research tools available within a selected genomic region.

Tracks are aligned to the current version of the Reference Genome Assembly (GRCz10) and link to data-specific pages at ZFIN or Ensembl. In addition, the previous genome assembly version (Zv9) is also available from the "Data Source" drop-down menu. Different assembly versions are provided to allow users to identify mutations, transgenics, or conserved nongenic elements (CNEs), that have been mapped only to the older Zv9 assembly. ZFIN does not provide "lift-over" of annotations from prior to later assembly versions, but instead relies on the original data providers to ensure that annotations are referenced to the most recent assembly. This ensures that data are consistent with the original annotations and minimizes the likelihood of errors and curatorial effort required to maintain the data annotations. Tracks that are not remapped onto the newest assembly will not be available in the default genome browser view, which shows the most current assembly, but will be available if the build version is changed to the previous assembly using the "Data Source" pull down menu located in the upper left corner of the GBrowse interface. The GBrowse implementation also supports the upload of custom tracks in a variety of formats. A tutorial is provided from the Custom Tracks Tab to guide the user through this process https://zfin.org/gbrowse2/annotation_help.html.

7 Conclusion

Over the past 25 years, genetic research of all kinds has benefited from the rapid evolution of powerful new tools and techniques in both genetics and computer science. Together, these advances have greatly accelerated the rate at which new genetic data can be gathered. As a result, the ZFIN web site and data have grown in diversity, volume, and complexity over the years. Facilitating the most effective use of these data remains a top priority at ZFIN. Curation and integration of various data types, maintenance of high quality control standards, and prioritization of the data and services most important to the zebrafish research community remain central to the ZFIN mission. In this chapter, many of the most important and heavily utilized ZFIN web pages and data types have been reviewed. Gene pages are the hub around which many of the data revolve, and are extensively hyperlinked to related data in both ZFIN and external resources. In many cases, collaboration allows partner resources to provide links back to ZFIN, often linking to a gene page. Searching for data in ZFIN is a critical feature of the web site. Basic searches such as gene symbol or anatomical structure are thoroughly supported. Data exploration now also has substantial support, particularly in the single box search interface. Exploration of more complex searches such as "which genes have curated expression in CaP and MiP motor neurons and produce a phenotype in cardiac muscle cells in the atrium?" (Answer: *isl1*) is encouraged. Data are available from ZFIN in tab delimited text files as well as from the ZebrafishMine, which also supports external use of ZFIN data through an API for several programming languages. Proper interpretation of these data is important for drawing correct conclusions or deriving new hypotheses. The ZFIN staff is always available (zfinadmn@zfin.org) to answer questions about data collection, the zfin.org web site, or the data provided. Expert curation, integration, and sharing of zebrafish genetic and genomic data will continue to provide the most useful resource possible for the zebrafish and greater biomedical research communities.

Acknowledgments

The ZFIN staff who build, maintain, and add data to ZFIN: Ken Frazer, Holly Paddock, Amy Singer, Sabrina Toro, Patrick Kalita, Prita Mani, Ryan Martin, Sierra Moxon, Christian Pich, Kevin Schapher, Xiang Shao, Ruben Lancaster, Anne Eagle, Jonathan Knight; who helped format the figures for this publication. Funding provided to Monte Westerfield, ZFIN's Principal Investigator by National Human Genome Research Institute of the National Institutes of Health [HG002659].

References

1. Westerfield M, Doerry E, Kirkpatrick AE et al (1997) An on-line database for zebrafish development and genetics research. Semin Cell Dev Biol 8:477–488. https://doi.org/10.1006/scdb.1997.0173
2. Howe DG, Bradford YM, Eagle A et al (2017) The Zebrafish Model Organism Database: new support for human disease models, mutation details, gene expression phenotypes and searching. Nucleic Acids Res 45:D758–D768. https://doi.org/10.1093/nar/gkw1116
3. Howe DG, Bradford YM, Conlin T et al (2013) ZFIN, the Zebrafish Model Organism Database: increased support for mutants and transgenics. Nucleic Acids Res 41:D854–D860. https://doi.org/10.1093/nar/gks938
4. Sprague J, Bayraktaroglu L, Bradford Y et al (2008) The Zebrafish Information Network: the zebrafish model organism database provides expanded support for genotypes and phenotypes. Nucleic Acids Res 36:D768–D772. https://doi.org/10.1093/nar/gkm956
5. Van Slyke CE, Bradford YM, Westerfield M, Haendel MA (2014) The zebrafish anatomy and stage ontologies: representing the anatomy and development of Danio rerio. J Biomed Semantics 5:12. https://doi.org/10.1186/2041-1480-5-12
6. Howe DG, Bradford YM, Eagle A et al (2016) A scientist's guide for submitting data to ZFIN. Methods Cell Biol 135:451–481. https://doi.org/10.1016/bs.mcb.2016.04.010
7. Degrave, Agnes, Fürthauer, Maximilian, Heyer, Vincent, Loppin, Benjamin, Obrecht-Pflumio, Sophie, Steffan, Tania, Thisse, Bernard, Thisse, Christine, Woehl R (2001) Expression of the zebrafish genome during embryogenesis (NIH R01 RR15402). ZFIN on-line Publ.
8. Barrett T, Wilhite SE, Ledoux P et al (2013) NCBI GEO: archive for functional genomics data sets--update. Nucleic Acids Res 41:D991–D995. https://doi.org/10.1093/nar/gks1193
9. Schriml LM, Mitraka E (2015) The Disease Ontology: fostering interoperability between biological and clinical human disease-related data. Mamm Genome 26:584–589. https://doi.org/10.1007/s00335-015-9576-9
10. Schriml LM, Arze C, Nadendla S et al (2012) Disease Ontology: a backbone for disease semantic integration. Nucleic Acids Res 40:D940–D946. https://doi.org/10.1093/nar/gkr972
11. Bradford YM, Toro S, Ramachandran S et al (2017) Zebrafish models of human disease: gaining insight into human disease at ZFIN. ILAR J 58(1):4–15. https://doi.org/10.1093/ilar/ilw040
12. Gaudet P, Livstone MS, Lewis SE, Thomas PD (2011) Phylogenetic-based propagation of

functional annotations within the Gene Ontology consortium. Brief Bioinform 12:449–462. https://doi.org/10.1093/bib/bbr042
13. Huntley RP, Sawford T, Mutowo-Meullenet P et al (2015) The GOA database: gene Ontology annotation updates for 2015. Nucleic Acids Res 43:D1057–D1063. https://doi.org/10.1093/nar/gku1113
14. Chibucos MC, Mungall CJ, Balakrishnan R et al (2014) Standardized description of scientific evidence using the Evidence Ontology (ECO). Database (Oxford) 2014:bau075. https://doi.org/10.1093/database/bau075
15. Chibucos MC, Siegele DA, JC H, Giglio M (2017) The evidence and conclusion ontology (ECO): supporting GO annotations. Methods Mol Biol 1446:245–259. https://doi.org/10.1007/978-1-4939-3743-1_18
16. Finn RD, Attwood TK, Babbitt PC et al (2017) InterPro in 2017-beyond protein family and domain annotations. Nucleic Acids Res 45:D190–D199. https://doi.org/10.1093/nar/gkw1107
17. Sigrist CJA, de Castro E, Cerutti L et al (2013) New and continuing developments at PROSITE. Nucleic Acids Res 41:D344–D347. https://doi.org/10.1093/nar/gks1067
18. Finn RD, Coggill P, Eberhardt RY et al (2016) The Pfam protein families database: towards a more sustainable future. Nucleic Acids Res 44:D279–D285. https://doi.org/10.1093/nar/gkv1344
19. Artimo P, Jonnalagedda M, Arnold K et al (2012) ExPASy: SIB bioinformatics resource portal. Nucleic Acids Res 40:W597–W603. https://doi.org/10.1093/nar/gks400
20. Kamens J (2015) The Addgene repository: an international nonprofit plasmid and data resource. Nucleic Acids Res 43:D1152–D1157. https://doi.org/10.1093/nar/gku893
21. Csályi K, Fazekas D, Kadlecsik T et al (2016) SignaFish: a Zebrafish-specific signaling pathway resource. Zebrafish 13:541–544. https://doi.org/10.1089/zeb.2016.1277
22. Howe K, Clark MD, Torroja CF et al (2013) The zebrafish reference genome sequence and its relationship to the human genome. Nature 496:498–503. https://doi.org/10.1038/nature12111
23. Ruzicka L, Bradford YM, Frazer K et al (2015) ZFIN, The zebrafish model organism database: updates and new directions. Genesis 53:498–509. https://doi.org/10.1002/dvg.22868
24. Nasevicius A, Ekker SC (2000) Effective targeted gene "knockdown" in zebrafish. Nat Genet 26:216–220. https://doi.org/10.1038/79951
25. Hwang WY, Fu Y, Reyon D et al (2013) Efficient genome editing in zebrafish using a CRISPR-Cas system. Nat Biotechnol 31:227–229. https://doi.org/10.1038/nbt.2501
26. Sander JD, Cade L, Khayter C et al (2011) Targeted gene disruption in somatic zebrafish cells using engineered TALENs. Nat Biotechnol 29(8):697. https://doi.org/10.1038/nbt.1934
27. Langmead B, Trapnell C, Pop M, Salzberg SL (2009) Ultrafast and memory-efficient alignment of short DNA sequences to the human genome. Genome Biol 10:R25. https://doi.org/10.1186/gb-2009-10-3-r25
28. Sprague J, Bayraktaroglu L, Clements D et al (2006) The Zebrafish Information Network: the zebrafish model organism database. Nucleic Acids Res 34:D581–D585. https://doi.org/10.1093/nar/gkj086
29. Ashburner M, Ball CA, Blake JA et al (2000) Gene ontology: tool for the unification of biology. The Gene Ontology Consortium. Nat Genet 25:25–29. https://doi.org/10.1038/75556
30. Westerfield M (2007) The zebrafish book : a guide for the laboratory use of zebrafish (Danio rerio), 5th edn. University of Oregon Press, Eugene, OR
31. Westerfield M (2000) The zebrafish book: a guide for the laboratory use of zebrafish (Danio rerio), 4th edn. University of Oregon Press, Eugene, OR
32. Sullivan J, Karra K, Moxon SAT et al (2013) InterMOD: integrated data and tools for the unification of model organism research. Sci Rep 3:1802. https://doi.org/10.1038/srep01802
33. Lyne R, Sullivan J, Butano D et al (2015) Cross-organism analysis using InterMine. Genesis 53:547–560. https://doi.org/10.1002/dvg.22869
34. Smith B, Ashburner M, Rosse C et al (2007) The OBO Foundry: coordinated evolution of ontologies to support biomedical data integration. Nat Biotechnol 25:1251–1255. https://doi.org/10.1038/nbt1346
35. Stein LD, Mungall C, Shu S et al (2002) The generic genome browser: a building block for a model organism system database. Genome Res 12:1599–1610. https://doi.org/10.1101/gr.403602
36. Bradford, Yvonne, Van Slyke, Ceri, Toro, Sabrina, Ramachandran, Sridhar (2016) The Zebrafish Experimental Conditions Ontology Systemizing Experimental Descriptions in ZFIN. CEUR Workshop Proceedings 1747. http://ceur-ws.org/Vol-1747/IP25_ICBO2016.pdf

Chapter 12

EchinoBase: Tools for Echinoderm Genome Analyses

Gregory A. Cary, R. Andrew Cameron, and Veronica F. Hinman

Abstract

The echinoderms are a phylum of invertebrate deuterostome animals that constitute important research models for a number of biological disciplines. EchinoBase (www.echinobase.org) is an echinoderm-specific genome database and web information system that provides a platform for the interrogation and exploration of echinoderm genomic data. This chapter outlines the datasets available on EchinoBase; from assembled genomes and genome annotations, to spatial and quantitative expression data, as well as functional genomics datasets. We also highlight the bioinformatic tools available on the website to facilitate rapid inquiries using these data (genome browsers, precompiled BLAST databases, etc.), and suggest optimized strategies for performing these inquiries. We conclude with a perspective on how one could integrate various genomic resources to predict putative noncoding regulatory regions. The available datasets and analyses they permit provide the basic components required for developing an understanding of how echinoderm genomes are regulated, especially during early development, and provides a platform for comparative genomic inquiries among species in this phylum.

Key words Echinoderm, Sea urchin, Sea star, Sea cucumber, Brittle star, Genome, Expression, Gene regulatory network (GRN)

1 Introduction

Species within the phylum Echinodermata have served as important research models for centuries. The biochemist, physiologist, and developmental biologist have all found ample reason to prefer these animals for experimental work. These animals utilize external fertilization and embryogenesis, many embryos and larvae are transparent facilitating visualizations, they are relatively easy to maintain and rear in a lab setting. What's more echinoderms along with hemichordates make up the Ambulacraria superphylum of invertebrate deuterostomes, the sister clade to chordates.

Dedication
This chapter is dedicated to the memory of Dr. Eric Davidson. A pioneer in understanding the genomic controls of developmental processes, he was a vigorous catalyst for the establishment and curation of echinoderm genome resources, including EchinoBase and its predecessor SpBase.

Martin Kollmar (ed.), *Eukaryotic Genomic Databases: Methods and Protocols*, Methods in Molecular Biology, vol. 1757, https://doi.org/10.1007/978-1-4939-7737-6_12, © Springer Science+Business Media, LLC, part of Springer Nature 2018

A significant contribution of echinoderm research in recent decades has been the generation of a detailed gene regulatory network description of sea urchin embryogenesis [1, 2]. These networks depict the hierarchical regulatory interactions that underpin development. Such a systems-level compendium demands understanding the developmental regulatory genes (i.e., transcription factors and signaling molecules) as well as how they control the precise spatiotemporal expression of other genes within the network. This includes identifying and interrogating noncoding cis-regulatory control sequences. Increasingly, the data supporting such models are obtained by genome-wide queries of sequence function and interspecies comparisons of regulatory interactions [3].

As with many areas of biological research, the deluge of genomic data is transforming how biologists approach research questions using these animals. The purpose of EchinoBase, the echinoderm genome database, is as a resource platform for echinoderm researchers to access, query, and streamline genomic inquiries to facilitate research projects involving or related to echinoderms. The site hosts data related to 4 of the 5 extant echinoderm classes (i.e., Echinoidea, Holothuria, Asteroidea, and Ophiuroidea), providing an exceptional platform for comparative evolutionary inquiries around the regulation of genes for developmental and cell biologists. In this chapter, we describe the data available on EchinoBase, as well as provide basic guidance for how users may access and work with the hosted datasets. The available data and pertinent methods presented in this chapter reflect a snapshot of the database from June 2017. The site is under active development and improvement; we welcome any and all feedback related to the website and underlying data. Such feedback can be provided through the web page contact form, or by contacting the authors directly.

2 Echinoderm Genome and Transcriptome Data Resources

There are several annotated echinoderm genome datasets available at EchinoBase. The first echinoderm genome to be fully sequenced was that of the sea urchin *Stronglycentrotus purpuratus* [4]. Since then, four additional genome resources have been generated and hosted on EchinoBase: *Lytechinus variegatus*, *Patiria miniata*, *Parastichopus parvamensis*, and *Ophiothrix spiculata* [5] (Fig. 1). Important features of the genome sequences (locations of genes, UTRs, repetitive elements, etc.) have been annotated. EchinoBase also presents transcriptome data for several species (Fig. 1). Furthermore, there are several hundred echinoderm transcriptome projects available in the GEO and SRA databases at NCBI (primarily as sequencing reads). We provide access to these data sets through links under the "Species > Other Species" menu.

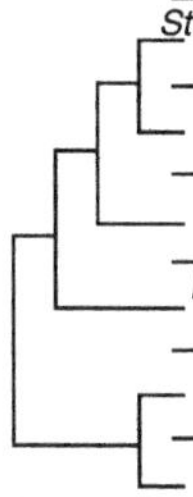

Species	Abbreviations & Annotation Prefix	Genome Assembly Version	Assembled Scaffold N50 (kb)	Transcriptome Datatypes	Annotated Genes
Strongylocentrotus purpuratus (purple sea urchin)	Sp | SPU	v3.1	402	S / Q / D	29,948
Lytechinus variegatus (green sea urchin)	Lv | LVA	v2.2	46	D	22,105
Eucidaris tribuloides (slate pencil urchin)	Et	v1.0	39	D	-
Parastichopus parvimensis (Warty sea cucumber)	Pp | PPA	v1.0	89	D	17,379
Patiria miniata (bat star)	Pm | PMI	v2.0	76	D	30,399
Ophiothrix spiculata (spiny brittle star)	Os	v1.0	72	-	-

Fig. 1 Echinoderm species with data hosted on EchinoBase. The resolved phylogenetic relationship of the species is shown by the dendrogram and the common name, species abbreviations, and annotation prefixes used are indicated. For each species the current version of the genome assembly is indicated along with the Scaffold N50 assembly statistic. There are also various transcript expression datasets hosted and the nature of the available datasets is indicated for each species (D = descriptive, sequence only; S = spatial expression patterns; Q = quantitative expression data; *see* Subheading 5). Finally, the number of annotated genes is indicated for all species for which gene annotation pipelines have been run. The annotation pipeline for *S. purpuratus* involved a GLEAN annotation pipeline whereas the others are based on MAKER2-based annotations

The most complete and detailed datasets are those pertaining to *Strongylocentrotus purpuratus*, the purple sea urchin. Many of the analyses described below are, at present, only available for this species. However, as data become available for other species these will be hosted and accessible in similar ways.

3 Home Page

3.1 Home Page Content

Typical users will arrive via the EchinoBase home page (Fig. 2). Here we describe how users may navigate through the site using menus and links found on this home page. Users can return to this home page by clicking on the EchinoBase logo found on the top left of any page on the site. There is also a website search box located in the top right corner of the home page that can be used to search the text on EchinoBase pages to locate specific datasets, tools, resources, or other information.

The home page is subdivided into four general regions (Fig. 2). At the top of the page are the navigation menus, which are described in more detail under Subheading 3.2. The second region is the species portal panel consisting of a looping slideshow of images and names of the various species for which data are hosted on EchinoBase. Users may click on any of the species depicted to navigate to the relevant species sub-page. The third region contains general information about EchinoBase and the Echinoderm phylum. The fourth section at the bottom of the home page is a location for reporting the most recent publications as well as news and events of interest to the community. There is also a panel of images indicating the current and historical sources of financial support for the development and maintenance of the site and underlying data.

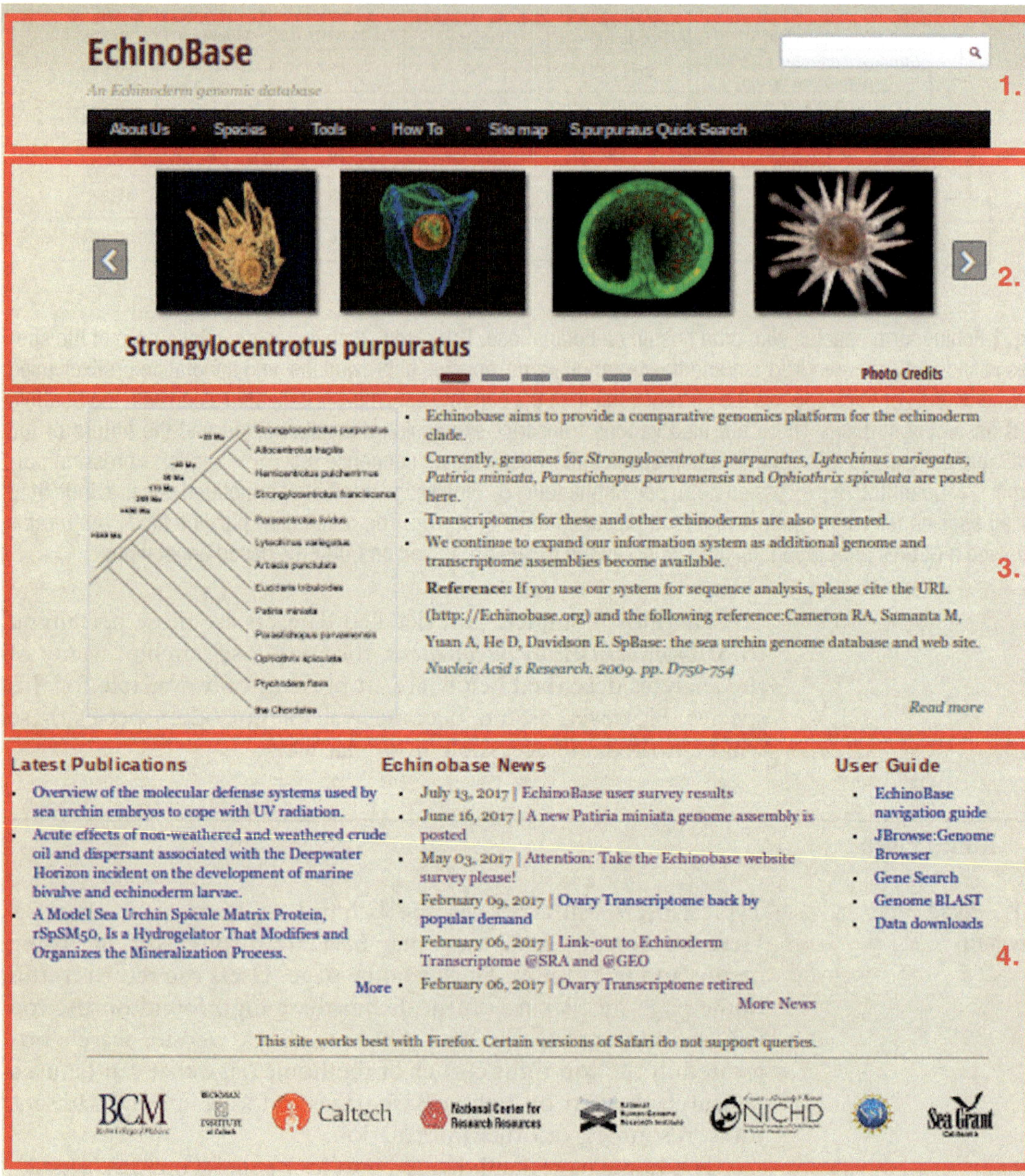

Fig. 2 Overview of the EchinoBase home page. At the top of the page (1) is the EchinoBase logo, which serves as a link back to the home page from any other page on the site, the website search box, the site navigation menus (i.e., About Us, Species, Tools, How To, Site Map), and a quick link to the *S. purpuratus* gene search tool. The next section (2) is the species portal, which consists of a rotating slideshow of images of each species with data hosted. Click on the radio buttons at the bottom or the left/right carrot buttons to flip between species and click on the name or images to navigate to species-specific information pages. The next section (3) presents general information about EchinoBase including the mission, nature of the hosted data, general information on echinoderms, and citation information. The section at the bottom of the page (4) contains community information including a list of recent echinoderm publications and news highlights of relevance to the community (both updated regularly) as well as user guides for various aspects of the site and a list of funding sources

3.2 Navigation Menus

The series of expandable menus along the top of the page allow a user to navigate to the data or resource most appropriate to their visit. These include navigation to individual species, hosted bioinformatic tools, information describing hosted data ("how to"), and an overall site map. Because these menus are available from any page on the site, and allow easy transition between data hosted on different parts of the site, we now describe the menus in more detail.

The "About Us" menu points to information about the people and policies that guide the EchinoBase web resource, our mission, and ways to both contact us and reference the data hosted.

The "Species" menu lists all species for which data is hosted on EchinoBase. A typical user may wish to interact with data for a single species and this menu may provide an efficient means of navigating to that portion of the site. Each listed species expands further to provide links to general information about the organism as well as the data, resources, methods, and tools available for that organism. The available tools may include genome and transcriptome database searches, blast sequence searches, genome browser, literature searches, and downloadable data (each of these functions are discussed in more detail below).

The "Tools" menu provides access to the same functions and data available under the "Species" menu, but is not organized by species. Instead, the individual tools available on EchinoBase are listed (e.g., gene search, transcriptome search, blast, and JBrowse) and the linked pages highlight the species for which these tools are available. Also available under the "Tools" menu is a catalog of bacterial artificial chromosomes (BAC) that have been characterized, including some that have been engineered to drive reporter gene expression. Finally, the "Tools" menu also links to the image and video gallery, a literature search portal (textpresso, [6]), and a link to data download pages.

The "How To" menu provides access to information on the genome assemblies and an in-depth guide of how to use the tools and data hosted on EchinoBase.

Finally the "Site Map" provides direct links to all static pages on the site (i.e., those not dynamically generated as a result of a database search), arranged hierarchically in reflection of the structure outlined above.

Given the depth of information available for the *S. purpuratus* genome, and given our perception that the majority of EchinoBase users are seeking information pertaining to this species, we have provided a rapid access point to the *S. purpuratus* gene search function via the "*S. purpuratus* Quick Search" link in the menu bar. Clicking this link takes the user directly to the gene search function for *S. purpuratus*, discussed in Subheading 4.1.

4 Finding Genes

4.1 Finding Genes by Name

For the four of the five species with hosted genome assemblies (i.e., *S. purpuratus*, *L. variegatus*, *P. miniata*, and *P. parvimensis*), EchinoBase also provides a database of annotated genes. These gene annotations have been generated through various gene prediction and annotation pipelines (i.e., GLEAN and MAKER2) [7, 8]. More detailed information about the approach used for each assembly can be found under the "How To" menu and navigating to the species of interest. Overall there are about 30,000 annotated genes for each species (29,891 Sp, 28,094 Lv, 29,697 Pm) except *P. parvimensis*, which is considerably lower (17,379 Pp).

These genes are contained within a searchable database and the "Gene Search" tool for each species allows a user to access these database records and retrieve the relevant data. The search tool is a means of retrieving records from the database relating to the terms input into the search field. The search function is greedy, meaning that any record containing the alphanumeric characters entered, even as a fragment or subset, will be returned as a result. Furthermore, there are two wild card characters available for searching. The first is the asterisk (*) which denotes any alphanumeric text, including nothing. Thus, searching for "sp-*br" will return, among other results, "Sp-Bra", "Sp-Brca1", "Sp-Ebr1", and "Sp-Gabbr1". The second wild card is the question mark (?) which denotes one alphanumeric character. In this case, searching for "sp-?br" returns, among other results, "Sp-Ebr1", "Sp-Ubr1", and "Sp-Tbrg4".

Users can restrict the search to particular attributes of annotated genes records, or can search all attributes simultaneously. The specific attributes available to search depend on which database is being queried, but in general the attributes include the official gene ID, gene name, gene synonyms, and genomic scaffold. Query text need not include the species-specific prefix for official gene ID or gene names, and fragment text will match to longer names within the database (e.g., searching "myc" will return both "Sp-Myc" and "Sp-Mycbp", a Myc binding protein). More information on nomenclature standards (i.e., the structure of official gene IDs and gene names) can be found under Subheading 7.1.

As the naming of individual genes may be quite variable, we have attempted to collect synonymous identifiers for each gene annotated in the databases and these can be explicitly searched via the "Synonym" field. For example, *Sp-Onecut* is also referred to as "onecut homeobox", "Onecut1/2/3-like", and "Sp-Hnf6". The search of the synonym field also matches text fragments (as with the official gene name) and so a search for "hnf" will return a list of genes including *Sp-Onecut* (a.k.a. "Hnf6"), *Sp-FoxA* (a.k.a. "Hnf3"), as well as *Sp-Hnf1*.

Searching the database for a genomic scaffold identifier will return all genes annotated to the scaffold matching the given identifier; these genes and their arrangement on the given scaffold can be investigated further by visual examination of the scaffold using the JBrowse tool (described in Subheading 6.1 below). A critical consideration when searching genomic scaffolds must be which version of the genome assembly one is considering, as the genes predicted to be linked may be variable between assemblies. The search field for every database indicates the assembly version that is currently searchable and this corresponds to the version currently accessible via JBrowse.

Finally, users may also search the *S. purpuratus* database by WHL ID and PubMed ID. The WHL identifiers originate from comprehensive transcriptomic analyses of *S. purpuratus* corresponding to surveys of gene expression from ten embryonic stages, six feeding larval and metamorphosed larval stages, and six adult tissues [9, 10]. The gene models generated from these data were given identifiers prefixed with WHL and were integrated with SPU gene models. Searching the *S. purpuratus* database with a PubMed identifier will return all of the genes examined by the indicated publication, although this is limited to those publications that have been annotated to particular genes in the database. The PubMed ID (aka PMID) is an 8-digit number and must be an exact match to the PMID annotated in the database (i.e., wildcard characters are not permissible).

The result of a database search performed on any or all of the attributes described above will yield a list of all database entries that match the search. These results will be presented in a table reporting, for each match, the official gene symbol (e.g., "SPU_…"), the official gene name (e.g., "Sp-…"), any synonyms and the manner in which each match was annotated (e.g., manually or electronically). By default only the first 10 matches are shown, but the results table can be expanded to show as many as 100 entries on one page. Additional matches can be reviewed by using the navigation links at the bottom of the table (i.e., "previous" and "next"). Furthermore, the result table may be sorted by any of the displayed columns by clicking on the column headers; clicking multiple times will change the sort order (i.e., from ascending to descending).

4.2 Finding Genes by Sequence Comparison

Frequently, users may not know *a priori* the appropriate database identifier to use to retrieve the desired sequence or record from the database. Therefore we also include BLAST interfaces to search the database based on sequence similarity. BLAST is a very common tool used for calculating sequence similarity, and the algorithm and parameters have been described extensively elsewhere [11, 12]. Therefore, in this chapter, we focus on the details of the BLAST implementation hosted on EchinoBase, the echinoderm specific sequence databases available to search, and how to interpret and effectively use the results.

The databases available vary by species and for more information on the sequence databases available for each species click on the "database" hyperlink on the blast interface. In general the available sequence databases will include genome assemblies as scaffolds (containing assembly gaps), where available, as well as ungapped contigs. In some cases RNA-seq transcriptome and predicted peptide sequences will also be available. For *S. purpuratus*, the GLEAN annotated gene and peptide sequences and BAC clone sequences are also available as searchable sequence databases. These databases can be searched using the common set of BLAST programs including blastn, blastp, blastx, tblastn, and tblastx, depending on query and database sequence type (click on the "program" hyperlink for descriptions of each search program).

We here envision a typical approach to finding a gene of interest within an EchinoBase database based on an existing sequence from another organism; this can and should be modified to suit a user's specific informational needs. First, using a protein sequence query of the gene of interest, search protein databases (using blastp) or nucleic acid databases (using tblastn), including transcriptome and/or genome databases. However, searching the genome with a protein sequence will incur gap penalties due to introns in the database whereas searching the protein or transcript data sets will incur no such penalties and will only incur penalties due to sequence dissimilarity. For searches against annotation databases (e.g., *S. purpuratus*), the results are linked to records within EchinoBase directly and the relevant record can be retrieved by clicking on the link. Otherwise, one may need to use the resultant sequence as a query in a subsequent BLAST search against a genome database to retrieve the relevant genomic coordinates. These coordinates can then be entered into JBrowse to direct the genome browser to the appropriate locus (*see* below, Subheading 6) from which the annotated gene can be identified and then retrieved by a database search.

4.3 Gene Information Page

Clicking on a database search item or link from JBrowse gene annotation propagates the database record as an html file (Fig. 3). This page presents all information pertaining to the given gene contained within the database including general gene information, expression data, functional annotation, associated GO terms, sequence information, reagent data, pubmed references, and much more information.

Fig. 3 (continued) FASTA records for each sequence can be viewed by clicking on the link or downloaded by clicking on the button at the right. The reagents section (6) presents experimental reagents specific to the annotated gene along with links to publications in which the reagent is developed and discussed. The comments and references sections (7) at the bottom of the page present additional information about the gene as well as links to publications in which the gene is referred, respectively

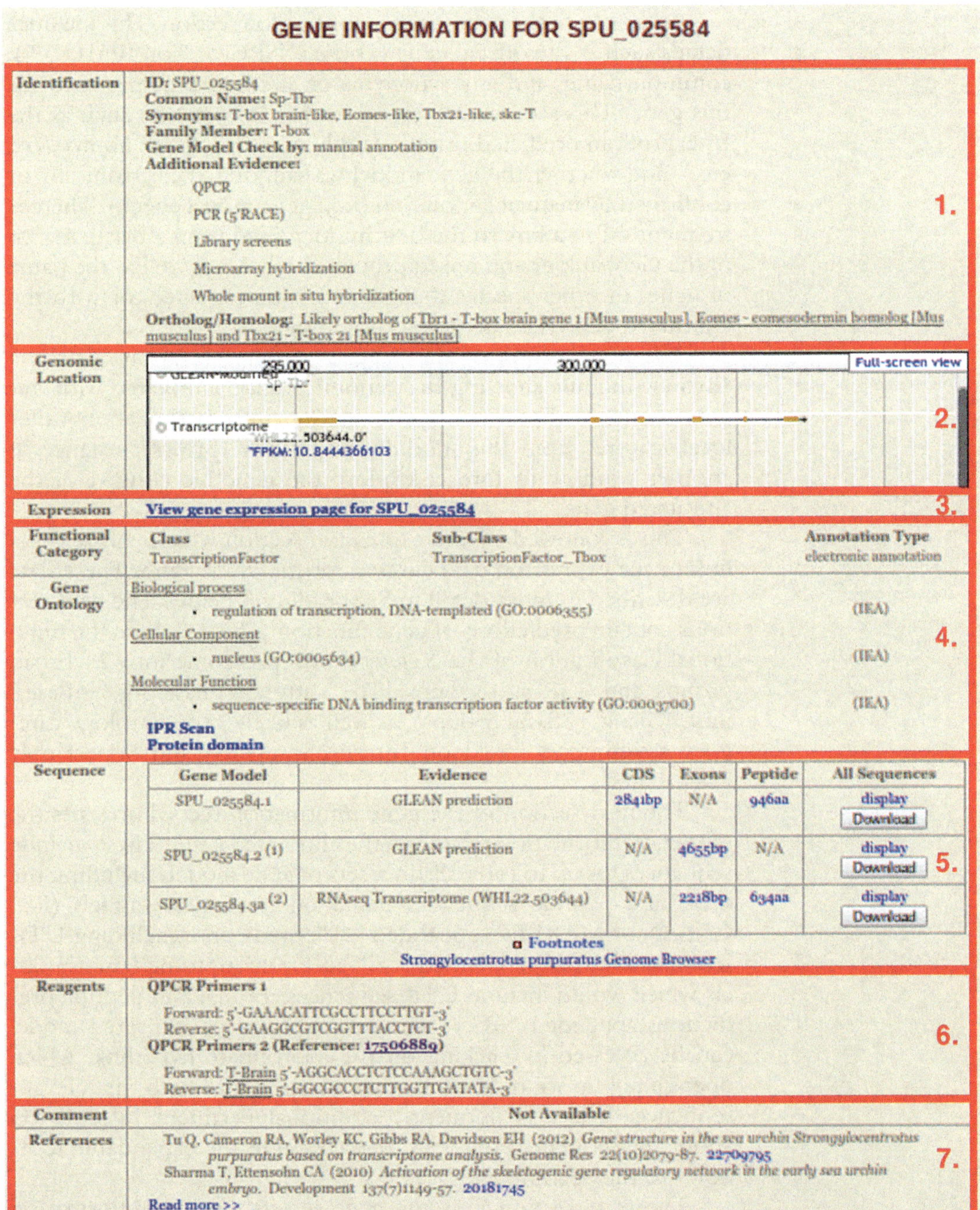

Fig. 3 Example gene information page. The top section (1) presents identifying information for the annotated gene including the official ID, common name, synonyms, annotation method and evidence, as well as gene family and orthology predictions. The next section (2) is a snapshot of the genomic locus including the annotated gene and transcriptome tracks; click on the browser window to launch the full version of the genome browser centered at this locus. The next section (3) is a link to a separate page containing the various expression data collected for this gene. Below the expression link are a set of gene function predictions (4) including both functional category assignments and gene ontology predictions. The next section (5) presents the sequence data associated with this gene including all the various gene models described in the text.

At the top of the page is the identification section. This includes details such as the official gene ID (i.e., "SPU_…" or "PMI_…"), common name, and any synonyms or alternative names used for this gene. This section also reports other information such as the InterProScan predicted protein family type (forkhead, homeobox, etc.), and whether the gene model was annotated electronically or confirmed by manual annotation (i.e., gene model check). There is frequently also a link to the best hit identified from a blastp search of the Genbank nr and refseq protein databases as well as the name of genes in other species that share a common ancestor with the annotated gene (i.e., ortholog/homolog).

The next section of the page is genomic location. This is a JBrowse module showing the genomic region annotated with this gene sequence. The user may obtain more information about other local features from this module, or open a JBrowse instance in another window to further explore the genomic context of the annotated gene.

This is followed by the expression section which consists of a link to the expression data curated for the given gene. These data are described in more detail in Subheading 5 below. The next sections contain predictions of gene function. These include the functional classification of the *S. purpuratus* proteome into 24 broad groups and 138 sub-groups [10] (http://spbase.org/SpBase/misc/Qiang-138-subgroups), as well as likely gene ontology category assignments based on InterProScan [13] and Superfamily [14] analyses.

The next section of the gene information record presents the sequence information related to the annotated gene. These include sequences for up to three distinct sets of gene models including the CDS and peptide sequences based on prediction models (i.e., GLEAN or MAKER2) appended ".1", predictions including UTR sequences (exons) appended ".2", and transcriptome based models which would include UTR sequences (exons) and peptide predictions appended ".3". The sequence for each indicated model can be accessed by clicking on the appropriate hyperlink, which opens a popup window containing a fasta record with the relevant sequence. There is frequently additional pertinent information about the sequences that can be accessed in the expandable footnotes to this section.

The next section presents gene-specific reagent information mined from the literature. Frequently these reagents include primer sets for real time qPCR, reverse transcriptase RT-PCR primer sets, primers for generating whole mount in situ hybridization probes (WMISH), the probe sequence itself, or sequences for either translation blocking or splice blocking morpholino antisense oligonucleotides (MASO). The publication source for these reagent sequences is indicated as a PubMed ID and a hyperlink.

The final sections include a comments section and a references section. The comments section would indicate if the gene is a duplicate or isomer of another gene, if it overlaps ribosomal RNA, or if it is derived purely from RNA-seq transcriptome data, among other pertinent information. The references section presents a list of published papers in which the gene is referred.

5 Expression Data

Probably one of the most frequent inquiries about the regulation of genes in the genome is when and where a given gene is expressed. To begin to address these questions for echinoderm genomes EchinoBase hosts spatiotemporal expression data in a variety of formats, from low throughput to high-density. These data are accessible by clicking on the link in the "expression" section of gene information page (Subheading 4.3). The information below is provided to orient you toward these datasets and how best to extract the most relevant information.

5.1 Expression Domains

The spatial pattern of gene expression is critical information toward understanding developmental gene regulation. For regulatory genes, overlapping expression domains suggests the potential for regulation and nonoverlapping patterns would indicate repression. For a few hundred *S. purpuratus* genes these data have been curated into tables for presentation on EchinoBase. The curation of spatial expression data at EchinoBase is ongoing and continuing to improve and currently these data may be found in one of two formats.

For some genes, such as *Sp-Hox11/13b* (SPU_002631), the spatial expression is reported in a table with two columns showing developmental time, ranging from 0 h to adult, and the set of expression domains defined for each stage (Fig. 4A). For example at 0 h the only expression domain is the egg, but by 6 h the three germ layers are distinct from one another. If a gene expressed within a particular domain at a given time point, the domain name is highlighted in green and bold font. This is a purely binary measure of gene expression (i.e., off or on). For other genes, such as *Sp-FoxA* (SPU_006676) there is a matrix display (i.e., "Early Embryo Spatial and Temporal Expression") (Fig. 4B). Each cell of this matrix indicates a given expression domain (egg, micromeres, etc.; rows) at a particular embryological time (0–30 h; columns). If a region is not defined at a given time it is indicated by grey shading; for example, the region "egg" has no meaning past 0 hpf and so it is grey. A region in which a gene is not expressed is a blank cell and if no data is present for the given time window it is indicated by a dash ("-"). If, however, the gene in question is expressed in a

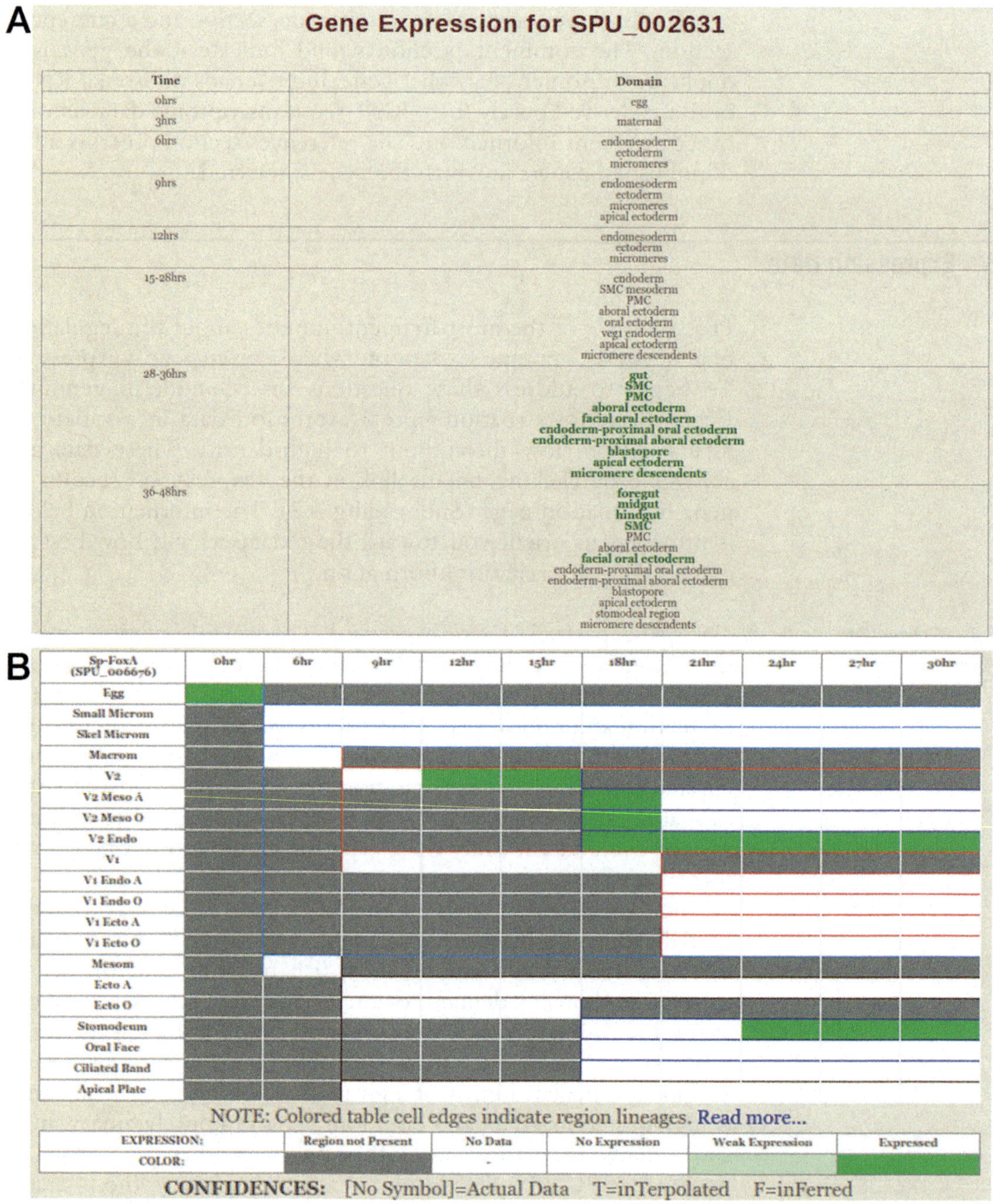

Fig. 4 Two representations of spatial expression data found on EchinoBase. For genes with curated spatial expression patterns, the patterns may be reported in one of two ways. In the first (A) the expression of *Sp-Hox11/13b* (SPU_002631) is reported in various developmental windows (rows). If the gene is expressed in some set of cells in the window, the domain of expression is represented in green colored, bolded text. In the second presentation (B) the expression of *Sp-FoxA* (SPU_006676) is reported in a matrix where the rows are different anatomical domains or regions and the columns represent different developmental timepoints. The coloration of each cell of the matrix indicates whether the region is not present (dark grey), whether there is no expression data ('-'), whether there is no expression (white), weak expression (light green), or expression (dark green)

region at the specified time, the cell is shaded in green corresponding to the level of expression (i.e., weak or strong). Glancing at either table can provide the user with a quick visual assessment of in which embryological regions and at what time a particular gene is expressed. Some degree of familiarity with sea urchin embryology is required to decipher the domain names, but there are guides available [15–17].

5.2 Quantitative Transcriptome Analysis

While spatial expression data is critical to fostering an understanding of developmental function, whole embryo quantitative assessments of gene expression are frequently obtained as these data are more readily generated and provide a snapshot of developmental expression dynamics. Such quantitative measures have been generated using a variety of technologies including quantitative real-time PCR (qPCR) [18], Nanostring nCounter [19], as well as high-throughput tiling microarray [20, 21] and RNA-seq transcriptome measurements [9, 10]. In general these data are presented as transcript counts per whole embryo at various stages and are presented in tabular or graphical format for each dataset.

The most interactive query of gene expression patterns is the quantitative developmental transcriptomes tool (Fig. 5). This tools enables users to enter gene names or symbols and display the

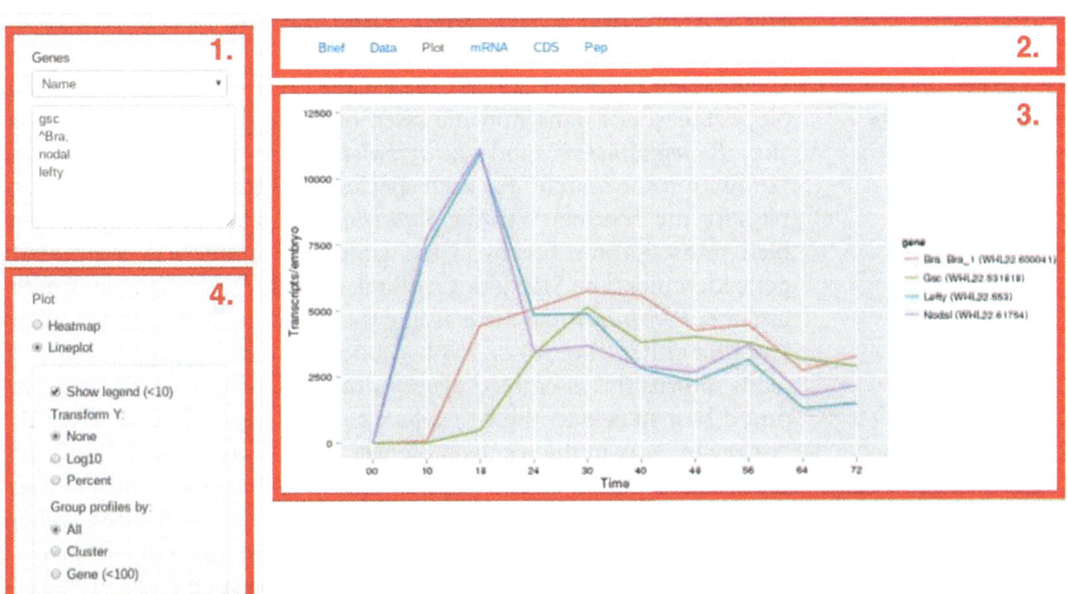

Fig. 5 Quantitative developmental transcriptomes tool. The names or official gene IDs of genes of interest are input in the box in the upper left (1). This populates information and data under the various tables (2); the entered names can be optimized based on the results returned under "brief" to ensure the data retrieved represent the genes of interest. The information under each table (3) may include a table of genes, data table, sequences, or plots of expression data. The presentation of the plots may be changed using the settings on the left (4) including to how to scale the axes, whether to include a legend, or to plot the data as a heatmap vs. a lineplot

quantitative expression profiles obtained by RNA-seq transcriptome measurements [9]. The user can enter several gene names (followed by a return) in the "genes" box and then explore the data in the right hand side of the page. The data returned includes a table of genes returned from the search based on the input, the official gene identifiers (which link to the relevant gene information page), transcriptome identifiers (e.g., "WHL_..."), as well as the functional classifications for the gene. In separate tabs users may access the sequences for the associated genes including mRNA, CDS, and peptide. Finally, the quantitative expression data may be examined either in tabular form (under "data" tab), in which the developmental timepoints in hours are listed along the top and the number of transcripts per embryo are indicated for each gene below, or visualized in a graphical format (under the "plot" tab). The display of the data under the "plot" tab can be further controlled by the settings on the left side of the page. Data can be presented as a heatmap, where colors indicate abundance at each timepoint, or as a lineplot. In the lineplot visualization, users may select to present the abundance values as total transcripts per embryo or transform these values by $\log_{10}$ or percent maximum expression. Finally, the plots may be plotted all on the same graph, grouped by assigned coexpression clusters ("cluster"), or plotted individually ("gene"). Checking the "legend" checkbox can be very helpful when plotting several expression profiles on the same graph.

5.3 Other Transcriptome Data

Other transcriptome datasets are available on EchinoBase. For example, databases of transcriptome assemblies for *P. miniata*, *L. variegatus*, *P. parvimensis*, and *E. tribuloides* are available under the "transcriptome search" for each species. This tool queries the assemblies for the best match to the *S. purpuratus* database for each assembled transcriptome based on the gene name, official *S. purpuratus* gene identifier (i.e., SPU), or GenBank identifier. For example, searching the *P. miniata* database using the transcriptome query tool by entering either "FoxA" or "SPU_006676" in the appropriate search fields returns the assembled *P. miniata* transcript that is the precomputed best match to the *S. purpuratus FoxA* gene. Clicking on the "query id" link in the results table will load the assembled *P. miniata* transcript sequence for further analysis. Still other transcriptome datasets are available via data repositories (i.e., GEO and SRA). Researchers willing to make their transcriptome data available through EchinoBase should contact us to discuss options for hosting such data.

6 Genomic Data

6.1 JBrowse Navigation and Available Data Tracks

Hosted genomes are available for exploration using JBrowse [22], a web-based genome browser (Fig. 6). In these genome browser instances users can visualize various genome tracks, or positionally mapped genomic annotations. Users can navigate to assembled scaffold sequences or genes by typing the gene name, official gene ID, or

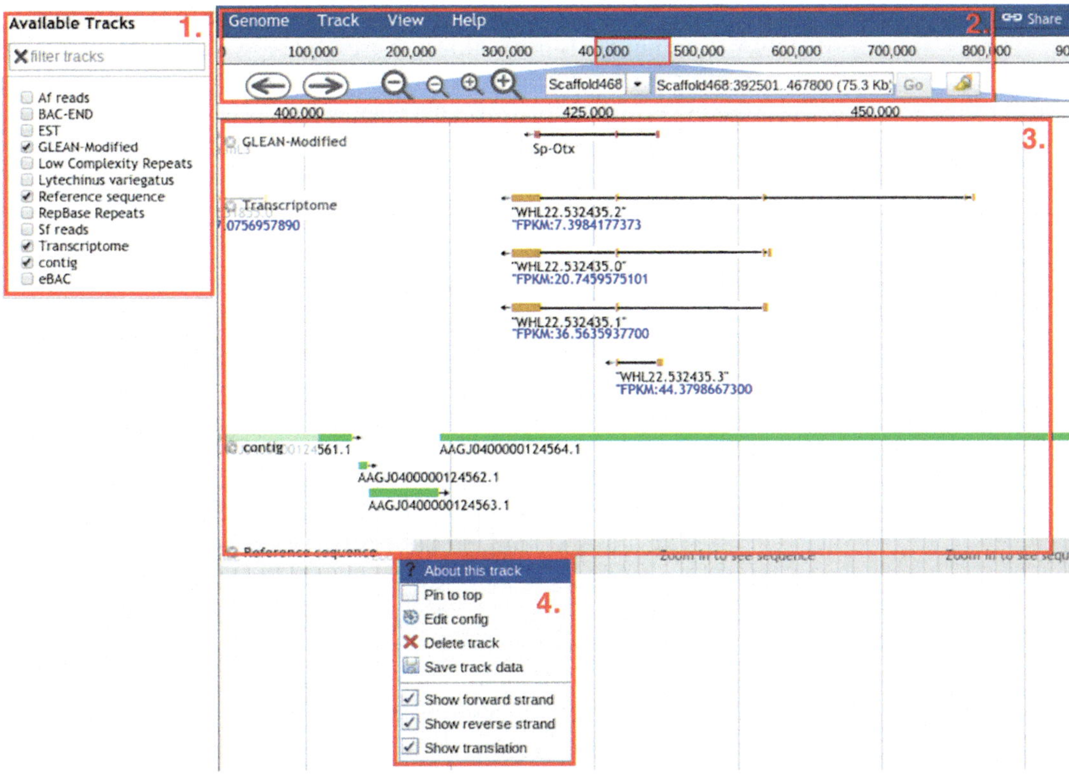

Fig. 6 JBrowse genome browser showing available tracks and manipulations. The tracks displayed in the genome browser window can be controlled by the panel on the left (1); click on the check box next to a track name to make that track visible in the browser. To navigate the genome browser use the controls at the top (2), including the text box to search for genes or directly enter the genomic position of interest (e.g., Scaffold468:3925091..467800), clicking the left/right arrows, and the magnifying glasses to zoom in and zoom out. Additional tracks can be loaded from the "track" menu. The tracks themselves (3) will display various pieces of information and to obtain more information on any feature simply click on the feature and a dialog box will open displaying additional information about that feature. To modify the display of a track, or to download data, click on the arrow to the right of the track name to access a track control menu (4)

genomic coordinates into the text entry box and clicking "go". Using the magnifying glass icons, users can zoom in to explore annotated features in closer detail or zoom out to get a broader sense of a genomic region. Users can also navigate along the displayed scaffold by either clicking on the right/left arrow icons or clicking on the reference ruler at the top of the page. Additional tracks can be added to the display by clicking on the appropriate checkbox under "available tracks" on the left side of the page and loaded tracks can be reordered by clicking and dragging on the track name.

The available sequence tracks vary by genome instance. For example, not all genomic analyses (and resultant browser tracks) available for *S. purpuratus* are available for, or relevant to, the *P. miniata* genome assembly. In general the available tracks include assembly contig, gene models (GLEAN or MAKER2), and predictions of repetitive sequence (inhouse, repbase, and low complexity). The assembly contig track shows the position of assembled contigs (comprised of continuously assembled from overlapping DNA fragment clones) along a scaffold, which may contain sequence gaps between assembled contigs. The gene model tracks display the positions of annotated genes based on predictions made in each species (i.e., GLEAN for *S. purpuratus* and MAKER2 for *L. variegatus*, *P. miniata*, and *P. parvimensis*). Tracks displaying repetitive sequence predictions are based on our inhouse repeat pipeline and low complexity repeats (using a custom repeatmasker pipeline), and Repbase transposable element detection [23]. The data available for *S. purpuratus* includes several additional genome browser tracks. "EST" and "Transcriptome" tracks display expression data as the position of expressed sequence tags or RNA-seq transcriptome sequences, respectively. There are two tracks that display information about BAC sequences including those enriched in shotgun sequencing analyses of the genome sequencing project (i.e., "eBACs") or those showing the positions of the sequenced ends of BAC clones (i.e., "BAC-END"). Finally, there are three tracks that address interspecies sequence homology. "AF reads" and "SF reads" display the positions of long reads from *Allocentrotus fragilis* and *Strongylocentrotus franciscanus* genomes, respectively, when mapped the *S. purpuratus* genome. At this evolutionary distance, regulatory sequences are statistically depleted of large indels (>20 bp), so examining the sequencing reads that align between the species is one way to highlight putative regulatory sequence. The "Lytechinus variegatus" track shows the location of patches of conserved noncoding sequences between *L. variegatus* and *S. purpuratus.* All three measures are potentially useful in the detection of functional noncoding sequence elements (*see* example below, Subheading 6.5).

6.2 Obtaining Sequence Data

Users can click on annotated features within the genome browser to generate a popup window that conveys further information about the feature (e.g., official gene ID, feature genomic coordinates, and sequence). For annotated genes these sequences will include the entire genomic sequence of the gene (including introns) as well as exonic sequences, which can be combined to predict the spliced transcript sequence. The sequence in unannotated regions of interest is available by sizing the JBrowse window to the desired region, selecting the "reference sequence" track from the left side panel, clicking on the track data icon menu, and selecting "save track data". A popup window will allow the user to select which part of the sequence to save (e.g., the entire scaffold or only the visible region) as well as how to access the data (e.g., save to disk or to view in a new window).

6.3 Uploading Genome Tracks

Users also have the option to upload their own genome browser tracks into the JBrowse application. These may be in any number of a variety of different genome browser formats (i.e., GFF3, GTF,

BigWig, BAM, FASTA, VCF, and tabix). These files may come from the user's own dataset or be from publically available datasets such as those available through the NCBI gene expression omnibus. However, one must be certain that the version of the genome used to generate the custom track corresponds to the genome build in use by JBrowse or the data will be incompatible.

Click on the "Track" menu item at the top of the page and then on the "Open track file or URL" menu item. A popup window then allows for the selection of local files (i.e., from the user's computer) or files hosted at a third party website indicated via URL. Then indicate the type of file being loaded (if not automatically detected) and modify how the track should be displayed in the browser (i.e., track name and whether to display the data as alignments, plots, variant calls, etc). This will generate a new track that can be viewed by clicking on the checkbox next to the track name.

6.4 ATAC-Seq Browser

There is a separate JBrowse instance available for *S. purpuratus* that presents data generated by a set of ATAC-seq experiments (GEO accession GSE95651). The ATAC-seq assay is used to identify regions of open chromatin [24] and was performed on whole embryos at several different embryologic timepoints (18, 24, 30, 39, 50, 60, and 70 hpf). These measurements were conducted in biological triplicate and the significant transposase-sensitive regions detected for each replicate are shown as separate tracks. For convenience, a user may wish to combine replicates into a single track by clicking on "add combination track" under the tracks menu and dragging multiple tracks onto the combination track. After the second track is added, the user will be asked how to combine the track data (by intersection of genomic regions, union of regions, subtraction of regions, and so on). The ATAC browser also contains several other tracks available in the other *S. purpuratus* genome browser instance.

6.5 Prediction of Noncoding Regulatory Regions

A common goal with such functional genomics datasets is to detect regulatory sequences in noncoding regions proximal to genes of interest. At present, the best approach to generating these predictions is to integrate several pieces of data, which may include ATAC-sensitivity data and sequence conservation blocks that are provided on EchinoBase. This of course may be augmented to include user-supplied data such as ChIP-seq datasets of transcription factor binding or histone post-translational modifications, but here we just focus on the data available through EchinoBase.

First, one needs to know when and where the gene of interest is expressed. For this purpose, the expression tools described under Heading 5 can be a useful starting point, if these data aren't otherwise available. For this example, we will consider the gene *Sp-SoxE* which is weakly expressed in the small micromeres starting at 9 hpf. Robust expression of *SoxE* begins at 30 hpf (Fig. 7, top).

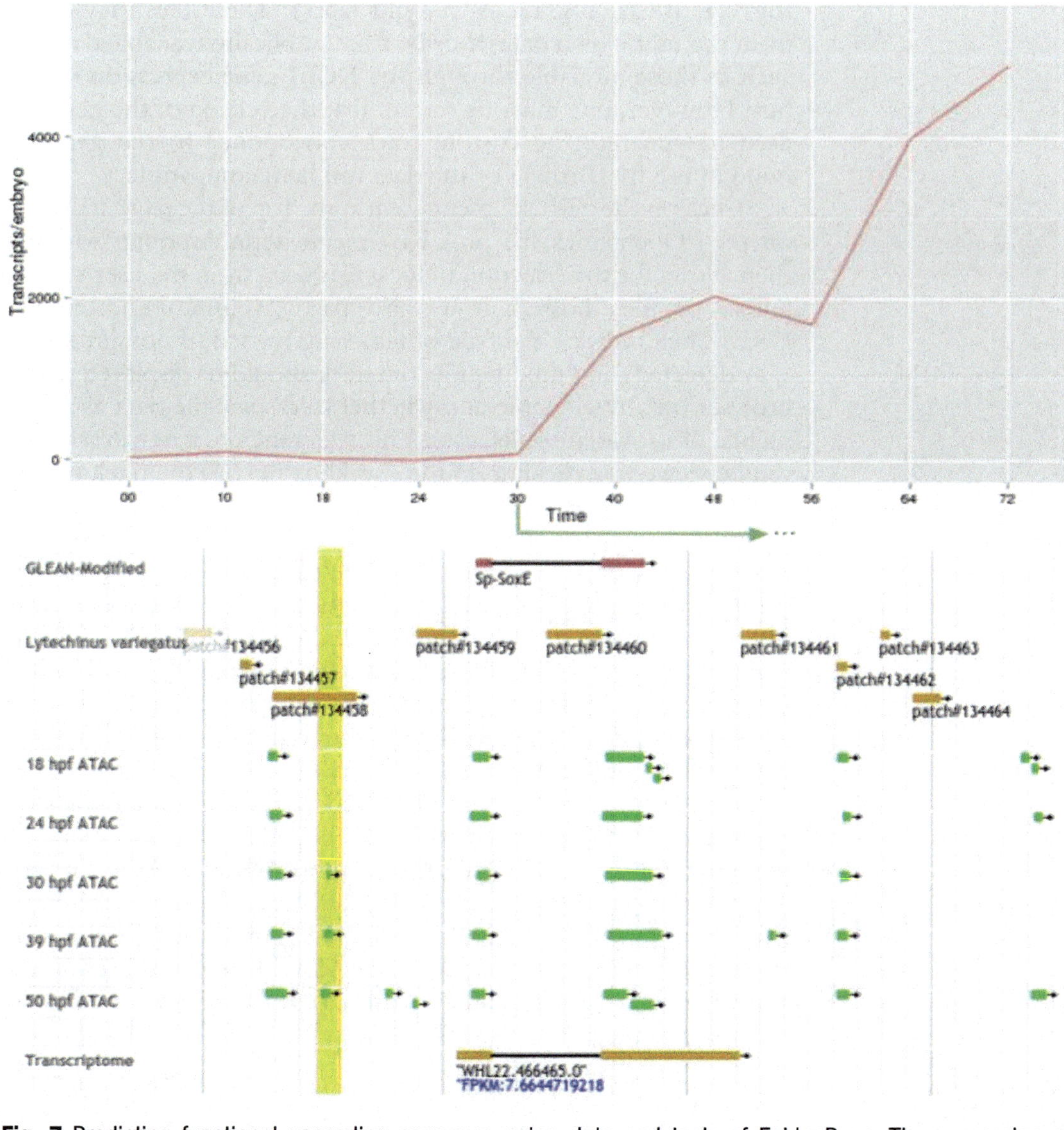

Fig. 7 Predicting functional noncoding sequence using data and tools of EchinoBase. The expression of *Sp-SoxE* obtained from the quantitative developmental transcriptome tool is shown (top). The region surrounding the *Sp-SoxE* locus (i.e., Scaffold375:577740..603419) is shown from the ATAC-seq browser (bottom) with tracks showing patches of noncoding sequence conservation between *Lytechinus variegatus* and *S. purpuratus* (gold) and ATAC-seq peaks at various developmental timepoints (green). The region highlighted in yellow shows a conserved region that becomes ATAC sensitive beginning at 30 hpf, which corresponds to when *Sp-SoxE* expression is activated (green arrow, top panel)

Start by navigating to the *Sp-SoxE* locus (Scaffold375:587477..593502) in the ATAC-seq browser and make certain the following tracks are turned on for viewing: GLEAN-modified, Transcriptome, and *Lytechinus variegatus*. We'll also need to turn on the ATAC-seq data. For simplicity, show the intersecting regions from the three replicates for each timepoint as a single combination track (described in Subheading 6.4). Focus on ATAC data from informative timepoints; for *Sp-SoxE* we might be particularly interested in detecting anything that activates the robust expression beginning at 30 hpf, so we would create a combination track for these replicates. Also generate combination tracks for preceding (i.e., 18 hpf and 24 hpf) and subsequent (i.e., 39 hpf and 50 hpf) timepoints to try to capture the transition. Once these combination tracks have been generated, the genome browser window should look similar to what is shown in Fig. 7.

Note that ATAC-seq peaks strongly overlap annotated exons; however these are not informative to us in our efforts to detect noncoding regulatory sequences. There are several ATAC-seq peaks that overlap patches of conserved noncoding sequence between *S. purpuratus* and *L. variegatus* (displayed in the "*Lytechinus variegatus*" track) and these are especially promising because they show regions of chromatin accessibility in highly conserved noncoding sequences over 50 Myr of evolution. This evolutionary distance is a "sweet spot" for detecting functional noncoding sequence conservation among these species, meaning that in the time since these species diverged only functional sequences are likely to be maintained [25]. One ATAC-seq peak that overlaps a patch of sequence conservation is absent in the early samples (i.e., 18 hpf and 24 hpf) but present beginning at 30 hpf onward (yellow highlight, Fig. 7). This is a promising candidate region for follow up reporter gene assay to validate activity as well as transcription factor binding site predictions and analyses.

A straightforward approach to assessing the activity of putative regulatory sequence is to identify a BAC clone that contains the sequence of interest. Such clones may already be known and this can be determined by examination of the BAC Table on EchinoBase (from the menus, see "Tools > BAC Table"), by identifying appropriately oriented BAC-ENDs in the genome browser that would encapsulate the desired sequence, or by screening the BAC library. An in-frame fusion of a reporter gene can be engineered into the locus of interest to assess gene expression and modifications of putative regulatory sequences would demonstrate functionality. In the case of *Sp-SoxE*, a recombinant *SoxE::GFP* BAC is already present in the collection, enabling rapid interrogation of the putative regulatory sequence.

7 Notes

7.1 Gene Naming Conventions

Each annotated gene has an official gene ID which takes the form of a three letter species code (e.g., "SPU" or "PMI") followed by an underscore and 6-digit numeric code. Thus examples for *S. purpuratus*, *L. variegatus*, and *P. miniata* would resemble "SPU_000001",

"LVA_000001" and "PMI_000001", respectively. The SPU identifiers will, in general, have three distinct versions. Version one, appended ".1", is the gene model predicted by GLEAN (i.e., CDS). Version two, appended ".2", is the GLEAN model extended to include predicted UTR sequence (i.e., exons). Version three corresponds to models emerging from the RNA-seq transcriptome measurements (i.e., WHL genes) and may have several identified transcript isoforms; thus version three IDs are appended with ".3" along with a subversion letter code (i.e., ".3a", ".3b", etc.).

The official gene name also consists of a species-specific prefix, in this case a two letter code (i.e., "Sp-", "Lv-", or "Pm-"), which precedes the gene name. The gene name is generally 3–5 alphanumeric characters, although additional characters are occasionally required. Gene names begin with an uppercase letter followed by all lowercase letters except in the case where previous usage dictates otherwise (e.g., "FoxA"). The use of punctuation is limited in gene names and is only used to separate adjacent numbers (e.g., "Nkx2-1").

7.2 Data Availability and Download

All datasets hosted on EchinoBase, as well as the entirety of the annotation databases and web code, are freely available to download from the website. There are also historical versions of genome assemblies and gene annotation sets. These facilitate users performing large-scale analyses of the genomic data locally, which allows more control over how data are manipulated and analyzed. To access these data download sites, simply navigate from the menus "Tools > Downloads" and select the species of interest.

Acknowledgments

EchinoBase was developed, in part, with support from the NIH (NIH1P41HD071837).

References

1. Davidson EH, Rast JP, Oliveri P et al (2002) A genomic regulatory network for development. Science 295:1669–1678
2. Peter IS, Davidson EH (2011) A gene regulatory network controlling the embryonic specification of endoderm. Nature 474:635–639
3. Cary GA, Hinman VF (2017) Echinoderm development and evolution in the post-genomic era. Dev Biol 427(2):203–211
4. Sea Urchin Genome Sequencing Consortium, Sodergren E, Weinstock GM et al (2006) The genome of the sea urchin Strongylocentrotus purpuratus. Science 314:941–952
5. Cameron RA, Kudtarkar P, Gordon SM et al (2015) Do echinoderm genomes measure up? Mar Genomics 22:1–9
6. Müller H-M, Kenny EE, Sternberg PW (2004) Textpresso: an ontology-based information retrieval and extraction system for biological literature. PLoS Biol 2:e309
7. Holt C, Yandell M (2011) MAKER2: an annotation pipeline and genome-database management tool for second-generation genome projects. BMC Bioinformatics 12:491
8. Kieras DE, Wood SD, Abotel K, et al. (1995) GLEAN: a computer-based tool for rapid

GOMS model usability evaluation of user interface designs. Proceedings of the 8th annual ACM symposium on user interface and software technology - UIST'95. ACM Press, New York, pp 91–100
9. Tu Q, Cameron RA, Davidson EH (2014) Quantitative developmental transcriptomes of the sea urchin Strongylocentrotus purpuratus. Dev Biol 385:160–167
10. Tu Q, Cameron RA, Worley KC et al (2012) Gene structure in the sea urchin Strongylocentrotus purpuratus based on transcriptome analysis. Genome Res 22:2079–2087
11. Camacho C, Coulouris G, Avagyan V et al (2009) BLAST+: architecture and applications. BMC Bioinformatics 10:421
12. National Center for Biotechnology Information (US) BLAST® Help, https://www.ncbi.nlm.nih.gov/books/NBK1762/
13. Jones P, Binns D, Chang H-Y et al (2014) InterProScan 5: genome-scale protein function classification. Bioinformatics 30:1236–1240
14. Wilson D, Pethica R, Zhou Y et al (2009) SUPERFAMILY–sophisticated comparative genomics, data mining, visualization and phylogeny. Nucleic Acids Res 37:D380–D386
15. Gilbert SF (2000) The early development of sea urchins. In: Developmental biology. Sinauer Associates, Sunderland (MA)
16. Li E, Cui M, Peter IS et al (2014) Encoding regulatory state boundaries in the pregastrular oral ectoderm of the sea urchin embryo. Proc Natl Acad Sci U S A 111:E906–E913
17. Davidson EH, Cameron RA, Ransick A (1998) Specification of cell fate in the sea urchin embryo: summary and some proposed mechanisms. Development 125:3269–3290
18. Howard-Ashby M, Materna SC, Brown CT et al (2006) High regulatory gene use in sea urchin embryogenesis: implications for bilaterian development and evolution. Dev Biol 300:27–34
19. Materna SC, Nam J, Davidson EH (2010) High accuracy, high-resolution prevalence measurement for the majority of locally expressed regulatory genes in early sea urchin development. Gene Expr Patterns 10:177–184
20. Samanta MP, Tongprasit W, Istrail S et al (2006) The transcriptome of the sea urchin embryo. Science 314:960–962
21. Wei Z, Angerer RC, Angerer LM (2006) A database of mRNA expression patterns for the sea urchin embryo. Dev Biol 300:476–484
22. Buels R, Yao E, Diesh CM et al (2016) JBrowse: a dynamic web platform for genome visualization and analysis. Genome Biol 17:66
23. Kapitonov VV, Jurka J (2008) A universal classification of eukaryotic transposable elements implemented in Repbase. Nat Rev Genetics 9:411–412. author reply 414
24. Buenrostro JD, Giresi PG, Zaba LC et al (2013) Transposition of native chromatin for fast and sensitive epigenomic profiling of open chromatin, DNA-binding proteins and nucleosome position. Nat Methods 10:1213–1218
25. Cameron RA, Davidson EH (2009) Flexibility of transcription factor target site position in conserved cis-regulatory modules. Dev Biol 336:122–135

Chapter 13

A Multi-Omics Database for Parasitic Nematodes and Trematodes

John Martin, Rahul Tyagi, Bruce A. Rosa, and Makedonka Mitreva

Abstract

Helminth.net (www.helminth.net) is a web-based resource that was launched in 2000 as simply "Nematode.net" to host and investigate gene sequences from nematode genomes. Over the years it has evolved to become the moniker for a collection of databases: Nematode.net and Trematode.net. These databases host information for 73 nematode (roundworms) and 17 trematode (flatworms) species and serve as backbone for a number of tools that allow users to query slices of the data for multifactorial combinations of species-omics properties. Recent focus has been on inclusion of gene and protein expression data, population genomics and cross-kingdom interactions (metagenomics datasets). This chapter describes the website, the available tools and some of the new features.

Key words Nematodes, Trematodes, Database, Search, Genome browser, BLAST, Functional annotation, Transcriptome, Proteome, Metagenome

1 Introduction

The website Nematode.net was established in 2000 as an outgrowth of the parasitic nematode transcriptomic project at Washington University's Genome Institute (WUGI). Over the next decade it developed into a specialty repository that made accessible the rapidly expanding nucleotide sequence data and related resources from species across the phylum Nematoda. Seventeen years later, it is a moniker for a collection of databases, Helminth.net, for the two main metazoan parasitic phyla Nematoda (roundworms) and Platyhelminthes (flatworms). Updates on development and improvements have been provided in 2004 [1], 2008 [2], 2012 [3], and 2015 [4] communicating to the community information needed to facilitate dissemination of diverse datasets in a useful and usage-friendly manner. Bioinformatics tools employed to access and/or analyze the omics datasets have also been developed (NemaPath [5], HelmCoP [6], modDFS [7], and TL-microbiome [8]) and are discussed in the appropriate sections

Martin Kollmar (ed.), *Eukaryotic Genomic Databases: Methods and Protocols*, Methods in Molecular Biology, vol. 1757,
https://doi.org/10.1007/978-1-4939-7737-6_13, © Springer Science+Business Media, LLC, part of Springer Nature 2018

below. A very detailed bioinformatics training protocol we have developed as part of our "Bioinformatics Workshop for Helminth Genomics" that was held at the Washington University's Genome Institute in 2015 and an "Introduction to Nematodes" lecture in six languages, are both provided in the left drop-down menu under "Education." The drop-down menu on the left also provides access to species hub pages that summarize the data available for each species. The Sitemap (Figs. 1 and 2) includes a summary of all

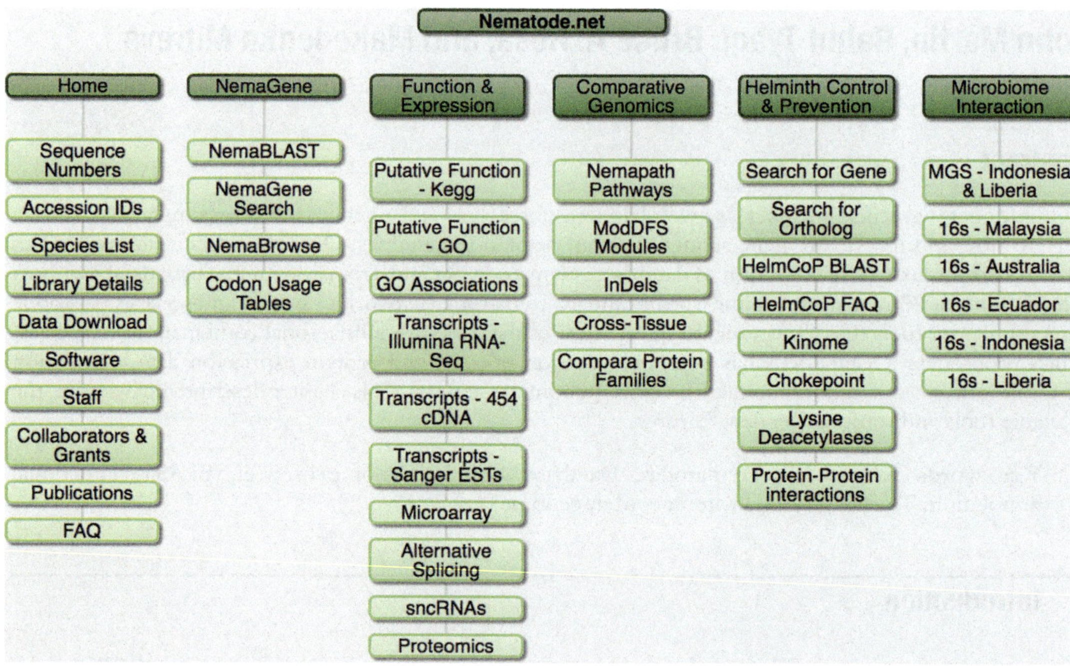

Fig. 1 Sitemap of Nematode.net. Shown are the tabs of the main navigation bar on the main page and the subcategories revealed by clicking on these tabs

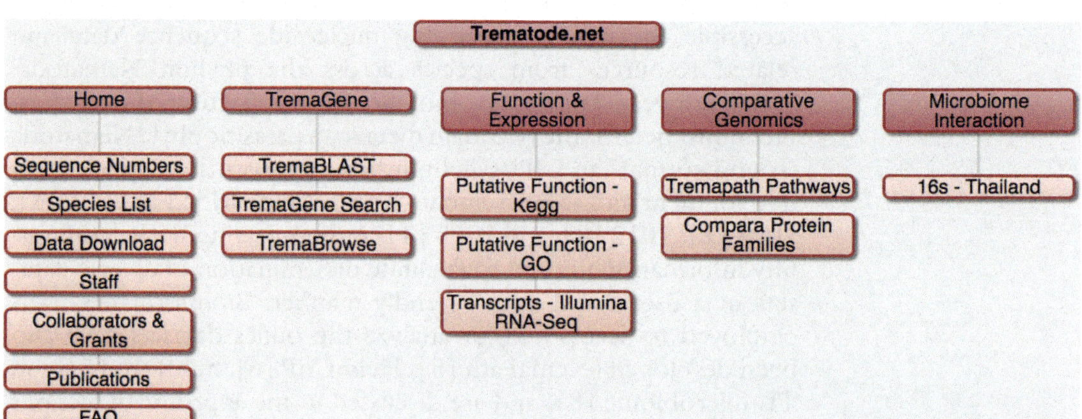

Fig. 2 Sitemap of Trematode.net. Shown are the tabs of the main navigation bar on the main page and the subcategories revealed by clicking on these tabs

main pages that can be accessed from the grey bar on the top for each of the two databases. This chapter provides detailed methods that outline the steps involved in navigating the website and in retrieving information from the database.

2 Materials

The following is needed to access, navigate, and extract information from the website:

- Device: Computer (desktop or laptop), tablet or smartphone.
- Internet access.
- Internet browser (designed to work on Firefox, Safari, Internet Explorer, and Chrome).

3 Methods

3.1 NemaGene and TremaGene

NemaGene (http://nematode.net/NemaGene.html) and TremaGene (http://trematode.net/TremaGene.html) are collections of transcript assembly contigs and genes produced and annotated at the McDonnell Genome Institute (MGI) or published by other researchers. Functional annotations are assigned by sequence similarity searches using InterProScan (software version 4.8, InterPro database release 32.0) [9, 10] and WU-BLAST 2.0 [11] and include annotation with InterPro (IPR) IDs, Gene Ontology (GO) terms [12], and KEGG Orthology IDs (KO) [13]. NemaGene currently hosts 1,456,372 entries spanning 73 species and TremaGene holds 253,360 entries spanning 12 species (7 are in progress and will become available in the near future: *Fasciola gigantica*, *Fasciolopsis buski*, *Opisthorchis viverrini*, *Paragonimus westermani*, *P. kellicotti*, *P. miyazaki*, and *P. heterotremus*).

Access to NemaGene/TremaGene frequently comes from other tools within the Helminth.net sites such as the contig links from NemaPath/TremaPath [5] which directly jump to the details pages that are the end point of a NemaGene/TremaGene search, or from external sites. But the NemaGene/TremaGene Search tool can also be used to extract custom slices of our database using available annotations as filters. This tool is also very useful for retrieving the full protein or nucleotide fasta of our genesets, or of a user-defined slice.

NemaGene can be searched using InterPro, GO, and/or KO ID filters. NemaGene is accessed via the link provided above, or via NemaGene Search available from the NemaGene menu. First click on the [+] Expand label for the Species selection section and select 1 or more species to start your query. After selecting the species of interest, expand the sections below to set specific filters you would like to apply. You are able to request a specific gene name (or comma-delimited list of gene names), orthologous protein families, InterPro ID, GO term, and/or KO ID. Comma-delimited lists of any of those IDs are also allowed as input. Filtering on multiple IDs of a single type will return genes/transcripts annotated by any of those IDs (i.e., a union set), but when setting filters using 2 or more ID types (i.e., InterPro ID + KO ID) each gene or transcript returned will be required to have at least one ID from each of the supplied lists (Fig. 3).

This will then lead to a page showing the slice of resulting filtered data retrieved from NemaGene (Fig. 4A). The Query Definition section now displays the query that was made to extract the results shown. Below this, the Data Download section

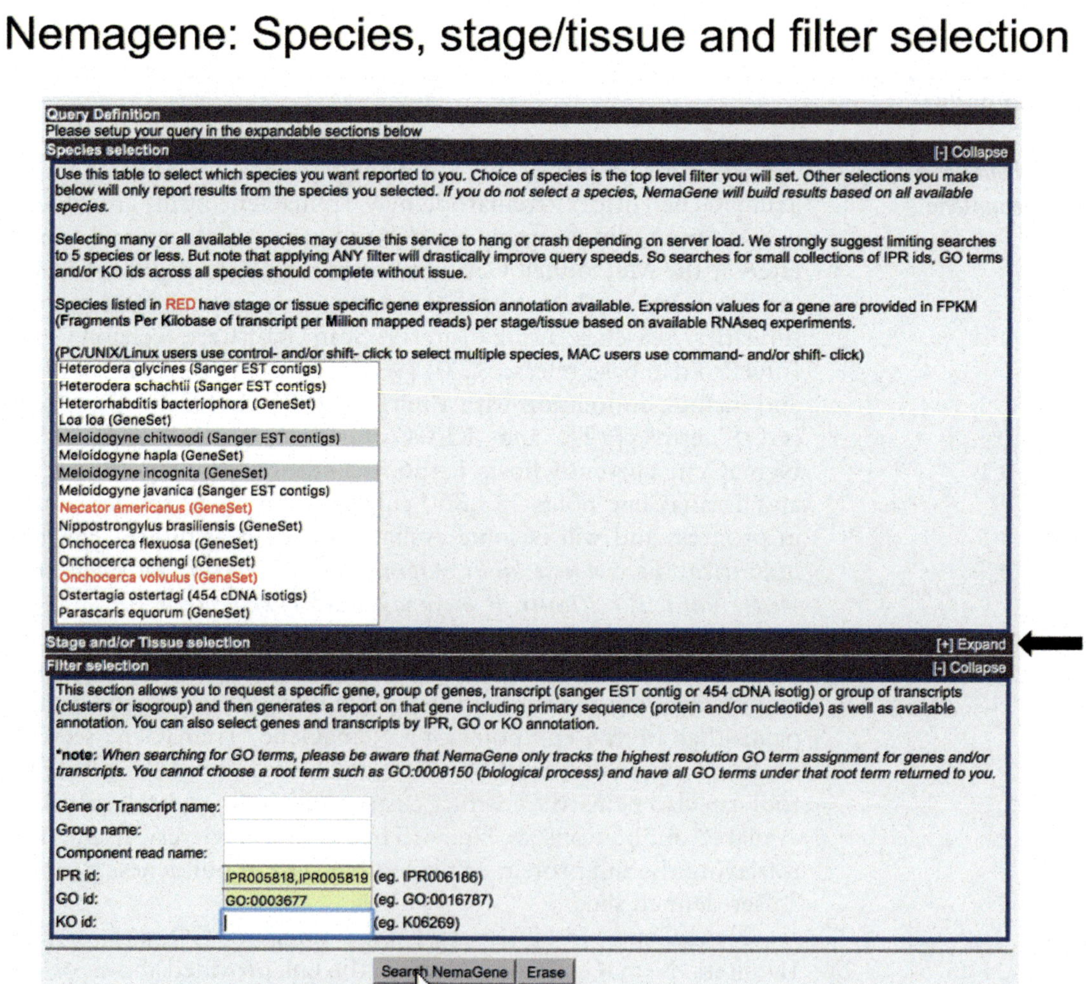

Fig. 3 Input selection for NemaGene. On the main NemaGene page, the user can select one of more species, stage, and/or tissue, and different combinations of filters. Clicking on "Search NemaGene" button (shown by a white arrow pointer) submits the input to NemaGene

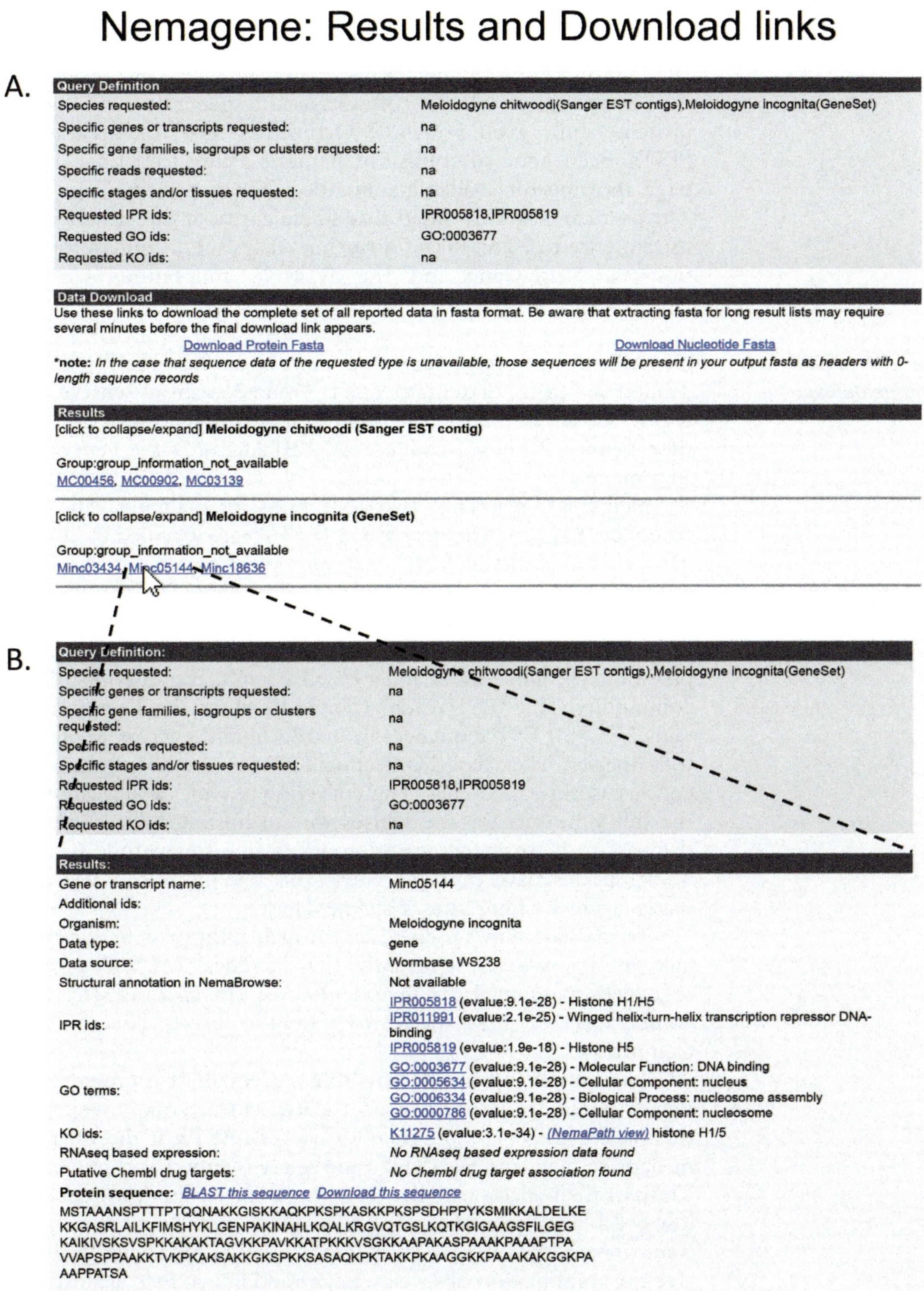

Nemagene: Results and Download links

A.

Query Definition

Species requested:	Meloidogyne chitwoodi(Sanger EST contigs),Meloidogyne incognita(GeneSet)
Specific genes or transcripts requested:	na
Specific gene families, isogroups or clusters requested:	na
Specific reads requested:	na
Specific stages and/or tissues requested:	na
Requested IPR ids:	IPR005818,IPR005819
Requested GO ids:	GO:0003677
Requested KO ids:	na

Data Download

Use these links to download the complete set of all reported data in fasta format. Be aware that extracting fasta for long result lists may require several minutes before the final download link appears.

Download Protein Fasta Download Nucleotide Fasta

***note:** *In the case that sequence data of the requested type is unavailable, those sequences will be present in your output fasta as headers with 0-length sequence records*

Results

[click to collapse/expand] **Meloidogyne chitwoodi (Sanger EST contig)**

Group:group_information_not_available
MC00456, MC00902, MC03139

[click to collapse/expand] **Meloidogyne incognita (GeneSet)**

Group:group_information_not_available
Minc03434, Minc05144, Minc18636

B.

Query Definition:

Species requested:	Meloidogyne chitwoodi(Sanger EST contigs),Meloidogyne incognita(GeneSet)
Specific genes or transcripts requested:	na
Specific gene families, isogroups or clusters requested:	na
Specific reads requested:	na
Specific stages and/or tissues requested:	na
Requested IPR ids:	IPR005818,IPR005819
Requested GO ids:	GO:0003677
Requested KO ids:	na

Results:

Gene or transcript name:	Minc05144
Additional ids:	
Organism:	Meloidogyne incognita
Data type:	gene
Data source:	Wormbase WS238
Structural annotation in NemaBrowse:	Not available
IPR ids:	IPR005818 (evalue:9.1e-28) - Histone H1/H5 IPR011991 (evalue:2.1e-25) - Winged helix-turn-helix transcription repressor DNA-binding IPR005819 (evalue:1.9e-18) - Histone H5
GO terms:	GO:0003677 (evalue:9.1e-28) - Molecular Function: DNA binding GO:0005634 (evalue:9.1e-28) - Cellular Component: nucleus GO:0006334 (evalue:9.1e-28) - Biological Process: nucleosome assembly GO:0000786 (evalue:9.1e-28) - Cellular Component: nucleosome
KO ids:	K11275 (evalue:3.1e-34) - *(NemaPath view)* histone H1/5
RNAseq based expression:	*No RNAseq based expression data found*
Putative Chembl drug targets:	*No Chembl drug target association found*

Protein sequence: *BLAST this sequence* *Download this sequence*

MSTAAANSPTTTPTQQNAKKGISKKAQKPKSPKASKKPKSPSDHPPYKSMIKKALDELKE
KKGASRLAILKFIMSHYKLGENPAKINAHLKQALKRGVQTGSLKQTKGIGAAGSFILGEG
KAIKIVSKSVSPKKAKAKTAGVKKPAVKKATPKKKVSGKKAAPAKASPAAAKPAAAPTPA
VVAPSPPAAKKTVKPKAKSAKKGKSPKKSASAQKPKTAKKPKAAGGKKPAAAKAKGGKPA
AAPPATSA

Fig. 4 NemaGene results. (A) The first result page on submitting input (Fig. 3) shows the details of the submitted query, the download links in both protein and nucleotide sequences of the results, and links to detail pages for each of the resulting genes(s). (B) The detail page of the results showing sequence and annotation for the selected gene

provides links to download the full fasta sequence file for all the genes/transcripts that were requested. The Results section will list all the resulting genes and/or transcripts organized by species and then by group if available (software used to generate orthologous protein family groups include OrthoMCL [14] or InParanoid [15]). Each gene or transcript name is a link to a final detail page showing the available annotations for that entity (Fig. 4B). The user can also download that single entity or directly forward its sequence to NemaBlast for further analysis. For more information on NemaGene and the available annotations see the NemaGene FAQ.

3.2 NemaBlast and TremaBlast

NemaBlast (http://nematode.net/NemaBlast.html) and TremaBlast (http://trematode.net/TremaBlast.html) enable visitors to search for a sequence of interest against a custom database they define. Both services use WU-BLAST 2.0 for generating alignments.

NemaBlast maintains two collections for mapping. The first comprises Expressed Sequence Tag (EST) reads grouped by library. This set represents all EST reads produced for species we have sequenced, grouped by species and sequencing library, allowing the user to mix and match in the creation of their search space. Assembled EST contigs are not hosted by NCBI therefore we are providing the annotated Sanger based assembled transcripts to the community. The "EST reads grouped by library" collection contains 275,850 EST sequences from 132 libraries across 31 nematode species. The second collection contains assembled transcript contigs, isotigs, and genes. This collection provides a view closer to the full gene sets for the species we are hosting. The complete database to blast search against contains all 73 nematode species. Some species have multiple entries because users can select to search against a transcript or a gene dataset.

TremaBlast allows users to search against the protein sets available in TremaGene. Currently this includes 221,003 protein sequences spanning 12 trematode species. The user can select any combination of these species to form the search space for this alignment.

An example illustrating how to use NemaBlast is presented in Figs. 5 and 6. To use NemaBlast, the user enters a query sequence, then selects the blast program to use—BLASTX if the query is nucleotide data, and BLASTP if the query is amino acid sequence. The user then selects whether they would like to filter the query for low-complexity sequence using SEG [16] and also whether they want to mask the query using RepeatMasker [17]. Then they indicate the combination of species they would like to map against and click the "BLAST Search" button. Search results in the form of standard WU-BLAST 2.0 text output will be emailed to the address provided in the form.

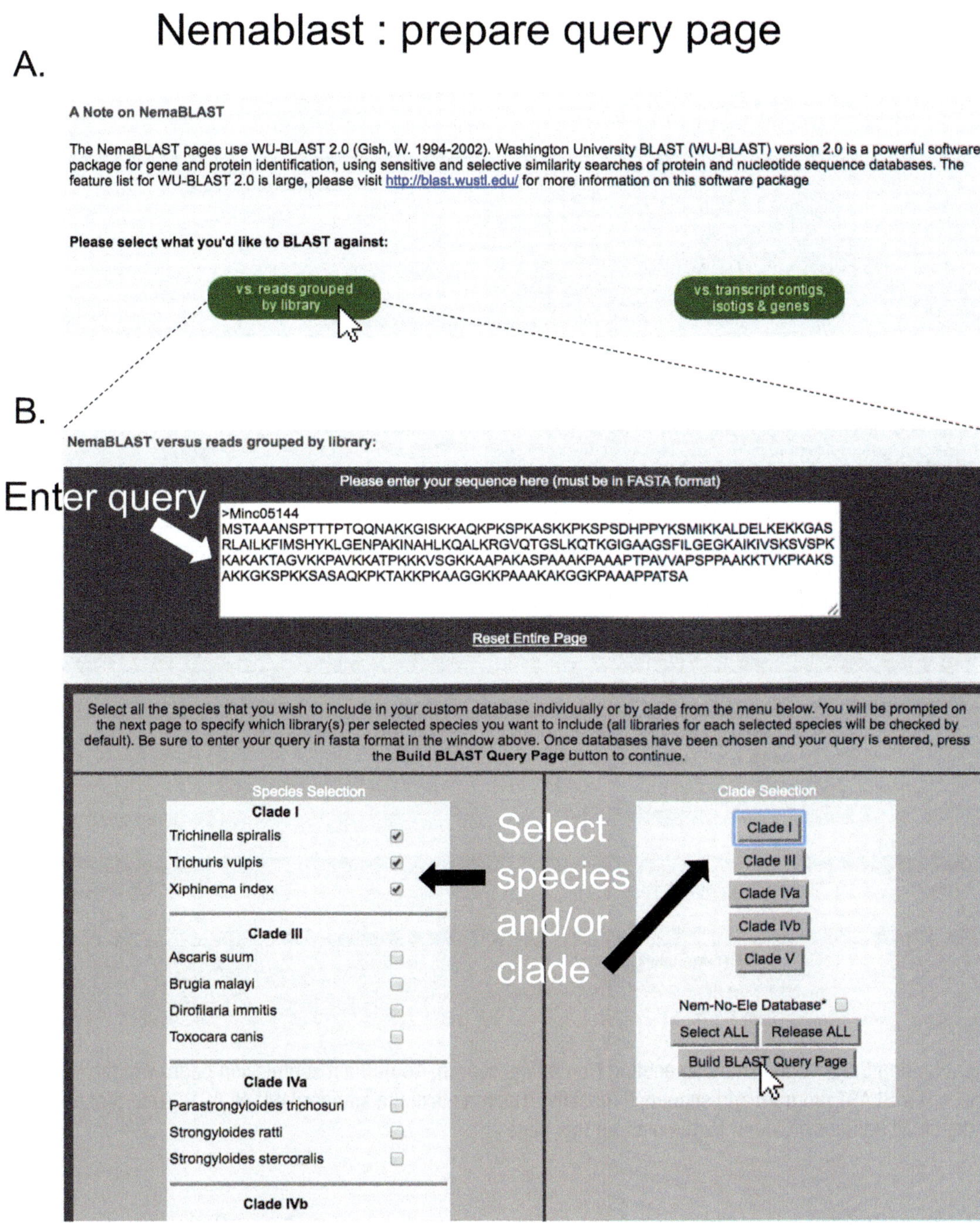

Fig. 5 Building NemaBlast query page. (A) On the main NemaBlast page, the user indicates whether they want to search the reads database or the transcript database. The illustrated example shows a search in the reads database (indicated by a white arrow pointer on the corresponding button). This leads to an intermediate page (B) where the user enters the query sequence and selects the species/clades of interest that will form the search database

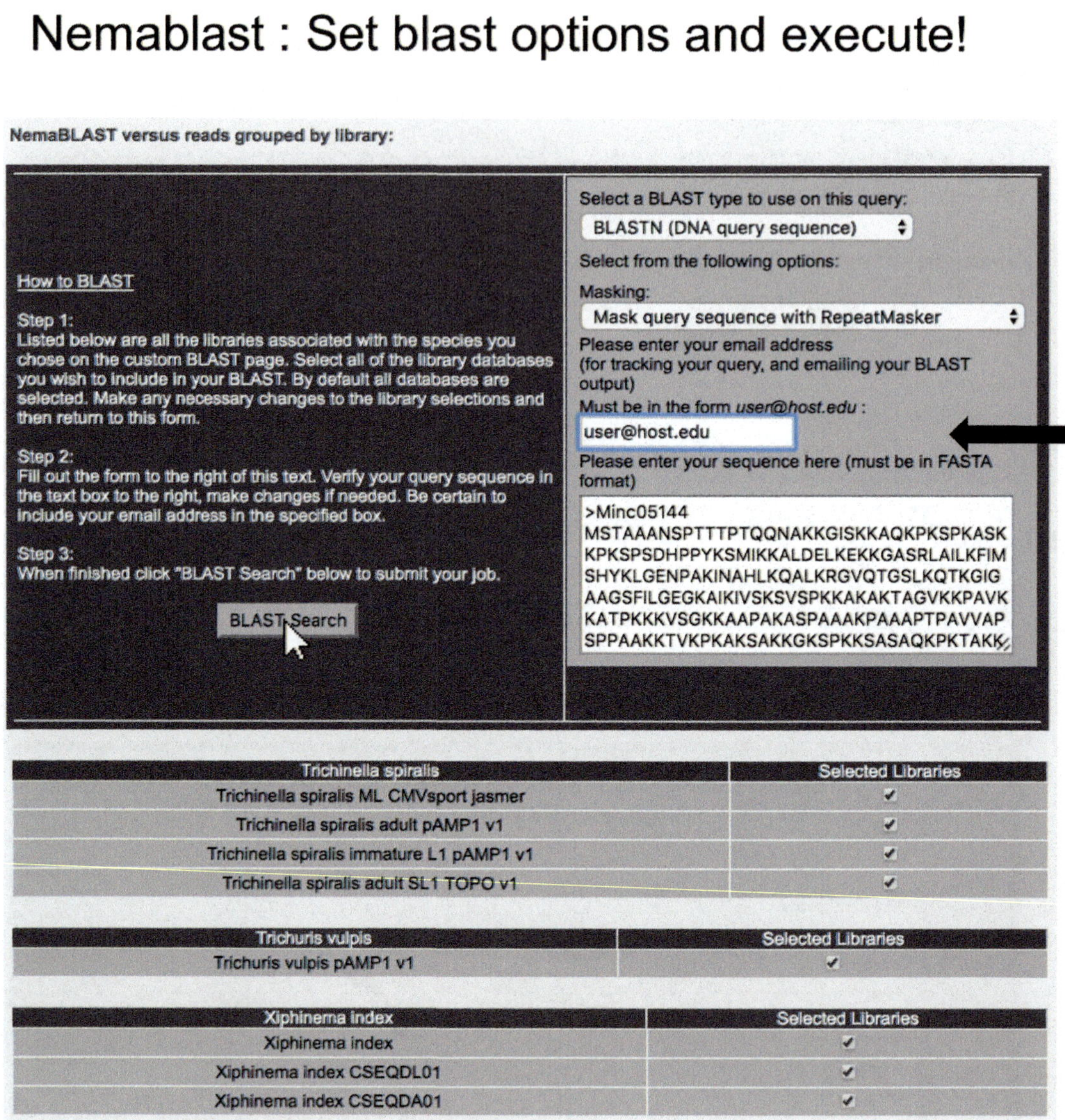

Fig. 6 Setting BLAST options and executing NemaBlast search. NemaBlast submission page with the relevant choices for BLAST program and sequence masking. The results in the standard WU-BLAST 2.0 format are sent to the email address provided by the user on this page

3.3 NemaBrowse and TremaBrowse

NemaBrowse (http://nematode.net/NemaBrowse.html) and TremaBrowse (http://trematode.net/TremaBrowse.hmtl) use GMOD's GBrowse [18] to offer a visualization of gene annotations and variants mapped on top of genomic assemblies. These provide a view of sometimes in-progress nematode and trematode genomes and, where variant calls on specific lab and field isolates are available, offer a useful comparative view.

Visualized annotations typically include Maker protein coding gene and RNA gene predictions [19, 20], tRNAs predicted by tRNAscan [21] and Single Nucleotide Polymorphism (SNP) loci predicted using the Genome Analysis ToolKit (GATK) [22] and annotated using SnpEff [23]. Exceptions to the typical annotation tools are noted on the entry portals. NemaBrowse currently hosts annotations for ten species and TremaBrowse hosts one species. Our recent focus has been to provide tracks for SNPs and SNP annotations in the browser. At present there are genetic variants called from 27 strains of the river blindness agent *Onchocerca volvulus* [24], 9 strains for the lungworm *Dictyocaulus viviparus* [25], a susceptible and a resistant strain of *Trichostongylus circumcincta* (unpublished), and 7 strains of *Fasciola hepatica* [26].

At the entry portal page into NemaBrowse, the user selects their species of interest and clicks the "Gene list" link (Fig. 7A). This leads to the gene list interface, which provides links directly to all the annotated gene features for that organism (Fig. 7B). Minimal annotation is made available in the gene list using either final gene product information as annotated by the BER pipeline, or information derived from WU-BLAST 2.0 mappings to the NCBI nonredundant (NR) database using a postalignment cutoff of 35 bits + 55% identity. Clicking on a Gene annotation link will take the user to the GBrowse view (Fig. 8). Once inside the GBrowse view the user can directly navigate along the reference and make use of the standard functions of the GBrowse environment.

Depending on the datasets available, the GBrowse tracks can include plottable information more than the basic tracks such as gene annotation, GC percent, and 6-frame translation. For the species with available variant information, these include tracks for SNP loci that can be selected for plotting. These loci are marked and identified according to their SNPEff annotation: noncoding SNPs (i.e., intronic or intergenic), synonymous coding SNPs and nonsynonymous coding SNPs. For any genes of interest, users can navigate the GBrowse for such information and can also provide their dataset of interest to be included as additional tracks.

3.4 Functional Annotations and Related Tools

3.4.1 NemaPath and TremaPath

NemaPath/TremaPath is a tool for visualizing the presence, absence, and overall coverage of enzymatic pathways in species based on the KO annotations of genes [13]. In addition to the single organism view NemaPath/TremaPath supports comparative views between pairs of species. This allows users to visually see and explore enzymatic pathway differences between these entities based on actual transcriptomic data. This tool can assist users in various ways, from identifying potential drug targets to helping understand the differences between species utilizing different survival strategies. NemaPath has 1,103,786 annotated genes and transcripts spanning 63 species, while TremaPath is populated by 204,647 proteins spanning 11 trematodes.

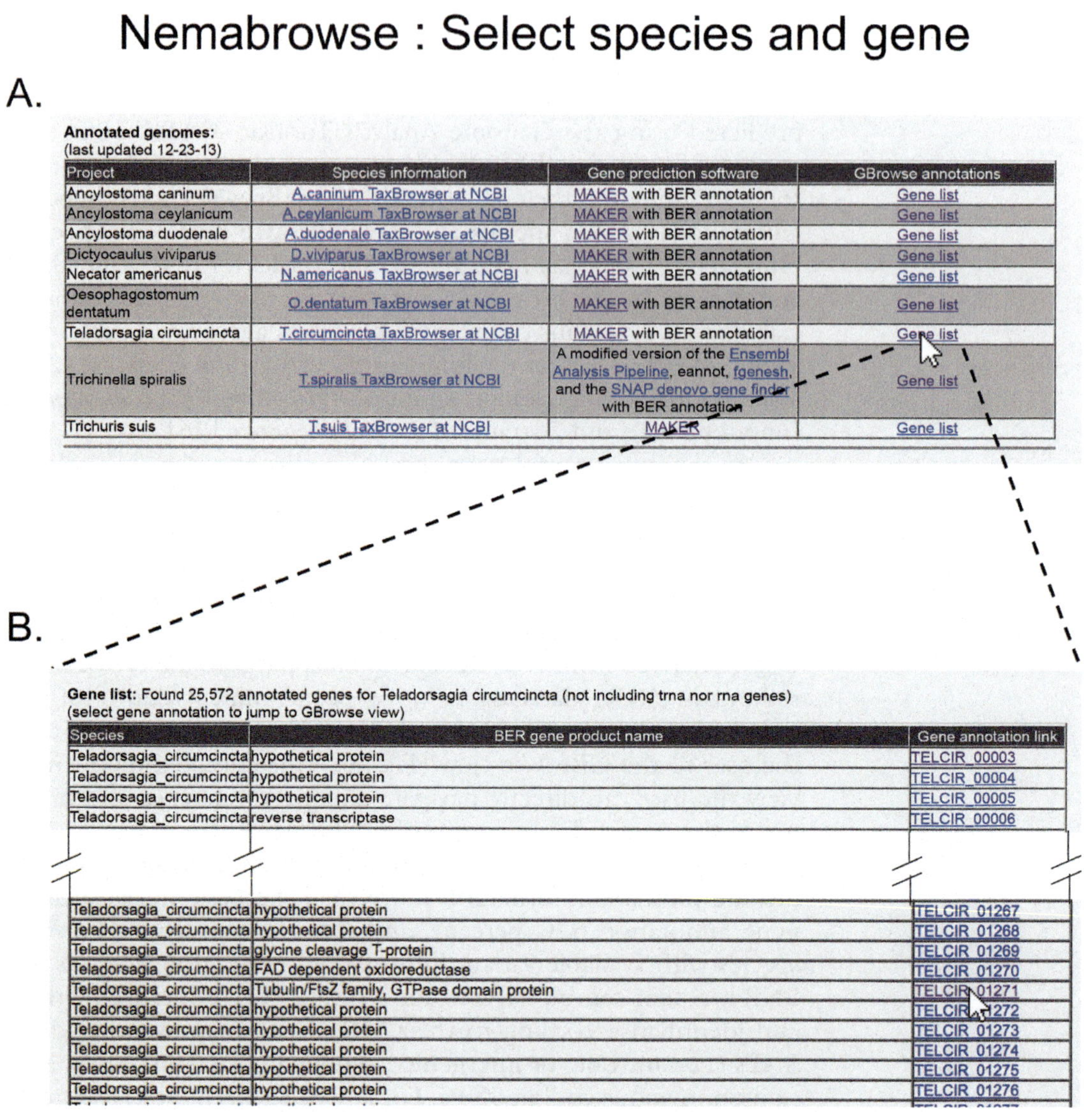

Fig. 7 Selecting NemaBrowse input. (A) On the main NemaBrowse page, the user can select one of the species for which NemaBrowse view is available. Clicking on "Gene list" link (shown by a white arrow pointer) takes the user to a page (B) that lists all the genes in the database for the species of interest

KO annotations are assigned to genes using WU-BLAST 2.0 alignments against the KEGG genes database (release 68.0). The KO IDs of the subjects hit are used to relate helminth sequences to the KEGG pathways. Enzymatic nodes of the KEGG pathway maps are painted to indicate the number of supporting genes found from each species.

Users first select a species (Fig. 9A) and are then provided a graphical distribution of the number of KO hits with varying *E*-value confidence scores for their chosen species (Fig. 9B). The user must then set an alignment strength threshold to assign KOs

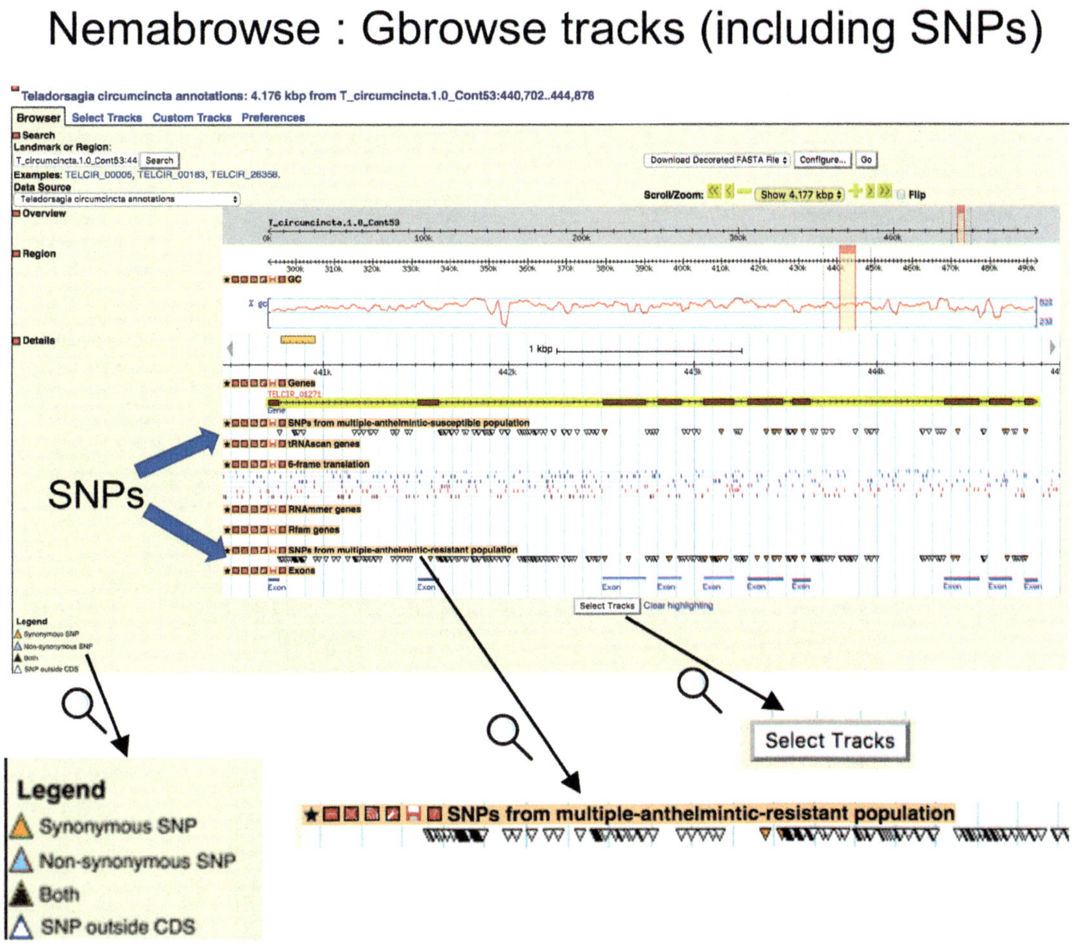

Fig. 8 GBrowse tracks in NemaBrowse. The GBrowse view centered on the gene selected as NemaBrowse input (Fig. 7) with some default tracks plotted. The user can select any of the available tracks using the "Select Tracks" button at the bottom. As an example, the two SNP tracks available for *T. circumcincta* are shown, with the SNP loci indicated by colored triangles, colored according to SNPEff annotation of the SNP. Zoomed in images of some parts are shown at the bottom for clarity

to genes. Only homologies whose alignment strength meets that cutoff are considered when populating the view. Choosing a less stringent score will be more sensitive, but can introduce false positive mappings. After confirming their choices (Fig. 9C) users are then presented with a menu of pathways supported by NemaPath/TremaPath (Fig. 10A). After pathway selection, a graphic displaying the compounds and reactions of that pathway for their species of choice is shown, with populated enzymes colored green, and darker shading indicating multiple genes annotated (Fig. 10B). This figure also includes more details about the enzymes that show up as mouse-over menu for each enzyme (Fig. 10C). The user can then optionally choose a second species for comparison using a drop-down menu near the top right of the page (Fig. 11A), map-

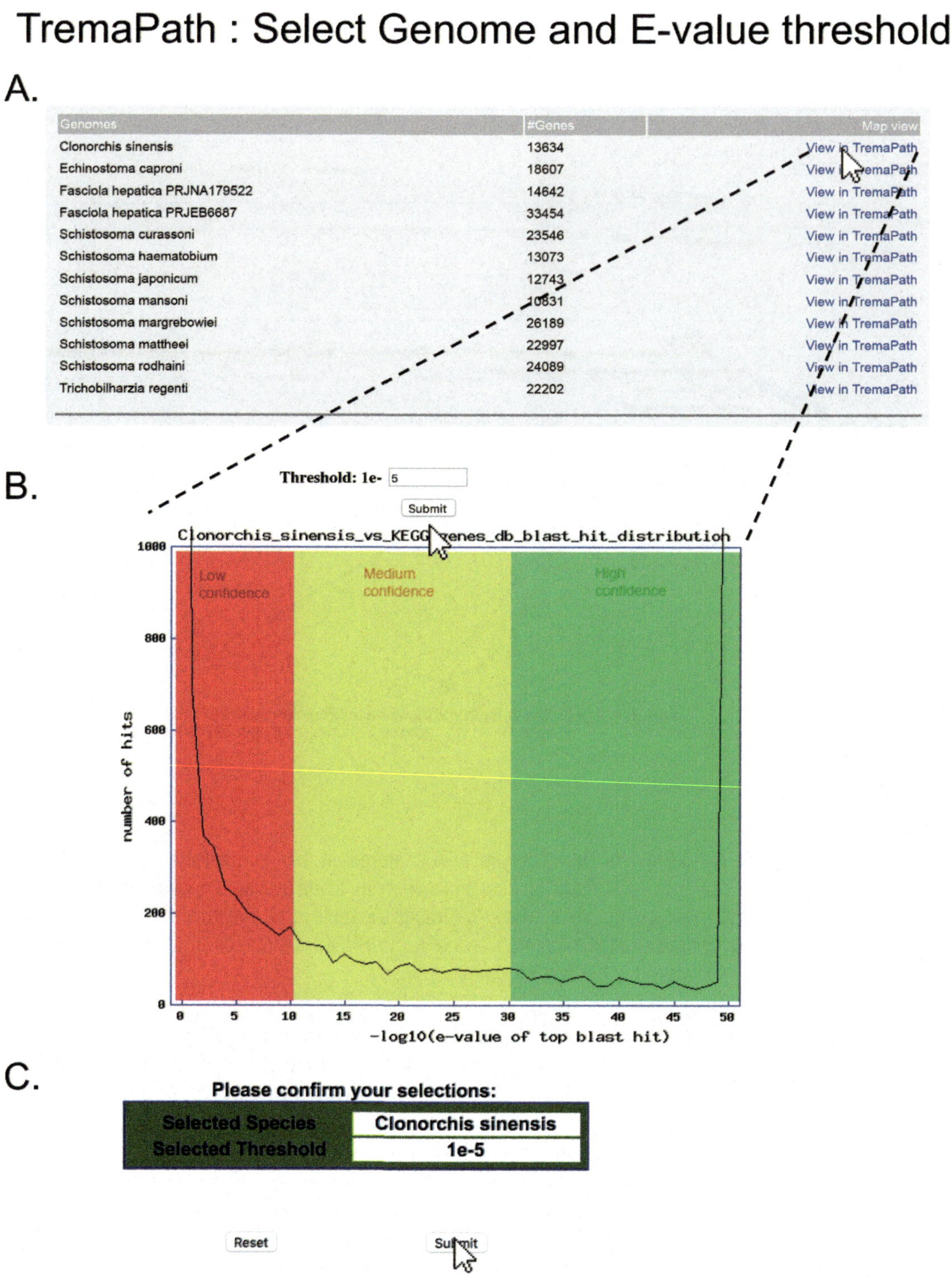

Fig. 9 Selecting genome and annotation threshold for TremaPath usage. On the main TremaPath page, clicking on the "Species-specific TremaPath comparisons" link takes the user to a page (A) where the user can select the species and assembly of interest. Clicking on "View in TremaPath" link leads to a page (B) that shows the distribution of KOs assigned at different *E*-value thresholds, with colors distinguishing low, medium and high confidence annotation. The user can indicate the *E*-value threshold they want by filling in the exponent and clicking the Submit button and confirming their input at the next page (C)

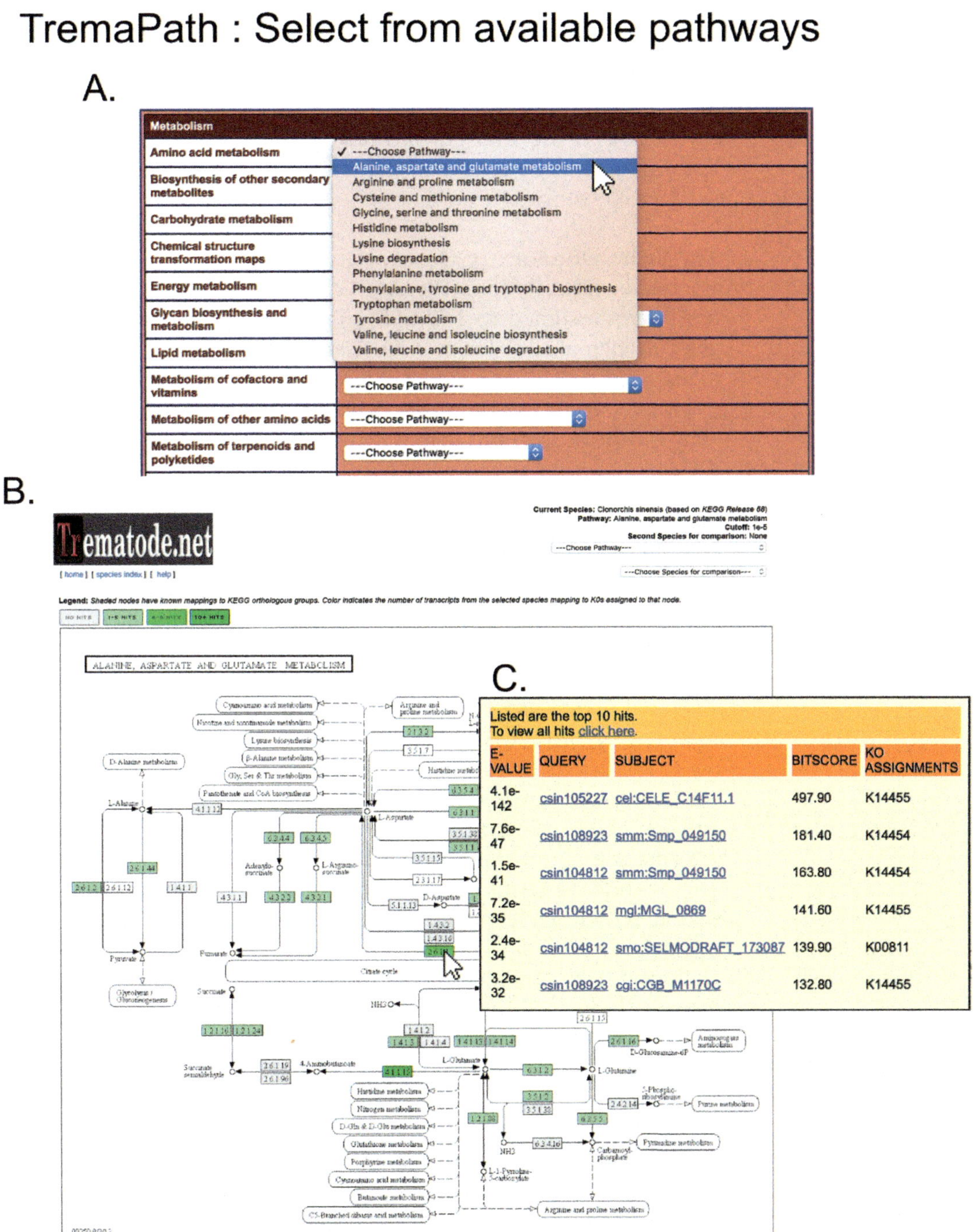

Fig. 10 Visualizing a metabolic pathway for a single species. (A) After confirming the species and annotation threshold (Fig. 9), the user selects the pathway of interest by drop-down menus grouped under major categories of KEGG pathways. (B) The resulting page shows KEGG pathways with the enzyme nodes shown in shades of green, representing the number of genes with the corresponding annotation. The user can mouse-over on any of these nodes to peek at the details of such genes and BLAST hits (C)

ping genes onto the same pathway and highlighting differences in pathway usage between the species (Fig. 11B). Information about the genes mapping to each node are available on mouse-over including the KEGG target(s) that sponsored the assignment to the node and the strength of the alignment(s). The query names link into NemaGene/TremaGene and the subject names link to the KEGG website.

3.4.2 AmiGO

The Gene Ontology (GO) association page (nematode.net/GO_associations.html) hosts the AmiGO tool [27] for viewing gene ontologies assigned to 172,505 NemaGene genes and transcripts from 31 nematode species. GO classifications, assigned by homologies detected using InterProScan, are loaded into the AmiGO software's backend database. AmiGO then provides a graphical view in which users can search NemaGene transcripts by ontological class.

First, the user selects the organism they want to explore. This will present an AmiGO overview showing the number of genes and transcripts annotated under each GO category. Categories expand to display child categories along with their assignment counts. The pie images next to each GO category are clickable and expand into views showing counts in all child categories beneath the level at which the user clicked. Genes and transcripts of the selected species assigned a specific GO term of interest can be explored by entering the GO ID into the "Search GO" field, leaving the "Terms" box checked, and clicking. This leads to a detailed view corresponding to the GO term. This view lists the transcripts and/or genes assigned to the current term and provides more information about the term itself. Note that if the user clicks on one of the GO terms on the original AmiGO overview, it would also directly lead them to this detailed view.

3.4.3 Transcriptomics Data

Nematode.net provides access to a large collection of transcript data including Illumina RNA-Seq (Table 1; nematode.net/IlluminaTranscripts.html), cDNA transcript assemblies (Table 2; nematode.net/cDNA454.html), and Sanger ESTs clusters (Table 3; nematode.net/SangerESTs.html). Trematode.net currently provides access to transcript data (Table 1; trematode.net/IlluminaTranscripts.html). All transcript expression data originate from either whole organism, developmental stage, gender, or tissue-specific RNA populations. Normalized gene expression values, for a subset of species, are also included on the gene pages.

The Illumina RNA-Seq table provides links to the experiment IDs (SRX ids) organized by species and annotated with stage, tissue, and sequencing platform information. The links point to the corresponding record in NCBI's Sequence Read Archive (SRA) [28]. This same layout is used in both the Nematode.net and Trematode.net Illumina transcript data tables.

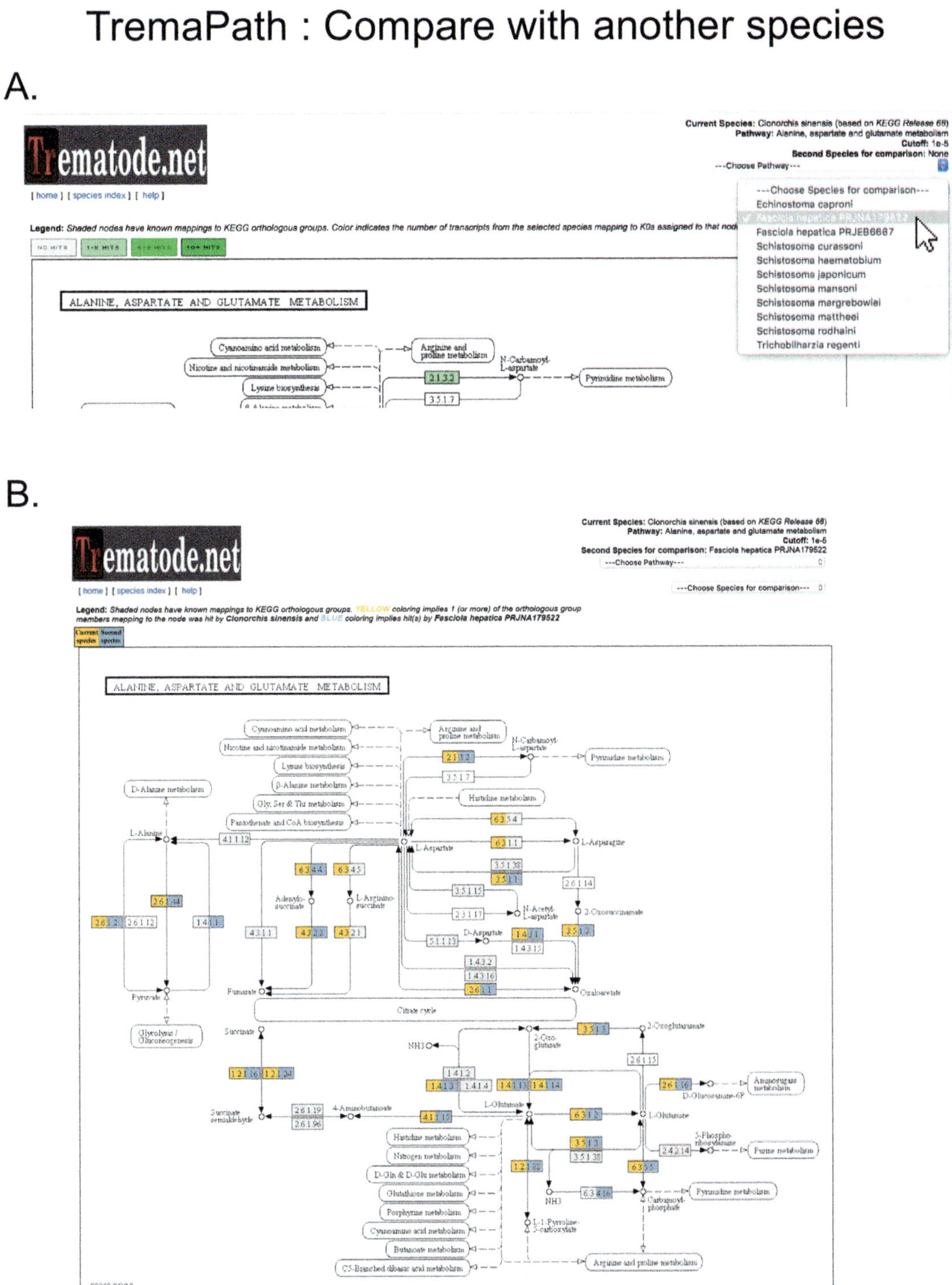

Fig. 11 Comparing a metabolic pathway for two species. (A) After visualizing the pathway of interest for one species (Fig. 10) in TremaPath, a second species can be selected on a drop-down menu on top right (selection shown with a white arrow). This leads to a pathway view (B) with the two species painted in different colors, juxtaposing the presence of enzymes from that pathway in the two species

Table 1
Available Illumina RNAseq reads

Nematode species	# RNAseq reads	Stages/tissues	Accession ids
Ancylostoma caninum	401157883	L2, L3 (nonactivated), L3 (untreated), female, male, oesophagus, gut	SRX1971542, SRX1971543, SRX1971544, SRX1971545, SRX1971546, SRX1971547, SRX1971548
Ancylostoma ceylanicum	1410548651	L3 (nonactivated), L3 (activated), L3 (infective), 48 h L3, 72 h L3, 72 h L4, female, male, L4 8 day female, 24 h small intestine, 24 h stomach, gut	SRX1116899, SRX1116900, SRX1116901, SRX1116902, SRX1116903, SRX1116904, SRX1116905, SRX1116906, SRX1116907, SRX1116908, SRX1116909, SRX1116910, SRX1116911, SRX1116912, SRX1116913, SRX1116915, SRX1116916, SRX1116917, SRX1116918, SRX1116919, SRX1116920, SRX1116921, SRX1116922, SRX1116923, SRX1127457
Ascaris suum	2259230956	Female head, male head, female pharynx, male pharynx, female intestine, male intestine, anterior intestine, mid intestine, posterior intestine, repro-associated unattached intestine, whole intestine, 24 h anterior intestine (treatment: hsiRNA2), 24 h anterior intestine (treatment: hsiRNA5), 24 h posterior intestine (treatment: hsiRNA2), 24 h posterior intestine (treatment: hsiRNA5), ovary, uterus, seminal vesicle, testis, Whole worm	SRX1013923, SRX1013925, SRX1013926, SRX1013928, SRX1013929, SRX1013930, SRX1013931, SRX1013932, SRX1013933, SRX1013934, SRX1013935, SRX1013936, SRX1013937, SRX1013938, SRX1013939, SRX1013940, SRX1013941, SRX1013942, SRX1013943, SRX1013944, SRX1013945, SRX1013946, SRX1013948, SRX1013949, SRX1013950, SRX1013951, SRX1013953, SRX1013954, SRX1013956, SRX1013957, SRX157781, SRX278110, SRX278111, SRX278113, SRX278114, SRX278115, SRX278116, SRX278117, SRX278118, SRX278119, SRX278120, SRX278121, SRX278122, SRX278123, SRX278124, SRX278125, SRX278126, SRX278127, SRX278128, SRX278129, SRX278130, SRX278131, SRX278133, SRX278134, SRX278135, SRX278136, SRX278137, SRX278138, SRX278139, SRX278140,SRX278141, SRX278142, SRX278143, SRX278144, SRX278151, SRX278152, SRX278153, SRX278154, SRX278155, SRX278156, SRX278157, SRX278158, SRX278159, SRX278160, SRX278161, SRX278162, SRX278163, SRX278164, SRX278165, SRX278166

(continued)

Table 1
(continued)

Nematode species	# RNAseq reads	Stages/tissues	Accession ids
Dictyocaulus viviparus	821,515,897	L1, L2, L3, L4, L5 (mixed), L5 (female), L5 (male), female, male, egg, hypobiotic larvae	SRX371002, SRX693266, SRX693267, SRX371003, SRX693295, SRX868541, SRX371004, SRX371005, SRX371006, SRX371007, SRX371008, SRX693298, SRX693301, SRX371010, SRX371011, SRX371012, SRX693296, SRX693299, SRX693297, SRX371009, SRX693300, SRX371413, SRX693302, SRX693304, SRX693303
Haemonchus contortus	184431397	Female intestine, male intestine	SRX736496, SRX736495
Necator americanus	37157105	L3, adult	SRX202018, SRX202022
Oesophago-stomum dentatum	654919328	L2, L3, L4, female, male, total RNA (drug-resistant population), total RNA (drug-susceptible population)	SRX2085121, SRX2085118, SRX2085106, SRX2085107, SRX2085122, SRX2085126, SRX2085123, SRX2085109, SRX2085108, SRX2085110, SRX2085112, SRX2085113, SRX2085125, SRX2085127, SRX2085111, SRX2085124, SRX2085119, SRX2085115, SRX2085116, SRX2085114, SRX2085117, SRX2085120
Teladorsagia circumcincta	226574759	Total RNA (anthelmintic resistant), total RNA(anthelmintic susceptible)	SRX1507696, SRX1507697, SRX1507698
Trichuris suis	693989167	10 day larvae, 16 day larvae, 17 day larvae, 21 day larvae, 28 day larvae, 5 day larvae, 42 day larvae, female, male, adult (mixed), female anterior body, male anterior body, ovary, testis	SRX1838979, SRX1838981, SRX1838978, SRX1838982, SRX1838974, SRX736471, SRX736470, SRX1838977, SRX1838980, SRX1838975, SRX1838976, SRX1838983, SRX1838985, SRX1838984, SRX1838986
Trematode species			
Fasciola hepatica	415324163	Egg, metacercariae, adult	SRX1037419, SRX1037418, SRX1037417, SRX1037421, SRX1037416, SRX1037420, SRX1798471, SRX1798472, SRX1037423, SRX1037422, SRX1798475, SRX1798474, SRX1798473
Paragonimus skrjabini	49816749	Adult	SRX1507709
Paragonimus westermani	46468226	Adult	SRX1507710

cDNA transcript assemblies are listed in an expandable table with a section for each of the nine nematode species providing various information. Expanding any of the rows provides information on the numbers and platform types of reads used in each assembly. Also shown are the numbers of isotigs (putative transcripts), isogroups (isotig group putatively representing all the expressed isoforms for a gene locus), and numbers of reads per stage if that information is available. For each assembly download links are provided for:

Isotig nucleotide fasta: Isotigs refer to alternatively spliced isoforms of genes.

Isotig protein translations: These are protein translations (made using prot4EST [29]) of the isotigs produced by the assembler.

Isogroup membership file: Isogroup refers to the grouping of isotigs that putatively represent multiple isoforms for a gene locus. This file lists the detected isogroups by the assembler and provides the list of member isotigs per isogroup. This file is generated by local perl scripts.

Read membership file: The read membership file lists the read members for each isotig in the isotig file.

Our currently hosted transcript assemblies are from both Roche/454 data (using the Newbler assembler (v2.5) [30] which generated the isotig and isogroup information) and Sanger EST data (which is clustered using the Phred/Phrap/Consed suite of analysis tools [31–33]), and the consensus of these clusters is provided on the site. In addition to the cluster consensus sequence, Nematode.net also provides translated protein sequence built using the prot4EST program [29].

3.4.4 Noncoding Small RNAs

Noncoding small RNAs have been published only for a very few nematodes and trematodes. While most of the publications have their miRNAs in miRBase [34], predictions of miRNA targets are not presented, primarily owing to the difficulty in reliably predicting miRNA–target relationship based on in silico bioinformatics methods. Nevertheless, some data related to miRNA, miRNA abundances, and potential mRNA targets of miRNAs have started to emerge and Nematode.net has now begun to host such data (nematode.net/smallRNAs.html). This is a new feature and as an example of such data hosting, we currently have included *Ascaris suum* intestinal miRNAs and their predicted targets in the database. Based on interest from the community we will expand this to other available nematode and trematode miRNA and target information.

3.4.5 Proteomics

Protein expression data are, at present, hosted in derivative tables that provide spectra abundance per protein per species (nematode.net/Proteomics.html). We are in the process of including this information directly on the gene pages.

Table 2
Available Roche/454 cDNA assembled transcripts

Species	# Isogroups	# Isotigs	Stages/tissues
Ancylostoma caninum	19,277	23,388	L3(infective), L3 (serum stimulated), adult female, adult male
Cooperia oncophora	na	30,025	L3, adult female, adult male
Dictyocaulus viviparus	20,529	36,626	L1, L3, L5, adult female, adult male, egg
Heterorhabditis bacteriophora	7310	7857	na
Necator americanus	9253	9693	na
Oesophagostomum dentatum	16,788	30,030	L2, L3, L4, adult female, adult male
Ostertagia ostertagi	na	34,871	L3, L4
Teladorsagia circumcincta	29,991	33,148	Adult
Trichostrongylus colubriformis	19,833	27,615	Adult female, adult male

4 Helminth Control and Prevention

4.1 HelmCoP (Nematode.net)

HelmCoP (Helminth Control and Prevention; nematode.net/HelmCoP.html) [6] is a database of integrated functional, structural, and comparative genomics data from plant, animal and human parasitic nematodes and trematodes, as well as model organisms and several host organisms (18 species) all capped by a query interface that allows users to ask complex questions of these data. HelmCoP's primary goal is to assist researchers in the process of building a list of candidate drug, pesticide, and vaccine targets in helminthes. HelmCoP has the versatility to enable users to search for drug targets for specific parasite species or for a group of species of interest and also to allow the user to search for broad-spectrum drug targets that span multiple taxonomic groups or phyla.

Querying HelmCoP is done using one of two forms. One is for users interested in building gene-based custom queries (Fig. 12A) and the other for users wanting to search the database using ortholog based queries. The upper half of both the gene and ortholog based search forms are used to build the set of genes that will be tested according to the user's filter selections. For the gene based search you can either enter a specific gene name or you can select a combination of species to define the gene space to which the filters you select will be applied.

If you are building an ortholog based query (i.e., results are returned by ortholog if any gene within that ortholog meets the

Table 3
Available Sanger EST assembled transcript clusters (i.e., genes)

Species	# EST clusters
Ancylostoma caninum	5484
Ancylostoma ceylanicum	4953
Ascaris suum	5137
Brugia malayi	1609
Caenorhabditis remanei	12,334
Dirofilaria immitis	2534
Ditylenchus africanus	5214
Globodera pallida	2973
Globodera rostochiensis	9482
Heterodera glycines	12,313
Heterodera schachtii	1595
Haemonchus contortus	9842
Meloidogyne arenaria	3356
Meloidogyne chitwoodi	5880
Meloidogyne hapla	11,193
Meloidogyne incognita	9107
Meloidogyne javanica	5165
Meloidogyne paranaensis	2263
Nippostrongylus brasiliensis	4532
Onchocerca flexuosa	1665
Ostertagia ostertagi	4794
Parastrongylus trichosuri	4923
Pratylenchus penetrans	488
Pristionchus pacificus	2654
Radopholus similus	5551
Strongyloides ratti	5237
Strongyloides stercoralis	3479
Toxocara canis	2082
Trichinella spiralis	5958
Trichuris muris	3735
Xiphinema index	5485
Zeldia punctata	202

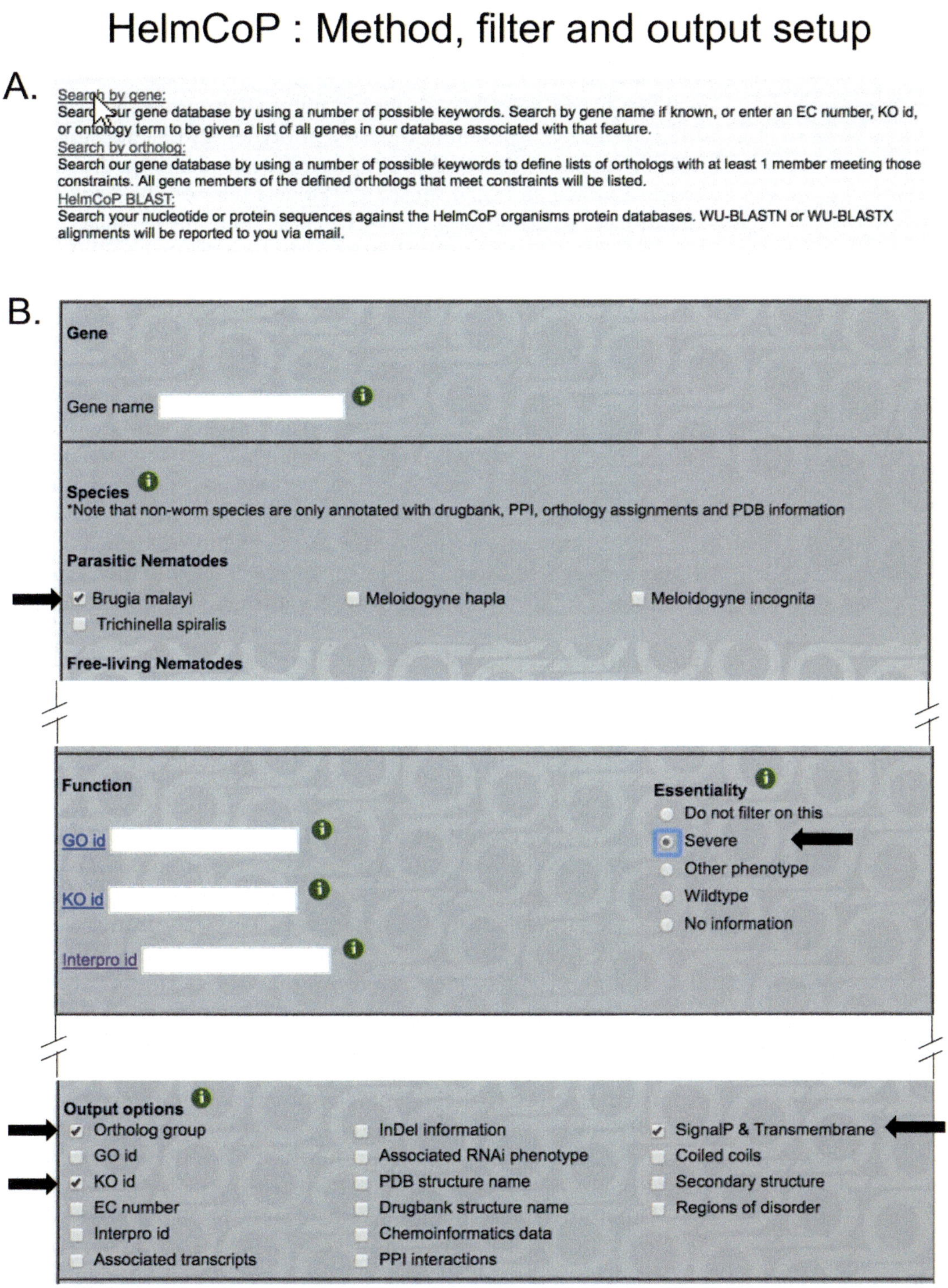

Fig. 12 HelmCoP options setup. (A) On the main HelmCoP page, the user indicates whether they want to search by gene, ortholog, or using the HelmCoP BLAST. The illustrated example shows a search by gene (indicated by a white arrow pointer on the corresponding link). (B) The main input page, where the user can select the gene(s) or species of interest (if any), the filters selecting any properties of interest (e.g., "Essentiality" required to be "Severe" as shown here), and the columns that the user wants to populate the results table. The example here shows a query for all genes in *Brugia malayi* that are annotated with "severe" essentiality, reporting their ortholog group, KO IDs, and any SignalP and Transmembrane annotation

criteria you define) you use the ortholog search page. Most users will not have a specific ortholog in mind when using this page, so the typical user will define their set of orthologs by choosing species to include or exclude (or ignore) when defining the search space. Leaving an organism set to "Do not filter on this" will cause that species to not be considered when defining the result set. Your choices will result in an initial set of orthologs that have at least one gene from any of the "included" species. But having even one gene member from an "excluded" species will remove the entire ortholog from the return set.

For both the gene and ortholog based searches, after defining the set of species to query (and thus defining the gene space), the next task is to apply desired filters to limit your output down to only those genes or orthologs of interest to you (Fig. 12B). The user can also request specific output columns in the result table at this step.

Several annotations are specific to the HelmCoP database. The **Function** section allows you to set specific GO, KO and/or IPR IDs that you require to be present in returned genes, or at least in one of the members of returned orthologous groups. The **Structure** based filters allow you to limit your return set by requiring them to have shown sequence similarity to the PDB ID you enter [35]. This section also allows you to filter your results based on the presence or absence of a detected signal peptide. The search for signal peptides is performed using the Phobius program [36]. The **Drugs** section allows you to filter on genes with homology to targets in DrugBank [37]. It also allows you to filter based on whether or not your returned gene (or at least one of the genes per each returned orthologous group) is considered a "Hopkins druggable target" [38]. The putative **Vaccine candidates** section allows the user to filter based on the presence of various structural based hints that may imply epitopes that are vaccine candidates. You can pick and choose specific traits or just turn on the "Show all vaccine candidates" switch to apply them all. Finally the user needs to set their **Output options**. These allow the user to customize the output to include only information of interest to them. Some columns will always be returned, such as gene and ortholog name as well as species of origin, but otherwise the user needs to select the other information they want reported.

Due to technical limitations, the output of a HelmCoP query is limited to 20,000 rows displayed in HTML. This limitation means that it is possible that ortholog-based search results may be truncated mid-ortholog. In other words you are not guaranteed to consistently have every gene member of filter-passing orthologs reported to you in the HTML display. The full and complete results are made available as a downloadable text file that is presented to the user on the results page (Fig. 13). To be assured of getting the full results the user should always download the provided full results text file.

Another consideration when using this tool is that due to the size of some return sets, selecting many or all of the available outputs can cause your query to take a long time to build. In some cases this may even cause the website to time out before the information can be fed back to user's browser. When a query is submitted we do first run a query manager that tries to estimate if your query is completeable within an amount of time that should avoid this timeout but the manager is not infallible. If the manager deems that the query would not complete within the time limits we will suggest which are the most time-intensive options for your search. Users can then reconstruct their query with fewer requested outputs or define a smaller starting search space (i.e., choosing fewer species or a more restrictive set of orthologs in the species selection).

Fig. 13 shows the HTML results of a HelmCoP search. The output table is gene-based with ortholog information provided as well for ortholog-based search returns (or in the case that the user selects to see orthologous group annotation for a gene based search

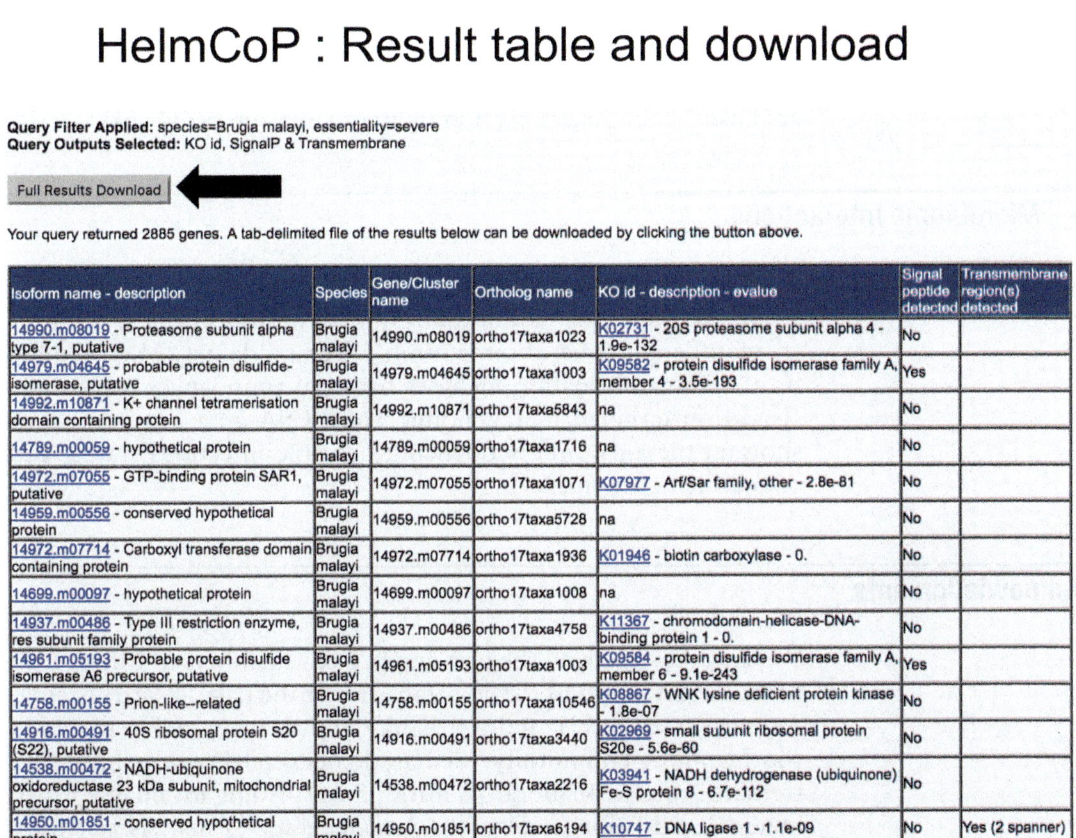
HelmCoP : Result table and download

Query Filter Applied: species=Brugia malayi, essentiality=severe
Query Outputs Selected: KO id, SignalP & Transmembrane

Full Results Download

Your query returned 2885 genes. A tab-delimited file of the results below can be downloaded by clicking the button above.

Isoform name - description	Species	Gene/Cluster name	Ortholog name	KO id - description - evalue	Signal peptide detected	Transmembrane region(s) detected
14990.m08019 - Proteasome subunit alpha type 7-1, putative	Brugia malayi	14990.m08019	ortho17taxa1023	K02731 - 20S proteasome subunit alpha 4 - 1.9e-132	No	
14979.m04645 - probable protein disulfide-isomerase, putative	Brugia malayi	14979.m04645	ortho17taxa1003	K09582 - protein disulfide isomerase family A, member 4 - 3.5e-193	Yes	
14992.m10871 - K+ channel tetramerisation domain containing protein	Brugia malayi	14992.m10871	ortho17taxa5843	na	No	
14789.m00059 - hypothetical protein	Brugia malayi	14789.m00059	ortho17taxa1716	na	No	
14972.m07055 - GTP-binding protein SAR1, putative	Brugia malayi	14972.m07055	ortho17taxa1071	K07977 - Arf/Sar family, other - 2.8e-81	No	
14959.m00556 - conserved hypothetical protein	Brugia malayi	14959.m00556	ortho17taxa5728	na	No	
14972.m07714 - Carboxyl transferase domain containing protein	Brugia malayi	14972.m07714	ortho17taxa1936	K01946 - biotin carboxylase - 0.	No	
14699.m00097 - hypothetical protein	Brugia malayi	14699.m00097	ortho17taxa1008	na	No	
14937.m00486 - Type III restriction enzyme, res subunit family protein	Brugia malayi	14937.m00486	ortho17taxa4758	K11367 - chromodomain-helicase-DNA-binding protein 1 - 0.	No	
14961.m05193 - Probable protein disulfide isomerase A6 precursor, putative	Brugia malayi	14961.m05193	ortho17taxa1003	K09584 - protein disulfide isomerase family A, member 6 - 9.1e-243	Yes	
14758.m00155 - Prion-like--related	Brugia malayi	14758.m00155	ortho17taxa10546	K08867 - WNK lysine deficient protein kinase - 1.8e-07	No	
14916.m00491 - 40S ribosomal protein S20 (S22), putative	Brugia malayi	14916.m00491	ortho17taxa3440	K02969 - small subunit ribosomal protein S20e - 5.6e-60	No	
14538.m00472 - NADH-ubiquinone oxidoreductase 23 kDa subunit, mitochondrial precursor, putative	Brugia malayi	14538.m00472	ortho17taxa2216	K03941 - NADH dehydrogenase (ubiquinone) Fe-S protein 8 - 6.7e-112	No	
14950.m01851 - conserved hypothetical protein	Brugia malayi	14950.m01851	ortho17taxa6194	K10747 - DNA ligase 1 - 1.1e-09	No	Yes (2 spanner)

Fig. 13 HelmCoP results. The result page for HelmCoP contains a table showing the properties selected by the user (Fig. 12) for the genes satisfying the filters of interest. A download link is also provided for the entire result table (since the shown table may be truncated, depending on how many genes are part of the result)

return). Every requested output will be a column, so requesting many outputs can result in your data spanning multiple page-widths. Where appropriate, the HTML view provides link-outs to the various resources each output type is based upon. For orthologous group output, it aggregates cases where multiple members of an ortholog report the same information.

Toward the top of the page, users will be shown the query filters and output requests that they have specified which have resulted in the provided report. The button for accessing a full, tab-delimited text version of the output table is also available here (*it is strongly recommended that users download the full report text file to avoid the issues mentioned above*).

4.2 Other Function Specific Candidate Drug Targets

Much research has been conducted on inhibitors for different gene families leading to a wealth of compounds that target them that have potential to be lead anthelmintic drugs. Nematode.net provides several derivative tables that provide information for specific gene families or specific functions that have been obtained as a result of genome-driven knowledge based drug target discovery. We host information on kinases, metabolic chokepoints, lysine deacetylases and protein-protein interactions as targets. These candidate targets are linked to inhibitors via homologous targets in drugbanks.

5 Microbiome Interactions

Both of the Helminth.net sites host information derived from the study of microbial communities in helminth-infected subjects (nematode.net/Microbiome.html; trematode.net/Microbiome.html). This information includes bacterial abundances per sample (based on targeted metagenomic 16S rRNA gene sequencing or shotgun metagenomic sequencing), sample infectious status, and cohort demographics.

Acknowledgments

We sincerely thank all the past and present members of Mitreva lab for their contribution to the database over the past 17 years (nematode.net/staff.html), and we thank the numerous collaborators in the helminth community (nematode.net/collaborators.html and trematode.net/collaborators.html), for providing invaluable worm material and being involved in data generation/analysis activities, and the dedicated members of the production group at The McDonnell Genome Institute (http://genome.wustl.edu/) for the library construction and sequencing. Helminth.net is funded by National Institutes of Health [AI081803 and GM097435] and NIFA [2013-01109].

References

1. Wylie T, Martin JC, Dante M, Mitreva MD, Clifton SW, Chinwalla A, Waterston RH, Wilson RK, McCarter JP (2004) Nematode.net: a tool for navigating sequences from parasitic and free-living nematodes. Nucleic Acids Res 32(Database issue):D423–D426. https://doi.org/10.1093/nar/gkh010
2. Martin J, Abubucker S, Wylie T, Yin Y, Wang Z, Mitreva M (2009) Nematode.net update 2008: improvements enabling more efficient data mining and comparative nematode genomics. Nucleic Acids Res 37(Database issue):D571–D578. https://doi.org/10.1093/nar/gkn744
3. Martin J, Abubucker S, Heizer E, Taylor CM, Mitreva M (2012) Nematode.net update 2011: addition of data sets and tools featuring next-generation sequencing data. Nucleic Acids Res 40(Database issue):D720–D728. https://doi.org/10.1093/nar/gkr1194
4. Martin J, Rosa BA, Ozersky P, Hallsworth-Pepin K, Zhang X, Bhonagiri-Palsikar V, Tyagi R, Wang Q, Choi YJ, Gao X, McNulty SN, Brindley PJ, Mitreva M (2015) Helminth.net: expansions to Nematode.net and an introduction to Trematode.net. Nucleic Acids Res 43(Database issue):D698–D706. https://doi.org/10.1093/nar/gku1128
5. Wylie T, Martin J, Abubucker S, Yin Y, Messina D, Wang Z, McCarter JP, Mitreva M (2008) NemaPath: online exploration of KEGG-based metabolic pathways for nematodes. BMC Genomics 9:525. https://doi.org/10.1186/1471-2164-9-525
6. Abubucker S, Martin J, Taylor CM, Mitreva M (2011) HelmCoP: an online resource for helminth functional genomics and drug and vaccine targets prioritization. PLoS One 6(7):e21832. https://doi.org/10.1371/journal.pone.0021832. PONE-D-11-02640 [pii]
7. Tyagi R, Rosa BA, Lewis WG, Mitreva M (2015) Pan-phylum comparison of nematode metabolic potential. PLoS Negl Trop Dis 9(5):e0003788. https://doi.org/10.1371/journal.pntd.0003788
8. Torbati ME, Mitreva M, Gopalakrishnan V (2016) Application of taxonomic modeling to microbiota data mining for detection of helminth infection in global populations. Data (Basel) 1(3):19. https://doi.org/10.3390/data1030019
9. Jones P, Binns D, Chang HY, Fraser M, Li W, McAnulla C, McWilliam H, Maslen J, Mitchell A, Nuka G, Pesseat S, Quinn AF, Sangrador-Vegas A, Scheremetjew M, Yong SY, Lopez R, Hunter S (2014) InterProScan 5: genome-scale protein function classification. Bioinformatics 30(9):1236–1240. https://doi.org/10.1093/bioinformatics/btu031
10. Hunter S, Jones P, Mitchell A, Apweiler R, Attwood TK, Bateman A, Bernard T, Binns D, Bork P, Burge S, de Castro E, Coggill P, Corbett M, Das U, Daugherty L, Duquenne L, Finn RD, Fraser M, Gough J, Haft D, Hulo N, Kahn D, Kelly E, Letunic I, Lonsdale D, Lopez R, Madera M, Maslen J, McAnulla C, McDowall J, McMenamin C, Mi H, Mutowo-Muellenet P, Mulder N, Natale D, Orengo C, Pesseat S, Punta M, Quinn AF, Rivoire C, Sangrador-Vegas A, Selengut JD, Sigrist CJ, Scheremetjew M, Tate J, Thimmajanarthanan M, Thomas PD, Wu CH, Yeats C, Yong SY (2012) InterPro in 2011: new developments in the family and domain prediction database. Nucleic Acids Res 40(Database issue):D306–D312. https://doi.org/10.1093/nar/gkr948
11. Gish W (1996–2003) http://blast.wustl.edu
12. Blake JA, Dolan M, Drabkin H, Hill DP, Li N, Sitnikov D, Bridges S, Burgess S, Buza T, McCarthy F, Peddinti D, Pillai L, Carbon S, Dietze H, Ireland A, Lewis SE, Mungall CJ, Gaudet P, Chrisholm RL, Fey P, Kibbe WA, Basu S, Siegele DA, McIntosh BK, Renfro DP, Zweifel AE, Hu JC, Brown NH, Tweedie S, Alam-Faruque Y, Apweiler R, Auchinchloss A, Axelsen K, Bely B, Blatter M, Bonilla C, Bouguerleret L, Boutet E, Breuza L, Bridge A, Chan WM, Chavali G, Coudert E, Dimmer E, Estreicher A, Famiglietti L, Feuermann M, Gos A, Gruaz-Gumowski N, Hieta R, Hinz C, Hulo C, Huntley R, James J, Jungo F, Keller G, Laiho K, Legge D, Lemercier P, Lieberherr D, Magrane M, Martin MJ, Masson P, Mutowo-Muellenet P, O'Donovan C, Pedruzzi I, Pichler K, Poggioli D, Porras Millán P, Poux S, Rivoire C, Roechert B, Sawford T, Schneider M, Stutz A, Sundaram S, Tognolli M, Xenarios I, Foulgar R, Lomax J, Roncaglia P, Khodiyar VK, Lovering RC, Talmud PJ, Chibucos M, Giglio MG, Chang H, Hunter S, McAnulla C, Mitchell A, Sangrador A, Stephan R, Harris MA, Oliver SG, Rutherford K, Wood V, Bahler J, Lock A, Kersey PJ, McDowall DM, Staines DM, Dwinell M, Shimoyama M, Laulederkind S, Hayman T, Wang S, Petri V, Lowry T, D'Eustachio P, Matthews L, Balakrishnan R, Binkley G, Cherry JM, Costanzo MC, Dwight SS, Engel SR, Fisk DG, Hitz BC, Hong EL, Karra K, Miyasato SR, Nash RS, Park J, Skrzypek MS, Weng S, Wong ED, Berardini TZ, Huala E, Mi H, Thomas PD, Chan J, Kishore R, Sternberg P, Van Auken K, Howe D, Westerfield M, Consortium GO (2013) Gene

Ontology annotations and resources. Nucleic Acids Res 41(Database issue):D530–D535. https://doi.org/10.1093/nar/gks1050

13. Kanehisa M, Goto S, Sato Y, Kawashima M, Furumichi M, Tanabe M (2014) Data, information, knowledge and principle: back to metabolism in KEGG. Nucleic Acids Res 42(Database issue):D199–D205. https://doi.org/10.1093/nar/gkt1076
14. Fischer S, Brunk BP, Chen F, Gao X, Harb OS, Iodice JB, Shanmugam D, Roos DS, Stoeckert CJ, Jr. (2011) Using OrthoMCL to assign proteins to OrthoMCL-DB groups or to cluster proteomes into new ortholog groups. Curr Protoc Bioinformatics Chapter 6:Unit 6 12 11–19. doi:https://doi.org/10.1002/0471250953.bi0612s35
15. Sonnhammer EL, Ostlund G (2015) InParanoid 8: orthology analysis between 273 proteomes, mostly eukaryotic. Nucleic Acids Res 43(Database issue):D234–D239. https://doi.org/10.1093/nar/gku1203
16. Wootton JC, Federhen S (1993) Statistics of local complexity in amino acid sequences and sequence databases. Comput Chem 17(2):149–163
17. Bedell JA, Korf I, Gish W (2000) MaskerAid: a performance enhancement to RepeatMasker. Bioinformatics 16(11):1040–1041
18. Stein LD (2013) Using GBrowse 2.0 to visualize and share next-generation sequence data. Brief Bioinform 14(2):162–171. https://doi.org/10.1093/bib/bbt001
19. Cantarel BL, Korf I, Robb SM, Parra G, Ross E, Moore B, Holt C, Sanchez Alvarado A, Yandell M (2008) MAKER: an easy-to-use annotation pipeline designed for emerging model organism genomes. Genome Res 18(1):188–196. https://doi.org/10.1101/gr.6743907
20. Lagesen K, Hallin P, Rodland EA, Staerfeldt HH, Rognes T, Ussery DW (2007) RNAmmer: consistent and rapid annotation of ribosomal RNA genes. Nucleic Acids Res 35(9):3100–3108. https://doi.org/10.1093/nar/gkm160
21. Lowe TM, Eddy SR (1997) tRNAscan-SE: a program for improved detection of transfer RNA genes in genomic sequence. Nucleic Acids Res 25(5):955–964
22. McKenna A, Hanna M, Banks E, Sivachenko A, Cibulskis K, Kernytsky A, Garimella K, Altshuler D, Gabriel S, Daly M, DePristo MA (2010) The Genome Analysis Toolkit: a MapReduce framework for analyzing next-generation DNA sequencing data. Genome Res 20(9):1297–1303. https://doi.org/10.1101/gr.107524.110
23. Cingolani P, Platts A, Wang le L, Coon M, Nguyen T, Wang L, Land SJ, Lu X, Ruden DM (2012) A program for annotating and predicting the effects of single nucleotide polymorphisms, SnpEff: SNPs in the genome of Drosophila melanogaster strain w1118; iso-2; iso-3. Fly 6(2):80–92. https://doi.org/10.4161/fly.19695
24. Choi YJ, Tyagi R, McNulty SN, Rosa BA, Ozersky P, Martin J, Hallsworth-Pepin K, Unnasch TR, Norice CT, Nutman TB, Weil GJ, Fischer PU, Mitreva M (2016) Genomic diversity in Onchocerca volvulus and its Wolbachia endosymbiont. Nat Microbiol 2:16207. https://doi.org/10.1038/nmicrobiol.2016.207
25. McNulty SN, Strube C, Rosa BA, Martin JC, Tyagi R, Choi YJ, Wang Q, Hallsworth Pepin K, Zhang X, Ozersky P, Wilson RK, Sternberg PW, Gasser RB, Mitreva M (2016) Dictyocaulus viviparus genome, variome and transcriptome elucidate lungworm biology and support future intervention. Sci Rep 6:20316. https://doi.org/10.1038/srep20316
26. McNulty SN, Tort JF, Rinaldi G, Fischer K, Rosa BA, Smircich P, Fontenla S, Choi YJ, Tyagi R, Hallsworth-Pepin K, Mann VH, Kammili L, Latham PS, Dell'Oca N, Dominguez F, Carmona C, Fischer PU, Brindley PJ, Mitreva M (2017) Genomes of Fasciola hepatica from the Americas reveal colonization with Neorickettsia Endobacteria related to the agents of potomac horse and human sennetsu fevers. PLoS Genet 13(1):e1006537. https://doi.org/10.1371/journal.pgen.1006537
27. Carbon S, Ireland A, Mungall CJ, Shu S, Marshall B, Lewis S, the AmiGO Hub, the Web Presence Working Group (2009) AmiGO: online access to ontology and annotation data. Bioinformatics 25(2):288–289. https://doi.org/10.1093/bioinformatics/btn615
28. Leinonen R, Sugawara H, Shumway M, International Nucleotide Sequence Database Collaboration (2011) The sequence read archive. Nucleic Acids Res 39(Database issue):D19–D21. https://doi.org/10.1093/nar/gkq1019
29. Wasmuth JD, Blaxter ML (2004) prot4EST: translating expressed sequence tags from neglected genomes. BMC Bioinformatics 5:187. https://doi.org/10.1186/1471-2105-5-187
30. Margulies M, Egholm M, Altman WE, Attiya S, Bader JS, Bemben LA, Berka J, Braverman MS, Chen YJ, Chen Z, Dewell SB, Du L, Fierro JM, Gomes XV, Godwin BC, He W, Helgesen S, Ho CH, Irzyk GP, Jando SC, Alenquer ML, Jarvie TP, Jirage KB, Kim JB, Knight JR, Lanza JR, Leamon JH, Lefkowitz SM, Lei M, Li J,

Lohman KL, Lu H, Makhijani VB, McDade KE, McKenna MP, Myers EW, Nickerson E, Nobile JR, Plant R, Puc BP, Ronan MT, Roth GT, Sarkis GJ, Simons JF, Simpson JW, Srinivasan M, Tartaro KR, Tomasz A, Vogt KA, Volkmer GA, Wang SH, Wang Y, Weiner MP, Yu P, Begley RF, Rothberg JM (2005) Genome sequencing in microfabricated high-density picolitre reactors. Nature 437(7057):376–380. https://doi.org/10.1038/nature03959

31. Ewing B, Hillier L, Wendl MC, Green P (1998) Base-calling of automated sequencer traces using phred. I. Accuracy assessment. Genome Res 8(3):175–185
32. Ewing B, Green P (1998) Base-calling of automated sequencer traces using phred. II. Error probabilities. Genome Res 8(3):186–194
33. Gordon D, Abajian C, Green P (1998) Consed: a graphical tool for sequence finishing. Genome Res 8(3):195–202
34. Kozomara A, Griffiths-Jones S (2011) miRBase: integrating microRNA annotation and deep-sequencing data. Nucleic Acids Res 39(Database issue):D152–D157. https://doi.org/10.1093/nar/gkq1027
35. Berman HM, Westbrook J, Feng Z, Gilliland G, Bhat TN, Weissig H, Shindyalov IN, Bourne PE (2000) The Protein Data Bank. Nucleic Acids Res 28(1):235–242
36. Kall L, Krogh A, Sonnhammer EL (2004) A combined transmembrane topology and signal peptide prediction method. J Mol Biol 338(5):1027–1036. https://doi.org/10.1016/j.jmb.2004.03.016
37. Knox C, Law V, Jewison T, Liu P, Ly S, Frolkis A, Pon A, Banco K, Mak C, Neveu V, Djoumbou Y, Eisner R, Guo AC, Wishart DS (2011) DrugBank 3.0: a comprehensive resource for 'omics' research on drugs. Nucleic Acids Res 39(Database issue):D1035–D1041. https://doi.org/10.1093/nar/gkq1126
38. Hopkins AL, Groom CR (2002) The druggable genome. Nat Rev Drug Discov 1(9):727–730. https://doi.org/10.1038/nrd892

Chapter 14

Using WormBase: A Genome Biology Resource for *Caenorhabditis elegans* and Related Nematodes

Christian Grove, Scott Cain, Wen J. Chen, Paul Davis, Todd Harris, Kevin L. Howe, Ranjana Kishore, Raymond Lee, Michael Paulini, Daniela Raciti, Mary Ann Tuli, Kimberly Van Auken, Gary Williams, and The WormBase Consortium

Abstract

WormBase (www.wormbase.org) provides the nematode research community with a centralized database for information pertaining to nematode genes and genomes. As more nematode genome sequences are becoming available and as richer data sets are published, WormBase strives to maintain updated information, displays, and services to facilitate efficient access to and understanding of the knowledge generated by the published nematode genetics literature. This chapter aims to provide an explanation of how to use basic features of WormBase, new features, and some commonly used tools and data queries. Explanations of the curated data and step-by-step instructions of how to access the data via the WormBase website and available data mining tools are provided.

Key words Data mining, Nematodes, Genomics, Genetics, *Caenorhabditis elegans*, Model organism database, Ontologies, User guide

1 Introduction

Since its inception in March 2000, WormBase has provided the nematode research community with an online resource for gene, genome, and other biological information about *Caenorhabditis elegans* and related nematodes [1, 2]. WormBase offers abundant gene-centric information including gene structure models, gene homology data, gene expression data, gene-affiliated phenotypes, gene ontology annotations, and gene interactions (physical, regulatory, and genetic) as well as an anatomy ontology, a life stage ontology, human disease relevance, publication information, and *C. elegans* researcher information. WormBase provides its users

The members of the WormBase Consortium are listed in the Acknowledgements.

Martin Kollmar (ed.), *Eukaryotic Genomic Databases: Methods and Protocols*, Methods in Molecular Biology, vol. 1757, https://doi.org/10.1007/978-1-4939-7737-6_14, © Springer Science+Business Media, LLC, part of Springer Nature 2018

with a number of tools including a genome browser (GBrowse and, more recently, JBrowse), BLAST/BLAT tools, an electronic PCR tool (ePCR), a genetic map browser, and a number of data mining tools including WormMine and SimpleMine.

This chapter serves as a user guide to the current version of WormBase (at the time of writing, WS257 release, April 2017) but should remain a relevant reference for years to come. Briefly, the chapter will begin by providing guidance on the basic mechanics of using the WormBase website, followed by an explanation of how to access genomic data for the multitude of nematode species now supported by WormBase, including how to use the available genome browsers. This is followed by a discussion of homology data, at the level of genomes, genes, and proteins and then an explanation of ontologies at WormBase, specifically how to use the Ontology Browser tool and how to navigate Gene Ontology data. The chapter then explores gene expression data, both small scale and large scale, and how to access it via gene report pages as well as via tools such as SPELL. Next is a discussion of how to access and interpret gene interaction data in WormBase, for physical, genetic, and regulatory interactions, followed by an explanation of phenotype data and where to find it. Reagents such as strains, transgenes, and RNAi clones are then reviewed followed by a discussion of integrated data views such as human disease models in nematodes and anatomy function data. After this is a review of the data mining tool WormMine, the WormBase instance of the Intermine biological data warehouse, a summary of the many data files available via the FTP site, and basic instructions for use of the WormBase RESTful API. Other available tools are then discussed, including BLAST/BLAT, the SimpleMine gene batch query tool, a new gene set enrichment analysis tool, and a description of our community annotation forms. We then close the chapter with a brief discussion of community resources, highlighting the many ways users can keep updated on WormBase and community activities.

2 The WormBase Website

The WormBase website provides quick and convenient access to the most important information for conducting research using *C. elegans* as a model system. The site presents a wide array of data both curated from the scientific literature and submitted to the consortium directly by users. These data range from whole genome sequences of *C. elegans* and a number of related nematodes and genomic-scale datasets down to the sequence of individual variations. Gene function and perturbation, expression, anatomy, phenotypes, literature, and researcher history and more are all directly accessible from a single search of the site.

2.1 The Home Page

The WormBase home page provides quick access to the most popular elements of the site, information about upcoming meetings, and quick glances of the WormBase forums. Figure 1 outlines the primary features of the WormBase home page.

The home page is illustrative of the organization of all pages across the site. The main navigation bar (across the top of any WormBase page) contains a class-specific and global search field (discussed below) and a series of drop-down menus. The "About" menu provides links to the WormBase mission statement and frequently asked questions (FAQs), as well as to lists of our advisory

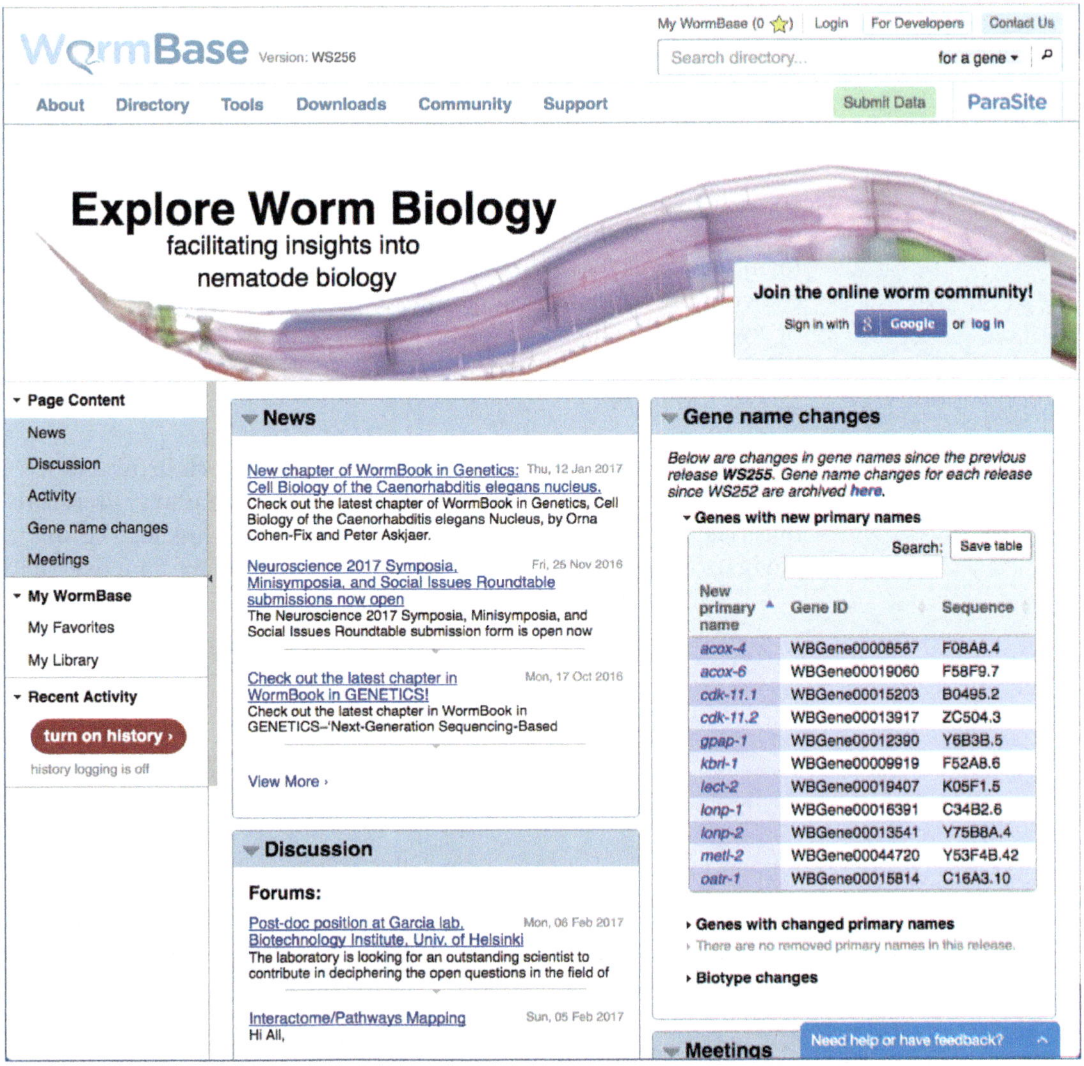

Fig. 1 The core elements of the WormBase home page. Each page on WormBase consists of three core areas. At the top of the page is a main navigation bar. On the left-hand side of the page is a page-specific navigation bar that controls the content displayed on a per-class basis. The majority of each page consists of the main view area where content specific to that page is displayed

board members and staff. The "Directory" menu provides links to basic and advanced search, to the genome browser for various nematode species, to resources such as external databases and methods sites, as well as to the WormBase schema, tree display and our "Submit Data" page. The "Tools" menu provides links to our commonly used tools such as GBrowse, BLAST/BLAT, electronic PCR (ePCR) search, genetic map, SPELL, and the WormBase ontology browser as well as links to data mining and batch query tools like WormMine, SimpleMine, and a gene set enrichment analysis tool. The "Downloads" menu provides links to download genomic and protein FASTA files as well as genomic annotation (GFF) files for all available species, and a link to the WormBase FTP site. The "Community" menu links to meeting information, the worm community forum, a "Submit Data" page, and to many other resources for the *C. elegans* and nematode research community. The "Support" menu links to a user guide, nomenclature explanation, FAQs, how-to videos and documentation for developers.

Underneath the main navigation bar, users will find a vertical navigation bar situated on the left hand side of the screen and a main view area to the right. This left side navigation bar controls the content that is displayed on the page in self-contained panels, or "widgets." This includes page-specific content widgets such as "Overview" and "Sequences," and "Tools" such as aligners and model browsers.

2.2 Logging in to WormBase

WormBase allows users to log in to the site to track browsing history, save favorite objects, and create rudimentary literature libraries. To log in to the site, visit the Login link in the main navigation bar (Fig. 2). We have adopted standard practice that

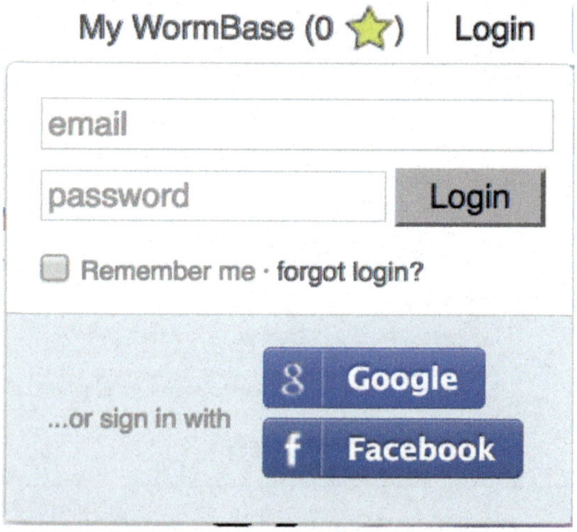

Fig. 2 Logging in to WormBase. From the main navigation bar, click "Login." Once you have authenticated using Google or Facebook, the navigation bar will change to read "Welcome (Your Name)" to notify you that you are logged in

authenticates users with their Facebook or Google accounts. When we do this, we request access to your profile, storing only your publicly provided name and email, if they have been provided. This allows us to identify you personally on the website and nothing more. No additional information is sent to either Google or Facebook.

Once a user is logged in, three new features become available on the site: “My Favorites,” “My Library;” and “My History.” First, by clicking on stars shown on every report page and in various search displays (Fig. 3), users can save often-accessed objects for quick subsequent access. Favorite objects are displayed on the “My WormBase” page (Fig. 4). Second, users can “star” literature references in a manner similar to favorites—by clicking on stars next to literature titles. These favorite references will be displayed in the “My Library” section on the “My WormBase” report page (Fig. 4, bottom). The third feature, recorded browsing history, is an opt-in only feature. To enable, log-in to the site. On the home page, in the “Activity” widget, look for a button reading “Turn On History.” Once enabled you will see not only your browsing history on the site but also items popularly saved across all users that have opted in. Your browsing history and saving preferences will also be included in this history.

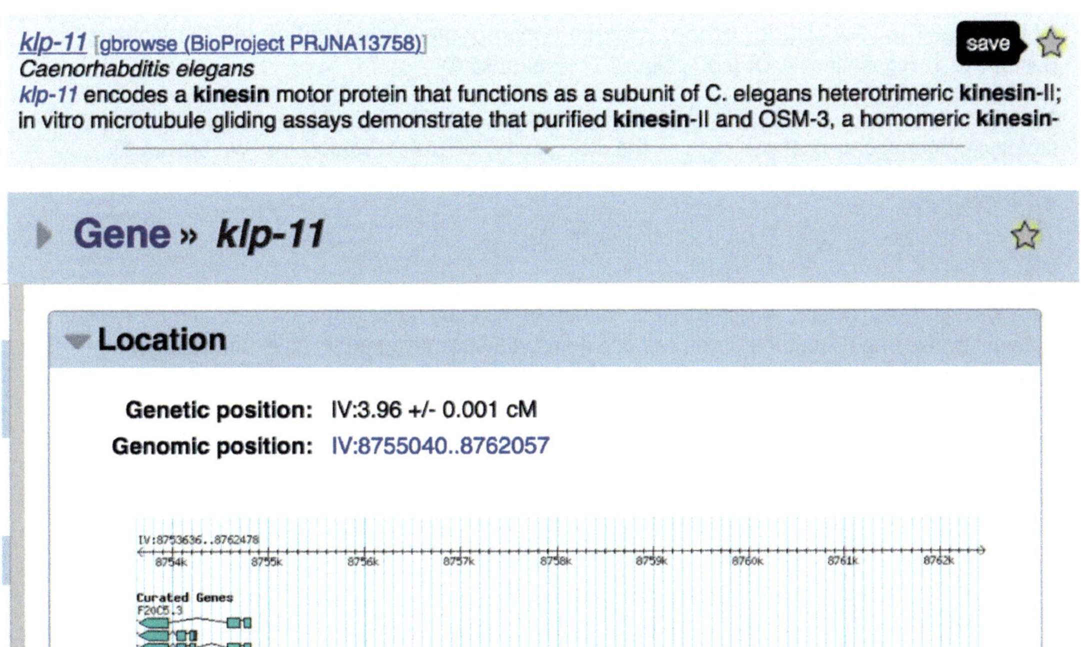

Fig. 3 Add items to your list of favorites by clicking on stars found in search results (top) and on report pages (bottom). You can remove an item from your favorites by clicking on the highlighted star

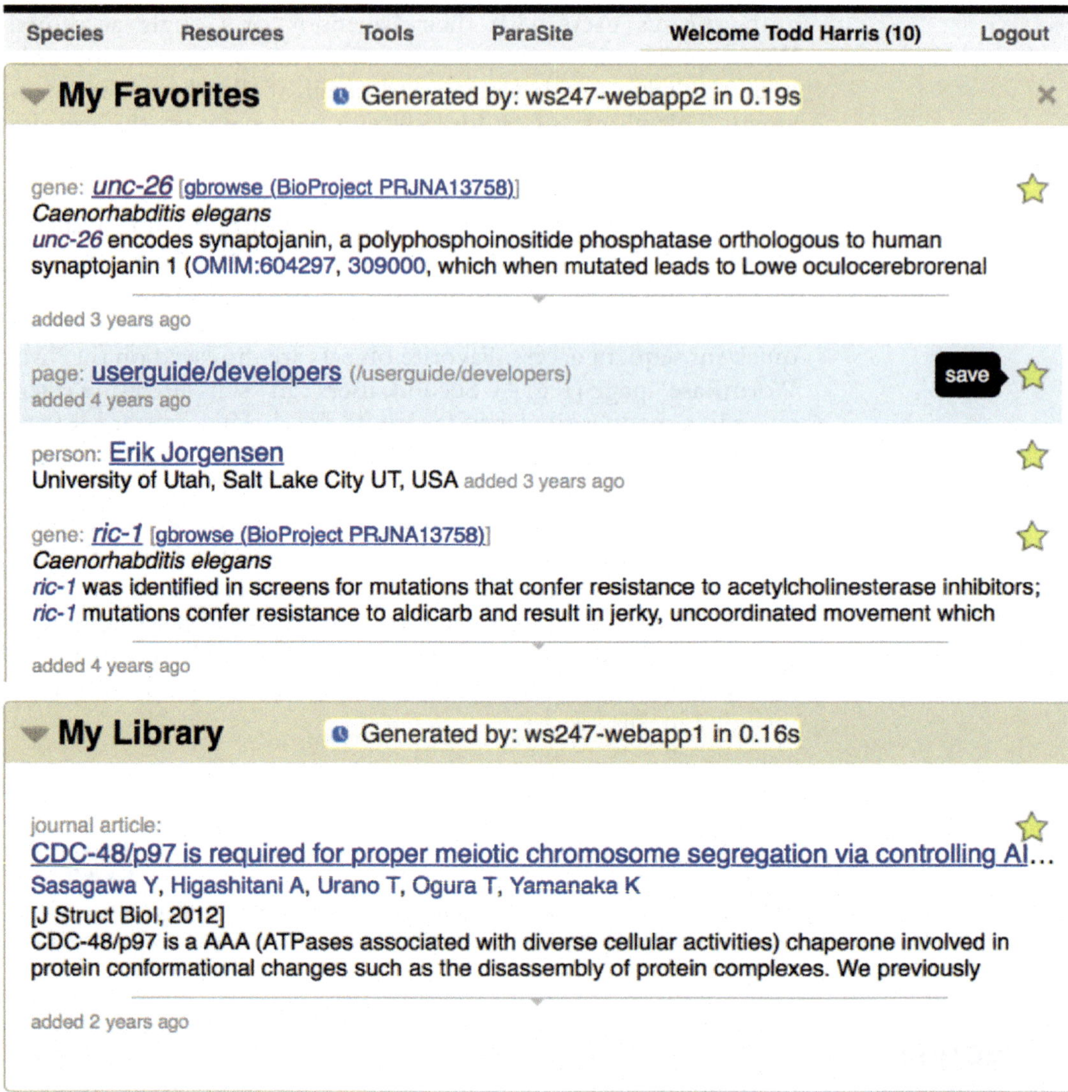

Fig. 4 Once logged in, click on "Welcome (Your Name)" in the main navigation bar located on every page (top) to visit your profile page. From there, you will find items you have favorited (middle) and your library of saved references (bottom)

2.3 Basic Searches

Basic searches are available from the main navigation bar presented at the top right hand side across the site (Fig. 5). By default, searches are constrained to the gene class. We do this for two reasons. First, the gene report pages are informationally rich, very popular landing pages, containing a great number of crosslinks to other data types. Thus, they act as important portals into the data contained in WormBase. Second, we wish to provide suitable performance and quick access to what most users are searching for. If you would like to change this behavior, simply select the desired

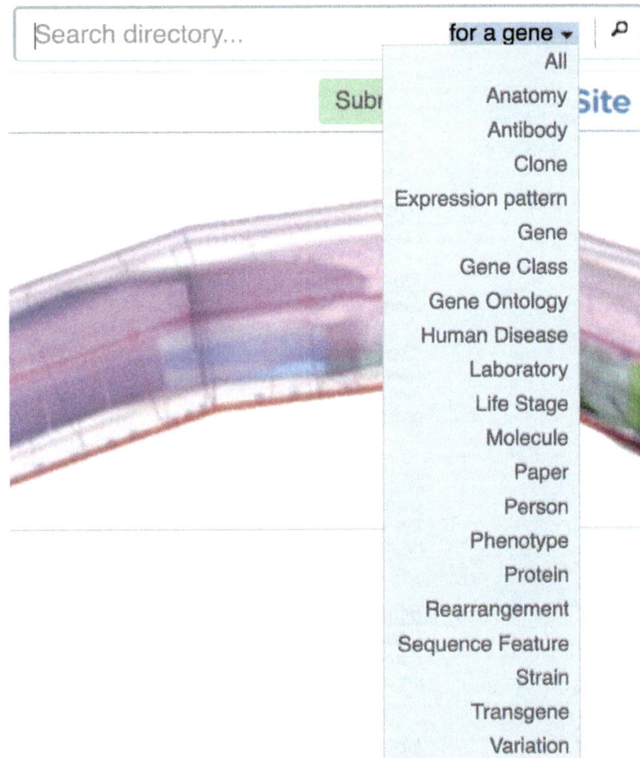

Fig. 5 Basic and global searches of WormBase are available from the top of every page. By default, the search is constrained to genes. This can be changed by selecting from available classes via the embedded drop-down menu, or by choosing "All" to search the entire database

class from the drop-down menu as shown in Fig. 5. To search for additional data classes, leave the search box empty and click on the magnifying glass icon to the right and select from the list of data classes on the subsequent page.

The search accepts a wide range of queries. Users can search for simple things like gene or sequence names (e.g., "lin-29"; "JC8.10a"), types of gene products (e.g., "kinesin"), or even related items associated with the item of interest (e.g., such as finding genes by searching with variation names of a given gene, like "e205").

If a search returns a single hit, users will be taken directly to the report page for that item. If multiple hits are returned, a disambiguation display allows one to select from search results, to download results, or to further constrain your search by class or species through a faceted results display. In the list view of results, we present multiple options for each result. For example, searches that result in gene hits show links to the gene report page, but also directly to the genome browser view (Fig. 6).

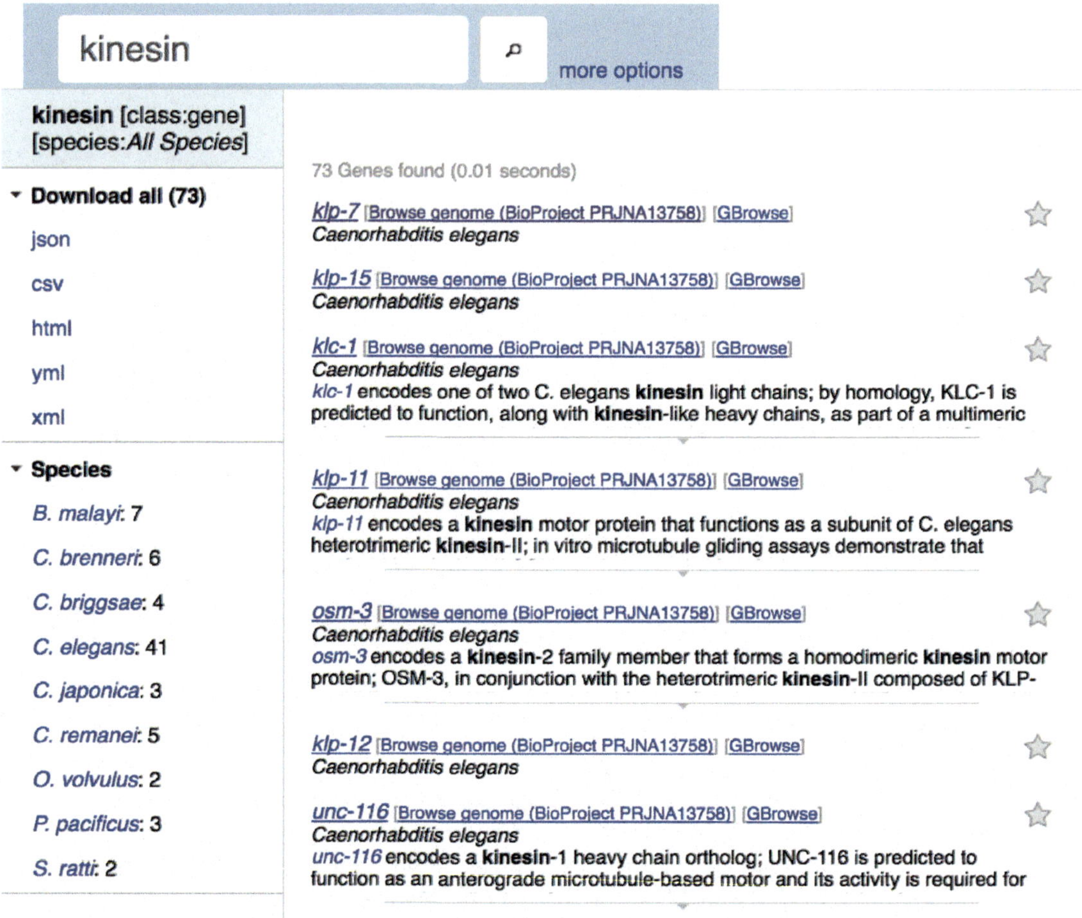

Fig. 6 The search results disambiguation screen showing several features of note. First, on the left hand side, you will find faceted results, broken down by class and species. Second, users may download search results in a variety of formats. In the main view, search results provide quick links that are class-specific. For example, genes (as shown here) provide direct links to gene report pages or to the genome browser. Search results can be directly added to your favorites list from this display too

2.4 Report Pages

Report pages are consistently structured across the site. On the left-hand side of the page is the vertically oriented navigation bar that controls which content is displayed. The main viewport for content is located on the right hand side. There are report pages for most data classes in WormBase, such as genes, proteins, anatomy terms, researchers, strains, transgenes, publications, and genetic variations.

2.4.1 Controlling the Display of Widgets

The left side navigation bar allows users to enable or disable specific widgets or turn on analysis tools specific to the type of data currently displayed (Fig. 7). If a widget is already open, clicking on its name in the left side navigation bar will scroll the display to the

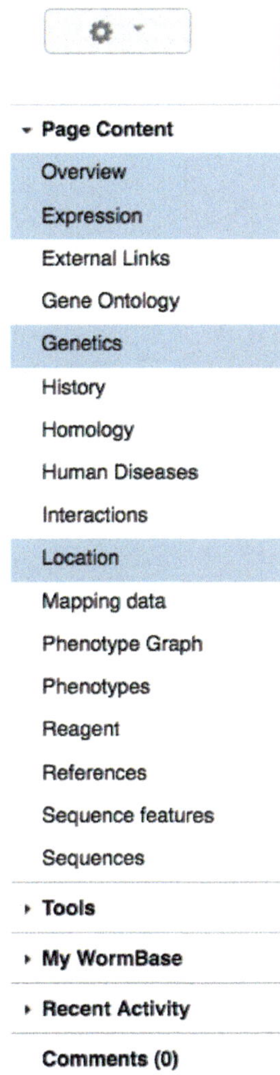

Fig. 7 The report page left-side navigation bar. This navigation bar contains titles of available widgets that can be enabled or disabled on a per-class (e.g., gene) basis. Clicking on a widget title will open it if it is not already opened and scroll the view to that widget. If it is already opened, the view will scroll to that position, an easy way to find a widget if you have many open. The navigation bar is subdivided into various sections including the primary widgets at the top, "Tools" for interacting with the current data, shortcuts to "My WormBase" and your "Recent Activity" on the site

location of that widget. If it is not open, it will be opened and the view scrolled to that widget, providing a convenient mechanism to find content if you have many widgets open.

2.4.2 Customizing Page Display

Every report page can be individually customized to suit user preferences. These preferences will persist from other pages of the same class (e.g., gene). Thus, it is possible to create a

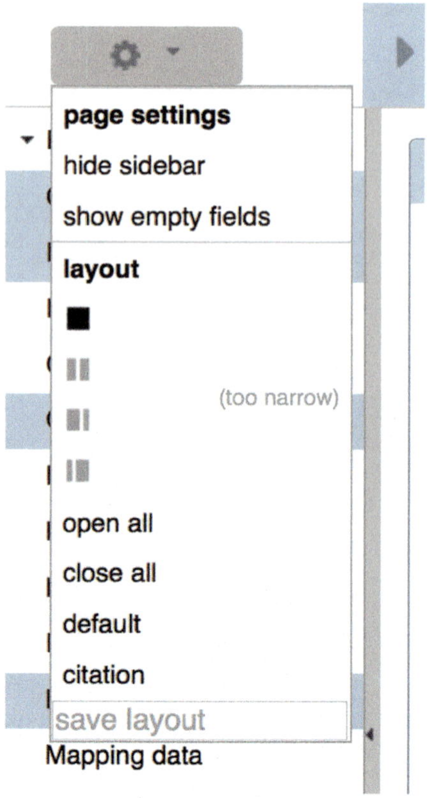

Fig. 8 The "layout" drop-down menu provides options for controlling the appearance of the main view of a report page. Users can select, for example, a single column, two columns of even width, or two columns of varying width. This menu also provides options for quickly opening or closing all widgets

sequence-oriented view of genes, or an anatomy-oriented view of expression data. Available layouts are specified from the drop-down "layout" menu at the top of the left side navigation bar (Fig. 8).

The display of report pages can be customized in additional ways besides enabling specific widgets, for example by setting a one- or two-column layout (Fig. 9). The order of widgets can be specified by dragging-and-dropping widgets to a specific position on the page. Selections will persist from one report page to the next. Instead of enabling or disabling widgets, clicking on the disclosure triangle will collapse the widget to a single title bar, temporarily minimizing it to reduce visual clutter (Fig. 10). To dismiss a widget altogether, simply click on the "X" that appears at the upper right corner of the widget when hovering the cursor over it or click on the "X" to the left of the widget name in the left side navigation bar. The widget can be reenabled at any time from the left side navigation bar.

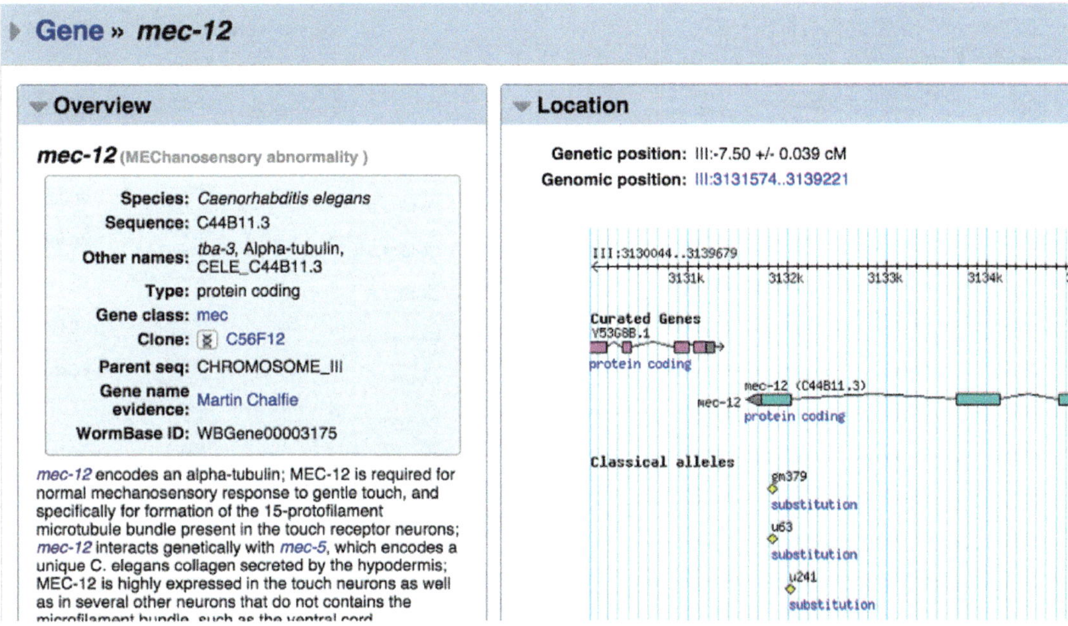

Fig. 9 Customizing the view of a gene report page. Here, a two-column layout makes it easy to compare the "Overview" and "Location" widgets side-by-side. Preferences will be saved and used from one report page to the next

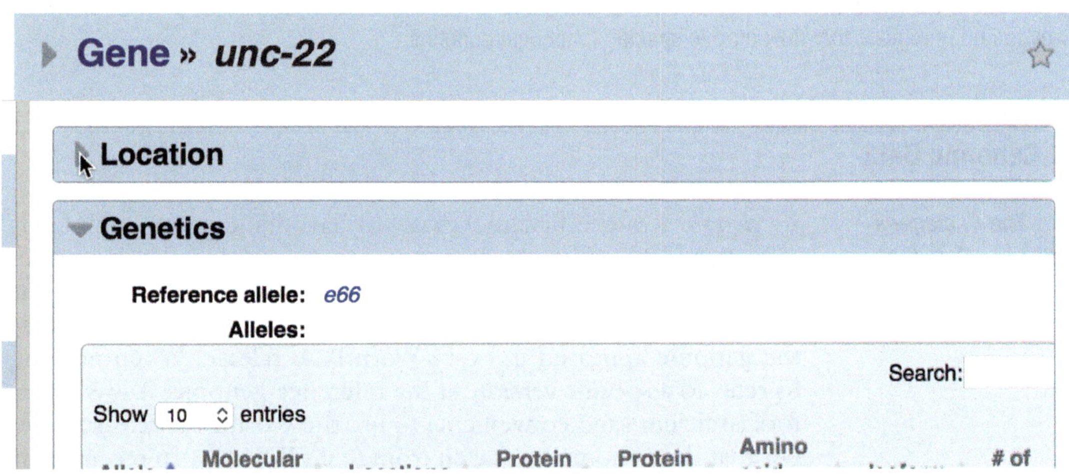

Fig. 10 Widget contents can be collapsed to a single title bar view to quickly hide their contents. Simply click on the disclosure triangle to open or close as necessary

2.4.3 Working with Tabular Data

Much of the data at WormBase is presented in a tabular format. We have standardized the display of tables with many features that make it easy to sort, search, and download data. For example, every table can be searched for its contents or sorted in ascending or descending order on any column (Fig. 11).

Search: Missense Save table

Show 10 entries

Allele	Molecular change	Locations	Protein effects	Protein change	Amino acid position	Isoform	# of Phenotypes	Method	Strain
gk170196	Substitution	Coding exon	Missense	P -> S	32	C44B11.3	0	Million mutation	VC40061
gk170195	Substitution	Coding exon	Missense	S -> N	50	C44B11.3	0	Million mutation	VC20161
gk636747	Substitution	Coding exon	Missense	R -> H	60	C44B11.3	0	Million mutation	VC40436
u76	Substitution	Coding exon	Missense	D -> N	69	C44B11.3	2	Substitution allele	
gk672907	Substitution	Coding exon	Missense	S -> F	140	C44B11.3	0	Million mutation	VC40513
gk600523	Substitution	Coding exon	Missense	G -> E	142	C44B11.3	0	Million mutation	VC40370
e1607	Substitution	Coding exon	Missense	G -> S	144	C44B11.3	6	Substitution allele	
gk583647	Substitution	Coding exon	Missense	L -> F	152	C44B11.3	0	Million mutation	VC40340
gk915672	Substitution	Coding exon	Missense	E -> K	196	C44B11.3	0	Million mutation	VC40989
gk854211	Substitution	Coding exon	Missense	P -> L	307	C44B11.3	0	Million mutation	VC40864

Showing 1 to 10 of 15 entries (filtered from 73 total entries) 1 2

Fig. 11 Working with tables. Tables can be sorted in ascending or descending order on any column, constrained by a search, or downloaded in any number of file formats. Here, we have filtered the *mec-12* alleles table by "Missense" and then sorted the table by the position of the mutation. This makes it easy to quickly see the location of variations in relation to their position in the predicted conceptually translated protein isoform as a way to find mutations that may map to specific functional domains

3 Genomic Data

3.1 The C. elegans Reference Genome

C. elegans was the first metazoan to have its genome completely sequenced [3], and since then the genome sequence has been subject to active curation and improvement. In the early days of the project, updates were frequent and consequently a new version of the genome appeared in every WormBase release. When needing to refer to a specific version of the reference genome, it was therefore sufficient (and convenient) to use the WormBase release number that the genome was taken from (e.g., WS118). In recent years however, the reference genome has stabilized, and is updated infrequently. The same version of the reference sequence often persists for many WormBase releases. In acknowledgment of this, in 2010 WormBase began assigning official names to new versions of the reference sequence itself, distinct from WormBase release names. When we submit an updated reference sequence to the International Nucleotide Sequence Database Collaboration (INSDC) [4], care is taken to label it with this version name, such that users should be able to obtain the genome from WormBase or the INSDC and know exactly what version they are using. This has not always been the case (*see* Table 1).

Table 1
Selected versions of the *C. elegans* reference genome

Genome release date	WormBase releases	WormBase assembly name	INSDC assembly name	UCSC assembly name
Mar 2004	WS120-WS122	–	–	WS120/ce2
May 2005	WS142-WS145	–	WS144	–
Dec 2006	WS169-WS176	–	–	WS170/ce4
Aug 2007	WS180-WS185	–	–	–
Mar 2008	WS189-WS193	–	WS190	WS190/ce6
Sep 2008	WS194-WS196	–	WS195	–
Apr 2009	WS202-WS214	–	–	WS210/ce8
May 2010	WS215-WS234	WBcel215	WS215	WS220/ce10
Nov 2012	WS235-date	WBcel235	WBcel235	WBcel235/ce11

The recent history of the reference genome can also be viewed at WormBase on the "Genome Assemblies" widget of the *C. elegans* landing page (http://www.wormbase.org/species/c_elegans#03--10)

A table with links to recent versions of the genome can be viewed in the "Genome Assemblies" widget of the *C. elegans* landing page (http://www.wormbase.org/species/c_elegans). The table provides a link to the file containing the reference genome sequence. Most users however obtain the genome sequence from our FTP site, which also includes a version of the sequence in which the repetitive regions have been masked (*see* Subheading 11.2).

3.2 Genes, CDSs, Transcripts and Proteins

WormBase annotates the reference genome with a variety of different types of feature, via a combination of analysis pipelines and data integration from both large and small scale studies. These features include sites associated with transcription (transcription initiation and termination sites, *cis* and *trans* splice sites), repetitive regions (including transposable elements), regulatory regions (enhancers, silencers, promoters, transcription-factor binding sites), and regions/sites that vary from the reference in mutant or wild isolate strains of *C. elegans.*

The most widely used annotations we produce are gene models, which are the exon/intron structures of transcribed regions. For protein-coding genes in particular, we initially manually annotate the coding sequence (CDS), which is the part of the transcript that is translated. Full-length transcript structures are created by software that extends the curated CDSs using alignments of transcriptome data (ESTs, cDNAs, and RNA-Seq reads), and other evidence (e.g., experimentally confirmed trans-splice sites, transcription initiation and termination sites). Identifiers are assigned to CDSs and full-length transcripts according to strict nomenclature (Fig. 12).

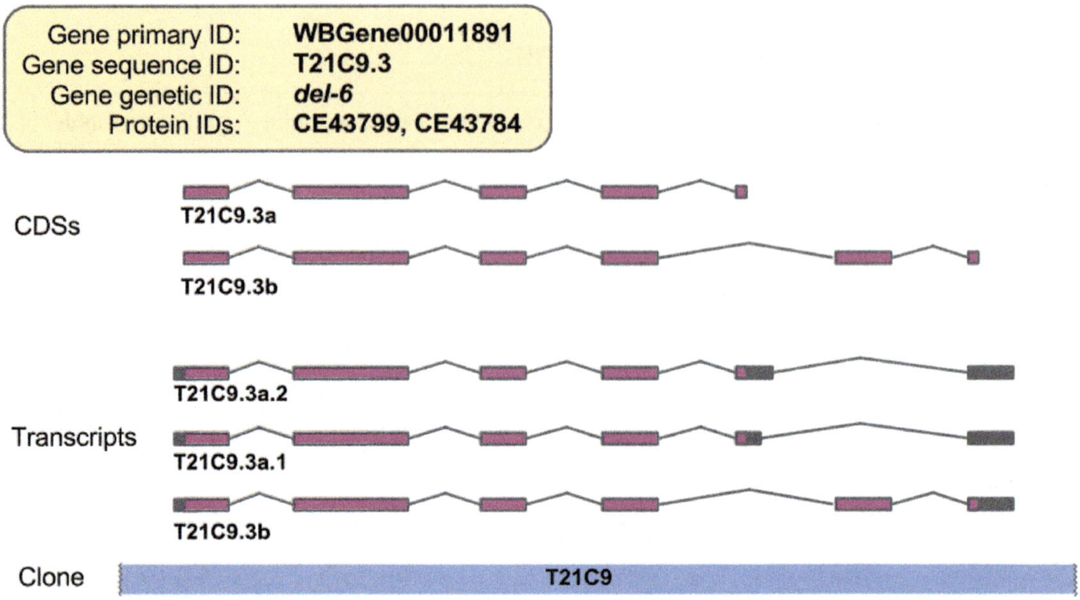

Fig. 12 The *C. elegans del-6* gene locus, showing curated CDS structures, and the predicted full-length transcript structures. The various different types of identifier associated with the gene and its transcripts are shown. Initially, a new protein-coding gene model comprises a single CDS, and that CDS is assigned a CDS sequence identifier to match the sequence identifier of the gene to which it belongs. If there is evidence for one or more additional isoforms of a CDS at a locus, then they are distinguished by giving them letters after their name. By default, the predicted full-length transcript structure is given the same sequence identifier as the curated CDS that it extends. However, if there is evidence for alternative splicing in the 5′ or 3′ untranslated region (UTR) of the transcript, then multiple full-length transcripts sharing the same CDS are created. These transcript isoforms are distinguished by the addition of a dot and numbers after the sequence name of the CDS

The CDS, transcript, and protein sequences associated with a gene can be viewed and downloaded from the "Sequences" widget of the gene report page (Fig. 13). Obtaining sequences for many genes at once can be achieved through our FTP site (Subheading 11.2) or via the WormBase ParaSite BioMart tool (available under the "Tools" menu).

3.3 The Genome Browser

Genome browsers at WormBase are among the most heavily used features of the web site. The older genome browser was originally designed for WormBase and later became GBrowse [5], a part of the Generic Model Organism Database project (GMOD, http://www.gmod.org/). It has reached end of life and is no longer being actively developed, so a new genome browser, JBrowse [6] has been chosen to take its place. JBrowse, also a GMOD software project, is in use at hundreds of sites worldwide. Since JBrowse is written entirely in JavaScript and is executed in the web browser rather than on a server, it provides a very fast and fluid interface for users. While GBrowse is still available at WormBase, all future genome browser development at WormBase will be devoted to JBrowse.

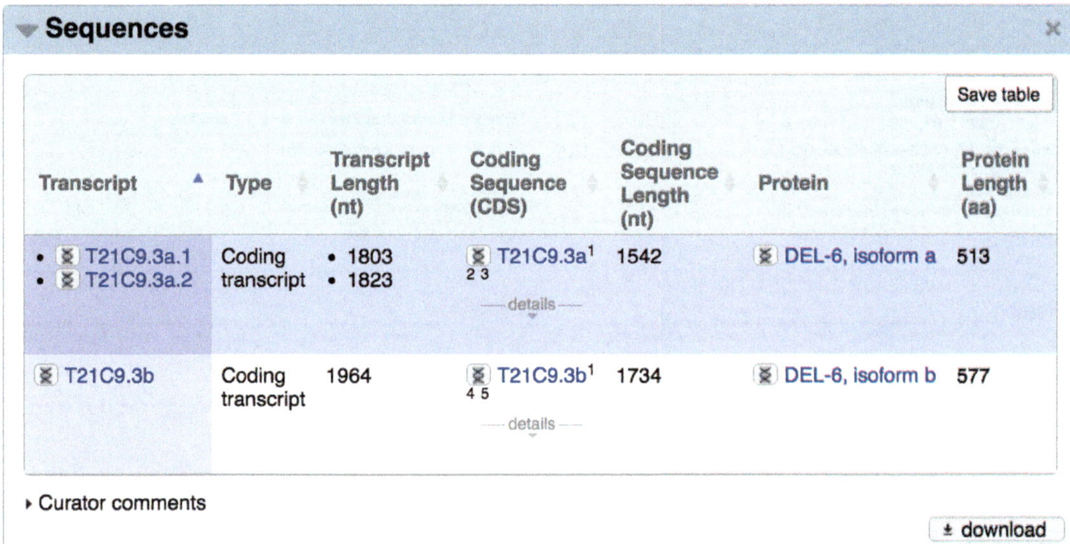

Fig. 13 The sequences widget from the *del-6* gene report page. The identifier for each CDS, transcript, and protein is hyperlinked to a page describing more information about that entity, and clicking on the double helix icon to the left of any identifier gives rise to a context-specific pop-up window showing the sequence. For example, for transcript sequences in particular, the pop-up shows spliced and unspliced sequences, with color-coding used to highlight the exonic regions

To access a genome browser, mouse over "Tools" and from the drop down, select either "GBrowse" or "JBrowse" from the menu. Alternatively, many report pages contain inline images of discrete genomic regions. Clicking on these will open the genome browser with the same coordinates of the image. From the genome browser, one can navigate to different regions by searching for coordinates or feature names. For example, when browsing the *C. elegans* genome, searching using the format *chromosome:start..stop* will open a view of the genome corresponding to the chromosome with a width of the start to the stop.

For users already familiar with GBrowse, the user interface of JBrowse will look familiar, with buttons for zooming and panning, as well as a search field and a menu for switching genomes (*see* Figs. 14a, b). One advantage JBrowse has compared to GBrowse is the visualization of the tracks themselves: the data tracks take up the majority of the web page, optionally using the full width of the screen, giving users the option of a large work space to view their data. The WormBase instance of JBrowse provides two methods for selecting tracks. The default method is via a list of track names, organized by category, on the left side of the page. When the user initially loads this page, there are approximately 65 tracks available in *C. elegans*, with over 1200 more tracks in a "collapsed" set of tracks from modENCODE [7]. Since finding specific tracks in such a large set can be a daunting task, JBrowse also provides a "track selector" button to switch to a faceted track selector, where

a

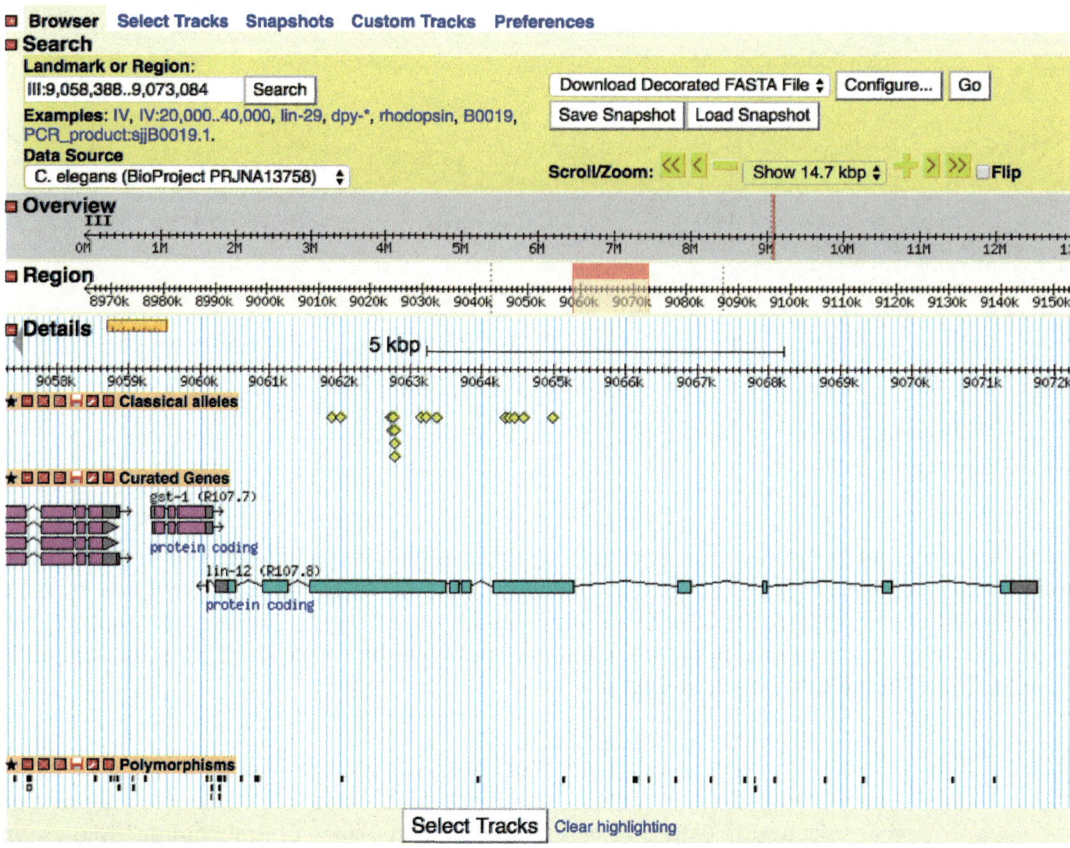

Fig. 14 (**a**) The graphical user interface of GBrowse showing the "Classical alleles," "Curated Genes," and "Polymorphism" tracks for *C. elegans* chromosome III (*lin-12* locus) (**b**) The graphical user interface of the same data in JBrowse

a track list is available by clicking on a tab in the upper left corner of the page and a drawer with all of the tracks available slides out from the left side of the page. In Fig. 14b, the pull out tab for JBrowse track selection is shown. The tracks can be selected en masse by category or the descriptions of the tracks can be searched to help narrow down the list. When using the faceted track selector, the list of checkboxes on the left is removed, and the track data can fill the full width of the screen.

JBrowse allows users to incorporate their own data so that it can be visualised alongside WormBase data. Users are not required to upload their data to render it; instead, users can supply either a local file or a URL specifying the location of a remote data file, and JBrowse will render the data by processing it locally. This is made possible due to the "in browser" execution of JBrowse; all of the software to process the data and display it is included in the JavaScript that is downloaded when the user first goes to a JBrowse page.

b

Fig. 14 (continued)

JBrowse has the ability to make high resolution screenshots. When the "Screen Shot" button is clicked, users are presented with a dialog box that lets them modify the view that will be imaged, for instance removing navigation or track selection portions of the page, as well as letting users specify options specific for each track, like the height of the track or placement of axes for quantitative tracks. Also the type of image (png, jpeg or pdf) as well as the size and resolution can be specified.

4 Comparative Genomics

4.1 Other Nematode Genomes

WormBase contains genomic data from many nematode species beyond *C. elegans* (http://www.wormbase.org/species/all). The genomes of some species are deeply integrated into WormBase and treated in a similar manner to *C. elegans*, with manual curation of transcript structures, assignment of WormBase identifiers to genes and other sequence features, systematic tracking of changes and maintenance of a full change history. We refer to these as WormBase "core" species. Historically, the set of core species was restricted to close relatives of *C. elegans*. Recently however, we have also added the genomes of selected parasitic nematodes to the core set.

Outside of the core species, WormBase imports other nematode genomes from the INSDC, or from direct submission, and provides a genome browser with annotation tracks displaying data provided by the authors (either via annotations on the INSDC records themselves, or by direct GFF3 submission to WormBase). We do not curate or assign WormBase identifiers to the annotations for these noncore species, but display them exactly as submitted/published. The original authors maintain ownership of the reference sequence and annotation for these genomes. When we become aware of a new genome, or an update to a genome we already have, we endeavor to incorporate the data into WormBase as quickly as possible. Selected genomes of particular relevance to the study of *C. elegans* are made available via the main WormBase website. The complete set of all nematode genomes (as well as platyhelminth genomes) can be found in our sister resource, WormBase ParaSite (*see* other chapter in this issue).

4.2 Protein-Level Homology and Domains

We align reference protein sets from a variety of organisms to both nematode protein sequences (protein-to-protein alignments) and nematode genome sequences (protein-to-genome alignments), using BLAST+ [8, 9]. The protein-to-genome alignments can be visualised on the genome browser (via the "Sequence Similarity"/"Proteins" track group), and the protein-to-protein results can be viewed in the Homology section on the protein and gene report pages of the website.

Another view of the protein similarity data is the "Protein Aligner" widget accessible from the protein report pages. This is a global multiple alignment of the protein of interest with the closest similar protein from each other species (by *p*-value), using MUSCLE [10]. The alignments are precalculated and colored in a way to reveal common properties conserved between the proteins.

While conserved regions of proteins can be apparent from viewing the color-coded protein alignments directly, a more sensitive fine-grained view of protein evolution can be achieved by considering that proteins comprise conserved domains.

WormBase annotates each protein with its domain architecture using InterProScan [11]. This applies a number of established resources and tools for domain annotation and additionally integrates the results into higher-level InterPro domain annotations. Because each InterPro entry has associated functional annotation (both textual, and by using terms from the Gene Ontology [12]), InterPro domain analysis thus provides automatic functional annotation for gene products. Gene Ontology annotations from InterProScan can be seen on the "Gene Ontology" widget of the gene report pages, alongside manually curated annotations (*see* Subheading 5.2 on Gene Ontology). In addition to the InterPro annotations, we also show active site annotations imported from Pfam [13], as well as phosphorylation sites based on submitted mass-spectrometry data. Functionally annotated clusters of related proteins made by the eggNOG project [14] are included in WormBase and shown in the "Homology Groups" section of the "Homology" widget of protein report pages.

4.3 Gene-Level Homology: Orthologs and Paralogs

WormBase stores and displays orthologous and paralogous relationships between pairs of genes, integrating data from a variety of resources and methods. Orthologs and paralogs can be seen in the respective ortholog and paralog sections of the "Homology" widget on gene report pages, and can be downloaded as tables. The "Method" column of the table shows the methods and resources that defined the relationship. This is an important feature as orthology predictions are very dependent on the underlying gene/protein sequences and algorithms used. Combining the results of multiple methods provides the user with an estimate of the prediction quality (i.e., orthologies predicted by multiple methods can be seen as being more reliable). Following the "Method" link provides more information about the method, including database versions used, as well as papers and who conducted the analysis. For WormBase genes included in TreeFam [15], an interactive TreeFam tree is shown.

As part of the preparation for each release, WormBase deploys the EnsemblCompara [16] software to compute orthologs and paralogs using the current WormBase protein set, and the most complete set of nematode proteomes. This is in contrast to the other imported sources of orthology which are based on snapshots of the proteome taken at some point in the past.

5 Ontologies at WormBase

5.1 The WormBase Ontology Browser

WormBase extensively uses the C. elegans Anatomy Ontology, Human Disease Ontology, Gene Ontology, Life Stage Ontology, and Phenotype Ontology to annotate genes [17–20]. Because these ontologies consist of sets of terms that are hierarchically related to each other, it is useful and convenient to peruse them graphically. Thus, we provide three standard graphical views for each ontology: a stand-alone hierarchy browser (Fig. 15a) that allows top-down, layer-by-layer expanded viewing of the whole ontology, a graph viewer (Fig. 15b) that illustrates a focus term and its related terms in the graph form, and an inference tree viewer (Fig. 15c) which also shows focus term relationships but in a tree form.

The expandable hierarchy browser (WormBase Ontology Browser or "WOBr") is accessible as a standalone tool from the WormBase "Tools" menu, and allows root-to-leaf drill down browsing of ontologies' directed graphs. Each term is a node and branch nodes can be toggled to expand or collapse with a click. The graph viewer is in the "Ontology Browser" widget on each ontology term page, which shows as a comprehensive ontology subgraph all relationships connecting the focus term to ontology roots. There is also an inset that provides quick access to sibling terms of the focus term. The inference tree viewer provides a summary of the focus term's direct relationship with its "child" terms and inferable relationships with its "ancestors" via chains of transitive relationships, up the hierarchy to root terms. We use information in annotation files, combined with inferred relationships, to provide quick access to lists of genes directly, and by inference, annotated with the focus term. We further make use of the available information to provide results of "pre-canned" complex queries. For example, some users are interested in knowing what genes may be specifically expressed in a specific tissue; and it would require several simple queries and steps to combine the results to answer this question. WormBase precomputes a list of genes that may be specifically expressed in a cell or tissue and displays the results at the bottom of the "Ontology Browser" widget on the anatomy term page (for example, see the "neuron" anatomy term report page: http://www.wormbase.org/species/all/anatomy_term/WBbt:0003679#03--10).

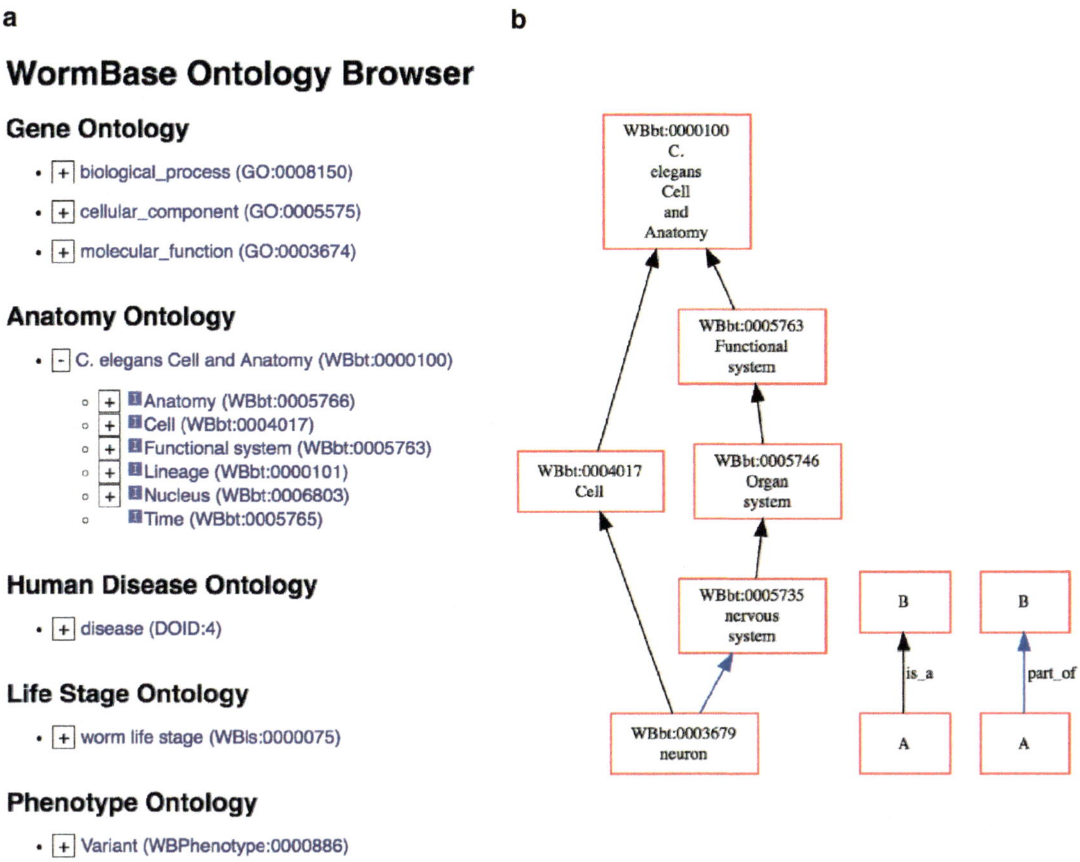

Fig. 15 Example views of the WormBase Ontology Browser (WOBr). (**a**) The hierarchy browser, (**b**) graph view for the anatomy term "neuron," and (**c**) inference tree view for the "neuron" term

5.2 Gene Ontology Data at WormBase

The Gene Ontology (GO) is a controlled vocabulary designed to describe three central aspects of gene function: (1) the Biological Processes (BP) in which a gene product is involved; (2) the Molecular Function (MF) or "activity" that is enabled by a gene product; and (3) the Cellular Component (CC), or subcellular location, where that function occurs [17].

In the GO, biological concepts are represented by GO "terms" that consist of a term name, textual definition describing the meaning of the term, and a unique, numerical identifier. Additional GO term information may include, for example, synonyms or free-text comments on term usage. Within the GO, terms are related to one

c

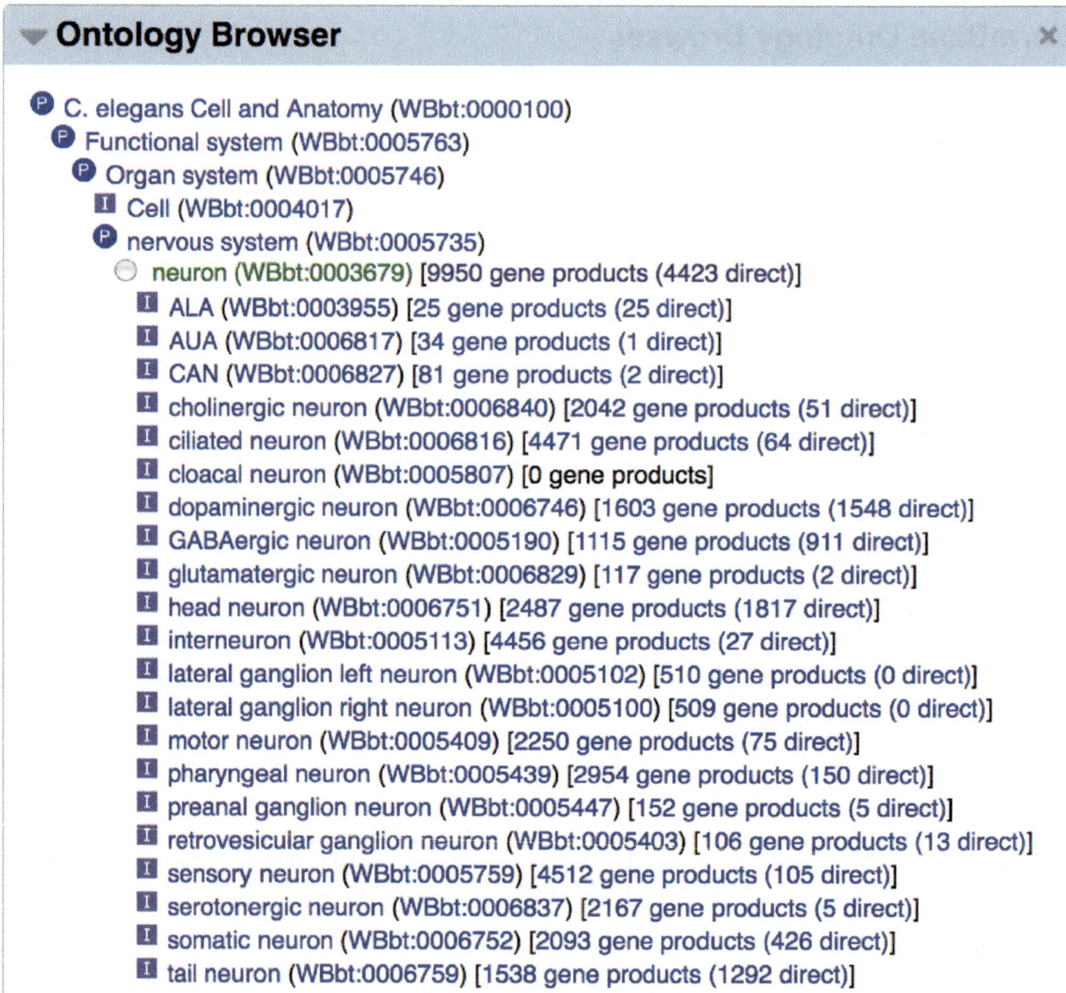

Fig. 15 (continued)

another via specific parent-child relationships. These relations include, but are not limited to, is_a, e.g., "plasma membrane" is_a "membrane"; part_of, e.g., the "nuclear envelope" is part_of the "nucleus"; and regulates, e.g., "regulation of G1/S transition of mitotic cell cycle" regulates the "G1/S transition of mitotic cell cycle." This formal representation of biological knowledge allows not only for a standardized view of gene function, but also for computational reasoning that forms one of the cornerstones of gene set analysis.

5.2.1 Gene Ontology Annotations in WormBase

GO annotations are associations between GO terms and WormBase genes. Although derived from a number of different curation pipelines, the basic GO annotation consists of a GO term, an evidence code indicating the type of experiment or analysis used to make the association, and a reference in which the primary data, or details about the experiment or analysis, may be found. Additional annotation fields may include evidence code-specific details, such as the interacting partner for an annotation inferred from a genetic interaction, annotation qualifiers such as "contributes_to", which is used, for example, to describe the role of noncatalytic members of multisubunit enzymes, and annotation extensions that provide additional contextual information such as the cell or tissue type in which a BP or MF occurs [21].

5.2.2 Annotations on Individual Gene Report Pages

On the WormBase gene report pages, GO annotations are visible under the "Gene Ontology" widget (Fig. 16). Annotations are listed in three separate tables, one for each branch of the ontology. Two display options are available. The default, "Summary view" provides a basic annotation display showing which GO terms have been associated with the gene and, if present, annotation extensions that provide specific contextual information for a term. The "Full view" (Fig. 17) additionally shows the evidence code used for the association, evidence code-specific details (also known as the "With" or "From" column), and a details menu that when opened lists the date on which the annotation was last updated, a brief citation and link for the associated reference, and the database that contributed the annotation to WormBase. A summary of the evidence codes used for GO curation at WormBase and the additional information associated with them is presented in Table 2. Table 3 lists the types of annotation extensions used for WormBase GO annotations and some examples of how they serve to qualify the respective GO term.

The GO display on gene report pages lists all GO annotations, regardless of the method used to make the association. Thus, it is not uncommon to see either redundant annotations with different supporting evidence or annotations to both parent and child GO terms listed for a single gene. As a general rule, though, automated methods for assigning GO terms result in annotations to less specific GO terms than manual methods that use published, experimental findings as supporting evidence and typically strive to annotate genes to the most granular GO term possible. A summary of the major GO annotation pipelines at WormBase is presented in Table 4 along with the associated evidence codes for each, and specific examples for the *gcy-8* receptor guanylate cyclase.

All nematode species represented in WormBase are assigned GO annotations via the InterPro2GO [24] automated annotation pipeline. This pipeline derives annotations from analysis of conserved processes, functions, and localizations associated with

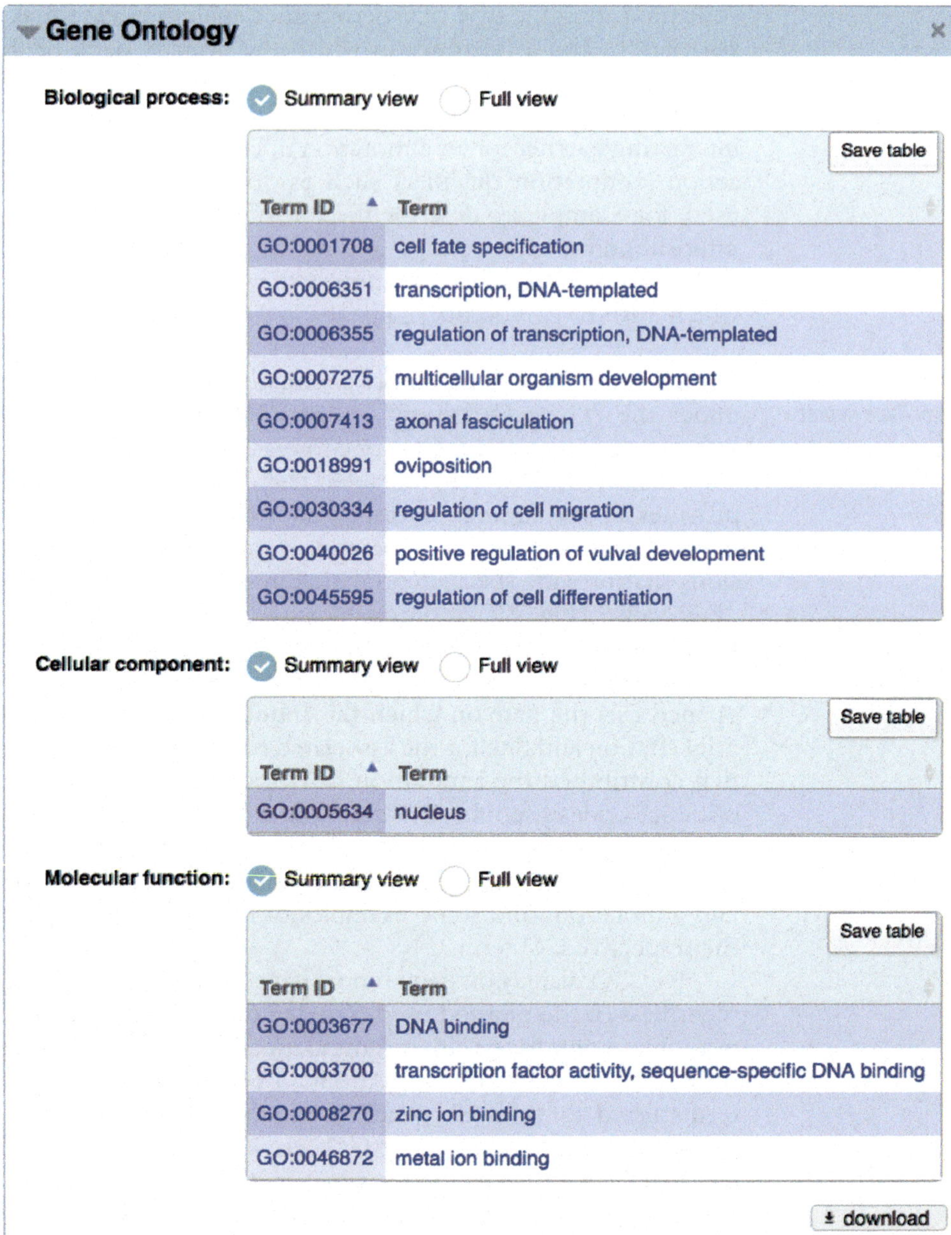

Fig. 16 Summary view of Gene Ontology data for the *lin-11* gene. The summary view displays GO IDs and term names from the three GO branches (BP, CC, and MF) annotated to *lin-11*. Depending upon experimental data, or the method used to create the annotation, a gene may be associated with GO terms of differing granularity, e.g., GO:0003677 DNA binding and GO:0003700, transcription factor activity, sequence-specific DNA binding

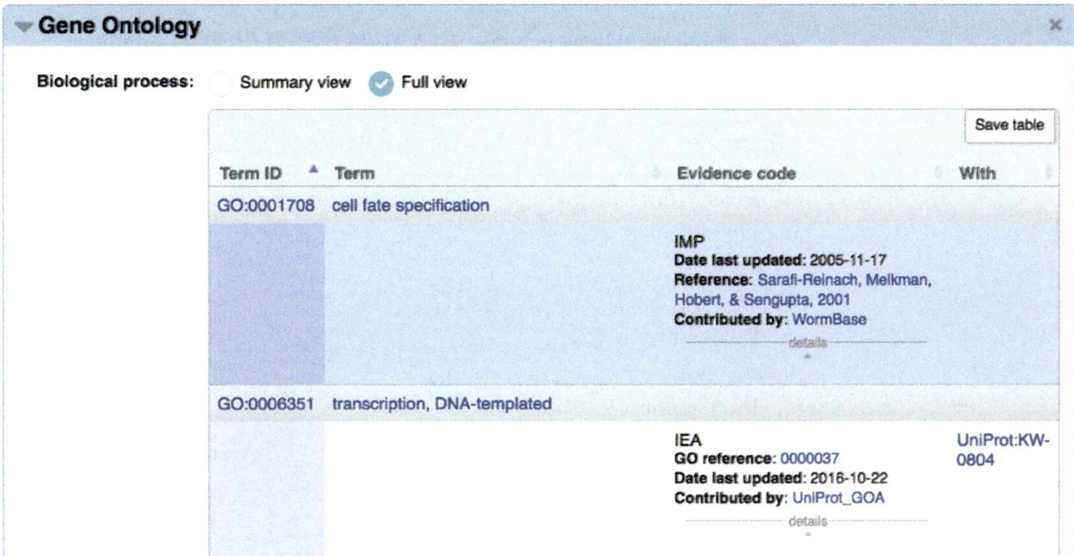

Fig. 17 Full view of Gene Ontology data for representative BP annotations to *lin-11*. The full view displays GO IDs, term names, three-letter evidence codes, the date the annotation was last updated, the associated reference, contributing annotation group, and additional supporting information in the "With" column

protein domains and families as catalogued by the InterPro database. Note, however, that manual annotations are largely limited to *C. elegans* with a few annotations also assigned to *C. briggsae*.

5.2.3 Searching and Browsing the GO and Associated Annotations

Browsing the GO is one of the best ways to learn about the variety of terms in the ontology and their relationships to one another. In WormBase, there are several entry points to find GO term information. From the search menu at the top right of each page, users can search for specific GO terms by typing the term name or unique GO identifier and then selecting "Gene Ontology" from the drop-down menu. If a specific GO term name is not known, then typing a few letters for the biological concept of interest will bring up an autocomplete menu that suggests possible matches. Given the complexity of the GO, rather than trying to find an exact term match, it can be useful to select a related term and then use the ontology browser (described above) to navigate the ontology.

Table 2
Gene Ontology evidence codes used in WormBase GO annotations

Evidence	Three-letter code	Supporting information
Inferred from Mutant Phenotype	IMP	Variations, RNAi Experiments, Phenotypes
Inferred from Genetic Interaction	IGI	Genes
Inferred from Physical Interaction	IPI	Genes, Proteins
Inferred from Direct Assay	IDA	na
Inferred from Sequence or Structural Similarity	ISS	Genes, Proteins
Inferred from Expression Pattern	IEP	na
Inferred from Curator	IC	GO term IDs
Inferred from Sequence Model	ISM	na
Inferred from Biological aspect of Ancestor	IBA	Panther Tree Nodes[a]
Inferred from Key Residues	IKR	Panther Tree Nodes[a]
Traceable Author Statement	TAS	na
Nontraceable Author Statement	NAS	na
No biological Data found	ND	na
Inferred from Electronic Annotation	IEA	InterPro entries[b,c], UniProtKB Keywords[c], UniProtKB Subcellular Localization[c]

[a]Gaudet et al., 2011, PMID:21873635 [22]
[b]Mitchell et al., 2015, PMID:25428371 [11]
[c]Huntley et al., 2015, PMID:25378336 [23]

Selecting a term from the search menu leads users to a GO term report page. Here the "Overview" widget displays the GO term name, associated definition, branch of the ontology to which the term belongs, and its unique GO ID. Two additional widgets on the GO term page allow users to see all associations (annotations) made to that GO term and placement of the term in the ontology. Like on the gene report pages, in the "Associations"

Table 3
Representative annotation extensions for WormBase GO annotations

Gene	GO term	Annotation extension	Type of contextual information
mom-4	protein serine/ threonine kinase activity	*has_input*: *lit-1*	Enzymatic activity—substrate
atf-7	nucleus	*part_of*: intestinal cell	Cellular component—cell type
let-381	mesodermal cell fate specification	*results_in_specification_of*: coelomocyte	Biological process—cell type

Table 4
Major GO annotation pipelines at WormBase[a]

Annotation pipeline	Associated evidence codes	Sample *gcy-8* annotation
Manual, literature-based	IMP, IGI, IDA, IPI, ISS, IEP, IC, ISM, ND, TAS, NAS	Thermotaxis
Phylogenetic based on PANTHER families[b]	IBA, IKR	Signal transduction
UniProt Keyword (KW) Mappings[c]	IEA	cGMP biosynthetic process
InterPro2GO Mappings[c,d]	IEA	Cyclic nucleotide biosynthetic process
Enzyme Commission (EC) Mappings[c]	IEA	Guanylate cyclase activity
UniProt Subcellular Localization (SL)[c]	IEA	Plasma membrane

[a]Other annotation pipelines include UniPathway and UniRule (Huntley et al., 2015, PMID:25378336 [23])
[b]Gaudet et al., 2011, PMID:21873635 [22]
[c]Huntley et al., 2015, PMID:25378336 [23]
[d]Mitchell et al., 2015, PMID:25428371 [11]

widget users can see a "Summary view" or "Full view" of the annotations associated with that GO term. Where applicable, the "Associations" widget also lists the InterPro motifs associated with specific GO terms; these mappings provide the basis for the

InterPro2GO annotations noted above. The "Ontology Browser" widget allows the user to see the GO term in the overall context of the GO. Two views are presented: the inferred tree view at the top of the widget and the graph view just below the tree view.

5.2.4 Downloading GO Annotations

The most common use of GO annotations is for gene set enrichment analysis. To perform such analyses, you can use the "Gene Set Enrichment Analysis" tool available under the "Tools" menu or download the complete set of GO annotations and perform the analysis yourself. The full set of *C. elegans* GO annotations is available as a Gene Association File (GAF) at the GO web site under the "Downloads" menu: http://www.geneontology.org/page/download-annotations. The GAF is a 17-column, tab-separated file that contains all of the information related to a GO annotation, including GO term ID, reference, evidence, and annotation extensions, as well as metadata about the WormBase gene, such as synonyms. Full details on the format of the GAF may be found at: http://www.geneontology.org/page/go-annotation-file-gaf-format-21.

Users may also download the table view of annotations on individual gene report pages by clicking on the "Save table" button in the upper right of each GO table or by clicking on the "Download" link at the bottom of the widget and selecting one of the four available formats.

6 Gene Expression Data

WormBase strives to maintain an up-to-date collection of gene expression descriptions extracted from the literature and directly submitted by individual laboratories. Gene expression data in WormBase include conventional expression pattern analysis, e.g., reporter gene analysis, antibody staining, in situ hybridization (ISH), single molecule fluorescent in situ hybridization (smFISH), RT-PCR, qPCR, Northern blots, Western blots (what we refer to as small-scale expression data), as well as RNA-Seq, microarray, and DNA tiling array data (large-scale expression data).

Expression data in WormBase can be accessed on any gene report page by turning on the "Expression" widget in the left side navigation bar.

6.1 Navigating Small-Scale Gene Expression Data

Small scale gene expression data can be found at the top of the "Expression" widget on a gene report page. Three main tables summarize the curated expression data: (1) the "Expressed in" table (Fig. 18) lists all the anatomical structures in which the gene product has been detected; (2) the "Expressed during" table lists the life stages in which the gene is expressed and (3) the "Subcellular localization" table contains the list of the subcellular components

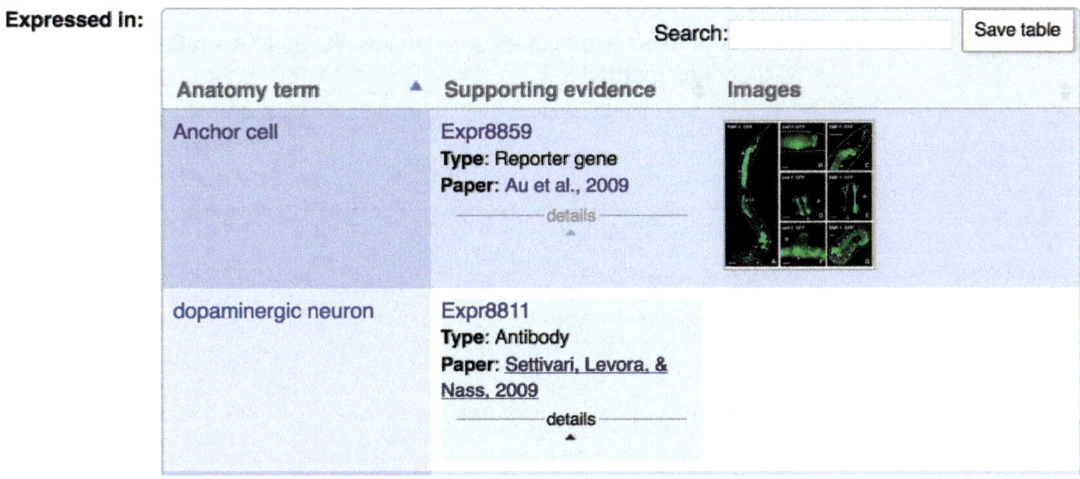

Fig. 18 Expression data can be accessed through the "Expression" widget on the gene report page. Once the "Expression" widget has been turned on, you can access all the available expression data for the gene

in which the gene product localizes. Note that each of these tables only appears if there is data present for the table.

The annotation to a specific cell/tissue is always made to the most granular term of the anatomy ontology. For example, if authors describe expression in HSN neurons, the annotation is made to HSNL and HSNR. This is especially important to know if you are browsing expression data from the anatomy page in the "Associations" widget as the correct input of the search is HSNL instead of HSN.

The "Supporting evidence" column specifies the type of experiment that has been used to determine the expression. Specifically, it will list if it was a reporter fusion analysis ("Reporter gene"), an in situ hybridization experiment ("In situ"), an Immunolocalization study ("Antibody"), or if the expression is driven by a *cis*-regulatory element ("Cis regulatory element"). Below the experimental type, we provide a reference to the paper from which the evidence was extracted.

WormBase release WS259 contains over 12,000 expression patterns determined by reporter gene fusions, 500 in situ hybridization experiments and over 1000 experiments for localization using commercially available antibodies or antibodies generated by individual laboratories (Table 5). Additional information on the transgene or construct used to determine expression—such as reporter, backbone vector, primers, or the antibody used to determine localization—can be found on the expression pattern report page, which can be accessed by clicking the "Expr####" listed in the "Supporting evidence" column.

Whenever possible, pending journal copyright permissions, we incorporate high-quality annotated images of gene expres-

Table 5
Number of expression patterns in version WS259 of WormBase grouped by detection method

Method	Number of expression patterns in WS259
Reporter Gene Fusions	12,791
Immunohistochemistry	1,137
In situ hybridization	565
RT PCR	330
Northern Blotting	356
Western blotting	96
Genome Editing	14
Cis Regulatory element	79
Total	15,368

Expression pattern for smf-1

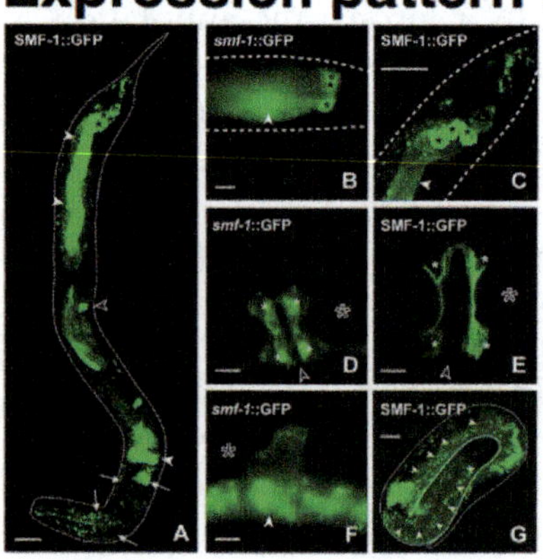

Reprinted from PLoS One, Au et al., 2009. PLoS 2009.

See original image

A, SMF-1::GFP strongly localizes to the anterior and posterior intestine (solid white arrowheads), to the anchor cell (hollow arrowhead) and to head neurons (white arrows). smf-1::GFP and SMF-1::GFP reveal expression of smf-1 gene in rectal gland cells (B,C, black asterisks), in the uterus (uv1, uv2, utse syncytium, D,E, solid white asterisks) as well as in the adult spermatheca (F) and the L1 hyp7 epidermis (G, white arrowheads). Dotted lines outline the cuticle of the worm. Hollow asterisks indicate position of fertilized embryos. Hollow arrowheads indicate position of the vulva. Scale bars are 5 um.

Expression pattern: Expression pattern for smf-1

Anatomical features: uterine seam cell
uterus
hyp7 syncytium
intestine
rectal gland cell
head neuron
spermatheca
Anchor cell
uv2
uv1

Paper: Au et al., 2009

Fig. 19 Images of gene expression. Expression images can be accessed through the 'view image' links in the "Expression pattern images" section of the "Expression" widget on the gene report page

sion directly submitted by individual laboratories or extracted from publications. In the gene report page "Expression" widget, clicking on the "view images" icon will display a pop-up window containing the image, the figure caption, spatiotemporal information, along with a link to the WormBase page for the original publication (Fig. 19). WormBase currently contains over 13,000 curated images.

6.1.1 Accessing Gene Expression Data

Depending on the scope of your search, conventional gene expression data can be accessed in different ways:

1. If you want to check the expression of a specific gene you can do so via the "Expression" widget on the gene report page.
2. If you wish to see which genes are expressed in a specific cell or tissue you can access the data on the anatomy report page by turning on the "Associations" widget in the left side navigation bar (Fig. 20).
3. An alternative way to find gene expression data is to use the WormBase Ontology Browser (WOBr) that provides an efficient way to browse anatomy terms and navigate the hierarchy (*see* Subheading 5.1). WOBr can be accessed from the anatomy report page by clicking the "Ontology Browser" widget on the left side navigation bar (Fig. 21). Next to each term you can see a number; this indicates how many genes have been directly—or indirectly—assigned to that particular anatomy term. With WOBr users can find specific anatomy terms without previous knowledge of the structure of the anatomy ontology.
4. Expression data may also be browsed by using WormMine (*see* Subheading 11.1). WormMine is an integrated search tool of WormBase data built with the Intermine data warehouse platform and can be accessed via the WormBase home page in the "Tools" menu. In the "Expression" tab on the WormMine home page are listed a few pre-canned (template) queries to browse gene expression. For instance, by clicking on the "Gene → Expression Pattern" query you are redirected to a template search page where you can simply retrieve all the expression patterns described for a particular gene—and export them in a table in your favorite format. The power of WormMine though, lies in the ability to construct complex queries that can be executed on single entities or lists. By clicking the "Edit Query" button you are now redirected to the Query Builder page, where you can navigate the WormBase data model. Here you can decide which columns you want to add in your output table.

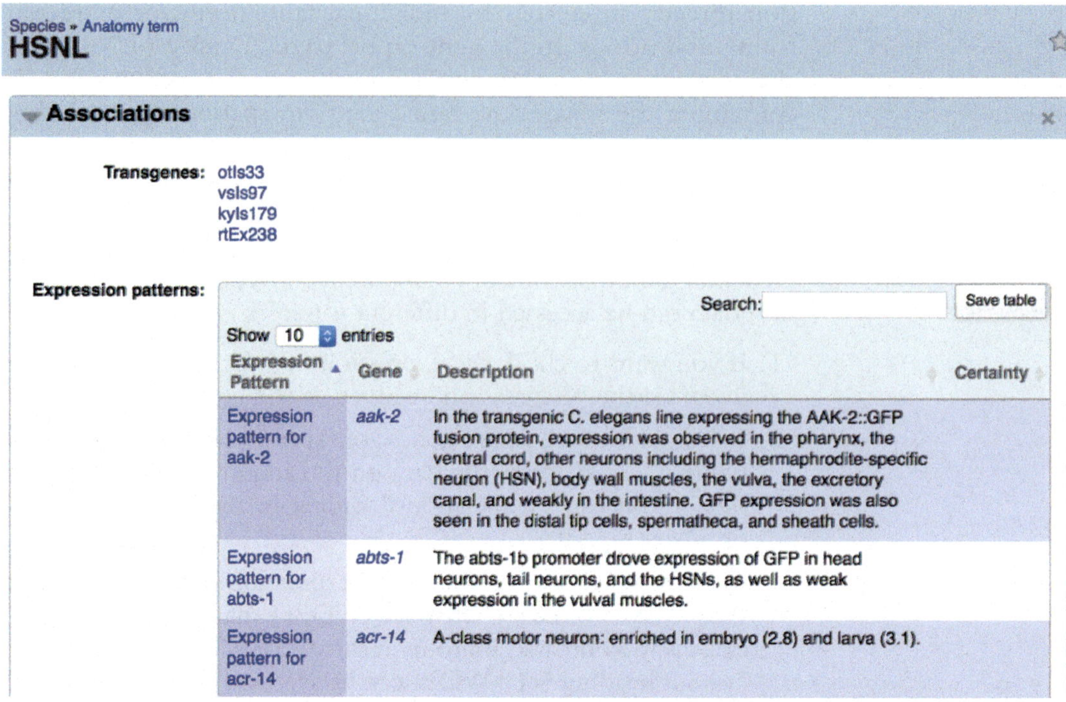

Fig. 20 On an anatomy term report page, users can retrieve a list of genes expressed in a specific cell/tissue through the "Associations" widget

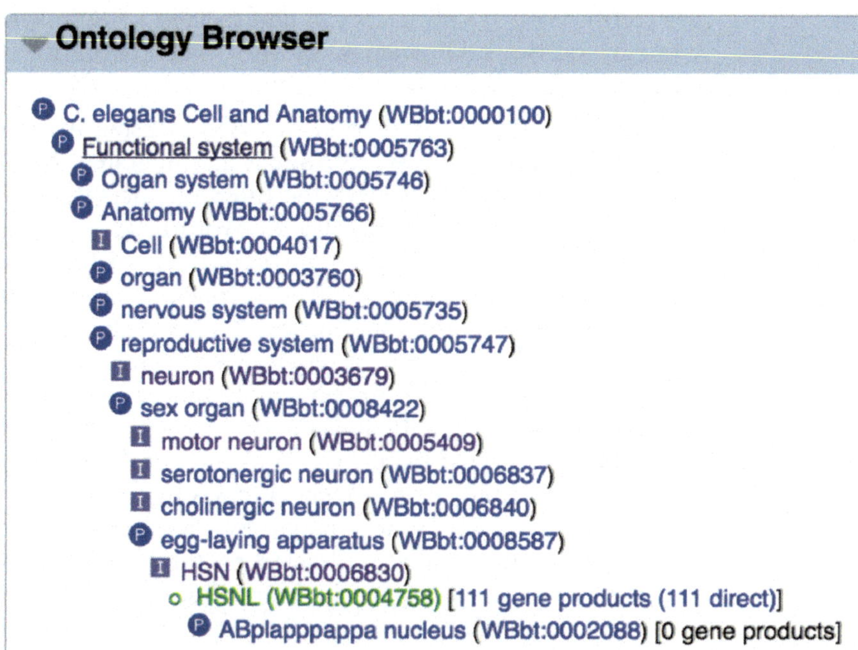

Fig. 21 By turning on the "Ontology Browser" widget on the anatomy term report page one can access the tree-like structure of the ontology, see how many genes have been found to be expressed in that specific cell/tissue, and easily navigate through the anatomy ontology without previous deep knowledge of the anatomy

6.1.2 Submitting Unpublished Expression Data: Micropublications

If you have unpublished expression data you can now "micropublish" it on WormBase by filling out the micropublication form (http://tazendra.caltech.edu/~azurebrd/cgi-bin/forms/expr_micropub.cgi).

The rationale behind the project is that not all data generated by publicly funded research is incorporated in the scientific literature. This information often includes high quality novel findings and is unfortunately not readily available to the scientific community. This knowledge can instead be shared with the public in the form of an open-access micropublication. Once you submit this data to WormBase, it will be reviewed by one or more experts in the field. If approved, your data will be assigned a stable digital object identifier (DOI), will be available on WormBase, and can be cited by traditional citation methods.

6.2 Navigating Large-Scale Gene Expression Data

In addition to small-scale expression data, WormBase provides a number of views of large-scale expression data, including RNA-seq data, microarray data, and expression clusters. This data is visible in the "Expression" widget on gene report pages below the small-scale expression data, on our genome browsers in special tracks, and via the SPELL tool (*see* Subheading 6.2.3).

6.2.1 RNA-Seq Expression Data

Short-read transcript data produced from coding transcript sequences (RNA-seq data) can be used to estimate the relative expression of loci. This is done by collecting the short-read data produced under selected conditions or life-stages and counting what proportion of reads are seen that align to the locus in question in comparison to the reads that align to the genome as a whole, normalized for the length of the locus. The expression is measured in Fragments Per Kilobase of transcript per Million fragments mapped (FPKM), as reported by the Cufflinks software [25, 26] and other packages.

The reads used to calculate the FPKM values come from the NCBI Sequence Read Archive (http://www.ncbi.nlm.nih.gov/Traces/sra/). Reads that originate from transcribed sequences are used but any experimental techniques that would cause biased results, such as selection for short sequences, ChIP-seq, ribosome fingerprinting and other unusual protocols are excluded. The Study ID and Experiment ID of each read library extracted from the NCBI Sequence Read Archive are noted and form part of the description used to annotate the results of the read libraries.

1. *modENCODE graph*: FPKM expression graphs from selected modENCODE [7] libraries are displayed as bar charts (Fig. 22) in the "Expression" widget. The bar chart titled "FPKM expression data from selected modENCODE libraries" displays the expression for each of various sets of life-stages using libraries which have been manually selected as representing the

baseline expression at each life-stage. In the first set, there is a time-series of measurements taken every 30 min during embryonic development. In the second are the "classical" life-stages with the first two "Early Embryo" and "Late Embryo" stages roughly dividing in two the time-series from the first set of data. Next are two male life stages, then expression from somatic cells in L4 and then dauer life-stages. The data have been produced from libraries made by using two different sets of protocol ("polyA+" selection and "Ribozero" selection). The data are quite variable, so to reduce the appearance of scattered points, the median value of the libraries at each stage has been calculated and is plotted as a grey bar.

2. *Mean and Median Values*: In the table "Aggregate expression estimates" (Fig. 23) further processing of the FPKM values for genes has been done by taking all available RNA-seq data and identifying those which are a control or which have been obtained when no particular experimental condition has been described. The life stages of these control data have been simplified to reduce them to the stages: "Embryo," "L1," "L2," "L3," "L4," "Dauer," "Adult," and "all stages" (where the life-stage has not been specified). The mean and median values of the FPKM data in each of these classes have been calculated. In species other than *C. elegans*, the appropriate life-stages are used. A final, "total over all stages," overall mean and median value for each gene has also been calculated by combining all of these control data values. These values are derived from baseline (control) expression data found in all available RNA-seq experiments and not just the selected modENCODE data that the bar chart (Fig. 22) represents. They are therefore available in species where there is no equivalent of the modENCODE data used to make the bar charts displayed in the *C. elegans* gene report page "Expression" widget. If these baseline expression values are required for all genes in the database, then they can be obtained from in a single file on the FTP site (Subheading 11.2).
3. *Box-plots of RNA-seq Study Data*: Each RNA-seq study that has an identifiable reference is summarised as a set of box-plots in the "FPKM expression" section of the "Expression" widget. Clicking on the various studies in the list on the left displays a set of box-plots of the data split by the study's independent variable (usually the life-stage) (Fig. 24). Below this is the FPKM value of every experiment in every study for those who wish to download and investigate the RNA-seq expression data for this gene in detail.
4. *RNA-seq Data in the Genome Browser*: A number of genome browser tracks are derived from RNA-seq data. The primary aim of these tracks is to highlight areas where the gene struc-

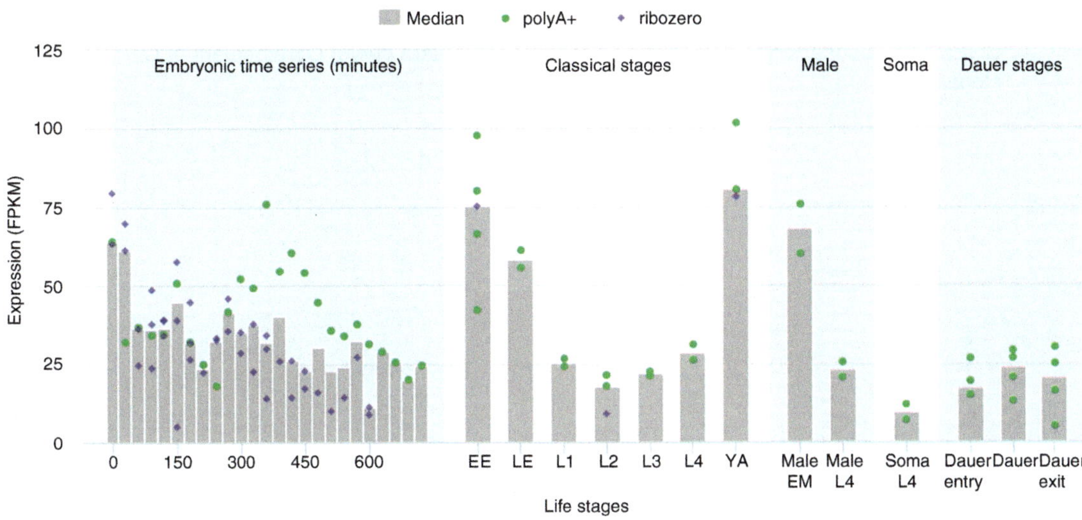

Fig. 22 Bar-chart of the FPKM expression values of a gene during various life-stages. The values are produced from RNA-seq data of selected modENCODE libraries. The dots represent individual libraries, the grey bars indicate the median values

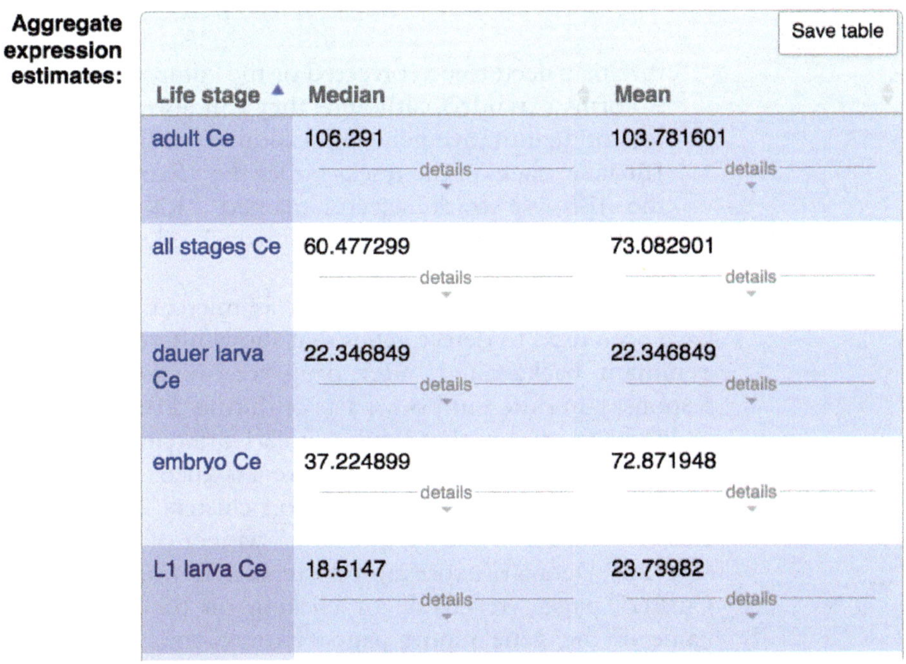

Aggregate expression estimates:

Save table

Life stage	Median	Mean
adult Ce	106.291 details	103.781601 details
all stages Ce	60.477299 details	73.082901 details
dauer larva Ce	22.346849 details	22.346849 details
embryo Ce	37.224899 details	72.871948 details
L1 larva Ce	18.5147 details	23.73982 details

Fig. 23 Mean and median of the baseline (no special experimental conditions) FPKM expression values of a gene during various life-stages

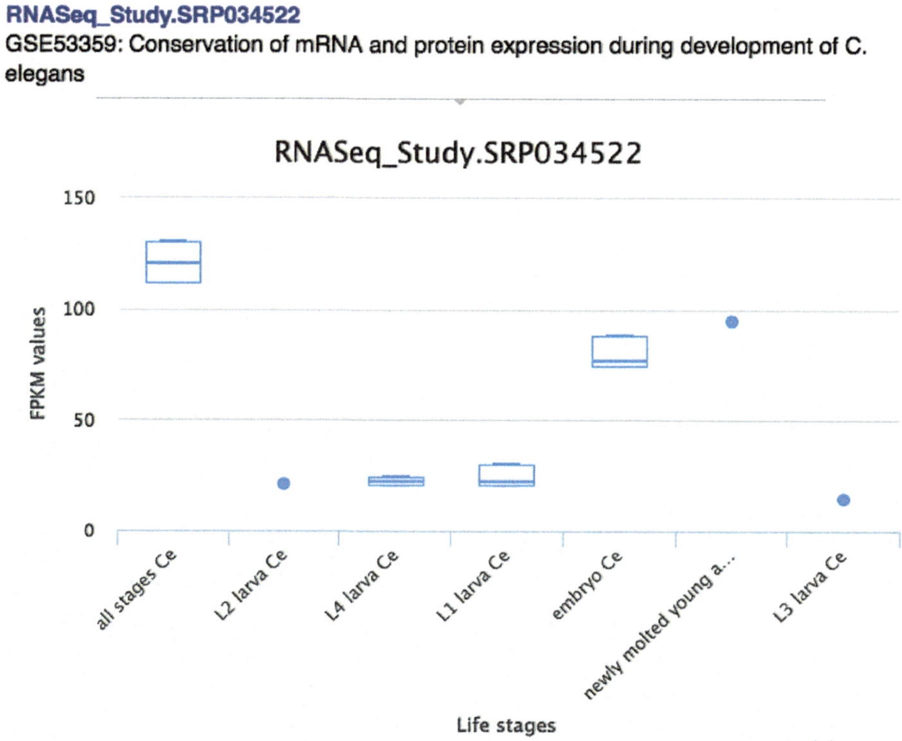

Fig. 24 Box-plots of FPKM expression values of experiments in a study, split by the life-stages

ture may need to be corrected or the complement of transcript isoforms extended, although they can also be used as indicators of quantitative gene expression. To display these tracks in JBrowse, click on the tracks under the "Expression" section of the JBrowse track selector named "RNASeq," "RNASeq Asymmetries," and "RNASeq Introns" (Fig. 25).

6.2.2 Expression Clusters

Genomic expression studies, such as microarray and RNA-seq, have been used to detect genes that show differential expression in a mutant background, after drug treatments, during immune responses, in different body parts or during different developmental life stages. Genes that exhibit similar differential expression profiles under the same condition are assigned to an "expression cluster." Users can access expression clusters in the "Expression cluster" section of the "Expression" widget of gene report pages (Fig. 26). Details of expression clusters can be found on "Expression Cluster" pages (reachable by clicking on the expression cluster name in the gene report page "Expression" widget), including regulation by genes, molecules or treatments, tissue- or life stage-specific information, and algorithms used to draw conclusions. WormBase expression clusters are generated from microarray, tiling array, RNA-Seq, proteomic analysis, quantitative PCR, and large-scale quantitative reporter gene analysis.

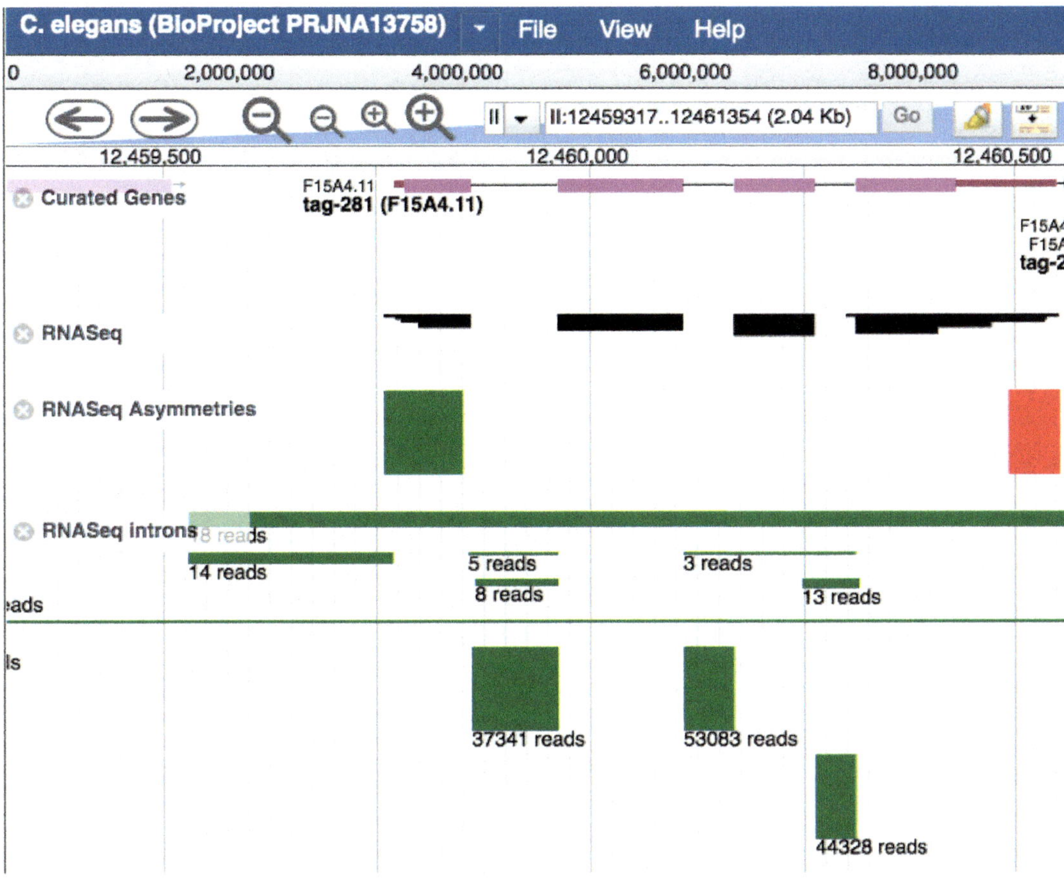

Fig. 25 Tracks in the genome browser JBrowse showing RNA-seq information for the gene *tag-281*. The "RNASeq" track shows the alignment of the RNA-seq reads against the genome. The block's height is proportional to the number of reads per library and so gives an indication of the level of expression. The "Asymmetries" track is based on a signal found by the modENCODE project where the ends of transcribed regions of the genome are often characterized by a preponderance of the aligned forward (green block) or reverse (red block) sense RNA-seq reads. The "Introns" track shows where RNA-seq reads span a region that is assumed to be an intron. The number of reads is indicated beneath the green blocks marking the introns. When two very similar genes occur near each other, reads can be aligned such that half of the read is in one gene and the other half is in the second gene, producing a spurious "intron" linking the two genes; these are artifacts and should be ignored

6.2.3 WormBase SPELL

SPELL (Serial Pattern of Expression Levels Locator) is a search engine to display, sort and download genomic expression data. It can also be used for clustering or GO enrichment analysis. SPELL can be accessed from the WormBase "Tools" menu. WormBase collects and displays multiple types of genomic expression data: microarray, tiling array, RNA sequencing, qPCR, and mass spectrometry proteomics studies. Gene Expression Omnibus (GEO), ArrayExpress and Sequence Read Archive (SRA) are the main sources from which we obtain microarray and RNA-Seq data; while tiling array and proteomics data mostly come from direct

Expression Cluster:

Search: [ ] Save table

Show 10 entries

Expression clusters	Description
[cgc5767]:expression_class_ME	Maternal-embryonic class (ME): genes that are in the intersection of the maternal and embryonic classes.
[cgc5767]:expression_class_MET	Maternal-embryonic transient class (MET): genes that are in the intersection of the maternal and embryonic transient classes.
[cgc6390]:Cluster_E	Germline-enriched and sex-biased expression profile cluster E.
[cgc6390]:oogenesis-enriched	oogenesis-enriched
cgc4386_cluster_4_3	Developmentally modulated gene cluster.
cgc4489_group_9	Genome-wide analysis of developmental and sex-regulated gene expression profile.
WBPaper00005124:progesterone_10-7M_regulated	Genes up or down regulated by 10e-07M of progesterone. The normalized values used were G/R ratio > 2.6 for up-regulation and G/R ratio < 0.38 for down-regulation, which corresponds to 1.39 and -1.39 log(base2) G/R ratio, respectively.
WBPaper00005124:progesterone_10-9M_regulated	Genes up or down regulated by 10e-09M of progesterone. The normalized values used were G/R ratio > 2.6 for up-regulation and G/R ratio < 0.38 for down-regulation, which corresponds to 1.39 and -1.39 log(base2) G/R ratio, respectively.
WBPaper00025032:PAL-1_target_genes	Potential PAL-1 target genes.
WBPaper00025141:N2_Expressed_Genes	Genes expressed in N2.

Showing 11 to 20 of 72 entries   1 2 3 4 5 ... 8

Fig. 26 Expression clusters displayed on a gene report page. The "Expression clusters" column displays the name of the expression cluster and the "Description" column displays a brief description of the expression cluster

author submission. The SPELL interface (Fig. 27) has three major functions listed on the left side menu. The "New Search" function is intended to query for clusters of genes with similar expression profiles to the query gene; it also displays biological pathways associated to the clustered genes. The "Dataset Listing" function allows users to browse and download specific datasets. The "Show Expression Levels" function gives an overview of expression levels across all experiments.

1. *Dataset Listing and Download*: When using "Dataset Listing," users may browse datasets according to biological topic, species, or experimental approach. In the WS257 release of WormBase, SPELL contained data from nine nematode species: *C. elegans*, *C. briggsae*, *C. brenneri*, *C. remanei*, *C. japonica*, *P. pacificus*, *B. malayi*, *O. volvulus*, and *S. ratti*. Each dataset is annotated to topics according to biological pathways that have been studied. To turn on the topic filter, click on

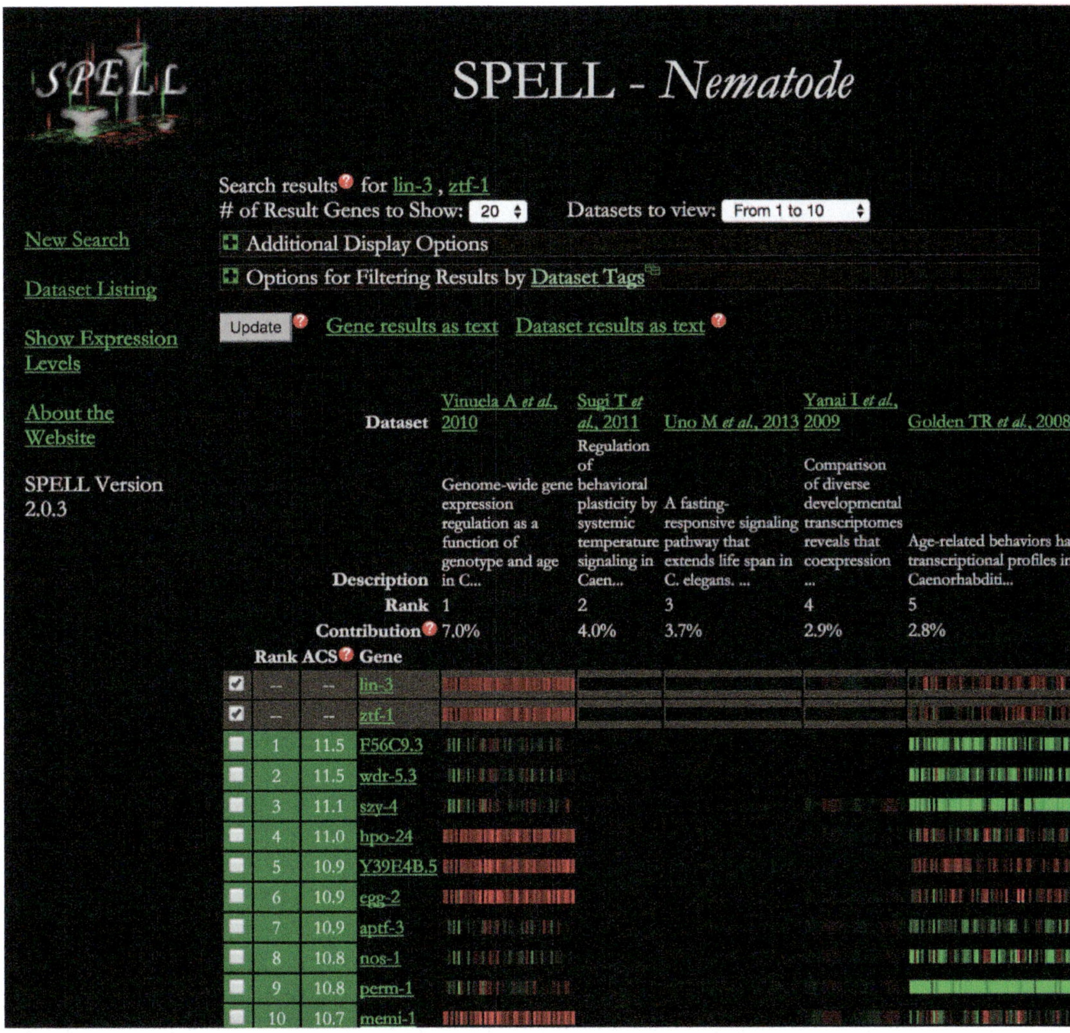

Fig. 27 SPELL search results showing genes with similar expression profiles across all datasets. Datasets are ranked according to their relevance to the queried genes

"Options for Filtering Results by Dataset Tags." If the dataset came from GEO, the dataset IDs and platform IDs are shown and linked back to the Gene Expression Omnibus site. At the end of each dataset entry, users can click on "details" to obtain more information about the study and names of the experiments. Each dataset entry contains a link to a tab-delimited file (e.g., "WBPaper12345678.ce.mr.csv") that contains the most up-to-date gene-centric data. Users can use the topic page to browse datasets of interest. One can also download all datasets with one click of the "Download All Datasets" option or download the original probe centric data from the "Download Other Files" option located under the SPELL title.

2. *Clustering and GO enrichment analysis*: The "New Search" option enables identification of new genes with similar expression profiles to a queried gene across all platforms. The search result will display each gene's expression profile across all experimental conditions in every dataset, ranked according to their relevance to the query. Users can provide a set of query genes that they believe have correlated expression. The search engine determines a relevance weight for each dataset based on how well correlated the query genes are in each dataset. Datasets in which the query genes are largely coexpressed receive a high weight, while datasets in which the query genes are not coexpressed are given a low weight. Negative correlations are treated as no correlation during score calculations. A multigene query, assuming the genes analyzed have good expression correlation, will generate best quality clustering results, because poor quality or irrelevant datasets will receive less weight. If only one query gene is entered, all datasets will get equal weight; users will still get clustering results. SPELL performs GO enrichment analysis on the clustering results. GO terms related to biological processes are displayed at the bottom of the result page.

7 Gene Interactions

WormBase curates four types of gene–gene interactions: physical, genetic, regulatory, and predicted. Physical interactions represent direct, physical, molecular interactions between genes and gene products and may be protein–protein interactions, protein–DNA interactions, or protein–RNA interactions. Genetic interactions represent phenotypic outcomes of double mutants (or other genetic perturbations) with respect to single mutant phenotypes and the control phenotype. Regulatory interactions represent how perturbation of one gene or gene product may affect the expression of a gene or localization of a gene product. Predicted interactions represent in silico predictions of genetic interactions between genes, based on a variety of criteria [27–29]. WormBase curates interactions between genes, sequence features (e.g., DNA binding sites, promoters, and enhancers), and occasionally molecules/chemicals, for example when a drug suppresses the effect of a mutation (genetic interactions) or if a chemical induces expression of a gene (regulatory interaction), and treatment conditions, like exposure to gamma irradiation or magnetic fields.

7.1 Gene Report Page "Interactions" Widget

On a gene report page, one may find interaction data in the "Interactions" widget. The first visual element at the top of the widget is the Cytoscape network viewer, which displays a graphic summary of all interactions with a gene (Fig. 28). If there are a large number of interactions the network graph may be collapsed

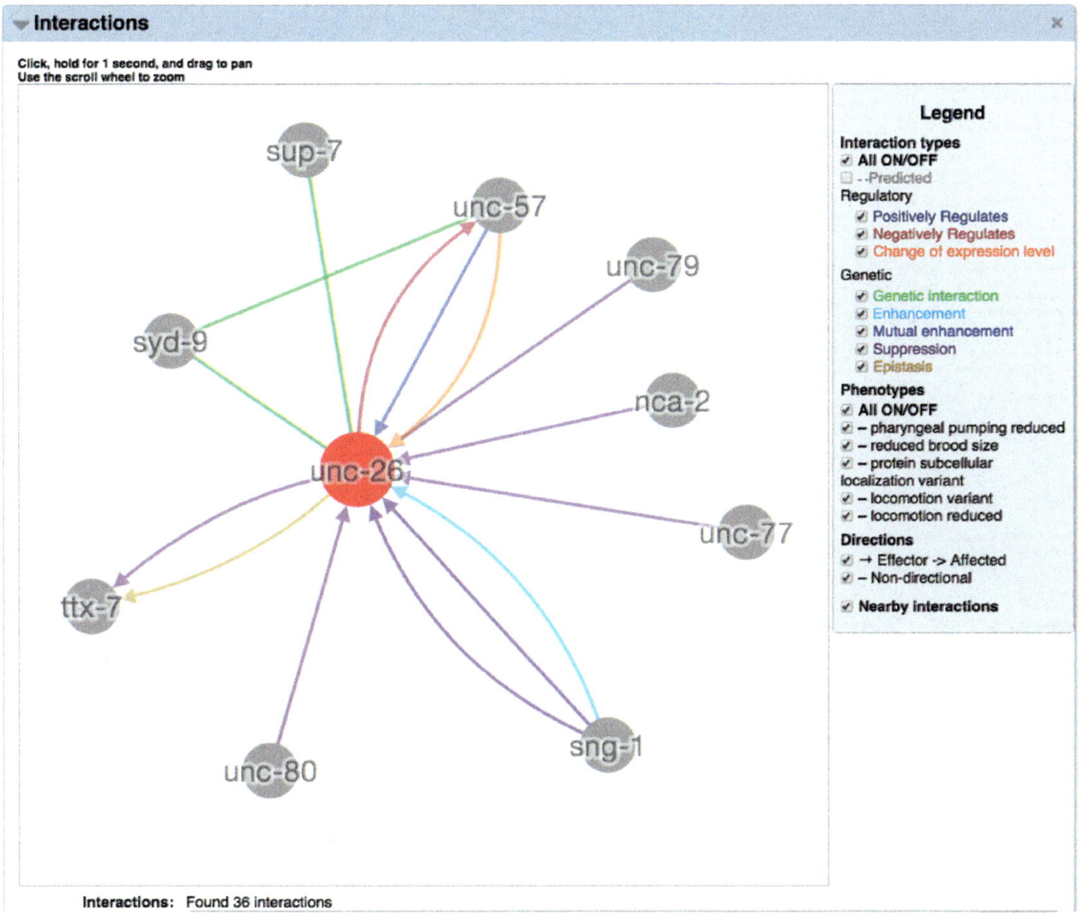

Fig. 28 The Cytoscape interaction network viewer. In the top portion of the "Interactions" widget on gene report pages is the network view of interactions for the focus gene as rendered in Cytoscape. The network view can be zoomed in and out (using the mouse scroll feature) and panned left/right/up/down (by clicking and dragging). Individual nodes can be clicked on and rearranged to customize the network view. Interaction types and interactor types can be toggled on or off by clicking the checkboxes in the network view legend to the right

by default to keep the widget operating optimally. To view the graph in this case, click on "View Interaction Network," and the Cytoscape network will load, but may require a few seconds to complete the loading process. Because of the large number of predicted interactions in WormBase and because of our priority to display interactions with experimental evidence first, predicted interactions for a gene are not displayed by default. To toggle on predicted interactions, click on the checkbox to the left of "Predicted" in the network viewer legend at the right.

The Cytoscape network viewer legend provides the ability to toggle on and off different interaction types as well as genetic interactions based on particular phenotypes, directional and non-directional interactions, nearby interactions (interactions between interactors of a focus gene), and different interactor node types (if more than one interactor type is present, e.g., DNA elements or molecules). For convenience there is also a "All ON/OFF" toggle for interaction types and for genetic interaction phenotypes, which can be used to quickly turn off all interactions when there are too many to visualize at once or to quickly turn on all interactions when there are a manageable number of interactions in total and you would like to see all interactions at a glance.

Below the Cytoscape interaction network viewer is a table of all interactions for the focus gene (Fig. 29). The table has seven columns: "Interactions," which name the interaction by the interacting genes and hyperlink to the individual interaction page; "Interaction Type" which displays the type and subtype of the interaction; "Effector" which displays the first interactor(s) in the interaction, which may play the role of the effector (e.g., suppressor or enhancer) in a directional genetic interaction, the role of the "prey" or "target" for a physical interaction, the role of regulator in a regulatory interaction, or may simply be a nondirectional

Interactions: Found 36 interactions

Search: Save table

Show 10 entries

Interactions	Interaction Type	Effector	Affected	Direction	Phenotype	Citations
apa-2 : unc-26	Predicted	*unc-26*	*apa-2*	non-directional		Zhong & Sternberg, 2006
aps-3 : unc-26	Predicted	*unc-26*	*aps-3*	non-directional		Zhong & Sternberg, 2006
chc-1 : unc-26	Predicted	*chc-1*	*unc-26*	non-directional		Lee et al., 2008
dbn-1 : unc-26	Predicted	*dbn-1*	*unc-26*	non-directional		Zhong & Sternberg, 2006
dyn-1 : unc-26	Predicted	*unc-26*	*dyn-1*	non-directional		Zhong & Sternberg, 2006
ehs-1 : unc-26	Predicted	*unc-26*	*ehs-1*	non-directional		Zhong & Sternberg, 2006
hipr-1 : unc-26	Predicted	*hipr-1*	*unc-26*	non-directional		Zhong & Sternberg, 2006
inf-1 : unc-26	Predicted	*unc-26*	*inf-1*	non-directional		Zhong & Sternberg, 2006
nca-2 : unc-26	Suppression	*nca-2*	*unc-26*	Effector->Affected		Jospin et al., 2007
rab-11.1 : unc-26	Predicted	*unc-26*	*rab-11.1*	non-directional		Zhong & Sternberg, 2006

Showing 1 to 10 of 36 entries 1 2 3 4

Fig. 29 The interactions table for a gene is displayed in the lower half of the "Interactions" widget on the gene report page. Interaction names in the leftmost column can be clicked to open up the web page for that particular interaction. The "Search" box in the table can be used to filter the table's content to specific interactors or interaction types

interactor; "Affected" which displays the second interactor(s) in the interaction, which may play the role of the affected gene (e.g., that which is suppressed or enhanced) in a directional genetic interaction, the role of "bait" for a physical interaction, the role of the regulated entity in a regulatory interaction, or also simply the role of a nondirectional interactor; "Direction" which displays the directionality of the interaction; for example, if a genetic perturbation of one gene (the effector) suppresses the phenotype of a genetic perturbation in another gene (the affected) in a directional genetic interaction; "Phenotype" which displays the relevant phenotype for genetic interactions; and "Citations" which displays links to WormBase paper pages for the articles from which the interaction data originated.

7.2 Interaction Page

By clicking on any interaction name in the table (usually listed by gene names concatenated by a colon), the user is directed to the WormBase interaction page for that interaction. The page has four widgets, including the "Overview" widget displaying interaction details and curator comments, an "External Links" widget to link out to view the interaction at an external database or website, an "Interactors" widget with a layout identical to the "Interactions" widget on a gene report page, and a "References" widget to display the primary research article reporting the interaction.

7.3 Interactions on Process and Pathway Pages

WormBase curates papers and interactions affiliated with certain biological topics, like signaling pathways and developmental processes. To see a process page, click on the magnifying glass icon next to the WormBase search box at the upper right corner of any WormBase page. This will direct you to the advanced search options page. Once there, click on "Process&Pathway" under "Classes" and then type in the name of a process, like "programmed cell death" and hit ENTER. By opening the "Interactions" widget on the process page (also identical in layout to the gene report page "Interactions" widget), users can see all interactions that have been annotated to the process. Note that these are not pathway diagrams, but rather the total network of gene interactions (physical, regulatory, genetic) that have been annotated as pertaining to the process.

8 Phenotype Data

Phenotypes are the observable traits of an organism, resulting from the organism's genotype interacting with its environment, and may manifest as gross phenotypes like body morphology defects or as more subtle phenotypes like changes in gene expression or metabolic throughput. WormBase organizes nematode phenotype terms according to an ontology (the Worm Phenotype Ontology [20]) which can be browsed using the WormBase Ontology

Browser (WOBr, *see* Subheading 5.1). Phenotype data are most commonly accessed via the gene report page, but are also accessible on variation (allele) pages and transgene pages, as well as on dedicated pages for each phenotype term. The following examples explore some common use cases and describe how to query for phenotype information.

8.1 Finding Phenotypes Associated with a Gene

Perhaps the most common query for phenotype information is to lookup all phenotypes attributed to a gene. Navigate to a gene report page and turn on the "Phenotypes" widget. The widget (Fig. 30) first displays all phenotypes resulting from alleles or RNAi experiments, followed by phenotypes NOT observed for alleles and RNAi, followed by interaction-based phenotypes, followed by overexpression phenotypes.

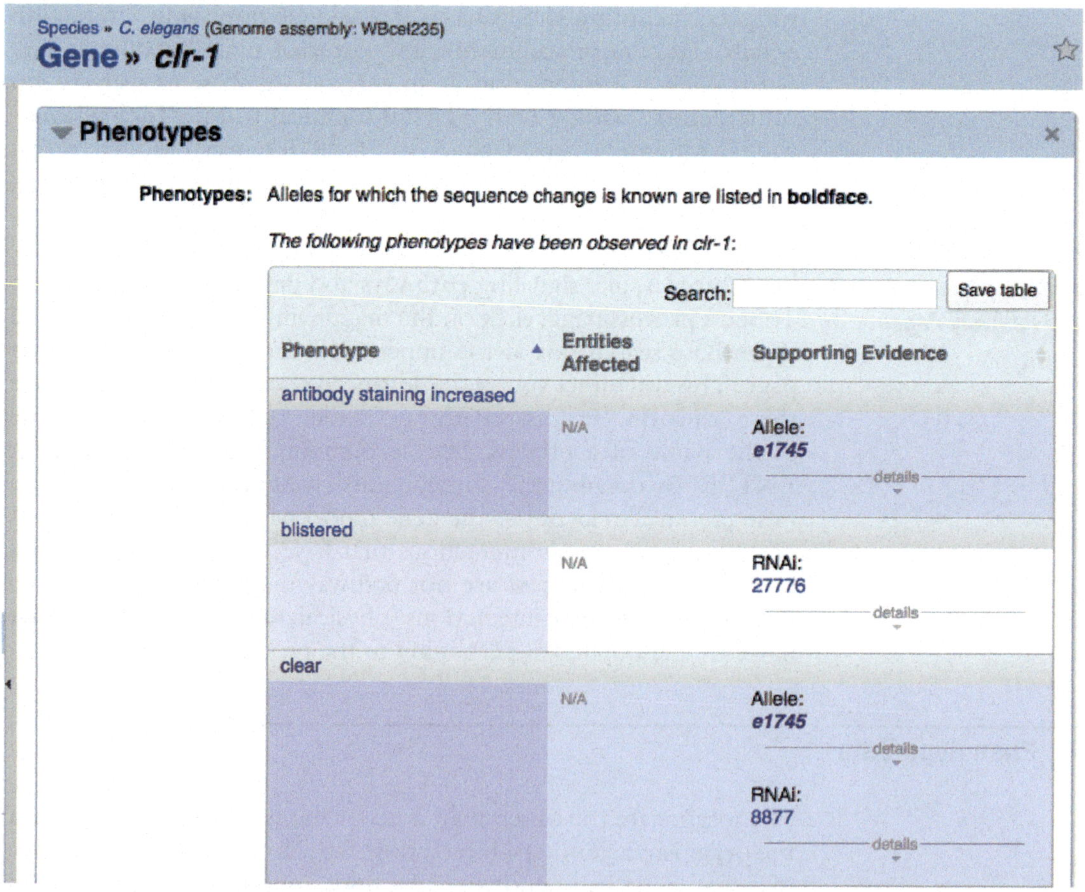

Fig. 30 The "Phenotypes" widget on a gene report page displays the phenotypes resulting from various perturbations of the focus gene. The topmost table displays observed phenotypes resulting from allele or RNAi perturbations, followed by a table of phenotypes assayed for but not observed for alleles or RNAi experiments. Below this is a table of interaction-based phenotypes and a table of overexpression phenotypes

For allele and RNAi-based phenotypes (the top-most table in the "Phenotypes" widget), the table presents three columns: "Phenotype," "Entities Affected," and "Supporting Evidence." The "Phenotype" column displays the name of the phenotype (with a hyperlink to the phenotype term page). The "Entities Affected" column lists all anatomy terms, life stages, and Gene Ontology terms that are affected as part of the phenotype as well as the manner in which each entity is affected, the default being "abnormal." The "Supporting Evidence" column displays the name of the allele or RNAi experiment annotated to the phenotype as well as additional information about the experiment, including the source (paper reference or personal communication), a remark about the phenotype result, and additional meta data. The table for phenotypes not observed has an identical layout to the allele and RNAi-based phenotypes table but represents phenotypes assayed for but not observed for the indicated genetic perturbation.

For interaction-based phenotypes, the phenotypes reported are those that are affected as part of a genetic interaction. This table has four columns. The "Phenotype" column displays the phenotype affected in the genetic interaction; the "Interactions" column lists the genes involved in the genetic interaction with a hyperlink to the interaction page; the "Interaction Type" column displays the type of genetic interaction; and the "Citations" column lists the papers from which the genetic interactions were annotated. For overexpression phenotypes, the table simply has two columns, "Phenotype" and "Supporting Evidence." The "Phenotype" column displays the phenotype name and the "Supporting Evidence" lists relevant meta data including paper reference and possibly a remark about the experiment.

Oftentimes exploring all phenotypes for a gene in tabular format is not ideal, as one would like to get a quick overall sense as to the nature of the phenotypes attributed to a gene without having to identify each phenotype by name in a (possibly large) list of phenotypes in alphabetical order. The "Phenotype Graph" widget provides a compact network graph view of all phenotypes annotated to a gene, clustered according to which branches of the Worm Phenotype Ontology are represented in the gene's annotations (Fig. 31). It provides a summary view of all affected phenotypes with their annotation types and counts. The graph is highly interactive. One can pan around the graph (by clicking and dragging), easily zoom in (using the mouse scroll function) to reveal more details, zoom out to have a broader view, or mouse-over or click on a node to show annotation counts and direct term connections. The Phenotype Graph offers an unweighted view in which each phenotype term node of the graph is of equal size, as well as a weighted view in which node sizes are proportional to the number of independent annotations to each term, with ancestor nodes

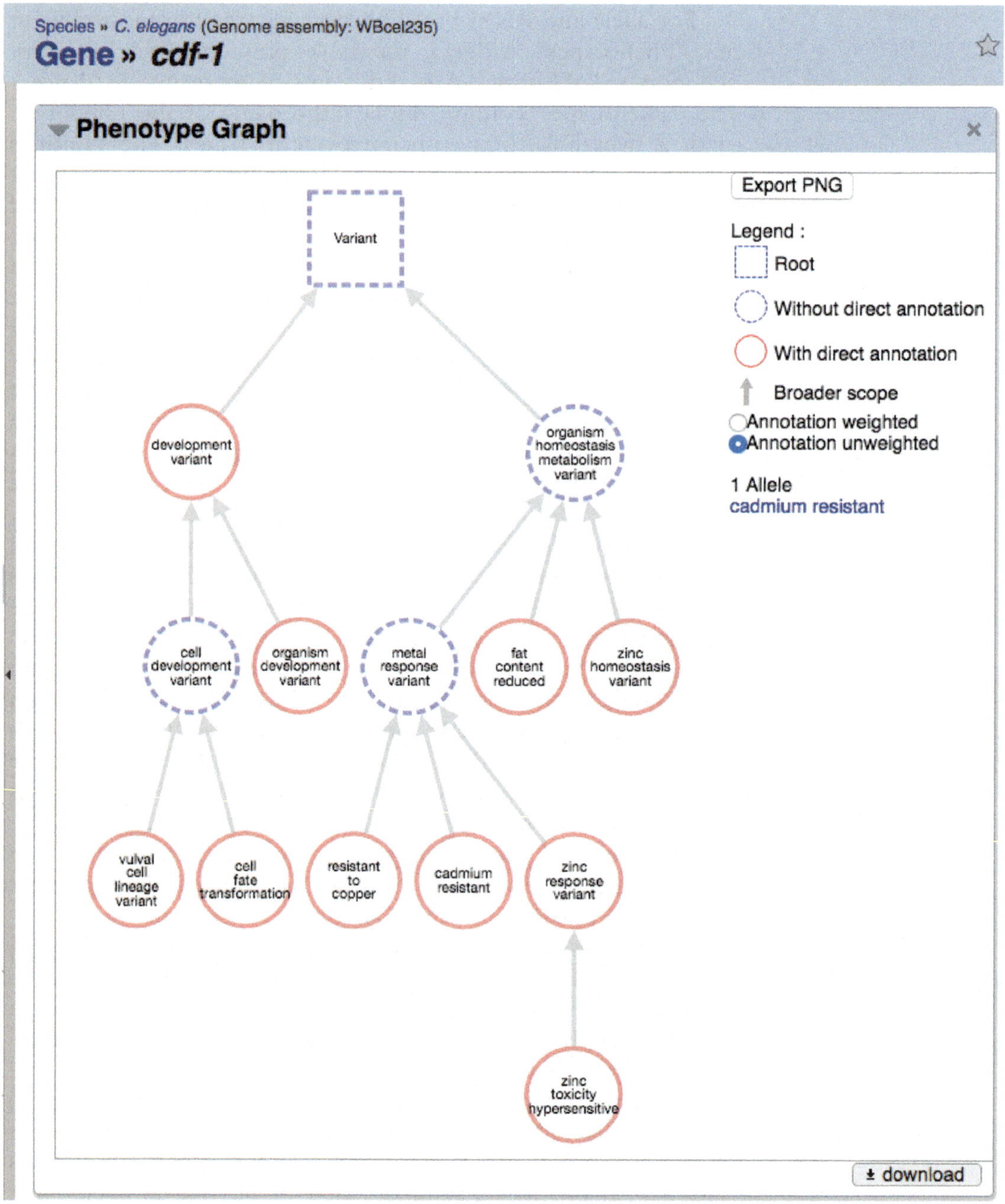

Fig. 31 The "Phenotype Graph" widget on gene report pages provides a concise ontology-guided graph view of phenotype annotations to enable more efficient perusal of phenotype terms annotated to the focus gene. The "Annotation weighted" view displays phenotypes as nodes whose size is proportional to the number of annotations to that term; the "Annotation unweighted" view displays each phenotype term as a node of the same size. Hovering the cursor over each node reveals the number and type of annotations to each term. The "Export PNG" button can be used to generate a PNG graphic file depicting the entire graph

inheriting annotations from descendant nodes. The entire graph may be exported in PNG format. To make sure labels are visible in the PNG, one should zoom in close enough first before exporting.

8.2 Finding All Genes Annotated to a Phenotype

Another common task is to lookup all genes annotated to a particular phenotype term as well as to any of the phenotype term's ontological descendants. For example, a researcher may wish to determine all essential genes, but searching for genes annotated to just the "lethal" phenotype will not include genes specifically annotated to the "larval lethal" or "embryonic lethal" phenotypes. Because the "larval lethal" and "embryonic lethal" phenotype terms are ontological descendants of the "lethal" phenotype term, we want a mechanism that allows users to find genes annotated to the "lethal" phenotype term and/or any of its "is_a" ontological descendants. The "Ontology Browser" widget on a phenotype term page provides this functionality and allows users to find all genes directly annotated to a phenotype term or indirectly annotated to that term via any of its ontological descendants. For example, you can look at the "lethal" phenotype term page by searching in the search box "for a phenotype" with the term "lethal," and selecting the autosuggest term that appears or pressing the ENTER/RETURN key. Once at the "lethal" phenotype term page, turn on the "Ontology Browser" widget. At the top of the widget you will see a tree representation of the phenotype ontology and the terms that lead to the "lethal" phenotype term (Fig. 32). To the right of the focus term ("lethal (WBPhenotype:0000062)" in this case) you will see the total number of genes annotated to this term or any of its descendants (4713 genes as of the WS257 release) and to the

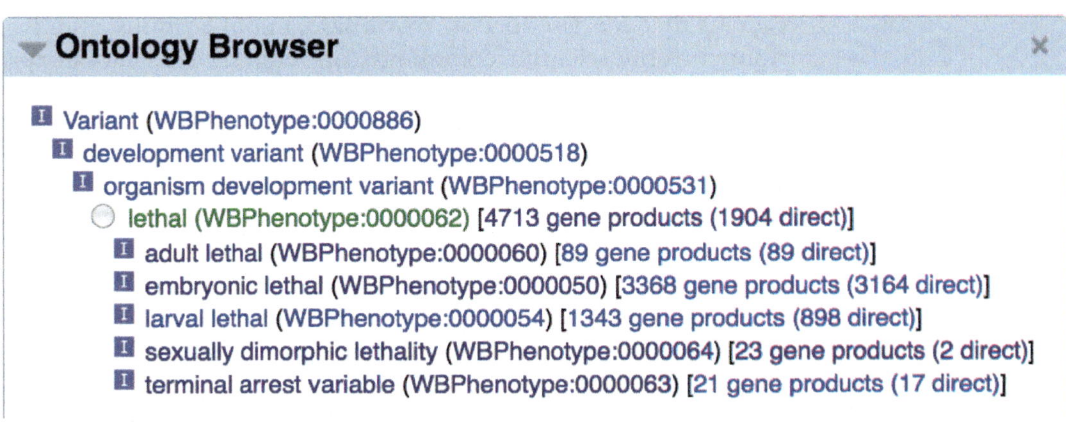

Fig. 32 The "Ontology Browser" widget on a phenotype term page is a good location to find out how many genes are annotated to that phenotype term or any of its ontological descendants. The "lethal" phenotype has 4713 genes annotated to it or any of its ontological descendant terms (e.g., "embryonic lethal") and 1904 genes annotated directly to it (as of WormBase release WS257). Clicking on any of the numbers will link to a page listing the names of all genes in that category, subdivided according to the nature of the annotation (direct or indirect annotation, phenotypes observed via alleles or RNAi experiments)

right of that you will see the number of genes directly annotated to the "lethal" term (1904 genes as of WS257). Clicking on either number will direct you to the list of genes, separated according to direct versus total annotations, and according to RNAi-based or allele-based phenotype associations. On the phenotype term page, one can also browse the list of RNAi experiments, alleles, and transgenes that are associated with the phenotype term, in the "RNAi," "Variation," and "Transgene" widgets, respectively.

9 Reagents

9.1 Strains

9.2.1 Strain Nomenclature

A strain is a set of individuals of a particular genotype with the capacity to produce more individuals of the same genotype. Strains are given nonitalicized names consisting of two or three uppercase letters followed by a number. WormBase has a long-standing collaboration with the *Caenorhabditis* Genetics Center (CGC), at the University of Minnesota. The CGC collects, maintains, and distributes stocks of *C. elegans.*

Strains can and should be preserved as frozen stocks at −70 °C or ideally in liquid nitrogen, in order to ensure long-term maintenance and to avoid drift or accumulation of modifier mutations.

WormBase assigns specific identifying codes to each laboratory engaged in dedicated long-term genetic research on *C. elegans.* Each laboratory is assigned a laboratory/strain code for naming strains, and an allele code for naming genetic variation (e.g., mutations) and transgenes. These designations are assigned to the laboratory head/PI who is charged with supervising their organization in laboratory databases and their associated biological reagents that are described in WormBase, in publications, and distributed to the scientific community on request. The laboratory/strain code is used (a) to identify the originator of community-supplied information on WormBase, which in addition to attribution facilitates communications between the community/curators and the originator if an issue related to the information should arise at a later date, and (b) provide a tracking code for activities at the CGC. The laboratory/strain designation consists of 2–3 uppercase letters while the allele designation has 1–3 lowercase letters. The final letter of a laboratory code should not be an "O" or an "I" so as not to be mistaken for the numbers "0" or "1" respectively. Additionally, allele designations should also not end with the letter "l" which could also be mistaken for the number "1." These codes are listed at the CGC and in WormBase. Investigators generating strains, alleles, transgenes, and/or defining genes require these designations and should apply for them at genenames@wormbase.org.

Examples: CB1833 is a strain of genotype *dpy-5(e61) unc-13(e51)*, originally constructed by S. Brenner at the MRC Laboratory of Molecular Biology (strain designation CB, allele designation *e*), and MT688 is a strain of genotype *unc-32(e189) +/+ lin-12(n137) III; him-5(e1467) V*, constructed in the laboratory of H.R. Horvitz at M.I.T. (strain designation MT, allele designation *n*).

Bacterial strain names employ the two or three letter Laboratory/Strain designation, followed by "b"; for example, CBb#. This facilitates distinguishing nematode strains from bacterial strains.

9.1.1 Accessing Strain Information via the Website

Strains carrying a gene of interest can be found within the "Genetics" widget on the gene report page. As well as a Venn diagram indicating strains which carry only the gene of interest and strains which are available to order from the CGC, the widget contains a table listing the genotype of all strains. The strain report page contains more detailed information about the strain, e.g., mutagen, who it was made by and when, from where it is available and which alleles, genes and transgenes it carries. Additionally, the alleles are listed in a table, which gives details of the molecular lesion.

9.2 Transgenes and Constructs

Transgenes in WormBase represent (at least partially) heritable, exogenously introduced DNA fragments in the nematode organism that generally tend to confer some functional or potentially functional property to the organism. Constructs, on the other hand, represent the DNA construct that exists independently of the organism (in a test tube, for instance). Thus, transgenes can be generated from constructs, but not the other way around.

Transgenes can be searched for in the main WormBase search box (after specifying "for a transgene") using conventional *C. elegans* transgene names. You can also type the name of a gene and see suggested transgene names (that contain the gene or a part of the gene) with the autosuggest feature of the search. On transgene report pages, the "Overview" and "Construction Details" widgets provide details of how the transgene was constructed and introduced into the nematode. The "Expression" widget displays expression patterns affiliated with the transgene and the "Phenotype" widget displays phenotypes resulting from overexpression of a gene from the transgene. To discover transgenes affiliated with a particular gene, navigate to the gene report page and open the "Reagents" widget. There are two tables that list the transgenes that the focus gene is a component of. The "Drives Transgenes" table lists the transgenes for which a regulatory region (e.g., promoter, enhancer) of the focus gene has been included and is used to drive expression. The "Expressed in Transgenes" table lists the transgenes in which the focus gene is the gene that is expressed.

Constructs are discoverable from transgene pages in the "Construction Details" widget, if available. Clicking on a construct name will redirect you to the construct page, which provides some complementary information about the construct not immediately apparent from the respective transgene report page.

9.3 RNAi Clones

WormBase users often like to know which RNAi reagent to use for effective knockdown of their gene of interest, usually from one of a few commercially available libraries of RNAi clones. These RNAi clones are represented in WormBase both as clone objects as well as PCR product objects, and usually have names that reflect their source. The ORFeome library RNAi clones (from the laboratory of Marc Vidal) [30] are stored in WormBase with names prefixed with "mv_" and are listed on a gene report page in the "Reagents" widget under the label "ORFeome Primers." To find the location of the clone in the library, it is best to search for the clone name (following the "mv_" prefix) at the WORFDB website (http://worfdb.dfci.harvard.edu/index.php?page=searchwm). This will direct you to the clone's web page in WORFDB listing the clone's plate and well location. Currently, these ORFeome RNAi clones can be purchased through Dharmacon/GE Life Sciences (http://dharmacon.gelifesciences.com) or Source BioScience (http://www.sourcebioscience.com). The RNAi library clones from the laboratory of Julie Ahringer [31] are stored in WormBase with names prefixed with "sjj_" and are listed in the "Reagents" widget of the gene report page as well, but under the "Primer pairs" label. Clicking on the PCR product name (with the "sjj_" prefix) will direct you to the PCR product page for that clone. The "Overview" widget of that page lists the Source BioScience location for the clone in the "Reagent" field, usually of the form <chromosome>-<plate><well row><well column>, for example V-4N08. Alternatively, you can find the Source BioScience clone location on the "sjj_*" clone page (search for a clone from the search box) in the "External Links" widget.

10 Integrated Views of Data in WormBase

10.1 Nematode Models of Human Disease

C. elegans has proven to be an effective, low-cost, preclinical model organism to study the genetics and interactions of human disease-causing genes. It has been used to study human disease gene orthologs, to model human disease and drug–disease interactions, study host-pathogen interactions and bacterial biofilms, and to screen for novel drugs and drug-targets [32–34]. With orthologs to ~50% or more of human disease genes, *C. elegans* is amenable to live animal drug and bioactive compound screening. *C. elegans* has the following advantages over tissue and cell culture systems or more expensive mammalian models: lower costs, genetics allowing the identification of effectors/interactors (aiding the identification of drug effector pathways), and experiments that can be done in a physiological whole animal context [35].

A review of the *C. elegans* literature corpus indicates that the worm has been used as a genetic model system for several diseases; examples include neuromuscular diseases like Amyotrophic Lateral Sclerosis (ALS) and Duchenne Muscular Dystrophy [36], complex neurological diseases like Parkinson's and Alzheimer's [37], ciliary diseases like Polycystic kidney disease (PKD) and Bardet-Biedl syndrome [38, 39] and premature aging syndromes like Werner syndrome [40]. *C. elegans* has also been used to study obesity [41] and prion diseases (modeled in *C. elegans* via the transgenic expression of the prion protein) [42].

10.1.1 Disease Vocabularies

WormBase uses a simple controlled vocabulary that allows the annotation of *C. elegans* genes as either "Experimental" or "Potential" models for a human disease based on manual curation of published papers, or based on orthology to a human gene implicated in disease, respectively.

A *C. elegans* gene is associated with a human disease via the use of a Disease Ontology (DO) term. The Disease Ontology project (www.disease-ontology.org) represents and organizes common and rare human diseases into an ontology, providing a knowledgebase of disease terms from several biomedical repositories as well as cross-references to clinical disease vocabularies like Online Mendelian Inheritance in Man (OMIM; www.ncbi.nlm.nih.gov/omim) [43, 44]. The disease ontology file that consists of disease ontology (DO) terms, synonyms, definitions and cross-references is imported into WormBase, for use in the annotation of genes to disease terms.

10.1.2 Manual Curation of Disease Relevant Data

This curation is based on reading of the published *C. elegans* literature that describes *C. elegans* models of disease. Data that curators look for include one or more of the following: data showing orthology between the nematode and human disease-causing gene(s), similarity between nematode and disease phenotypes, similar processes in nematodes and humans underlying the abnormal phenotypes, transgenic rescue of nematode phenotypes by the human gene, transgenic expression of the *C. elegans* gene in human cell lines causing phenotypes, similarity of genetic and physical interactions between nematode and human proteins, etc. Manual curation results in the annotation of *C. elegans* genes as "Experimental models" of disease. In addition, a "Disease relevant description" is sometimes written that allows a text description of the nematode model for disease, called "Human disease relevance," similar in style to the current "Overview" widget at the top of any WormBase gene report page, which describes gene function.

10.1.3 Automated Orthology Based Curation

This type of curation uses orthology between nematode genes and human genes, as predicted by both a number of external methods and internally using the EnsemblCompara method [16] (*see* Subheading 4.3). Human ortholog information is then cross-referenced with human gene and disease data in the Online Mendelian Inheritance in Man (OMIM) database (www.omim.org) [44]. OMIM disease IDs and their causative genes referenced in the disease ontology file (*see* explanation of Disease Ontology above) are used as a way to link *C. elegans* genes to DO terms. This automated orthology-based curation results in the annotation of *C. elegans* genes as "Potential Models" of disease.

10.1.4 Display of Human Disease Relevant Data

WormBase displays both manually curated disease data from the *C. elegans* literature and automated orthology-based human disease-related data for genes on the gene report page, in the "Overview" widget and in the "Human Diseases" widget.

1. *Disease Data in the "Overview" Widget*: When a gene has been curated as an experimental model for human disease, the "Overview" widget of the gene report page features a collapsible subheading, "Human disease relevance," which consists of a textual description of the model for human disease.

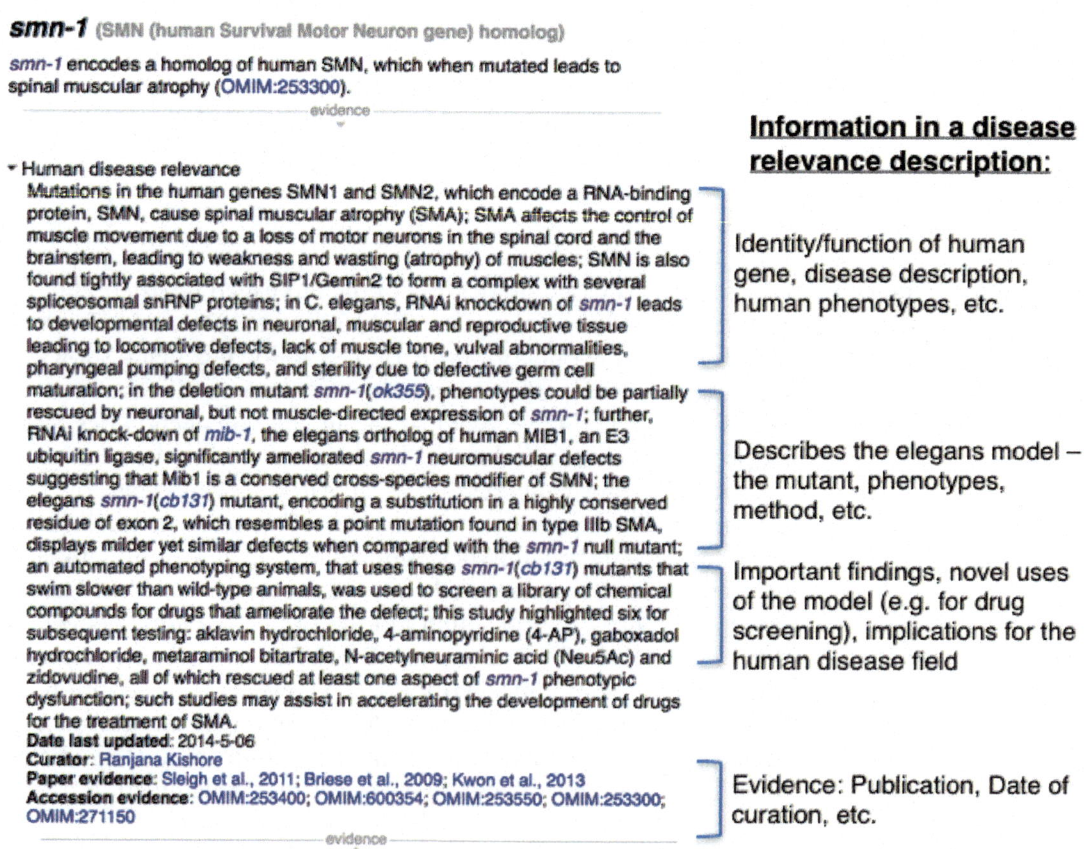

Fig. 33 "Human disease relevance" description (when curated) appears in the "Overview" widget on the gene report page. It is a free-text description describing the human disease and how it is modeled in *C. elegans*

This complements the description of normal gene function in *C. elegans* (Fig. 33).

2. *"Human Diseases" Widget*: When the "Human Diseases" widget on a gene report page is turned on, the page scrolls to the human disease relevant data for that gene; both manually curated and automated orthology-based data are presented here. Manually curated data include the annotation of a gene to the controlled vocabulary term "Experimental model" for a human disease (described using a Disease Ontology (DO) term) based on experimental evidence from the manually curated literature, or a "Potential model" for a human disease/DO term, based on orthology with a gene(s) in the Online Mendelian Inheritance for Man (OMIM). Also included in the "Human Diseases" widget is a list of the orthologous OMIM genes and diseases with hyperlinks, providing easy access to the OMIM resource. Both experimental and potential model data are supported with various evidence like literature references, date of annotation, and the curator responsible for the annotation (Fig. 34).
3. *Searching and Querying for Human Disease Data*: Human disease-relevant information for *C. elegans* genes may be queried for by typing in the name of a disease in the search box on the top right-hand side of the WormBase home page, with the search context "Human Disease" selected from the drop-down menu. The autocomplete function suggests a corresponding DO term for the disease and returns a page with all of the relevant information for the human disease including: DO term definition, synonyms, cross-references, genes curated as experimental models and/or potential models for the disease, the disease model descriptions, and the orthologous OMIM genes and hyperlinks. The data is presented in tabular form for ease of viewing for example "Parkinson's disease" (Fig. 35).
4. *Browsing Disease Data Using the Ontology Browser*: On a given disease term report page, e.g., Alzheimer's disease, one can also browse the data via the "Ontology Browser" widget. The Ontology Browser depicts the hierarchical structure of the ontology, showing the placement of the term "Alzheimer's disease," parent terms and the relationship between the different terms. The browser also shows the number of *C. elegans* genes annotated with this disease term (Fig. 36).
5. *Human Disease Relevant Data Files*: Files with *C. elegans* gene associations to human diseases (DO terms) are available for download via the WormBase FTP site, organized by WormBase release number. For example, the files for WS250 are available at: ftp://ftp.wormbase.org/pub/wormbase/releases/WS262/

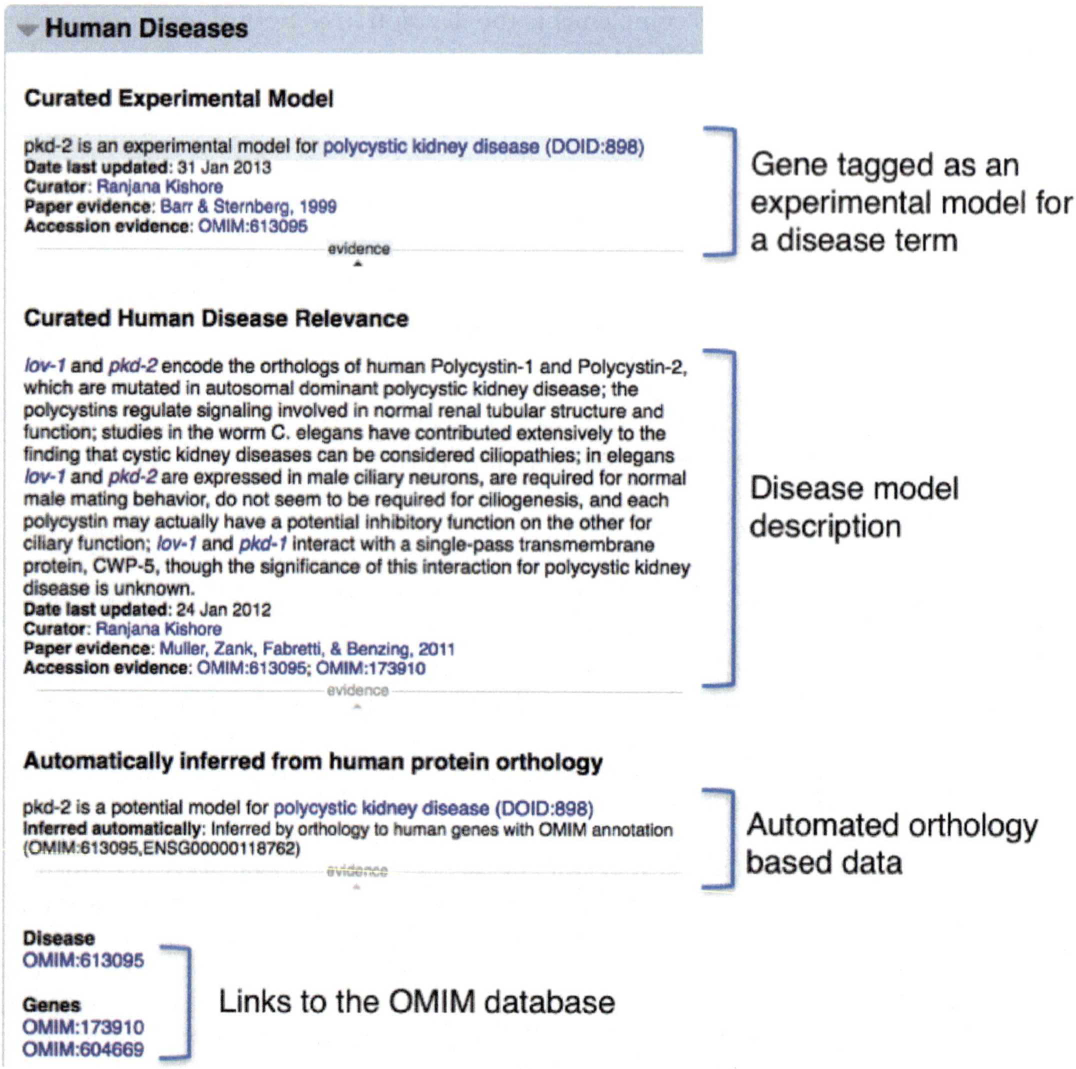

Fig. 34 When the "Human Diseases" widget title is clicked on in the left side navigation bar of the gene report page, the page scrolls to all of the disease-relevant data for that gene, both manually annotated and automatically generated

DISEASE/ and ftp://ftp.wormbase.org/pub/wormbase/releases/WS262/ONTOLOGY/.

10.2 Anatomy Function

Anatomy function is inferred from the observed phenotypic consequences when cells and tissues of interest are specifically affected by physical operations or genetic perturbations like genetic or laser ablation, genetic or expression mosaics, blastomere isolation, and optogenetics. Depending on the treatment, one can infer whether a body part is either necessary or sufficient to support a normal

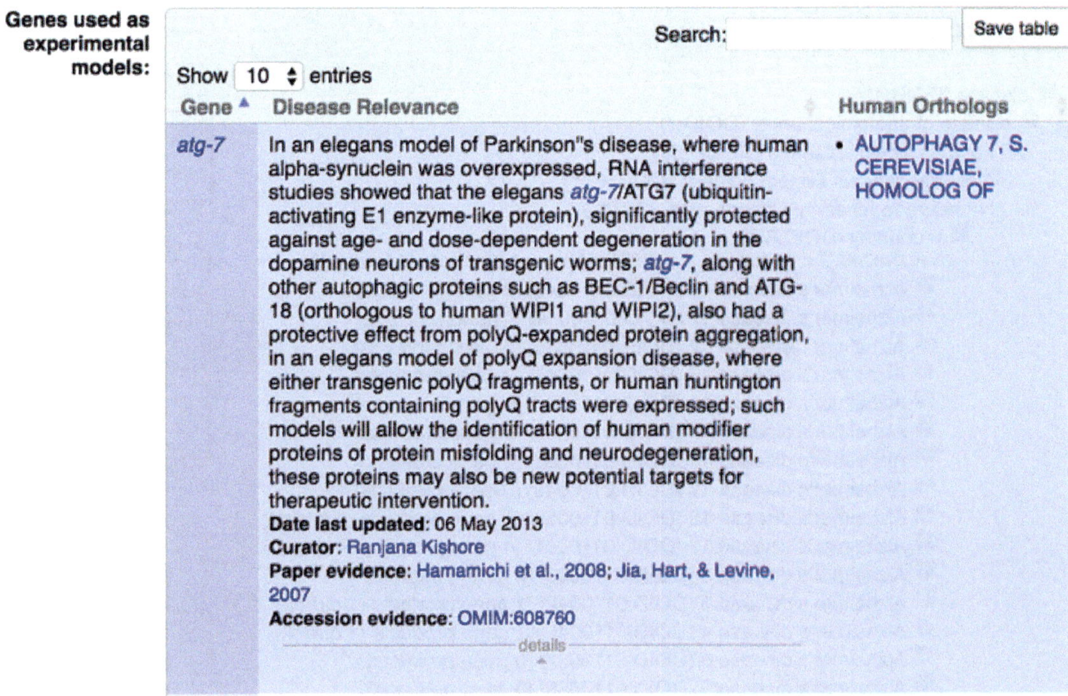

Genes used as experimental models:

Search: | Save table

Show 10 entries

Gene	Disease Relevance	Human Orthologs
atg-7	In an elegans model of Parkinson"s disease, where human alpha-synuclein was overexpressed, RNA interference studies showed that the elegans *atg-7*/ATG7 (ubiquitin-activating E1 enzyme-like protein), significantly protected against age- and dose-dependent degeneration in the dopamine neurons of transgenic worms; *atg-7*, along with other autophagic proteins such as BEC-1/Beclin and ATG-18 (orthologous to human WIPI1 and WIPI2), also had a protective effect from polyQ-expanded protein aggregation, in an elegans model of polyQ expansion disease, where either transgenic polyQ fragments, or human huntington fragments containing polyQ tracts were expressed; such models will allow the identification of human modifier proteins of protein misfolding and neurodegeneration, these proteins may also be new potential targets for therapeutic intervention. **Date last updated:** 06 May 2013 **Curator:** Ranjana Kishore **Paper evidence:** Hamamichi et al., 2008; Jia, Hart, & Levine, 2007 **Accession evidence:** OMIM:608760 details	• AUTOPHAGY 7, S. CEREVISIAE, HOMOLOG OF

Fig. 35 A portion of the "Overview" widget on the disease ontology term page showing the 'Genes used as experimental models' table for Parkinson's disease. This page can be obtained from a context-dependent search using the Human disease term "Parkinson's disease" from the top-right hand corner search box in WormBase or from a gene report page in the "Human diseases" widget relevant to "Parkinson's disease," where this term is hyperlinked

physiological function. When it is available, an anatomy function table can be found on specific phenotype report pages (in the "Associated Anatomy" widget) and anatomy report pages (in the "Associations" widget).

For example, on the "BAG" anatomy term page (http://www.wormbase.org/species/all/anatomy_term/WBbt:0006825#01--10), users can find reference to the anatomy function WBbtf0434 in the "Associations" widget. Alternatively, users can find reference to the same information in the "Associated Anatomy" widget on the "omega turns variant" phenotype term page (http://www.wormbase.org/species/all/phenotype/WBPhenotype:0000551#01--10) under the "BAG" neuron header in the "Body Parts Involved" column. By expanding all details under the table columns, the user can best comprehend the full annotation. Specifically, Bretscher and coworkers reported "Coablation of AFD and BAG abolished the suppression of reversals and omega turns following a fall in CO2…. These data suggest that together

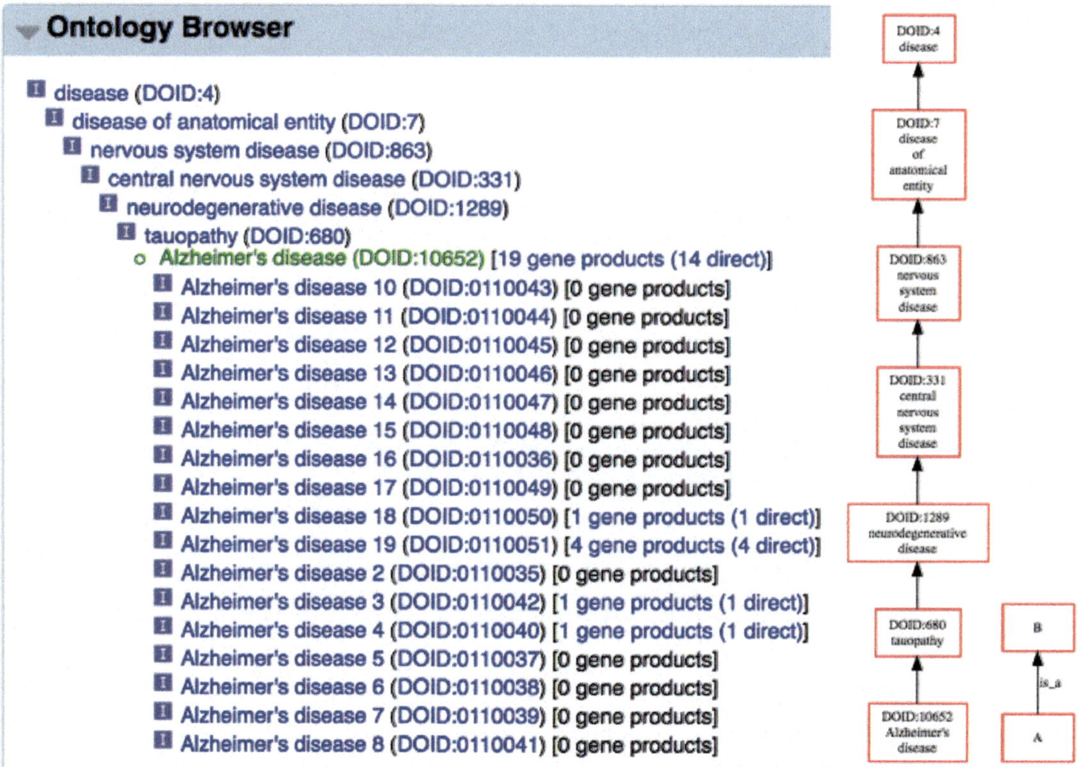

Fig. 36 Ontology Browser view of the disease term "Alzheimer's disease" in the context of the full ontology showing relationships between terms. Also shows number of genes directly or indirectly annotated to this disease term

BAG and AFD act to suppress reversals and omega turns when CO2 decreases" [45].

In addition to the annotation of anatomy function using worm phenotype ontology terms, we have recently begun to more fully represent the affected phenotypes with a specifically constructed phrase built from controlled vocabularies from the OBO ontologies (http://www.obofoundry.org/). For example, in anatomy function WBbtf0434, the observed phenotype is represented by the phrase "ENTITY:WBbt:0007833(organism) | GO:0040011(locomotion) | GO:0035178(turning) | CHEBI:16526(carbon dioxide) QUALITY:PATO:0000460(abnormal)". Via this phrase, users can make more atomized associations between anatomy and aspects of biology. For example, one can infer that BAG affects locomotion.

11 Bulk Data Analysis and Downloads

11.1 WormMine

The Intermine biological data warehouse (http://intermine.org/) [46, 47] is a powerful tool to perform a variety of queries of biological databases and to manage and manipulate lists of biological entities. This section discusses the use of WormMine, the WormBase instance of Intermine. For the lab biologist, the power of WormMine lies in the ability to generate custom queries and share these queries efficiently with other users, ready-made template queries for common data requests, list editing and comparison, and user login for storing custom queries and tables of results within the context of a high-performance user-friendly web-based interface. Intermine instances have already been established for FlyBase (FlyMine) [48], the Mouse Genome Informatics (MGI) database (MouseMine) [49], and several other model organism databases (MODs) and data sets [50–52], providing biologists across different disciplines a uniform and standard way to access biological data. For a simpler batch gene query, try the SimpleMine tool (*see* Subheading 12.2) from the WormBase "Tools" menu.

11.1.1 WormMine Data Mining: The Phenotype Data Use-Case

The WormBase instance of the InterMine biological data warehouse, which we refer to as "WormMine" (available at http://www.wormbase.org/tools/wormmine/begin.do), is a great tool with which to perform large scale data queries of WormBase data. WormMine can be reached via the "WormMine" link under the "Tools" menu at the top of any WormBase page. Many types of data can be retrieved through WormMine. By way of example to illustrate the power of WormMine, what follows here is specific for querying and retrieving phenotype specific data.

11.1.2 WormMine Lists

Some phenotype data in WormMine are available in the form of pregenerated lists. Lists in WormMine are named "list" objects that hold a list of WormMine-recognized entities. The power of WormMine lists are that they can be used as a filtering criteria when performing queries (e.g., when you want to know all phenotypes attributed to all genes in a list; see below) and that they can be used with Boolean operators to compare and contrast different lists of items with the same object type.

WormMine has several pregenerated lists of phenotypes with all of their respective ontological descendant terms as well as lists of genes that are annotated to those phenotypes. For example, WormMine has a list containing the "lethal" phenotype plus all of the ontological descendant terms of the "lethal" phenotype, including "larval lethal" and "embryonic lethal." To get to precomputed lists from the WormMine front page (Fig. 37), click on the gray "Lists" tab at the top of the page to open the list entry page. Then, click on the "View" option at the top to the right of the "Upload" label to arrive at the WormMine lists view page. If we click on the "Life span variant and descendant phenotype terms" list, we can view, in table form, the list of WB phenotypes including the generic "life span variant" phenotype plus all of its ontological descendant terms. This table view represents the gen-

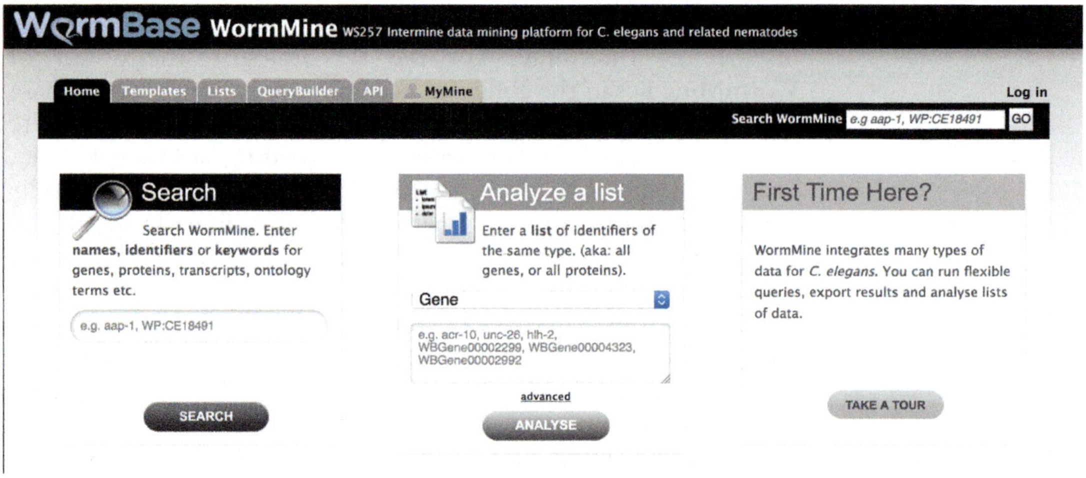

Fig. 37 The WormMine home page provides a search box, a list upload box and a link to a basic tour of the WormMine tool on the WormBase Wiki site. Tabs along the top of the page provide links to template queries (also available via the menu at the bottom of the home page), list upload and viewing, the "QueryBuilder" tool, information about the WormMine API, and the "MyMine" tab for managing saved lists and queries once a user has logged in

eral layout of all WormMine query results and provides a number of functions for manipulating lists or query results (*see* below).

To upload a list of items into WormMine, click on the "Lists" tab at the top of any WormMine page and click on "Upload" at the top left of the panel if it is not already selected. Next to the "Select type" field, select the data type of your list items from the drop-down menu. Paste your list of names/identifiers into the available space or use the "Browse…" button to upload a list from a text file. Once you have pasted or uploaded your list of items, click on "Create List" at the bottom right of the panel. WormMine will then process your list, making sure that it can find matches to each item in your list. A confirmation page will appear indicating the list of items that WormMine recognizes and a list of items that are not found. If there are any ambiguous names or identifiers, the user will be asked to specify which WormMine objects were intended for the list. Note that deprecated object names (e.g., pseudogene names or names of dead genes) will not be recognized, will be displayed in a list of objects not found, and will be omitted from the final WormMine list. Once you have confirmed the set of items for your list and provided a list name in the field provided, click on the green button "Save a list of # <items>". You will then be directed to a WormMine table listing your items along with some basic information for each item in your list. You can then find your list under the Lists tab by selecting the "View" option to the right of the "Upload" option. If you are not logged in, the list will be saved for you during your web browser session, but will be removed once you have quit out of your web browser application. If you are logged in, you can permanently save your list into your personal lists for viewing or manipulation at a later date.

Once you have a series of lists that contain the same object type, you may perform list operations such as finding the union, intersection, subtraction, or asymmetric difference of two lists. For example, to find the intersection of two lists, click on the "Lists" tab, select the "View" option, and click the checkbox to the left of each of the two lists you would like to perform the operation on. Note that to perform a list operation, the two lists must contain the same type of object (e.g., both have a list of genes). Once you have selected the two lists, click on "Intersect" at the top of the panel and you will be prompted to enter a name for the resulting intersection list. Once you have entered a name for the new list, click on "Save" and the resulting intersection list will be generated and stored in the list of WormMine lists. When performing an

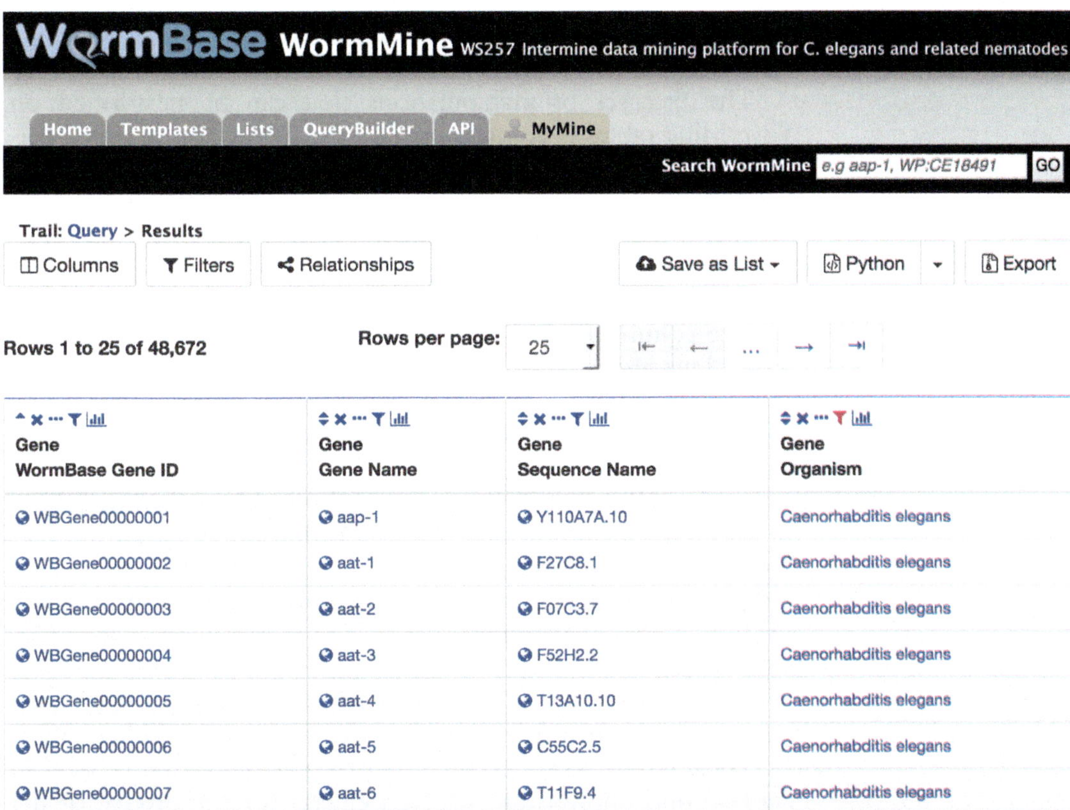

Fig. 38 WormMine tables display the results of a query or display default information for a list of objects. Users can manage table columns (add or remove columns, specify sort order) and manage filters from the buttons at the upper left. Users can save a list of objects represented in the table (e.g., save a list of anatomy terms from a query for tissues where a gene is expressed) and export the table in a variety of formats using the buttons at the upper right of the table. Each column header also has several buttons that allow individual manipulation of columns, e.g., sorting, filtering, or hiding

"asymmetric difference" operation, in addition to being prompted for a new list name, you will be prompted to indicate whether the resulting list should be List 1 minus List 2 or vice versa.

11.1.3 WormMine Tables

WormMine tables enable a user to perform a variety of tasks for analyzing or manipulating a list or table of results. An example view of a WormMine table is presented in Fig. 38. The table view appears whenever viewing a list or the results of a query. When viewing a list, the table that appears represents the list of objects plus basic (default) information about each object in the list. For example, gene lists by default display the WormBase gene ID, the gene public name, the gene sequence name and the organism to which the gene belongs (Fig. 38). Other data types similarly have default columns displayed for lists. Once a table is loaded, one can specify the number of rows to show per page or navigate to subsequent or previous pages using the arrow buttons to the right of the "Rows per page" selection. Clicking on the ellipsis ("…") button in the middle of the arrow buttons allows a user to specify a page number to view.

The first set of manipulations that can be performed on WormMine tables is column operations. At the top of every column is a series of icons representing (from left to right) the sort function (the triangle icons), the remove column function (the "X" icon), the toggle visibility function (the ellipsis icon), the filter function (the funnel-shaped icon), and the column summary function (the bar chart icon). Clicking on the sort function icon will sort the column first in ascending order or, by clicking again, in descending order. Clicking on the "X" remove column icon will remove the column entirely from the table. Clicking on the ellipsis icon ("…") will collapse the column, resulting in each column entry being replaced by an ellipsis. Subsequently clicking on the expand icon ("↔") will reopen the column for viewing. Clicking on the filter icon will present a pop-up window with a list of existing filters for that column and will present the option to apply a new filter. Clicking on the column summary icon will display a summarized list of all entries that exist in that column. This is particularly useful when there are a small number of options in that column and you would like to see how many entries exist for each type or what entries you might like to filter on. The display offers checkboxes next to each entry so that a filter can be applied in place to only show those selected entries (or the inverse of the selection, i.e., show everything except what was selected). Note that whenever an operation is applied to a WormMine table, an "undo" icon will appear allowing the user to undo their last operation (or series of operations).

Next, a user can manage the columns presented by clicking on the "Manage Columns" or "Columns" button. Clicking on this button will present a pop-up window allowing a user to reorder the columns, remove columns (by clicking on the red (–) icon to the right of the column name or by dragging the column to the trash icon), adjust the sort order (for sequential sorting of columns), or add one or more columns by clicking on the green "Add a column" button. When adding columns, the user will be redirected to a model browser allowing the user to select attributes to add to the table. Adjacent to the "Manage Columns"/"Columns" button is the "Manage Filters"/"Filters" button, which allows a user to add a new filter or edit an existing filter. Adjacent to the "Filters" button is the "Manage Relationships"/"Relationships" button, which allows a user to specify whether a particular attribute is required or optional for viewing an entry.

To the right of these buttons is the "Save as List" button, allowing a user to create a new list of objects based on the objects represented in the table being viewed (with all filters and constraints applied) or add objects in the table to an existing list. To save a list, click on the "Save as List" button, choose "Create List" and then click on the data type for which you would like to create a list. This will prompt the user for a name for the new list which will then be saved with other WormMine lists (see above). Rather than select all entities of a particular data type, one can choose the "Pick items from the table" option which will allow for selecting individual items for the destination list. To add items to an existing list, click on "Save as List" and select "Add to List" and the items you would like to add. At this point you will be prompted to select an existing list to add the new items to.

Finally, one can export the table in a variety of formats by clicking on the "Export" button. Specify a filename in the "File name" field and the format of the file in the drop-down menu to the right of the "File name" field. Format options include tab-separated values (.tsv), comma-separated values (.csv), XML, and JSON formats. In addition, the left side panel of the "Export" pop-up window provides options to specify which columns and rows to include, whether to apply compression to the downloaded file, and whether to include column headers in the file. Once parameters for the exported file are set, one can view the expected output in the "Preview" tab. To complete the download, click on "Download file."

11.1.4 WormMine Queries

The preexisting WormMine lists of phenotypes and genes with those phenotypes are informative, but the real power of WormMine lies in the ability to query the data in a variety of ways. To get a first look at querying phenotype data with WormMine, we can take advantage of two template queries for phenotype data, available under the "PHENOTYPES" template query tab located at about the middle (vertically) of the WormMine home page along with the "GENOMICS," "PROTEINS," etc. tabs. Here we can see two template queries that are available: "Genes → Phenotypes" and "Phenotype → Genes". If we click on the "Genes → Phenotypes" query, we will see a description of the query and some options. As the description states, this template query "returns a list of all phenotypes attributed to a gene or a list of genes." The default gene input for the query is *unc-26*. If we click on the green "Show Results" button, we arrive at the query results page, displayed in the table format as described above. What these results depict are all observed variation (allele) induced phenotypes (RNAi phenotypes are not yet available in WormMine, as of this writing) for the gene *unc-26*, along with the alleles of unc-26 which conferred the phenotypes. This is the same information that can be gleaned from the "Phenotypes" widget on the *unc-26* gene page, so this may not seem to add much query power to our existing approaches. However, WormMine also allows querying by using lists instead of just an individual gene. If we click on the "Query" link at the upper left corner of the results table (after the "Trail" header), we can go back to our query options page to perform the same query on a LIST of genes, instead of just unc-26. Click on the checkbox to the left of the text "constrain to be IN" and this will turn on the ability to select one of a list of gene lists in WormMine that may be used as input for the query. Select the list "*C. elegans* genes with a cell cycle variant - or descendant - allele phenotype…" and click on the green "Show Results" button. The results are, again, displayed in the WormMine table format. This query allows a user to see what other phenotypes are associated with genes that are already known to have a "cell cycle variant" or descendant phenotype. This type of query can, of course, be performed with any list of genes, as long as the list has been generated and saved within WormMine.

The second available template query for phenotype data in WormMine is the "Phenotype → Genes" query. Both this query and the template query described above ("Genes → Phenotypes") can be found under the "Phenotypes" tab of the WormMine front page (as described above), but also under the gray "Templates" tab at the top of the WormMine interface. Once selected, we can see the description of the "Phenotype → Genes" template query: "Return all genes annotated with a particular phenotype. Select either observed or not observed." The default phenotype to search with is "transgene expression variant." Clicking on "Show Results" brings us to the results table simply displaying all genes that have been annotated (via alleles) with the "transgene expression variant" phenotype. As with the first template query, this query can be modified by going back to the query options page (click on the "Query" at the upper left of the results table) and either typing in a different single phenotype to search with or selecting a list of phenotypes. As before, we can check the checkbox to the left of "constrain to be IN" and then select a registered list of phenotypes to query with. Select the "Cell cycle variant and descendant phenotype terms…" list and click "Show Results" to see the list of all genes annotated to the phenotype "cell cycle variant" or any of its ontological descendant terms.

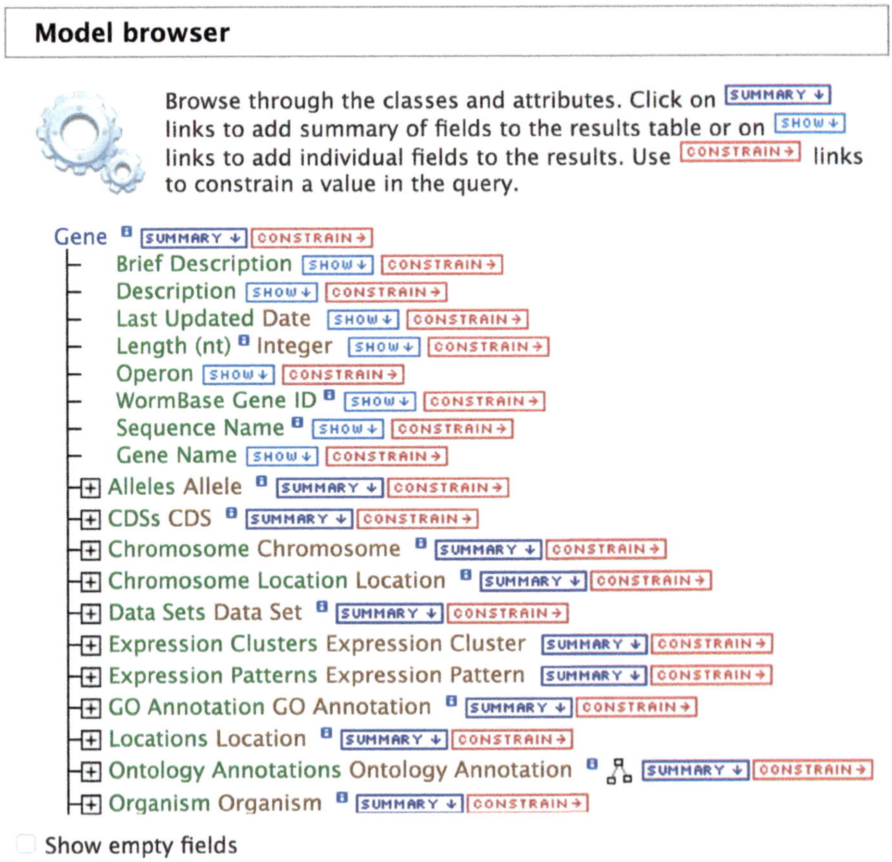

Fig. 39 The WormMine "QueryBuilder" tool displays a model browser for the data type that is being queried. Individual tags/attributes may be selected to be shown (by clicking on "SHOW") or constrained on (by clicking on "CONSTRAIN"). Clicking on "SUMMARY" selects a set of default attributes to be shown for that category of data. Any selections or modifications made in the model browser will immediately be displayed in the "Query Overview" panel to the right of the model browser

11.1.5 The WormMine QueryBuilder

To get a better understanding of how queries in WormMine are constructed, and to enable you to perform your own queries, let us take a closer look at the first template query "Genes → Phenotypes" discussed earlier. Once we have selected this template query from the "PHENOTYPES" tab on the WormMine front page and arrived at the template query options page, click on the "Edit Query" button at the lower right of the options panel. This will take you to the template query's QueryBuilder page, outlining the underlying construction of this query. The upper left quadrant of the QueryBuilder page displays the "Model browser" (Fig. 39) where a user can specify what data is to be displayed and/or constrained on for a query. The upper right quadrant of the page displays the "Query Overview" which summarizes the data chosen to be displayed and constrained on using the model browser. The bottom "Columns to Display" panel of the QueryBuilder page displays the arrangement of output columns for the results table, where the order in which the columns will be displayed in the output table can be rearranged and columns can be removed, given a column header title/description, or set to be sorted.

The query overview panel indicates the data to be displayed as well as where the data may be found in the model browser. We can see that the "WB Gene ID," "Sequence Name," and "Gene Name" have been selected for display from under the "Gene" heading and this can be seen directly in the model browser to the left (note the highlighted tags in the model browser). The query overview also indicates that a constraint has been set for the gene name to be equal to "unc-26" (thereby only returning data relevant to the gene unc-26). We can also see that the "WormBase ID" and "Public Name" have been selected under the "Alleles" header and subsequently "Identifier" and "Name" under the nested "Phenotypes Observed" header. We can verify this in the model browser panel by expanding the relevant nodes by clicking on the "+" sign, first to the left of the "Alleles" header and then to the left of the now exposed "Phenotypes Observed" header under the "Alleles" header. Note the additional highlighted tags for "WormBase ID," "Public Name," "Identifier," and "Name" in the model browser.

To set any data to display in the query results, click on the "Show" button immediately to the right of any desired data attribute title in the model browser and the query overview and columns-to-display panels should update immediately. To remove any data or constraints from the query, click on the red and white "X" icon next to the relevant data type or constraint in the query overview panel or the bottom columns-to-display panel (cannot remove constraints from the bottom panel). To edit an existing constraint (the "Gene Name = unc-26" constraint in this case) click on the blue edit icon immediately to the right of the red and white "X" cancellation button. This brings up the constraint options dialogue box which is the same dialogue box that would appear when applying a new constraint (which can be set by clicking on the red "CONSTRAIN" text next to the data attribute in the model browser). From here we can change the constraint to query using a different gene or a list of genes. We may also use the drop-down list of operators to perform a more generalized query. For example, we may select the "not equal to" option "!=" or use the "CONTAINS" option with the text "unc" to search for all phenotypes for any genes whose names contain the "unc" prefix. Note that the less-than, less-than-or-equal-to, greater-than, and greater-than-or-equal-to options only apply meaningfully to numerical data.

The above description of WormMine is not intended to cover all of its features, but provides an adequate first glance at the tool and its capabilities with respect to phenotype data. A complementary guide to WormMine may be found on the WormBase Wiki WormMine user guide page (http://wiki.wormbase.org/index.php/User Guide:WormMine). Our look at the QueryBuilder focused only on querying from the "Gene" context, but the same general principles apply when building queries in other contexts. The QueryBuilder tool can also be used in the context of "Alleles" and "Phenotype" to extract phenotype data, although the "Alleles" context provides no more than the "Gene" context does (the entire "Alleles" model browser is nested inside the "Gene" model browser) and querying from the "Phenotype" context only provides the list of alleles (not the affiliated genes) annotated to an indicated phenotype or phenotypes. To build a query from scratch, click on the gray "QueryBuilder" tab on the WormMine home page to arrive at the QueryBuilder starting page. From this page a user may view recent query history, import a query from XML (exported from the QueryBuilder previously) or select a data type to begin a query from scratch.

11.2 FTP Site

To complement WormMine, we also provide a collection of precalculated files on our FTP site. Many of these are data files required as input for genome analysis software, e.g., the reference genome sequence (in FASTA format) and the genome annotations (in GFF3 format), and we provide them for convenience. Others capture the results of queries that multiple users have requested in the past.

Many of these files are linked to directly from various different pages on the WormBase site. However, visiting the FTP site itself allows comprehensive access to all the files (ftp://ftp.wormbase.org/pub/wormbase). The files are organized into two parallel directory structures. The first ("releases") organizes the files first by release, then by species, and finally by genome project. This structure is convenient for downloading all (or many) files for a single WormBase release, or single genome. The second structure ("species") organizes the files first by species, and then by data type, with all files for a particular species/data-type, across all releases and genome projects, located in the same folder. This structure is useful for retrieving the latest file for a species/data-type of interest, and to retrieve multiple versions of that file.

In both structures, files are named according to the scheme *G_SPECIES.BIOPROJECT.RELEASE.SUFFIX*, for example: c_elegans.PRJNA13758.WS257.genomic.fa.gz. Note that the second element of the filename is the INSDC BioProject identifier for the genome project, which is our main way of disambiguating between multiple genome projects for the same species. Some of the main files we provide for each species are listed in Table 6.

11.3 RESTful API

The WormBase website implements a simple yet powerful Application Programming Interface (API) that follows the RESTful design pattern. Each widget on the website corresponds to a unique API endpoint using a generic URI structure:

http://api.wormbase.org/rest/widget/[CLASS]/[ID]/[WIDGET]

Table 6
FTP site file names and their contents

Category	Suffix	Description
Genome	.genomic.fa.gz	Reference genome sequence
	.genomic_masked.fa.gz	Genome sequence with repettitive sequence masked with Ns
	.genomic_softmasked.gz	Genome sequence with repetitive sequence made lower-case
Sequences for genomic features	.mRNA_transcripts	Full length protein-coding transcripts
	.CDS_transcripts.fa.gz	CDS portion of protein-coding transcripts
	.proteins.fa.gz	Reference proteome
	.ncRNA_transcripts.fa.gz	Non-coding transcripts
	.transposons.fa.gz	Transposable elements
	.transposon_transcripts.fa.gz	Transcripts associated with transposable elements
	.intergenic_sequences.fa.gz	Sequences between adjacent genes
Genome annotations	.annotations.gff3.gz	Genome annotations in GFF v3
	.annotations.gff2.gz	Genome annotations in GFF v3
	.canonical_transcripts.gtf.gz	The canonical gene set in GTF
Misc	.xrefs.txt.gz	Cross-references between WormBase annotations and other resources such as INSDC and UniProt

For example, to fetch the gene report "Overview" widget for *unc-26* (WBGene00006763) in JSON, send the following curl request:

curl -H content-type:application/json\ http://api.wormbase.org/rest/widget/gene/WBGene00006763/overview

The API is described in more complete detail on the WormBase website (http://www.wormbase.org/about/userguide/for_developers/api-rest#10--10).

12 Tools

12.1 BLAST/BLAT Tool

The BLAST/BLAT tools [53, 54] at WormBase are the standard way to compare a protein or nucleotide query sequence with all protein and genome sequences at WormBase, whether you are looking for an exact match (e.g., by BLASTing a *C. elegans* sequence) or if you are looking for the best nematode BLAST match for a foreign sequence (e.g., BLASTing a human protein sequence). The tool can be reached via the "Tools" menu on any WormBase page. Once the tool has opened, paste in your list of protein or nucleotide sequences in FASTA format. Next, select your query tool, BLAST or BLAT, noting that BLAT tends to work more efficiently with exact matches (it is best not to use BLAT when querying with a nonnematode sequence). Next, select the type of query (blastn, blastp, blastx, or tblastn). There is a "Filter" checkbox, checked by default, that enables the tool to filter out low complexity regions of the query sequence. Finally, select the WormBase version/release, the sequence type (genome or EST for nucleotide queries), the species, and, if applicable, the BioProject ID you would like to BLAST or BLAT against and click the Submit button. Results will be displayed for each sequence submitted (one result page on top of the next), listing the best matches at top with links to the corresponding entities. For nucleotide queries, click on the expandable "+" in the box to the left of the query hit to see a genome view, outlining where the sequence aligns to the genome.

12.2 SimpleMine

SimpleMine is a simple bulk data download tool. Users can submit a list of gene names to get a tab-delimited file containing gene IDs from various databases, phenotypes from alleles and RNAi studies, anatomical and developmental life stages of expression from individual and genomic studies, as well as summarized descriptions about their functions. SimpleMine can be accessed from the WormBase "Tools" menu. To begin a query, type or paste in a list of gene names or identifiers (or upload a list of genes from a file with the "Browse…" button), choose your results format ("download" to download a tab-delimited file; "html" to see the results in your web browser), and click on the "query list" button. Please note that when processing a tab-delimited file of gene names with a spreadsheet program, gene names are often automatically converted into dates, so make sure that all gene names are read-in only as text.

12.3 Gene Set Enrichment Analysis

If a user provides a list of genes, say from an RNA-seq analysis, the Gene Set Enrichment Analysis Tool (available under the WormBase Tools menu) may find tissues, Gene Ontology terms or phenotypes that are over-represented regarding gene annotation frequency. This tool uses the most up to date WormBase gene expression, Gene Ontology, and phenotype data, respectively, and applies a hypergeometric statistical model [55]. The input can be of any format of WormBase gene names, either entered in the provided box or in a file on the user's computer. The output will inform the user if any of the input genes are not recognized or for which there is no available data and thus excluded from the analysis. A successful analysis will yield a table and a graph of enriched terms. Both the table and the graph may be exported for further analysis or presentations.

12.4 Community Annotation Forms

As the rate at which nematode research articles are published continues to grow, the demand for manual curation is overwhelming the capacity of the WormBase literature curation team. In order to stay up-to-date and current with the literature, WormBase needs the support and curation effort of the entire nematode research community. To that end, WormBase has begun development of a series of web-based, user-friendly community annotation forms designed for easy and efficient submission of data. Links to these forms are available in the "Community Curation" widget of the "Submit Data" page (http://www.wormbase.org/about/userguide/submit_data#01--10). The "Submit Data" page can be accessed via the "Submit Data" link directly below the search box on any WormBase page or in the "Community" and "Support" drop-down menus in the main navigation bar.

Currently, data for four different data types can be submitted via the community curation forms: phenotype, allele sequence, expression data micropublication, and gene descriptions. Each of the forms shares common features designed to expedite the data submission process: (1) personal recognition: once any form has been filled out and data submitted, the form will remember users based on their IP address, making it a bit easier to submit data on subsequent visits; (2) autocomplete functionality: wherever a field requires a controlled vocabulary, the field will provide matching options as you type so as to ease the term lookup process, reduce errors, and save time; (3) term information: also for controlled vocabulary fields, term information boxes appear at the upper right corner of the screen to provide additional information about the term and often provide links to the relevant WormBase page; (4) in-line help: green question marks adjacent to entry fields may be clicked to provide help information about that field in the term information box at the upper right corner of the screen; (5) clearly marked mandatory fields to make clear what entries are required for submission.

13 Community Resources

WormBase offers a number of services to support the *C. elegans* research community. These include tools to interact directly with WormBase curators and developers, to request assistance with specific problems you may have, to publish brief research missives, and to stay up-to-date with items of interest to the community. In brief, your best approach for staying up-to-date.

13.1 Help Desk

WormBase provides a responsive help desk service to assist with problems you may have using the website or interpreting data. Need help with a data mining query? Looking for information on a specific gene or have new data to submit? The help desk is appropriate for any and all queries that you may have.

There are two options for submitting queries to the help desk. Most directly, you can send an email to *help@wormbase.org*. Alternatively, on the website look for a small tab at the bottom of every window that reads "Need help or have feedback?" (Fig. 1, bottom right). Click on the tab to expand it. If this is during normal working hours (typically within the range of 6 AM–8 PM GMT −5), you may have the option of chatting directly with a WormBase curator or developer. If no members of WormBase staff are available, you will be provided with a form to enter your name and email (optional) and a brief description of your query. Do note that if you do not provide your email, we will not be able to follow up with you on your query.

Regardless of how you contact us, we will quickly triage your issue and make sure that the most appropriate curator or developer handles your query. You can track progress on resolution of your query on our public issue tracker at http://github.com/wormbase/website/issues. We aim to respond to every query within 24 h, and often do so much more quickly than that. As mentioned above, the website has a built-in chat feature. During normal working hours, you can text chat directly with WormBase staff for quick resolution of your query.

13.2 The Worm Community Forum

The Worm Community Forum is a joint resource sponsored by WormBase and Worm Atlas [56] (www.wormatlas.org). Here, you can submit queries to an audience beyond WormBase staff, to include a wide range of *C. elegans* and nematode researchers. The Forum includes a variety of sections including those for new lab announcements, job postings, and help with specific experimental techniques. You must register in order to post and reply to topics, but you may browse without registering.

13.3 The WormBase Blog

WormBase maintains a blog (http://blog.wormbase.org) as a home for longer format narratives discussing major new features, meeting announcements and so on. These are generally longer narratives that will not fit within the Twitter confines of 140 characters. You can always stay up-to-date with the WormBase blog by visiting the home page periodically. Items posted to the blog will appear in the "News" section. You can always subscribe by either email or to the RSS feed by visiting the WormBase Blog directly.

13.4 Twitter @ WormBase

At WormBase, we use Twitter to inform users of breaking service status issues, as an aggregator of various other outreach channels (such as the blog), and as a means to contact us and ask us questions. We are @wormbase on Twitter (http://twitter.com/wormbase).

13.5 The Worm Breeder's Gazette

The Worm Breeder's Gazette (http://wbg.wormbook.org) is an extension of the long-running print version of the newsletter. Operated under the auspices of WormBook [57], the new online version of the Worm Breeder's Gazette lets authors publish brief (approximately one printed page) research findings, methods, or announcements of general relevance to the community. Submissions are handled directly online. We aim to publish submitted articles within a week after submission, following a brief editorial period to correct typographical errors and insert links to relevant resources.

13.6 The WormBase YouTube Channel

WormBase also manages a YouTube channel (https://www.youtube.com/user/WormBaseHD) that provides users with brief instructional videos on the basics of how to use the WormBase website and carry out common query tasks.

Acknowledgments

WormBase is supported by grant #U41 HG002223 from the National Human Genome Research Institute at the US National Institutes of Health, the UK Medical Research Council and the UK Biotechnology and Biological Sciences Research Council. At the time of writing, the WormBase Consortium included Paul W. Sternberg, Paul Kersey, Matthew Berriman, Lincoln Stein, Tim Schedl, Todd Harris, Scott Cain, Sibyl Gao, Paulo Nuin, Adam Wright, Kevin Howe, Bruce Bolt, Paul Davis, Michael Paulini, Faye Rodgers, Matthew Russell, Myriam Shafie, Gary Williams, Juancarlos Chan, Wen J. Chen, Christian Grove, Ranjana Kishore, Raymond Lee, Hans-Michael Müller, Cecilia Nakamura, Daniela Raciti, Gary Schindelman, Mary Ann Tuli, Kimberly Van Auken, Daniel Wang, and Karen Yook.

References

1. Harris TW, Baran J, Bieri T et al (2014) WormBase 2014: new views of curated biology. Nucleic Acids Res 42:D789–D793. https://doi.org/10.1093/nar/gkt1063
2. Howe KL, Bolt BJ, Cain S et al (2016) WormBase 2016: expanding to enable helminth genomic research. Nucleic Acids Res 44:D774–D780. https://doi.org/10.1093/nar/gkv1217
3. *C. elegans* Sequencing Consortium (1998) Genome sequence of the nematode *C. elegans*: a platform for investigating biology. Science 282:2012–2018.
4. Nakamura Y, Cochrane G, Karsch-Mizrachi I, International Nucleotide Sequence Database Collaboration (2013) The International Nucleotide Sequence Database Collaboration. Nucleic Acids Res 41:D21–D24. https://doi.org/10.1093/nar/gks1084
5. Stein LD, Mungall C, Shu S et al (2002) The generic genome browser: a building block for a model organism system database. Genome Res 12:1599–1610. https://doi.org/10.1101/gr.403602
6. Skinner ME, Uzilov AV, Stein LD et al (2009) JBrowse: a next-generation genome browser. Genome Res 19:1630–1638. https://doi.org/10.1101/gr.094607.109
7. Gerstein MB, ZJ L, Van Nostrand EL et al (2010) Integrative analysis of the *Caenorhabditis elegans* genome by the modENCODE project. Science 330:1775–1787. https://doi.org/10.1126/science.1196914
8. Altschul SF, Madden TL, Schäffer AA et al (1997) Gapped BLAST and PSI-BLAST: a new generation of protein database search programs. Nucleic Acids Res 25:3389–3402
9. Camacho C, Coulouris G, Avagyan V et al (2009) BLAST+: architecture and applications. BMC Bioinformatics 10:421. https://doi.org/10.1186/1471-2105-10-421

10. Edgar RC (2004) MUSCLE: multiple sequence alignment with high accuracy and high throughput. Nucleic Acids Res 32:1792–1797. https://doi.org/10.1093/nar/gkh340
11. Mitchell A, Chang H-Y, Daugherty L et al (2015) The InterPro protein families database: the classification resource after 15 years. Nucleic Acids Res 43:D213–D221. https://doi.org/10.1093/nar/gku1243
12. Gene Ontology Consortium (2015) Gene Ontology Consortium: going forward. Nucleic Acids Res 43:D1049–D1056. https://doi.org/10.1093/nar/gku1179
13. Finn RD, Bateman A, Clements J et al (2014) Pfam: the protein families database. Nucleic Acids Res 42:D222–D230. https://doi.org/10.1093/nar/gkt1223
14. Powell S, Forslund K, Szklarczyk D et al (2014) eggNOG v4.0: nested orthology inference across 3686 organisms. Nucleic Acids Res 42:D231–D239. https://doi.org/10.1093/nar/gkt1253
15. Li H, Coghlan A, Ruan J et al (2006) TreeFam: a curated database of phylogenetic trees of animal gene families. Nucleic Acids Res 34:D572–D580. https://doi.org/10.1093/nar/gkj118
16. Vilella AJ, Severin J, Ureta-Vidal A et al (2009) EnsemblCompara GeneTrees: complete, duplication-aware phylogenetic trees in vertebrates. Genome Res 19:327–335. https://doi.org/10.1101/gr.073585.107
17. The Gene Ontology Consortium (2017) Expansion of the Gene Ontology knowledgebase and resources. Nucleic Acids Res 45:D331–D338. https://doi.org/10.1093/nar/gkw1108
18. Lee RYN, Sternberg PW (2003) Building a cell and anatomy ontology of *Caenorhabditis elegans*. Comp Funct Genomics 4:121–126. https://doi.org/10.1002/cfg.248
19. Schriml LM, Arze C, Nadendla S et al (2012) Disease Ontology: a backbone for disease semantic integration. Nucleic Acids Res 40:D940–D946. https://doi.org/10.1093/nar/gkr972
20. Schindelman G, Fernandes JS, Bastiani CA et al (2011) Worm Phenotype Ontology: integrating phenotype data within and beyond the *C. elegans* community. BMC Bioinformatics 12:32. https://doi.org/10.1186/1471-2105-12-32
21. Huntley RP, Harris MA, Alam-Faruque Y et al (2014) A method for increasing expressivity of Gene Ontology annotations using a compositional approach. BMC Bioinformatics 15:155. https://doi.org/10.1186/1471-2105-15-155
22. Gaudet P, Livstone MS, Lewis SE, Thomas PD (2011) Phylogenetic-based propagation of functional annotations within the Gene Ontology consortium. Brief Bioinform 12:449–462. https://doi.org/10.1093/bib/bbr042
23. Huntley RP, Sawford T, Mutowo-Meullenet P et al (2015) The GOA database: gene Ontology annotation updates for 2015. Nucleic Acids Res 43:D1057–D1063. https://doi.org/10.1093/nar/gku1113
24. Burge S, Kelly E, Lonsdale D et al (2012) Manual GO annotation of predictive protein signatures: the InterPro approach to GO curation. Database (Oxford) 2012:bar068. https://doi.org/10.1093/database/bar068
25. Trapnell C, Williams BA, Pertea G et al (2010) Transcript assembly and quantification by RNA-Seq reveals unannotated transcripts and isoform switching during cell differentiation. Nat Biotechnol 28:511–515. https://doi.org/10.1038/nbt.1621
26. Trapnell C, Roberts A, Goff L et al (2012) Differential gene and transcript expression analysis of RNA-seq experiments with TopHat and Cufflinks. Nat Protoc 7:562–578. https://doi.org/10.1038/nprot.2012.016
27. Zhong W, Sternberg PW (2006) Genome-wide prediction of *C. elegans* genetic interactions. Science 311:1481–1484. https://doi.org/10.1126/science.1123287
28. Lee I, Lehner B, Crombie C et al (2008) A single gene network accurately predicts phenotypic effects of gene perturbation in *Caenorhabditis elegans*. Nat Genet 40:181–188. https://doi.org/10.1038/ng.2007.70
29. Lee I, Lehner B, Vavouri T et al (2010) Predicting genetic modifier loci using functional gene networks. Genome Res 20:1143–1153. https://doi.org/10.1101/gr.102749.109
30. Rual J-F, Ceron J, Koreth J et al (2004) Toward improving *Caenorhabditis elegans* phenome mapping with an ORFeome-based RNAi library. Genome Res 14:2162–2168. https://doi.org/10.1101/gr.2505604
31. Kamath RS, Fraser AG, Dong Y et al (2003) Systematic functional analysis of the *Caenorhabditis elegans* genome using RNAi. Nature 421:231–237. https://doi.org/10.1038/nature01278
32. Culetto E, Sattelle DB (2000) A role for *Caenorhabditis elegans* in understanding the function and interactions of human disease genes. Hum Mol Genet 9:869–877
33. Artal-Sanz M, de Jong L, Tavernarakis N (2006) *Caenorhabditis elegans*: a versatile platform for drug discovery. Biotechnol J 1: 1405–1418. https://doi.org/10.1002/biot.200600176
34. Giacomotto J, Ségalat L (2010) High-throughput screening and small animal models, where are we?

Br J Pharmacol 160:204–216. https://doi.org/10.1111/j.1476-5381.2010.00725.x

35. O'Reilly LP, Luke CJ, Perlmutter DH et al (2014) *C. elegans* in high-throughput drug discovery. Adv Drug Deliv Rev 69–70:247–253. https://doi.org/10.1016/j.addr.2013.12.001
36. Li J, Le W (2013) Modeling neurodegenerative diseases in *Caenorhabditis elegans*. Exp Neurol 250:94–103. https://doi.org/10.1016/j.expneurol.2013.09.024
37. Alexander AG, Marfil V, Li C (2014) Use of *Caenorhabditis elegans* as a model to study Alzheimer's disease and other neurodegenerative diseases. Front Genet 5:279. https://doi.org/10.3389/fgene.2014.00279
38. O'Hagan R, Wang J, Barr MM (2014) Mating behavior, male sensory cilia, and polycystins in *Caenorhabditis elegans*. Semin Cell Dev Biol 33:25–33. https://doi.org/10.1016/j.semcdb.2014.06.001
39. Blacque OE, Sanders AAWM (2014) Compartments within a compartment: what *C. elegans* can tell us about ciliary subdomain composition, biogenesis, function, and disease. Organogenesis 10:126–137. https://doi.org/10.4161/org.28830
40. Lee S-J, Gartner A, Hyun M et al (2010) The *Caenorhabditis elegans* Werner syndrome protein functions upstream of ATR and ATM in response to DNA replication inhibition and double-strand DNA breaks. PLoS Genet 6:e1000801. https://doi.org/10.1371/journal.pgen.1000801
41. Zheng J, Greenway FL (2012) *Caenorhabditis elegans* as a model for obesity research. Int J Obes (Lond) 36:186–194. https://doi.org/10.1038/ijo.2011.93
42. Park K-W, Li L (2011) Prion protein in *Caenorhabditis elegans*: distinct models of anti-BAX and neuropathology. Prion 5:28–38
43. Kibbe WA, Arze C, Felix V et al (2015) Disease Ontology 2015 update: an expanded and updated database of human diseases for linking biomedical knowledge through disease data. Nucleic Acids Res 43:D1071–D1078. https://doi.org/10.1093/nar/gku1011
44. Amberger JS, Bocchini CA, Schiettecatte F et al (2015) OMIM.org: Online Mendelian Inheritance in Man (OMIM®), an online catalog of human genes and genetic disorders. Nucleic Acids Res 43:D789–D798. https://doi.org/10.1093/nar/gku1205
45. Bretscher AJ, Kodama-Namba E, Busch KE et al (2011) Temperature, oxygen, and salt-sensing neurons in *C. elegans* are carbon dioxide sensors that control avoidance behavior. Neuron 69:1099–1113. https://doi.org/10.1016/j.neuron.2011.02.023
46. Smith RN, Aleksic J, Butano D et al (2012) InterMine: a flexible data warehouse system for the integration and analysis of heterogeneous biological data. Bioinformatics 28:3163–3165. https://doi.org/10.1093/bioinformatics/bts577
47. Kalderimis A, Lyne R, Butano D et al (2014) InterMine: extensive web services for modern biology. Nucleic Acids Res 42:W468–W472. https://doi.org/10.1093/nar/gku301
48. Lyne R, Smith R, Rutherford K et al (2007) FlyMine: an integrated database for *Drosophila* and *Anopheles* genomics. Genome Biol 8:R129. https://doi.org/10.1186/gb-2007-8-7-r129
49. Motenko H, Neuhauser SB, O'Keefe M, Richardson JE (2015) MouseMine: a new data warehouse for MGI. Mamm Genome 26:325–330. https://doi.org/10.1007/s00335-015-9573-z
50. Balakrishnan R, Park J, Karra K et al (2012) YeastMine--an integrated data warehouse for *Saccharomyces cerevisiae* data as a multipurpose tool-kit. Database (Oxford) 2012: bar062. https://doi.org/10.1093/database/bar062
51. Contrino S, Smith RN, Butano D et al (2012) modMine: flexible access to modENCODE data. Nucleic Acids Res 40:D1082–D1088. https://doi.org/10.1093/nar/gkr921
52. Rhee DB, Croken MM, Shieh KR et al (2015) toxoMine: an integrated omics data warehouse for *Toxoplasma gondii* systems biology research. Database (Oxford) 2015:bav066. https://doi.org/10.1093/database/bav066
53. Altschul SF, Gish W, Miller W et al (1990) Basic local alignment search tool. J Mol Biol 215:403–410. https://doi.org/10.1016/S0022-2836(05)80360-2
54. Kent WJ (2002) BLAT--the BLAST-like alignment tool. Genome Res 12:656–664. https://doi.org/10.1101/gr.229202
55. Angeles-Albores D, N Lee RY, Chan J, Sternberg PW (2016) Tissue enrichment analysis for *C. elegans* genomics. BMC Bioinformatics 17:366. https://doi.org/10.1186/s12859-016-1229-9
56. WormAtlas, Altun ZF, Herndon LA, Wolkow CA, Crocker C, Lints R, Hall DH (eds) (2002–2017). http://www.wormatlas.org. Accessed 10 Apr 2017
57. Greenwald I (2016) WormBook: WormBiology for the 21st Century. Genetics 202:883–884. https://doi.org/10.1534/genetics.116.187575

Chapter 15

Using WormBase ParaSite: An Integrated Platform for Exploring Helminth Genomic Data

Bruce J. Bolt, Faye H. Rodgers, Myriam Shafie, Paul J. Kersey, Matthew Berriman, and Kevin L. Howe

Abstract

WormBase ParaSite (parasite.wormbase.org) is a comprehensive resource for the genomes of parasitic nematodes and flatworms (helminths). It currently includes genomic data for over 100 helminth species, adding value by way of consistent functional annotation, gene comparative analysis and gene expression analysis. We provide several ways of exploring the data including a choice of genome browsers, genome and gene summary pages, text and sequence searching, a query wizard, bulk downloads, and programmatic interfaces. WormBase ParaSite is released three to six times per year, and is developed in collaboration with WormBase (www.wormbase.org) and Ensembl Genomes (www.ensemblgenomes.org).

Key words Genome browser, Comparative genomics, Functional genomics, Helminths, WormBase, Ensembl, Parasitology

1 Introduction

1.1 Background

Parasitic nematodes and flatworms, collectively known as helminths, infect at least 25% of the human population globally [1] and are responsible for long term, chronic diseases that result in pain, social stigma, physical and mental disabilities, and in the worst cases, death. Helminth infections also have a devastating impact on agriculture, causing high morbidity in farm animals and significant annual reduction in crop yields.

The study of helminth biology, with a view to understanding and controlling helminth infection, is thus an active field of research. In recent years, the genomes of numerous helminth species have become the subject of genome sequencing and annotation projects, which are the cornerstone of modern large-scale approaches to systems biology. The aim of WormBase ParaSite is to systematically integrate the results of these efforts and present them consistently via a single set of interfaces and exploratory tools.

Martin Kollmar (ed.), *Eukaryotic Genomic Databases: Methods and Protocols*, Methods in Molecular Biology, vol. 1757, https://doi.org/10.1007/978-1-4939-7737-6_15, © Springer Science+Business Media, LLC, part of Springer Nature 2018

1.2 Infrastructure and Release Cycle

WormBase ParaSite is a sister project to the WormBase [2] and Ensembl Genomes [3] projects, leveraging infrastructure and expertise from both and adding data and functionality of specific utility to researchers engaged in helminth genomics. Releases of WormBase ParaSite are loosely coupled with WormBase. For the species in common with WormBase (e.g., *C. elegans* and other free-living nematodes), we synchronise the data with a specific release of WormBase. For example, WormBase ParaSite release 9 was synchronized with, and made public as close as possible to, WormBase release WS258. This means that users can navigate between WormBase ParaSite and WormBase and be confident that the underlying data is consistent.

1.3 Sources of Data

We aim to include all publicly available nematode and platyhelminth genomes, including those of free-living (nonparasitic) species, as these are useful for comparative analysis and the study of the evolution of parasitism.

Our primary source for genome sequences is the International Nucleotide Sequence Database Collaboration (INSDC) resources [4]. Where multiple genome assemblies exist for the same species (e.g., different genome projects for *Haemonchus contortus* [5, 6]), we aim to include all. Different genomes from the same species are distinguished by their INSDC BioProject identifier [7] which is linked to an archive record describing the provenance of the genome project.

Release 9 of WormBase ParaSite (April 2017) included 134 genomes from 114 species, from 24 genome project data providers. We add value to these primary data by way of a number of computational pipelines for functional annotation, for example: protein domains and Gene Ontology [8] terms using InterProScan [9], gene comparative analysis using Ensembl Compara [10], and alignment of life-stage-specific RNA-Seq data sets to the genome. Access to these analyses is described below; the analyses themselves are described in more detail elsewhere [11].

2 Website Overview

2.1 Home Page and General Navigation

The WormBase ParaSite home page (Fig. 1) is the main point of access to the resource. As well as providing convenient entry points to all of our data and tools for the current release, it also displays news and information on topics relevant to the helminthology community (for example, recent articles from our blog; *see* Subheading 8).

A panel of pictograms on the left-hand side of the home page provides links to the most common starting points for using WormBase ParaSite, such as the Genome List, BioMart (Subheading 5) and BLAST (Subheading 6.1). The header bar also includes

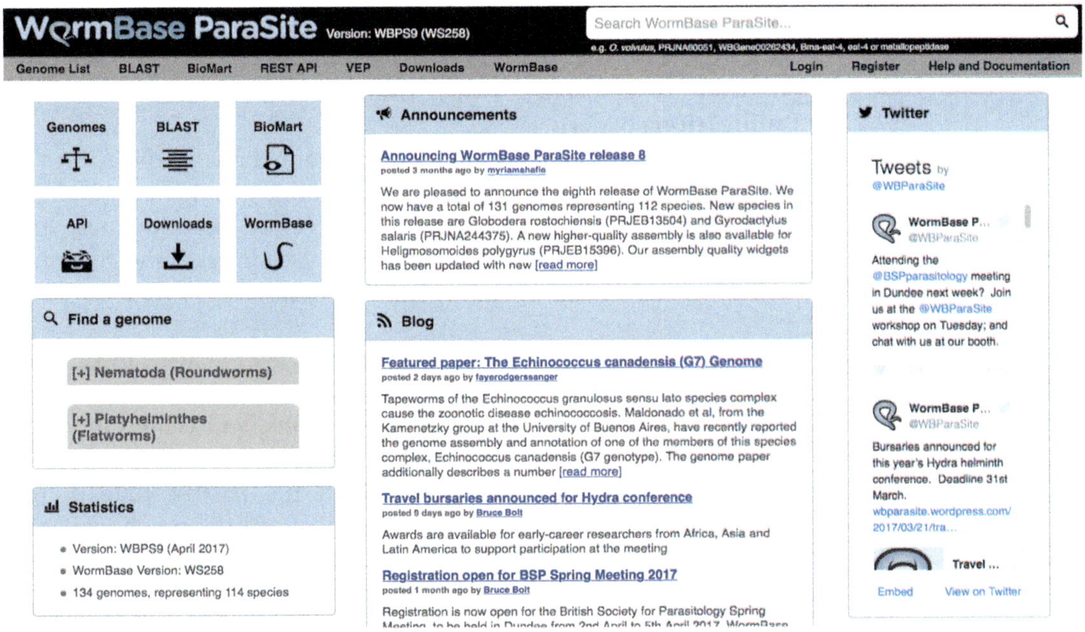

Fig. 1 WormBase ParaSite home page (release 9)

these links and is present on every page of the site. The header also includes a search box, and links to our Help and user-account login pages (Subheading 2.5).

2.2 Finding a Genome

There are three main ways to find a genome of interest: (a) the Genome table (accessible via the **Genomes** home page pictogram or the **Genome List** link in the header); (b) the **Find a genome** panel on the home page, where genomes are arranged taxonomically according to the species of origin; and (c) typing the first few letters of the name of the species or genus of interest into the search bar, which will auto-complete.

The **Genome List** table contains columns for basic information about the genome including species of origin, the genome project identifier, the data provider, and (for nematodes) the clade to which the species belongs. Also included are a set of popular metrics used to assess assembly quality, including N50 [12], CEGMA [13], and BUSCO [14] scores. The table can be sorted by any of these fields, by clicking the column name in the table header.

2.3 Genome Overview Pages

The genome overview pages collect together a variety of information and onward links for a single genome of interest. These include: a short description about the species of origin (in the **About** panel); information about the source and method for the genome assembly and annotation (in the **Genome Assembly and Annotation** panel); shortcut links to example genes and regions in the genome (via the **Navigation** panel); links to a collection of files associated with the genome (the **Downloads** panel); and selected publications associated with the genome project (the **Key Publications** panel).

The **Assembly Statistics** panel contains basic statistics on the genome assembly and annotation, including genome size and the number of coding genes, noncoding genes, pseudogenes, and transcripts in the current annotation. It also presents the common metrics used to assess assembly quality graphically (Fig. 2).

2.4 Searching

The search bar located on the top right of every page can be used to search for genes by several criteria, including gene name, product name, protein domain names/accessions, and Gene Ontology accessions. While typing in the search bar, an autocomplete menu will appear listing common search terms having that prefix. Additionally, the name of tools or services offered on the website can be entered into the search box, for quick navigation to other parts of the website. It can also be used to access a species directly; any species name that is a partial match to the user input will appear on top of the list of suggested search terms.

The search results page is divided into several sections. For each gene matching the search term, the gene name, primary identifier, description, species, and location in the genome are displayed. If the gene has any orthologues in *C. elegans*, a link to the WormBase gene page is also shown.

All searches are performed both in WormBase ParaSite and in WormBase. The number of results for both websites are displayed in the top-left menu. To refine the results, the user can select a species on the pulldown menu at the top of the search results.

2.5 User Accounts

User accounts allow the saving of configuration and custom data tracks between sessions and computers. Users can create an account via the **Register** link in the header bar, providing their name and email address. An email is sent to this address to verify it and give the user the possibility to set a password. The account can then be accessed through the **Login** link in the right of the header bar.

Once logged in to their account, users can create or join a group to share data with other users. They can also create and save genome browser configurations (e.g., attached custom tracks). Finally, BLAST and VEP results (Subheading 6) can be saved permanently by linking them to an account. Otherwise, these results are deleted after 7 days. After saving genome browser configurations and tool results against a user account, they become available to the user through any computer from which they login to the website.

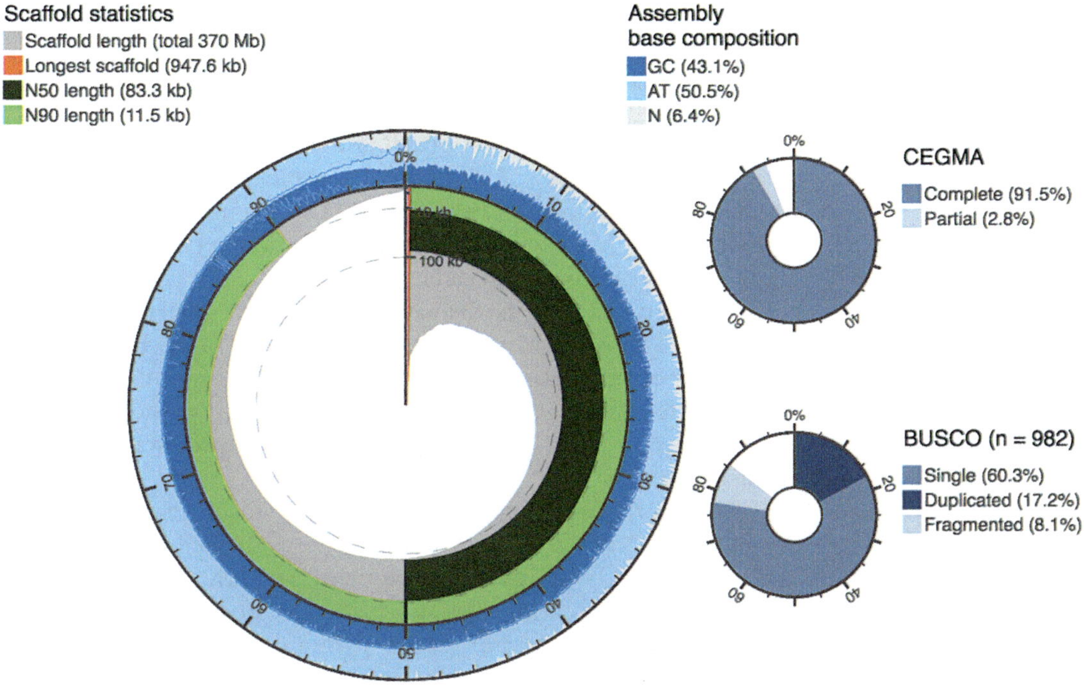

Fig. 2 Genome overview Assembly Statistics panel for the *Haemonchus contortus* genome (project PRJEB506). Left: the large circle represents the length of the genome, with the scaffolds ordered from longest to shortest. The longest scaffold is indicated with an orange line. The grey inner segment plots the cumulative length as a proportion of the total assembly size, and at each point shows what proportion of the assembly is contained within scaffolds of at least that length. The N50 and N90 scores are shown graphically by the dark and light green radial arcs. Base composition is plotted in blue along the outer rim of the circle. Right: assembly completeness using two popular methods (CEGMA and BUSCO) are shown graphically

3 Gene Report Pages

Each gene in WormBase ParaSite has a report page, which can be accessed by searching for the gene (by any of its identifiers), or selecting the gene structure in the genome browser (Subheading 4). The content of the page is divided into two sections. The first, top section is an overview showing the gene name and primary identifier, the biotype (e.g., protein-coding), a short description of the function (where available), genomic location, the number of transcripts that the gene has, the number of homologues, and the source/provider of the transcript structures. A table summarising the transcripts of the gene is shown at the bottom of the overview section.

The second, lower section of the page is context-dependent and is controlled by selecting links from the left-hand navigation menu. For example, the **External references** link populates the lower part of the page with the table showing cross-references to other resources for the gene, and the **Literature** link performs a search of the gene identifier against Europe PubMed Central [15] and displays the results in the lower part of the page. The other functions are described in more detail below.

3.1 Summary View

The Summary view is shown by default when first navigating to the gene report page (Fig. 3). It contains a **Genomic context** panel which shows the transcript structures of the gene, and those of its 5′ and 3′ neighbours, with the gene of interest highlighted in green. For protein-coding genes in particular, the coding part of each transcript (CDS) is displayed in red, and the untranslated regions (UTRs) are transparent. The panel can be customized via the **Configure this page** button, where a limited number of tracks can be added. The complete set of tracks, including gene expression tracks, are available via the main genome browser views (*see* Subheading 4).

Two alternative views of the transcript structures of the gene can be obtained by selecting submenu items in the left-hand navigation panel. The **Splice Variants** option shows the transcript structures aligned and overlaid with the positions of protein domains and features. For genes with multiple transcripts, the **Transcript comparison** option shows a multiple alignment of all transcript isoforms, which reveals which sequence segments are shared by multiple transcripts, and which are unique to a single transcript.

3.2 Sequence View

The **Sequence** menu item provides annotated sequence of the gene and surrounding genomic region. The **Download sequence** button can be used to obtain a multientry FASTA file with any or all of the following sequences: cDNA, CDS, peptide, 5′ UTRs, 3′ UTRs, exons, introns, or genomic sequence, with a user-specified length of flanking DNA. Alternatively, these can be downloaded in RTF format, where additional options can be selected, such as displaying alternating exons as upper and lower case, showing line numbering and replacing ambiguous bases with Ns. Additionally, by highlighting text sequence using the left mouse button, it is possible to send a user selected section of the sequence to BLAST (Fig. 4).

3.3 Gene Ontology View

The three options underneath **Gene Ontology** menu item show functional annotation for the gene, via associations with terms in the Gene Ontology (GO) [8]. There is a separate table for annotations to each of the biological process, molecular function and cellular component aspects of the GO. The evidence, source and specific transcripts associated with the GO term are shown in the table for each annotation. The **view associated genes** link in the final column of the table displays a complete list of genes in the genome associated with that GO term, and their positions in the genome. If a karyotype image is available (i.e., if the genome is

Fig. 3 Gene summary page for the *Brugia malayi* gene Bm3298. An overview of the gene and its transcripts are displayed on all pages related to this gene

assembled into chromosomes), the positions of genes annotated with the GO term will also be available as an image. By clicking **Search BioMart**, the user will be taken to ParaSite BioMart where the relevant GO term will have been autofilled as a query filter (*see* Subheading 5).

3.4 Comparative Genomics View

The **Comparative Genomics** submenu can be used to explore the relationship between the gene of interest and homologous genes in helminths, human, and model organisms. Homologues are predicted using the Ensembl Compara pipeline [10], which produces an evolutionary tree for each gene family, and uses the tree to determine orthologues (a pair of genes in different species related by a speciation event) and paralogues (a pairs of genes in the same species related by a duplication event).

The Compara tree for the gene of interest can be explored by following the **Gene tree** menu option. The tree is shown alongside a pictorial representation of the multiple alignment of the proteins in the family (Fig. 5). The alignment is coloured according to the degree of conservation, with highly conserved regions represented in darker green. The full amino acid alignment of the sequences in

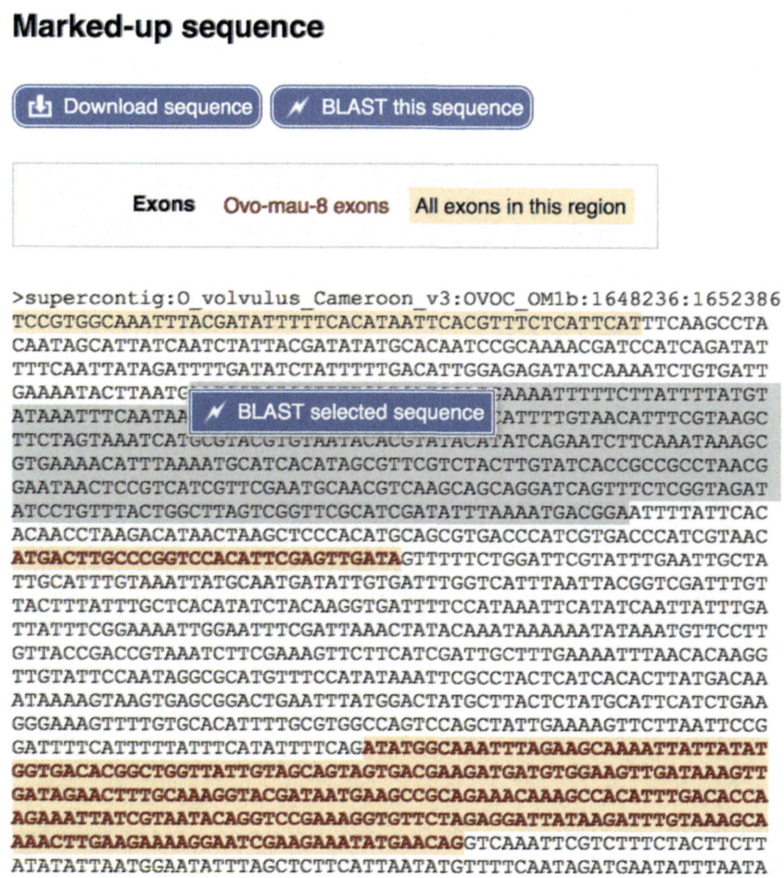

Fig. 4 Part of the marked-up sequence for the *Schistosoma mansoni* gene Smp_110730. Exons are marked in red text. A section of the sequence has been selected, with the option of sending this selection directly to BLAST

a subtree can be explored interactively by clicking a node of interest and selecting **View in Wasabi** [16].

The default view for the tree is for the gene of interest to be highlighted in red and for portions of the tree to be collapsed. Individual parts of the tree can be expanded and collapsed as required by clicking on the parent node of the subtree and choosing the appropriate option from the resulting pop-up menu. The tree display can be further customized via the **Configure this page** button on the left followed by **Display options**, for example to collapse or hide all platyhelminth or all nematode genes, or genes with only low coverage in the protein multiple alignment. The icons along the top of the image allow the image of the tree and the underlying data to be exported in a variety of formats.

Tables of predicted **Orthologues** and **Paralogues** can be viewed by selecting the corresponding menu item. For each orthologue, the table shows a number of properties, including the type

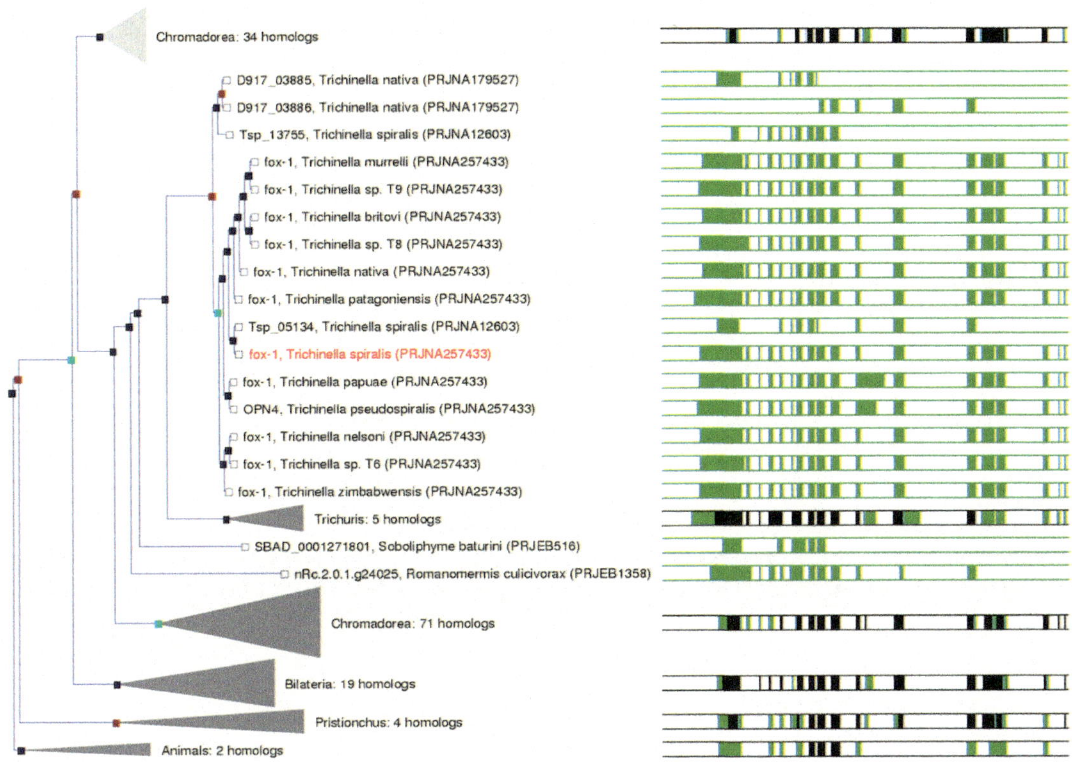

Fig. 5 Part of a Compara tree showing the evolution of *fox-1* in *Trichinella spiralis*, a gene involved in sex determination. In this view, the *Trichinella* subtree has been expanded for additional details

of relationship (one-to-one or one-to-many), percentage identity, and links to pairwise protein and cDNA alignments.

3.5 Variation and Expression Views

For genomes with variation data present in the European Variation Archive (EVA) [17], for example *Strongyloides ratti*, a Variation menu item is present. Two main views are provided: a **Variation Table** showing all variants that have been identified as affecting the current gene/transcript; and a **Variation Image** showing the variants within the context of the gene, exons and protein domains. Each variant is linked to a page showing genotypes across all samples/strains (Fig. 6). These variation-specific pages can also be reached by selecting from the Variations track on the genome browser (Subheading 4.1).

For genomes with data in Gene Expression Atlas [18], for example *Schistosoma mansoni*, an **Expression** menu item is present. This populates the lower part of the page with a display showing normalized expression levels of the gene in various assayed conditions.

3.6 Transcripts and Proteins

Various elements of the gene-report pages refer to specific transcripts of the gene. Following a linked transcript identifier opens

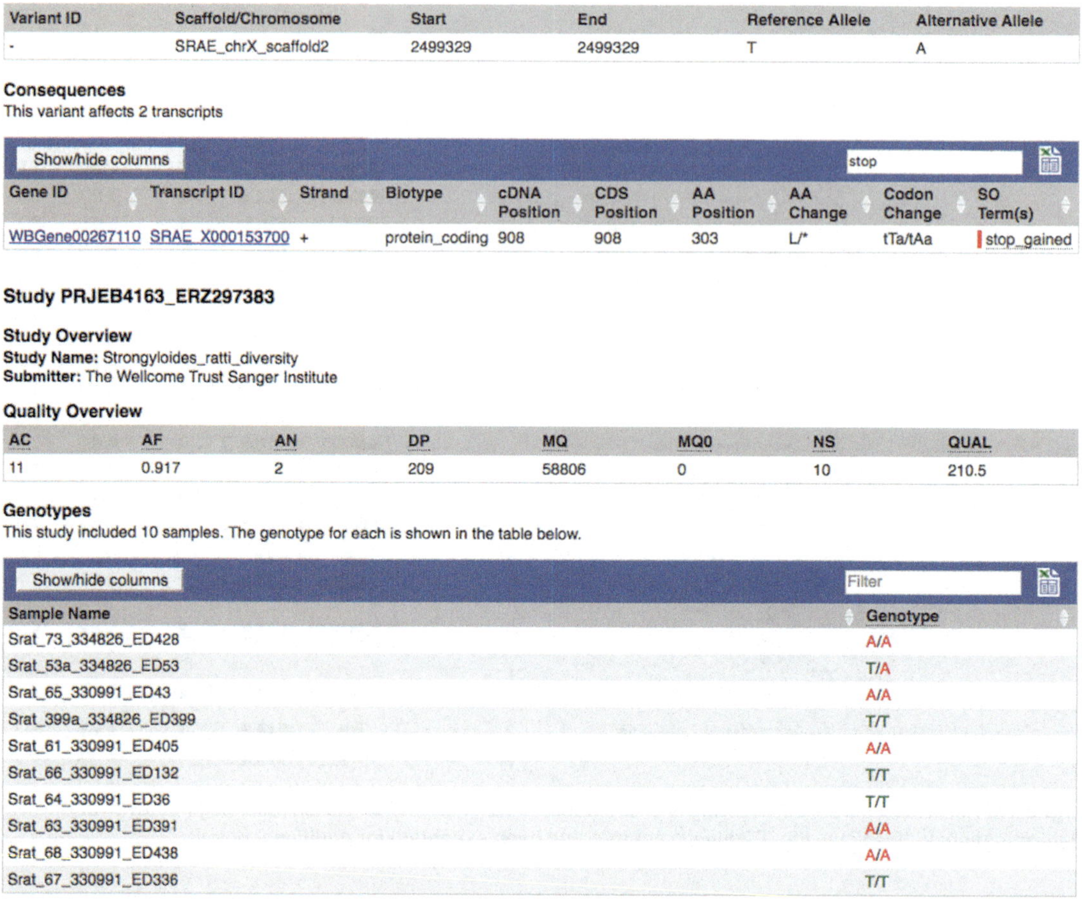

Fig. 6 Summary view of a single variant in the *Strongyloides ratti* genome, showing its effect on the reference annotation, data associated with the confidence of the variant call, and genotypes in ten strains. Six of the ten strains carry a genotype that gives rise to a nonsense mutation in the gene

up a page with information specific to that transcript, with a transcript-oriented left-hand navigation menu.

The **Sequence** submenu can be used to view sequences associated with the transcript. The **Exons** option displays a table showing basic information about each exon, including genomic location and sequence. For protein-coding genes the **cDNA** option provides a graphical view of aligned cDNA, CDS and peptide sequence, and the **Protein** option shows the peptide sequence with alternating exons marked in black and blue. As with genes, users can **Download sequence** for transcripts in a variety of formats.

For protein-coding genes, the **Protein information** submenu contains links to pages that display features and domains from the InterPro [19] member resources. The **Protein Summary** option displays the features in graphical form, in relation to the exon structure of the transcript; **Domains and features** shows the same information in tabular form.

4 Genome Browsers

For every genome in WormBase ParaSite, we provide two ways of browsing the genome: (a) integrated browser panels, embedded into the main pages of the site; and (b) a pop-out standalone browser using the JBrowse [20] platform.

4.1 Integrated Ensembl Genome Browser

The integrated genome browser panels are provided by the Ensembl framework [21]. There are a number of access points to these views: from the species information page (via the **Example Region** button); from any gene report page (via the **Location** tab) and from the search results page (via the **Location** link).

The browser comprises three panels, each showing a different level of detail. The upper panel shows the full length of the chromosome or scaffold; the middle panel provides an overview of a 500 kilobase part of that chromosome or scaffold and shows the gene structures; and the lower panel shows a smaller subregion (e.g., around a single gene) in more detail. For each of the upper two panels, a red box shows which region is being viewed in the panel immediately below. Clicking and dragging in each of these panels affords precise control over which region is being displayed in each. Depending on which of the **Drag** or **Select** controls is active, this will either select a region for display, or scroll the whole panel upstream or downstream.

A large range of configuration options are provided to customise the lower panel. Tracks can be added or removed using the **Configure tracks** button, and can be reordered by dragging the handle located on the left-hand side. A number of tracks are switched on for each genome by default, for example gene structures, %GC content and repetitive regions. Any customization made to this section will be saved against the user account (*see* Subheading 3.4) while logged in.

For a number of species, we have aligned transcriptomic data from RNA-Seq studies found in the public nucleotide archives. Where available, these tracks can be added to the browser via the **Add RNA-Seq tracks** button located in the left-hand menu of any genome browser view (Fig. 7). A summary description is available for each study, which provides more information about the origin of the data and links to the original publication.

Where variation data for a species are available in the European Variation Archive (EVA), we display these as a genome browser track. Each variant is colour-coded to correspond with the most severe predicted consequence this variant will have on the affected gene(s). Each variant is linked to a page describing further information (Subheading 3.5).

Users can create tracks from their own data via the **Add custom tracks** button on the left-hand menu. A number of file formats

Fig. 7 Integrated Ensembl genome browser showing a zoomed region on chromosome 1 of *Schistosoma mansoni*. Top: the length of the chromosome showing the zoomed area in red. Middle: interactive genome browser showing genes, pseudogenes, and RNA genes. Bottom: track browser showing gene models (including exon structure), predicted noncoding RNAs, expression in two life-stages and variations

are supported, including the UCSC BED/Wiggle format and their large-file extensions [22], UCSC Track Hubs [23], BAM [24] and VCF [25]. Small text files (typically less than 20 MB) can be uploaded directly to WormBase ParaSite, while larger files must be hosted on a remote web or FTP server, from where they can be attached directly to the WormBase ParaSite browser.

4.2 ParaSite JBrowse

To complement the embedded Ensembl genome browser, we also provide a JBrowse for each species. JBrowse is the browser used by WormBase and a number of other model organism databases. All tracks available via the embedded Ensembl browser are also available in JBrowse.

JBrowse enables a degree of on-the-fly analysis by allowing users to create new tracks by combining existing ones using arithmetic and set operations. These "combination tracks" can be

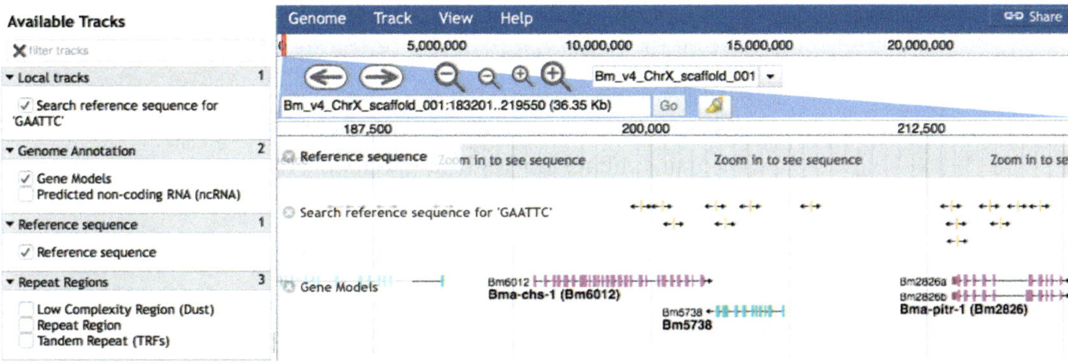

Fig. 8 JBrowse genome browser showing a region on chromosome X of *Brugia malayi*. Left: selector showing all available and currently available tracks. Right: length of the chromosome is depicted along the top, with the red box marking the part of the chromosome that is shown in detail below. The main panel shows the gene models and the location of *Eco*RI restriction sites (GAATTC)

used as input for subsequent combination tracks, allowing complex analysis tracks to be built up iteratively. Additionally, custom tracks can be attached to the genome browser without needing to be uploaded to WormBase ParaSite; the web-browser can instead be pointed to a local file or URL specifying the location of a remote file, and JBrowse will read, process and display the data completely locally within the web-browser. This allows tracks to be rendered and manipulated very quickly. JBrowse also offers various other features, including searching for specific short nucleotide or protein motifs in the genome sequence (Fig. 8).

5 ParaSite BioMart

WormBase ParaSite provides an advanced search and data export facility to enable the generation of custom data tables and sequence files, using a customized version of the BioMart platform [26, 27]. All species within WormBase ParaSite are available for query and export, plus a number of additional nonworm comparator species, including human, mouse, and zebrafish.

There are two main steps to using ParaSite BioMart. Firstly, a set of criteria are defined which the genes, transcripts, or proteins must conform to in order to be included in the results—these are referred to as *Query Filters*. Next, the data-types to include in the output list are defined—these are *Output Attributes*.

5.1 Query Filters

A variety of filters are provided to narrow down the set of genes or sequences that will ultimately appear in the output. These include:

- Species: genes/sequences from a single species, or taxonomic clade
- Genomic region: genes/sequences from a custom-defined list of name/start-end regions

- Gene Ontology: genes annotated with a specified GO term, or one of its descendants in the ontology
- Homologues: genes having (or alternatively, lacking) an orthologue in the specified genome or set of genomes
- Domains: genes having the specified protein domains, or set of domains.

5.2 Output Attributes and Results

There are two output modes: **Create a data table** and **Retrieve sequences**. For data tables, the selection of output attributes determines which columns will appear in the table. For sequences, it determines which sequences associated with the filtered set of genes will be output, and which additional attributes will be added as decorations to the FASTA header.

Once output attributes have been selected, clicking the **Results** button presents the first ten results of the query. This provides a fast preview of the requested data matching the criteria, and acts as a quick check that all the required information is present and correct before downloading the full file. The preview can be extended to include additional rows by selecting a new value from the **View 10 rows** drop-down menu. The query can also be edited at this stage by returning to the **Query Filters** and/or **Output Attributes** page.

Results of the query can be downloaded for use offline and further processing. The **Export all results to** drop-down menu is used to define the output format for the export. Both uncompressed and compressed text files options are provided, which can be imported into most statistical packages including R. Additionally, by selecting "XLS" format, results may be opened in Microsoft Excel.

For queries returning a large number of results, it is recommended that the **Compressed web file (notify by email)** option is selected. This avoids timeout issues by running the query on the WormBase ParaSite server in the background, then sending an email when the entire file is ready for download. There is no requirement to remain on the BioMart website while the file is processing and the download may be started from any computer.

5.3 Example: Finding Human Orthologues and Annotating with Function

Numerous use-cases for ParaSite BioMart exist, including extracting the 5′ flanking sequences of genes, and converting between different types of gene identifier. A common use-case is to add additional attributes to a list of helminth genes that have been obtained elsewhere (e.g., from a publication that reports the set as having differential expression under a certain experimental condition). Examples of such attributes would be Gene Ontology annotations and human orthologues.

The task is made simple by ParaSite BioMart. First, the list of gene IDs is uploaded as a Query Filter. This can be done by copy-and-paste from another window, or by uploading a file containing the identifiers. Second, Output Attributes relating to GO annotations and human orthologues are selected. The process is summarized in Fig. 9.

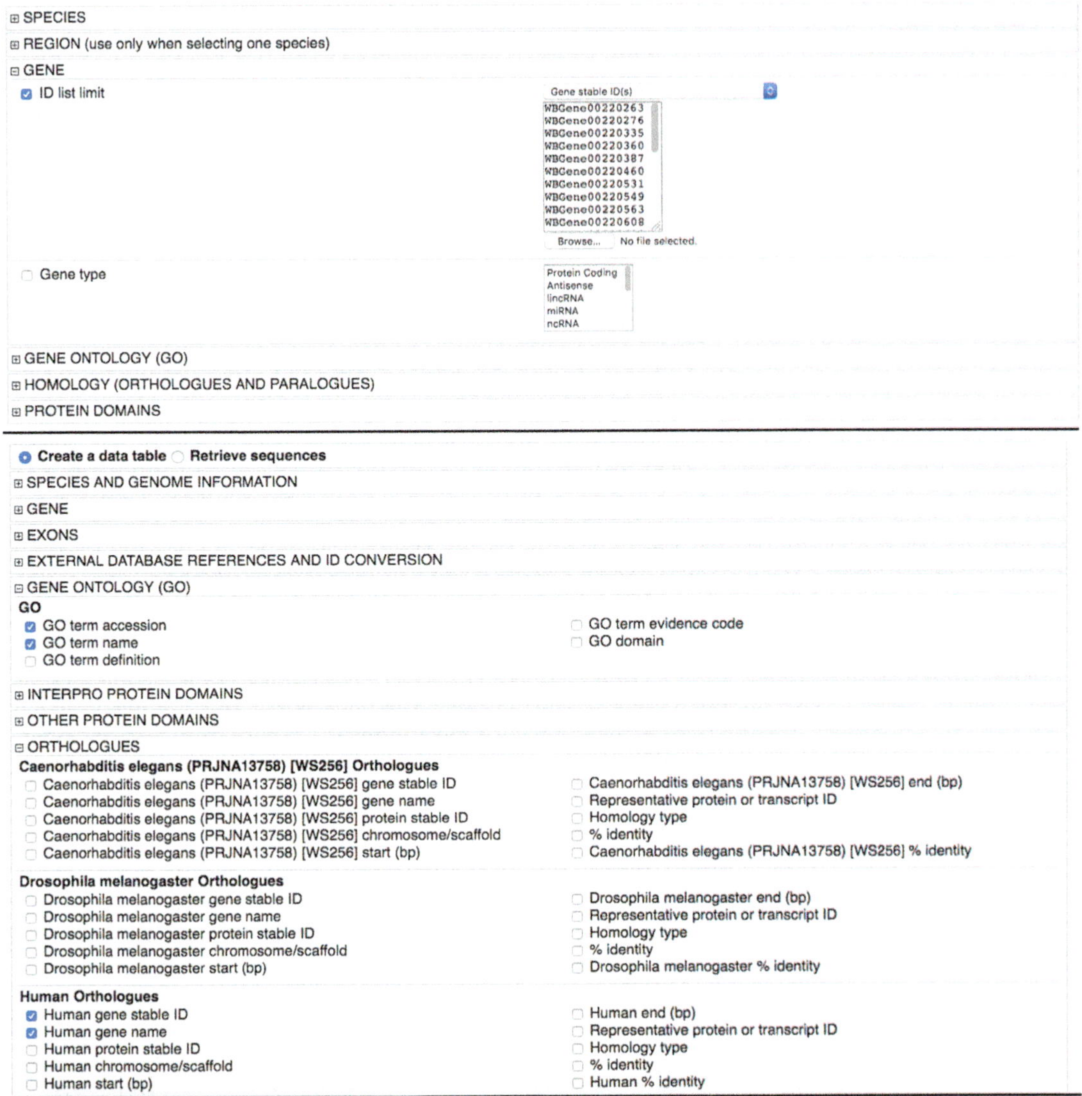

Gene stable ID	Transcript stable ID	GO term accession	GO term name	Human gene stable ID	Human gene name
WBGene00220276	Bm15a	GO:0006233	dTDP biosynthetic process	ENSG00000168393	DTYMK
WBGene00220276	Bm15a	GO:0004798	thymidylate kinase activity	ENSG00000168393	DTYMK
WBGene00220276	Bm15a	GO:0046939	nucleotide phosphorylation	ENSG00000168393	DTYMK
WBGene00220276	Bm15a	GO:0005524	ATP binding	ENSG00000168393	DTYMK
WBGene00220276	Bm15c	GO:0006233	dTDP biosynthetic process	ENSG00000168393	DTYMK
WBGene00220276	Bm15c	GO:0004798	thymidylate kinase activity	ENSG00000168393	DTYMK
WBGene00220276	Bm15c	GO:0046939	nucleotide phosphorylation	ENSG00000168393	DTYMK
WBGene00220276	Bm15c	GO:0005524	ATP binding	ENSG00000168393	DTYMK
WBGene00220276	Bm15c	GO:0016021	integral component of membrane	ENSG00000168393	DTYMK
WBGene00220276	Bm15c	GO:0016020	membrane	ENSG00000168393	DTYMK
WBGene00220335	Bm74	GO:0043565	sequence-specific DNA binding	ENSG00000157557	ETS2
WBGene00220335	Bm74	GO:0003700	transcription factor activity, sequence-specific DNA binding	ENSG00000157557	ETS2
WBGene00220335	Bm74	GO:0006355	regulation of transcription, DNA-templated	ENSG00000157557	ETS2
WBGene00220335	Bm74	GO:0005634	nucleus	ENSG00000157557	ETS2
WBGene00220335	Bm74	GO:0003677	DNA binding	ENSG00000157557	ETS2
WBGene00220335	Bm74	GO:0043565	sequence-specific DNA binding	ENSG00000134954	ETS1
WBGene00220335	Bm74	GO:0003700	transcription factor activity, sequence-specific DNA binding	ENSG00000134954	ETS1

Fig. 9 Using the WormBase ParaSite BioMart to annotate a list of gene identifiers. Top panel: setting the query filters by entering a list of gene IDs. Middle panel: selecting the output attributes. Bottom panel: preview of the results

6 Other Tools

6.1 Sequence Searching with BLAST

Our BLAST service (using NCBI-BLAST+ [28]) allows users to search their DNA or protein sequences for similarity to the genomes, transcriptomes, and proteomes in WormBase ParaSite. The computation required for a BLAST analysis can sometimes take a few minutes, and it is often useful to perform several analyses and compare the results. We therefore provide a "job" system, whereby the user can create one or more BLAST jobs, and return later to explore the results.

The query for BLAST can be supplied by pasting in a raw nucleotide or protein sequences, or uploading a file of sequences. To select which genomes, transcriptomes or proteomes to search against, shortcut buttons are available to select common groups of species; for example, all species, all nematodes or all platyhelminths. A **custom species list** button opens an interactive taxonomic tree where single species as well as entire clades can be selected.

Once a BLAST job has been created, the user is presented with a list of all of their current jobs, including those already completed. By default, completed BLAST jobs are available for review for 7 days. However, if the user logs in and saves the results to their account, they will be kept permanently. From the job overview table, results can be download and existing jobs can be edited and resubmitted, or deleted.

The **view results** link leads to a new page showing a table of the results. By default, the matches are ordered by score, from highest to lowest. The table can be customized to show more or fewer columns and the results can be ordered according to any of the parameters in the table. Columns include "Genomic-location," which provides a link to the genome-browser view of the hit, and "%ID", which shows the percentage identity of the alignment, and also links to a text display of the alignment itself. Below the results table, a visualization shows the distribution of High-scoring Segment Pairs (HSPs) on the query sequence as a heat-map, with high-scoring matches shown in darker colours.

6.2 Variant Effect Predictor

The Ensembl Variant Effect Predictor (VEP) [29] takes a list of genomic variations (from, for example, a genome resequencing project) and calculates the putative effect that those variations would have on the structure and function of the reference genes and transcripts. We provide an instance of the VEP for use with all genomes in WormBase ParaSite. As with BLAST, VEP analyses are performed as "jobs."

VEP can be reached from anywhere on the site via the top menu bar. The starting point is to select a genome to work with, and then pasting or uploading the list of genomic variants, in one of the supported formats. The **Identifiers and frequency data** panel can be used to select which types of gene/protein identifiers are displayed on the results page. The **Filtering options** panel can be used to (for example) restrict to variants that are in coding region or to show only certain consequences of each variant.

The VEP results page is divided into two sections. A tabular and graphical overview of the data is shown at the top (Fig. 10). A full table of annotated variants is shown underneath. If the table is very large, it can be navigated through (by using the arrow buttons under the **Navigation** heading) or filtered (using the **Filters** panel). The entire and the filtered results set can be downloaded in tab-delimited plain text (equivalent to the results table), VEP or VCF format.

6.3 Data Downloads

Data from the current and all previous releases of WormBase ParaSite are available to download from the FTP server (linked from the **Downloads** button located in the header of every page). A number of files are available, representing the genome assembly, genome annotation, transcriptome, and proteome (Table 1). A fully searchable and sortable table summarizes the downloads available for each genome in the current release.

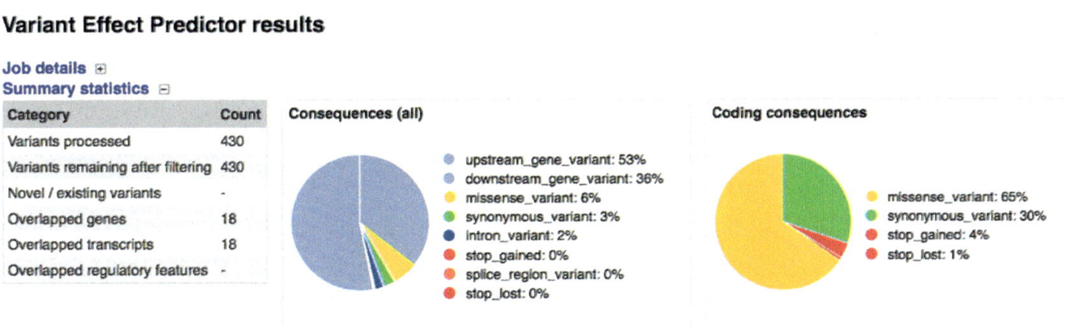

Fig. 10 Example of VEP summary view after analysis of a set of 430 variants in *Strongyloides ratti*

Table 1
Summary of the file formats available for download from the WormBase ParaSite FTP site (ftp://ftp.ebi.ac.uk/pub/databases/wormbase/parasite)

Category	File suffix	Format	Description
Genome sequence	genomic.ga	FASTA	Genome sequence
	genomic_masked.fa	FASTA	Genome with repetitive sequence replaced with Ns
	genomic_softmasked.fa	FASTA	Genome with repetitive sequence in lower-case
Gene sequence	mRNA_transcripts.fa	FASTA	Sequences of spliced, full-length transcript
	CDS_transcripts.fa	FASTA	Sequence of the CDS-portion of the spliced full-length protein-coding transcripts
	proteins.fa	FASTA	Protein sequences of canonical protein-coding transcripts
Annotations	annotations.gff3	GFF3	all annotations
	canonical_geneset.gtf	GTF	The canonical gene set

7 Programmatic Access

Programmatic access to data within WormBase ParaSite is possible using our Application Programming Interface (API) which is built around the Ensembl REST API [30]. This service provides access to a number of different data types via "end points" which can be invoked using any programming language. The **REST API** header link displays the complete catalogue of endpoints and example code for each data type given in Perl, Python, Ruby and Java.

For downloading tables of data to use in the statistical programming environment R, it is recommended to make use of the package "biomaRt," which forms part of the Bioconductor library [31]. This package allows the entire database of all genes to be filtered to include only those matching specific criteria, with the results stored as an R data structure containing the custom attribute data types selected during the query. Queries are formed in a similar methodology to the web interface (Subheading 5), but specified within the R script. The query is carried out on the WormBase ParaSite server, with only the results being transferred to the user computer, reducing bandwidth and memory requirements.

8 Community

We maintain a blog (http://wbparasite.wordpress.com) with how-to articles, website updates and detailed information about papers of interest. We also have a presence on Twitter, under the username @WBParaSite. Users can follow the account to receive news on website and data updates, papers of interest to the community and details of upcoming meetings. The latest Tweets are shown in a panel on the home page.

Questions, comments and feedback are always welcome to our email helpdesk (http://parasite.wormbase.org/Help/Contact). We endeavour to respond to all questions within 24 h.

Acknowledgments

WormBase ParaSite is funded by the UK Biotechnology and Biological Sciences Research Council [BB/K020080]. We are thankful to Alessandra Traini, Eleanor Stanley, and Jane Lomax for their previous work on WormBase ParaSite. We would also like to thank members of the WormBase and WTSI Parasite genomics groups for useful advice and feedback.

References

1. Hotez PJ, Brindley PJ, Bethony JM, King CH, Pearce EJ, Jacobson J (2008) Helminth infections: the great neglected tropical diseases. J Clin Invest 118(4):1311–1321. https://doi.org/10.1172/JCI34261
2. Howe KL, Bolt BJ, Cain S, Chan J, Chen WJ, Davis P, Done J, Down T, Gao S, Grove C, Harris TW, Kishore R, Lee R, Lomax J, Li Y, Muller HM, Nakamura C, Nuin P, Paulini M, Raciti D, Schindelman G, Stanley E, Tuli MA, Van Auken K, Wang D, Wang X, Williams G, Wright A, Yook K, Berriman M, Kersey P, Schedl T, Stein L, Sternberg PW (2016) WormBase 2016: expanding to enable helminth genomic research. Nucleic Acids Res 44(D1):D774–D780. https://doi.org/10.1093/nar/gkv1217
3. Kersey PJ, Allen JE, Armean I, Boddu S, Bolt BJ, Carvalho-Silva D, Christensen M, Davis P, Falin LJ, Grabmueller C, Humphrey J, Kerhornou A, Khobova J, Aranganathan NK, Langridge N, Lowy E, McDowall MD, Maheswari U, Nuhn M, Ong CK, Overduin B, Paulini M, Pedro H, Perry E, Spudich G, Tapanari E, Walts B, Williams G, Tello-Ruiz M, Stein J, Wei S, Ware D, Bolser DM, Howe KL, Kulesha E, Lawson D, Maslen G, Staines DM (2016) Ensembl Genomes 2016: more genomes, more complexity. Nucleic Acids Res 44(D1):D574–D580. https://doi.org/10.1093/nar/gkv1209
4. Cochrane G, Karsch-Mizrachi I, Takagi T, International Nucleotide Sequence Database C (2016) The International Nucleotide Sequence Database Collaboration. Nucleic Acids Res 44(D1):D48–D50. https://doi.org/10.1093/nar/gkv1323
5. Schwarz EM, Korhonen PK, Campbell BE, Young ND, Jex AR, Jabbar A, Hall RS, Mondal A, Howe AC, Pell J, Hofmann A, Boag PR, Zhu XQ, Gregory T, Loukas A, Williams BA, Antoshechkin I, Brown C, Sternberg PW, Gasser RB (2013) The genome and developmental transcriptome of the strongylid nematode Haemonchus contortus. Genome Biol 14(8):R89. https://doi.org/10.1186/gb-2013-14-8-r89
6. Laing R, Kikuchi T, Martinelli A, Tsai IJ, Beech RN, Redman E, Holroyd N, Bartley DJ, Beasley H, Britton C, Curran D, Devaney E, Gilabert A, Hunt M, Jackson F, Johnston SL, Kryukov I, Li K, Morrison AA, Reid AJ, Sargison N, Saunders GI, Wasmuth JD, Wolstenholme A, Berriman M, Gilleard JS, Cotton JA (2013) The genome and transcriptome of Haemonchus contortus, a key model parasite for drug and vaccine discovery. Genome Biol 14(8):R88. https://doi.org/10.1186/gb-2013-14-8-r88
7. Barrett T, Clark K, Gevorgyan R, Gorelenkov V, Gribov E, Karsch-Mizrachi I, Kimelman M, Pruitt KD, Resenchuk S, Tatusova T, Yaschenko E, Ostell J (2012) BioProject and BioSample databases at NCBI: facilitating capture and organization of metadata. Nucleic Acids Res 40(Database issue):D57–D63. https://doi.org/10.1093/nar/gkr1163
8. The Gene Ontology Consortium (2017) Expansion of the Gene Ontology knowledgebase and resources. Nucleic Acids Res 45(D1):D331–D338. https://doi.org/10.1093/nar/gkw1108
9. Jones P, Binns D, Chang HY, Fraser M, Li W, McAnulla C, McWilliam H, Maslen J, Mitchell A, Nuka G, Pesseat S, Quinn AF, Sangrador-Vegas A, Scheremetjew M, Yong SY, Lopez R, Hunter S (2014) InterProScan 5: genome-scale protein function classification. Bioinformatics 30(9):1236–1240. https://doi.org/10.1093/bioinformatics/btu031
10. Vilella AJ, Severin J, Ureta-Vidal A, Heng L, Durbin R, Birney E (2009) EnsemblCompara GeneTrees: complete, duplication-aware phylogenetic trees in vertebrates. Genome Res 19(2):327–335. https://doi.org/10.1101/gr.073585.107
11. Howe KL, Bolt BJ, Shafie M, Kersey P, Berriman M (2016) WormBase ParaSite—a comprehensive resource for helminth genomics. Mol Biochem Parasitol. https://doi.org/10.1016/j.molbiopara.2016.11.005
12. Earl D, Bradnam K, St John J, Darling A, Lin D, Fass J, Yu HO, Buffalo V, Zerbino DR, Diekhans M, Nguyen N, Ariyaratne PN, Sung WK, Ning Z, Haimel M, Simpson JT, Fonseca NA, Birol I, Docking TR, Ho IY, Rokhsar DS, Chikhi R, Lavenier D, Chapuis G, Naquin D, Maillet N, Schatz MC, Kelley DR, Phillippy AM, Koren S, Yang SP, Wu W, Chou WC, Srivastava A, Shaw TI, Ruby JG, Skewes-Cox P, Betegon M, Dimon MT, Solovyev V, Seledtsov I, Kosarev P, Vorobyev D, Ramirez-Gonzalez R, Leggett R, MacLean D, Xia F, Luo R, Li Z, Xie Y, Liu B, Gnerre S, MacCallum I, Przybylski D, Ribeiro FJ, Yin S, Sharpe T, Hall G, Kersey PJ, Durbin R, Jackman SD, Chapman JA, Huang X, DeRisi JL, Caccamo M, Li Y, Jaffe DB, Green RE, Haussler D, Korf I, Paten B (2011) Assemblathon 1: a competitive assessment of de novo short read assembly methods. Genome Res 21(12):2224–2241. https://doi.org/10.1101/gr.126599.111
13. Parra G, Bradnam K, Korf I (2007) CEGMA: a pipeline to accurately annotate core genes in eukaryotic genomes. Bioinformatics 23(9):1061–1067. https://doi.org/10.1093/bioinformatics/btm071

14. Simao FA, Waterhouse RM, Ioannidis P, Kriventseva EV, Zdobnov EM (2015) BUSCO: assessing genome assembly and annotation completeness with single-copy orthologs. Bioinformatics 31(19):3210–3212. https://doi.org/10.1093/bioinformatics/btv351
15. Consortium EP (2015) Europe PMC: a full-text literature database for the life sciences and platform for innovation. Nucleic Acids Res 43(Database issue):D1042–D1048. https://doi.org/10.1093/nar/gku1061
16. Veidenberg A, Medlar A, Loytynoja A (2016) Wasabi: an integrated platform for evolutionary sequence analysis and data visualization. Mol Biol Evol 33(4):1126–1130. https://doi.org/10.1093/molbev/msv333
17. Cook CE, Bergman MT, Finn RD, Cochrane G, Birney E, Apweiler R (2016) The European Bioinformatics Institute in 2016: data growth and integration. Nucleic Acids Res 44(D1):D20–D26. https://doi.org/10.1093/nar/gkv1352
18. Petryszak R, Keays M, Tang YA, Fonseca NA, Barrera E, Burdett T, Fullgrabe A, Fuentes AM, Jupp S, Koskinen S, Mannion O, Huerta L, Megy K, Snow C, Williams E, Barzine M, Hastings E, Weisser H, Wright J, Jaiswal P, Huber W, Choudhary J, Parkinson HE, Brazma A (2016) Expression Atlas update—an integrated database of gene and protein expression in humans, animals and plants. Nucleic Acids Res 44(D1):D746–D752. https://doi.org/10.1093/nar/gkv1045
19. Finn RD, Attwood TK, Babbitt PC, Bateman A, Bork P, Bridge AJ, Chang HY, Dosztanyi Z, El-Gebali S, Fraser M, Gough J, Haft D, Holliday GL, Huang H, Huang X, Letunic I, Lopez R, Lu S, Marchler-Bauer A, Mi H, Mistry J, Natale DA, Necci M, Nuka G, Orengo CA, Park Y, Pesseat S, Piovesan D, Potter SC, Rawlings ND, Redaschi N, Richardson L, Rivoire C, Sangrador-Vegas A, Sigrist C, Sillitoe I, Smithers B, Squizzato S, Sutton G, Thanki N, Thomas PD, Tosatto SC, Wu CH, Xenarios I, Yeh LS, Young SY, Mitchell AL (2017) InterPro in 2017—beyond protein family and domain annotations. Nucleic Acids Res 45(D1):D190–D199. https://doi.org/10.1093/nar/gkw1107
20. Buels R, Yao E, Diesh CM, Hayes RD, Munoz-Torres M, Helt G, Goodstein DM, Elsik CG, Lewis SE, Stein L, Holmes IH (2016) JBrowse: a dynamic web platform for genome visualization and analysis. Genome Biol 17:66. https://doi.org/10.1186/s13059-016-0924-1
21. Aken BL, Achuthan P, Akanni W, Amode MR, Bernsdorff F, Bhai J, Billis K, Carvalho-Silva D, Cummins C, Clapham P, Gil L, Giron CG, Gordon L, Hourlier T, Hunt SE, Janacek SH, Juettemann T, Keenan S, Laird MR, Lavidas I, Maurel T, McLaren W, Moore B, Murphy DN, Nag R, Newman V, Nuhn M, Ong CK, Parker A, Patricio M, Riat HS, Sheppard D, Sparrow H, Taylor K, Thormann A, Vullo A, Walts B, Wilder SP, Zadissa A, Kostadima M, Martin FJ, Muffato M, Perry E, Ruffier M, Staines DM, Trevanion SJ, Cunningham F, Yates A, Zerbino DR, Flicek P (2017) Ensembl 2017. Nucleic Acids Res 45(D1):D635–D642. https://doi.org/10.1093/nar/gkw1104
22. Kent WJ, Zweig AS, Barber G, Hinrichs AS, Karolchik D (2010) BigWig and BigBed: enabling browsing of large distributed datasets. Bioinformatics 26(17):2204–2207. https://doi.org/10.1093/bioinformatics/btq351
23. Raney BJ, Dreszer TR, Barber GP, Clawson H, Fujita PA, Wang T, Nguyen N, Paten B, Zweig AS, Karolchik D, Kent WJ (2014) Track data hubs enable visualization of user-defined genome-wide annotations on the UCSC Genome Browser. Bioinformatics 30(7):1003–1005. https://doi.org/10.1093/bioinformatics/btt637
24. Li H, Handsaker B, Wysoker A, Fennell T, Ruan J, Homer N, Marth G, Abecasis G, Durbin R, Genome Project Data Processing S (2009) The Sequence Alignment/Map format and SAMtools. Bioinformatics 25(16):2078–2079. https://doi.org/10.1093/bioinformatics/btp352
25. Danecek P, Auton A, Abecasis G, Albers CA, Banks E, DePristo MA, Handsaker RE, Lunter G, Marth GT, Sherry ST, McVean G, Durbin R, Genomes Project Analysis G (2011) The variant call format and VCFtools. Bioinformatics 27(15):2156–2158. https://doi.org/10.1093/bioinformatics/btr330
26. Smedley D, Haider S, Ballester B, Holland R, London D, Thorisson G, Kasprzyk A (2009) BioMart—biological queries made easy. BMC Genomics 10:22. https://doi.org/10.1186/1471-2164-10-22
27. Zhang J, Haider S, Baran J, Cros A, Guberman JM, Hsu J, Liang Y, Yao L, Kasprzyk A (2011) BioMart: a data federation framework for large collaborative projects. Database (Oxford) 2011:bar038. https://doi.org/10.1093/database/bar038
28. Camacho C, Coulouris G, Avagyan V, Ma N, Papadopoulos J, Bealer K, Madden TL (2009) BLAST+: architecture and applications. BMC Bioinformatics 10:421. https://doi.org/10.1186/1471-2105-10-421
29. McLaren W, Gil L, Hunt SE, Riat HS, Ritchie GR, Thormann A, Flicek P, Cunningham F (2016) The Ensembl Variant Effect Predictor. Genome Biol 17(1):122. https://doi.org/10.1186/s13059-016-0974-4

30. Yates A, Beal K, Keenan S, McLaren W, Pignatelli M, Ritchie GR, Ruffier M, Taylor K, Vullo A, Flicek P (2015) The Ensembl REST API: Ensembl data for any language. Bioinformatics 31(1):143–145. https://doi.org/10.1093/bioinformatics/btu613

31. Durinck S, Moreau Y, Kasprzyk A, Davis S, De Moor B, Brazma A, Huber W (2005) BioMart and Bioconductor: a powerful link between biological databases and microarray data analysis. Bioinformatics 21(16):3439–3440. https://doi.org/10.1093/bioinformatics/bti525

Chapter 16

Using FlyBase to Find Functionally Related *Drosophila* Genes

Alix J. Rey, Helen Attrill, Steven J. Marygold, and The FlyBase Consortium*

Abstract

For more than 25 years, FlyBase (flybase.org) has served as an online database of biological information on the genus *Drosophila*, concentrating on the model organism *D. melanogaster*. Traditionally, FlyBase data have been organized and presented at a gene-by-gene level, which remains a useful perspective when the object of interest is a specific gene or gene product. However, in the modern era of a fully sequenced genome and an increasingly characterized proteome, it is often desirable to compile and analyze lists of genes related by a common function. This may be achieved in FlyBase by searching for genes annotated with relevant Gene Ontology (GO) terms and/or protein domain data. In addition, FlyBase provides preassembled lists of functionally related *D. melanogaster* genes within "Gene Group" reports. These are compiled manually from the published literature or expert databases and greatly facilitate access to, and analysis of, established gene sets. This chapter describes protocols to produce lists of functionally related genes in FlyBase using GO annotations, protein domain data and the Gene Groups resource, and provides guidance and advice for their further analysis and processing.

Key words FlyBase, *Drosophila*, *D. melanogaster*, Database, Functionally related genes, Gene Ontology, Protein domain, Gene group

1 Introduction

FlyBase gathers genetic, genomic, and functional information on *Drosophila* by manual curation of the research literature and computational incorporation of data from relevant sources [1, 2]. Data are partitioned into separate classes (e.g., gene, transcript, allele) to enable entity-specific searching and display, with much of the data being presented on individual gene reports on the website. While this approach has many benefits, it is often desirable to search for and view groups of genes whose products are related in some way, such as their known or predicted function. For example, a list of

*The members of the FlyBase Consortium are given in the Acknowledgments.

Martin Kollmar (ed.), *Eukaryotic Genomic Databases: Methods and Protocols*, Methods in Molecular Biology, vol. 1757, https://doi.org/10.1007/978-1-4939-7737-6_16, © Springer Science+Business Media, LLC, part of Springer Nature 2018

functionally related genes may provide the starting point for a genetic/molecular screen, or be the basis for in silico analyses using associated data (phenotypes, reagents, genomic data, etc.), or allow comparison with equivalent gene sets in other species. FlyBase provides three main ways to search for functionally related *Drosophila* genes: via Gene Ontology (GO) annotations, protein domain information, and our Gene Group resource.

The GO is a widely used controlled vocabulary aimed at labeling gene products with biological attributes [3]. It is divided into three aspects: "Molecular Function" describes the molecular activity being carried out—for example, protein kinase activity or ubiquitin-protein transferase activity; "Biological Process" describes the context in which the gene product acts—for example, protein ubiquitination or Wnt signaling pathway; and "Cellular Component" describes where it acts—either a subcellular region, such as the cytosol, or a macromolecular complex, such as the anaphase-promoting complex. The GO is arranged in a hierarchical structure, with more specific child terms nested under higher-level parent terms. For example, "protein kinase activity" is a child of "kinase activity."

GO annotations may be added manually by a curator based either on experimental data in published research (e.g., direct assay, genetic interaction) or from predictions/assertions based on sequence, such as similarity to a characterized gene. Alternatively, GO annotations may be added computationally via automated [4] or curator-reviewed pipelines [5]. This combinatorial approach results in good coverage of GO annotation data over the *D. melanogaster* genome: 73% of sequence-localized genes and 88% of protein-coding genes have associated GO terms (FlyBase release FB2017_02).

The intimate relationship between structure and function can be exploited to find genes encoding proteins with particular functional attributes. For example, the BAR (Bin-Amphiphysin-Rvs)-domain is characteristic of proteins involved in promoting membrane curvature in intracellular trafficking, while proteins containing an RNA recognition motif (RRM) are associated with single-stranded RNA binding. Thus the possession of common motifs or domains can be used as a handle to search and retrieve protein-coding genes of shared function. In FlyBase, protein-coding genes are linked to UniProtKB accessions, and this relationship is used to associate these genes with domain data from InterPro [4]. InterPro aggregates data from many sources to produce integrated protein signatures classified as domains, families, repeats, and sites. (InterPro uses the "signature" to describe these collective terms, but in this text InterPro domain and signature should be considered interchangeable.) Thus, in contrast to GO data, protein domain annotations are derived entirely computationally and are applied to all protein-coding genes in an unbiased fashion. InterPro domains are associated with 82% of *D. melanogaster* protein-coding genes (FB2017_02).

Despite the strengths of using GO and/or protein domain annotations to identify functionally related genes, these approaches are not always straightforward or even appropriate. Take, for example, the seemingly simple query: "which genes encode the general transcription factors of *D. melanogaster*?" These genes are not defined by a single GO term; combinatorial queries using advanced tools may find candidates, but the accuracy of the results would depend on an in-depth knowledge of the GO and the subject area, and would be limited by annotation coverage. Similarly, a protein domain query is unsuited to this task as the individual subunits do not share a common sequence motif. Ultimately, many familiar functional grouping terms used within the scientific literature and research community fall beyond the scope of the GO and/or cannot be defined by protein domains.

The FlyBase "Gene Group" resource was established to fill this gap, allowing users to easily access lists of functionally related *D. melanogaster* genes [6]. Gene Groups are manually curated based on published research papers, reviews, and online databases. The resource includes genes whose products share a function based on their evolutionary history (gene families, e.g., actins, odorant receptors), contribution to a macromolecular complex (e.g., ribosome or proteasome subunits), or a common molecular function (e.g., deubiquitinases or tRNAs). Gene Groups are arranged in a hierarchical fashion to allow users to drill-down to specific subsets. For example, the "protein kinase" group is divided into 11 main subgroups, which are further subdivided. In contrast to GO or protein domain data, there is no automated compilation pipeline for Gene Groups—this ensures their integrity and utility, though limits their number and genome coverage. The Gene Groups resource currently comprises over 612 groups (FB2017_02), covering over 21% of all sequence-localized genes and 24% of protein-coding genes. Many areas of biology have been covered in depth, such as intracellular transport, autophagy and cytoskeletal groups (Table 1).

In this chapter, we provide step-by-step protocols to find functionally related *D. melanogaster* genes in FlyBase using GO annotations, protein domain information and the Gene Groups resource. We also describe methods to build combinatorial queries in order to retrieve sets of genes satisfying multiple conditions, and to download gene lists for further analysis/processing. Finally, we discuss the relative merits of the three main approaches described herein, including guidance on which approach to use in different situations (*see* **Notes 1–4**).

Table 1
An overview of the Gene Groups in FlyBase release FB2017_02

Theme	Number of genes	Example gene groups
Gene expression	1227	General transcription factors Ribosomal proteins Spliceosomal complexes Transfer RNAs Translation factors
Post-translational modification	919	Protein kinases Protein phosphatases Ring finger domain proteins Ubiquitination enzymes
Receptor and receptor signaling	395	Chemoreceptors G protein coupled receptors Neuropeptides Odorant binding proteins
Metabolism	404	Glutathione S-transferases Oxidative phosphorylation complexes Proteasome subunits
Transmembrane transport	326	ATP-binding cassette transporter-like Ion channels Nuclear pore complex Vacuolar ATPase subunits
Intracellular transport	217	Bloc complexes Intracellular transport groups SNAREs Tethering factors
Small GTPase signaling	201	RAS GTPase superamily RAS superfamily GAPs RAS superfamily GEFs
Chromatin organization	191	Chromatin modifying complexes Chromatin remodeling complexes Polycomb group complexes SMC complexes
Cytoskeletal	180	Actins Dynein subunits Kinesins Myosins Tubulins
Cell–cell communication and adhesion	86	Beat, Side families Cadherins Integrins
Apoptosis and autophagy	52	Autophagy-related complexes Autophagy-related genes Caspases

(continued)

Table 1
(continued)

Theme	Number of genes	Example gene groups
Cell cycle	30	Anaphase-promoting complex Chromosomal passenger complex Origin recognition complex
Immunity	30	Drosomycins Nimrod genes Peptidoglycan recognition proteins
Other	108	Heat shock proteins Tetraspanins

To provide a summary, the Gene Groups have been divided into major biological themes. The number of genes for each theme is shown, together with a small selection of example Gene Groups. Note, as some genes belong to more than one group, the number of genes for each theme does not represent the number of unique genes. There are 3789 unique genes within the Gene Group collection (FB2017_02)

2 Methods

2.1 Using the GO to Find Functionally Related Genes

A list of genes annotated with a particular GO term, or any of the children of that term, can be obtained via a Term Report page. Term Reports themselves can be queried/accessed by either of two FlyBase search tools: QuickSearch or Vocabularies. QuickSearch is located in the center of the FlyBase homepage and allows rapid querying of almost all data in FlyBase via a tabbed interface [7]. In each QuickSearch tab, a link to specific documentation is provided via the question mark icon. Additionally, YouTube video tutorials are available for many data types, including the GO. These can be accessed by clicking the YouTube icon, where present, after selecting the relevant QuickSearch tab. Vocabularies is a dedicated tool for browsing and searching all the controlled vocabularies used in FlyBase to annotate data with standardized terms [8]. Additional documentation is shown in the section at the foot of the Vocabularies page, which includes a link to a YouTube video tutorial.

2.1.1 Searching the GO Using QuickSearch

1. From the FlyBase homepage, click on the QuickSearch “GO” tab. Alternatively, from any FlyBase page, click on the “Tools” menu from the Navigation Bar (NavBar) and select “Query Tools and Portals,” then “QuickSearch.” Either route takes you to the “QuickSearch Search Page” (Fig. 1a).
2. From the “Data Field” drop-down menu select “all GO terms,” or chose “molecular function,” “biological process,” or “cellular component,” to restrict the search to a particular aspect of the GO. Type your query into the “Enter term” field. Valid entries are GO terms, synonyms (e.g., “smoothened signaling pathway,”

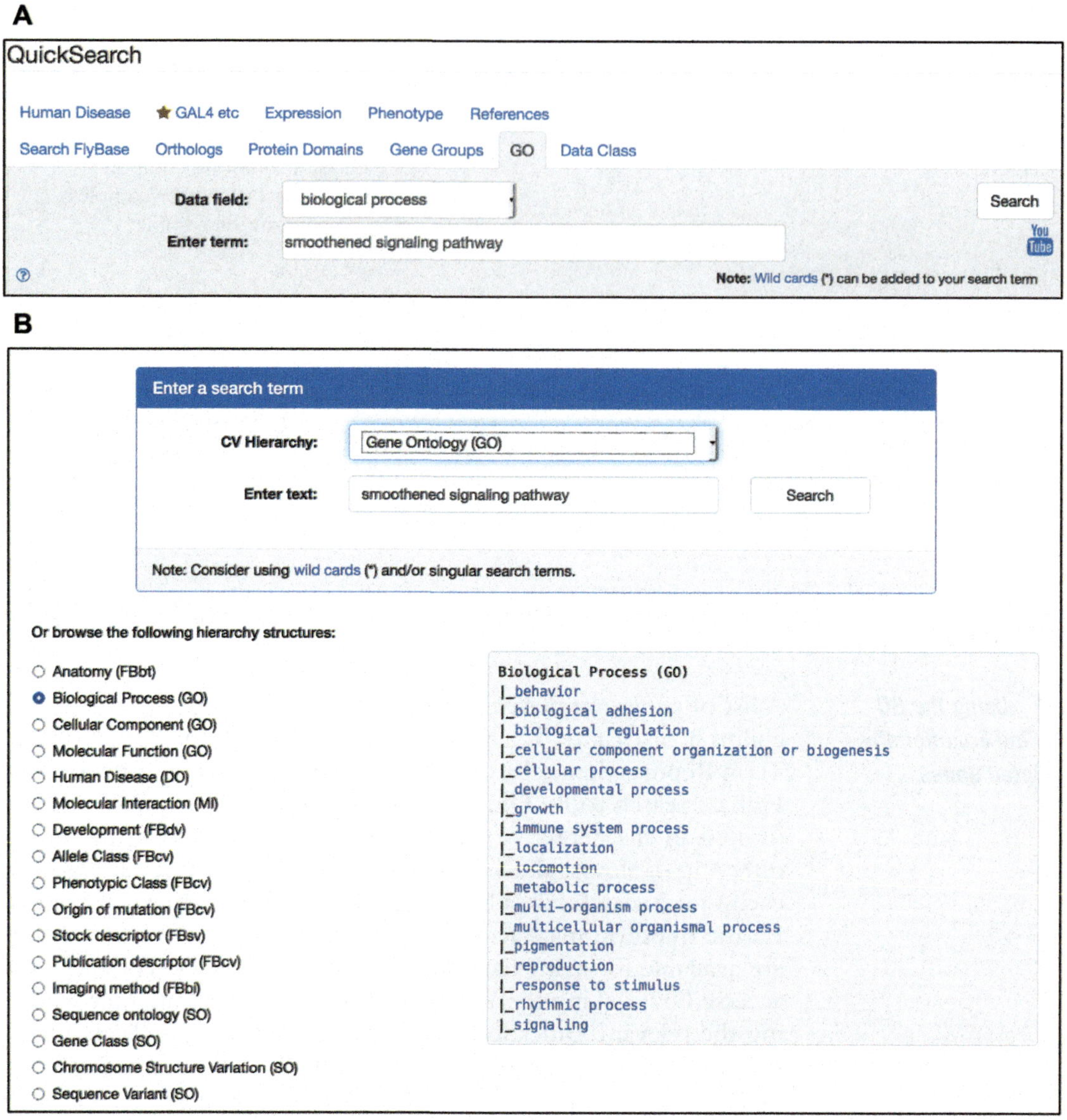

Fig. 1 (**a**) The Gene Ontology (GO) tab of the QuickSearch tool. (**b**) The Vocabularies tool page, which offers two search options: a search term box (top) and a browsing window to select top-level GO terms (bottom)

"hedgehog signaling pathway"), or GO identifiers (e.g., "GO:0007224"). GO terms that match the entered text appear in a drop-down list when typing and can be clicked to populate the field. The search is case-insensitive and a wildcard (*) can be added to search for matches to partial terms.

3. Click the "Search" button or press "enter." This takes you to a hit-list of "Matching CV terms" listing similar terms.

4. From this list, select a term by clicking on it—choosing a general, high-level term is a good starting point; more specific terms may be selected in subsequent steps. This takes you to a Term Report page for this GO term—*see* Subheading 2.1.3 for details.

2.1.2 Searching the GO Using Vocabularies

1. From the FlyBase homepage, click on the "Vocabularies" icon located near the top of the page. Alternatively, from any FlyBase page, click on the "Tools" menu from the Navigation Bar (NavBar) and select "Query Tools and Portals," then "Vocabularies." Either route takes you to the "Vocabularies Search Page" (Fig. 1b).
2. Select "Gene Ontology (GO)" from the drop-down menu under "CV Hierarchy" to restrict the search to GO terms. Type your query into the "Enter text" field. Valid entries are GO terms/synonyms or GO identifiers. GO terms that match the entered text appear in a drop-down list when typing and can be clicked to populate the field. The search is case-insensitive and a wildcard (*) can be added to search for matches to partial terms.

   (Alternatively, select an aspect of the GO from the "Or browse the following hierarchy structures" section. Selected top-level GO terms will be displayed in the right-hand panel (Fig. 1b). As in **step 4**, clicking on a GO term will open its Term Report.)
3. Click the "Search" button or press "enter." This takes you to a hit-list of "Matching CV terms" listing similar terms.
4. Clicking on a GO term name opens its Term Report—*see* Subheading 2.1.3 for details.

2.1.3 Viewing GO Annotations in a Term Report and Hit-List

A Term Report (Fig. 2a) displays information and data associated with a controlled vocabulary term and is the destination page for GO queries via the QuickSearch or Vocabularies tools. The "General Information" section at the top contains the term name, ID, definition, and synonyms. Further down the report, the GO hierarchy is shown in a tree view, centered on the chosen term. The number of genes associated with each term and its children is displayed to the right-hand side of each term name. (Where no number is shown, there are no annotations to this term.) Below the tree, a "Spanning Tree View Settings" panel allows the user to adjust the number of levels shown for parents and children. Clicking on a term name within the tree generates the corresponding Term Report.

The "Annotations" section of the Term Report shows two relevant numbers. The first, displayed in a table under the "Records" column, is the number of genes annotated with the exact GO term only. The second, shown in a prominent box, is the number of

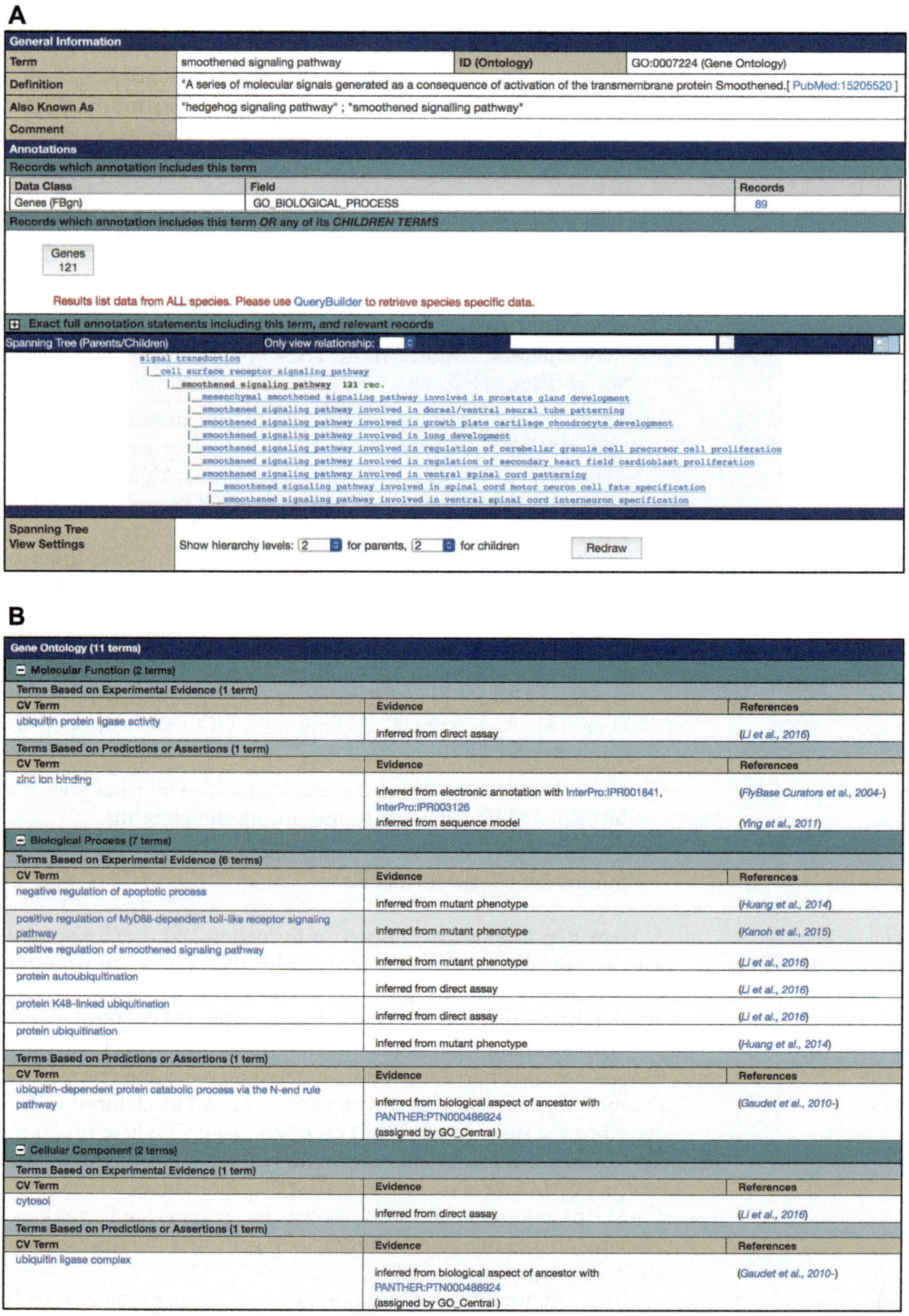

Fig. 2 (**a**) A GO Term report, using the term "smoothened signaling pathway" as an example. (**b**) The GO section of a Gene report. The gene *Ubr3* is shown here as an example

genes annotated with the GO term or its children, which is usually what is desired. Clicking on either number returns those genes in the form of a hit-list, representing the list of Drosophila genes that are related to each other by virtue of sharing a common GO annotation. FlyBase hit-lists can be sorted, analyzed or exported in several ways—*see* Subheadings 2.4 and 2.5.

Clicking on an individual gene in a hit-list takes the user to the corresponding Gene report. Here, all GO annotations associated with the specified gene are displayed within the "Gene Ontology (GO)" section (Fig. 2b). Clicking on a GO term within this section takes the user to the corresponding Term Report, thus providing an alternative route to generate a list of genes annotated with a particular GO term and its children.

2.2 Using Protein Domain Data to Find Functionally Related Genes

A list of *D. melanogaster* genes whose product(s) contain a specified protein domain (as defined by InterPro signatures) can be obtained by using the "Protein Domains" tab of the QuickSearch tool. Additional documentation may be obtained by clicking the question mark within the interface.

2.2.1 Searching Protein Domains Using QuickSearch

1. From the FlyBase homepage, click on the QuickSearch "Protein Domains" tab (Fig. 3a). Leave the 'Species' box unchecked to restrict the query to *D. melanogaster*.

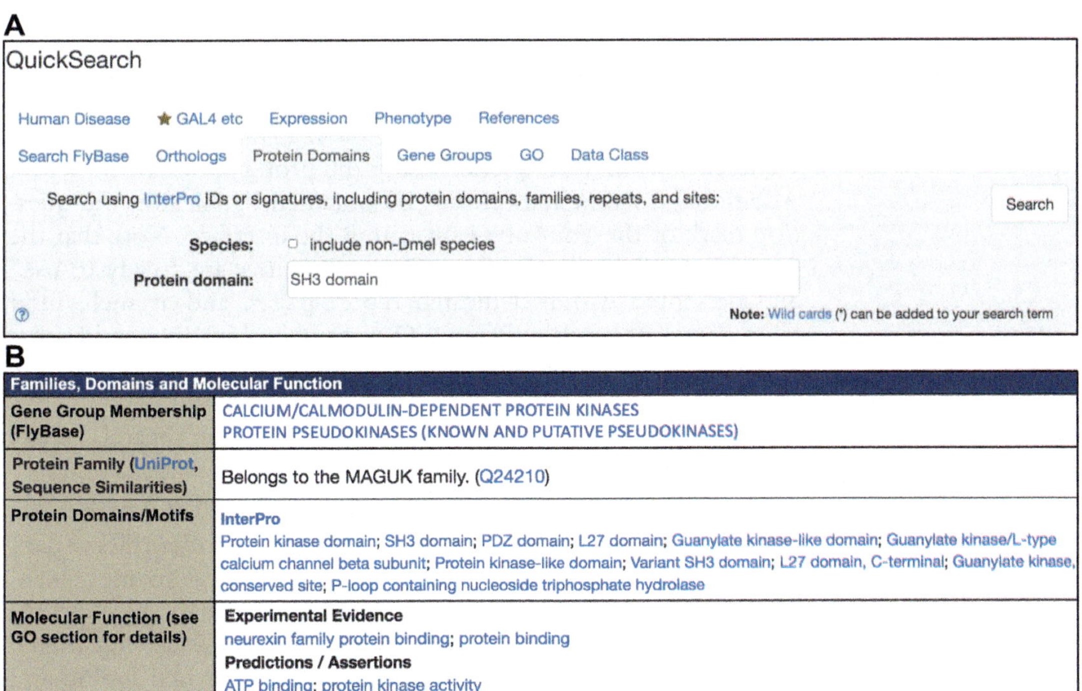

Fig. 3 (**a**) The Protein Domains tab of the QuickSearch tool. (**b**) The "Families and Domains and Molecular Function" section of a Gene report, which includes information on protein domains and Gene Group membership. The gene *CASK* is shown as an example

2. Type your query into the search box. Valid entries are InterPro terms (e.g., "SH3 domain" or "WD40 repeat") or InterPro identifiers (e.g., IPR001452). InterPro signatures that match the entered text appear in a drop-down list when typing and can be clicked to populate the field. The search is case-insensitive and a wildcard (*) can be added to search for matches to partial terms.
3. Click the "Search" button or press "enter." Genes that match the query are displayed in a hit-list, representing the list of *D. melanogaster* genes that are related to each other by virtue of sharing a common protein domain. FlyBase hit-lists can be sorted, analyzed or exported in several ways—*see* Subheadings 2.4 and 2.5.
4. Clicking on an individual gene in a hit-list takes the user to the corresponding Gene report. Here all InterPro signatures associated with the gene are displayed in the "Protein Domains/Motifs" subsection of the "Families, Domains and Molecular Function" section (Fig. 3b), as well as the "Polypeptide Data" subsection of the "Gene Model and Products" section (not shown). Clicking on a signature term takes the user to the corresponding page at InterPro, which contains detailed information on the domain.

2.3 Using Gene Groups to Find Functionally Related Genes

A list of *D. melanogaster* genes contained within a manually curated Gene Group can be obtained by using the "Gene Groups" tab of the QuickSearch tool (Fig. 4a). The "browse" link can be used to view all current Gene Groups as a nested hierarchy, where a specific group can be selected by clicking on it (Fig. 4b). Alternatively, Gene Groups may be queried using the protocol described below. Additional documentation may be obtained by clicking the question mark or the YouTube icon within the interface. Note that the gene lists available via the Gene Groups resource are "ready-to-use" and presented within dedicated report pages, and as such differ from gene lists resulting from GO or protein domain searches that are generated "on-the-fly" from gene-associated annotation data.

2.3.1 Searching Gene Groups Using QuickSearch

1. From the FlyBase homepage, click on the QuickSearch "Gene Groups" tab (Fig. 4a).
2. Type your query into the "Enter text" field. Valid entries are Gene Group names/symbols, synonyms or identifiers (e.g., ACTINS, FBgg0000184), or the symbols/names, synonyms, or identifiers of any member genes (e.g., Act42A, CG12051, FBgn0000043). Gene Group names that match the entered text appear in a drop-down list when typing and can be clicked to populate the field. The search is case-insensitive and a wildcard (*) can be added to search for matches to partial terms.

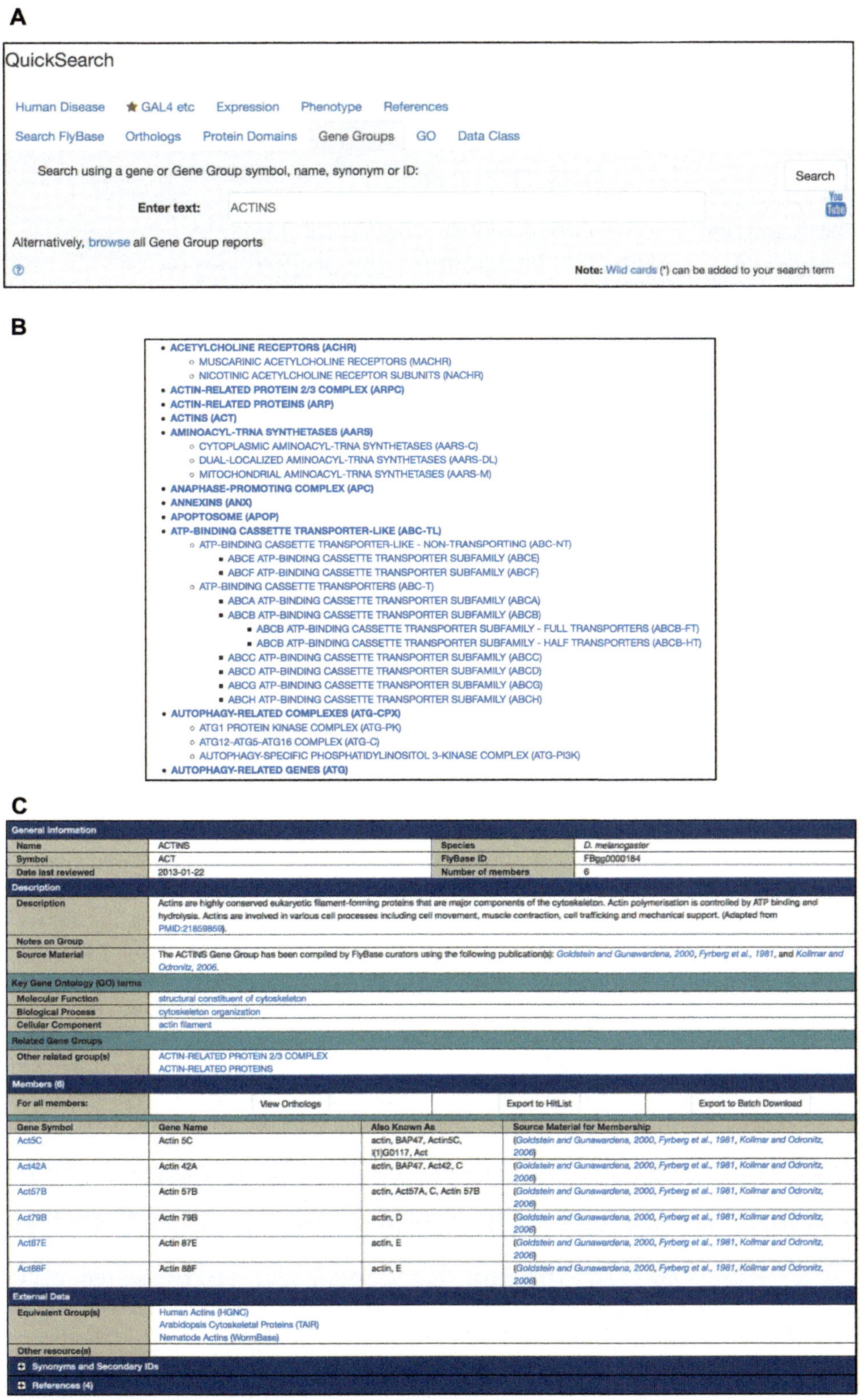

Fig. 4 (**a**) The Gene Groups tab of the QuickSearch tool. (**b**) Gene Groups are available as a browsable list. Groups are displayed as a nested hierarchy, with the top-level groups arranged in alphabetical order. The top section of the list is shown from ACETYLCHOLINE RECEPTORS to AUTOPHAGY-RELATED GENES. (**c**) A Gene Group report, using the ACTINS Gene Group as an example

3. Click the "Search" button or press "enter." Groups that match the query are displayed in a hit-list.
4. Click on a Gene Group to open the corresponding Gene Group report page.

2.3.2 Viewing Gene Lists in Gene Group Reports

Gene Group reports contain the list of member genes together with additional information organized into sections (Fig. 4c) [6]. The "Description" section gives an important overview of the criteria used to compile a particular group, and the "Notes on Group" field may contain justification for the inclusion or exclusion of particular genes. This section also displays "Key Gene Ontology (GO) terms"—these terms, or their children, are associated with most/all of the member genes and are typical, though not necessarily diagnostic, of that group. Clicking on a key GO term takes the user to a Term Report, where other genes annotated with this term or its children can be found (*see* Subheading 2.1.3).

Gene Groups are constructed in a hierarchical fashion, with only the terminal groups populated with genes. The "Related Gene Groups" subsection displays the groups immediately above or below the group (called "Parent group(s)" or "Component group(s)," respectively) and clicking on these links displays the corresponding Gene Group page. Any nonhierarchical but functionally relevant relationships (e.g., receptor–ligand groups such as Frizzled-type receptors and Wnts) are displayed as "Other related group(s)."

The "Members" section contains all genes belonging to the group (displayed under their terminal group heading) with gene symbols hyperlinked to their Gene report page (where Gene Group membership is displayed in the "Families, Domains and Molecular Function" section (Fig. 3b), thereby providing an alternative entry into Gene Group reports). The attribution for membership of an individual gene to a particular group is shown in the "Source Material for Membership" column of the table. At the top of the "Members" table are three export buttons, provided to facilitate further analysis of the group. The "View Orthologs" button runs the gene list through the "QuickSearch-Orthologs" tab [1] to retrieve the predicted orthologs of each *D. melanogaster* gene in humans and model organisms, powered by the DRSC Integrative Ortholog Prediction Tool (DIOPT) [9]. The "Export to HitList" and "Export to Batch Download" buttons export the genes in the members table to these tools for further analyses (*see* Subheadings 2.4 and 2.5).

The "External Data" section of a Gene Group report includes links to equivalent gene collections at other databases to facilitate cross-organism analyses, notably human gene families at the HGNC, which are also manually compiled and verified [10]. Indeed, the reciprocal links that exist between HGNC gene families and FlyBase Gene Groups should be the primary method to compare related gene sets between humans and *D. melanogaster*, rather than using the "View Orthologs" option described above. Other expert/specialized databases are also listed in the "External Data" section where relevant, for example the Heat Shock Protein Information Resource [11] or the Ribosomal Protein Gene Database [12] for the HEAT SHOCK PROTEINS and RIBOSOMAL PROTEINS Gene Groups, respectively.

2.4 Combinatorial Queries

The methods above describe how to find a set of functionally related *Drosophila* genes based on a single GO term (and its children), a single protein domain, or a specific Gene Group (and its subgroups). It is sometimes useful to combine searches of multiple terms within or between any one of the three classifications to define a gene set based on additional criteria. For example, the subset of genes from the ION CHANNEL Gene Group that also have GO annotation under "sensory perception" (to identify ion channels known/predicted to be involved in perception of sensory stimuli), or a list of genes annotated with an "EF-hand domain" and the GO term "synaptic signaling" (to identify candidate Ca^{2+}-binding proteins involved in signaling at the synapse). Simple intersections can be achieved using the Analysis tools available from a gene hit-list using protocol in Subheading 2.4.1 below. More complex queries require export of the hit-list to the QueryBuilder tool [8], as described in protocol in Subheading 2.4.2 below. (A detailed description of the use of QueryBuilder is beyond the scope of this chapter—additional information, templates, and examples are available online.)

2.4.1 Hit-List Analysis Tools

1. Generate an initial hit-list of genes from a GO or protein domain search, or directly from a Gene Group, as described in Subheadings 2.1.3, 2.2.1, and 2.3.2.
2. From a hit-list of genes, click on the "Analyze" button (Fig. 5a). From the drop-down menu, select one of "Molecular function (GO)," "Biological Process (GO)," "Cellular Component (GO)," or "InterPro Domains" (Fig. 5a). This generates a second hit-list showing the distribution of the most frequent GO term or protein domain annotations associated with the genes in the first hit-list (Fig. 5b). Note that for GO term refinements, the numbers shown correspond to genes with annotations to that exact term—that is, annotations to more specific child terms are not included in the given counts.
3. Click on the number in the "Related records" column to produce a third hit-list. This contains the subset of genes from the initial list that are also associated with the additional GO term/protein domain selected in **step 1**—that is, the intersection of the two criteria.
4. If desired, repeat the steps above to define finer level intersections of the list.

2.4.2 QueryBuilder

1. Generate an initial hit-list of genes from a GO or protein domain search, or directly from a Gene Group, as described in Subheadings 2.1.3, 2.2.1, and 2.3.2.
2. Click on the "Export" button (Fig. 5a). From the drop-down menu, select "QueryBuilder".
3. A new QueryBuilder session appears with the first segment of the query populated with the genes from **step 1**. Click on the "+" button to add a query segment, then select a data class from the drop-down menu and a specific field/term to query within that class. For example, choose the "Controlled Vocabularies" data class and select a specific GO term, or choose the "Genes" data class and select a term from the InterPro Domains field (Fig. 5c).
4. Repeat **step 3** to include additional query legs.
5. Combine individual query segments using Boolean operators (AND, OR, BUT NOT) in order to generate lists that combine or exclude the given criteria.
6. Once the query is assembled, click the "Run query" button.
7. From the results page (Fig. 5c), click on the "Genes" box to generate a hit-list of genes matching the search criteria.

2.5 Downloading Lists of Functionally Related Genes

The hit-list of genes obtained via one of the approaches described above, together with associated data if desired, can be easily downloaded using the Batch Download tool (Fig. 6). Bulk files listing all FlyBase GO annotations and Gene Group data are also available. Both these options enable further processing/analysis of lists of genes offline or using other web-based tools. Protocols to obtain these files are given below. (Detailed descriptions of the use of Batch Download and the contents of the bulk files are beyond the scope of this article—additional information is available online.)

2.5.1 Batch Download

1. From a gene hit-list, click on the "Export" button (Fig. 5a). From the drop-down menu, select "Batch Download". Alternatively, from a Gene Group report, simply click the "Export to Batch Download" button at the top of the "Members" table (Fig. 4c).
2. A new Batch Download session appears with the data entry box populated with the genes from **step 1** (Fig. 6a).
3. Choose the "Output format" as "HTML table" or "tab-separated file" as required. Then choose to "Send results" to "Browser" or "File" as desired.
4. Click on the "Continue to Select Fields" button to be directed to a template resembling a FlyBase Gene report page and check the boxes corresponding to the data of interest (Fig. 6b). These may be directly relevant to the original search (e.g., gene

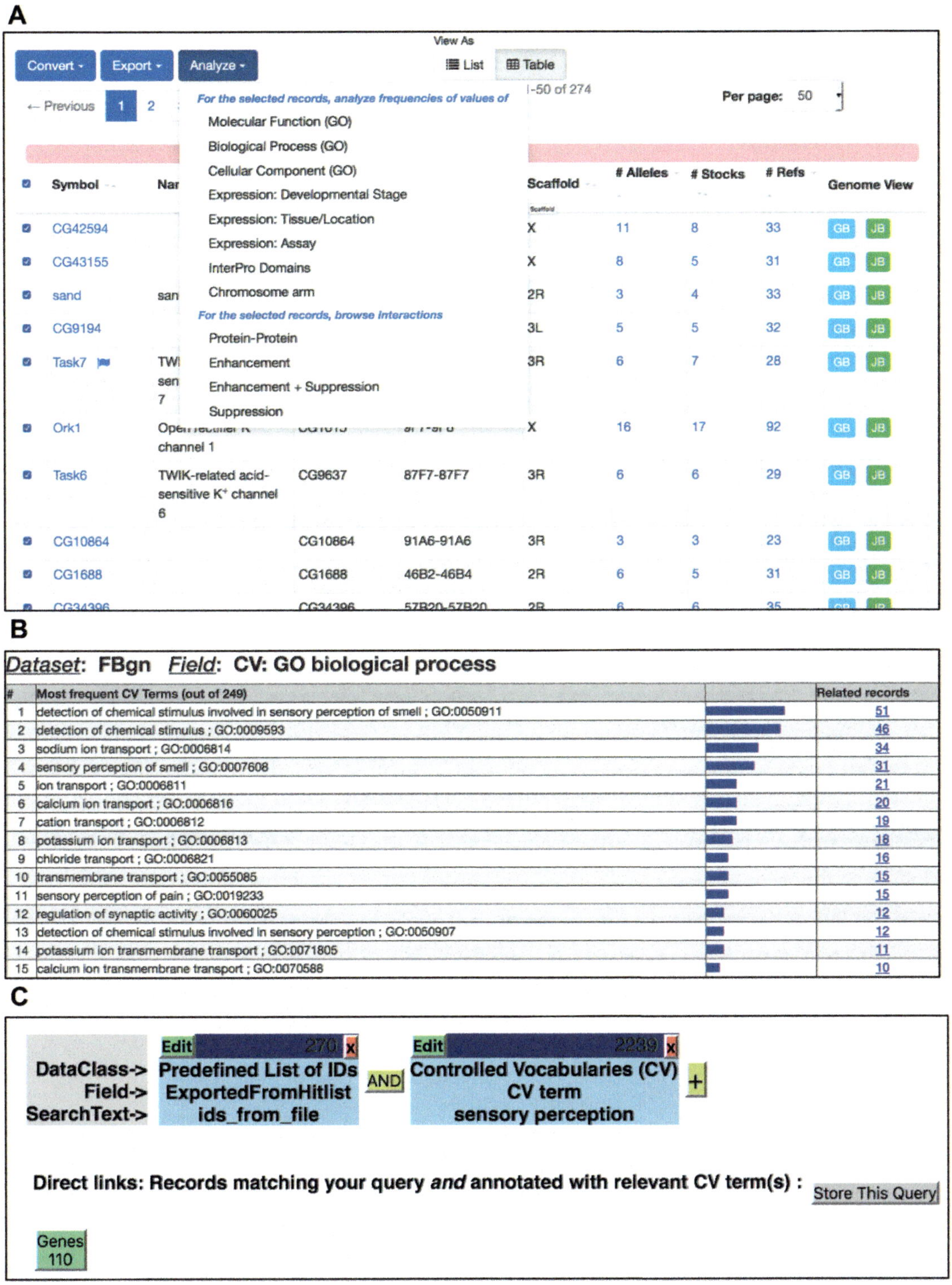

Fig. 5 (**a**) A hit-list of genes. In this example, the hit-list is populated with genes exported from the ION CHANNELS Gene Group. The "Analyze" drop-down menu is shown. (**b**) A results analysis of individual Biological Process GO terms associated with genes from the ION CHANNELS Gene Group. (Only the top 15 most frequently associated GO terms are shown). (**c**) A QueryBuilder results page, showing a 2-leg query (top). First query: IDs imported from the ION CHANNELS Gene Group; second query: GO term search for "sensory perception." A results button, with the number of genes returned from the query, is displayed at the bottom—clicking this generates a new gene hit-list

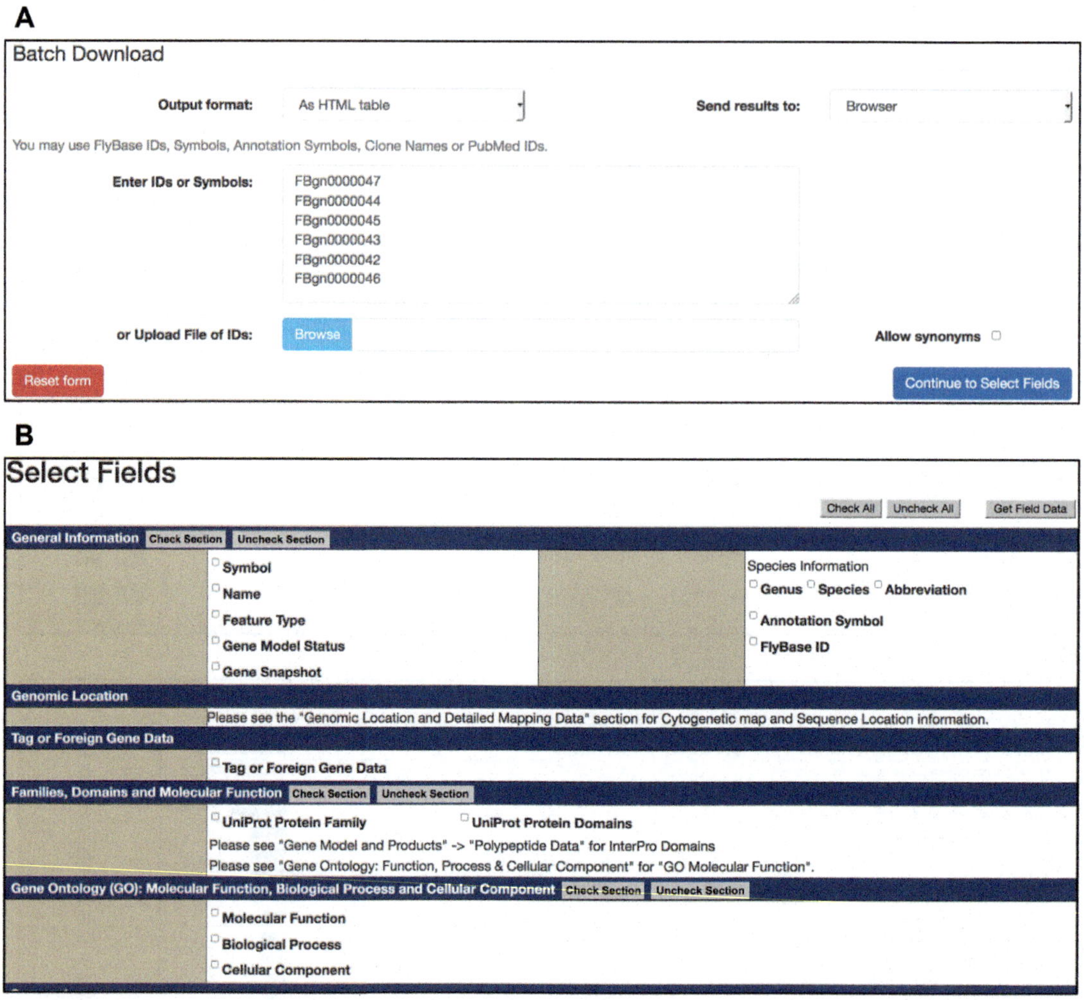

Fig. 6 (**a**) The FlyBase Batch Download interface, using the FlyBase Gene IDs (FBgns) exported from the ACTINS Gene Group as an example. (**b**) The Batch Download interface for selecting data fields for download (only the top section is shown)

symbols and synonyms, GO annotations, InterPro domains) or a different type of data (e.g., genomic data, expression data, physical interactions, available reagents) to be analyzed in the context of the given gene list.

5. Finally, click on the "Get Field Data" button to retrieve the data in the method and format selected in **step 3**.

2.5.2 Bulk Files

FlyBase bulk files can be accessed from any page by clicking on "Current release" from the "Downloads" menu in the NavBar. For GO data, the "gene_association.fb.gz" file within the "Genes" section contains all GO annotations for *D. melanogaster* genes within FlyBase in the standardized GO Annotation File (GAF) format.

For Gene Groups data, two files are available within the "Gene Groups" section of the Downloads page. The first (gene_group_data_fb_*.tsv.gz) includes the symbol, name and ID of every group, any parent/child relationships between groups, and the symbol and ID of all member genes. The second file (gene_groups_HGNC_fb_*.tsv.gz) lists just the groups themselves together with any corresponding HGNC gene family IDs.

3 Notes

1. Three distinct, but overlapping, approaches to finding functionally related genes in FlyBase are presented in this chapter and it is important to consider the advantages and limitations of each method. For established and/or evolutionary conserved gene sets, the Gene Groups resource should be the first place to look, benefiting from manual curation from expert sources and supplemented with explanatory notes for edge cases and/or atypical members. However, the genomic coverage of Gene Groups is relatively limited and its scope does not extend to broad biological phenomena or to predicted/uncharacterized gene sets. Thus, if a list of candidate genes involved in a process/pathway is required or the property sought is not confined to particular protein classes, then querying GO annotations is the most appropriate route, benefitting from high genomic coverage and a high degree of manual verification. Protein domain data are also worth consulting where there is good structure–function correlation: while they are not manually validated, they have a similar genomic coverage as GO annotations and are particularly useful when wanting to cast a wider net. For example, a search for "SH2 domain" retrieves many candidate phosphotyrosine-binding proteins involved in receptor tyrosine-kinase signaling that GO annotation may not capture. Of course, the results of some queries using the three approaches will overlap significantly. For example, the PROTEIN KINASE Gene Group comprises 243 genes, of which 219 are annotated with the GO term "protein kinase activity" or its child terms, and 220 have a "Protein kinase domain" (Fig. 7). In this case, where there is well-defined structure–function relationship, the Gene Group presentation provides the complete and accurate picture and differences in overlap with protein domain signature and GO annotation arise from either sequence divergence or the presence of pseudokinases. Ultimately, the approach taken to identify a group of functionally related genes depends on the details of the query itself and the accuracy/scope required in the answer. It will often be informative to experiment with all three methods, combining or refining the results with additional criteria as necessary.

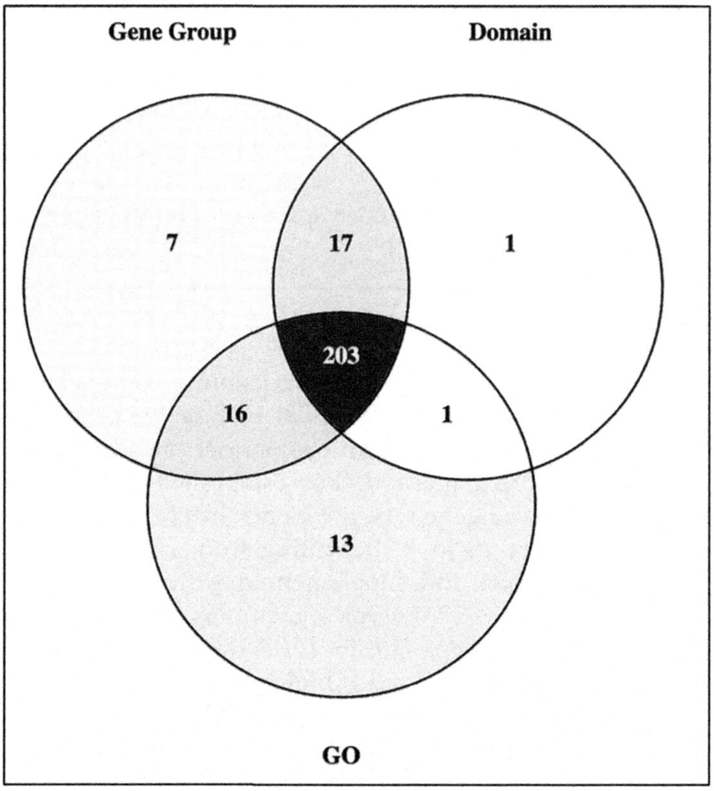

Fig. 7 A Venn diagram showing the overlap between genes annotated with the GO term 'protein kinase activity' (GO:0004672) or its child terms, the InterPro term 'Protein kinase domain' (IPR000719) and the 'PROTEIN KINASE' Gene Group. The diagram was generated using Venny 2.1

2. It is worth noting that a subset of GO annotations in FlyBase are computationally derived from InterPro domain associations via "InterPro2GO" mapping [4, 13], and that GO annotations associated with members of a Gene Group are reviewed and improved during the compilation of a group. Both of these pipelines act to increase the overlap in results obtained when querying using different methods.
3. For some species (e.g., humans [10]), genes belonging to particular families/groups are given symbols/names with identical prefixes or "root symbols", meaning that functionally related genes can be retrieved/classified by their nomenclature to some extent. This approach should not be used to identify *D. melanogaster* gene sets—gene nomenclature is generally not as systematic in this species with many genes given an esoteric symbol/name based on their mutant phenotype. Notable exceptions are genes encoding ncRNAs, whose symbols have a systematic prefix ("tRNA:", "snoRNA:", etc.). (*See* the "Nomenclature" link under the "Help" menu on the NavBar of any FlyBase page.)

4. The chapter focuses on methods to identify functionally related genes within FlyBase, taking advantage of GO annotations, protein domain associations, and membership of Gene Groups. Of course, there are several other methods, tools, and resources within FlyBase to identify other kinds of "related gene sets" based on these and other criteria. For example, FlyBase compiles sets of genes within experimentally derived datasets, such as protein–protein interaction sets or gene expression clusters, while any number of de novo sets could be constructed based on phenotype, expression, genomic data, etc. The protocols described herein are readily expandable/transferable to encompass a wider scope of data within FlyBase.

Acknowledgments

FlyBase is funded by the National Human Genome Research Institute at the US National Institutes of Health (#U41HG000739, PI N. Perrimon) and the UK Medical Research Council (#MR/N030117/1, PI N.H. Brown). At the time of writing, the FlyBase Consortium included: Norbert Perrimon, Julie Agapite, Kris Broll, Madeline Crosby, Gilberto dos Santos, David Emmert, Sian Gramates, Kathleen Falls, Beverley Matthews, Susan Russo Gelbart, Christopher Tabone, Pinglei Zhou, Mark Zytkovicz; Nicholas Brown, Giulia Antonazzo, Helen Attrill, Silvie Fexova, Phani Garapati, Tamsin Jones, Aoife Larkin, Steven Marygold, Gillian Millburn, Alix Rey, Vitor Trovisco, Jose-Maria Urbano; Thomas Kaufman, Bryon Czoch, Josh Goodman, Gary Grumbling, Victor Strelets, Jim Thurmond; Richard Cripps, Maggie Werner-Washburne, Phillip Baker.

References

1. Gramates LS, Marygold SJ, Santos GD, Urbano JM, Antonazzo G, Matthews BB, Rey AJ, Tabone CJ, Crosby MA, Emmert DB, Falls K, Goodman JL, Hu Y, Ponting L, Schroeder AJ, Strelets VB, Thurmond J, Zhou P, FlyBase Consortium (2017) FlyBase at 25: looking to the future. Nucleic Acids Res 45(D1):D663–D671. https://doi.org/10.1093/nar/gkw1016
2. Marygold SJ, Crosby MA, Goodman JL, FlyBase Consortium (2016) Using FlyBase, a database of Drosophila genes and genomes. Methods Mol Biol 1478:1–31. https://doi.org/10.1007/978-1-4939-6371-3_1
3. The Gene Ontology C (2017) Expansion of the Gene Ontology knowledgebase and resources. Nucleic Acids Res 45(D1):D331–D338. https://doi.org/10.1093/nar/gkw1108
4. Finn RD, Attwood TK, Babbitt PC, Bateman A, Bork P, Bridge AJ, Chang HY, Dosztanyi Z, El-Gebali S, Fraser M, Gough J, Haft D, Holliday GL, Huang H, Huang X, Letunic I, Lopez R, Lu S, Marchler-Bauer A, Mi H, Mistry J, Natale DA, Necci M, Nuka G, Orengo CA, Park Y, Pesseat S, Piovesan D, Potter SC, Rawlings ND, Redaschi N, Richardson L, Rivoire C, Sangrador-Vegas A, Sigrist C, Sillitoe I, Smithers B, Squizzato S, Sutton G, Thanki N, Thomas PD, Tosatto SC, Wu CH, Xenarios I, Yeh LS, Young SY, Mitchell AL (2017) InterPro in 2017-beyond protein family and domain annotations. Nucleic Acids Res 45(D1):D190–D199. https://doi.org/10.1093/nar/gkw1107
5. Gaudet P, Livstone MS, Lewis SE, Thomas PD (2011) Phylogenetic-based propagation of

functional annotations within the Gene Ontology consortium. Brief Bioinform 12(5):449–462. https://doi.org/10.1093/bib/bbr042

6. Attrill H, Falls K, Goodman JL, Millburn GH, Antonazzo G, Rey AJ, Marygold SJ, FlyBase Consortium (2016) FlyBase: establishing a Gene Group resource for Drosophila melanogaster. Nucleic Acids Res 44(D1):D786–D792. https://doi.org/10.1093/nar/gkv1046
7. Marygold SJ, Antonazzo G, Attrill H, Costa M, Crosby MA, Dos Santos G, Goodman JL, Gramates LS, Matthews BB, Rey AJ, Thurmond J, FlyBase Consortium (2016) Exploring FlyBase data using QuickSearch. Curr Protoc Bioinformatics 56(1):31 31–31 31 23. https://doi.org/10.1002/cpbi.19
8. St Pierre SE, Ponting L, Stefancsik R, Mcquilton P, FlyBase Consortium (2014) FlyBase 102—advanced approaches to interrogating FlyBase. Nucleic Acids Res 42(Database issue):D780–D788. https://doi.org/10.1093/nar/gkt1092
9. Hu Y, Flockhart I, Vinayagam A, Bergwitz C, Berger B, Perrimon N, Mohr SE (2011) An integrative approach to ortholog prediction for disease-focused and other functional studies. BMC Bioinformatics 12:357. https://doi.org/10.1186/1471-2105-12-357
10. Yates B, Braschi B, Gray KA, Seal RL, Tweedie S, Bruford EA (2017) Genenames.org: the HGNC and VGNC resources in 2017. Nucleic Acids Res 45(D1):D619–D625. https://doi.org/10.1093/nar/gkw1033
11. Ratheesh Kumar R, Nagarajan NS, PA S, Sinha D, Veedin Rajan VB, Esthaki VK, D'Silva P (2012) HSPIR: a manually annotated heat shock protein information resource. Bioinformatics 28(21):2853–2855. https://doi.org/10.1093/bioinformatics/bts520
12. Nakao A, Yoshihama M, Kenmochi N (2004) RPG: the Ribosomal Protein Gene database. Nucleic Acids Res 32(Database issue):D168–D170. https://doi.org/10.1093/nar/gkh004
13. Tweedie S, Ashburner M, Falls K, Leyland P, McQuilton P, Marygold S, Millburn G, Osumi-Sutherland D, Schroeder A, Seal R, Zhang H, FlyBase Consortium (2009) FlyBase: enhancing Drosophila Gene Ontology annotations. Nucleic Acids Res 37(Database issue):D555–D559. https://doi.org/10.1093/nar/gkn788

Chapter 17

Hymenoptera Genome Database: Using HymenopteraMine to Enhance Genomic Studies of Hymenopteran Insects

Christine G. Elsik, Aditi Tayal, Deepak R. Unni, Gregory W. Burns, and Darren E. Hagen

Abstract

The Hymenoptera Genome Database (HGD; http://hymenopteragenome.org) is a genome informatics resource for insects of the order Hymenoptera, which includes bees, ants and wasps. HGD provides genome browsers with manual annotation tools (JBrowse/Apollo), BLAST, bulk data download, and a data mining warehouse (HymenopteraMine). This chapter focuses on the use of HymenopteraMine to create annotation data sets that can be exported for use in downstream analyses. HymenopteraMine leverages the InterMine platform to combine genome assemblies and official gene sets with data from OrthoDB, RefSeq, FlyBase, Gene Ontology, UniProt, InterPro, KEGG, Reactome, dbSNP, PubMed, and BioGrid, as well as precomputed gene expression information based on publicly available RNAseq. Built-in template queries provide starting points for data exploration, while the QueryBuilder tool supports construction of complex custom queries. The List Analysis and Genomic Regions search tools execute queries based on uploaded lists of identifiers and genome coordinates, respectively. HymenopteraMine facilitates cross-species data mining based on orthology and supports meta-analyses by tracking identifiers across gene sets and genome assemblies.

Key words Hymenoptera, *Apis mellifera*, Genome, Database, Data mining, Orthology, Pathway, Gene expression, Single nucleotide polymorphism, InterMine

1 Introduction

The Hymenoptera Genome Database (HGD; http://hymenopteragenome.org) is an informatics resource for data associated with sequenced genomes of hymenopteran insects [1]. HGD currently includes genomes of eleven bee species, ten ant species, and the parasitoid jewel wasp (Table 1). Goals of HGD have been to (1) support species genome consortia with genome annotation tools, (2) provide access to data via genome browsers, BLAST and data download, (3) add value to the genome data by integrating it with external data sources in a data mining warehouse (HymenopteraMine) and (4) maintain the value of the genome consortia's published work by porting gene annotations to

Martin Kollmar (ed.), *Eukaryotic Genomic Databases: Methods and Protocols*, Methods in Molecular Biology, vol. 1757, https://doi.org/10.1007/978-1-4939-7737-6_17, © Springer Science+Business Media, LLC, part of Springer Nature 2018

Table 1
Species in the hymenoptera genome database

HGD division	Species	Common name or group	Genome/gene set reference(s)
BeeBase	*Apis mellifera*	European honey bee	[2, 3]
	Apis dorsata	Giant honey bee	
	Apis florea	Dwarf honey bee	
	Bombus impatiens	Common Eastern bumble bee	[4]
	Bombus terrestris	Buff-tailed bumble bee	[4]
	Eufriesea mexicana	Orchid bee	[5]
	Dufourea novaeangliae	Sweat bee	[5]
	Habropoda laboriosa	Southeastern blueberry bee	[5]
	Lasioglossum albipes	Sweat bee	[6]
	Megachile rotundata	Alfalfa leafcutting bee	[5]
	Melipona quadrifasciata	Stingless bee	[5]
Ant Genomes Portal	*Acromyrmex echinatior*	Panamanian leaf cutter ant	[7]
	Atta Cephalotes	Leaf cutter ant	[8]
	Camponotus floridanus	Florida carpenter ant	[9]
	Cardiocondyla obscurior		[10]
	Cerapachys biroi	Clonal raider ant	[11]
	Harpegnathos saltator	Jumping ant	[9]
	Linepithema humile	Argentine ant	[12]
	Pogonomyrmex barbatus	Red harvester ant	[13]
	Solenopsis invicta	Red fire ant	[14]
	Wasmannia auropuncata	Little fire ant	
NasoniaBase	*Nasonia vitripennis*	Parasitoid jewel wasp	[15, 16]

upgraded genome assemblies and providing identifier cross-references for updated gene sets.

Most hymenopteran insect genome sequencing projects have been carried out by small research consortia. HGD's initial contributions have focused on supporting hymenopteran insect genome consortia in the genome annotation and analysis process. More recently, with the availability of easy-to-deploy annotation pipelines, such as Maker2 [17] and the web-based Apollo annotation platform [18], the need for annotation support from HGD has decreased. Although we still provide Apollo annotation tools, we have shifted our efforts to support the use of the genomics data in downstream analyses. To make HGD more effective for post-genome-sequencing analyses, we have integrated the genome assemblies with other sources of biological data and developed data mining tools that support complex queries across species.

In addition to providing data mining and browsing tools, an important role of HGD is mapping identifiers across alternate datasets of the same species. Following publications of several genome projects, many of the genome assemblies and gene sets available at NCBI have been upgraded. To preserve the information from the original consortium publications and provide users with the most-up-to-date information, we provide resources for both the original consortium gene sets and the updated RefSeq gene sets and assemblies, along with references between gene identifiers across the gene sets. For entry into HGD, we require either a published consortium gene set or availability of a RefSeq gene set.

2 Methods

2.1 Website Navigation

HGD is divided into three major divisions (BeeBase, NasoniaBase, and the Ant Genomes Portal) to facilitate species-specific data download and genome browsing (Fig. 1). However, data mining (HymenopteraMine) and sequence search (BLAST) tools are unified across HGD. The navigation bar of the HGD home page includes links to the individual divisions and to the pages that are common across divisions (Hymenoptera Home, Hymenoptera Mine, BLAST, Genome Consortium Publications, Data Usage Policy, How to Cite). Within each division, the navigation bars include the links common to all divisions, as well as links that expand to species-specific pages (JBrowse and Data Sets).

2.2 JBrowse and Apollo

Genome browsing is provided for each species using JBrowse [19] as implemented by Apollo [18]. From a user's perspective, the main differences between JBrowse and Apollo are the gene editing functions and the user annotation pane that are available only when logged into Apollo. The evidence tracks are identical across the browsers. All HGD users can access JBrowse, while only users registered for annotation can access Apollo. An Apollo registration link is provided in the navigation bar of a division only when there is an active annotation project for a species. Currently, active annotation is available for the three *Apis* species.

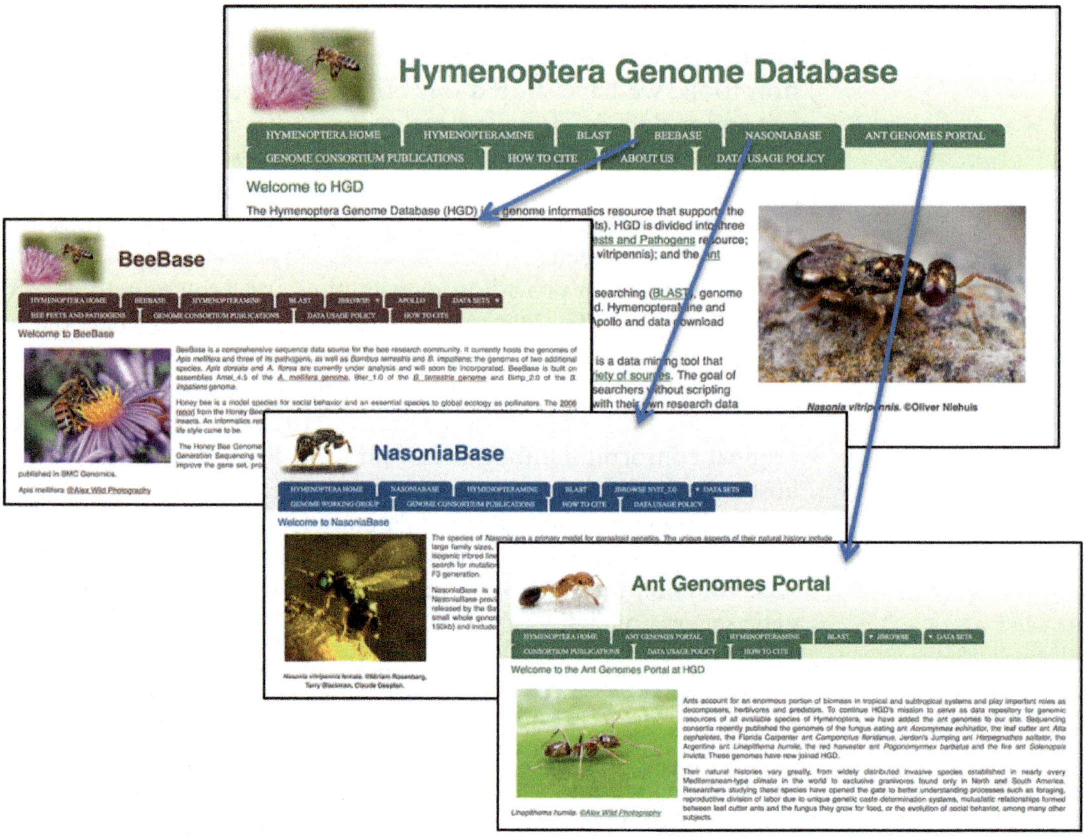

Fig. 1 The Hymenoptera Genome Database (HGD) includes three divisions—BeeBase, NasoniaBase, and the Ant Genomes Portal. The navigation bar on the HGD home page provides access to each of the divisions, as well as to resources available to all three divisions (e.g., HymenopteraMine and BLAST). The navigation bar for each division provides tabs to division specific resources (e.g., JBrowse and data download pages) as well as links the HGD shared resources and the HGD home page

2.3 BLAST Search

The HGD BLAST search interface leverages the SequenceServer platform [20], modified to make dataset selection easier. Datasets for all species are provided in a single interface. The name of each dataset indicates whether it is a consortium or NCBI dataset. You can select any combination of either protein or nucleotide datasets for a single search. When the search database is a genome assembly, BLAST hits are linked to JBrowse viewers based on matched coordinates. When the search database is coding sequence, transcript or peptide, BLAST hits are linked to a JBrowse location based on the hit identifier. SequenceServer also provides downloadable tab-delimited or BLAST XML reports and graphical overviews of the matches.

2.4 HymenopteraMine

HymenopteraMine is the data mining resource for HGD. It leverages the InterMine platform [21] to integrate biological data from a variety of sources (Table 2). Combining genomes for multiple hymenopteran species in a single data mining warehouse allows

Table 2
HymenopteraMine external data sources

Data source	Reference
BioGrid	[22]
dbSNP	[23]
FlyBase	[24]
FlyReactome	[25]
Gene Ontology	[26]
InterPro	[27]
KEGG	[28]
OrthoDB	[29]
PubMed	[30]
Reactome	[25]
RefSeq	[31]
SRA	[32]
UniProt	[33]
UniProt-GOA	[34]

users to leverage cross-species information using orthologue relationships from OrthoDB [29]. The complexity of data in HymenopteraMine makes query construction challenging, so the search tools support a range of user skills, such as simple keyword search and predefined template queries for new users and the QueryBuilder for power users.

2.4.1 HymenopteraMine Navigation, Tutorials, and MyMine

The HymenopteraMine home page is accessible from the navigation bars of HGD, BeeBase, NasoniaBase, and the Ant Genomes Portal. HymenopteraMine has its own navigation bar with HymenopteraMine-specific links (Home, MyMine, Templates, Lists, QueryBuilder, Regions, Data Sources, Data Model, Help, API), as well as a link to HGD BLAST (Fig. 2). The HGD home page is accessible by clicking "Hymenoptera Genome Database" in the header. The Help Tab leads to a HymenopteraMine tutorial that includes a link to a YouTube channel with HymenopteraMine videos. The Data Model tab provides helpful information about the interconnection of data types in HymenopteraMine; it opens a new browser tab showing data network diagrams and links to tables indicating which types of identifiers are needed for specific data sets.

HymenopteraMine maintains user accounts allowing you to save your work after ending a session. Clicking "Log in" to the right of the navigation bar leads to an option to create an account.

The MyMine tab in the navigation bar leads to a history of queries performed during the current session. If you are logged in, lists created, queries performed, and template queries are saved for future use.

2.4.2 Quick Search and Report Page

The HymenopteraMine home page provides basic search tools (Fig. 2). The Quick Search tool is used to perform a full text search of all datasets loaded in HymenopteraMine, and supports the use of wild cards. Data input types for Quick Search include gene identifiers, transcript identifiers, protein identifiers, gene symbols, gene names, functional annotation terms, and species names. Quick Search is a good place to start to explore the data before performing more complex queries. For example, searching a species name will provide a list of all datasets for that species. A faceted search tool in the search result page allows users to filter the results by category before selecting an entity to access a report page (Fig. 3).

A report is provided for each entity in HymenopteraMine. Each report is divided into sections appropriate for the data class. Users may find the most familiar reports to be those for genes,

Fig. 2 The HymenopteraMine navigation bar is available on all HymenopteraMine pages. The HymenopteraMine home pages provides the Quick Search and Quick List tools, and access to categorized template queries

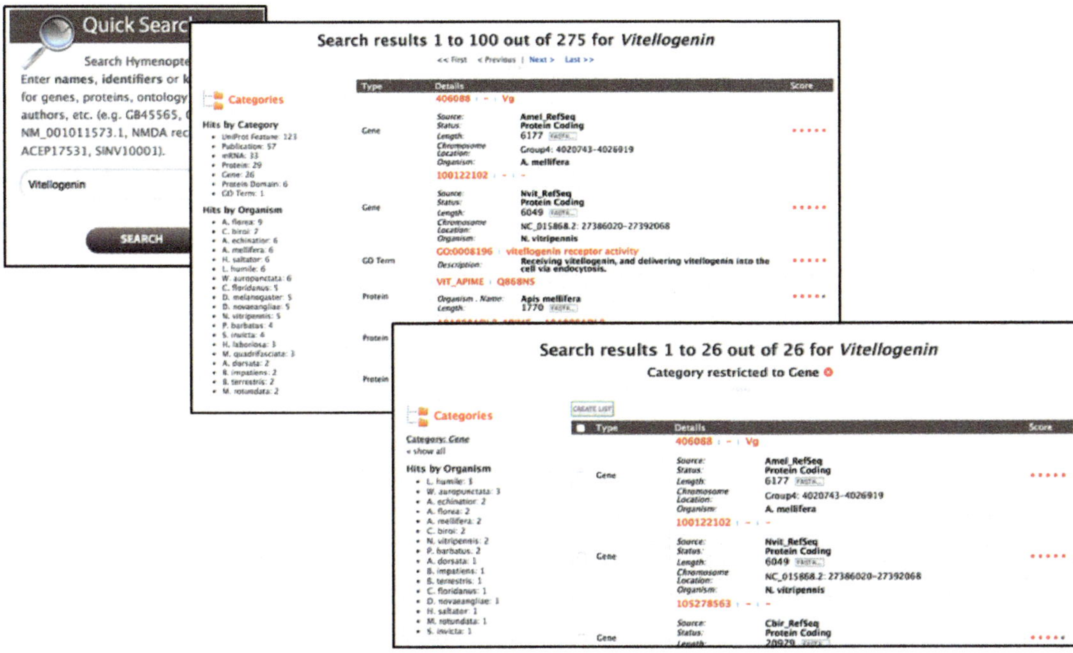

Fig. 3 The Quick Search tool performs a full text search and retrieves all data objects containing the searched term. A faceted selector to the left of the result list allows you to filter results by data class or organism

transcripts and proteins. These contain information similar to that found in gene pages of other model organism databases. However, most of the information in a HymenopteraMine report is provided in the form of tables that can be customized and downloaded in various formats, or saved as lists for further HymenopteraMine analyses. Sequences can be downloaded from reports in fasta format. The Function section of a Gene Report provides GO annotations and may also include pathways. The Transcripts section of a Gene Report gives a visual representation of the transcripts highlighting gene structure with links to JBrowse. Transcript identifiers are linked to Transcript Reports. For *A. mellifera*, Transcript Reports include a Gene Expression section that provides various forms of expression values (raw read counts, normalized read counts, FPKM and RPKM) for RNAseq data with metadata from the Sequence Read Archive. The Protein section of the Gene Report lists protein identifiers that link to Protein Reports with more information including protein domains, UniProt keywords, and curated notes from UniProt.

2.4.3 List Analysis

The ability to upload and analyze lists of identifiers is one of the most important HymenopteraMine features because it allows you to gather a variety of functional annotation information associated with your own data. The Quick List tool provided on the home page is a slimmed-down version of the List Tool (Fig. 2). As opposed to Quick Search, which performs a full text search of keywords and supports wildcards, Quick List searches only gene and protein datasets based on gene identifiers (ids and symbols), transcript identifiers and protein identifiers.

The full List Tool is available by clicking the "Lists" tab in the navigation bar or the word "advanced" in the Quick List box. Clicking the "Lists" tab brings you to either an upload interface or a view of existing lists. You can move from one to the other by clicking "Upload" or "View" in the brown bar below the main navigation bar. HymenopteraMine provides users with premade gene lists, for entire gene sets, that may be used as background populations in enrichment widgets (described in Subheading 2.4.7). The List upload menu of the List tool allows users to select from a pull-down menu of many data classes, to limit the search to a particular species, and to upload a list of identifiers.

Both Quick List and the full List tool perform a database lookup to validate the identifiers and may prompt the user to select from duplicates due to their presence in multiple datasets. For example, entering the gene symbol "Nmdar1" returns results for both *A. mellifera* and *D. melanogaster*. A green button saying "Save a list of 0 Genes" indicates that you must click the "Add" button to the right of a gene to save it to the list. The "Add" button does not appear if there are no duplicate identifiers. Before saving the final list, you may wish to enter a name for it. The list is saved by clicking the green "Save a List of X Genes" (where "X" is a number), and the result is a table with preset column output of associated information, such as gene identifier, gene secondary identifier, gene name, gene symbol, gene source (i.e., the gene set), gene status (i.e., the type of gene), chromosome and coordinate location, and organism. As with all table outputs in HymenopteraMine, the columns can be rearranged or deleted; column management tools (described in Subheading 2.4.8) can be used to add additional information; and the table can be exported. Lists of the additional data types, such as organisms or chromosomes and coordinates, can be saved using the "Save as List" pull-down menu above the table.

Saved lists can be retrieved by clicking "View" in the brown bar that is displayed under the navigation bar when the Lists tab is selected. Here you can perform set operations (union, intersection, subtraction, and asymmetric difference) to create new lists. Once you have saved a list, the predefined template queries and the QueryBuilder (described in Subheading 2.4.5) automatically provide the option to use it as long as the query is based on the appropriate data class. Lists are automatically deleted upon ending a session or accidental server disconnection unless you are logged in to MyMine, so working while logged in is recommended.

2.4.4 Template Queries

The complicated network of data in HymenopteraMine and the presence of alternative identifiers (Tables 3 and 4, and discussed in Subheading 2.4.10) can make it difficult to construct a query. HymenopteraMine provides predefined template queries to serve as starting points for data exploration. The complete list of templates is available via the Templates tab in the main navigation bar. Templates are also divided into categories in the home page template menu, which has tabs for GENES, GENE EXPRESSION, PROTEINS, HOMOLOGY, FUNCTION, and VARIATION. In addition, the ALIAS AND DBXREF tab provides templates that convert identifiers between gene sets for an organism, and the ENTIRE GENE SET tab lists templates for queries that output all genes or proteins for an organism.

If you click a template name, you will access a query form that may already be prepopulated with example identifiers, and may include pull-down menus. Some templates include options for numerical operations. An example of a simple template is Gene ID → Alias ID, listed under the ALIAS AND DBXREF category. Clicking on the template name opens the template interface where you can enter a gene ID. An example of a complex template query is A. mellifera Transcript → Expression and MetaData, under the GENE EXPRESSION category, in which you enter a transcript ID and you have options to constrain the output by expression levels and metadata values. If you have already saved a list with the appropriate data class, an option is provided to constrain the search to the list of identifiers rather than a single identifier. You obtain results by clicking "Show Results." Alternatively, you can click "Edit Query" to access the QueryBuilder (described in Subheading 2.4.5) if you wish to modify the query.

2.4.5 QueryBuilder

The QueryBuilder is the most flexible and sophisticated search tool of HymenopteraMine. However, use of the QueryBuilder is not intuitive; becoming a power user requires practice. On the other hand, for users without scripting skills, mastering the use of QueryBuilder to create a large complex data set would likely be more efficient than learning a scripting language to compile the data from the original sources. Before trying to build a query with QueryBuilder, it may be helpful to investigate the structure of some of the template queries using the "Edit Query" button available in the template query menus. After clicking "Edit Query" you will see the constructed template query in the same interface that is used to build the query from scratch.

A detailed example using the QueryBuilder is provided below. However, first we will provide an overview of the features. Clicking QueryBuilder in the navigation bar leads to the entry page for construction. The "QueryBuilder" box on the left includes options to browse the HymenopteraMine data model, import a query from

Table 3
Gene set and alias data source names in HymenopteraMine

Species	Primary gene set data source name(s)	Alias source name(s) (G or T)
A. dorsata	Ador_RefSeq	Ador_OrthoDB (G)
A. echinatior	aech_OGSv3.8_HGD, Aech_RefSeq	aech_OGSv3.8_C (T)
A. florea	Aflo_RefSeq	Aflo_OrthoDB (G)
A. mellifera	amel_OGSv3.2, Amel_RefSeq	amel_OGSv1 (G)
B. impatiens	bimp_OGSv1.0, Bimp_RefSeq	Bimp_OrthoDB (G)
B. terrestris	Bter_RefSeq	Bter_OrthoDB (G)
C. biroi	armyant.OGS.V1.8.6_HGD, Cbir_RefSeq	armyant.OGS.V1.8.6_C (T)
C. floridanus	cflo_OGSv3.3_HGD, Cflo_RefSeq	cflo_OGSv3.3_C (T)
C. obscurior	cobs_OGSv1.4	
D. novaeangliae	Dufourea_novaeangliae_v1.1_HGD, Dnov_RefSeq	Dufourea_novaeangliae_v1.1_C, Dnov_OrthoDB (T)
E. mexicana	Eufriesea_mexicana_v1.1_HGD	Eufriesea_mexicana_v1.1_C, Emex_OrthoDB (T)
H. laboriosa	Habropoda_laboriosa_v1.2_HGD	Habropoda_laboriosa_v1.2_C, Hlab_OrthoDB (T)
H. saltator	hsal_OGSv3.3_HGD, Hsal_RefSeq	hsal_OGSv3.3_C (T)
L. albipes	lalb_OGSv5.42_HGD	lalb_OGSv5.42_C (T)
L. humile	lhum_OGSv1.2, Lhum_RefSeq	
M. rotundata	Megachile_rotundata_v1.1_HGD, Mrot_RefSeq	Megachile_rotundata_v1.1_C, Mrot_OrthoDB (T)
M. quadrifasciata	Melipona_quadrifasciata_v1.1_HGD	Melipona_quadrifasciata_v1.1_C, Mqua_OrthoDB (T)
N. vitripennis	nvit_OGSv1.2, Nvit_EviGene, Nvit_RefSeq	
P. barbatus	pbar_OGSv1.2, Pbar_RefSeq	
S. invicta	sinv_OGSv2.2.3_HGD, Sinv_RefSeq	sinv_OGSv2.2.3_C (T)
W. auropuncata	Waur_RefSeq	Waur_OrthoDB (G)

"HGD" and "C" (for Consortium) indicate alternative identifiers for official gene sets. OrthoDB aliases may change in the future. "G" or "T" indicates whether alias IDs are for genes (G) or transcripts (T)

XML, and login to view saved queries. Selecting "Browse the Data Model" leads to a tree-like model of the data classes (Fig. 4). Clicking a plus sign in the tree reveals subclasses, while clicking the name of a data class opens the Model Builder (described below) at that class. The tree is useful for seeing the numbers of data objects within different data classes. Mousing over the "i" symbol next to

Table 4
Primary gene or protein identifiers used in data sets for different species

Data class	Data set	Species: *Acromyrmex echinatior*	*Apis dorsata*	*Apis florea*	*Apis mellifera*	*Bombus impatiens*	*Bombus terrestris*	*Camponotus floridanus*	*Cardiocondyla obscurior*	*Cerapachys biroi*	*Drosophila melanogaster*	*Dufourea novaeangliae*	*Eufriesea mexicana*	*Habropoda laboriosa*	*Harpegnathos saltator*	*Lasioglossum albipes*	*Linepithema humile*	*Megachile rotundata*	*Melipona quadrifasciata*	*Nasonia vitripennis*	*Pogonomyrmex barbatus*	*Solenopsis invicta*	*Wasmannia auropunctata*
Gene	BioGRID										F												
Gene	Evidential Gene Set																			E			
Gene	GO	R	R		O,R	O,R	R	R		R	F,R	R			R		R			O,R	R	R	R
Gene	KEGG				R						F									R			
Gene	Official Gene Set	O			O	O		O	O	O		O	O	O	O	O	O	O	O	O	O	O	
Gene	OrthoDB	O	R	R	O	O	R	O	O	R	F	O	O	O	O	O	O	O	O	E	O	O	R
Gene	Publications	R	R	R	O,R	R	R	R		R	F,R				R		R	R		O,R	R	R	
Gene	RefSeq	R	R	R	R	R	R	R		R		R			R		R	R		R	R	R	R
Gene	Gene Symbols	R	R	R	R	R	R	R		R	F	R			R		R	R		R	R	R	R
Protein	Fly Reactome										U												
Protein	InterPro	U	U	U	U		U	U		U	U	U			U		U	U	U		U	U	
Protein	Reactome										U												
Protein	UniProt	R	R	R	O,R	O,R	R	R		R	F	R			R		R	R		O,R	R	O,R	
Transcript	Gene Expression				O																		
Sequence Variant	dbSNP				R																		

E = Evidential Gene Set (*N. vitripennis*), F=FlyBase, O=OGS, R = RefSeq, U=UniProt.

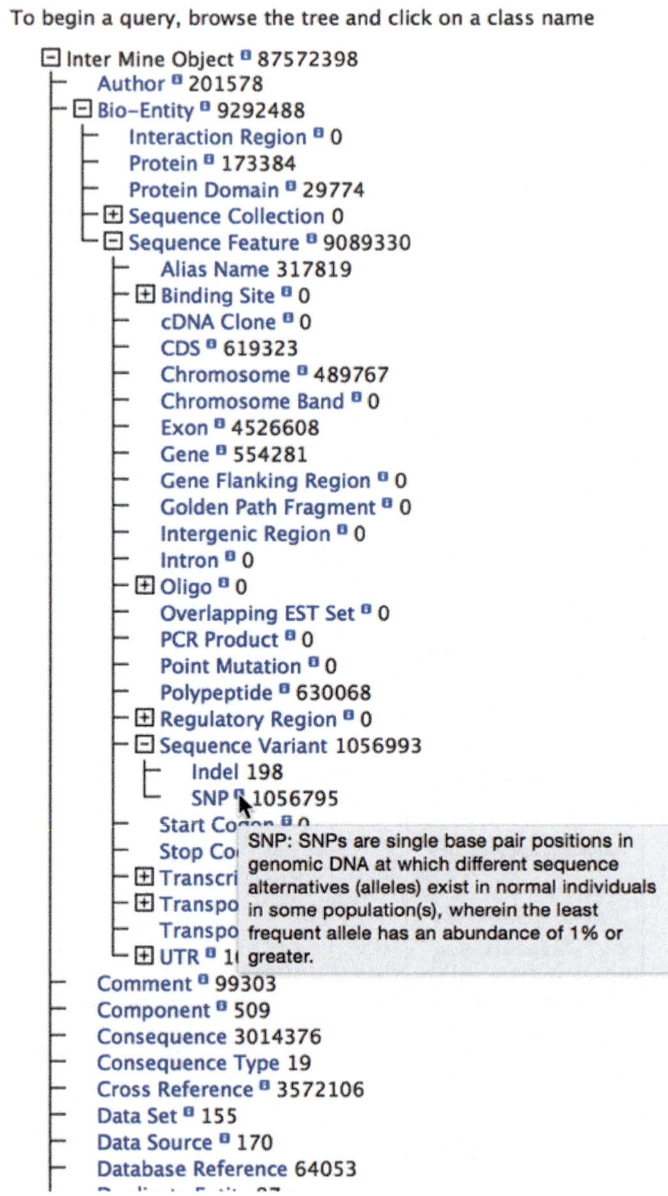

Fig. 4 The hierarchical data model tree, accessible by clicking "Browse Data Model" on the QueryBuilder page, shows the data classes in HymenopteraMine and numbers of data objects for each class. Clicking an "i" symbol provides a definition for a data class

a class provides a description. The classes that will be of most interest to HymenopteraMine users are listed under the "Sequence Feature" class, which is directly under "BioEntity." Although exploring the data with this tree is useful for developers, it is not intuitive for most users, and the structure is subject to change.

Instead of selecting "Browse the Data Model," selecting a data class from the pull-down menu under "Select a Data Type to Begin a Query" on the right side of the main QueryBuilder page leads to the same Model Builder mentioned above; the advantage of starting with the pull-down menu is you do not need to find the correct class in the data model tree. The Model Browser shows the hierarchical data structure, similar to the data tree described above, but is zoomed into the selected data class. The Model Browser also shows relationships between data classes. To the right of each data class is the word CONSTRAIN, and either the word SUMMARY or SHOW. Clicking any of these words initiates query construction. You add constraints to the query using "CONSTRAIN," and you select output data classes with SUMMARY or SHOW. SUMMARY is provided for any class with a "+" sign next to it, indicating that it can be divided into more than one subclass or attribute; SUMMARY is used to select a collection of the attributes as output. SHOW is provided for individual attributes that cannot be further subdivided (e.g., an identifier).

As you construct a query using the Model Browser, query building blocks are shown in the Query Overview Panel on the right side of the page. If you initiate query construction by clicking the word "CONSTRAIN" in the Model Browser next to the data class on which the search will be performed, a box appears allowing you to enter a constraint identifier. Once you have made selections in the constraint box, the constraint appears in the Query Overview. You add query output by clicking "SHOW" next to an attribute or "SUMMARY" next to a class. When the word "collection" appears next to a data class name in the Query Overview, it indicates that there is a collection of attributes related to this class. Clicking on the data class name next to "collection" in the Query Overview allows you to easily navigate to that data class in the Model Browser tree for further modifications. In an iterative fashion, you can add additional constraints and outputs to build a complex query. You can remove constraints and outputs using the red "X," and you can edit constraints by clicking the pencil symbol. Once you have completed query construction, you can view and rearrange the output column order in the "Fields selected for output" section below the Model Browser. You can export the query using "Export XML" at the bottom of the page to save the query locally. If you are logged into MyMine, the "Start building a template query" button at the bottom of the page allows you to make the query into a template that will be saved in your MyMine account. Finally, to run the query, you would click the green "Show results" button. After running the query, it is automatically saved in your MyMine query history. From there you can rerun it or edit it. Clicking edit returns you to the QueryBuilder interface, which provides another opportunity to export or create a template query if you did not already do so.

QueryBuilder Example: Say you would like more information about a gene, "BIMP17180" from the *B. impatiens* official gene set (bimp_OGSv1.0). In this example, you will build a query to retrieve the *A. mellifera* homologue(s) of the *B. impatiens* gene, and pathways of those homologues. You will also retrieve the gene symbols for both the *B. impatiens* and *A. mellifera* genes. As you follow the instructions to build the query, note how the Query Viewer displays each step of query construction. The last few steps in this example provide instructions for turning your constructed query into a template query if you are logged into MyMine. An advantage of creating a template query is it makes it convenient to rerun the query with different constraint values or with a premade list of identifiers.

1. It is advisable to log into MyMine before you start building a query so that you will be able to save your work.
2. Click the QueryBuilder tab in the HymenopteraMine navigation bar.
3. Select "Gene" in the "Select a Data Type to Begin a Query" box. This brings you to the Model Browser, starting with the "Gene" data class.
4. The first query building block will be to constrain the gene ID to "BIMP17180." Click CONSTRAIN next to "DB identifier" under "Gene" in the model browser, enter the gene ID "BIMP17180" in the box, and click "Add to Query" (Fig. 5A). Notice that the constraint has been added to the Query Viewer on the right. The pencil symbol allows you to edit the constraint in case you decide to use a different ID. Since this is a unique identifier that is only found in the bimp_OGSv1.0 gene set, there is no need to constrain the species or dataset.
5. We would like the output to include our entered gene ID, so click "Show" next to "DB identifier" under "Gene" in the Model Browser. Notice that in the Query Overview, "DB

Fig. 5 (continued) font shown next to the data class in the Query Overview. Here, clicking "Gene" next to "Cross Reference" allows you to navigate to the correct area of the Model Browser to add the symbol of the cross reference gene. (C) The first subclass under "Homologues Homologue" is "Gene," with a red up arrow, indicating that it is a reference to the parent "Gene" class, rather than the homologous gene. Look further down to see "Homologue Gene." Clicking "Show" next to "DB identifier" under "Homologue Gene" adds an output column for the gene ID of the homologue. (D) The "DB identifier" selected to show here is for the "Cross Reference Gene" within the "Db Cross References x Ref" collection that is a subclass of "Homologue Gene" rather than the "Db Cross References x Ref" collection that is a subclass of "Gene". (E) This final Query Overview shows two query constraints and eight output columns. (F) After query construction is complete, you can rearrange columns in the "Columns to Display" section. At the bottom of the page are options to save, export and run the query. The "Start building a template query" button shows up if you are logged into MyMine

Fig. 5 The Model Browser and Query Overview are used in query construction. (A) Query construction is initiated by clicking CONSTRAIN. A pop-up menu allows entry of a constraint, such as an identifier. After clicking "Add to Query" the constraint is shown in the Query Overview. (B) During query construction, to easily navigate to the correct subtree to add additional attributes related to a particular data class, click on the word in brown

identifier" is now shown in a light blue box to indicate it will be included in the output.

6. The next step is determine what kind of identifier is needed in order to output a gene symbol and homologues using Table 4 in this chapter, or the "Identifier Relationship Table" available by clicking the HymenopteraMine "Data Model" tab. Notice that for *B. impatiens*, Table 4 indicates "R" (RefSeq) for gene symbol and "O" (OGS) for OrthoDB.
7. We will first take steps to retrieve the gene symbol. We will need to use a database cross reference relationship to retrieve the RefSeq ID for the gene, so the next query building block to add is the database cross reference ID. Use the scroll bar to the right of the Model Browser to scroll down until you see "Db Cross References x Ref." Click the "+" sign next to "Db Cross References" to show subclasses. Look down a couple lines for "Cross Reference Gene" and click the "+" sign to see its attributes. Click SHOW next to "DB identifier" to output the database cross reference identifier. Notice that several lines have been added to the Query Viewer, and these are indented to indicate that this information is a subclass of "Gene." The line with "Db Cross References x Ref collection" was added because you selected an attribute that is part of the collection of attributes for the relationship between the Db Cross Reference dataset and the Gene dataset. More specifically, you selected the attribute "DB identifier" of the "Cross Reference Gene."
8. The next query building block is to add the gene symbol for the database cross reference as output. You will notice that the Model Browser view has automatically been reset to the top of the tree. In order to output the gene symbol for the database cross reference gene, rather than the original gene, you must be sure to navigate back to the correct part of the tree. To easily do so, click the word "Gene" within "Cross Reference Gene" in the Query Overview. This causes the Model browser to adjust so that "Cross Reference Gene" and its attributes show in the central area of the view (Fig. 5B). Click SHOW next to "Symbol" to add it as output to the query.
9. The objective of the next few steps is to add the homologous *A. mellifera* gene (or genes) to the output. Remember that Table 4 indicated that the *B. impatiens* OGS ID is required to retrieve relationships to OrthoDB data. Therefore you will go back to the original "Gene" as a starting point when looking at the Model Browser, because you have already constrained the gene based on an OGS ID.
10. Use the Model Browser scroll bar to scroll down. You should click the "-" sign next to "Db Cross References x Ref" to close

that part of the tree to avoid confusion. You will see "Homologues Homologue" a few lines down. Click the plus sign to see the subclasses. Many data classes include a subclass that refers back to the parent data class (in this case the initial "Gene"). These are recognized with a red arrow pointing up. Under Homologues, the first subclass listed is "Gene" with the red up arrow (Fig. 5C). This is not the homologous gene, but is a reference to the parent "Gene" class. Do not select this. Instead, look further down the list of subclasses until you see "Homologue Gene." This is where you find attributes for the homologous genes. Click the "+" sign next to "Homologue Gene."

11. Under "Homologue Gene" select SHOW next to "DB identifier" so the gene ID of the homologue will be included in the output (Fig. 5C). Notice that several lines have been added to the Query Viewer. The line with "Homologues Homologue Collection" was added because you selected an attribute that is part of the collection of attributes for homologue relationships to genes. More specifically, you selected the attribute "DB identifier" that is part of the "Homologue Gene," i.e., the homologous gene itself.
12. The next building block to add is a constraint specifying that you want only *A. mellifera* homologues. Click the word "Gene" within "Homologue Gene" in the Query Viewer to jump back to the appropriate part of the Model Browser tree.
13. Scroll down using the Model Browser scroll bar to find the word "Organism," keeping an eye on the inner line on the left side to ensure that you stay next to the line that descends from "Homologue Gene" (i.e., make sure that you stay in the "Homologue Gene" subtree).
14. Click the "+" next to "Organism" to show its attributes.
15. Click CONSTRAIN next to "Short Name," and then use the pull-down menu to select "A. mellifera" and click "Add to Query." The homologue organism constraint will show up in the Query Viewer.
16. The next step is to determine which kind of identifier is used for *A. mellifera* homologues, and also which kind of identifier is needed for *A. mellifera* gene symbols and pathways. Notice in Table 4 that for *A. mellifera* there is an "O" (OGS) in the OrthoDB cell, and "R" (RefSeq) for both Gene Symbol and KEGG. This means that the *A. mellifera* homologues that will be retrieved will have OGS IDs, so a database cross reference is needed to connect the OGS IDs to the RefSeq IDs before KEGG pathways can be retrieved.

17. To add the building block for the database cross reference ID of the *A. mellifera* homologue, click "Gene" in "Homologue Gene" to return to the correct subtree in the Model Browser.
18. Follow the line that descends from "Homologue Gene" until you find "Db Cross References x Ref" (i.e., make sure that you select the "Db Cross References x Ref" that is a subclass of "Homologue Gene" rather than the one that is a subclass of the root "Gene") (Fig. 5D).
19. Similar to what you did in **step 7**, click the "+" sign next to "Db Cross References" to show subclasses under "Db Cross References." Look down a couple lines for "Cross Reference Gene" and click the "+" sign to see its attributes. Click SHOW next to "DB identifier" to output the database cross reference identifier.
20. The next building block to add is to output the gene symbol for the *A. mellifera* database cross reference (RefSeq) gene. Click the word "Gene" within "Cross Reference Gene" in the Query Viewer, making sure that you are looking at the "Cross Reference Gene" under "Homologue Gene" to jump to the correct subtree of the Model Browser.
21. Under "Cross Reference Gene" in the Model Browser, click SHOW next to "Symbol" to add the gene symbol for the *A. mellifera* RefSeq gene to the output. A line with the word "Symbol" will appear under "DB identifier" under "Cross Reference Gene."
22. The next step is to output pathway information for the database cross reference of the *A. mellifera* homologous gene. To make sure that you select the correct pathway data class, again click the word "Gene" within "Cross Reference Gene" in the Query Viewer, making sure that you are looking at the "Cross Reference Gene" under "Homologue Gene" to jump to the correct subtree of the Model Browser.
23. Scroll down using the Model Browser scroll bar to find the word "Pathways," again keeping an eye on the inner line on the left side to ensure that you stay in the correct subtree.
24. Click the "+" next to "Pathways" to open the subtree.
25. Click SHOW next to "Identifier" to include the pathway identifier in the output. Notice that several more lines have been added to the Query Viewer, because you have added information from another data collection.
26. The last building block to add to the query is to output the name of the pathway. To navigate back to the correct Pathway subtree in the Model Browser, click the word "Pathway" in the line "Pathways Pathway Collection" in the query viewer.
27. Under "Pathways," click SHOW next to "Name" to output the name of the homologue pathway.

28. At this point, query construction is complete (Fig. 5E). Scroll down to the "Fields Selected for Output" area of the page. You will see blocks representing the output columns (Fig. 5F). The blocks appear in the order you added them to the query, not necessarily the order they are shown in the Query Overview. You may rearrange the columns by dragging the blocks.
29. If you are not logged in to MyMine, and you wish to save this query, you can click "Export XML" to save it locally. You will be able to import it in a future HymenopteraMine session.
30. You can either click "Show Results" at this point to run the query, or if you are logged into MyMine, you can click "Start building a template query" to create a template query saved in your MyMine Template collection. The only additional steps required to create the template are to name the template and click "Save Template." Before saving, you also have the options of providing a title, description and comments. You will have the opportunity to run the query after you complete the template.
31. Whether or not you have created a template, if you are logged into MyMine the query will automatically be saved in your query history after you run it. From your query history list in MyMine, you can rerun the query or edit it. However, the query history is temporary, and its preservation across sessions is not guaranteed. Therefore, it is advisable to name your query and click "Save query" at the bottom of the QueryBuilder page before you end the session to save the query in your MyMine Queries list. If you have already run the query before saving, simply find it in your query history to save it.

2.4.6 Regions Search

The Genomic Region search tool allows you to perform a coordinate based search for genomic features. The list of available features depends on the species selected. Gene, mRNA, coding regions (CDS), exons, and polypeptides are available for all species. Some species include additional features, such as miRNA, tRNA, indels and SNPs. You perform a search by selecting desired output features, providing a list of regions, and optionally entering a distance in bases to extend the regions. You can either paste lists of locations that include the scaffold or chromosome identifier and the start and end coordinates, or upload the locations as a text file. The search result page provides options to download data for individual regions or all regions at once in tab, csv, gff3, fasta, or bed formats. A pull-down menu allows selection of a feature type for creating a list that you can further augment using template queries or "Manage Columns" (described in Subheading 2.4.8). The example in Subheading 2.4.11 includes a Regions search.

2.4.7 Enrichment Widgets

After saving or viewing a list of genes you will automatically be presented with widgets showing results of enrichment analyses for gene ontology terms, pathways and publications. Each widget provides options for test correction (Holm-Bonferroni, Benjamini Hochberg, and Bonferroni) and *p*-value cutoff. The default background population is all genes in the organism that have annotations of the type being calculated. Therefore, since HymenopteraMine contains multiple gene sets with annotations per organism, it is recommended that you do not use the default background population. Rather, you can click "Change" to select a background population from among your saved lists. Premade lists of gene IDs for each organism's gene sets are provided. The List Tool makes it easy to create more refined background populations for specific questions. For example, you can create lists of all expressed genes to use as the background to test for enrichment in differentially expressed genes. The example in Subheading 2.4.11 includes the Gene Ontology Enrichment Widget.

2.4.8 Manipulating Outputs and Augmenting Queries Through Column Management

HymenopteraMine table outputs can be manipulated in many ways, from minor changes like sorting rows or deleting columns, to extensive changes, like adding new columns and filters. The example in Subheading 2.4.11 includes column management.

Rearranging columns: You can rearrange column order in two ways. First, you can reorder the columns when building a query with QueryBuilder, as described above. Second, you can use the "Manage Columns" button appearing above the output table. With "Manage Columns," you can use up and down arrows next to the column list to rearrange the order. The red circle next to each column name lets you delete the column.

Using column headers: Each column header in a table contains symbols for table management. Arrows are for sorting in descending or ascending order; the "X" is for column deletion, the "…" symbol is to hide a column in order to make other columns more visible; the funnel-shaped symbol is to filter the rows based on values; the histogram symbol is to show counts of individual values, and to provide an alternate filter interface. An example use of the filter in the histogram interface is to determine which gene source has the largest number of output genes, and apply a filter to select only that gene source. This assumes that gene source was included as output in the original query. If it was not include in the original query, it can be added using Manage Columns as described below.

Manage columns to alter query output: The Manage Columns interface provides yet another mechanism to build a query. It does not allow the incorporation of new query constraints, but allows the addition of new output columns. The starting point for building a query with Manage Columns is any table; the table may have originated from the List Tool, a Regions Search, a table within a report page, or a query output. You may find that using a simple template query followed by Manage Columns is easier than using the QueryBuilder to construct a complex query. Although you cannot add new query constraints with Manage Columns, you can use the column filters described above or "Manage Filters" described below to further constrain the final output. Upon clicking "Manage Columns," the "Selected Columns" interface automatically appears, listing the columns; the "+ Add a Column" button opens an interactive data model tree organized hierarchically, similar to the tree in the Model Browser. The tree is automatically rooted with the appropriate data class. Tag symbols within the tree delineate columns that you can add to the table. Plus symbols indicate data subclasses that can be opened to reveal additional tags or subclasses. You select a column to add as output by clicking to highlight in blue any attribute with a tag symbol. Once you have selected the desired columns, clicking "Apply Changes" brings you back to the column list, where you can use the arrows to rearrange the column order. Clicking "Apply Changes" in this window provides the modified table.

Manage filters to alter query constraints: Manage Filters provides an interface complementary to the Manage Columns interface. While Manage Columns can be used to alter query outputs, Manage Filters can be used to alter query constraints. Although the word "filter" is used in this interface, you can relate each filter to a constraint from the original query. After clicking Manage Filters, an interface is provided for editing current filters or adding new filters. For example, if the original query was constrained using a particular gene identifier, clicking on the corresponding filter in Manage Filters allows you to change that identifier. To create a new filter, click "Define a New Filter." An interactive data model tree, similar to the one described under "Manage Columns" will appear. From the tree you can select attributes to use as filters. An important difference between adding a filter with the Manage Filters interface versus incorporating the constraint in the original query is that multiple values for the selected attribute must exist in the current table in order for that attribute to be selected as a filter. If the current table has only one value for a particular attribute, you will receive a message "There is only one possible value… You might want to remove this constraint."

Sorting with manage columns: In addition to the simple ascending or descending sort you can perform on a single column using the sort symbol in the header, you can create a prioritized list of sort attributes using the Sort Order tab in Manage Columns. In the Sort Order interface, the primary output column on which you can sort is listed on the left side. The panel on the right lists all data classes that are available as sort criteria, and can be selected by clicking the plus sign. In addition to the existing columns, all attributes directly connected to each data class can be used as sort criterion without requiring the attribute itself to be present in the table. For example, gene IDs can be sorted based on gene length without adding a gene length column to the output.

Manage relationships: The default result for a complex query is to show only rows for which values exist in all related data classes. For example, if you wish to generate a table containing *A. mellifera* gene IDs, *D. melanogaster* homologues to the *A. mellifera* genes and pathways for the *D. melanogaster* homologues, the default output will not include any *A. mellifera* gene for which there is a *D. melanogaster* homologue but no pathway. You can use "Manage Relationships" to change the behavior such that all *A. mellifera* genes that have *D. melanogaster* homologues are shown regardless of whether pathway information exists. The Manage Relationships interface lists all data class relationships present in your original query, with buttons to switch any of them from "Required" to "Optional." Selecting "Optional" not only causes rows lacking results to be included in the output, but also modifies the format for the affected column, such that the output is shown as nested sub-tables within the main results set. However, when the table is exported as tab or comma separated values, the format reverts to the inline layout.

2.4.9 Exporting Query Outputs and Queries

The "Export" button at the top of each table provides many download options. Tables can be downloaded as tab separated values (TSV), comma separated values (CSV), JSON or XML; sequences associated with identifiers in the tables can be downloaded in fasta format; coordinate information associated with the identifiers can be downloaded in GFF3 or BED format.

The Export form includes a text box to enter a file name. To the right of the text box the default .tsv file extension opens a pull-down menu allowing you to modify the file type. The tabs listed on the left side of the Export form depend on the file type selected, and allow you to modify default settings. For TSV, CSV, JSON, and XML, the "All Columns" tab allows you to select which columns to download, the "All Rows" tab allows you to decrease the number of rows for export. For TSV and CSV, the "No Column Headers" tab allows you to add column headers. By default, none of the file formats are compressed, but the "No Compression" tab allows you to set the option to gzip or zip compression. A preview of the first three rows is provided for TSV, CSV, JSON, and XML formats.

The queries themselves can be downloaded as XML, allowing them to be shared with other users or imported back into your private MyMine account. An advantage of saving your query after you are satisfied with the table content and arrangement is that manipulations on the table after running a query, such as sorting or filtering, will be maintained in the XML code.

2.4.10 Gene Aliases and Database Cross References in HymenopteraMine

A HymenopteraMine feature that is particularly important for the species in HGD is the maintenance of gene and transcript alias identifiers and database cross reference identifiers. Table 3 lists the data sources of alias identifiers and database cross references, and Table 4 indicates which type of identifier is used for each data set.

We define a "Database cross reference" as an identifier for an equivalent gene locus in an alternative primary gene set. For example, for most HymenopteraMine organisms, the consortium OGS (with HGD identifiers) contains gene identifiers that cross reference gene identifiers in the RefSeq gene set and vice versa. Any database cross reference ID is also the primary database ID of another primary gene set, and therefore may be connected to other data classes, such as homologues and pathways (Table 4). We use database cross references for gene sets that have been generated using different methods, and sometimes using different assembly releases. As a result, the cross-referencing gene sets are not equivalent to each other, so any gene in one gene set may not have a database cross reference in another gene set, and some genes may have multiple cross references due to disagreement in gene boundaries. Rather than eliminating database cross references for genes with ambiguous relationships, we keep the full list of cross references for all genes so you will be aware when there is a disagreement between gene sets (e.g., you will see a gene in one gene set that maps to two different genes in the other). Several species have database cross references between the OGS and RefSeq gene sets. For *Nasonia vitripennis*, there are cross references between the original published consortium OGS (nvit_OGSv1.2) [29], the Evidential Gene Set [35], and RefSeq.

When working with a primary gene set identifier, you may need to use a database cross reference in order to retrieve a specific data set. For example, to retrieve KEGG pathways for *A. mellifera* genes, you would need RefSeq gene IDs rather than OGS gene IDs. Table 4, also available via the HymenopteraMine Data Model tab, indicates the type of identifier required to retrieve information from each data set for each species. If necessary, you can convert identifiers using the Gene ID → Database Cross Reference ID template query.

An alias ID is an alternative identifier for a gene or transcript, and is not connected directly to any data class other than the primary gene. Alias identifiers occur due to (1) HGD identifiers being assigned to an OGS, (2) the existence of an upgraded OGS (e.g., *A. mellifera* OGSv1 and OGSv3.2), or (3) an alternative identifier used by OrthoDB. The assignment of HGD identifiers is the major source of aliases. We assign gene and transcript identifiers to official gene sets provided by consortia because many of the original gene sets provide only transcript identifiers that do not indicate which transcripts belong to the same gene locus. The OGS are loaded into HymenopteraMine using the HGD ID as the primary ID and the consortium ID as the alias. We also assign aliases between old and new consortium gene sets in a similar way that we compute database cross references. Finally, we provide aliases when the OrthoDB website displays an alternate identifier for a species, so that you can use the OrthoDB website in downstream analyses. Table 3 indicates the names of alias identifier sources for each species, and also indicates whether the alias identifiers are assigned to transcripts or genes. Whether an alias is associated with a transcript or gene determines which type of template query to use for conversion to primary IDs. Each template query that involves an alias includes a description that indicates which species it is suited for.

If you are unsure whether an identifier is an alias or an ID from a primary gene set, you can quickly check the identifier using the List Tool. Use the pull-down menu to select "Gene" or "Alias Name." If the identifier is verified in the database when "Gene" is selected, it is an identifier from a primary gene set. If it is verified when "Alias Name" is selected, it is an alias. If the identifier is not found with either choice, it is not a gene or transcript identifier in HymenopteraMine. Once you have determined that your identifier is an alias, you can use a template to convert the alias to a primary identifier.

2.4.11 Step-By-Step Meta-Analysis Example Including Regions Search, List Tool, Enrichment Widget, Template Query, and Column Management

This example will combine two studies on honey bee scouting behavior. The first study used microarray analysis to identify genes that were differentially expressed between foragers that scout for food sources and foragers called recruits, which do not scout for food sources [35]. The second study used genome-wide association analysis (GWAS) to identify genome variants associated with behavioral differences between scouts and recruits [36]. We would like to identify differentially expressed genes from the first study within 50 kb of single nucleotide polymorphisms (SNPs) identified in the second study.

1. The first step is to create a list of SNP coordinates to use in a regions search. You have the choice of following instructions for downloading the data from the original journal source and using Excel or OpenOffice-Calc to format them or using preformatted SNP coordinates provided in Table 5 of this chapter.

   Method using original journal source plus Excel: For this example we will use the supplemental data file from the GWAS study [36]. Go to the publicly available journal article webpage (http://journals.plos.org/plosone/article?id=10.1371/journal.pone.0146430). Scroll down to the supporting information section. Click "S2 Table." After downloading the file, change the file extension to ".txt". Open the file using Excel/OOCalc. Although the file was originally labeled "CSV," check both tab

Table 5
SNP locations from [36] for use in the example described in Subheading 2.4.11

Group1:2097573.0.2097573	Group6:16625187.0.16625187	Group12:2253248.0.2253248
Group1:2097586.0.2097586	Group7:9560908.0.9560908	Group12:2253284.0.2253284
Group1:2179980.0.2179980	Group8:12005663.0.12005663	Group12:2253301.0.2253301
Group1:7106109.0.7106109	Group9:4087742.0.4087742	Group12:2253308.0.2253308
Group1:9605960.0.9605960	Group9:7003402.0.7003402	Group12:2253502.0.2253502
Group1:9612560.0.9612560	Group9:11035994.0.11035994	Group12:2253524.0.2253524
Group1:9976934.0.9976934	Group10:5450230.0.5450230	Group12:2253528.0.2253528
Group2:3079339.0.3079339	Group10:8609127.0.8609127	Group12:2253589.0.2253589
Group2:7060217.0.7060217	Group10:8624631.0.8624631	Group12:2253812.0.2253812
Group2:7064150.0.7064150	Group12:1118838.0.1118838	Group12:2253824.0.2253824
Group2:8478663.0.8478663	Group12:1137567.0.1137567	Group12:2278672.0.2278672
Group2:9476451.0.9476451	Group12:1219747.0.1219747	Group12:2282983.0.2282983
Group2:14317970.0.14317970	Group12:1294097.0.1294097	Group12:2302941.0.2302941
Group4:5736174.0.5736174	Group12:1325665.0.1325665	Group12:2383470.0.2383470
Group4:5812780.0.5812780	Group12:1421540.0.1421540	Group12:2425407.0.2425407
Group4:5816424.0.5816424	Group12:1473174.0.1473174	Group12:2484094.0.2484094
Group4:5823271.0.5823271	Group12:1473195.0.1473195	Group12:2618272.0.2618272
Group4:6418141.0.6418141	Group12:1640082.0.1640082	Group12:2872625.0.2872625
Group4:6472951.0.6472951	Group12:1744256.0.1744256	Group12:2872631.0.2872631
Group5:471817.0.471817	Group12:1770057.0.1770057	Group12:2996797.0.2996797
Group5:471818.0.471818	Group12:1805764.0.1805764	Group12:3535567.0.3535567
Group5:10104581.0.10104581	Group12:1814177.0.1814177	Group12:3535855.0.3535855
Group5:10232114.0.10232114	Group12:1814180.0.1814180	Group12:3540115.0.3540115
Group5:11006033.0.11006033	Group12:1820238.0.1820238	Group12:3607075.0.3607075
Group5:11012508.0.11012508	Group12:1842196.0.1842196	Group12:3614603.0.3614603
Group5:11012515.0.11012515	Group12:1842255.0.1842255	Group12:3638525.0.3638525
Group5:11012516.0.11012516	Group12:1842386.0.1842386	Group12:3682798.0.3682798
Group5:11012538.0.11012538	Group12:1842411.0.1842411	Group12:3845071.0.3845071
Group5:11048772.0.11048772	Group12:1842525.0.1842525	Group12:4531251.0.4531251
Group5:11062231.0.11062231	Group12:1859484.0.1859484	Group12:4636018.0.4636018
Group5:11063028.0.11063028	Group12:1859516.0.1859516	Group12:5139164.0.5139164
Group5:11095696.0.11095696	Group12:1882625.0.1882625	Group12:5500706.0.5500706

(continued)

Table 5
(continued)

Group5:11104858.0.11104858	Group12:1916728.0.1916728	Group12:9498320.0.9498320
Group5:11134225.0.11134225	Group12:2160939.0.2160939	Group13:9587246.0.9587246
Group5:11286749.0.11286749	Group12:2195252.0.2195252	Group15:6478761.0.6478761
Group6:1062671.0.1062671	Group12:2196250.0.2196250	Group16:1670311.0.1670311
Group6:4544720.0.4544720	Group12:2206117.0.2206117	Group16:1670313.0.1670313
Group6:12559697.0.12559697	Group12:2215036.0.2215036	Group16:6118876.0.6118876

and space as column delimiter. Like many supplemental datasets, this file is not optimally formatted for use in meta-analyses. The first step is to delete all unnecessary columns. We need only the chromosome identifier and SNP coordinate, so delete all columns except A and C. Now delete any extra rows, including the first two rows and the rows at the end with "Unplaced" scaffolds. We will ignore any scaffold that was not assigned to a chromosome. We need to create a list of chromosome locations using the original *A. mellifera* genome assembly "Group" identifiers, rather than identifiers like "LG1." Use the Find and Replace Function to replace "LG" with "Group." The next step is to format the chromosome location for pasting into the HymenopteraMine Regions text input box, which requires both a start and an end coordinate. For SNPs, you use the same coordinate for each. So in your Excel/OOCalc spreadsheet, copy the column of coordinates to the next column. Now you should have three columns: chromosome identifier, start, and end.

Alternative method: If you do not have access to the publication or Excel/OOCalc, you should follow the alternative instructions in **step 2**d below to use identifiers provided in Table 5 of this chapter.

2. Once you have properly formatted SNP locations, the next step is to perform a Regions Search to obtain genes within 50 kb of the SNP coordinates. Go to HymenopteraMine. It is recommended that you login to MyMine so you can save your work.

   (a) Click the Regions Tab in the HymenopteraMine navigation bar.

   (b) Select "A. mellifera" from the "1. Select Organism" pull-down menu.

   (c) Click the square next to "2. Select Feature Types" to uncheck all options and then click the box next to "gene" as the chosen feature option.

   (d) If you created a spreadsheet with 3 columns, highlight all columns simultaneously and copy them into the text box.

The columns entered will automatically be tab-delimited. If you choose to use the locations provided in Table 5 of this chapter, be aware that they are formatted differently than what was described for the Excel/OOCalc spreadsheet. Each cell in Table 5 contains an entire SNP location, with chromosome ID, start and end coordinate. Table 5 has three columns to save space within the chapter, but the columns must not be selected simultaneously. Highlight each column and paste it, one-at-a time, into the Regions search text box, so that you are always extending the list when you add the next column.

(e) Type "50 kb" into the text box of "4. Extend your regions at both sides."

(f) Click "Search" to run the Regions Search tool (Fig. 6A).

(g) After the Region Search is successfully run you are presented with an output page listing each of the regions and the numbers of genes identified within the regions (Fig. 6B). To save a list of all the genes, use the "Create List by feature type" button above the output after selecting "Gene" in the pull-down menu and click "Go." This action creates a new list and produces a List Analysis page.

(h) Click on the histogram in the "Gene Source" column header to see the column summary (Fig. 6C). You will notice that the list contains two gene sources, RefSeq and the *A. mellifera* official gene set (amel_OGSv3.2). Within the column summary pop-up box, filter the list by checking "amel_OGSv3.2", clicking the blue filter box, and selecting "Restrict table to matching rows." Now save the list of amel_OGSv3.2 genes using "Save as List" above the table, clicking "Gene (333 Genes)" and in the pop-up menu. Naming the list "OGSv3.2 genes within 50 kb of SNPs for scouting," and clicking "Create List." We will use the OGS gene list for the remainder of this example, but if you wish to also save the RefSeq genes, you could now click "Undo" above the table to remove the filter, then filter using the Gene Source column summary to restrict table to rows matching "Amel_RefSeq".

(i) While in the process of saving the lists, you may have noticed enrichment widgets appearing below the table. At this point it is not advisable to rely on these enrichments, because they were performed on the original list containing identifiers from two gene sets. The enrichment widgets do not alter the original list even after you filter it.

(j) To see a valid enrichment, click the Lists tab in the navigation bar. If you are presented with the Lists Upload page rather than the View page, click "View" in the brown bar

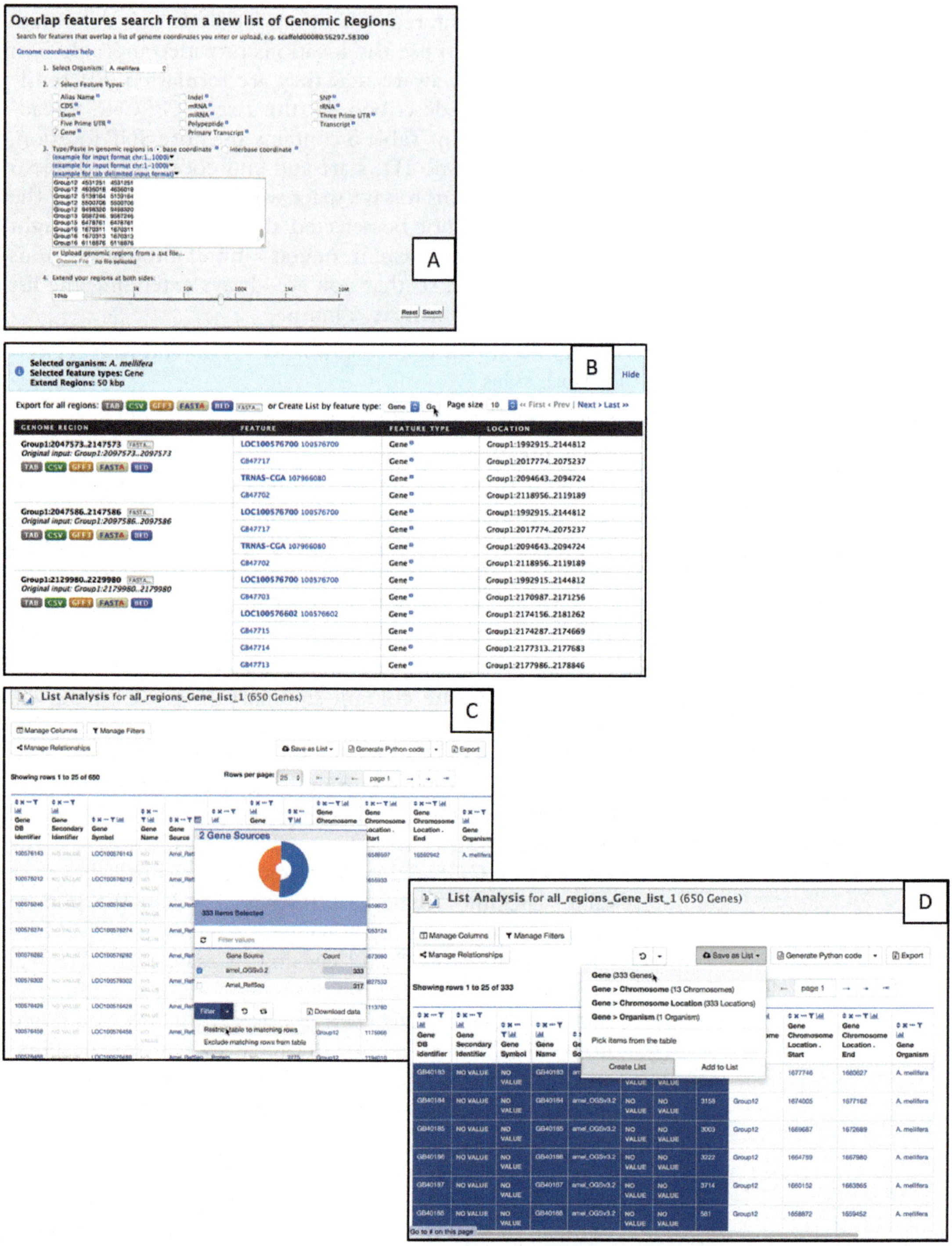

Fig. 6 (A) The Regions search interface takes chromosome coordinate input lists in several formats. (B) The output of a Regions search consists of pages showing results for each region. The "Create List by feature type" function allows you to create a list of identifiers for further use in HymenopteraMine. (C) After creating a list, the list is presented in a List Analysis page. Since this is a list of all *A. mellifera* genes within the searched regions, it includes genes from two gene sets. The table can be filtered for one gene set using the column summary tool in the column header. (D) A new list can be created from the filtered table

below the HymenopteraMine navigation bar. Once on the list view page, click your list name, "OGSv3.2 genes within 50 kb of SNPs for scouting." The Gene Ontology enrichment shows some significantly enriched GO terms; however this analysis is still not correct, because the default background population set is all genes in a species annotated with the appropriate data type. HymenopteraMine has two gene sets for *A. mellifera* annotated with GO terms, so two gene sets were used as the background population. To correct this, click "Change" below "Background Population." The resulting pull-down menu shows all of your lists as well as some premade lists. For this analysis, select "A. mellifera all amel_OGSv3.2 Genes (15314)" (Fig. 7).

3. To continue with the goal of identifying differentially expressed (DE) genes [35] within 50 kb of the SNPs associated with scouting behavior, the next step is to create a DE gene list. Go to the supporting online material webpage (http://science.sciencemag.org/content/suppl/2012/03/07/335.6073.1225.DC1?_ga=1.205106516.512133959.1337265718) and download Table S3B (http://science.sciencemag.org/highwire/filestream/593583/field_highwire_adjunct_files/0/1213962_SuppTable_S3B.xlsx). You will use the gene IDs listed in column E.

   (a) In HymenopteraMine, click the Lists tab in the navigation bar. If you are presented with a view of your lists rather than the Upload page, click "Upload" in the brown bar just below the navigation bar.

   (b) By reading the publication [35], we know that the gene IDs used in this study were from the old *A. mellifera* gene set (amel_OGSv1.0). Therefore, we need to create a list of alias identifiers and later we will convert them to amel_OGSv3.2 gene IDs. Choose "Alias Name" from the "Select Type" pull-down menu and "A. mellifera" from the "for Organism" menu (Fig. 8A).

   (c) Paste all the IDs from Column E of the publication supplemental table into the text box. There is no need to be concerned with the 260 non-amel_OGSv1.0 identifiers that are also found in this column.

   (d) Click "Create List." HymenopteraMine performs a lookup and removes any identifiers that are not found. Notice that 775 Alias Names of the 959 amel_OGSv1.0 identifiers entered are found in HymenopteraMine. The missing 260 identifiers are from other species, such as *Drosophila melanogaster*, or OGSv1 genes that did not map to a gene in the updated *A. mellifera* gene set (Fig. 8B).

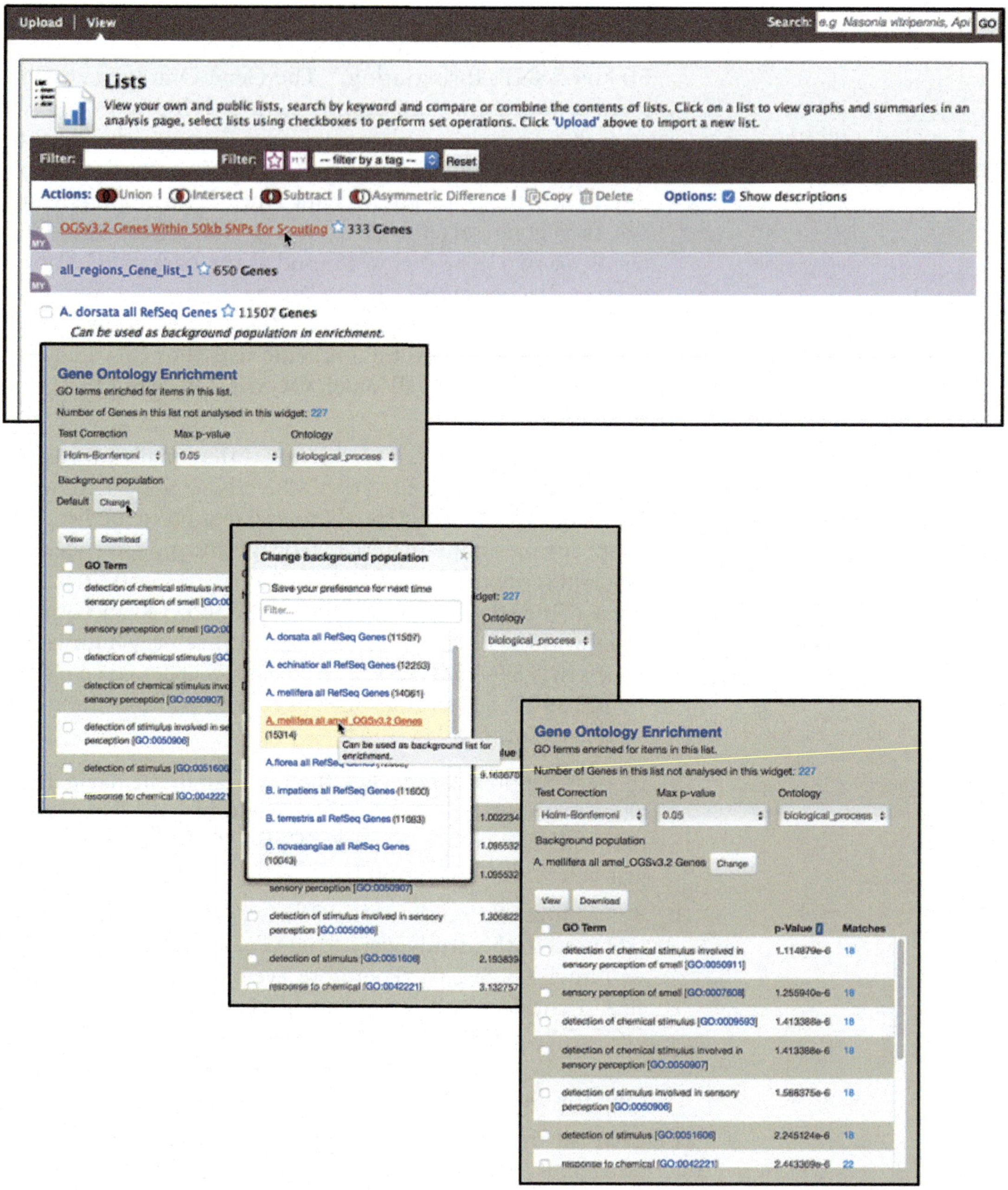

Fig. 7 Enrichment Widgets appear when you view any list of gene identifiers. However, the default background population gene list is usually not the appropriate dataset for the analysis. The background population can be changed and replaced with one of the premade gene lists for each species, or you can use the Lists tool to create a custom background population gene list for your study

(e) Enter a name for the list such as "Alias OGSv1 DE Genes Scout vs Recruit" and click the green "Save a list of 775 AliasNames" button. You are taken to a page showing the list you created. Notice that there are no enrichment widgets as this is not a Gene list (Fig. 8C).

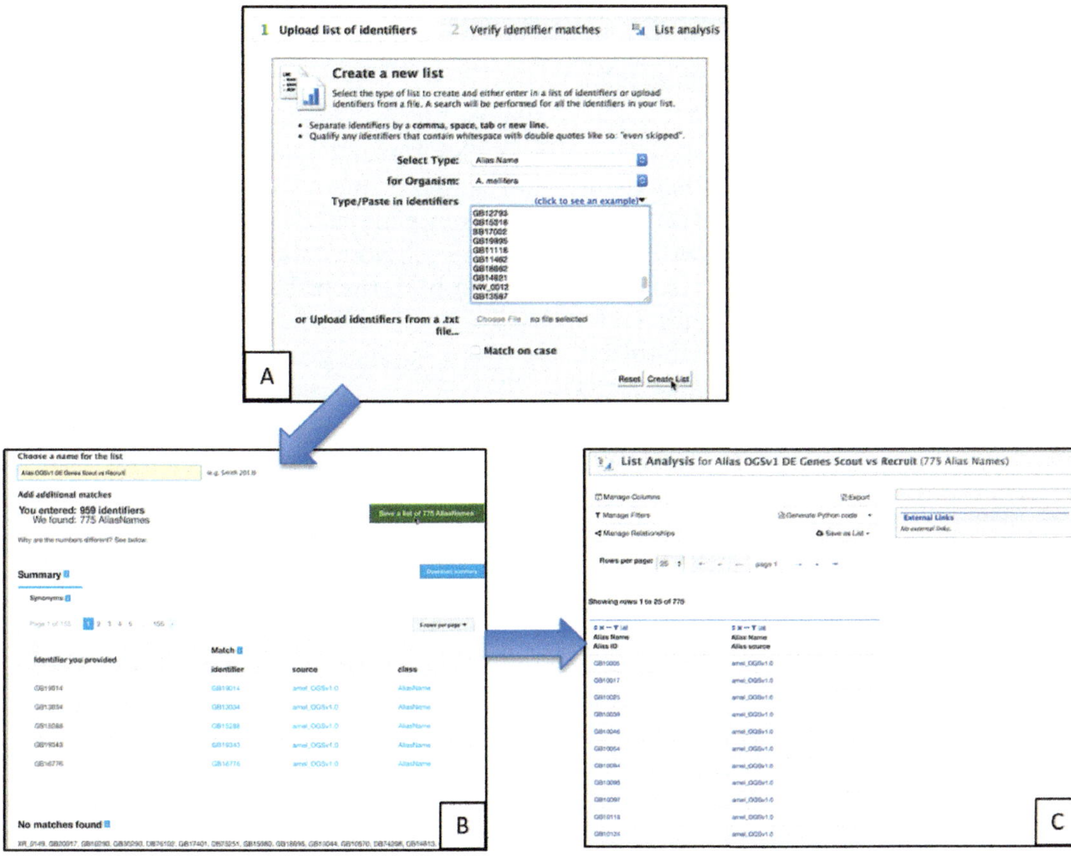

Fig. 8 (A) The Lists Upload page accepts any identifier, but the identifier will not be validated in the HymenopteraMine lookup step unless the correct data class is selected. Here "Alias Name" is selected, because the A. mellifera identifiers in the entered list are from the old (amel_OGSv1) gene set. (B) After performing a lookup, some identifiers are eliminated from the list because they are not found in HymenopteraMine. (C) Clicking "Save a list of 775 AliasNames" in the previous figure leads to this list of alias identifiers. Enrichment widgets are not provided since this is not a list of primary gene set identifiers

4. The next step is to convert your alias ID list to a gene ID list using a template query.
   (a) Click the Home tab in the navigation bar, and then click the ALIAS AND DBXREF tab in the template categories bar, halfway down the page.
   (b) Click the template "Alias ID → Gene ID", or if you cannot see that template listed, click "More Queries" and then click the template name. In the template query pop-up menu, you change the default alias ID to your list of IDs by checking "constrain to be," making sure that "IN" is selected, and then selecting your list name "Alias OGSv1 DE Genes Scout vs Recruit." Make sure that *A. mellifera* is selected under "Organism > Short Name". Toggle on the optional constraint to specify a gene source, otherwise

both RefSeq and amel_OGSv3.2 gene IDs will be included in the output. Make sure that "amel_OGSv3.2" is selected (Fig. 9A).

(c) Click "Show Results." The output is a table with 855 rows showing amel_OGSv1 IDs and amel_OGSv3.2 IDs. Note that there is not always a one-to-one correspondence in IDs due to changes in gene models in the updated gene set. For example, you will notice that the amel_OGSv1 ID GB11731 is listed in both the first and second rows, with cross references to amel_OGSv3.2 genes GB40009 and GB40010, due to the original gene being split in the new gene set (Fig. 9B).

(d) Now you must save a list of amel_OGSv3.2 gene IDs by clicking "Save as List" above the table, clicking "Gene (851 Genes)" and naming the list such as "OGSv3.2 DE Genes Scout vs Recruit". Notice that you will be saving 851 genes, even though the table has 885 rows. This is due to some of the amel_OGSv3.2 gene IDs being listed more than once, due to multiple genes of the old gene set being merged in the new gene set. Finally, click "Create List" (Fig. 9C).

5. The next step is to determine if any of the 851 genes in the list "OGSv3.2 DE Genes Scout vs Recruit" are found in the list "OGSv3.2 genes within 50 kb of SNPs for scouting." Click the Lists tab in the navigation bar. If necessary, switch from the Upload page to the View page by clicking "View" in the brown bar below the navigation bar. Check the boxes next to the lists you made, "OGSv3.2 DE Genes Scout vs Recruit" and "OGSv3.2 genes within 50 kb of SNPs for scouting." Click "Intersect" in the white "Actions" bar above your list of lists. Name the list for the intersection, such as "DE vs genes within 50 kb SNP intersection" and click "Save" (Fig. 10). Now you have a list of 22 DE genes that are located within 50 kb of the SNPs associated with scouting behavior.

Once you have your final list of genes, you would like to get more information about them. You have many options. We will show two approaches: using column manager to add gene descriptions and pathway information and using the list in a template query to retrieve GO terms.

6. The first step is to use the column manager on the table provided on the list analysis page.

(a) If necessary, return to the table by clicking on the name of the list "DE vs genes within 50 kb SNP intersection" shown on the Lists "View" page.

(b) Click "Manage Columns" above the table. First we will remove columns that are not of interest (Fig. 11A). Click

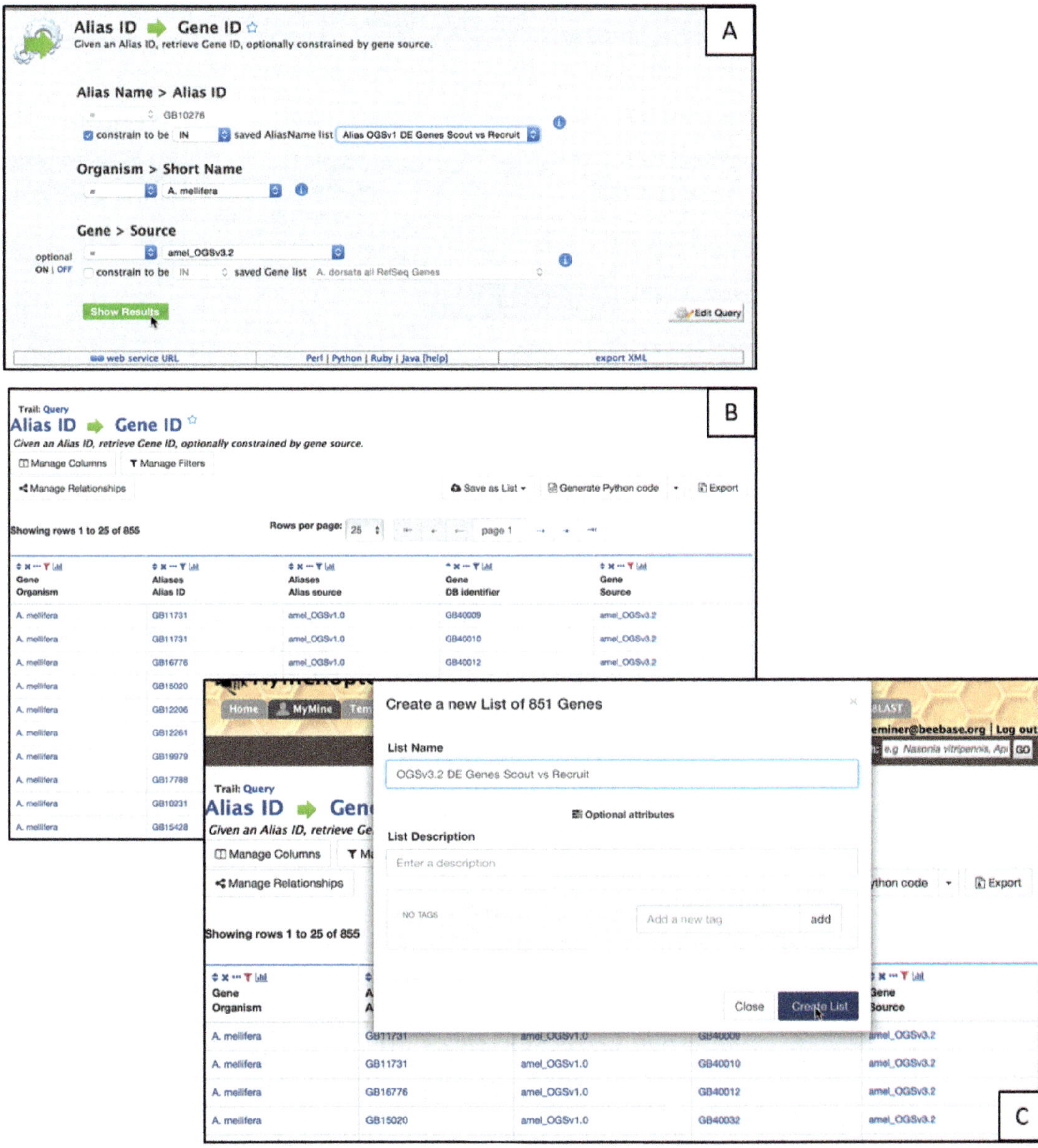

Fig. 9 (A) The list of alias IDs can be used in the "Alias ID → Gene ID" template query to convert the identifiers to primary gene set IDs. The optional constraint for Gene > Source is turned on, and amel_OGSv3.2 is selected so that only IDs from that gene set will be returned. (B) The output of the template query includes amel_OGSv1 IDs and amel_OGSv3,2 IDs. The first two rows demonstrate that there is not always a one-to-one correspondence in IDs due to changes in gene models in the updated gene set, as amel_OGSv1 ID GB11731 is listed in both rows, with cross references to amel_OGSv3.2 genes GB40009 and GB40010, due to the original gene being split in the new gene set. (C) Saving the final list of amel_OGSv3.2 gene IDs by clicking "Save as List" above the table, shows that 851 genes will be saved, even though the table has 885 rows. This is due to multiple genes of the old gene set being merged in the new gene set

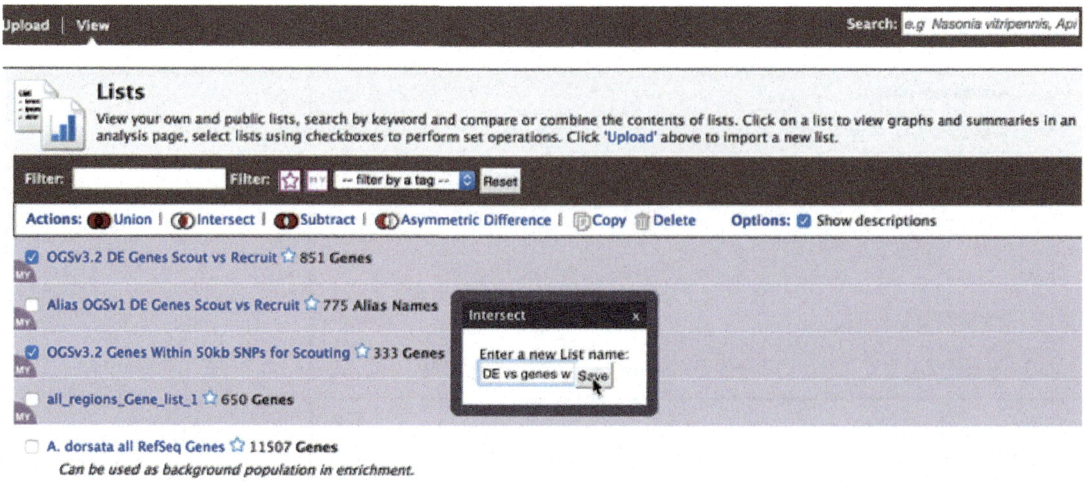

Fig. 10 A List intersection is performed on the Lists View page by selecting each list, clicking "Intersection," and providing a name for the new list

the red circle next to all columns except "Gene ≫ DB identifier". We are not interested in "Gene ≫ Symbol" or "Gene ≫ Name", because OGS gene sources are not assigned symbols or names.

(c) Now we would like to retrieve additional information, including gene symbols, descriptions, and pathways, associated with RefSeq genes (Fig. 11B). Click the green "+ Add a Column" button on the upper right. Scroll down the data model tree to find "Db Cross References" and click the "+" button. Along the way, you may wish to close the "Organism," "Chromosome Location," and "Chromosome" parts of the tree, which were open because their attributes were included in the original table. Within "Db Cross References," open the "Cross Reference" subtree. Under "Cross Reference," select "Symbol," "Description," and "DB identifier." Once selected, each will be highlighted in a blue bar. Scroll further down within the "Db Cross References" subtree and open the sub-subtree for "Pathways." Be sure that the "Pathways" you open is descended from "Db Cross References," rather than the one you could find further down the page directly descended from "Gene." Within the correct "Pathways," select "Identifier" and "Name" so that each is highlighted in a blue bar. Now all of the columns you will be adding are highlighted in blue.

(d) Click the green "Add five new columns" box. You are returned to the column list, which includes your new columns listed in the order of output. If you wish to change the output order, use the arrows next to the column name to move the position (Fig. 11C).

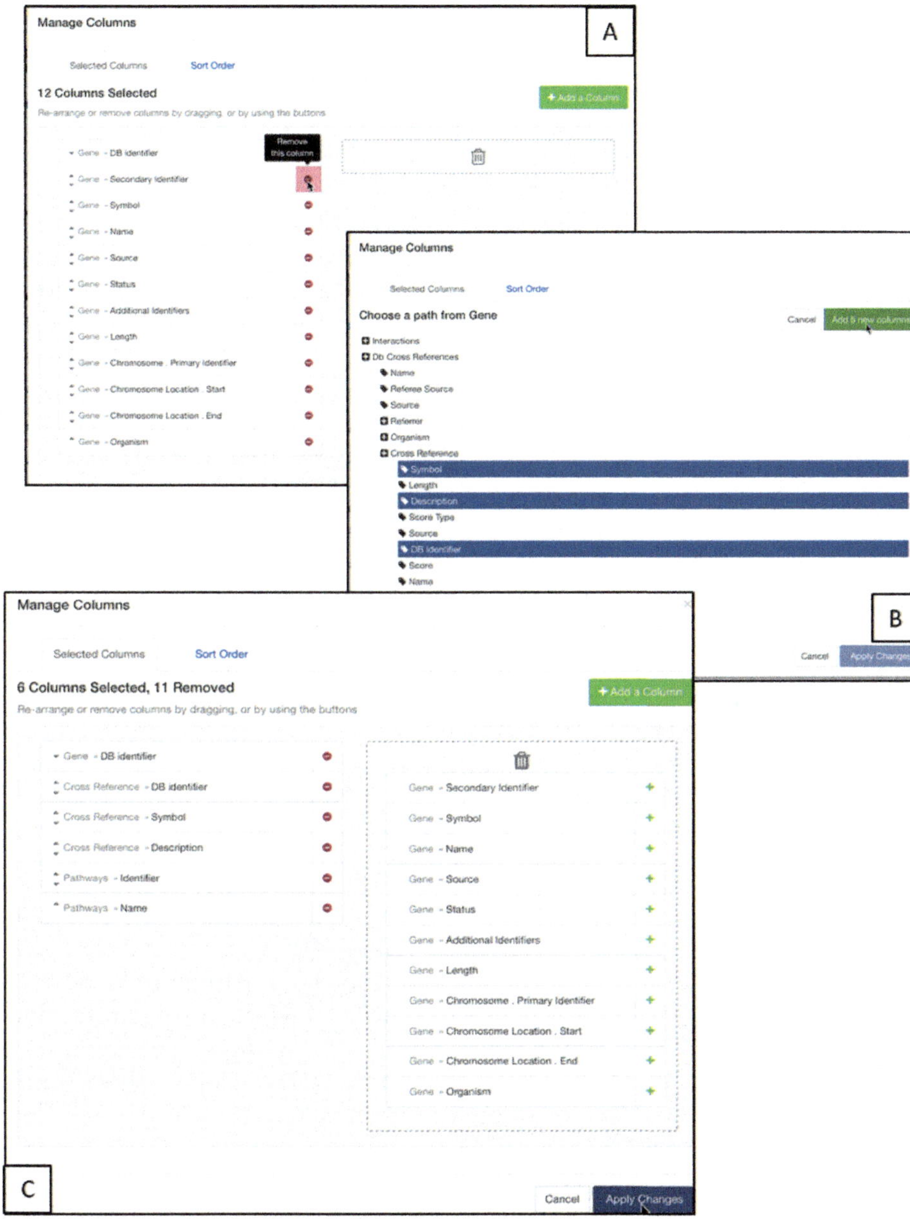

Fig. 11 (A) The "Manage Columns" button above the HymenopteraMine table leads to an interface displaying the columns in order. A column can be deleted by clicking the red circle, and the column position can be modified using the up or down arrows. (B) Clicking "+ Add a Column" in the previous menu leads to a hierarchical display of the data model, similar to the Model Browser in the QueryBuilder, but without the means for adding constraints. A column is selected by clicking the name so it is highlighted in a blue bar. (C) Once clicking "Apply Changes" after selecting columns, the column list is updated to include the new columns. Clicking "Apply Changes" once more leads to the updated table

(e) Click "Apply Changes." You will notice that you have only nine rows of output, even though you started with 22 gene IDs. This is due to the lack of pathway information for 13 genes. You would still like to see symbols and gene descriptions for all genes, so you can modify the default requirement that a relationship must be present in order for output to be provided.

(f) Click the "Manage Relationships" button. Toggle the relationship for "Gene ≫ Db Cross References ≫ Cross Reference Pathways" to be optional, and click "Apply Changes" (Fig. 12A). The table now has a row for each gene. When you make a relationship optional, the table format is modified so that rows contain embedded subtables, each indicating the number of results in that subtable. Clicking on a subtable opens it for viewing. Your output shows that most genes have zero or one pathway annotation. GB52164 (cyclin-dependent kinase 7) and GB52468 (cytochrome b-c1 complex subunit 8) are each annotated with two pathways. Click "2 pathways" in a cell to view the pathway names (Fig. 12B). If you choose to export the table using the "Export" button, and download all columns, the pathway information will no longer be shown as embedded rows. For this table, the exported file has 24 rows, since two genes each have two pathways, and 14 of the rows have no information in the pathway columns.

7. The final task is to use a template query to retrieve gene ontology terms for our intersection gene list. Click "Home" in the navigation bar and click "Function" in the template category bar on the home page.

   (a) Click the template name "Gene → GO Terms" (Fig. 13A). In the template pop-up menu, under "Gene > DB identifier" check "constrain to be," select "IN," and select your intersection list, "DE vs genes within 50 kb SNP intersection." There is no need to select the organism, since these identifiers are unique to *A. mellifera*, but if you are unsure, you could toggle on the "Organism → Short Name" constraint and select "A. mellifera".

   (b) Click "Show Results." The output provides GO identifiers and terms. Clicking on the histogram symbol to show the column summary for Gene DB Identifier shows that 19 of your 22 genes are annotated with GO terms (Fig. 13B). Mousing over either a GO term identifier or name within the table allows you to see a description of the term. The Namespace column indicates whether each term is part of the biological process, molecular function or cellular component ontology, and you can filter for any one of these using the column header tools. The Qualifier column shows whether any term is annotated with a term such as "NOT" or "Contributes to," which would indicate that the given function is not intended to describe the actual

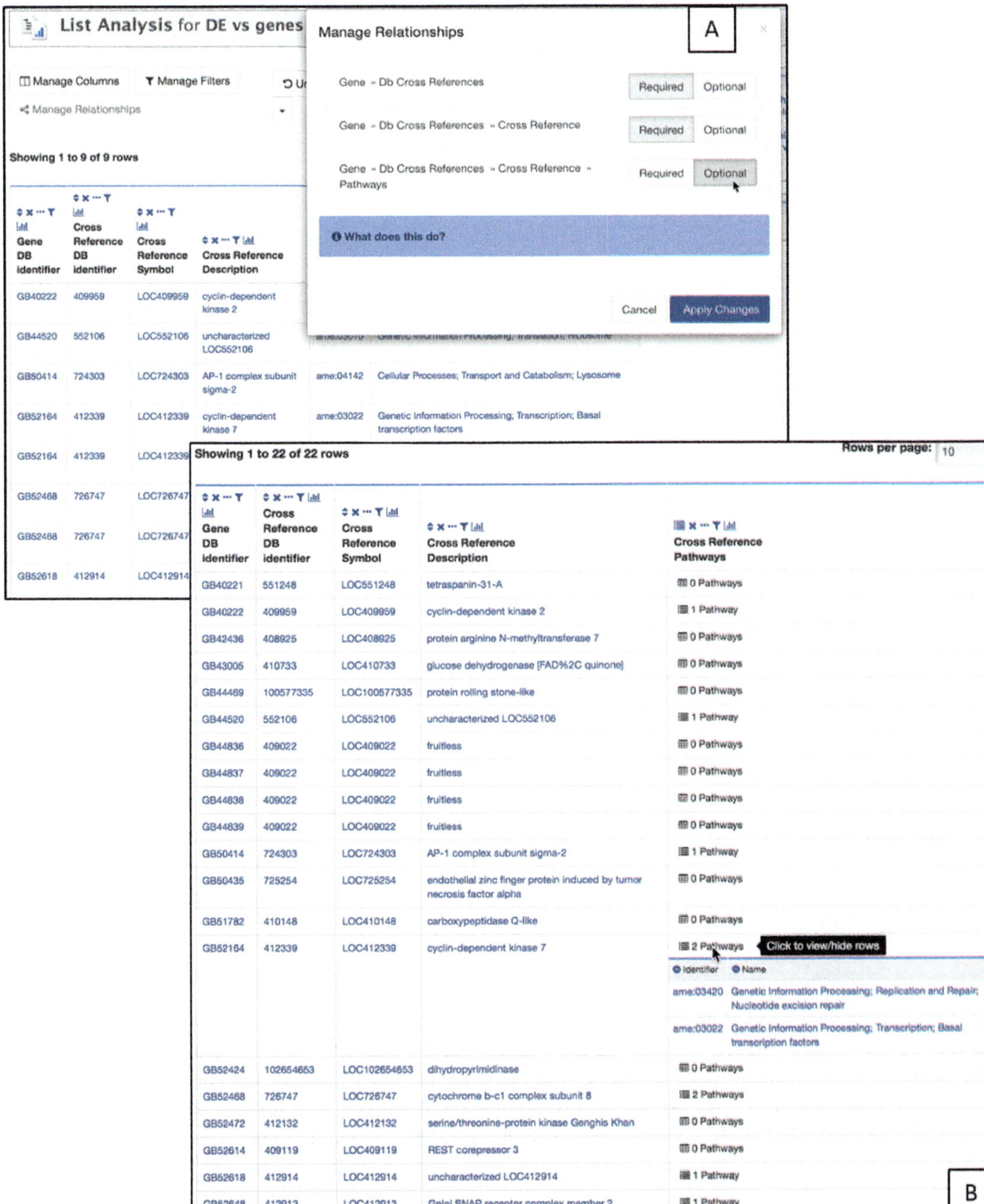

Fig. 12 (A) The Manage Relationships interface allows you to modify the default property that all relationships must be present in order for a row to be present in the output. In our example, we started with 22 genes but after adding new columns (Fig. 11), the number of genes in the table is reduced to 9. This is due to lack of pathway information for 13 genes. We can modify the relationship to remove the requirement for the existence of a pathway. (B) After modifying the pathway relationship, all 22 genes are included in the output table. Making the relationship optional also changes the format of the output, so that the column(s) included in the optional relationship are now shown as embedded subtables that can be clicked on to view. Exporting the table changes the format back to inline with missing values for some rows

Fig. 13 (A) Our gene list can be used in the Gene → GO Term template query. It is not necessary to constrain the organism since the identifiers are unique to *A. mellifera*. (B) The output of the template query shows that 19 of the genes are annotated with GO terms. The faded words "NO VALUE" in the Qualifier column are an important output in this table. "NO VALUE" is the result we are looking for. We may wish to filter the output for Qualifiers other than "NO VALUE" (such as "NOT") so that we interpret the GO Terms correctly

function of the gene. The grey "NO VALUE" result represents the usual case in which the gene is annotated to have the function listed. It is always a good idea to view the column summary in the Qualifier column to see if it contains any information needed to properly interpret a GO annotation, and perhaps filter out rows with "NOT" or keep only rows with "NO VALUE."

3 Notes

3.1 Hymeneptera Mine Release for This Chapter

The most current release of HymenopteraMine can be accessed from the main HGD navigation bar, or with the following URL: http://hymenopteragenome.org/hymenopteramine/. This chapter is based on HymenopteraMine release 1.2, which will be maintained here once it is no longer the current release: http://hymenopteragenome.org/hymenopteramine-release-1.2/. You can perform this example anonymously, but it is advisable to login to your MyMine account so that you can save your work.

3.2 Computing Database Cross References

We compute database cross references for gene identifiers based on overlapping coding exons on the same strand. When alternate assemblies are used for generating the primary gene sets, we first use the UCSC LiftOver tool [37] to map OGS genes to the newest RefSeq genome assembly. We do not attempt to compute database cross references for individual transcripts.

3.3 Computing Aliases

To assign HGD identifiers to consortium gene sets that do not have gene IDs, we identify transcripts originating from the same gene locus based on overlapping coding sequences. Each group of overlapping transcripts is assigned a gene identifier in the form of species code plus a set of digits (e.g., SINV10017). Transcripts are assigned identifiers that include the gene ID plus a suffix in the form of "-RA", where the last letter varies to distinguish transcript isoforms. There is a one-to-one relationship between consortium transcript identifiers and HGD transcript identifiers.

In addition to assigning HGD identifiers, we compute aliases for old and new gene sets of the same source (e.g., aliases between *A. mellifera* OGSv1 and OGSv3.2). In order to do so, we first map the old gene set to a newer assembly (if necessary) using the UCSC LiftOver tool, and then compute overlapping coding sequences on the same strand to identify coding transcripts from the same gene locus. We then assign as Aliases any pair of genes that have any coding sequence in common. Most gene IDs have one-to-one alias relationships, but some genes have no alias or multiple aliases due to disagreement in gene models.

3.4 Computing Gene Expression Levels and Variant Effects

We download fastq files for *A. mellifera* Illumina runs with reads of at least 100 bp from the SRA, trim for adaptors using Fastq-MCF (https://code.google.com/p/ea-utils/wiki/FastqMcf), trim for quality using DynamicTrim [38] and align reads to the *A. mellifera* genome assembly Amel_4.5 using TopHat2 [39]. We determine FPKM (Fragments Per Kilobase of transcript per Million mapped reads) and normalized read counts for each expression dataset for transcripts in the amel_OGSv3.2 and RefSeq gene sets using cuffquant and cuffnorm, which are part the Cufflinks package [40].

We also use HTSeq [41] to determine raw read counts per transcript, and use the raw counts to compute RPKM (Reads Per Kilobase of transcript per Million mapped reads). We compute *A. mellifera* SNP effects using SnpEff [42].

3.5 Orthologues

HymenopteraMine includes orthologues from OrthoDB [29], which identifies orthologous groups of genes that are descended from a single ancestral gene. We use the OrthoDB data set computed based on a common insect ancestor to allow the inclusion of *Drosophila melanogaster* (a Dipteran). This means that any orthologous group in HymenopteraMine can include duplicated genes that emerged after divergence from the common insect ancestor. All pairwise relationships within an orthologous group are called orthologues even if some might be classified as paralogues in an analysis of a smaller taxonomic group (e.g., the relationship between *A. mellifera* and *A. florea* genes since divergence from the *Apis* ancestor). Users may investigate gene lineages in OrthoDB to clarify relationships.

3.6 Drosophila melanogaster Data

In order to leverage orthology with well-annotated fly genes, we use the same FlyBase release that was used in the OrthoDB release, and it may not be the most recent FlyBase release.

3.7 Sequence Identifiers Used in BLAST Databases

Sequence identifiers used in BLAST databases of genome assemblies are the same identifiers used in JBrowse. RefSeq or GenBank chromosome and scaffold identifiers are used for all species except A. mellifera, for which the original consortium identifiers are used. The identifiers used for RNA or protein BLAST databases are either the RefSeq identifier or the HGD official gene set identifier. For species with new assigned HGD identifiers, the original consortium IDs are also provided in the BLAST output.

Acknowledgments

The authors would like to thank Colin M. Diesh for his contributions to the development of HymenopteraMine, HGD BLAST, and HGD JBrowse/Apollo. This work was supported by USDA National Institute of Food and Agriculture Hatch Project 1009273 and Agriculture and Food Research Initiative Competitive grant no. 2010-65205-20407.

References

1. Elsik CG, Tayal A, Diesh CM, Unni DR, Emery ML, Nguyen HN, Hagen DE (2016) Hymenoptera Genome Database: integrating genome annotations in HymenopteraMine. Nucleic Acids Res 44(D1):D793–D800. https://doi.org/10.1093/nar/gkv1208
2. Elsik CG, Worley KC, Bennett AK, Beye M, Camara F, Childers CP, de Graaf DC, Debyser

G, Deng J, Devreese B, Elhaik E, Evans JD, Foster LJ, Graur D, Guigo R, Teams HP, Hoff KJ, Holder ME, Hudson ME, Hunt GJ, Jiang H, Joshi V, Khetani RS, Kosarev P, Kovar CL, Ma J, Maleszka R, Moritz RF, Munoz-Torres MC, Murphy TD, Muzny DM, Newsham IF, Reese JT, Robertson HM, Robinson GE, Rueppell O, Solovyev V, Stanke M, Stolle E, Tsuruda JM, Vaerenbergh MV, Waterhouse RM, Weaver DB, Whitfield CW, Wu Y, Zdobnov EM, Zhang L, Zhu D, Gibbs RA, Honey Bee Genome Sequencing C (2014) Finding the missing honey bee genes: lessons learned from a genome upgrade. BMC Genomics 15:86. https://doi.org/10.1186/1471-2164-15-86

3. Honeybee Genome Sequencing Consortium (2006) Insights into social insects from the genome of the honeybee Apis mellifera. Nature 443(7114):931–949
4. Sadd BM, Barribeau SM, Bloch G, de Graaf DC, Dearden P, Elsik CG, Gadau J, Grimmelikhuijzen CJ, Hasselmann M, Lozier JD, Robertson HM, Smagghe G, Stolle E, Van Vaerenbergh M, Waterhouse RM, Bornberg-Bauer E, Klasberg S, Bennett AK, Camara F, Guigo R, Hoff K, Mariotti M, Munoz-Torres M, Murphy T, Santesmasses D, Amdam GV, Beckers M, Beye M, Biewer M, Bitondi MM, Blaxter ML, Bourke AF, Brown MJ, Buechel SD, Cameron R, Cappelle K, Carolan JC, Christiaens O, Ciborowski KL, Clarke DF, Colgan TJ, Collins DH, Cridge AG, Dalmay T, Dreier S, du Plessis L, Duncan E, Erler S, Evans J, Falcon T, Flores K, Freitas FC, Fuchikawa T, Gempe T, Hartfelder K, Hauser F, Helbing S, Humann FC, Irvine F, Jermiin LS, Johnson CE, Johnson RM, Jones AK, Kadowaki T, Kidner JH, Koch V, Kohler A, Kraus FB, Lattorff HM, Leask M, Lockett GA, Mallon EB, Antonio DS, Marxer M, Meeus I, Moritz RF, Nair A, Napflin K, Nissen I, Niu J, Nunes FM, Oakeshott JG, Osborne A, Otte M, Pinheiro DG, Rossie N, Rueppell O, Santos CG, Schmid-Hempel R, Schmitt BD, Schulte C, Simoes ZL, Soares MP, Swevers L, Winnebeck EC, Wolschin F, Yu N, Zdobnov EM, Aqrawi PK, Blankenburg KP, Coyle M, Francisco L, Hernandez AG, Holder M, Hudson ME, Jackson L, Jayaseelan J, Joshi V, Kovar C, Lee SL, Mata R, Mathew T, Newsham IF, Ngo R, Okwuonu G, Pham C, Pu LL, Saada N, Santibanez J, Simmons D, Thornton R, Venkat A, Walden KK, Wu YQ, Debyser G, Devreese B, Asher C, Blommaert J, Chipman AD, Chittka L, Fouks B, Liu J, O'Neill MP, Sumner S, Puiu D, Qu J, Salzberg SL, Scherer SE, Muzny DM, Richards S, Robinson GE, Gibbs RA, Schmid-Hempel P, Worley KC (2015) The genomes of two key bumblebee species with primitive eusocial organization. Genome Biol 16:76. https://doi.org/10.1186/s13059-015-0623-3
5. Kapheim KM, Pan H, Li C, Salzberg SL, Puiu D, Magoc T, Robertson HM, Hudson ME, Venkat A, Fischman BJ, Hernandez A, Yandell M, Ence D, Holt C, Yocum GD, Kemp WP, Bosch J, Waterhouse RM, Zdobnov EM, Stolle E, Kraus FB, Helbing S, Moritz RF, Glastad KM, Hunt BG, Goodisman MA, Hauser F, Grimmelikhuijzen CJ, Pinheiro DG, Nunes FM, Soares MP, Tanaka ED, Simoes ZL, Hartfelder K, Evans JD, Barribeau SM, Johnson RM, Massey JH, Southey BR, Hasselmann M, Hamacher D, Biewer M, Kent CF, Zayed A, Blatti C III, Sinha S, Johnston JS, Hanrahan SJ, Kocher SD, Wang J, Robinson GE, Zhang G (2015) Social evolution. Genomic signatures of evolutionary transitions from solitary to group living. Science 348(6239):1139–1143. https://doi.org/10.1126/science.aaa4788
6. Kocher SD, Li C, Yang W, Tan H, Yi SV, Yang X, Hoekstra HE, Zhang G, Pierce NE, Yu DW (2013) The draft genome of a socially polymorphic halictid bee, Lasioglossum albipes. Genome Biol 14(12):R142. https://doi.org/10.1186/gb-2013-14-12-r142
7. Nygaard S, Zhang G, Schiott M, Li C, Wurm Y, Hu H, Zhou J, Ji L, Qiu F, Rasmussen M, Pan H, Hauser F, Krogh A, Grimmelikhuijzen CJ, Wang J, Boomsma JJ (2011) The genome of the leaf-cutting ant Acromyrmex echinatior suggests key adaptations to advanced social life and fungus farming. Genome Res 21(8):1339–1348. https://doi.org/10.1101/gr.121392.111
8. Suen G, Teiling C, Li L, Holt C, Abouheif E, Bornberg-Bauer E, Bouffard P, Caldera EJ, Cash E, Cavanaugh A, Denas O, Elhaik E, Fave MJ, Gadau J, Gibson JD, Graur D, Grubbs KJ, Hagen DE, Harkins TT, Helmkampf M, Hu H, Johnson BR, Kim J, Marsh SE, Moeller JA, Munoz-Torres MC, Murphy MC, Naughton MC, Nigam S, Overson R, Rajakumar R, Reese JT, Scott JJ, Smith CR, Tao S, Tsutsui ND, Viljakainen L, Wissler L, Yandell MD, Zimmer F, Taylor J, Slater SC, Clifton SW, Warren WC, Elsik CG, Smith CD, Weinstock GM, Gerardo NM, Currie CR (2011) The genome sequence of the leaf-cutter ant Atta cephalotes reveals insights into its obligate symbiotic lifestyle. PLoS Genet 7(2):e1002007. https://doi.org/10.1371/journal.pgen.1002007
9. Bonasio R, Zhang G, Ye C, Mutti NS, Fang X, Qin N, Donahue G, Yang P, Li Q, Li C, Zhang P, Huang Z, Berger SL, Reinberg D, Wang J, Liebig J (2010) Genomic comparison of the ants Camponotus floridanus and Harpegnathos

saltator. Science 329(5995):1068–1071. https://doi.org/10.1126/science.1192428

10. Schrader L, Kim JW, Ence D, Zimin A, Klein A, Wyschetzki K, Weichselgartner T, Kemena C, Stokl J, Schultner E, Wurm Y, Smith CD, Yandell M, Heinze J, Gadau J, Oettler J (2014) Transposable element islands facilitate adaptation to novel environments in an invasive species. Nat Commun 5:5495. https://doi.org/10.1038/ncomms6495
11. Oxley PR, Ji L, Fetter-Pruneda I, McKenzie SK, Li C, Hu H, Zhang G, Kronauer DJ (2014) The genome of the clonal raider ant Cerapachys biroi. Curr Biol 24(4):451–458. https://doi.org/10.1016/j.cub.2014.01.018
12. Smith CD, Zimin A, Holt C, Abouheif E, Benton R, Cash E, Croset V, Currie CR, Elhaik E, Elsik CG, Fave MJ, Fernandes V, Gadau J, Gibson JD, Graur D, Grubbs KJ, Hagen DE, Helmkampf M, Holley JA, Hu H, Viniegra AS, Johnson BR, Johnson RM, Khila A, Kim JW, Laird J, Mathis KA, Moeller JA, Munoz-Torres MC, Murphy MC, Nakamura R, Nigam S, Overson RP, Placek JE, Rajakumar R, Reese JT, Robertson HM, Smith CR, Suarez AV, Suen G, Suhr EL, Tao S, Torres CW, van Wilgenburg E, Viljakainen L, Walden KK, Wild AL, Yandell M, Yorke JA, Tsutsui ND (2011) Draft genome of the globally widespread and invasive Argentine ant (Linepithema humile). Proc Natl Acad Sci U S A 108(14):5673–5678. https://doi.org/10.1073/pnas.1008617108
13. Smith CR, Smith CD, Robertson HM, Helmkampf M, Zimin A, Yandell M, Holt C, Hu H, Abouheif E, Benton R, Cash E, Croset V, Currie CR, Elhaik E, Elsik CG, Fave MJ, Fernandes V, Gibson JD, Graur D, Gronenberg W, Grubbs KJ, Hagen DE, Viniegra AS, Johnson BR, Johnson RM, Khila A, Kim JW, Mathis KA, Munoz-Torres MC, Murphy MC, Mustard JA, Nakamura R, Niehuis O, Nigam S, Overson RP, Placek JE, Rajakumar R, Reese JT, Suen G, Tao S, Torres CW, Tsutsui ND, Viljakainen L, Wolschin F, Gadau J (2011) Draft genome of the red harvester ant Pogonomyrmex barbatus. Proc Natl Acad Sci U S A 108(14):5667–5672. https://doi.org/10.1073/pnas.1007901108
14. Wurm Y, Wang J, Riba-Grognuz O, Corona M, Nygaard S, Hunt BG, Ingram KK, Falquet L, Nipitwattanaphon M, Gotzek D, Dijkstra MB, Oettler J, Comtesse F, Shih CJ, Wu WJ, Yang CC, Thomas J, Beaudoing E, Pradervand S, Flegel V, Cook ED, Fabbretti R, Stockinger H, Long L, Farmerie WG, Oakey J, Boomsma JJ, Pamilo P, Yi SV, Heinze J, Goodisman MA, Farinelli L, Harshman K, Hulo N, Cerutti L, Xenarios I, Shoemaker D, Keller L (2011) The genome of the fire ant Solenopsis invicta. Proc Natl Acad Sci U S A 108(14):5679–5684. https://doi.org/10.1073/pnas.1009690108
15. Werren JH, Richards S, Desjardins CA, Niehuis O, Gadau J, Colbourne JK, Nasonia Genome Working G, Werren JH, Richards S, Desjardins CA, Niehuis O, Gadau J, Colbourne JK, Beukeboom LW, Desplan C, Elsik CG, Grimmelikhuijzen CJ, Kitts P, Lynch JA, Murphy T, Oliveira DC, Smith CD, van de Zande L, Worley KC, Zdobnov EM, Aerts M, Albert S, Anaya VH, Anzola JM, Barchuk AR, Behura SK, Bera AN, Berenbaum MR, Bertossa RC, Bitondi MM, Bordenstein SR, Bork P, Bornberg-Bauer E, Brunain M, Cazzamali G, Chaboub L, Chacko J, Chavez D, Childers CP, Choi JH, Clark ME, Claudianos C, Clinton RA, Cree AG, Cristino AS, Dang PM, Darby AC, de Graaf DC, Devreese B, Dinh HH, Edwards R, Elango N, Elhaik E, Ermolaeva O, Evans JD, Foret S, Fowler GR, Gerlach D, Gibson JD, Gilbert DG, Graur D, Grunder S, Hagen DE, Han Y, Hauser F, Hultmark D, HCt H, Hurst GD, Jhangian SN, Jiang H, Johnson RM, Jones AK, Junier T, Kadowaki T, Kamping A, Kapustin Y, Kechavarzi B, Kim J, Kim J, Kiryutin B, Koevoets T, Kovar CL, Kriventseva EV, Kucharski R, Lee H, Lee SL, Lees K, Lewis LR, Loehlin DW, Logsdon JM Jr, Lopez JA, Lozado RJ, Maglott D, Maleszka R, Mayampurath A, Mazur DJ, McClure MA, Moore AD, Morgan MB, Muller J, Munoz-Torres MC, Muzny DM, Nazareth LV, Neupert S, Nguyen NB, Nunes FM, Oakeshott JG, Okwuonu GO, Pannebakker BA, Pejaver VR, Peng Z, Pratt SC, Predel R, Pu LL, Ranson H, Raychoudhury R, Rechtsteiner A, Reese JT, Reid JG, Riddle M, Robertson HM, Romero-Severson J, Rosenberg M, Sackton TB, Sattelle DB, Schluns H, Schmitt T, Schneider M, Schuler A, Schurko AM, Shuker DM, Simoes ZL, Sinha S, Smith Z, Solovyev V, Souvorov A, Springauf A, Stafflinger E, Stage DE, Stanke M, Tanaka Y, Telschow A, Trent C, Vattathil S, Verhulst EC, Viljakainen L, Wanner KW, Waterhouse RM, Whitfield JB, Wilkes TE, Williamson M, Willis JH, Wolschin F, Wyder S, Yamada T, Yi SV, Zecher CN, Zhang L, Gibbs RA (2010) Functional and evolutionary insights from the genomes of three parasitoid Nasonia species. Science 327(5963):343–348. https://doi.org/10.1126/science.1178028
16. Rago A, Gilbert DG, Choi JH, Sackton TB, Wang X, Kelkar YD, Werren JH, Colbourne JK (2016) OGS2: genome re-annotation of the jewel wasp Nasonia vitripennis. BMC Genomics 17:678. https://doi.org/10.1186/s12864-016-2886-9

17. Holt C, Yandell M (2011) MAKER2: an annotation pipeline and genome-database management tool for second-generation genome projects. BMC Bioinformatics 12:491. https://doi.org/10.1186/1471-2105-12-491
18. Lee E, Helt GA, Reese JT, Munoz-Torres MC, Childers CP, Buels RM, Stein L, Holmes IH, Elsik CG, Lewis SE (2013) Web Apollo: a web-based genomic annotation editing platform. Genome Biol 14(8):R93. https://doi.org/10.1186/gb-2013-14-8-r93
19. Buels R, Yao E, Diesh CM, Hayes RD, Munoz-Torres M, Helt G, Goodstein DM, Elsik CG, Lewis SE, Stein L, Holmes IH (2016) JBrowse: a dynamic web platform for genome visualization and analysis. Genome Biol 17:66. https://doi.org/10.1186/s13059-016-0924-1
20. Priyam A, Woodcroft BJ, Rai V, Munagala A, Moghul I, Ter F, Gibbins MA, Moon H, Leonard G, Rumpf W, Wurm Y (2015) Sequenceserver: a modern graphical user interface for custom BLAST databases. BioRxIV. https://doi.org/10.1101/033142
21. Smith RN, Aleksic J, Butano D, Carr A, Contrino S, Hu F, Lyne M, Lyne R, Kalderimis A, Rutherford K, Stepan R, Sullivan J, Wakeling M, Watkins X, Micklem G (2012) InterMine: a flexible data warehouse system for the integration and analysis of heterogeneous biological data. Bioinformatics 28(23):3163–3165. https://doi.org/10.1093/bioinformatics/bts577
22. Chatr-Aryamontri A, Oughtred R, Boucher L, Rust J, Chang C, Kolas NK, O'Donnell L, Oster S, Theesfeld C, Sellam A, Stark C, Breitkreutz BJ, Dolinski K, Tyers M (2017) The BioGRID interaction database: 2017 update. Nucleic Acids Res 45(D1):D369–D379. https://doi.org/10.1093/nar/gkw1102
23. Sherry ST, Ward MH, Kholodov M, Baker J, Phan L, Smigielski EM, Sirotkin K (2001) dbSNP: the NCBI database of genetic variation. Nucleic Acids Res 29(1):308–311
24. Gramates LS, Marygold SJ, Santos GD, Urbano JM, Antonazzo G, Matthews BB, Rey AJ, Tabone CJ, Crosby MA, Emmert DB, Falls K, Goodman JL, Hu Y, Ponting L, Schroeder AJ, Strelets VB, Thurmond J, Zhou P, the FlyBase C (2017) FlyBase at 25: looking to the future. Nucleic Acids Res 45(D1):D663–D671. https://doi.org/10.1093/nar/gkw1016
25. Fabregat A, Sidiropoulos K, Garapati P, Gillespie M, Hausmann K, Haw R, Jassal B, Jupe S, Korninger F, McKay S, Matthews L, May B, Milacic M, Rothfels K, Shamovsky V, Webber M, Weiser J, Williams M, Wu G, Stein L, Hermjakob H, D'Eustachio P (2016) The reactome pathway knowledgebase. Nucleic Acids Res 44(D1):D481–D487. https://doi.org/10.1093/nar/gkv1351
26. Gene Ontology Consortium (2015) Gene Ontology Consortium: going forward. Nucleic Acids Res 43(Database issue):D1049–D1056. https://doi.org/10.1093/nar/gku1179
27. Finn RD, Attwood TK, Babbitt PC, Bateman A, Bork P, Bridge AJ, Chang HY, Dosztanyi Z, El-Gebali S, Fraser M, Gough J, Haft D, Holliday GL, Huang H, Huang X, Letunic I, Lopez R, Lu S, Marchler-Bauer A, Mi H, Mistry J, Natale DA, Necci M, Nuka G, Orengo CA, Park Y, Pesseat S, Piovesan D, Potter SC, Rawlings ND, Redaschi N, Richardson L, Rivoire C, Sangrador-Vegas A, Sigrist C, Sillitoe I, Smithers B, Squizzato S, Sutton G, Thanki N, Thomas PD, Tosatto SC, Wu CH, Xenarios I, Yeh LS, Young SY, Mitchell AL (2017) InterPro in 2017-beyond protein family and domain annotations. Nucleic Acids Res 45(D1):D190–D199. https://doi.org/10.1093/nar/gkw1107
28. Kanehisa M, Goto S, Sato Y, Furumichi M, Tanabe M (2012) KEGG for integration and interpretation of large-scale molecular data sets. Nucleic Acids Res 40(Database issue):D109–D114. https://doi.org/10.1093/nar/gkr988
29. Zdobnov EM, Tegenfeldt F, Kuznetsov D, Waterhouse RM, Simao FA, Ioannidis P, Seppey M, Loetscher A, Kriventseva EV (2017) OrthoDB v9.1: cataloging evolutionary and functional annotations for animal, fungal, plant, archaeal, bacterial and viral orthologs. Nucleic Acids Res 45(D1):D744–D749. https://doi.org/10.1093/nar/gkw1119
30. Resource Coordinators NCBI (2017) Database resources of the National Center for Biotechnology Information. Nucleic Acids Res 45(D1):D12–D17. https://doi.org/10.1093/nar/gkw1071
31. O'Leary NA, Wright MW, Brister JR, Ciufo S, Haddad D, McVeigh R, Rajput B, Robbertse B, Smith-White B, Ako-Adjei D, Astashyn A, Badretdin A, Bao Y, Blinkova O, Brover V, Chetvernin V, Choi J, Cox E, Ermolaeva O, Farrell CM, Goldfarb T, Gupta T, Haft D, Hatcher E, Hlavina W, Joardar VS, Kodali VK, Li W, Maglott D, Masterson P, McGarvey KM, Murphy MR, O'Neill K, Pujar S, Rangwala SH, Rausch D, Riddick LD, Schoch C, Shkeda A, Storz SS, Sun H, Thibaud-Nissen F, Tolstoy I, Tully RE, Vatsan AR, Wallin C, Webb D, Wu W, Landrum MJ, Kimchi A, Tatusova T, DiCuccio M, Kitts P, Murphy TD, Pruitt KD (2016) Reference sequence (RefSeq) database at NCBI: current status, taxonomic expansion, and functional annotation. Nucleic

Acids Res 44(D1):D733–D745. https://doi.org/10.1093/nar/gkv1189

32. Kodama Y, Shumway M, Leinonen R, International Nucleotide Sequence Database Consortium (2012) The sequence read archive: explosive growth of sequencing data. Nucleic Acids Res 40(Database issue):D54–D56. https://doi.org/10.1093/nar/gkr854
33. The UniProt Consortium (2017) UniProt: the universal protein knowledgebase. Nucleic Acids Res 45(D1):D158–D169. https://doi.org/10.1093/nar/gkw1099
34. Huntley RP, Sawford T, Mutowo-Meullenet P, Shypitsyna A, Bonilla C, Martin MJ, O'Donovan C (2015) The GOA database: gene ontology annotation updates for 2015. Nucleic Acids Res 43(Database issue):D1057–D1063. https://doi.org/10.1093/nar/gku1113
35. Liang ZS, Nguyen T, Mattila HR, Rodriguez-Zas SL, Seeley TD, Robinson GE (2012) Molecular determinants of scouting behavior in honey bees. Science 335(6073):1225–1228. https://doi.org/10.1126/science.1213962
36. Southey BR, Zhu P, Carr-Markell MK, Liang ZS, Zayed A, Li R, Robinson GE, Rodriguez-Zas SL (2016) Characterization of genomic variants associated with scout and recruit behavioral castes in honey bees using whole-genome sequencing. PLoS One 11(1):e0146430. https://doi.org/10.1371/journal.pone.0146430
37. Hickey G, Paten B, Earl D, Zerbino D, Haussler D (2013) HAL: a hierarchical format for storing and analyzing multiple genome alignments. Bioinformatics 29(10):1341–1342. https://doi.org/10.1093/bioinformatics/btt128
38. Cox MP, Peterson DA, Biggs PJ (2010) SolexaQA: at-a-glance quality assessment of Illumina second-generation sequencing data. BMC Bioinformatics 11:485. https://doi.org/10.1186/1471-2105-11-485
39. Kim D, Pertea G, Trapnell C, Pimentel H, Kelley R, Salzberg SL (2013) TopHat2: accurate alignment of transcriptomes in the presence of insertions, deletions and gene fusions. Genome Biol 14(4):R36. https://doi.org/10.1186/gb-2013-14-4-r36
40. Trapnell C, Williams BA, Pertea G, Mortazavi A, Kwan G, van Baren MJ, Salzberg SL, Wold BJ, Pachter L (2010) Transcript assembly and quantification by RNA-Seq reveals unannotated transcripts and isoform switching during cell differentiation. Nat Biotechnol 28(5):511–515. https://doi.org/10.1038/nbt.1621
41. Anders S, Pyl PT, Huber W (2015) HTSeq—a Python framework to work with high-throughput sequencing data. Bioinformatics 31(2):166–169. https://doi.org/10.1093/bioinformatics/btu638
42. Cingolani P, Platts A, Wang le L, Coon M, Nguyen T, Wang L, Land SJ, Lu X, Ruden DM (2012) A program for annotating and predicting the effects of single nucleotide polymorphisms, SnpEff: SNPs in the genome of Drosophila melanogaster strain w1118; iso-2; iso-3. Fly (Austin) 6(2):80–92. https://doi.org/10.4161/fly.19695

Chapter 18

Navigating the i5k Workspace@NAL: A Resource for Arthropod Genomes

Monica F. Poelchau, Mei-Ju May Chen, Yu-Yu Lin, and Christopher P. Childers

Abstract

The i5k Workspace@NAL is a genome database tailored toward newly sequenced arthropod genomes and their research communities. With 56 arthropod genomes and counting, the i5k Workspace strives to facilitate public data access, visualization, and community curation across arthropod species. Any researcher with an arthropod genome project who would like to take advantage of the i5k Workspace facilities is encouraged to submit their data. In this chapter, we explain how to use the i5k Workspace@NAL to submit, find, and improve arthropod genomics data.

Key words i5k, Arthropods, Insects, Genomics, Database, Genome portal, Community annotation

1 Introduction

The i5k Workspace@NAL (https://i5k.nal.usda.gov/) is a genome portal designed to help arthropod researchers access, visualize, share, and curate data associated with arthropod genome assemblies. The i5k Workspace was established under the purview of the i5k initiative, which has tasked itself with coordinating the sequencing and assembly of 5000 arthropod genomes [1, 2]. The genome projects resulting from this initiative—and from other sources—can benefit from a centralized web portal where researchers and their communities can find, share, and improve genomic data. Many arthropod genome projects serve small scientific communities, but providing additional services within a community genome portal enhances their value for comparative genomics. To fill this need, the National Agricultural Library of the USDA's Agricultural Research Service developed the i5k Workspace@NAL, a genome portal for arthropod genomes. The i5k Workspace particularly welcomes arthropod genome projects that do not fit within the scope of other existing genome databases focused on particular taxonomic groups. As of April 2017, 56 arthropod genomes are

Martin Kollmar (ed.), *Eukaryotic Genomic Databases: Methods and Protocols*, Methods in Molecular Biology, vol. 1757, https://doi.org/10.1007/978-1-4939-7737-6_18, © Springer Science+Business Media, LLC, part of Springer Nature 2018

available. All of the hosted data is user-submitted, and we encourage new content submissions.

The arthropod genome database landscape comprises many valuable resources, including but not limited to VectorBase [3] for vector genomes; HymenopteraGenomeDatabase [4] for Hymenoptera; and InsectBase [5] for general access to insect genomes and transcriptomes. Within this context, the i5k Workspace@NAL is particularly suited toward genome projects that can benefit from our broad taxonomic scope, as well as the community curation activities at the i5k Workspace.

Previously, we introduced the then new i5k Workspace@NAL, its design and its functions [6]. In this publication, we provide guidance on how new and recurring users can (1) submit new data using our new data submission pages; (2) find data hosted by the i5k Workspace; and (3) manually annotate existing genomes. Any questions about the i5k Workspace can be directed to i5k@ars.usda.gov or our contact page: https://i5k.nal.usda.gov/contact.

2 Submitting Data to the i5k Workspace

The data hosted at the i5k Workspace is currently entirely user-submitted. Anyone with an appropriate dataset can submit their data to us in order to take advantage of our tools. In this section, we explain the basics behind submitting data to the i5k Workspace, and walk through an example data submission process.

What is an i5k Workspace Project? An i5k Workspace project is a collection of data centered on the genome assembly of an arthropod. This generally includes: a genome assembly, accessioned by the International Nucleotide Sequence Database Collaboration (INSDC); one or more sets of predicted gene models; and other data files, mapped to the genome assembly. These datasets allow the user to explore information in the context of the genome assembly.

What is a project coordinator? A project coordinator is the point of contact for questions related to an i5k Workspace project. Usually, the project coordinator's main responsibility is to approve or reject users who wish to use the Apollo manual annotation software for their project.

What does the i5k Workspace do with your data? If you are starting an i5k Workspace project, we generate the following resources for you:

1. An organism page, which can include any information that you like about the organism and the genome project. The page also includes links to pages that describe the assembly and other analyses; contact information for the project coordinator; summary statistics about the assembly and gene predictions; and links to download the data.

2. The JBrowse [7] genome browser, including the Apollo manual annotation tool [8].
3. The BLAST+ sequence search tool [9], using our custom BLAST interface [6].
4. HMMER [10, 11] and Clustal Omega [12]/ClustalW [13, 14] web applications.
5. A data downloads page to share with collaborators.

Considerations prior to submission. There are several things that a submitter should consider prior to submitting data to the i5k Workspace:

1. Currently, *all data submitted to the i5k Workspace will be publicly available.* Only manual annotations generated using the Apollo software are under password protection. Submitted genome assemblies need to already be accessioned by an INSDC member organization (e.g., NCBI, ENA, DDBJ). Almost all i5k Workspace resources are based on the genome assembly. Therefore, all of our products benefit from (1) the additional contaminant screens that these organizations require, and (2) the reduced ambiguity about the assembly origin when accession numbers are used.
2. The i5k Workspace is primarily designed for "orphaned" arthropod genomes—datasets that are not hosted by other genome databases, typically with a more focused taxonomic scope. If there is another genome database that may be a better fit for your dataset, we will get in touch with the database owners to help you determine whether your dataset is best hosted at the i5k Workspace or elsewhere. Submitters can always opt to host their projects with us if they feel that is the best fit for the project.
3. If you are not sure what i5k Workspace projects we host—we add new genomes regularly—view our organism overview page, which lists all projects by organism (https://i5k.nal.usda.gov/species). Within each individual organism page, the assembly used for this organism is listed. If you want to add a genome to an existing i5k Workspace project, get in touch with us.

How to get started. Here, we explain each step within the submission process.

2.1 Register for a Data Submission Account

Go to https://i5k.nal.usda.gov/register/project-dataset/account. Fill out each field to the best of your knowledge. Once you click "Submit," a confirmation message should display. We review all account requests. Once we have approved your account, you should receive an email including your username and a link for a one-time login. You may use this login link to generate your password.

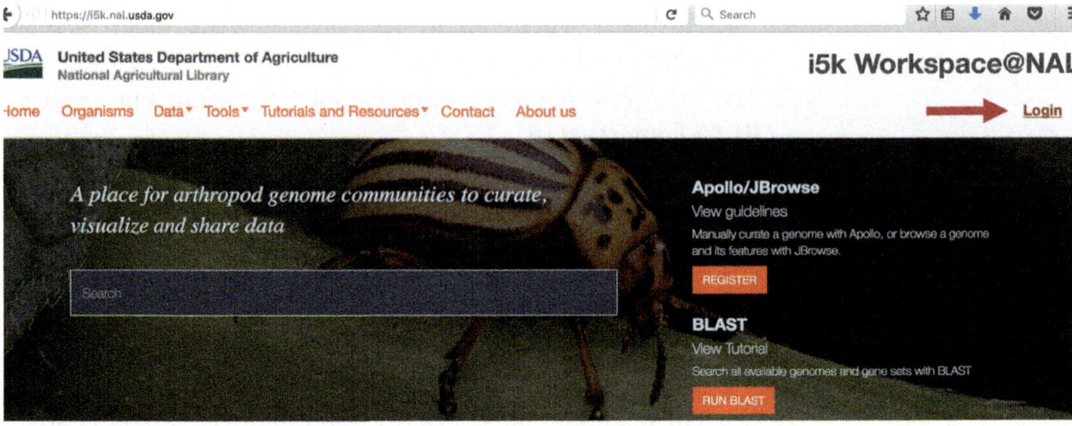

Fig. 1 Screenshot of the i5k Workspace@NAL homepage, highlighting the "Login" feature

2.2 Start an i5k Workspace Project

1. Log in to the i5k Workspace with your new data submission credentials (Fig. 1).
2. Under the menu bar, select "Data → Submit data → Request a new i5k Workspace Project". Fill out the form to the best of your knowledge:
   (a) *Genus and species*: the genus and species names for the i5k Workspace project that you would like to start.
   (b) *NCBI Taxonomy ID*: If you do not know the NCBI taxonomy ID, you can search for it at NCBI (https://www.ncbi.nlm.nih.gov/taxonomy).
   (c) *Common name*: the common name for your organism.
   (d) *Is the genome assembly already hosted at another genome portal, or is there another genome portal that would also be appropriate to host your dataset?* Select "yes" if this genome is already available at another taxon-specific genome portal, or if you are aware of another suitable genome portal.
   (e) *Have you submitted the genome assembly to NCBI, or another INSDC member?* Select "yes" if you have already submitted the genome assembly to an INSDC organization, such as NCBI, ENA, or DDBJ.
   (f) *Is this a re-assembly or a new assembly of an existing i5k Workspace organism?* Select "yes" if this organism is already hosted by the i5k Workspace. A list of all hosted organisms is available here: https://i5k.nal.usda.gov/species.
   (g) *Were you involved in the generation of this genome assembly?* Select "Yes" if you were, "No" if not.
   (h) *Briefly describe your plans for this genome project at the i5k Workspace.* Let us know a bit about why you would like to use the i5k Workspace.

(i) *Full name, email address*: Choose the contact information that you would like us to use to contact you.

3. Once you submit the request form, you should receive a confirmation email to the email address that you specified. We will contact you if we have any questions. You should receive an email once we have approved your project request.

2.3 Submit Your Genome Assembly

1. All information in this form, except for the provider's email address and the md5sum field, will be reformatted for display in the genome assembly's analysis page, which is visible to the public.
2. Under the menu bar, select "Data → Submit data → Submit a genome assembly". Complete the fields to the best of your ability.

   (a) *Full name, email address*: Enter the contact information of the project coordinator.

   (b) *Organism*: Select the organism for this assembly from the drop-down menu. The Common Name entry will auto-fill once an organism is selected.

   (c) *Project description to display in your organism page*. Enter any information that you would like that will convey to other users what this genome project is about. You can view examples of other project descriptions in our other organism pages.

   (d) *Image URL for your organism page*. If there is an image available online for your organism, enter the URL here. If you or someone from your group took the photo, we will ask you to fill out an image permission form (https://i5k.nal.usda.gov/image_permission/new). If not, we ask that the image that you provide can be shared openly under a creative commons license (https://creativecommons.org/licenses/).

   (e) *Will you manually curate this assembly using i5k Workspace tools?* Select "Yes" if you would like to open up this genome assembly for manual annotation using the Apollo software.

   - *Is the curation coordinator the same as the genome coordinator?* If the individual who should receive emails from us about new Apollo users is different than the genome coordinator, please enter their information here.
   - *Do you need assistance developing an Official Gene Set?* We are prototyping a system to develop official gene sets or consensus gene sets from manual annotations and a single, comprehensive gene set. If you would like to consider this service, enter "Yes."

- *Specify curation timeframe.* If you have a specific time frame in mind, enter the dates—otherwise, select "There is no set time frame for curation."

(f) *Geo location, tissues/life stage included, sex, strain, Other notes*: Enter this information if it is available; however, it is not required.

(g) *Sequencing platform and version*: Enter the sequencing technology and version used to generate the raw reads for the genome assembly.

(h) *Data source URL*: If there is a URL for the genome assembly, enter it here.

(i) *Assembly name and version*: If a name and version already exists for this assembly, please enter this here. Otherwise, enter the appropriate name and version of your choice.

(j) *NCBI/INSDC Genome Assembly accession #*: Enter the accession number for the genome assembly.

(k) *Analysis method*: Enter the program(s) and program version used to generate the assembly.

(l) *Is the assembly published?* If yes, please enter the citation; if no, please specify whether the Ft. Lauderdale [15] and Toronto [16] codes of conduct should apply.

(m) *Other notes*: Enter any other information that you would like to share.

(n) *File name*: The name of the assembly file(s).

(o) *Md5sum*: The md5 checksum of the assembly file. This is a 128-bit hash frequently used to verify file integrity.

3. Once you submit the form, you should receive a confirmation email to the email address that you specified. We will contact you if we have any questions.

2.4 Submit Gene Predictions or Data Mapped to the Genome Assembly

1. If you have files that you would like to visualize on the genome browser or share with your community, you may also submit these. Currently, we only accept files that relate to a genome assembly that we host. Examples of file types include .gff3, .bam, .vcf, and .gtf format—contact us if you have another format in mind. Make sure that the data is mapped to the *same* assembly that was submitted to the i5k Workspace.
2. All information in this form, except for the provider's email address and the md5sum field, will be reformatted for display in the gene prediction set analysis page, which will be visible to the public. Some of this information will also be displayed as track information in the genome browser.
3. Under the menu bar, select "Data → Submit data → Submit Gene Predictions" or "Data → Submit data → Submit a

Mapped Dataset". Below, we list information on some of the steps in these forms that may require additional clarification.

(a) Select the organism that your dataset belongs to. If you do not know which genome assembly is used for this organism, then visit the organism's page, where this information is available (https://i5k.nal.usda.gov/species).

(b) For the "Analysis Method" section, enter the program and its version that was used to generate the result file. Additional information or Methods can be entered if desired.

(c) Under the "Analysis provider" or "Data provider" section, enter the provider contact information.

(d) For gene predictions, under "Gene set information," list the name and version of the dataset. Include a descriptive track name for display in the genome browser—this should be a name that even individuals outside your lab can understand. Specify whether the gene set is an Official gene set (OGS).

(e) For mapped datasets, optionally enter some information about the collection location, tissues or life stages included, etc., so users can understand the conditions under which the data were generated.

(f) Finally, enter the file name and the md5sum of the file, so we can verify whether your file transferred correctly.

2.5 Send Us Your Files

The i5k Workspace data submission forms are currently only available for metadata; however, an update where file upload is possible is in progress. Therefore, once we have the metadata for your datasets, we will contact you with the best way to send us your files. Then, we will begin processing them and setting up i5k Workspace resources for your project. We will contact you when they are ready.

Maintaining your project. In principle, no other investment on the part of the data submitter is necessary. We will enroll you in our low-volume mailing list to keep you abreast of new developments and occasional service disruptions; you may opt out if desired.

2.5.1 Finding Data at the i5k Workspace

As of April 2017, the i5k Workspace hosts 56 arthropod genomes (Table 1); a current list can be found at (https://i5k.nal.usda.gov/species). As mentioned above, the i5k Workspace hosts genome assemblies, and any data that can be mapped to the assembly, including files in bam, sam, vcf, gff3, and bed format. Here, we provide several methods and examples for finding and retrieving data at the i5k Workspace.

1. *Website search for general content.* To search for general content at the i5k Workspace, such as organism or gene information, navigate to the home page (https://i5k.nal.usda.gov/)

Table 1
Organisms hosted at the i5k Workspace as of April 2017. Class, order, Name (genus and species), and common name are given

Class	Order	Name	Common name
Arachnida	Araneae	*Latrodectus hesperus*	Western black widow spider
Arachnida	Araneae	*Loxosceles reclusa*	Brown recluse spider
Arachnida	Araneae	*Parasteatoda tepidariorum*	Common house spider
Arachnida	Scorpiones	*Centruroides exilicauda*	Bark scorpion
Entognatha	Diplura	*Catajapyx aquilonaris*	Silvestri's Northern Forcepstail
Insecta	Blattodea	*Blattella germanica*	German cockroach
Insecta	Coleoptera	*Aethina tumida*	Small hive beetle
Insecta	Coleoptera	*Agrilus planipennis*	Emerald ash borer
Insecta	Coleoptera	*Anoplophora glabripennis*	Asian long-horned beetle
Insecta	Coleoptera	*Leptinotarsa decemlineata*	Colorado potato beetle
Insecta	Coleoptera	*Nicrophorus vespilloides*	Common Sexton Beetle
Insecta	Coleoptera	*Onthophagus taurus*	Bull-headed Dung beetle
Insecta	Coleoptera	*Tribolium castaneum*	Red flour beetle
Insecta	Diptera	*Bactrocera cucurbitae*	Melon fruit fly
Insecta	Diptera	*Bactrocera dorsalis*	Oriental fruit fly
Insecta	Diptera	*Bactrocera oleae*	Olive fruit fly
Insecta	Diptera	*Ceratitis capitata*	Mediterranean fruit fly
Insecta	Diptera	*Drosophila biarmipes*	NA
Insecta	Diptera	*Drosophila bipectinata*	NA
Insecta	Diptera	*Drosophila elegans*	NA
Insecta	Diptera	*Drosophila eugracilis*	NA
Insecta	Diptera	*Drosophila ficusphila*	NA
Insecta	Diptera	*Drosophila kikkawai*	NA
Insecta	Diptera	*Drosophila rhopaloa*	NA
Insecta	Diptera	*Drosophila takahashii*	NA
Insecta	Diptera	*Dufourea novaeangliae*	NA
Insecta	Diptera	*Mayetiola destructor*	Hessian fly
Insecta	Ephemeroptera	*Ephemera danica*	Mayfly
Insecta	Hemiptera	*Cimex lectularius*	Bed bug
Insecta	Hemiptera	*Diaphorina citri*	Asian Citrus Psyllid

(continued)

Table 1 (continued)

Class	Order	Name	Common name
Insecta	Hemiptera	*Gerris buenoi*	Water strider
Insecta	Hemiptera	*Halyomorpha halys*	Brown marmorated stink bug
Insecta	Hemiptera	*Homalodisca vitripennis*	Glassy-winged sharpshooter
Insecta	Hemiptera	*Oncopeltus fasciatus*	Milkweed bug
Insecta	Hemiptera	*Pachypsylla venusta*	Hackberry petiole gall psyllid
Insecta	Hymenoptera	*Athalia rosae*	Turnip sawfly
Insecta	Hymenoptera	*Cephus cinctus*	Wheat stem sawfly
Insecta	Hymenoptera	*Copidosoma floridanum*	NA
Insecta	Hymenoptera	*Diachasma alloeum*	Parasitoid wasp
Insecta	Hymenoptera	*Fopius arisanus*	NA
Insecta	Hymenoptera	*Habropoda laboriosa*	NA
Insecta	Hymenoptera	*Lasioglossum albipes*	White-footed sweat bee
Insecta	Hymenoptera	*Megachile rotundata*	Alfalfa leafcutting bee
Insecta	Hymenoptera	*Melipona quadrifasciata*	NA
Insecta	Hymenoptera	*Microplitis demolitor*	NA
Insecta	Hymenoptera	*Neodiprion lecontei*	Redheaded pine sawfly
Insecta	Hymenoptera	*Orussus abietinus*	Parasitic wood wasp
Insecta	Hymenoptera	*Trichogramma pretiosum*	Parasitic wasp
Insecta	Lepidoptera	*Amyelois transitella*	Navel orangeworm moth
Insecta	Lepidoptera	*Manduca sexta*	Tobacco hornworm
Insecta	Odonata	*Ladona fulva*	Scarce chaser
Insecta	Thysanoptera	*Frankliniella occidentalis*	Western flower thrips
Insecta	Trichoptera	*Limnephilus lunatus*	Caddisfly
Malacostraca	Amphipoda	*Hyalella azteca*	NA
Maxillopoda	Calinoida	*Eurytemora affinis*	Calanoid copepod
Maxillopoda	Harpacticoida	*Tigriopus californicus*	NA

and type a query, e.g., "*Anoplophora glabripennis*," into the main search bar. The results page will show all content at the i5k Workspace containing those search terms. Study the "Filter by content type" section on the right of the search results page, and select the content type of interest—e.g., Analysis (Fig. 2). This will display the three Analysis pages for

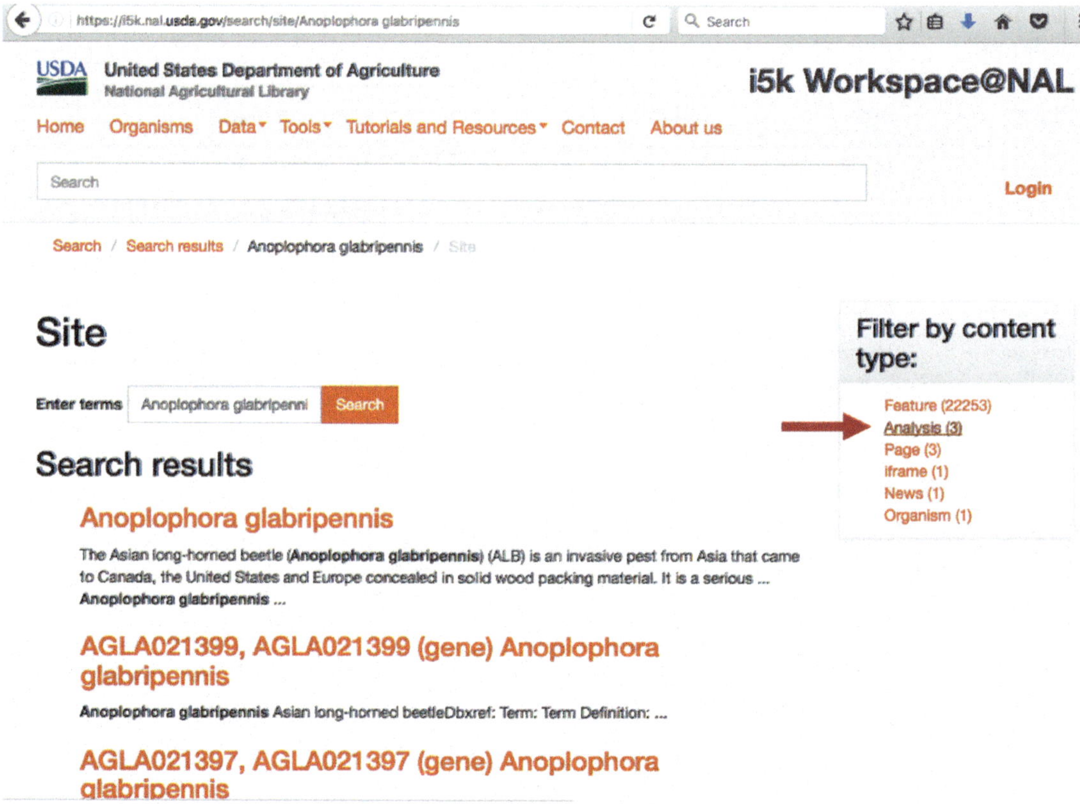

Fig. 2 Screenshot of the "Search Results" page for *Anoplophora glabripennis.* The link to the filter for the "Analysis" content type is highlighted by an arrow

Anoplophora glabripennis, including the genome assembly and two annotation datasets.

2. *Bulk data downloads for full files.* Users can find all publicly available files via the menu item "Data → Data Downloads", under https://i5k.nal.usda.gov/data/, or https://i5k.nal.usda.gov/content/data-downloads. These two sites host identical content, but provide separate user interfaces. Users should note that not all data are published in peer-reviewed journals, and that each genome projects' community contact should be contacted prior to using the data in any publication, respecting the Ft. Lauderdale [15] and Toronto agreements [16]. Datasets are organized by species name, then current vs. legacy genome assembly, then by Data type.
3. *Sequence search—BLAST+.* The i5k Workspace has several options for searching sequence datasets. Users can use BLAST+ version 2.2.29 [9] to search protein and nucleotide sequences of all genome assemblies and Official or Primary Gene Sets—navigate to https://i5k.nal.usda.gov/webapp/blast/, or select "Tools → BLAST" from the main menu. For a detailed description of the BLAST interface, we refer users to our pre-

vious publication [6] and our BLAST tutorial (https://i5k.nal.usda.gov/content/blast-tutorial). Also, for an annotation workflow using our BLAST interface, *see* Subheading 2.5.2 below.

4. *Sequence alignment—ClustalW and Clustal Omega*. ClustalW [17] and Clustal Omega [12] are multiple alignment programs for nucleotide and protein sequences. ClustalW is one of the oldest programs available for multiple sequence alignment, and is implemented at the i5k Workspace for users accustomed to this implementation. Clustal Omega is the latest version of Clustal. Users can select "Tools—Clustal" from the main menu bar, or navigate to https://i5k.nal.usda.gov/webapp/clustal/. There, the user can enter multiple sequences of a single sequence type (i.e., nucleotide or peptide) into the "Query Sequence" box (Fig. 3). The web application autodetects the sequence type. Users can change various parameter settings within the web application, or use the default settings. On the results page, users can view the full alignment in default or "colorful" mode (Fig. 4). Users can also download the alignment in aln format, as well as submission details. Results can be retrieved via the result URL for up to one week. Finally, the alignment can be exported directly to the hmmsearch

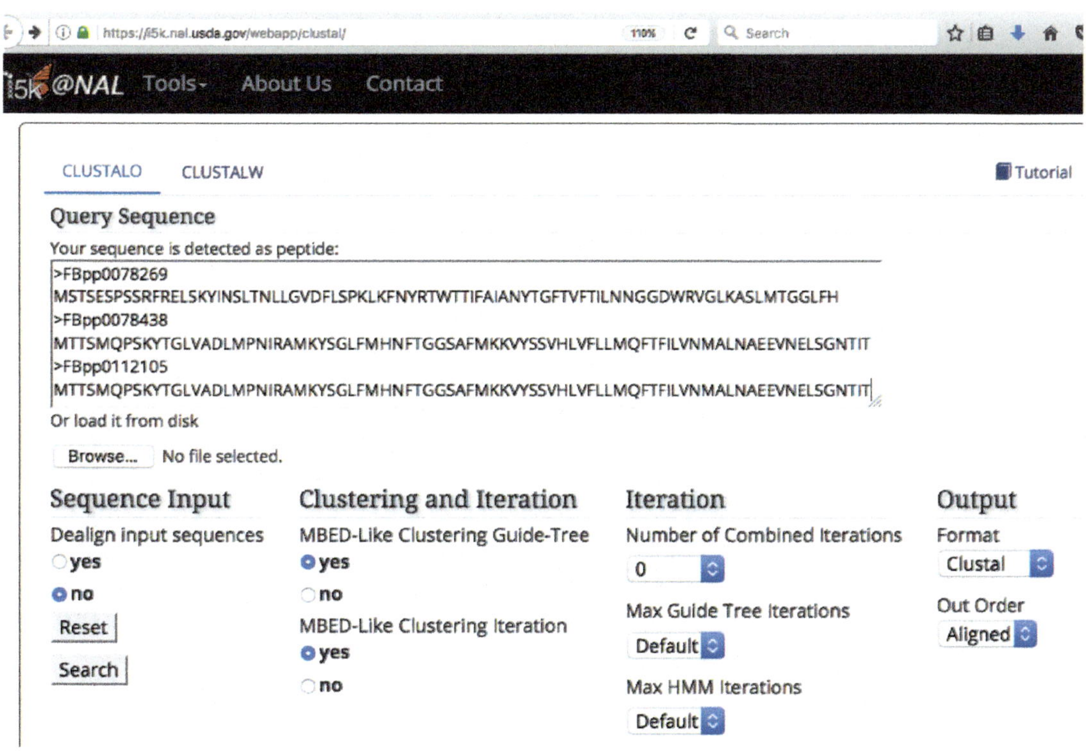

Fig. 3 Screenshot of an example Clustal Omega query

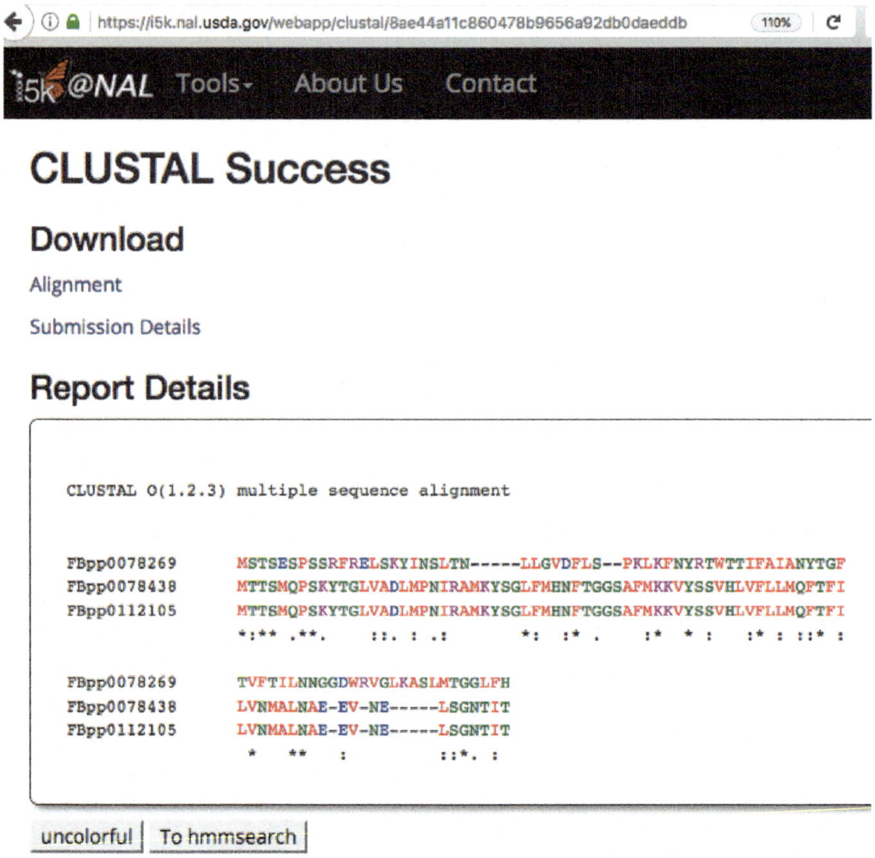

Fig. 4 Screenshot of an example Clustal Omega result

program (see below). A manual on how to use the i5k Workspace Clustal webapp is available here: https://i5k.nal.usda.gov/webapp/clustal/manual/.

5. *Sequence search—HMMER*. The i5k Workspace HMMER implementation is available at https://i5k.nal.usda.gov/webapp/hmmer/, or via "Tools—HMMER" on the main menu. HMMER [11] is a method to search sequence datasets and perform sequence alignments by generating a probabilistic model, or "profile HMM," of the query sequence, and searching the sequence database using this profile instead of the sequence itself. This method is especially useful when searching a sequence database for representatives of a gene family. In the i5k Workspace implementation, a user can search all i5k Workspace protein datasets using the phmmer (protein query, fasta format) or hmmsearch (multiple alignment query in alignment or fasta format) programs. For a full tutorial on how to use the i5k Workspace HMMER web application, see our manual: https://i5k.nal.usda.gov/webapp/hmmer/manual/.

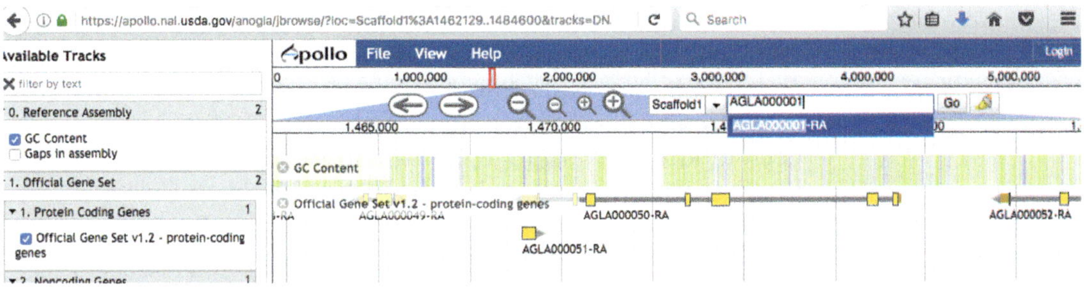

Fig. 5 Screenshot of a search for an mRNA ID within the JBrowse genome browser

6. *Genome browser*. I5k Workspace users can search and browse for gene IDs and sequences in the JBrowse genome browser [7]. Available browsers are listed under https://i5k.nal.usda.gov/available-genome-browsers, or the user can navigate to Tools → JBrowse/Apollo → JBrowse/Apollo Organisms. There are several methods for a user to search for the location of a particular gene in the genome browser. First, the user can perform a BLAST search on the desired genome sequence, which allows the user to view BLAST High-scoring Segment Pairs (HSPs) mapped onto the genome browser (*see* workflow in Subheading 2.5.2 below). Alternatively, a user can search within the genome browser by feature ID. For example, in the Asian longhorned beetle genome browser (*Anoplophora glabripennis*, https://apollo.nal.usda.gov/anogla/jbrowse), the user can enter the ID "AGLA000001-RA" into the search box at the top of the browser, and JBrowse will jump to the correct coordinates for this model (Fig. 5). The user will still need to select the appropriate track from the "Available tracks" menu on the left side of the browser. A detailed workflow for using the JBrowse genome browser within a community annotation context is below.

2.5.2 Improving Data at the i5k Workspace via Community Annotation

The majority of i5k Workspace projects take advantage of the Apollo manual curation software to improve annotation quality via community curation. Apollo is a plugin to the JBrowse genome browser [7] that allows registered users to collaboratively edit gene structures and gene functional information [8]. Computationally predicted gene models are not always correct [18], and manual inspection and corrections by community annotators can improve models of gene structure and function. For example, in the Asian longhorned beetle *Anoplophora glabripennis*, a community of annotators received one session of online training, and using a set of guidelines, modified 1234 transcript models, of which 75% had modifications to gene structure [19]. This demonstrates that with minimal training effort, communities can contribute effectively toward improving computationally predicted gene models.

Community annotation at the i5k Workspace is a collaborative effort among many individuals, often working together across different time zones and countries. As such, we encourage annotators to communicate with each other, and if there is disagreement on a gene model, to work together to find the best solution. The i5k Workspace works with each project coordinator to facilitate communication and collaboration whenever possible.

The use of Apollo for manual annotation is documented in detail elsewhere (http://genomearchitect.github.io/users-guide/). Here, we describe an example workflow for a new annotator interested in finding the gene alpha-catenin in the Colorado Potato Beetle *Leptinotarsa decemlineata*, using a "training" Apollo application. Note that for actual community annotation, you must first register for an account (https://i5k.nal.usda.gov/web-apollo-registration), and that the URLs for BLAST and Apollo access will differ slightly. Also note that the Apollo version used here is 1.0.4, which has a slightly different interface and different functions than Apollo2, the newer version.

1. Familiarize yourself with the process of manual annotation in Apollo (http://genomearchitect.github.io/users-guide/) and the i5k Workspace annotation guidelines (https://i5k.nal.usda.gov/content/rules-web-apollo-annotation-i5k-pilot-project).
2. One way to locate where your gene resides is to use the i5k Workspace BLAST+ interface. Locate the gene sequence from the "best" reference that you can find. It is preferable to use a sequence that has not been computationally inferred; otherwise, errors may inadvertently be propagated to the species that you are annotating. In this example, we will use the protein sequence of alpha-catenin from *Drosophila melanogaster* (CG17947/FBpp0070037, Supplemental File 1, http://flybase.org/cgi-bin/getseq.html?source=dmel&id=FBpp0070037&chr=3L&dump=PrecompiledFasta&targetset=translation).
3. For this exercise, we will use the i5k Workspace "Training" BLAST website, which provides link-outs to a "Training" Apollo application. Navigate to https://i5k.nal.usda.gov/training/blast/, or select "Tools → Training Tools → Training BLAST" from the menu. If you wish to annotate as part of an existing i5k Workspace project, you will need to register for an account (https://i5k.nal.usda.gov/web-apollo-registration; (Fig. 6)), and can annotate pending the genome coordinator's approval.
4. Select the organism(s) that you would like to search against, then choose the specific databases for these organisms. In the example, we chose the genome assembly for the Colorado Potato Beetle, *Leptinotarsa decemlineata* (Fig. 7).

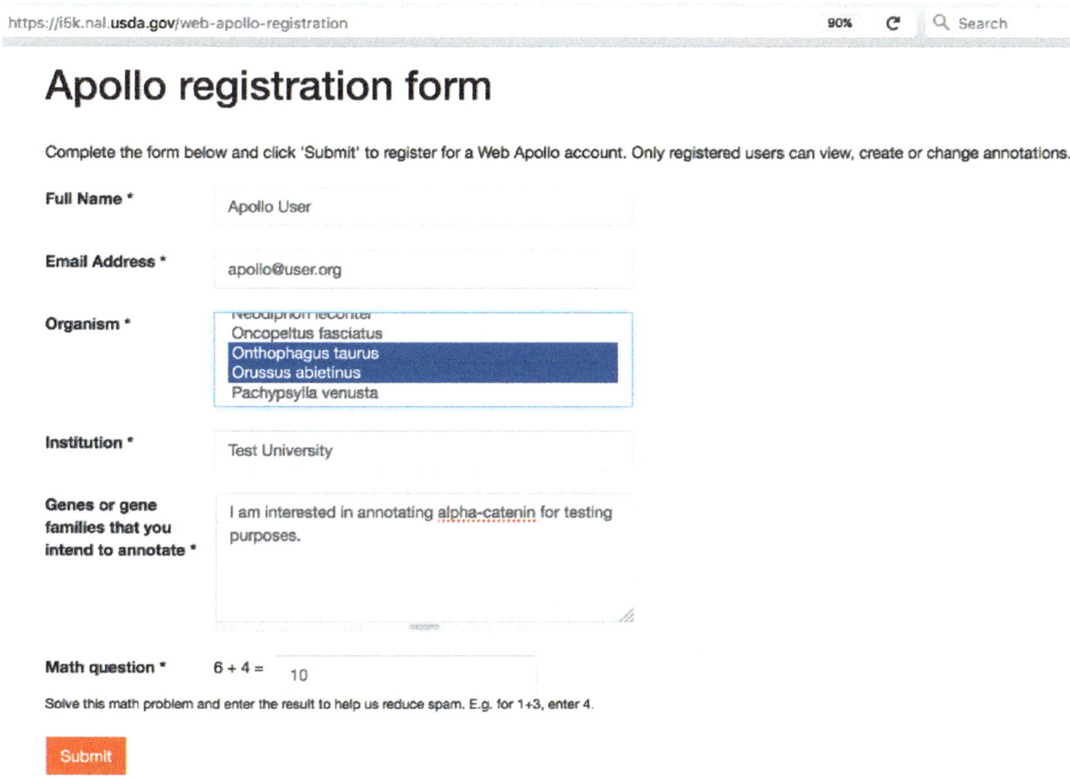

Fig. 6 Screenshot of the Apollo registration form with example registration content. Two organisms (*Onthophagus taurus* and *Orussus abietinus*) are selected for registration

5. Paste the query sequence (from Supplemental File 1, or http://flybase.org/cgi-bin/getseq.html?source=dmel&id=FBpp0070037&chr=3L&dump=PrecompiledFasta&targetset=translation) into the "Query Sequence" box. The BLAST program should be automatically selected—in this example, tblastn.
6. Press the "Search" button. A "Query Submission" page should display, noting the status of your submission.
7. Once the search is complete, a result page should display (Fig. 8). Details of the BLAST result page are documented in our tutorial (https://i5k.nal.usda.gov/content/blast-tutorial). View the result table in the bottom left panel. Scroll to the right of the table until you find the "evalue" column. You can sort the High-scoring Segment Pairs (HSPs) from the BLAST output by *e*-value using the arrows to the right of the column header. The HSPs with the highest *e*-values are on Scaffold1. To focus the result page on Scaffold1, type "Scaffold1" in the search box underneath the "sseqid" column. This will restrict the search results in the result table and both graphical panels to results on Scaffold1. Viewing the "Query coverage graph" panel on the top left, almost all of the

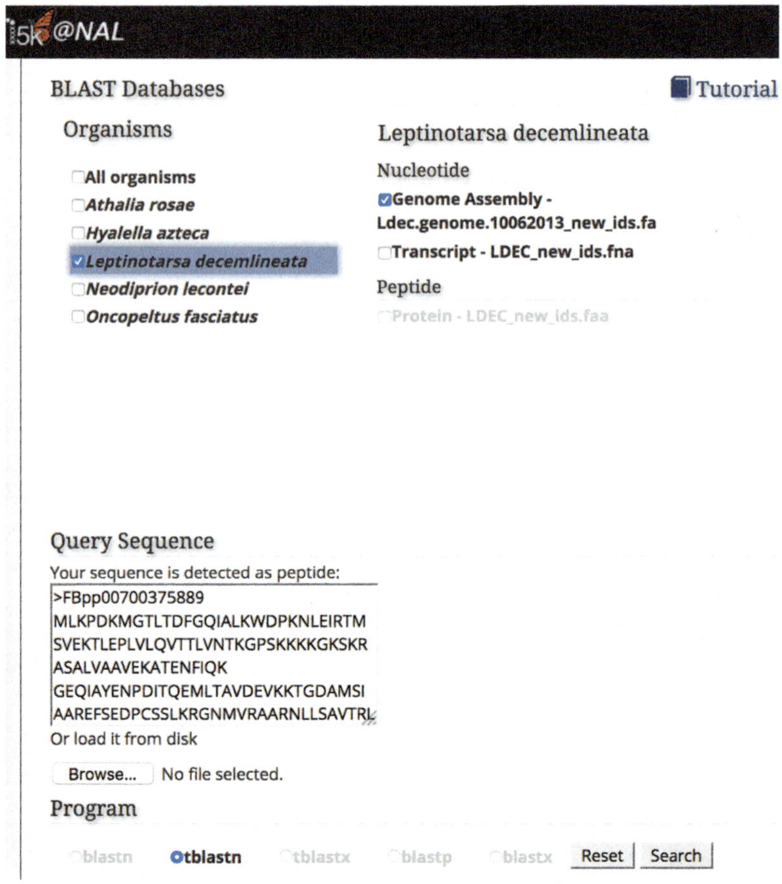

Fig. 7 Screenshot of the BLAST input page with an example query. The *Leptinotarsa decemlineata* genome assembly is selected as the database

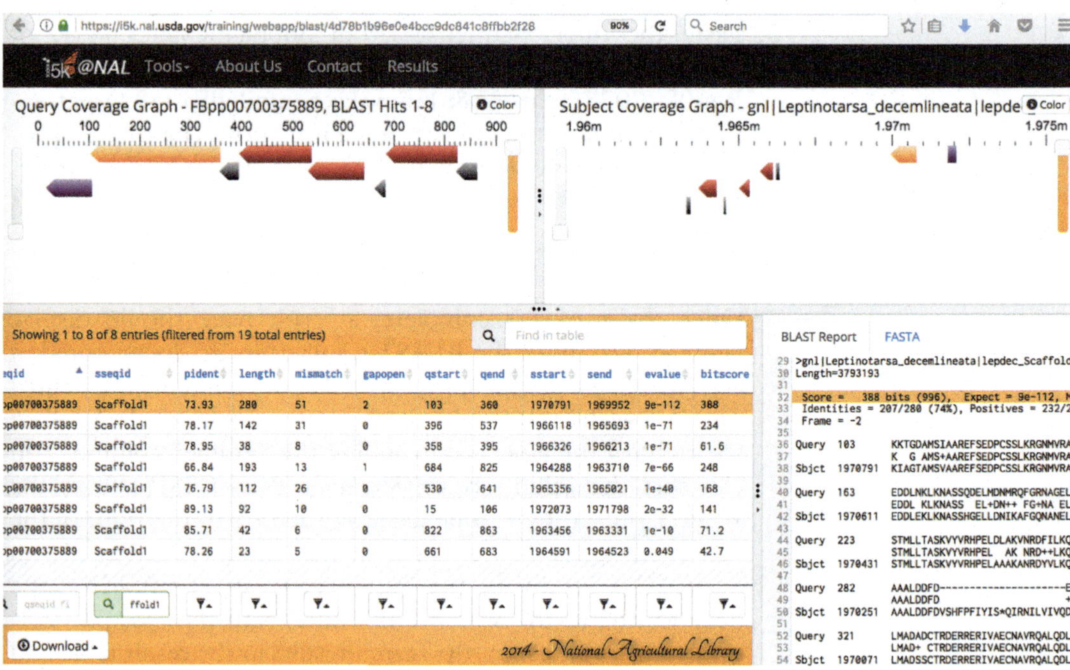

Fig. 8 Screenshot of an example BLAST result page. Please refer to the text for a detailed description of the results

query protein is covered by HSPs on Scaffold1. On the "Subject coverage graph" panel on the top right, all of the HSPs appear in the correct orientation. Note that the arrows representing the HSPs in the Query and Subject coverage graphs denote the orientation of the aligned sequence—if both query and subject align in the same orientation, then the arrow points to the right. If the query and subject sequences align in reverse orientation, then the arrow points to the left. The arrow direction in the query coverage graph should *not* be interpreted as the direction in which to read the query sequence. Hovering the mouse over each HSP highlights the corresponding information in each of the other three result panels—this allows us to see that each HSP on the Subject coverage graph is in the same consecutive order as the HSPs aligned against the Query.

8. To view an HSP in the genome browser, view the result table on the bottom left panel, and click on the blue box to the left of the best HSP result in the "blastdb" column (Fig. 9).
9. The JBrowse genome browser should open in a new window (Fig. 10). The BLAST+ HSP is displayed on the browser main

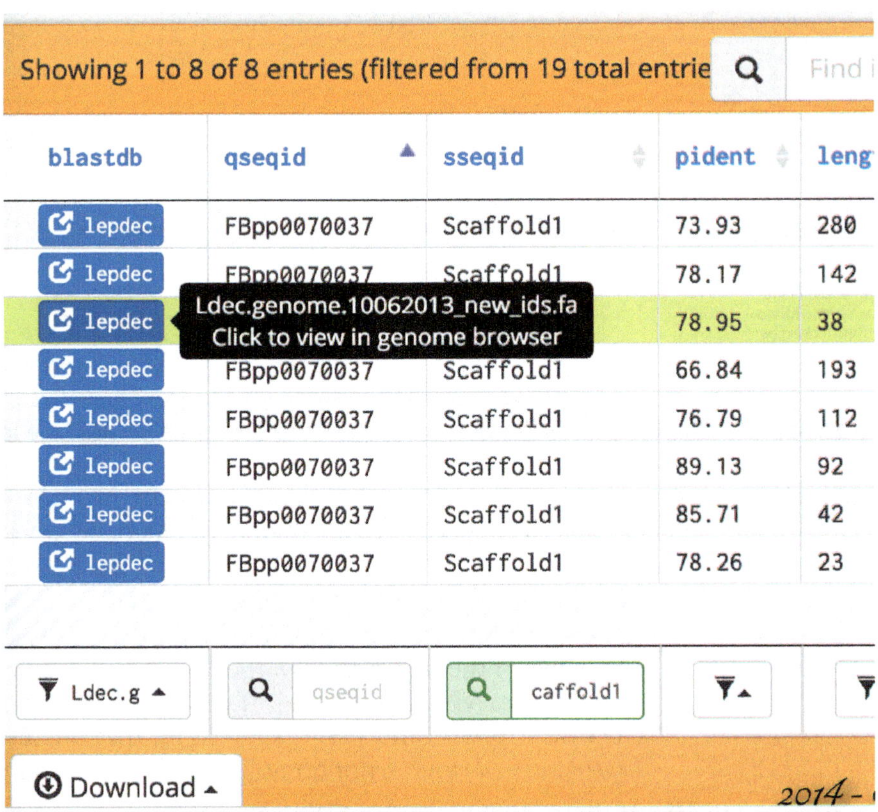

Fig. 9 A close-up of the icons that link from the BLAST result page out to the JBrowse genome browser when a genome assembly is selected as the database

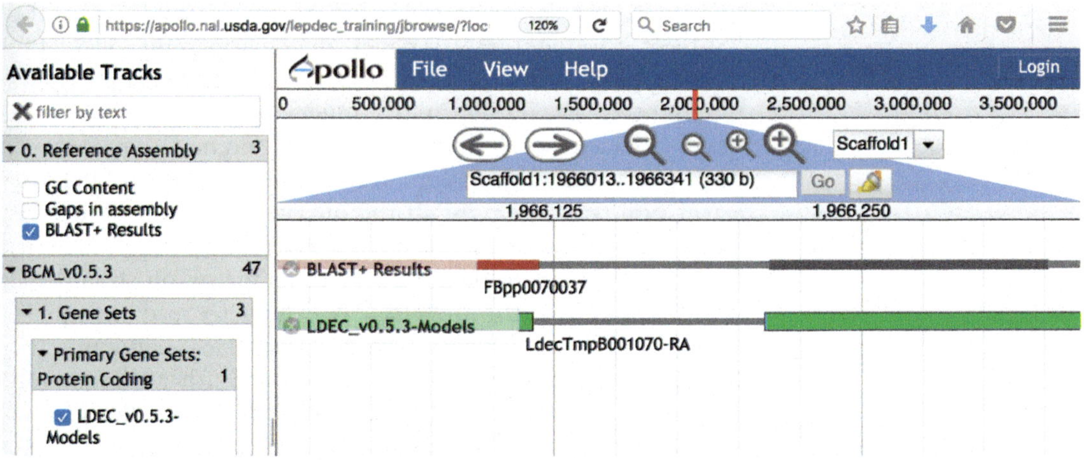

Fig. 10 Screenshot of a selected HSP from the BLAST search results, displayed as a track in the JBrowse genome browser, as well as two gene models from the *Leptinotarsa decemlineata* gene model track "LDEC_v0.5.3-Models"

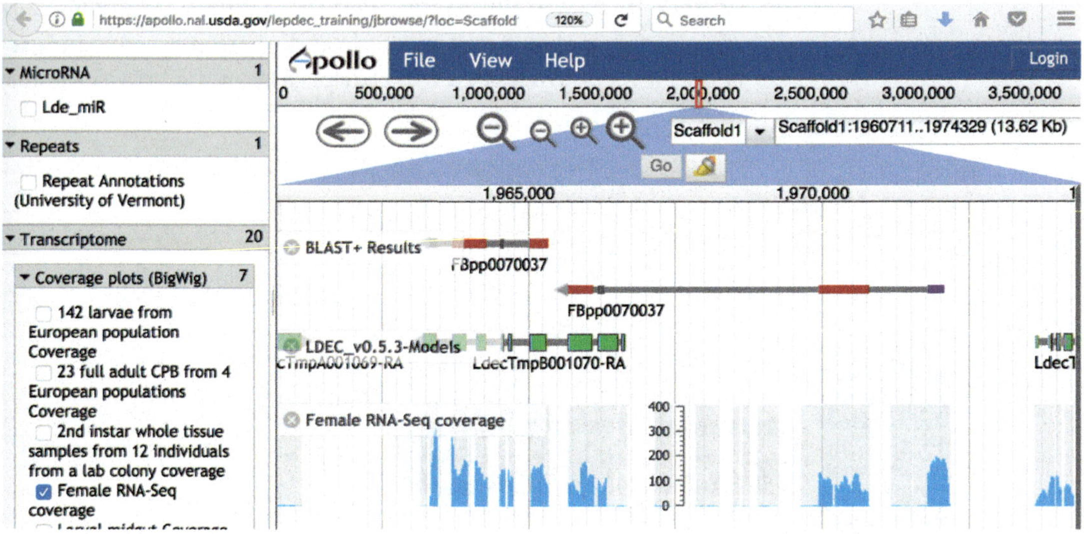

Fig. 11 Screenshot of the JBrowse genome browser for *Leptinotarsa decemlineata*, this time including a coverage plot of an aligned RNA-Seq track

window as a track under the name "BLAST+ Results". In the "Available Tracks" panel on the left side of the browser, select the gene prediction set of choice—in this case, the track "LDEC_v0.5.3-Models". Zoom out using the "-" icon in the top center of the screen to view the entire gene prediction and set of HSPs. The gene model "LdecTmpB001070-RA" has CDS overlap with the HSPs, suggesting that this model is a *Leptinotarsa decemlineata* homolog of alpha-catenin.

10. At the bottom of the "Available Tracks" panel on the left, under "Transcriptome," select a "Coverage Plots" track to view RNA-Seq support for this model (Fig. 11).

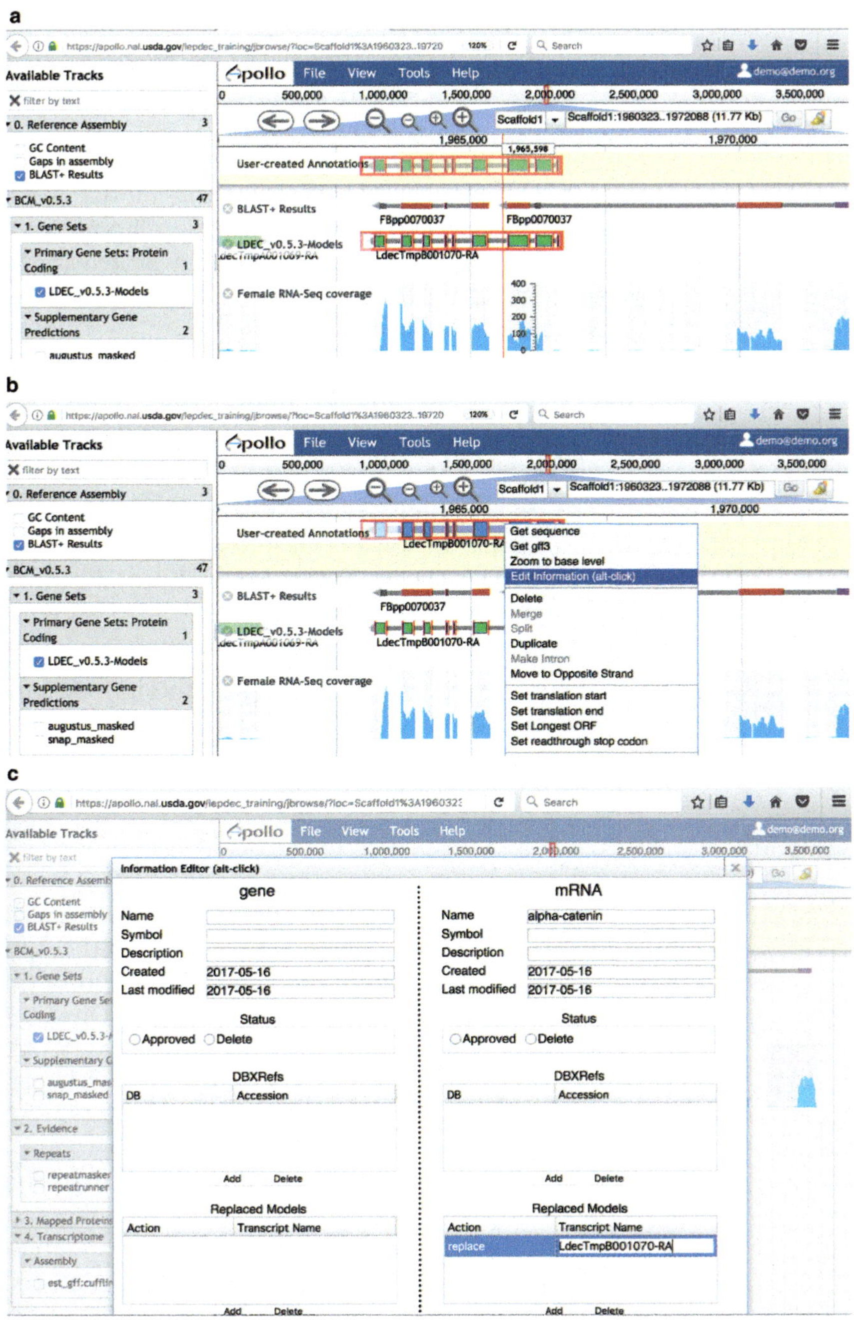

Fig. 12 Screenshot of the Apollo (v1.0.4) manual curation tool. This is the same view as in Fig. 11, after logging in to Apollo. (**a**) The gene model "LdecTmpB001070-RA" is being dragged into the User-created Annotations track toward the top of the browser; (**b**) the Information Editor, which is used to enter functional annotation information, can be opened via right-click (PC) or control-click (Mac); (**c**) text entry in the Information Editor

11. If manual annotation of this model is desired, log on to Apollo using the "Login" button at the top left of the screen. Use the username/password combination "demo/demo". Drag the model LdecTmpB001070-RA into the User-created Annotations track (Fig. 12a). Note that because this is a "training" Apollo

instance, other users may have already created a gene model here—feel free to ignore these. Also be aware that the NAL may clear out the user-created annotations track from time to time.

12. Following the guidelines in the Apollo documentation (http://genomearchitect.github.io/users-guide/), determine whether the protein sequence of this model and UTRs are complete. In this example, based on homology with the *Drosophila melanogaster* alpha-catenin protein and RNA-Seq support, two 5′ exons should be added, and the sequence of existing exons adjusted.
13. When you are satisfied with the structural annotation of this model, open the Apollo Information Editor by control-clicking (PC) or right-clicking (Mac) on the model in the User-created Annotations track (Fig. 12b). This will allow you to edit the functional information of this model.
14. In this example, we will give the mRNA the name "alpha-catenin", and will enter the gene model name "LdecTmpB001070-RA" into the "Replaced Models" field (Fig. 12c). This information will allow us to generate a consensus gene set downstream.
15. There is no need to save the model—this is done automatically by the Apollo software.

Acknowledgments

We would like to thank our users, the i5k coordinating committee, the National Agricultural Library leadership, and the Apollo development team for their support and encouragement of this project. We also thank the NAL Information Systems Division team for logistical support. Funding for this project is from the US Department of Agriculture—Agricultural Research Service; and the USDA-ARS Bee Research Laboratory.

EEO/Nondiscrimination Statement

The US Department of Agriculture (USDA) prohibits discrimination in all its programs and activities on the basis of race, color, national origin, age, disability, and where applicable, sex, marital status, familial status, parental status, religion, sexual orientation, genetic information, political beliefs, reprisal, or because all or part of an individual's income is derived from any public assistance program. (Not all prohibited bases apply to all programs.) Persons with disabilities who require alternative means for communication of program information (Braille, large print, audiotape, etc.) should contact USDA's TARGET Center at (202) 720-2600 (voice and TDD). To file a complaint of discrimination, write to USDA, Director, Office of Civil Rights, 1400 Independence Avenue, S.W., Washington, D.C. 20250-9410, or call (800) 795-3272 (voice) or (202) 720-6382 (TDD). USDA is an equal opportunity provider and employer.

References

1. i5K Consortium (2013) The i5K initiative: advancing arthropod genomics for knowledge, human health, agriculture, and the environment. J Hered 104:595–600. https://doi.org/10.1093/jhered/est050
2. Robinson GE, Hackett KJ, Purcell-Miramontes M et al (2011) Creating a buzz about insect genomes. Science 331:1386–1386. https://doi.org/10.1126/science.331.6023.1386
3. Giraldo-Calderón GI, Emrich SJ, MacCallum RM et al (2015) VectorBase: an updated bioinformatics resource for invertebrate vectors and other organisms related with human diseases. Nucleic Acids Res 43:D707–D713. https://doi.org/10.1093/nar/gku1117
4. Elsik CG, Tayal A, Diesh CM et al (2016) Hymenoptera Genome Database: integrating genome annotations in HymenopteraMine. Nucleic Acids Res 44:D793–D800. https://doi.org/10.1093/nar/gkv1208
5. Yin C, Shen G, Guo D et al (2016) InsectBase: a resource for insect genomes and transcriptomes. Nucleic Acids Res 44:D801–D807. https://doi.org/10.1093/nar/gkv1204
6. Poelchau M, Childers C, Moore G et al (2014) The i5k Workspace@NAL—enabling genomic data access, visualization and curation of arthropod genomes. Nucleic Acids Res. https://doi.org/10.1093/nar/gku983
7. Skinner ME, Uzilov AV, Stein LD et al (2009) JBrowse: a next-generation genome browser. Genome Res 19:1630–1638. https://doi.org/10.1101/gr.094607.109
8. Lee E, Helt GA, Reese JT et al (2013) Web Apollo: a web-based genomic annotation editing platform. Genome Biol 14:R93. https://doi.org/10.1186/gb-2013-14-8-r93
9. Camacho C, Coulouris G, Avagyan V et al (2009) BLAST+: architecture and applications. BMC Bioinformatics 10:421. https://doi.org/10.1186/1471-2105-10-421
10. Mistry J, Finn RD, Eddy SR et al (2013) Challenges in homology search: HMMER3 and convergent evolution of coiled-coil regions. Nucleic Acids Res 41:e121. https://doi.org/10.1093/nar/gkt263
11. Eddy SR (1998) Profile hidden Markov models. Bioinformatics 14:755–763. https://doi.org/10.1093/bioinformatics/14.9.755
12. Sievers F, Wilm A, Dineen D et al (2011) Fast, scalable generation of high-quality protein multiple sequence alignments using Clustal Omega. Mol Syst Biol. https://doi.org/10.1038/msb.2011.75
13. Higgins DG, Sharp PM (1988) CLUSTAL: a package for performing multiple sequence alignment on a microcomputer. Gene 73:237–244. https://doi.org/10.1016/0378-1119(88)90330-7
14. Higgins D, Thompson J, Gibson T et al (1994) CLUSTAL W: improving the sensitivity of progressive multiple sequence alignment through sequence weighting, position-specific gap penalties and weight matrix choice. Nucleic Acids Res. https://doi.org/10.1093/nar/22.22.4673
15. The Wellcome Trust (2003) Sharing data from large-scale biological research projects: a system of tripartite responsibility. Report of a meeting organized by the Wellcome Trust, 14–15 Jan 2003, Fort Lauderdale, USA
16. Toronto 2009 Data Release Workshop Authors (2009) Benefits and best practices of rapid pre-publication data release. Nature 461: 168–170. https://doi.org/10.1038/461168a
17. Larkin MA, Blackshields G, Brown NP et al (2007) Clustal W and Clustal X version 2.0. Bioinformatics 23:2947–2948. https://doi.org/10.1093/bioinformatics/btm404
18. Yandell M, Ence D (2012) A beginner's guide to eukaryotic genome annotation. Nat Rev Genet 13:329–342. https://doi.org/10.1038/nrg3174
19. McKenna DD, Scully ED, Pauchet Y et al (2016) Genome of the Asian longhorned beetle (*Anoplophora glabripennis*), a globally significant invasive species, reveals key functional and evolutionary innovations at the beetle–plant interface. Genome Biol 17:227. https://doi.org/10.1186/s13059-016-1088-8

INDEX

Martin Kollmar (ed.), *Eukaryotic Genomic Databases: Methods and Protocols*, Methods in Molecular Biology, vol. 1757, https://doi.org/10.1007/978-1-4939-7737-6, © Springer Science+Business Media, LLC, part of Springer Nature 2018

H

I

J

K

M

N

O

Printed by Printforce, the Netherlands